Theory of Beam-Columns

Volume 2:

Space Behavior and Design

TITLES IN THE SERIES

Architectural Acoustics
M. David Egan
ISBN 13: 978-1-932159-78-3, ISBN 10: 1-932159-78-9, 448 pages

Earth Anchors
By Braja M. Das
ISBN 13: 978-1-932159-72-1, ISBN 10: 1-932159-72-X, 242 pages

Limit Analysis and Soil Plasticity
By Wai-Fah Chen
ISBN 13: 978-1-932159-73-8, ISBN 10: 1-932159-73-8, 638 pages

Plasticity in Reinforced Concrete
By Wai-Fah Chen
ISBN 13: 978-1-932159-74-5, ISBN 10: 1-932159-74-6, 474 pages

Plasticity for Structural Engineers
By Wai-Fah Chen & Da-Jian Han
ISBN 13: 978-1-932159-75-2, ISBN 10: 1-932159-75-4, 606 pages

Theoretical Foundation Engineering
By Braja M. Das
ISBN 13: 978-1-932159-71-4, ISBN 10: 1-932159-71-1, 440 pages

Theory of Beam-Columns, Volume 1: In-Plane Behavior and Design
By Wai-Fah Chen & Toshio Atsuta
ISBN 13: 978-1-932159-76-2, ISBN 10: 1-932159-76-9, 513 pages

Theory of Beam-Columns, Volume 2: Space Behavior and Design
By Wai-Fah Chen & Toshio Atsuta
ISBN 13: 978-1-932159-77-6, ISBN 10: 1-932159-77-0, 732 pages

Theory of Beam-Columns

Volume 2: Space Behavior and Design

by Wai-Fah Chen and Toshio Atsuta

ISBN-10: 1-932159-77-0
ISBN-13: 978-1-932159-77-6

Printed and bound in the U.S.A. Printed on acid-free paper
10 9 8 7 6 5 4 3 2

This J. Ross Publishing edition, first published in 2008, is an unabridged republication of the work originally published by McGraw-Hill, Inc., New York, in 1977.

Library of Congress Cataloging-in-Publication Data

Chen, Wai-Fah, 1936–
Theory of beam-columns / by Wai-Fah Chen & Toshio Atsuta.
v. cm.
Reprint. Originally published: New York : McGraw-Hill, Inc., c1977.
Includes bibliographical references and index.
Contents: v. 1. In-plane behavior and design — v. 2. Space behavior and design.
ISBN 978-1-932159-76-9 (v. 1 : hardcover : alk. paper) — ISBN 978-1-932159-77-6 (v. 2 : hardcover : alk. paper)
1. Columns. I. Atsuta, Toshio, 1940- II. Title.
TA660.C6C45 2008
624.1'7725—dc22 2007044139

Direct all inquiries to J. Ross Publishing, Inc., 5765 N. Andrews Way, Fort Lauderdale, FL 33309.

Phone: (954) 727-9333
Fax: (561) 892-0700
Web: www.jrosspub.com

To
Lily Chen
Chika Atsuta

CONTENTS

PREFACE

This is the second volume of the two-volume treatise which presents systematically for the first time the most comprehensive theory on beam-columns. It covers virtually all the subjects and methods of analysis for solving problems of beam-columns in elastic or plastic, in-plane or in-space, short or long, of steel or of concrete.

Volume 2 focuses particular attention on the space behavior and design of beam-columns subjected to compression combined with biaxial bending moments acting at the ends in two perpendicular directions. Extensive theoretical studies on the behavior of such beam-columns have been made in recent years. As a result of these studies, several practical methods for the design of such beam-columns have been proposed. However, engineers who need to study the basic theory or who have need for guidance in design, often find it difficult to read or to locate or to use the available literature, recorded in various countries and languages. It seems that the time has come to bring the literature together and put it in the form of a book that will provide engineers and research workers the information needed to enable them to follow the latest developments in this field.

The book begins with a rigorous derivation of basic differential equations of biaxially loaded beam-columns and proceeds to offer a unified approach to the various approximations and numerical solution techniques. Various methods for which accurate solutions can be obtained are reported. While the emphasis is on the analytical methods, numerical and experimental results for various three-dimensional beam-column problems are presented. Some of the results obtained are presented here for the first time. In many cases existing results have been rederived to enable them to be presented in accordance with the unified viewpoint of the present treatment. While this volume focuses particular attention on the analysis and behavior of three-dimensional beam-columns, the development of improved procedures for the practical design of such beam-columns, is also discussed and illustrated. Further, the book treats the flexural-torsional buckling of laterally unsupported beam-columns as the special, or limiting case, of biaxially loaded beam-column problems and devotes an entire chapter to this particular subject.

Here, as in Vol. 1, most relationships given in the book are expressed in dimensionless form. For numerical examples, both the Customary (English) and SI (International System) units are used. The references are given in a separate list at the end of each chapter in alphabetical order by the last name of the senior author. Exercise problems are given at the end of each chapter and the answers to some of these are given at the end of the book.

The real application of some numerical procedures and their associated solution techniques require not only the mastery of the theory, but also a considerable experience in computer programming. With this in mind, three chapters have been written separately by three active investigators—Dr. N. S. Trahair of the University of Sydney (Chap. 3), Dr. S. Vinnakota of the Swiss Federal Institute of Technology (Chap. 10) and Dr. S. Rajasekaran of P.S.G. College of Technology (Chap. 12) who have contributed much to developments in this field. Here, an in-depth discussion on this subject together with their experience of computer solution is presented and it is hoped that this will be beneficial to those attempting to implement their own programs.

Much of the research on the behavior of beam-columns subjected to biaxial bending, conducted at Fritz Engineering Laboratory (Dr. L. S. Beedle, Director) Lehigh University, provided a background for the book and has been drawn on extensively. The book contains many results first presented in the form of Technical Reports, prepared under various phases of research projects on this subject. Those sponsoring this work include the American Iron and Steel Institute (J. A. Gilligan, Task Group Chairman), the Naval Ship Systems Command, the Naval Facilities Engineering Command, the National Science Foundation (M. P. Gaus and C. A. Babendreier, Program Directors), the American Petroleum Institute (L. A. Boston, Project Advisory Committee Chairman) and the Canadian Steel Industries Construction Council (J. Springfield, Project Coordinator).

Professor Chen wishes to thank his students S. Santathadaporn, N. Tebedge, M. T. Shoraka, G. P. Rentschler, D. A. Ross and J. McGraw and his collaborator, L. W. Lu, at Lehigh University without whose efforts this volume would not have been possible; and to the research teams of New York University (C. Birnstiel), the

University of Illinois (E. H. Gaylord), the Royal Military College of Canada (J. S. Ellis), the Swiss Federal Institute of Technology (S. Vinnakota) and the University of Cincinnati (B. C. Ringo) through the coordination of Task Group 3 on biaxially loaded beam-columns of the Structural Stability Research Council (J. Springfield. Task Group Chairman), who have exchanged information and contributed to the general development.

The authors also thank Dr. N. S. Trahair, Dr. S. Vinnakota and Dr. S. Rajasekaran who each contributed a chapter and who read portions of the manuscript and offered valuable suggestions.

We also express our sincere thanks to Miss Shirley Matlock for her rapid and careful typing of the manuscript.

May 1976 W. F Chen
Bethlehem, Pa. T. Atsuta

NOTATION

All notation will be specifically defined where it is first introduced. The following Table will serve as a reference and guide. In general, capital letters will be used for properly dimensioned physical quantities, and the corresponding lower-case letters will stand for the same quantities expressed in dimensionless form. Where more than one meaning has been assigned to a symbol, the correct use will be obvious from the context. As far as possible, subscripts will have consistent values; representative samples are given in the Table.

A	= area
B, D	= width and depth of a cross section
C_{mx}, C_{my}	= equivalent moment factors used in AISC Specification
E, E_t	= Young's modulus, tangent modulus
e	= end eccentricity
f_c'	= concrete compressive strength
G	= elastic shear modulus
I	= moment of inertia
K	= $\int \sigma a^2\, dA$ where a is defined as the distance from a point on the cross section to the shear center
K_T	= St. Venant torsion constant

L = member length
l = segment length
M = bending moment
M_ω = bimoment or warping moment
M_{pcx}, M_{pcy} = fully plastic moments of section about the x- and y-axis, respectively, reduced for the presence of axial load when $P \neq 0$
M_{px}, M_{py} = fully plastic moments of section about the x- and y-axis, respectively, when $P = 0$
M_{ucx}, M_{ucy} = ultimate bending moments of an axially loaded beam-column about the x- and y-axis respectively, when there is zero moment about the other axis ($P \neq 0$)
M_{ux}, M_{uy} = ultimate bending moments of a beam about the x- and y-axis, respectively, reduced for the presence of lateral torsional buckling, if necessary, when $P = 0$
M_{xl}, M_{yl} = moments applied at the end $z = L$ of a beam-column about the x- or y-axis, respectively
M_{x0}, M_{y0} = moments applied at the end $z = 0$ of a beam-column about the x- or y-axis, respectively
M_{ycx}, M_{ycy} = initial yield moments of section about the x- and y-axis, respectively, reduced for the presence of axial load, when $P \neq 0$
M_{yx}, M_{yy} = initial yield moments of section about the x- and y-axis, respectively, when $P = 0$
N = axial tensile force
P = axial compressive force
P_{ex}, P_{ey} = Euler buckling loads about x- and y-axis, respectively
P_u = ultimate load of an axially loaded column
P_y = axial load at full yield condition of a section
P_z = elastic torsional buckling load
r_0 = polar radius of gyration
T_{sv} = St. Venant torsional moment
T_w = warping torsional moment
t = thickness of plate or tubular wall
u_i, v_i, θ_i = initial deflection of shear center in x or y direction, and initial twisting angle
u, v, w = displacements in x, y or z directions
X, Y, Z = local coordinate system whose origin moves with the cross section and whose axes are always parallel to the x, y, z axes
x, y, z = global coordinate system fixed in space
x_0, y_0 = location of shear center
$\{ \}$ = vector
$[ ]$ = matrix
ξ, η, ζ = local coordinate system fixed to the cross section and deforming with cross section
θ = angle of twist
ε = axial strain
ν = Poisson's ratio

σ = normal stress
σ_r = residual stress
σ_w = warping normal stress
σ_y, τ_y = yield stresses in simple tension and shear respectively
τ = shearing stress
τ_w = warping shearing stress
Φ = curvature
ϕ = capacity reduction factor
ω = double sectorial area coordinate

1

INTRODUCTION

1.1 BEAM-COLUMN UNDER BIAXIAL LOADING

A three-dimensional space structure is often treated as a collection of two-dimensional planar structures; that is, structures with all their members lying in a single plane and with all the loads applied in the same plane. This procedure is equivalent to setting a number of secondary interaction bending moments and torques equal to zero. The beam-columns in a planar frame are, therefore, designed to resist bending moments acting in the plane of the frame. The theory of *in-plane* beam-columns is presented in Vol. 1 of this book.

While this idealization has resulted in satisfactory designs in the past, it does not necessarily represent the true loading condition existing in a space structure and may not give the optimum design. In an actual building framework, the beam-columns are frequently subjected to bending moments acting in two perpendicular directions in addition to an axial compression (commonly called *biaxial loading* or *biaxial bending*). The obvious example is a corner column in a space building frame. The biaxial moments may result from the space action of the entire framing system (Fig. 1.1a) or from an axial load biaxially located with respect to the principal axes of the beam-column cross section (Fig. 1.1b).

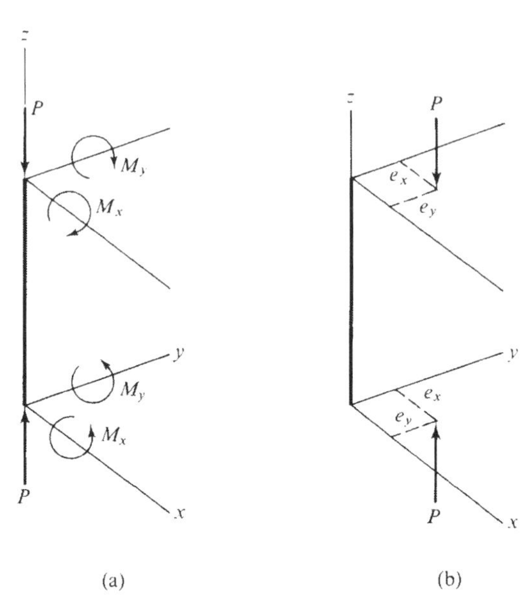

FIGURE 1.1
Beam-column under biaxial loading

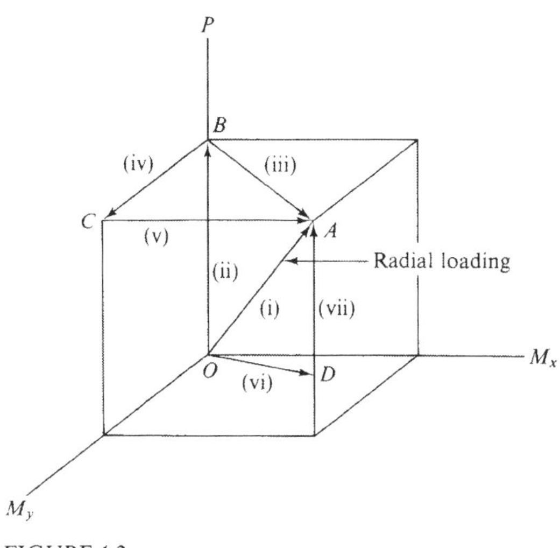

FIGURE 1.2
Various loading paths

Although the loading conditions shown in Fig. 1.1(a) and Fig. 1.1(b) are statically equivalent to each other and often are considered identical in terms of stress resultants, the plastic behavior of these two beam-columns may be quite different, depending on the detailed loading program acting at the ends of the beam-column in Fig. 1.1(a).

Plastic action is load path dependent and usually requires step-by-step calcula-

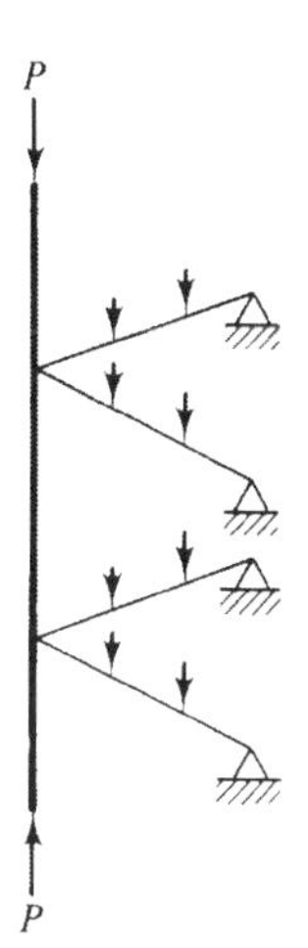

FIGURE 1.3
Elastically restrained column subjected to biaxial loading (center column)

tions that follow the history of loading. If P, M_x and M_y increase proportionally (or sometimes termed *radial loading*), the plastic behavior of the beam-column in Fig. 1.1(a) is equivalent to that of Fig. 1.1(b). The term radial comes from the plot of the loading path in a generalized stress space as shown diagrammatically in Fig. 1.2 [see path (i)]. Different loading paths for Fig. 1.1(a) are also shown in the figure. In the path O-B-A, as marked by (ii) and (iii), the beam-column is first loaded axially to point B and then the axial load P is held constant while the beam-column is loaded to failure by two end moments, M_x and M_y, which increase proportionally in magnitude from zero. In the path O-B-C-A, as marked by (ii), (iv) and (v), the beam-column is first loaded axially up to point B and then bent by M_y to C while keeping P constant and, finally, bent by M_x to failure while keeping P and M_y constant. Loading path O-D-A can be interpreted in a similar manner and was found especially convenient for the experimental investigations of an elastically restrained column which is also restrained against sway (Fig. 1.3).

Behavior

If an axial load is applied with an eccentricity in a plane of symmetry, a beam-column will deflect but remain untwisted at loads less than the buckling load. However, if a beam-column is loaded with biaxial eccentricity, it will usually deflect and twist at any load as illustrated in Fig. 1.4. Typical load versus midheight displacements of the beam-column for an elastic-plastic material are shown in Fig. 1.5. The essential feature of the beam-column due to space action is that the lateral displacement is always accompanied by a rotation of the beam-column sections. The importance of this twisting lies in the fact that the ultimate load carrying capacity of such beam-columns, especially beam-columns with open thin-walled sections that have small torsional rigidity, may be less than the maximum load carrying capacity for in-plane loading.

The state of the beam-column section under any axial load that is not in a

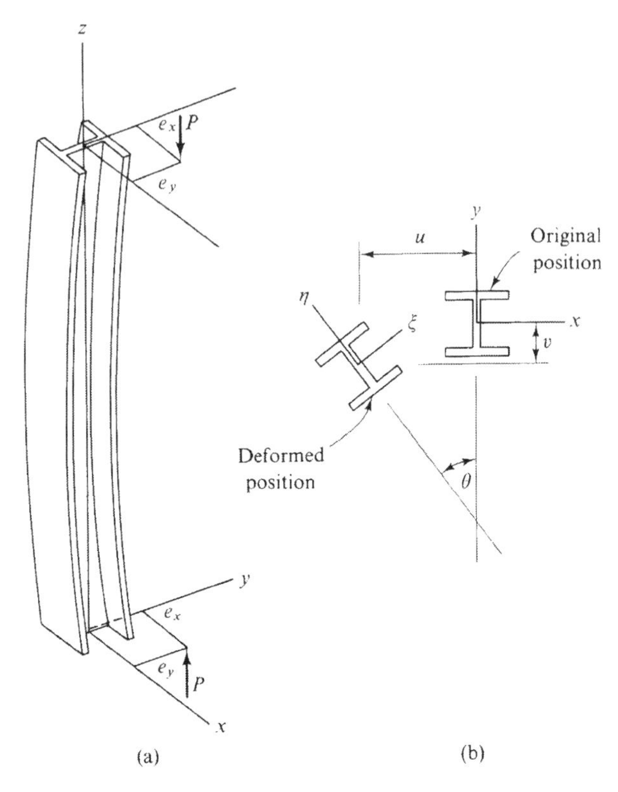

FIGURE 1.4
Isolated H-column under biaxial loading

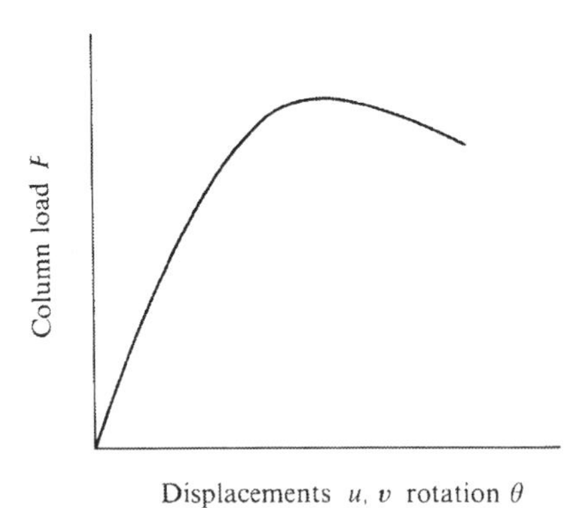

FIGURE 1.5
Load-displacements or load-rotation curve for beam-column subjected to biaxial loading.

plane of symmetry may be explained by considering a simple physical model (Fig. 1.6).

The biaxial load can be decomposed into four components as shown in this figure. The first three are statically equivalent to an axial force $4P$ and bending moments M_x and M_y about the two principal axes of the section. However, these

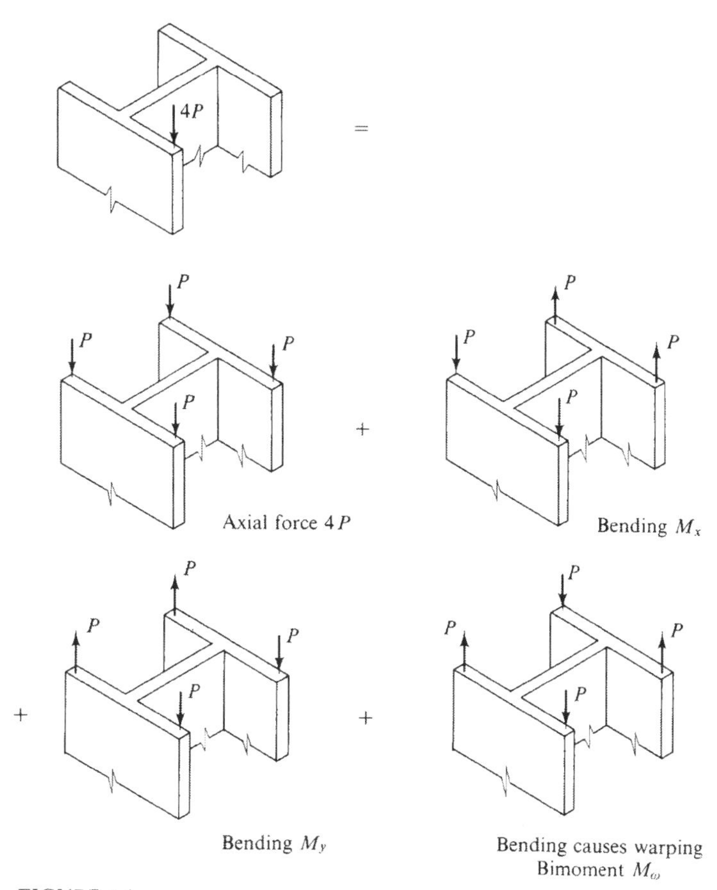

FIGURE 1.6
Decomposition of a biaxial loading

three equivalent systems do not produce the biaxial load $4P$. It is necessary to consider a fourth system which produces zero axial and bending moment resultants on the section. This fourth system causes the beam-column to warp or twist. The twisting effect is small for beam-columns with solid or closed wall sections but it is significant for beam-columns with open thin-walled sections because of their small torsional rigidity. Computations by Harstead, Birnstiel and Leu (1968) on the biaxially loaded elastic H-column show clearly that the twisting effect becomes so large that even the elastic analysis has substantial error introduced by the commonly used Goodier's approximate linearized formulation. Even greater error can be expected for the plastic case of biaxially loaded beam-columns.

Purpose

The recent development of the *limit state approach to design* has focused particular attention on two requirements: accurate information regarding the behavior of structures throughout the entire range of loading up to ultimate load and simple procedures to enable designers to assess this behavior. The purpose of this book is to present the complete theory which attempts to satisfy these requirements in the case of biaxially loaded beam-columns.

To this end, the book presents, systematically, the analysis and design methods of beam-columns under biaxial loading and offers a unified approach to the various analytical and numerical solution techniques. Both metal and reinforced concrete or composite beam-columns are treated. Various methods of solution are presented in separate chapters which are best suited for the types of beam-column problems considered. Both refined and simplified theory and design procedures, along with experimental results, are presented.

1.2 BASIC CONCEPTS

The solution of the biaxially loaded beam-column requires consideration of geometry or compatibility, equilibrium or dynamics and of the relation between stress and strain. Compatibility and equilibrium are independent of material properties and, hence, valid for elastic and elastic-plastic beam-columns. The differentiating feature is the relation between stress and strain. The extreme difficulty in obtaining an exact plastic analysis of beam-columns under biaxial loading, even with the aid of digital computers, is due mainly to the fact that the stress-strain relation in the plastic range is far more complicated than Hooke's law for linearly elastic materials.

Plastic behavior is extremely load path dependent and almost always requires step-by-step solutions that follow the history of loading. They are further complicated by the fact that the elastic-plastic boundary is moving and the stress-strain relationship for loading and unloading is different. Even without this complication, there are no solutions available for beam-columns under biaxial loading that consider nonlinear elasticity.

It is apparent that an exact elastic-plastic solution of the biaxially loaded beam-column is unlikely. Drastic simplifications and idealizations are essential for a reasonably approximate solution. The geometry or compatibility of the beam-column, the stress-strain relations and the equations of equilibrium must be idealized to accomplish a solution.

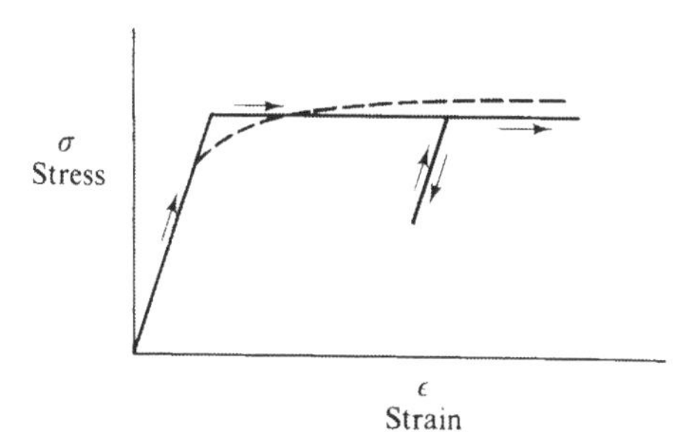

FIGURE 1.7
Idealization as elastic perfectly plastic

For example, the material may be idealized as perfectly plastic. This ignores work-hardening as symbolized by the stress-strain curve for simple tension (Fig. 1.7). This idealization is reasonable for materials with sharply defined yield strength such as mild structural steel. However, it may be interpreted as an approximation to the work-hardening material by using a suitably chosen yield stress (see dashed curve in Fig. 1.7). The adoption of this ideal material must not be thought of as neglecting work-hardening but, rather, as averaging its effect over the field of flow. In Vol. 1, Chap. 9, Chen and Santathadaporn (1969) have shown that the idealized material applied to eccentrically loaded columns which are assumed to fail by excessive bending in the plane of the applied moments, results in good agreement with the more precise analysis of von Kármán (1908, 1910) which utilizes the real stress-strain curve.

The difficulty of an exact analysis of even a perfectly plastic beam-column under biaxial loading has led to the approximate formulation of equilibrium equations in terms of generalized stresses and strain-rates, such as force and moment resultants and rates of extension and curvature of the beam-column section. In addition, the traction boundary conditions are not specified in detail, at least for end loading conditions. Normally, they are expressed in terms of the stress resultants and displacements. This approximate formulation of the equilibrium equations was generally adopted in the past.

Moreover, as a consequence of such an approximate formulation of equilibrium, there is one fundamental assumption about the strain distribution; that is, the plane cross sections remain plane after loading for each of the thin-wall flat plates of which the beam-column is composed. As remarked by Bleich (1952), this assumption appears to be justified because St. Venant's theory of torsion indicates that the longitudinal center line of the cross section of a thin flat plate remains in a plane during torsion. Warping can only vary slightly across the plate because of its small thickness and the entire cross section must remain approximately plane. This assumption enables one to obtain the stress distribution over the cross sections of the beam-column for a given stress-strain curve, provided the curvatures at each station along the beam-column are known.

1.3 GENERAL APPROACH

The methods of solution to the problem of biaxially loaded beam-columns may be categorized into two major groups: "short" beam-column and "long" or "intermediate" length beam-column (Table 1.1).

Short beam-column

All beam-columns deflect under loading but the term "short" beam-column used here implies that the effect of this lateral deflection upon the overall geometry can be ignored in an analysis. Hence, the strength of a short beam-column is limited only by full plastic yielding of the material of the cross section, provided, of course, the

Table 1.1 THE BIAXIALLY LOADED BEAM-COLUMN PROBLEM

Basic Equations—Chap. 2
Short Beam-Column
  Strength—Chap. 5 (Interaction Surface)
  Behavior—Chap. 6 (M-P-Φ)
Long or Intermediate Length Beam-Column
  Buckling (Eigenvalue) Approach—Chap. 3
  Load-Deflection Approach
    Elastic Analysis—Chap. 4
    Plastic Analysis { Approximate Methods—Chaps. 7 and 8
      Numerical Methods—Chaps. 9 and 10
      Finite Element Methods—Chaps. 11 and 12
Design Methods—Chap. 13

local buckling of metal or the crushing of concrete does not occur. Hence, the complex problem of biaxially loaded beam-columns can be approached from the *plastic hinge concept* by taking into account the effect of axial force and biaxial moments under the condition that the entire cross section will be fully plastic. This approach is exact only for beam-columns of zero length or for beam-columns with sufficient lateral bracing. This is described in Chap. 5.

The basic quantity required in the "long" or "intermediate" length beam-column analysis is the value EI which here can be considered as the slope of the relation between biaxial moment M_x and M_y and curvature Φ. This is the basic relationship which reflects the load-deformation behavior of a "short" beam-column. Moment-curvature-thrust relationships for a cross section (or the generalized stress-generalized strain relationships) are studied in Chap. 6.

Long or intermediate length beam-column

For a *long* or *intermediate* length beam-column, the *instability* of the member arising from the magnification of the primary moments by the axial load acting on the laterally deflected beam-column must be considered. The instability of a beam-column may be classified in accordance with its load-deflection characteristic as being the *bifurcation* or *non-bifurcation* type. For the bifurcation type of instability, the beam-column deforms in the direction of the applied load as the load is increased from zero until a *critical* (or *buckling*) *load* is reached and the beam-column has two states of equilibrium. The beam-column prefers a stable configuration and changes its deformation suddenly from the unstable to the stable state (see Chap. 1, Vol. 1 for further discussion). The flexural-torsional buckling of a beam or a beam-column subjected to loading in its plane is of the bifurcation type and the out-of-plane deformations remain zero until the critical loading condition is reached. Thus, the in-plane behavior of a beam-column up to the critical or buckling load can be analyzed independently of the out-of-plane buckling behavior. The solution from this in-plane analysis can then be substituted in the flexural-torsional equations which govern the out-of-plane buckling of the beam-column.

Elastic as well as plastic buckling analysis of beams and beam-columns are described in Chap. 3. For a *long* or purely elastic beam-column, the use of formal mathematics to obtain solutions from the governing differential equations is possible for some cases. However, for an intermediate length beam-column for which the flexural torsional buckling load usually occurs in the plastic or nonlinear range of the material, numerical methods such as *finite difference method* and *finite integral method* must be used to obtain solutions. In Chap. 3, a considerable volume of data for flexural-torsional buckling problems is presented which has been obtained by using an exact or a particular numerical method. Against the background of this information, design methods are derived and developed or reviewed.

Biaxially loaded beam-columns exhibit the non-bifurcation type of instability in which the deflection increases until a maximum load is reached, beyond which static equilibrium can only be sustained by decreasing the load for a plastic beam-column. The problem must, therefore, be approached from the standpoint of *load-deflection analysis.*

Any rigorous analysis which attempts to cover this behavior is complicated. For the case of a long beam-column for which elastic analysis may be applied, the governing differential equilibrium equations may be solved rigorously by the use of formal mathematics. Closed form solutions for simplified equations are possible for some cases. Elastic solutions of biaxially loaded beam-columns are presented in Chap. 4.

In the plastic or nonlinear range, the differential equations are often intractable and recourse must be made to numerical methods to obtain solutions. The numerical methods which have been used by various investigators include the *finite integral method* (Chap. 3), *numerical integration method* (Chap. 9) and the *finite difference method* (Chap. 10). The disadvantage of these methods is that they are efficient and successful only for isolated members or small structures.

For practical purposes, the differential equilibrium equations of a biaxially loaded plastic beam-column can be simplified considerably by introducing additional assumptions. The assumptions which have been used by various investigators include establishing equilibrium only at midheight of the beam-column or at a number of stations along the length of the member (Chaps. 7 and 8), assuming the displacements of a beam-column to be given by known simple functions (Chap. 7) and idealizing the shape of the biaxial moment-curvature-thrust relationships (Chap. 7). The accuracy of these assumptions can be checked by comparing them with more refined solutions. The approximate approaches are generally found quite accurate for many cases and, thus, can be used efficiently to generate a large volume of data from which practical design methods can be developed (Chap. 13).

Recently, there has been an increasing number of applications of the finite element method to flexural-torsional buckling and to three-dimensional nonlinear analysis of biaxially loaded thin-walled beam-columns. In this development, the beam-column is divided into a number of *finite segments* (Chap. 11) or *finite elements* (Chap. 12) and the assemblage of these elements is treated as a space structure. The *matrix stiffness method* is then applied to obtain solutions. The advantages of this method are that it can be used for large structures and it is systematic in obtaining solutions. Complex beam-column problems such as local buckling interaction and elastically restrained beam-columns subjected to biaxial loading as shown in Fig. 1.3, can all be solved on the basis of this method.

1.4 HISTORICAL DEVELOPMENTS

Solutions that describe the elastic behavior of beam-columns with various end conditions are the most highly developed aspect of beam-column research. These theories and important solutions are given in several texts as well as in Vol. 1 of this book. Here, the names of Timoshenko (1945), Timoshenko and Gere (1961), Bleich (1936, 1952, 1953), Vlasov (1961), Goodier (1941, 1942) and Johnston (1941), among others, can be mentioned but their work is also comparatively recent. The contribution made by Wagner in 1929 for the torsional buckling of thin-walled open sections to the currently well established solutions for torsion and flexure buckling gives an indication of the rapid progress in this area.

Analytical studies on elastic beam-columns with a thin-walled open section loaded biaxially, with respect to the principal axes of the beam-column cross section, are very extensive. Goodier (1941, 1942), following the related work on flexural-torsional buckling of Wagner (1936), Wagner and Pretschner (1936) and Kappus (1937), extended the governing differential equations to include beam-columns under biaxial bending with identical loading conditions at each end. Goodier's equations are simplified by the assumption that the twisting as well as the displacements of any cross section of the beam-column are small compared to the eccentricities of the loading. Excellent discussions of the theory are given by Bleich (1952), Timoshenko and Gere (1961) and Kollbrunner and Meister (1961).

Goodier's simplified equations have been solved exactly by Culver (1966) and approximately by Thürlimann (1953), Dabrowski (1961) and Prawel and Lee (1964), among others. Discussion of the elastic theory of biaxially loaded beam-columns is given in Chap. 4.

Harstead, Birnstiel and Leu (1968) have reported that Goodier's equations are not applicable at larger loads to elastic problems such as those selected by Culver (1966). This is due to the fact that, as the value of rotation of the beam-column cross section becomes larger, the error in Goodier's approximation becomes considerable. Refined elastic theory, along with approximate plastic solutions, is given in Chaps. 7 and 8. Elastic finite deformation solutions of biaxially loaded beam-columns have been reported by Soltis and Christiano (1972).

The post-buckling behavior of beam-columns bent out of the plane of the applied moment and twist at the same time is generally complicated for detailed solutions, even for the case of a double symmetrical section. Galambos (1959) defined failure of the beam-column as the load at which the beam-column begins to deflect laterally accompanied by twisting. This criterion does enable solutions to be obtained for the critical load but the resulting value is a conservative estimate of the ultimate load. Nevertheless, good correlation of this estimate with test results was observed. Elastic, as well as plastic buckling theory of beams and beam-columns, is presented in Chap. 3.

The plastic behavior of biaxially loaded beam-columns has been studied extensively in recent years by several research teams working with Birnstiel at New York University (1963–68), Chen at Lehigh University (1966–76), Gaylord at the University of Illinois (1965–73), Ellis at the Royal Military College of Canada (1964–71) and Vinnakota at the Swiss Federal Institute of Technology (1974–76), among others, but their work has been coordinated by the Task Group on biaxially loaded beam-columns of the U.S. Column Research Council since 1965. A state of the art report on this subject has been published (Chen and Santathadaporn, 1968). Since the combinations of single curvature about one axis and single or double curvature about the other present formidable problems, most of the research has concentrated on the case of symmetrical single curvature about both axes of wide flange sections. The ultimate strength of plastic biaxially bent beam-columns is evaluated by numerical procedures that determine, at successive increments of load and deflection, the load deflection curve up to and beyond maximum load.

In the early development, Pinadzhyan (1956) studied the ultimate load carrying capacity of beam-columns with H-shaped cross section under biaxial loading. Klöppel and Winkelmann (1962) have conducted experimental and analytical studies for isolated steel beam-columns of channel and H-shaped cross sections under biaxial loading. The solution was based on assuming polynomial expressions for the lateral displacements of the beam-column axis and satisfying equilibrium at a sufficient number of points so that the coefficients of a power series solution are determined. Twist was found later in a separate operation. A numerical procedure for the analysis of this problem was developed. However, the procedure is not described in their paper and the details are not available. An interaction formula for the maximum load carrying capacity of such beam-columns was proposed. It should be noted that the warping strains that result from the twisting of the cross section of the beam-column were neglected. Therefore, the results reported by Klöppel and Winkelmann are not "exact" but still provide useful information.

Birnstiel and Michalos (1963), following the related work of Johnston (1961), presented a general procedure for determining the ultimate load carrying capacity of beam-columns under biaxial loading. Warping strains due to nonuniform twist were considered. However, their procedure requires successive trials and corrections and needs considerable computational effort for a solution. In a later work, Harstead (1966) succeeded in reducing the laborious trial and correction procedure to a few cycles by solving a system of linear equations for the corrections at each station along the beam-column (Harstead, Birnstiel and Leu, 1968). The New York University team also conducted experiments on isolated H-columns subjected to biaxial loading (Birnstiel, 1968). The results of these tests and the effect of warping restraint at beam-column ends on the ultimate load-carrying capacity of the beam-column, and the effect of residual thermal strains on the behavior of the beam-column, are examined and compared by Harstead, Birnstiel and Leu (1968). The agreement between the numerical and experimental results appears to be satisfactory. A somewhat similar approach for the computation of failure loads of composite steel-concrete beam-columns in biaxial bending has recently been reported by Virdi and Dowling (1976).

Both methods are based on the determination of the actual deflected shape of the beam-column and, hence, give an almost "exact" solution to the problem. However, the labor involved in the numerical determination of the ultimate load carrying capacity of biaxially loaded beam-columns based upon the determination of the "exact" deflected shape makes its practical use unlikely. Therefore, simplifications and idealizations are essential if simple design formula are to be developed. The approximate solutions obtained by Ježek for eccentrically loaded in-plane beam-columns of rectangular cross section (Chap. 8, Vol. 1) indicates the possibility of extending Ježek's concept to the problem of biaxially loaded beam-column.

According to Ježek's concept, the solution of the system of governing differential equations can be simplified drastically by establishing equilibrium only at midheight of the beam-column, by idealizing the material as elastic-perfectly plastic (Fig. 1.7), and by assuming the deflected shape of the beam-column axis as the half wave of a sine curve. The solution based upon these assumptions leads to analytical expressions for the ultimate load-carrying capacity of eccentrically loaded in-plane beam-columns of rectangular cross section. A comparison of the results of the approximate method with results obtained from the "exact" theory and the results of tests, indicates that the idealization and simplification are reasonable.

The extension of Ježek's concept to biaxially loaded beam-columns proves to be equally successful. Sharma and Gaylord (1969) and Santathadaporn and Chen (1973) applied this concept and assumed, as well, an elastic-perfectly plastic idealization for the material and sinusoidal variations for the lateral displacements and rotation of the cross section of the deflected axis of the beam-column. The solution of the equations of equilibrium was simplified by only considering equilibrium at midheight of the beam-column between the applied force at the ends of the beam-column and the internal resistance of the midheight of the beam-column.

Despite these drastic idealizations and simplifications, however, no analytical expressions which define the relation between the slenderness ratio of the beam-column and the ultimate load-carrying capacity of the beam-column, as was obtained by Ježek for the case of eccentrically loaded in-plane beam-column, were obtained, hence, a large number of numerical solutions for the maximum load-carrying capacity of wide flange steel beam-columns under biaxial loading are given in the form of maximum strength interaction curves. Comparison with more refined solutions indicates that, for all practical purposes, the approximation is satisfactory. The approximate and the more refined theories are presented in Chaps. 7 and 8, respectively.

The Column Deflection Curve (CDC) concept, developed thoroughly in recent years for the two-dimensional in-plane analysis of beam-columns, has been discussed thoroughly in Chap. 8 of Vol. 1 of this book. The extension of this concept to three-dimensional biaxially loaded beam-column problems was reported by Ellis, Jury and

Kirk (1964) for the case of box sections and by Aglan (1972) for the case of H-sections. Here, as in the previous in-plane case, the CDCs in space are constructed by numerical integration of curvatures at each station along the length of beam-column. The particular applications of CDCs to box and H-sections are described in Chap. 9. The procedure presented can also be modified to handle beam-columns of general cross sections.

As an extension to the case of elastically restrained beam-columns, Milner (1965) appears to be the first to report the theoretical and experimental study of restrained and biaxially loaded H-section beam-columns. In his procedure, the governing differential equations of equilibrium are first expressed in terms of finite differences and a numerical integration procedure is used for the solution. His main purpose was to investigate the effect of the irreversible nature of plastic strains and, also, the effect of the residual stress upon the elastic-plastic behavior of the elastically restrained H-section beam-columns subjected to biaxial loading.

Milner's results indicated that the effect of unloading after yielding occurred in a biaxially loaded beam-column, was to strengthen the beam-column rather than weaken it. The effect of the order of load application upon the failure load was observed to be significant. Also, the effect of residual stress is less when the loading is eccentric and decreases as the eccentricity increases so that this effect may be small for restrained beam-columns. More recent developments and solutions of elastically restrained beam-columns under biaxial bending and torsion using finite difference method for numerical solutions have been reported by Vinnakota and his associates (1974). The applications of the finite difference method to biaxially loaded beam-columns are given in Chap. 10.

In recent years, there has been an increasing number of applications of the finite element method to biaxially loaded beam-column problems. In this approach, the actual beam-column is physically replaced by an assembly of finite segments. Here, as in the CDC method, load-deformation response of the segment throughout the entire range of loading up to the ultimate load must be known first before the assemblage of these segments can be solved. The behavior of beam-column segment under biaxial bending is studied thoroughly in Chap. 6. The systematic development and utilization of the finite segment method, along with its more refined counterpart—finite element method, are presented in Chaps. 11 and 12.

A different approach to the complex problem of biaxially loaded beam-columns is to extend the simple plastic hinge concept by taking into account the effect of axial force P and biaxial moment M_y on the fully plastic moment capacity M_x. This approach is correct only for beam-columns of short length or for beam-columns with sufficient lateral bracing, but it will not be applicable for the plastic analysis of space frames in which beam-column length is relatively long and the effect of the geometrical change of the beam-column on the ultimate strength of the beam-column becomes appreciable. Nevertheless, the concept of the interaction surface approach can be considered as a first step in extending the plastic planar structural analysis and design to more realistic plastic space frame analysis and design.

There is very extensive literature on the application of upper and lower bounds of limit analysis of plasticity to obtain interaction curves or surfaces for various loading combinations and various shapes of cross section. Various procedures and approaches that have been made to solve this problem are presented in Chap. 5. The most significant contribution in this aspect is the development of the method of superposition by Chen and Atsuta (1972). Using this method, exact interaction equations for commonly used structural shapes can be obtained directly by subtracting rectangular areas from the enclosing rectangular areas. The method is extremely powerful and efficient for computer solution.

As a result of this and allied developments, there exist, at present, several methods of design of biaxially loaded beam-columns. These are the linear interaction formulas currently used in the United States [Johnston (1976)], Sharma and Gaylord's method (1969), Horne's method (1964), Tebedge and Chen's method (1974), Young's method (1973) and Wood's method (1974), among others. Some of these design methods are developed further in Chap. 13 for general application and some practical limits are suggested for their use. Comparisons are also made in Chap. 13 with an extensive range of tests on biaxially loaded beam-columns. Some of these methods are found suitable for adoption in practice.

1.5 ORGANIZATION OF THE BOOK

The contents of this book are divided roughly into four parts as illustrated in Table 1.1. The first part, Chaps. 1 and 2, begins with the historical review of the subject, covers the basic theory of bending and torsion of thin-walled elastic beams, derives rigorously the basic differential equations of biaxially loaded beam-columns and proceeds to offer a unified approach to the various approximate and refined theories of beam-columns with which the literature abounds.

The second part, Chaps. 5 and 6, deals with the strength and behavior of a short beam-column segment for which the effect of lateral deflections upon the overall geometry of the beam-column can be ignored. Hence, the behavior and strength of the beam-column are determined almost entirely by the nonlinear properties of the material. The load-deformation behavior for a beam-column segment of unit length is the relation between axial thrust, biaxial moment and curvature. This relationship is required in later calculations for long or intermediate length beam-columns.

The third part deals with the analysis of long or intermediate length beam-columns. It contains a total of 8 chapters. The special cases of flexural-torsional buckling of laterally unsupported beam-columns are treated first in Chap. 3, followed by the presentation of elastic analysis of biaxially loaded beam-columns in Chap. 4. Elastic-plastic beam-column theory is presented in Chaps. 7 to 12, where the formal mathematical solution of the governing differential equations became intractable and various approximate and numerical methods are employed to obtain solutions.

The fourth part, Chap. 13, deals with practical design methods for the design of biaxially loaded beam-columns. Various methods proposed are reviewed and some of these are developed further and refined to achieve both simplicity in use and, as far as possible, a realistic representation of actual behavior. Extensive comparisons are also made with the results of tests on actual beam-columns, providing final confirmation of the validity of the proposed nonlinear interaction equation method. It is concluded that the nonlinear interaction equation is suitable for adoption in practice.

1.6 REFERENCES

Aglan, A. A., "The Ultimate Carrying Capacity of Beam Columns." Thesis presented to the University of Windsor, Windsor, Ontario, Canada, in partial fulfillment of the requirements for the degree of Doctor of Philosophy, (1972).

Birnstiel, C., "Experiments on H-Columns Under Biaxial Bending," *Journal of the Structural Division, ASCE*, Vol. 94, No. ST10, Proc. Paper 6186, pp. 2429–2448, October, (1968).

Birnstiel, C. and Michalos, J., "Ultimate Strength of H-Columns Under Biaxial Bending," *Journal of the Structural Division, ASCE*, Vol. 89, No. ST2, Proc. Paper 3505, pp. 161–197, April, (1963).

Bleich, F., *Buckling Strength of Metal Structures*, McGraw-Hill Book Co., Inc., New York, (1952).

Bleich, F. and Bleich, H., "Bending, Torsion and Buckling of Bars Composed of Thin Walls," Congress, International Association for Bridge and Structural Engineers, Berlin, English edn, p. 871, (1936).

Bleich, H., "Refinement of the Theory of Torsional Buckling of Thin-Walled Columns," *Proceedings, First Mid-Western Conference on Solid Mechanics*, April, (1953).

Chen, W. F. and Atsuta, T., "Interaction Equations for Biaxially Loaded Sections," *Journal of the Structural Division, ASCE*, Vol. 98, No. ST5, Proc. Paper 8902, pp. 1035–1052, May, (1972).

Chen, W. F. and Santathadaporn, S., "Review of Column Behavior Under Biaxial Loading," *Journal of the Structural Division, ASCE*, Vol. 94, No. ST12, Proc. Paper 6316, pp. 2999–3021, December, (1968).

Chen, W. F. and Santathadaporn, S., "Curvature and the Solution of Eccentrically Loaded Columns," *Journal of the Engineering Mechanics Division, ASCE*, Vol. 95, No. EM1, Proc. Paper 6382, pp. 21–40, February, (1969).

Culver, C. G., "Exact Solution of the Biaxial Bending Equations," *Journal of the Structural Division, ASCE*, Vol. 92, No. ST2, Proc. Paper 4772, pp. 63–83, April, (1966a).

Culver, C. G., "Initial Imperfections in Biaxial Bending," *Journal of the Structural Division, ASCE*, Vol. 92, No. ST3, Proc. Paper 4846, pp. 119–135, June, (1966b).

Dabrowski, R., "Dünnwandige Stäbe unter Zweiachsig Aussermittigem Druck," *Der Stahlbau*, Wilhelm Ernst and Son, Berlin-Wilmersdorf, Germany, December, (1961).

Ellis, J. S., Jury, E. J. and Kirk, D. W., "Ultimate Capacity of Steel Columns Loaded Biaxially," *Transactions, Engineering Institute of Canada*, Vol. 7, No. A-2, pp. 3–11, February, (1964).

Galambos, T. V., "Inelastic Lateral-Torsional Buckling of Eccentrically Loaded Wide-Flange Columns." Thesis presented to Lehigh University, Bethlehem, Pa., in partial fulfillment of the requirements for the degree of Doctor of Philosophy (University Microfilms, Inc., Ann Arbor), (1959).

Goodier, J. N., "Buckling of Compressed Bars by Torsion and Flexure," *Cornell University Engineering Experiment Station Bulletin* No. 27, Ithaca, N. Y., December, (1941).

Goodier, J. N., "Flexural-Torsional Buckling of Bars of Open Section Under Bending, Eccentric Thrust or Torsional Loads," *Cornell University Engineering Experiment Station Bulletin* No. 28, Ithaca, N. Y., January, (1942a).

Goodier, J. N., "Torsional and Flexural Buckling of Bars of Thin Walled Open Section Under Compressive and Bending Loads," *Journal of Applied Mechanics*, Vol. 9, No. 3, September, pp. A-103–A-107, (1942b).

Harstead, G. A., "Elasto-Plastic Behavior of Columns Subjected to Biaxial Bending." Dissertation presented to New York University at the Bronx, N. Y., in partial fulfillment of the requirements for the degree of Doctor of Philosophy, June, (1966).

Harstead, G. A., Birnstiel, C. and Leu, K. C., "Inelastic H-Columns Under Biaxial Bending," *Journal of the Structural Division, ASCE*, Vol. 94, No. ST10, Proc. Paper 6173, pp. 2371–2398, October, (1968).

Horne, M. R., *The Plastic Design of Columns*, British Constructional Steelwork Association Publication No. 23, London, (1964).

Johnston, B. G., "Lateral Buckling of I-Section Columns with Eccentric Loads in Plane of Web," *Journal of Applied Mechanics*, p. A-176, (1941).

Johnston, B. G., "Buckling Behavior Above the Tangent Modulus Load," *Journal of the Engineering Mechanics Division, ASCE*, Vol. 87, No. EM6, Proc. Paper 3019, pp. 79–99, December, (1961).

Johnston, B. G., Editor, *Guide to Stability Design Criteria for Metal Structures*, 3rd edn, John Wiley and Sons, New York, (1976).

Kappus, R., "Twisting Failure of Centrally Loaded Open-Section Columns in the Elastic Range," *Luftfahrtforschung*, Vol. 14, No. 9, September, pp. 444–457, (1937). Translated by J. Vanier, National Advisory Committee for Aeronautics, No. 851, (1938).

Klöppel, K. and Winkelmann, E., "Experimental and Theoretical Studies on the Load Carrying Capacity of Biaxially Eccentrically Compressed Steel Members," *Der Stahlbau*, Vol. 31, No. 2, p. 33, February–April, (1962).

Kollbrunner, C. F. and Meister, M., *Knicken, Biegedrillknicken, Kippen*, Springer, Berlin, (1961).

Milner, H. R., "The Elastic Plastic Stability of Stanchions Bent About Two Axes." Dissertation presented to the College at the University of London, in partial fulfillment of the requirements for the degree of Doctor of Philosophy, December, (1965).

Pinadzhyan, V. V., "Problem of the Ultimate Load Capacity of Short Eccentrically Compressed Columns of H-Shaped Cross Section in the Case of Double Axis Eccentricity of the Force of Application," *Dokladi Armenian Academy of Science*, Vol. 21, No. 2, Nauk (in Russian), (1956).

Prawel, S. P. and Lee, G. C., "Biaxial Flexure of Columns by Analog Computers," *Journal of the Engineering Mechanics Division, ASCE*, Vol. 90, No. EM1, Proc. Paper 3805, pp. 83–111, February, (1964).

Santathadaporn, S. and Chen, W. F., "Analysis of Biaxially Loaded H-Columns," *Journal of the Structural Division, ASCE*, Vol. 99, No. ST3, Proc. Paper 9621, pp. 491–509, March, (1973).

Sharma, S. S. and Gaylord, E. H., "Strength of Steel Columns with Biaxially Eccentric Load," *Journal of the Structural Division, ASCE*, Vol. 95, No. ST12, Proc. Paper 6960, pp. 2797–2812, December, (1969).

Soltis, L. A. and Christiano, P., "Finite Deformation of Biaxially Loaded Columns," *Journal of the Structural Division, ASCE*, Vol. 98, No. ST12, Proc. Paper 9407, pp. 2647–2662, December, (1972).

Tebedge, N. and Chen, W. F., "Design Criteria for Steel H-Columns Under Biaxial Loading," *Journal of the Structural Division, ASCE*, Vol. 100, No. ST3, Proc. Paper 10400, pp. 579–598, March, (1974).

Thürlimann, B., "Deformations of and Stresses in Initially Twisted and Eccentrically Loaded Columns of Thin-Walled Open Cross-Section," *Report No. 3* to the Column Research Council and the Rhode Island Department of Public Works, Graduate Division of Applied Mathematics, *Report Nos. E797-3, E696-3*, Brown University, Providence, Rhode Island, June, (1953).

Timoshenko, S. P., "Theory of Bending, Torsion and Buckling of Thin-Walled Members of Open Cross Section," *Journal of the Franklin Institute*, Philadelphia, Pa., Vol. 239, No. 3, March, p. 201; No. 4, April, p. 249; No. 5, May, p. 343, (1945).

Timoshenko, S. P. and Gere, J. M., "Theory of Elastic Stability," 2nd edn, McGraw-Hill Book Co., Inc., New York, (1961).

Vinnakota, S. and Aoshima, Y., "Inelastic Behavior of Rotationally Restrained Columns under Biaxial Bending," *The Structural Engineer*, Vol. 52, No. 7, pp. 245–255, London, England, July (1974a).

Vinnakota, S. and Aoshima, Y., "Spatial Behavior of Rotationally and Directionally Restrained Beam-Columns," International Association for Bridge and Structural Engineering, Publications, Vol. 34-II, pp. 169–194, Zurich, (1974b).

Vinnakota, S. and Aysto, P., "Inelastic Spatial Stability of Restrained Beam-Columns," *Journal of the Structural Division, ASCE*, Vol. 100, No. ST11, Proc. Paper 10919, pp. 2235–2254, November, (1974).

Virdi, K. S. and Dowling, P. J., "The Ultimate Strength of Biaxially Restrained Columns," *Proceedings, Institute of Civil Engineers*, Part 2, No. 61, pp. 41–58, March, (1976).

Vlasov, V. Z., "Thin-Walled Elastic Beams," National Science Foundation, Washington, D.C. and the Department of Commerce by the Israel Program for Scientific Translations, Jerusalem, 2nd edn, (1961).

von Kármán, T., "Die Knickfestigkeit Gerader Stäbe," *Physikalische Zeitschrift*, Vol. 9, p. 136, (1908).

von Kármán, T., "Untersuchungen Über Knickfestigkeit," Mitteilungen Über Forschungsarbeiten auf dem Gebiete des Ingenieurwesens, No. 81, Berlin, (1910).

Wagner, H., "Torsion and Buckling of Open Sections," 25th Anniversary Publication, Technische Hochschule, Danzig, 1904–1929, pp. 329–343, (1936); translated as Technical Memorandum No. 807, U.S. National Advisory Committee for Aeronautics.

Wagner, H. and Pretschner, W., "Torsion and Buckling of Open Sections," *Luftfahrtforschung*, Vol. 1, December, pp. 174–180; translated by J. Vanier, Technical Memorandum No. 784, U.S. National Advisory Committee for Aeronautics, (1936).

Wood, R. H., "A New Approach to Column Design," Building Research Establishment Report, Her Majesty's Stationery Office, (1974).

Young, B. W., "Steel Column Design," *The Structural Engineer*, Vol. 51, No. 9, pp. 323–336, London, England, September, (1973).

2

REVIEW OF BENDING AND TORSION

2.1 INTRODUCTION

Prior to the study of the space behavior of beam-columns, basic equations for biaxial bending and torsion of thin-walled elastic beam-columns are studied first in this chapter.

Assumptions adopted in the theory of bending and torsion of thin-walled members are (Fig. 2.1)

1. Elastic homogeneous material: E = Young's modulus, G = shearing modulus, ν = Poisson's ratio
2. Long prismatic member: $D/L, B/L < 0.1$
3. Thin-walled cross section: $t/D, t/B < 0.1$
4. No shear deformation
5. No distortion in cross section.

The distortion of a thin-walled cross section may be produced by the longitudinal bending stress and/or shear stress, but this is neglected in the present analysis.

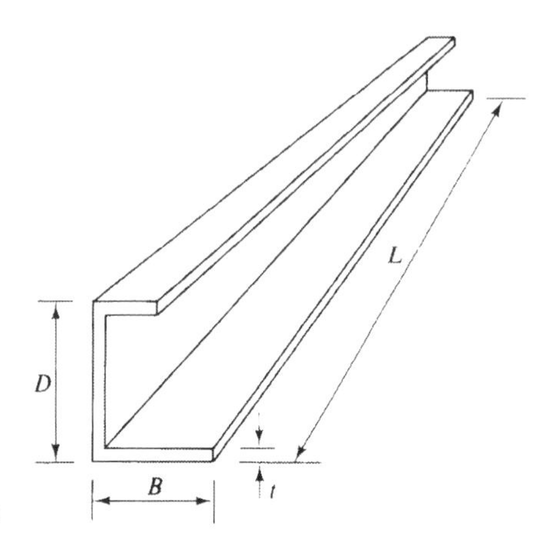

FIGURE 2.1
Thin-walled beam

In the following, the basic equations are derived for tension and bending, pure (uniform) torsion, and warping (nonuniform) torsion, separately. Since only elastic beam-columns are treated here, the basic equations of an actual member under combined bending and torsion can be obtained by superimposing these first order effects. The second order effect due to axial thrust is then added afterwards, because this effect is usually coupled with other deformations and results in nonlinear and second order terms in the basic governing equations.

As for tension and bending theory, equations derived previously for an in-plane beam-column in Vol. 1 of this book can be used. Emphasis is therefore placed here on biaxial bending about general axes which is quite different from the in-plane bending cases.

Theory of torsion consists of two parts, i.e., St. Venant (uniform) torsion and warping torsion (nonuniform torsion). The torsional response of a member differs depending on the degree of warping restraint at ends and the shape of a cross section. Saint Venant torsion is predominated for members with thin-walled closed section, while warping torsion is predominated for thin-walled open cross section. Therefore, equations of torsion are derived separately for these two cases.

Since the basic theory of bending and torsion of thin-walled elastic beams has been treated in detail in many books, such as Timoshenko and Goodier (1951), Kollbrunner and Basler (1966), Vlasov (1961) and Galambos (1968), we therefore review here only those basic equations which are needed in later chapters.

2.2 BIAXIAL BENDING OF BEAMS

Since it is assumed that there is no distortion of the shape of a cross section nor shear deformation, and further, warping deformation is to be considered separately later, it can be assumed here that the plane before deformation remains plane after deformation due to bending. Consider an arbitrary beam cross section as shown in Fig. 2.2. The forces acting on the cross section are:

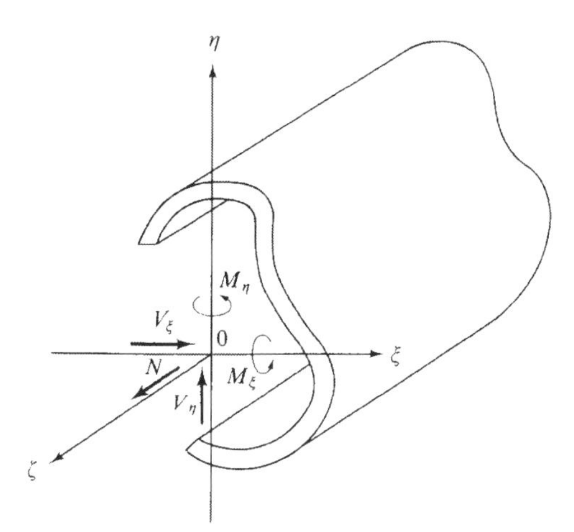

FIGURE 2.2
Forces acting on cross section

normal force (tension):	N
biaxial bending moments:	M_ξ, M_η
biaxial shear forces:	V_ξ, V_η

in which the bending moments are considered positive by the right-hand screw rule on the front face. Normal force and shear forces are considered positive in the positive direction of ξ, η and ζ. These moments and forces produce normal stress and shear stress in the thin wall of the beam. At this stage, the axes ξ and η can be arbitrary.

2.2.1 Normal Stress Due to Bending

From the assumption that plane remains plane, the normal strain is expressed in the form

$$\varepsilon = \varepsilon_0 + \Phi_\xi \eta - \Phi_\eta \xi = (1 \quad \eta \quad -\xi) \begin{Bmatrix} \varepsilon_0 \\ \Phi_\xi \\ \Phi_\eta \end{Bmatrix} \tag{2.1}$$

in which the notations ( ) and { } represent row and column vectors respectively and

Φ_ξ, Φ_η = curvatures about ξ and η axes, respectively
ε_0 = normal strain at the origin 0.

The normal stress is

$$\sigma = E\varepsilon = E(1 \quad \eta \quad -\xi) \begin{Bmatrix} \varepsilon_0 \\ \Phi_\xi \\ \Phi_\eta \end{Bmatrix} \tag{2.2}$$

The three generalized strains $(\varepsilon_0, \Phi_\xi, \Phi_\eta)$ of the cross section are determined from the three generalized stresses (N, M_ξ, M_η),

$$\begin{Bmatrix} N \\ M_\xi \\ M_\eta \end{Bmatrix} = \begin{Bmatrix} \int \sigma \, dA \\ \int \sigma\eta \, dA \\ -\int \sigma\xi \, dA \end{Bmatrix} = E \begin{bmatrix} A & S_y & -S_x \\ S_y & I_x & -I_{xy} \\ -S_x & -I_{xy} & I_y \end{bmatrix} \begin{Bmatrix} \varepsilon_0 \\ \Phi_\xi \\ \Phi_\eta \end{Bmatrix} \tag{2.3}$$

in which A, S_x, S_y, I_x, I_y and I_{xy} are section properties defined by

$$\begin{aligned} A &= \int dA, \qquad & S_x &= \int \xi \, dA, \qquad & S_y &= \int \eta \, dA \\ I_x &= \int \eta^2 \, dA, \qquad & I_y &= \int \xi^2 \, dA, \qquad & I_{xy} &= \int \xi\eta \, dA \end{aligned} \tag{2.4}$$

Solving Eq. (2.3) for the generalized strains $(\varepsilon_0, \Phi_\xi, \Phi_\eta)$ and making use of Eq. (2.2), one obtains the general expression for the normal stress σ due to bending and tension in terms of the generalized stresses (N, M_ξ, M_η),

$$\sigma = E\varepsilon = (1 \;\; \eta \;\; -\xi) \begin{bmatrix} A & S_y & -S_x \\ S_y & I_x & -I_{xy} \\ -S_x & -I_{xy} & I_y \end{bmatrix}^{-1} \begin{Bmatrix} N \\ M_\xi \\ M_\eta \end{Bmatrix} \tag{2.5}$$

As a special case, if the axes ξ and η are selected to be the centroidal coordinate of the section, i.e., $S_x = 0$ and $S_y = 0$, then Eq. (2.5) is simplified as

$$\begin{aligned} \sigma &= \frac{1}{A(I_x I_y - I_{xy}^2)} (1 \;\; \eta \;\; -\xi) \begin{bmatrix} I_x I_y - I_{xy}^2 & 0 & 0 \\ 0 & I_y A & I_{xy} A \\ 0 & I_{xy} A & I_x A \end{bmatrix} \begin{Bmatrix} N \\ M_\xi \\ M_\eta \end{Bmatrix} \\ &= \frac{N}{A} + \frac{I_y M_\xi + I_{xy} M_\eta}{I_x I_y - I_{xy}^2} \eta - \frac{I_x M_\eta + I_{xy} M_\xi}{I_x I_y - I_{xy}^2} \xi \end{aligned} \tag{2.6}$$

Further, if the axes ξ and η are the principal axes of the cross section, i.e., $I_{xy} = 0$, then Eq. (2.6) turns out to be the well-known expression for the normal stress due to bending and tension,

$$\sigma = \frac{N}{A} + \frac{M_\xi}{I_x} \eta - \frac{M_\eta}{I_y} \xi \tag{2.7}$$

The moment of inertia matrix becomes a tensor if it is written in the form given in Eq. (2.3), or

$$[I] = \begin{bmatrix} I_x & -I_{xy} \\ -I_{xy} & I_y \end{bmatrix} \tag{2.8}$$

then it follows the tensor transformation rule, i.e.,

$$[I'] = [R][I][R]^T \tag{2.9}$$

in which $[R]$ is the rotation matrix related with rotation of coordinate axes ξ and η by the angle θ (Fig. 2.3),

$$[R] = \begin{bmatrix} \cos\theta & \sin\theta \\ -\sin\theta & \cos\theta \end{bmatrix} \tag{2.10}$$

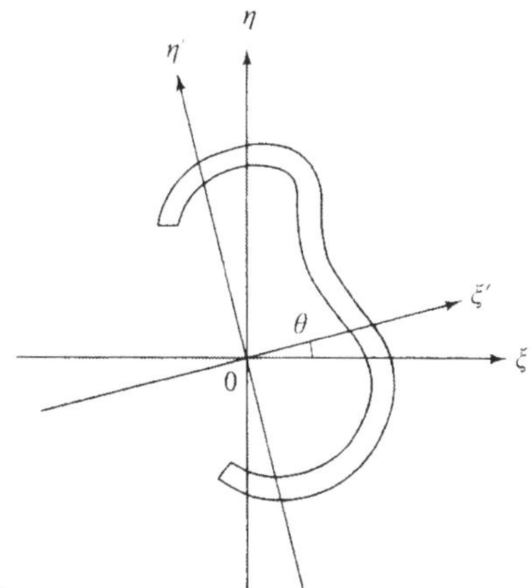

FIGURE 2.3
Rotation of coordinate axes

The principal moments of inertia I_x and I_y used in Eq. (2.7) are denoted by I_x^* and I_y^* which are the elements of the diagonalized inertia tensor by the transformation of Eq. (2.8),

$$\begin{bmatrix} I_x^* & 0 \\ 0 & I_y^* \end{bmatrix} = \begin{bmatrix} \cos\theta & \sin\theta \\ -\sin\theta & \cos\theta \end{bmatrix} \begin{bmatrix} I_x & -I_{xy} \\ -I_{xy} & I_y \end{bmatrix} \begin{bmatrix} \cos\theta & -\sin\theta \\ \sin\theta & \cos\theta \end{bmatrix} \tag{2.11}$$

From which the rotation angle required for the principal direction is computed by

$$2\theta^* = \tan^{-1} \frac{2I_{xy}}{I_x - I_y} \tag{2.12}$$

It is also known that the principal moments of inertia I_x^* and I_y^* are the eigenvalues of the inertia tensor, i.e., the solutions of the equation

$$\begin{vmatrix} I_x - I^* & -I_{xy} \\ -I_{xy} & I_y - I^* \end{vmatrix} = 0 \tag{2.13}$$

2.2.2 Shear Stress Due to Bending

Consider a thin-walled beam element dz shown in Fig. 2.4. The differences in bending moments at the ends give the shear forces,

$$V_\eta = \frac{\partial M_\xi}{\partial z} \quad \text{and} \quad V_\xi = -\frac{\partial M_\eta}{\partial z} \tag{2.14}$$

Taking the new coordinate s along the contour of the cross section starting from reference point A, a plate element $abcd$ can be cut out. Stresses acting on the element are the normal stresses σ on the transverse edges and shearing stresses τ on all edges.

From the equilibrium condition of the plate element *abcd*, one obtains,

$$\tau_{zs} = \tau_{sz} \qquad (= \tau \text{ hereafter})$$

$$\frac{\partial \tau t}{\partial z} = 0 \qquad (\tau t = \text{constant in } z) \tag{2.15}$$

$$\frac{\partial \tau t}{\partial s} + \frac{\partial \sigma t}{\partial z} = 0$$

Since the normal stress is given by Eq. (2.6), the shear stress is solved from the third equation of Eq. (2.15) as

$$\begin{aligned} \tau t &= -\int_0^s \frac{\partial \sigma t}{\partial z}\, ds + \tau_A t_A \\ &= -\frac{I_y V_\eta - I_{xy} V_\xi}{I_x I_y - I_{xy}^2} \int_0^s \eta t\, ds - \frac{I_x V_\xi - I_{xy} V_\eta}{I_x I_y - I_{xy}^2} \int_0^s \xi t\, ds + \tau_A t_A \end{aligned} \tag{2.16}$$

for which Eqs. (2.14) were used.

If the coordinate axes are selected in the principal direction of the cross section, $I_{xy} = 0$, Eq. (2.16) is reduced to

$$\tau t = -\frac{V_\eta}{I_x} \int_0^s \eta t\, ds - \frac{V_\xi}{I_y} \int_0^s \xi t\, ds + \tau_A t_A \tag{2.17}$$

The shear flow at the reference point A, i.e., $\tau_A t_A$, must be known first before the bending shear flow distribution along the contour can be calculated from Eq. (2.16) or Eq. (2.17). For the case of an open thin-walled cross section, it is convenient to

FIGURE 2.4
Equilibrium of beam-element

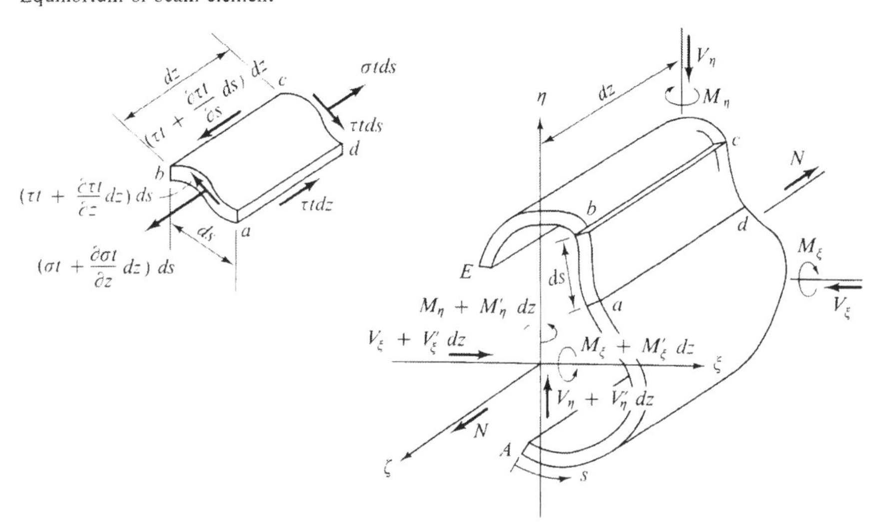

select the reference point A at one of the free edges where there is no shear stress $\tau_A = 0$. Thus

$$\tau t = -\frac{V_\eta}{I_x}\int_0^s \eta t\, ds - \frac{V_\xi}{I_y}\int_0^s \xi t\, ds \tag{2.18}$$

EXAMPLE 2-1 *Bending Shear Stress on an Open Cross Section* Determine the bending shear stress distribution on the antisymmetric channel section of constant thickness as shown in Fig. 2.5. The section is subjected to only a shear force $V_\eta = 10$ kips.

SOLUTION The centroidal coordinates are ξ and η shown in Fig. 2.5.

$$I_x = I_\xi = 114 \text{ in}^4, \quad I_y = I_\eta = 8.87 \text{ in}^4, \quad I_{xy} = I_{\xi\eta} = 11.6 \text{ in}^4$$

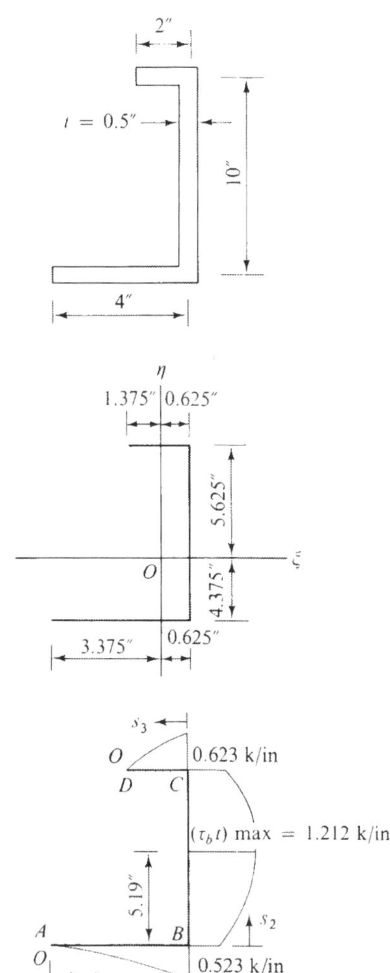

FIGURE 2.5
Bending shear distribution on open section (Example 2.1)

From Eq. (2.16) with $V_\xi = 0$,

$$\tau t = \frac{I_{\xi\eta} V_\eta t}{I_\xi I_\eta - I_{\xi\eta}^2} \int_0^s \xi \, ds - \frac{I_\eta V_\eta t}{I_\xi I_\eta - I_{\xi\eta}^2} \int_0^s \eta \, ds$$

$$= \frac{11.6 \times 10 \times 0.5}{877} \int_0^s \xi \, ds - \frac{8.87 \times 10 \times 0.5}{877} \int_0^s \eta \, ds$$

$$= 0.066 \int_0^s \xi \, ds - 0.0506 \int_0^s \eta \, ds$$

Start s at point A, $(\tau t)_A = 0$

$$\text{A-B:} \quad \tau t = (\tau t)_A + 0.066 \int_0^{s_1} (s_1 - 3.375) \, ds_1 - 0.0506 \int_0^{s_1} (-4.375) \, ds_1$$

$$= 0.033 s_1^2 - 0.00138 s_1$$

$$\text{B-C:} \quad \tau t = (\tau t)_B + 0.066 \int_0^{s_2} 0.625 ds_2 - 0.0506 \int_0^{s_2} (s_2 - 4.375) \, ds_2$$

$$= -0.0253 s_2^2 + 0.263 s_2 + 0.523$$

$$\text{C-D:} \quad \tau t = (\tau t)_C + 0.066 \int_0^{s_3} (0.625 - s_3) \, ds_3 - 0.0506 \int_0^{s_3} 5.625 \, ds_3$$

$$= -0.033 s_3^2 - 0.243 s_3 + 0.623$$

It is known that at point D ($s_3 = 2$) the shear flow vanishes $(\tau t)_D = 0$.

For the case of a closed cross section, one cannot find, in general, such a convenient reference point. In order to solve the unknown shear flow at a point, we need one more condition. The compatibility of normal displacement, w (warping displacement), around the contour of the cross section, provides this condition, i.e.,

$$\oint \frac{\partial w}{\partial s} \, ds = 0 \tag{2.19}$$

The symbol $\oint ds$ indicates the integration over the entire contour of the closed cross section. Shear strain and displacement are related by

$$\gamma_{sz} = \frac{\partial u_s}{\partial z} + \frac{\partial w}{\partial s} \tag{2.20}$$

in which u_s is the displacement in the direction of s. Using Eq. (2.20), Eq. (2.19) becomes

$$\oint \left[\gamma_{sz} - \frac{\partial u_s}{\partial z} \right] ds = 0 \tag{2.21}$$

Since twisting of the cross section will be considered separately and superimposed later, the only displacements of the section are translations $\bar{u}_\xi$ and $\bar{u}_\eta$ in the two coordinate directions. Thus, the displacement in the direction tangent to the contour u_s is given by

$$u_s = \bar{u}_\xi \cos(\xi, s) + \bar{u}_\eta \cos(\eta, s) \tag{2.22}$$

in which cos (,) indicates the directional cosine between the coordinate axes. Now Eq. (2.21) is reduced to

$$\begin{aligned}\oint \gamma_{sz}\, ds &= \oint \frac{\partial u_s}{\partial z}\, ds \\ &= \frac{d}{dz}\left[\bar{u}_\xi \oint \cos(\xi, s)\, ds + \bar{u}_\eta \oint \cos(\eta, s)\, ds\right] \\ &= \frac{d}{dz}\left(\bar{u}_\xi \oint d\xi + \bar{u}_\eta \oint d\eta\right) = 0\end{aligned} \tag{2.23}$$

for which the relations $d\xi = \cos(\xi, s)\, ds$ and $d\eta = \cos(\eta, s)\, ds$ and $\oint d\xi = 0, \oint d\eta = 0$ are used. Since the shear strain γ_{sz} is directly related to the shear stress τ by the linear stress-strain relationship $\tau = G\gamma_{sz}$, it follows from Eq. (2.16) and Eq. (2.23) that

$$\begin{aligned}G\oint \gamma_{sz}\, ds &= 0 = \oint \tau\, ds \\ &= -\frac{I_y V_\eta - I_{xy} V_\xi}{I_x I_y - I_{xy}^2} \oint \frac{\int_0^s \eta t\, ds}{t}\, ds - \frac{I_x V_\xi - I_{xy} V_\eta}{I_x I_y - I_{xy}^2} \oint \frac{\int_0^s \xi t\, ds}{t}\, ds + \tau_A t_A \oint \frac{ds}{t}\end{aligned} \tag{2.24}$$

Thus the shear flow at the arbitrary reference point A is solved:

$$\tau_A t_A = \left[\frac{I_y V_\eta - I_{xy} V_\xi}{I_x I_y - I_{xy}^2} \oint \frac{\int_0^s \eta t\, ds}{t}\, ds + \frac{I_x V_\xi - I_{xy} V_\eta}{I_x I_y - I_{xy}^2} \oint \frac{\int_0^s \xi t\, ds}{t}\, ds\right] \Big/ \oint \frac{ds}{t} \tag{2.25}$$

The shear flow distribution along the contour is now determined from Eq. (2.16), using Eq. (2.25)

$$\begin{aligned}\tau t &= \frac{I_y V_\eta - I_{xy} V_\xi}{I_x I_y - I_{xy}^2}\left(\oint \frac{\int_0^s \eta t\, ds}{t}\, ds \Big/ \oint \frac{ds}{t} - \int_0^s \eta t\, ds\right) \\ &\quad + \frac{I_x V_\xi - I_{xy} V_\eta}{I_x I_y - I_{xy}^2}\left(\oint \frac{\int_0^s \xi t\, ds}{t}\, ds \Big/ \oint \frac{ds}{t} - \int_0^s \xi t\, ds\right)\end{aligned} \tag{2.26}$$

When the coordinate axes are in the principal direction, $I_{xy} = 0$, Eq. (2.26) is reduced to

$$\tau t = \frac{V_\eta}{I_x}\left(\oint \frac{\int_0^s \eta t\, ds}{t}\, ds \Big/ \oint \frac{ds}{t} - \int_0^s \eta t\, ds\right)$$
$$+ \frac{V_\xi}{I_y}\left(\oint \frac{\int_0^s \xi t\, ds}{t}\, ds \Big/ \oint \frac{ds}{t} - \int_0^s \xi t\, ds\right) \tag{2.27}$$

Further, if the reference point A is selected at zero shear point, the shear flow Eq. (2.26) becomes equal to that of an open section given in Eq. (2.18).

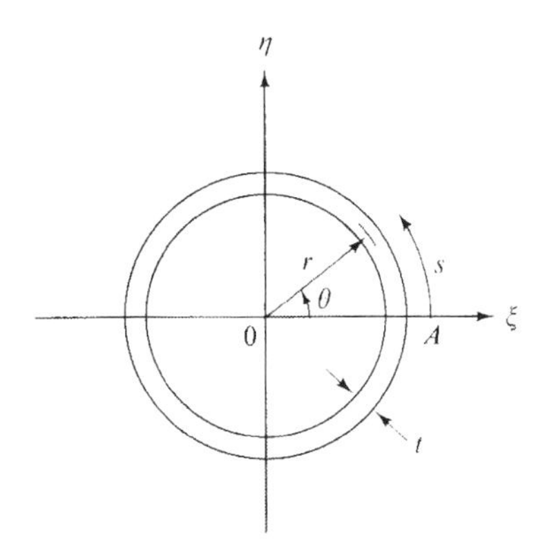

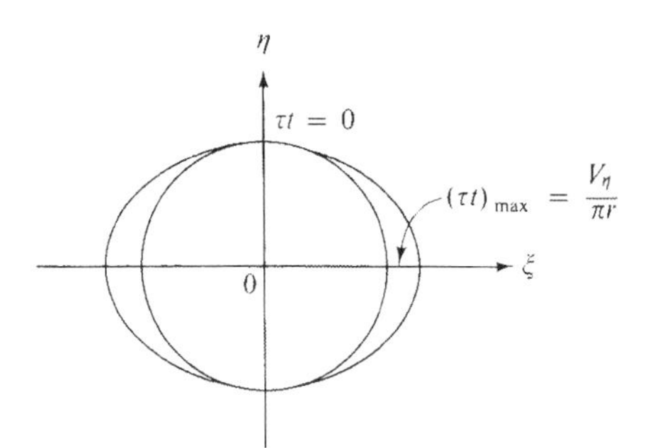

FIGURE 2.6
Bending shear distribution on closed section (Example 2.2)

EXAMPLE 2-2 *Bending Shear Stress on a Closed Cross Section* Determine the shear flow distribution in the circular tube shown in Fig. 2.6. Shear force due to bending is V_η only.

SOLUTION Select the reference point A on ξ axis. Take the polar coordinate,

$$\xi = r\cos\theta, \quad \eta = r\sin\theta, \quad s = r\theta$$

then

$$\int_0^s \xi t\, ds = r^2 t \int_0^\theta \cos\theta\, d\theta = r^2 t \sin\theta$$

$$\int_0^s \eta t\, ds = r^2 t \int_0^\theta \sin\theta\, d\theta = r^2 t(1 - \cos\theta)$$

$$I_x = I_\xi = \oint \eta^2 t\, ds = r^3 t \int_0^{2\pi} \sin^2\theta\, d\theta = \frac{1}{2} r^3 t \int_0^{2\pi} (1 - \cos 2\theta)\, d\theta = \pi r^3 t$$

$$I_y = I_\eta = \oint \xi^2 t\, ds = r^3 t \int_0^{2\pi} \cos^2\theta\, d\theta = \frac{1}{2} r^3 t \int_0^{2\pi} (1 + \cos 2\theta)\, d\theta = \pi r^3 t$$

$$I_{xy} = I_{\xi\eta} = \oint \xi\eta t\, ds = r^3 t \int_0^{2\pi} \sin\theta\cos\theta\, d\theta = \frac{1}{2} r^3 t \int_0^{2\pi} \sin 2\theta\, d\theta = 0$$

From Eq. (2.27), with $V_\xi = 0$, we have

$$\begin{aligned} \tau t &= \frac{V_\eta}{I_\xi}\left[\frac{1}{2\pi r}\oint\left(\int_0^s \eta t\, ds\right) ds - \int_0^s \eta t\, ds\right] \\ &= \frac{V_\eta}{\pi r^3 t}\left[\frac{1}{2\pi r} r^3 t \int_0^{2\pi} (1 - \cos\theta)\, d\theta - r^2 t(1 - \cos\theta)\right] \\ &= \frac{V_\eta}{\pi r} \cos\theta \end{aligned}$$

From which we see that the maximum shear flow is twice the average value

$$(\tau t)_{\text{max}} = 2(\tau t)_{\text{mean}} = 2\left(\frac{V_\eta}{2\pi r}\right)$$

2.2.3 Shear Center

So far we have assumed that the member only bends without twist. In order that this assumption holds true, resultant shear forces V on the cross section which is the integral of the shear stress over the area

$$\begin{aligned} V_\xi &= \int_0^E \tau \cos(\xi, s) t\, ds = \int_0^E \tau t\, d\xi \\ V_\eta &= \int_0^E \tau \cos(\eta, s) t\, ds = \int_0^E \tau t\, d\eta \end{aligned} \tag{2.28}$$

must pass through a specific point which is defined here as the *shear center* of the cross section. The symbol $\int_0^E ds$ indicates the integration over the entire open or closed cross section and $\cos(\xi, s)$ and $\cos(\eta, s)$ are the directional cosines of vector s. The location of the shear center $S(\xi_s, \eta_s)$ is determined from the condition

$$\int_0^E \tau\rho t\, ds = -V_\xi \eta_s + V_\eta \xi_s \tag{2.29}$$

in which ρ is the distance from the centroid of the cross section to the tangent of the contour ds (Fig. 2.7). Substituting τt of Eq. (2.26) into Eq. (2.29), one obtains

$$\begin{aligned}
&\frac{I_y V_\eta - I_{xy} V_\xi}{I_x I_y - I_{xy}^2}\left[\left(\int_0^E \rho\, ds\right)\left(\int_0^E \frac{\int_0^s \eta t\, ds}{t}\, ds \Big/ \int_0^E \frac{ds}{t}\right) - \int_0^E \rho\left(\int_0^s \eta t\, ds\right) ds\right] \\
&+ \frac{I_x V_\xi - I_{xy} V_\eta}{I_x I_y - I_{xy}^2}\left[\left(\int_0^E \rho\, ds\right)\left(\int_0^E \frac{\int_0^s \xi t\, ds}{t}\, ds \Big/ \int_0^E \frac{ds}{t}\right) - \int_0^E \rho\left(\int_0^s \xi t\, ds\right) ds\right] \\
&= -V_\xi \eta_s + V_\eta \xi_s
\end{aligned} \tag{2.30}$$

In order that Eq. (2.30) holds true for any combination of V_ξ and V_η, the shear center must be

$$\begin{aligned}
\xi_s &= \frac{I_y}{I_x I_y - I_{xy}^2}\left[\left(\int_0^E \rho\, ds\right)\left(\int_0^E \frac{\int_0^s \eta t\, ds}{t}\, ds \Big/ \int_0^E \frac{ds}{t}\right) - \int_0^E \rho\left(\int_0^s \eta t\, ds\right) ds\right] \\
&\quad - \frac{I_{xy}}{I_x I_y - I_{xy}^2}\left[\left(\int_0^E \rho\, ds\right)\left(\int_0^E \frac{\int_0^s \xi t\, ds}{t}\, ds \Big/ \int_0^E \frac{ds}{t}\right) - \int_0^E \rho\left(\int_0^s \xi t\, ds\right) ds\right] \\
\eta_s &= -\frac{I_x}{I_x I_y - I_{xy}^2}\left[\left(\int_0^E \rho\, ds\right)\left(\int_0^E \frac{\int_0^s \xi t\, ds}{t}\, ds \Big/ \int_0^E \frac{ds}{t}\right) - \int_0^E \rho\left(\int_0^s \xi t\, ds\right) ds\right] \\
&\quad + \frac{I_{xy}}{I_x I_y - I_{xy}^2}\left[\left(\int_0^E \rho\, ds\right)\left(\int_0^E \frac{\int_0^s \eta t\, ds}{t}\, ds \Big/ \int_0^E \frac{ds}{t}\right) - \int_0^E \rho\left(\int_0^s \eta t\, ds\right) ds\right]
\end{aligned} \tag{2.31}$$

It is to be noted that the elastic shear center is a sectional property which is independent of applied load. Let us now introduce a new coordinate ω defined by

$$\omega = \int_0^s \rho\, ds \qquad \text{or} \qquad d\omega = \rho\, ds \tag{2.32}$$

which is called *double sectorial area* or *unit warping* with respect to the centroid. From Fig. 2.7, it is known that $d\omega = \rho\, ds$ is equal to double of the triangular area ΔOPQ, thus, ω is double of the sectorial area OAP shown by the hatching in Fig. 2.7. For a closed section, the ω at the final point E is equal to double of the enclosed area A_0 of the cross section;

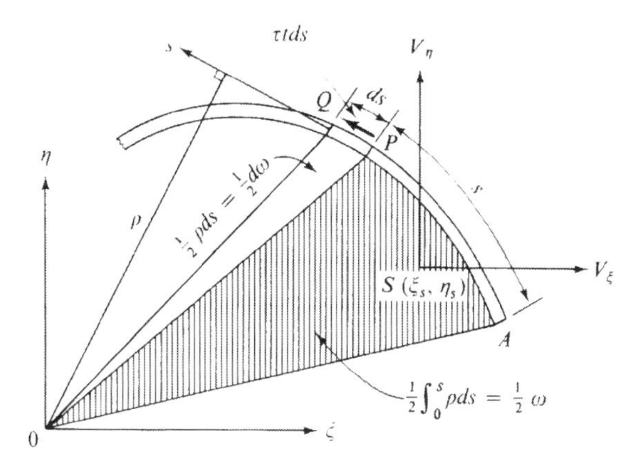

FIGURE 2.7
Unit warping and sectorial area

$$\omega_E = \int_0^E \rho \, ds = 2A_0 \tag{2.33}$$

The double sectorial area coordinate ω plays an important role in the analysis of torsion just as the rectangular coordinates ξ and η do in the bending analysis. Now we have three independent coordinates; ξ, η and ω. The warping products of inertia of section is defined by

$$\begin{aligned} I_{\omega y} &= \int_0^E \omega \eta t \, ds \\ I_{\omega x} &= \int_0^E \omega \xi t \, ds \end{aligned} \tag{2.34}$$

the last term in Eq. (2.31) can be integrated by parts and yields

$$\begin{aligned} \int_0^E \rho\left(\int_0^s \xi t \, ds\right) ds &= \int_0^E \left(\int_0^s \xi t \, ds\right) d\omega \\ &= \left(\int_0^E d\omega\right)\left(\int_0^E \xi t \, ds\right) - \int_0^E \omega \xi t \, ds \\ &= -I_{\omega x} \end{aligned}$$

in which

$$\int_0^E \xi t \, ds = 0$$

Similarly,

$$\int_0^E \rho\left(\int_0^s \eta t \, ds\right) ds = -I_{\omega y} \tag{2.35}$$

Thus the location of the shear center [Eq. (2.31)], is determined by

$$\xi_s = \frac{I_y}{I_x I_y - I_{xy}^2}\left[\left(\int_0^E d\omega\right)\left(\int_0^E \frac{\int_0^s \eta t\, ds}{t}\, ds \bigg/ \int_0^E \frac{ds}{t}\right) + I_{\omega y}\right]$$
$$- \frac{I_{xy}}{I_x I_y - I_{xy}^2}\left[\left(\int_0^E d\omega\right)\left(\int_0^E \frac{\int_0^s \xi t\, ds}{t}\, ds \bigg/ \int_0^E \frac{ds}{t}\right) + I_{\omega x}\right]$$
$$\eta_s = -\frac{I_x}{I_x I_y - I_{xy}^2}\left[\left(\int_0^E d\omega\right)\left(\int_0^E \frac{\int_0^s \xi t\, ds}{t}\, ds \bigg/ \int_0^E \frac{ds}{t}\right) + I_{\omega x}\right]$$
$$+ \frac{I_{xy}}{I_x I_y - I_{xy}^2}\left[\left(\int_0^E d\omega\right)\left(\int_0^E \frac{\int_0^s \eta t\, ds}{t}\, ds \bigg/ \int_0^E \frac{ds}{t}\right) + I_{\omega y}\right] \tag{2.36}$$

If the reference point A is selected at zero shear stress point, then Eq. (2.36) is reduced to

$$\xi_s = \frac{I_y I_{\omega y} - I_{xy} I_{\omega x}}{I_x I_y - I_{xy}^2}$$
$$\eta_s = -\frac{I_x I_{\omega x} - I_{xy} I_{\omega y}}{I_x I_y - I_{xy}^2} \tag{2.37}$$

Further, if the coordinate axes are in principal direction, $I_{xy} = 0$, then

$$\xi_s = \frac{I_{\omega y}}{I_x}, \quad \eta_s = -\frac{I_{\omega x}}{I_y} \tag{2.38}$$

EXAMPLE 2-3 *Shear Center of an Open Cross Section* Calculate the location of shear center for the anti-symmetric channel section studied in Example 2-1.

SOLUTION The double sectorial area $\omega = \int_0^s \rho\, ds$
Start s at point A, $\omega_A = 0$

$$\text{A-B:} \quad \omega = \omega_A + 4.375 s_1$$
$$\text{B-C:} \quad \omega = \omega_B + 0.625 s_2$$
$$\text{C-D:} \quad \omega = \omega_C + 5.625 s_3$$

The ω-distribution is shown in Fig. 2.8

$$I_{\omega x} = \int_A^D \omega \xi t\, ds = 39.2 \text{ in}^5$$
$$I_{\omega y} = \int_A^D \omega \eta t\, ds = 179 \text{ in}^5$$

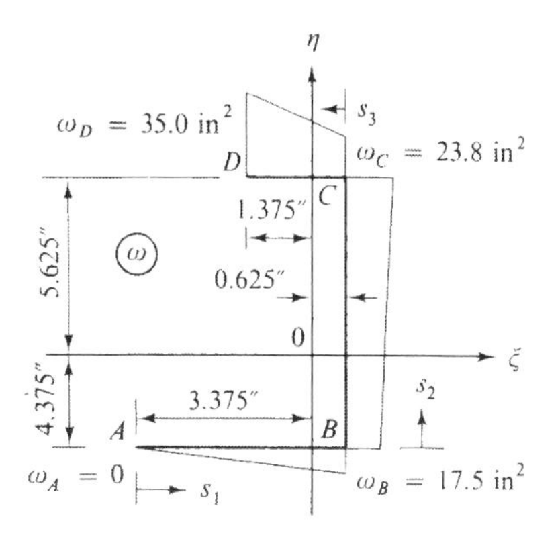

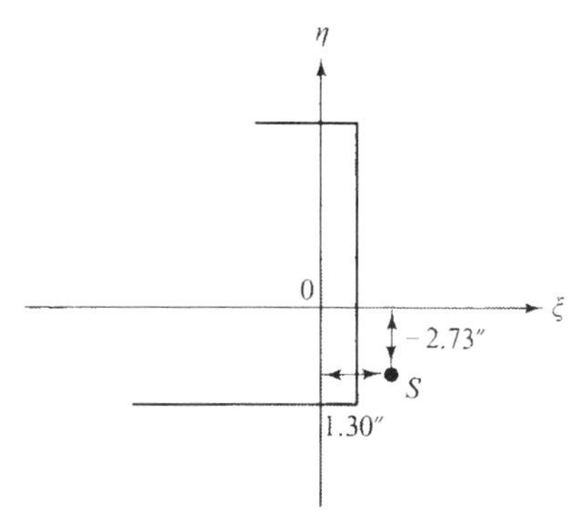

FIGURE 2.8
Calculation of shear center (Example 2.3)

The location of shear center from Eq. (2.37),

$$\xi_s = \frac{8.87 \times 179 - 11.6 \times 39.2}{114 \times 8.87 - (11.6)^2} = 1.30 \text{ in}$$

$$\eta_s = -\frac{114 \times 39.2 - 11.6 \times 179}{114 \times 8.87 - (11.6)^2} = -2.73 \text{ in}$$

2.3 UNIFORM TORSION

Consider a prismatic solid member which is subjected to equal twisting moment M_ζ at both ends (Fig. 2.9). Warping displacement of the cross section is assumed completely free so that the member may be twisted uniformly along the length with a constant twisting rate $\theta' = d\theta/d\zeta$ about the fixed point S_o which is defined here as the *torsion center* of the cross section.

2.3.1 Basic Equations

Take an arbitrary coordinate ξ and η on the cross section with its origin at the torsion center S_o as shown in Fig. 2.9. Since stresses do not alter along the longitudinal

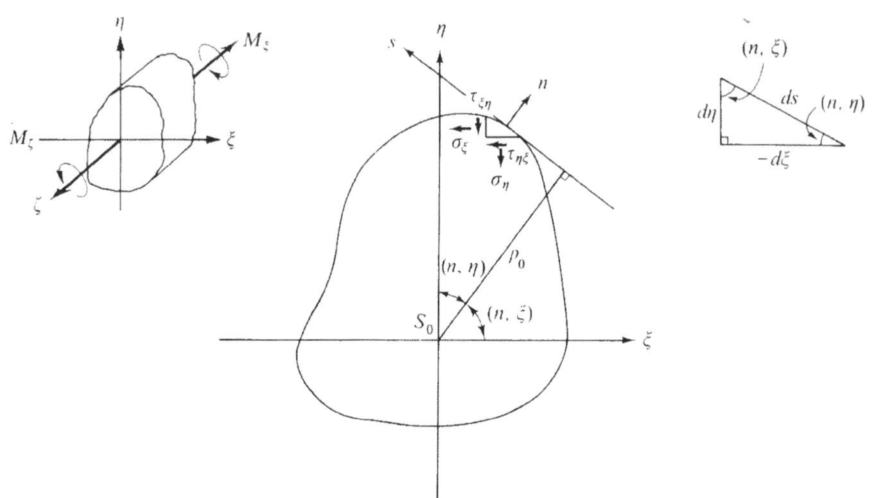

FIGURE 2.9
Boundary condition

axis, ζ, the basic equations in the theory of elasticity become as follows. The equilibrium equations are

$$\frac{\partial \sigma_\xi}{\partial \xi} + \frac{\partial \tau_{\xi\eta}}{\partial \eta} = 0$$

$$\frac{\partial \tau_{\xi\eta}}{\partial \xi} + \frac{\partial \sigma_\eta}{\partial \eta} = 0 \tag{2.39}$$

$$\frac{\partial \tau_{\zeta\xi}}{\partial \xi} + \frac{\partial \tau_{\zeta\eta}}{\partial \eta} = 0$$

Stress-strain relations are

$$\varepsilon_\xi = \frac{\partial u}{\partial \xi} = \frac{1}{E}[\sigma_\xi - \nu(\sigma_\eta + \sigma_\zeta)]$$

$$\varepsilon_\eta = \frac{\partial v}{\partial \eta} = \frac{1}{E}[\sigma_\eta - \nu(\sigma_\zeta + \sigma_\xi)]$$

$$\varepsilon_\zeta = \frac{\partial w}{\partial \zeta} = \frac{1}{E}[\sigma_\zeta - \nu(\sigma_\xi + \sigma_\eta)]$$

$$\gamma_{\xi\eta} = \frac{\partial u}{\partial \eta} + \frac{\partial v}{\partial \xi} = \frac{1}{G}\tau_{\xi\eta} \tag{2.40}$$

$$\gamma_{\zeta\eta} = \frac{\partial v}{\partial \zeta} + \frac{\partial w}{\partial \eta} = \frac{1}{G}\tau_{\zeta\eta}$$

$$\gamma_{\zeta\xi} = \frac{\partial w}{\partial \xi} + \frac{\partial u}{\partial \zeta} = \frac{1}{G}\tau_{\zeta\xi}$$

Boundary conditions on the surface are (Fig. 2.9)

$$\begin{aligned}
\sigma_\xi \cos(n,\xi) + \tau_{\xi\eta}\cos(n,\eta) &= 0 \\
\tau_{\xi\eta}\cos(n,\xi) + \sigma_\eta \cos(n,\eta) &= 0 \\
\tau_{\zeta\xi}\cos(n,\xi) + \tau_{\zeta\eta}\cos(n,\eta) &= 0
\end{aligned} \tag{2.41}$$

in which $\cos(n,\xi)$ and $\cos(n,\eta)$ are directional cosines of the normal vector n on the boundary surface.

2.3.2 St. Venant Stress Function

Since the displacement of the entire section is defined only by the twisting angle $\theta'\zeta$, about the torsion center S_o, displacements at an arbitrarily point $P(\xi,\eta,\zeta)$ on the cross section can be expressed in the form

$$\begin{aligned}
u &= -\theta'\zeta\eta \\
v &= \theta'\zeta\xi \\
w &= \theta'\phi(\xi,\eta)
\end{aligned} \tag{2.42}$$

in which $\theta' = d\theta/d\zeta$ and $\phi(\xi,\eta)$ is the *St. Venant stress function* related directly with the warping displacement w. Substituting Eq. (2.42) into Eq. (2.40) we obtain

$$\begin{aligned}
\tau_{\zeta\eta} &= G\theta'\left(\frac{\partial\phi}{\partial\eta} + \xi\right) \\
\tau_{\zeta\xi} &= G\theta'\left(\frac{\partial\phi}{\partial\xi} - \eta\right) \\
\sigma_\xi = \sigma_\eta &= \sigma_\zeta = \tau_{\xi\eta} = 0
\end{aligned} \tag{2.43}$$

From the last equation of Eq. (2.39)

$$\frac{\partial^2\phi}{\partial\xi^2} + \frac{\partial^2\phi}{\partial\eta^2} = \nabla^2\phi = 0 \tag{2.44}$$

The last boundary condition in Eq. (2.41) is

$$\left(\frac{\partial\phi}{\partial\xi} - \eta\right)\cos(n,\xi) + \left(\frac{\partial\phi}{\partial\eta} + \xi\right)\cos(n,\eta) = 0 \tag{2.45}$$

Since the directional cosines of the normal vector n are

$$\begin{aligned}
\cos(n,\xi) &= \frac{d\eta}{ds} \\
\cos(n,\eta) &= -\frac{d\xi}{ds}
\end{aligned} \tag{2.46}$$

thus

$$\left(\frac{\partial \phi}{\partial \xi} - \eta\right)\frac{d\eta}{ds} - \left(\frac{\partial \phi}{\partial \eta} + \xi\right)\frac{d\xi}{ds} = 0 \tag{2.47}$$

Now the elasticity problem is reduced to the following problem: find the stress function ϕ which satisfies the harmonic equation, Eq. (2.44), subjected to the boundary condition Eq. (2.47).

Once the stress function $\phi(\xi, \eta)$ is found, the corresponding torsional moment T can be determined directly by

$$\begin{aligned} T_{sv} &= \iint_A (\xi\tau_{\zeta\eta} - \eta\tau_{\zeta\xi})\, d\xi\, d\eta \\ &= G\theta' \iint_A \left(\xi^2 + \eta^2 + \xi\frac{\partial \phi}{\partial \eta} - \eta\frac{\partial \phi}{\partial \xi}\right) d\xi\, d\eta \end{aligned} \tag{2.48}$$

Defining the *St. Venant torsional constant* K_T by

$$K_T = \iint_A \left(\xi^2 + \eta^2 + \xi\frac{\partial \phi}{\partial \eta} - \eta\frac{\partial \phi}{\partial \xi}\right) d\xi\, d\eta \tag{2.49}$$

Eq. (2.48) can be written in the simple form

$$T_{sv} = GK_T\theta' \tag{2.50}$$

2.3.3 Prandtl's Stress Function

Prandtl's stress function Ψ is defined in relation with the shear stresses,

$$\tau_{\zeta\xi} = \frac{\partial \Psi}{\partial \eta}, \quad \tau_{\zeta\eta} = -\frac{\partial \Psi}{\partial \xi} \tag{2.51}$$

It is clear that this function satisfies the equilibrium condition, Eq. (2.39). Substituting Eq. (2.51) into Eq. (2.43), one obtains relationships between St. Venant stress function ϕ and Prandtl's stress function Ψ as

$$\begin{aligned} \frac{\partial \Psi}{\partial \eta} &= G\theta'\left(\frac{\partial \phi}{\partial \xi} - \eta\right) \\ -\frac{\partial \Psi}{\partial \xi} &= G\theta'\left(\frac{\partial \phi}{\partial \eta} + \xi\right) \end{aligned} \tag{2.52}$$

Eliminating ϕ from Eqs. (2.52), we obtain

$$\frac{\partial^2 \Psi}{\partial \xi^2} + \frac{\partial^2 \Psi}{\partial \eta^2} = \nabla^2 \Psi = -2G\theta' \tag{2.53}$$

Making use of Eqs. (2.52), the boundary condition, Eq. (2.47), turns out to be a simple expression

$$\frac{\partial \Psi}{\partial \eta}\frac{d\eta}{ds} + \frac{\partial \Psi}{\partial \xi}\frac{d\xi}{ds} = \frac{d\Psi}{ds} = 0 \tag{2.54}$$

This boundary condition implies that the stress function Ψ is a constant on the boundary

$$\Psi = \text{constant} \qquad \text{(on the boundary)} \tag{2.55}$$

Now the solution to the torsional problem is reduced to the following problem: find a function Ψ which satisfies Eq. (2.53) and takes a constant value on the given boundary.

The torsional moment is expressed by

$$\begin{aligned} T_{sv} &= \iint_A (\xi\tau_{\zeta\eta} - \eta\tau_{\zeta\xi})\, d\xi\, d\eta \\ &= -\iint_A \left(\xi\frac{\partial\Psi}{\partial\xi} + \eta\frac{\partial\Psi}{\partial\eta}\right) d\xi\, d\eta \end{aligned} \tag{2.56}$$

This surface integral can be transformed into a line integral by Green's theorem,

$$T_{sv} = -\oint \Psi[\xi \cos(n,\xi) + \eta\cos(n,\eta)]\, ds + 2\iint \Psi\, d\xi\, d\eta \tag{2.57}$$

Denoting the distance from the torsion center S_0 to the tangent of the boundary line ds by ρ_0, it is known from Fig. 2.9 that

$$\rho_0 = \xi\cos(n,\xi) + \eta\cos(n,\eta) \tag{2.58}$$

and from Eq. (2.46),

$$\rho_0\, ds = \xi\, d\eta - \eta\, d\xi \tag{2.59}$$

Now the torsional moment in Eq. (2.57) becomes

$$T_{sv} = -\oint \Psi\rho_0\, ds + 2\iint \Psi\, d\xi\, d\eta \tag{2.60a}$$

Since the stress function is a constant on a boundary,

$$T_{sv} = -2\Psi_0 A_0 + 2\iint \Psi\, d\xi\, d\eta \tag{2.60b}$$

in which A_0 is the area enclosed by the boundary, Ψ_0 is the value of Ψ on the boundary. Equation (2.60) implies that the torsional moment T_{sv} is equal to double of the volume surrounded by the surfaces $\zeta = \Psi$ and $\zeta = \Psi_0$ on the boundary.

When the cross section is composed of several closed contour lines as shown in Fig. 2.10, denote the constant stress functions as $\Psi_0, \Psi_1, \Psi_2 \ldots$ on these contours and the surrounded areas by $A_0, A_1, A_2 \ldots$ then Eq. (2.60) can be extended to this case in the form

$$T_{sv} = -2(\Psi_0 A_0 - \Psi_1 A_1 - \Psi_2 A_2 - \cdots) + 2\iint \Psi\, d\xi\, d\eta \tag{2.61}$$

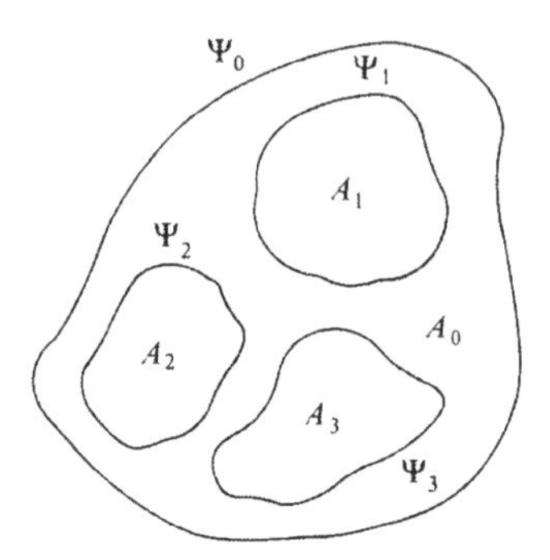

FIGURE 2.10
Constant stress functions on boundaries

The warping displacement is derived next. From the last two equations of Eq. (2.40)

$$\begin{aligned} w &= w_A + \int_A \left(\frac{\partial w}{\partial \xi} d\xi + \frac{\partial w}{\partial \eta} d\eta \right) \\ &= w_A + \int_A \left[\left(\frac{1}{G} \tau_{\zeta\xi} - \frac{\partial u}{\partial \zeta} \right) d\xi + \left(\frac{1}{G} \tau_{\zeta\eta} - \frac{\partial v}{\partial \zeta} \right) d\eta \right] \end{aligned} \tag{2.62}$$

in which w_A is the warping displacement at a reference point A. Substituting u, v from Eq. (2.42) and $\tau_{\zeta\xi}$, $\tau_{\zeta\eta}$ from Eq. (2.51), we find

$$w = w_A + \int_A \left[\frac{1}{G} \left(\frac{\partial \Psi}{\partial \eta} d\xi - \frac{\partial \Psi}{\partial \xi} d\eta \right) + \theta'(\eta \, d\xi - \xi \, d\eta) \right] \tag{2.63}$$

If the section has several closed boundaries, each value of $\Psi_0, \Psi_1, \ldots$ can be determined from the condition that the warping displacement is continuous on each boundary, i.e.,

$$\begin{aligned} 0 &= \oint \left[\frac{1}{G} \left(\frac{\partial \Psi}{\partial \eta} d\xi - \frac{\partial \Psi}{\partial \xi} d\eta \right) + \theta'(\eta \, d\xi - \xi \, d\eta) \right] \\ &= -\oint \left[\frac{1}{G} \left(\frac{\partial \Psi}{\partial \eta} \cos(n, \eta) \, ds + \frac{\partial \Psi}{\partial \xi} \cos(n, \xi) \, ds \right) + \theta' \rho_0 \, ds \right] \\ &= -\oint \left(\frac{1}{G} \frac{\partial \Psi}{\partial n} ds \right) - \theta' \oint \rho_0 \, ds \\ &= -\frac{1}{G} \oint \frac{\partial \Psi}{\partial n} ds - 2\theta' A_i \end{aligned} \tag{2.64}$$

in which Eq. (2.46) and Eq. (2.59) are used. Thus the value Ψ_i on the boundary C_i is determined by

$$\oint_{C_i} \frac{\partial \Psi_i}{\partial n} ds = -2\theta' G A_i \tag{2.65}$$

2.3.4 Membrane-Torsion Analogy

Consider the vertical deflection Z of a membrane which is subjected to an internal pressure p. Since tension R of the membrane is constant in all directions, the equilibrium equation of an element $d\xi\, d\eta$ in the direction perpendicular to the tangent plane is given by (Fig. 2.11)

$$R\, d\eta\left(-\frac{\partial^2 Z}{\partial \xi^2}\, d\xi\right) + R\, d\xi\left(-\frac{\partial^2 Z}{\partial \eta^2}\, d\eta\right) = p\, d\xi\, d\eta \tag{2.66}$$

or

$$\frac{\partial^2 Z}{\partial \xi^2} + \frac{\partial^2 Z}{\partial \eta^2} = -\frac{p}{R} \tag{2.67}$$

and the boundary condition is

$$Z = \text{constant} \qquad \text{(on the boundary)} \tag{2.68}$$

It is known that the differential equation and the boundary condition of the membrane problem, Eq. (2.67) and Eq. (2.68), are analogous to those of the torsion problem, Eq. (2.53) and Eq. (2.55). In this analogy, the parameters are compared to each other as follows:

Stress function Ψ to deflection Z

Twisting angle $2G\theta'$ to pressure-tension ratio p/R

Shear stress $\tau_{\zeta\xi} = \dfrac{\partial \Psi}{\partial \eta}$ to slope of membrane $\dfrac{\partial Z}{\partial \eta}$

Shear stress $\tau_{\zeta\eta} = -\dfrac{\partial \Psi}{\partial \xi}$ to slope of membrane $-\dfrac{\partial Z}{\partial \xi}$

Moment $T_{sv} = 2\displaystyle\iint \Psi\, d\xi\, d\eta$ to double of volume $2V = 2\displaystyle\iint Z\, d\xi\, d\eta$

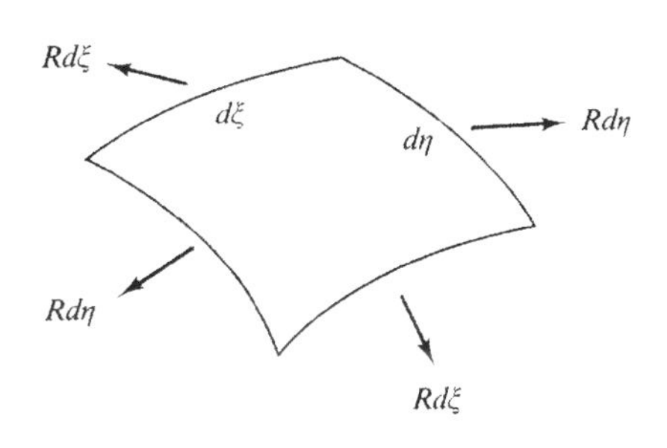

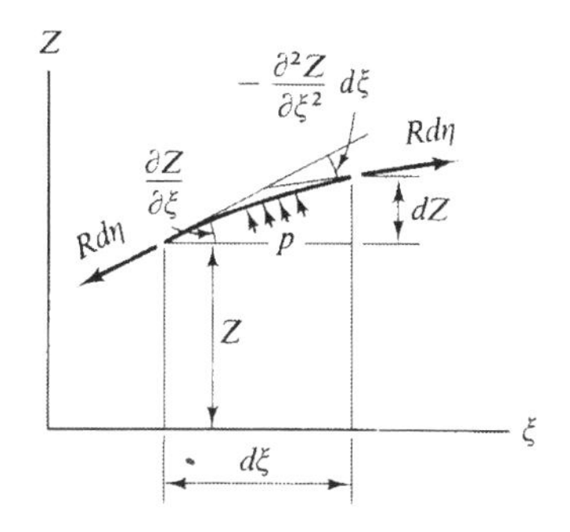

FIGURE 2.11
Equilibrium of membrane

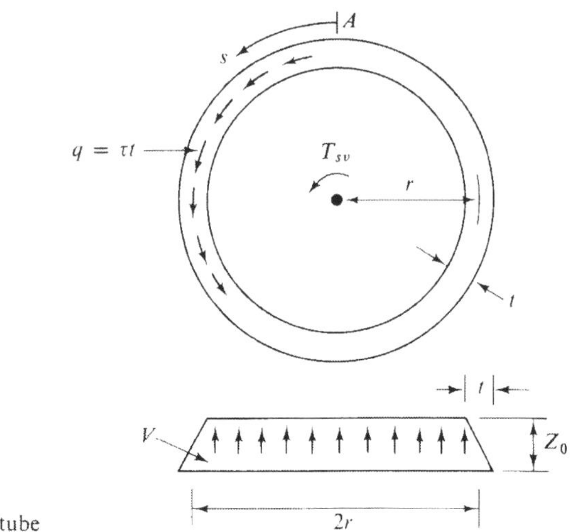

FIGURE 2.12
Torsion of Circular tube

2.3.5 Uniform Torsion of Thin-Walled Closed Cross Section

Assuming that the deflection of the membrane to be Z_0 which changes linearly through the thickness t as shown in Fig. 2.12, the shear stress, i.e., the slope of the membrane, becomes constant,

$$\tau = \frac{Z_0}{t} \tag{2.69}$$

The torsional moment is double of the volume under the membrane which is approximately given by

$$T_{sv} = 2V = 2A_0Z_0 \tag{2.70}$$

in which A_0 is the area enclosed by the middle surface. From Eqs. (2.69) and (2.70), the shear stress is directly obtained as

$$\tau = \frac{T_{sv}}{2A_0t} \tag{2.71}$$

from which it is known that shear flow along the contour has a constant value at all points,

$$q = \tau t = \frac{T_{sv}}{2A_0} = \text{constant} \tag{2.72}$$

The constant shear flow is known also from Eq. (2.15) since $\partial\sigma t/\partial z = \partial\sigma t/\partial\zeta = 0$, thus $\partial\tau t/\partial s = 0$. The torsional moment due to the shear flow q about the torsion center is

$$T_{sv} = \oint q\rho_0 \, ds = \oint q \, d\omega_0 \tag{2.73}$$

in which $d\omega_0 = \rho_0\, ds$ is the double sectorial area about the torsion center defined later in Eq. (2.91). Since the shear flow is constant, Eq. (2.73) has the value

$$T_{sv} = q \oint d\omega_0 = 2qA_0 \tag{2.74}$$

This gives the same results as Eq. (2.72).

The St. Venant torsional constant K_T can be obtained from energy consideration. The strain energy per unit length stored in the beam segment is

$$\begin{aligned} V_E &= \frac{1}{2}\int_V \tau\gamma\, dV \\ &= \frac{1}{2G}\oint \tau^2 t\, ds \\ &= \frac{q^2}{2G}\oint \frac{1}{t}\, ds \\ &= \frac{T_{sv}^2}{8GA_0^2}\oint \frac{1}{t}\, ds \end{aligned} \tag{2.75}$$

while the external work done is

$$W = \frac{1}{2} T_{sv}\theta' \tag{2.76}$$

Equating the strain energy to the external work, one obtains the torsional equation

$$T_{sv} = \frac{4A_0^2}{\oint \frac{1}{t}\, ds} G\theta' \tag{2.77}$$

from Eq. (2.50) the St. Venant torsional constant is obtained

$$K_T = 4A_0^2 \Big/ \int \frac{1}{t}\, ds \tag{2.78}$$

As an example, consider a circular tube section of mean radius r and thickness t (Fig. 2.12),

$$K_T = \frac{4(\pi r^2)^2}{\frac{1}{t}(2\pi r)} = 2\pi r^3 t \tag{2.79}$$

and the magnitude of shear stress is calculated from Eq. (2.71),

$$\tau = \frac{T_{sv}}{2(\pi r^2)t} = \frac{T_{sv}}{2\pi r^2 t} \tag{2.80}$$

2.3.6 Uniform Torsion of Thin-Walled Open Cross Section

Take a curvilinear coordinate system (s, n) as shown in Fig. 2.13. The tangential coordinate s starts from reference point A. Since the change of the function Ψ in s direction is small, Eq. (2.53) may be written as

$$\frac{\partial^2 \Psi}{\partial n^2} = -2G\theta' \tag{2.81}$$

On the boundary $n = \pm t/2$, $\Psi = 0$. Except near the end area, the stress function can be written as

$$\Psi = G\theta' \left(\frac{t^2}{4} - n^2 \right) \tag{2.82}$$

and the corresponding shear stress along the contour is

$$\tau_{zs} = -\frac{\partial \Psi}{\partial n} = -2G\theta' n \tag{2.83}$$

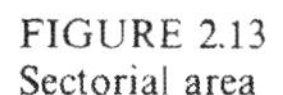
FIGURE 2.13
Sectorial area

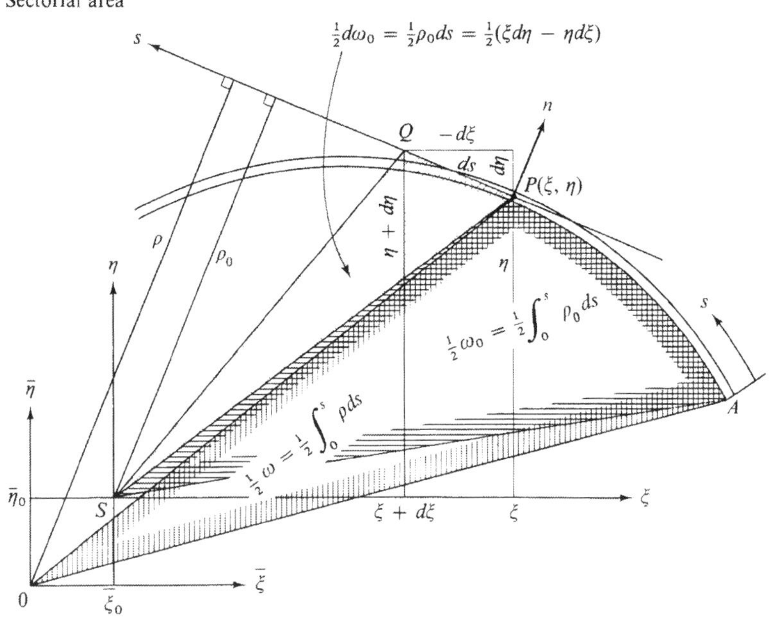

$$\tfrac{1}{2}d\omega_0 = \tfrac{1}{2}\rho_0 ds = \triangle SQP = \triangle SQP' - \triangle SP = \triangle SQ + \square QP - \triangle SP$$

$$= \tfrac{1}{2}(\xi + d\xi)(\eta + d\eta) + \tfrac{1}{2}(2\eta + d\eta)(-d\xi) - \tfrac{1}{2}\xi\eta$$

$$= \tfrac{1}{2}(\xi d\eta - \eta d\xi)$$

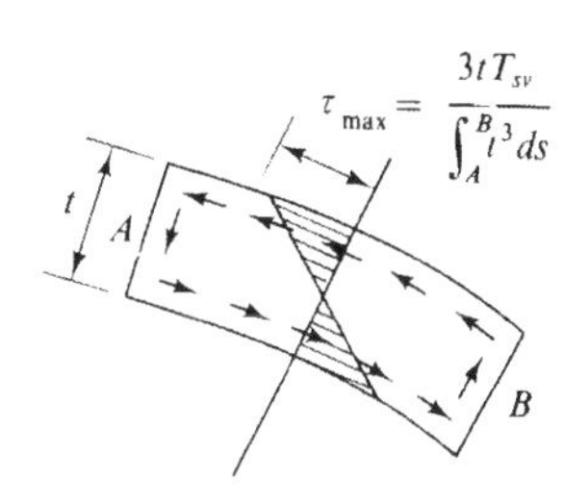

FIGURE 2.14
St. Venant shear stress

It is shown that the shear stress is linearly distributed through the wall thickness and the shear flow is shown in Fig. 2.14. The torsional moment is double the volume under the membrane

$$
\begin{aligned}
T_{sv} &= 2\int_0^E \int_{-t/2}^{t/2} \Psi \, dn \, ds = 2G\theta' \int_0^E \left[\int_{-t/2}^{t/2} \left(\frac{t^2}{4} - n^2 \right) dn \right] ds \\
&= \frac{1}{3} G\theta' \int_0^E t^3 \, ds
\end{aligned}
\tag{2.84}
$$

from which K_T is known as

$$
K_T = \frac{1}{3} \int_0^E t^3 \, ds \tag{2.85}
$$

The maximum shear stress on the surface is from Eq. (2.83) with $n = \pm t/2$,

$$
(\tau_{zs})_{\max} = \pm G\theta' t = \pm \frac{t}{K_T} T_{sv} \tag{2.86}
$$

For a cross section consisting of several flat, thin elements, the St. Venant torsional constant can be calculated by

$$
K_T = \sum \frac{b_i t_i^3}{3} \tag{2.87}
$$

in which b_i and t_i are the width and thickness of i-th plate element, respectively.

From Eq. (2.85), the St. Venant torsional constant for a circular tube section with a slit shown in Fig. 2.15 is

$$
K_T = \frac{1}{3} t^3 \oint ds = \frac{2\pi}{3} r t^3
$$

Comparing with the value of K_T for a closed tube given in Eq. (2.79), it is clear that St. Venant torsional rigidity GK_T of an open section is much less than that of a closed section. This fact is easily understood from the membrane analogy since the torsional rigidity is directly related to the enclosed area.

2.3.7 Warping of Open Cross Section

Warping deformation of closed sections are generally small and not important, therefore, only open cross sections are treated herein. Taking the local coordinate

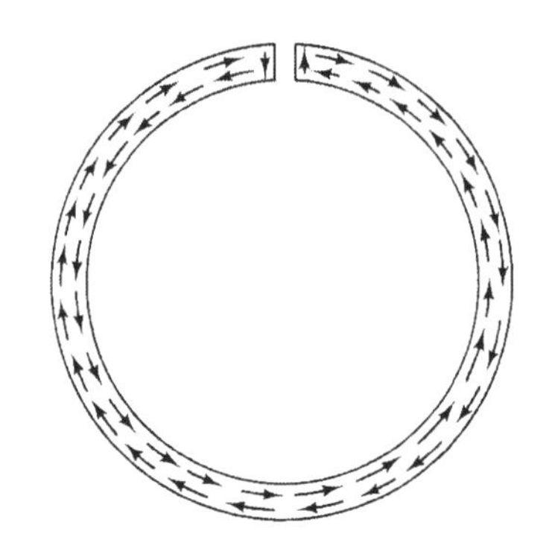

FIGURE 2.15
St. Venant shear stress on circular tube with a slit

system (s, n) as shown in Fig. 2.13 again, the warping expression, Eq. (2.63), is transformed into

$$w = w_A + \int_0^s \left[\frac{1}{G} \left(\frac{\partial \Psi}{\partial n} ds - \frac{\partial \Psi}{\partial s} dn \right) + \theta'(\eta \, d\xi - \xi \, d\eta) \right] \tag{2.88}$$

Substituting the stress function Ψ for a thin-walled open section given in Eq. (2.82), and making use of Eq. (2.59), Eq. (2.88) is reduced to

$$w = w_A - \theta' \int_0^s (2n \, ds + \rho_0 \, ds) \tag{2.89}$$

Since the n coordinate is in the direction of wall thickness and n-term can be neglected compared with ρ_0, the warping deformation of Eq. (2.89) may be written as

$$w = w_A - \theta' \int_0^s \rho_0 \, ds \tag{2.90}$$

Defining the new double sectorial area with respect to the torsion center, by

$$\omega_0 = \int_0^s \rho_0 \, ds \tag{2.91}$$

the warping is obtained by

$$w = w_A - \theta' \omega_0 \tag{2.92}$$

The two double sectorial areas ω_0 and ω are related. Using the centroidal coordinates, $\bar{\xi}$ and $\bar{\eta}$,

$$\xi = \bar{\xi} - \bar{\xi}_0 \qquad \text{and} \qquad \eta = \bar{\eta} - \bar{\eta}_0 \tag{2.93}$$

The relation between ω_0 and ω is

$$
\begin{aligned}
\omega_0 &= \int_0^s \rho_0 \, ds = -\int_0^s (\eta \, d\xi - \xi \, d\eta) \\
&= -\int_0^s [(\bar{\eta} - \bar{\eta}_0) \, d\bar{\xi} - (\bar{\xi} - \bar{\xi}_0) \, d\bar{\eta}] \\
&= \omega + \bar{\eta}_0\bar{\xi} - \bar{\eta}_0\bar{\xi}_A - \bar{\xi}_0\bar{\eta} + \bar{\xi}_0\bar{\eta}_A
\end{aligned}
\tag{2.94}
$$

in which $A(\bar{\xi}_A, \bar{\eta}_A)$ is the reference point and $S_0(\bar{\xi}_0, \bar{\eta}_0)$ is the torsion center which is identical to the shear center of the cross section. This will be shown later.

2.4 NONUNIFORM TORSION

In the case that warping deformation due to torsion is restrained, additional normal stress σ_w and shear stress τ_w are generated. Using the equations derived above, the warping normal strain is [Eq. (2.92)]

$$
\varepsilon_w = \frac{dw}{d\zeta} = w'_A - \omega_0\theta'' \tag{2.95a}
$$

Thus the warping normal stress is

$$
\sigma_w = E\varepsilon_w = E(w'_A - \omega_0\theta'') \tag{2.95b}
$$

in which prime (′) denotes differentiation with respect to ζ. Since there are no sectional forces applied except for torsional moment, the resultant normal force and bending moments with respect to the centroidal coordinates, $\bar{\xi}$ and $\bar{\eta}$ must be zero, i.e.,

$$
\begin{aligned}
N = 0 &= \int_0^E \sigma_w \, dA = E\int_0^E (w'_A - \omega_0\theta'') \, dA \\
M_{\bar{\xi}} = 0 &= \int_0^E \bar{\eta}\sigma_w \, dA = E\int_0^E (w'_A - \omega_0\theta'')\bar{\eta} \, dA \\
M_{\bar{\eta}} = 0 &= \int_0^E \bar{\xi}\sigma_w \, dA = E\int_0^E (w'_A - \omega_0\theta'')\bar{\xi} \, dA
\end{aligned}
\tag{2.96}
$$

Substituting ω_0 of Eq. (2.94) into Eq. (2.96) and using the sectional properties of

$$
\begin{aligned}
&\int_0^E t \, ds = A \\
&\int_0^E \bar{\xi}t \, ds = \int_0^E \bar{\eta}t \, ds = \int_0^E \bar{\xi}\bar{\eta}t \, ds = 0 \\
&\int_0^E \bar{\xi}^2 t \, ds = I_y, \quad \int_0^E \bar{\eta}^2 t \, ds = I_x \\
&\int_0^E \omega\bar{\xi}t \, ds = I_{\omega x}, \quad \int_0^E \omega\bar{\eta}t \, ds = I_{\omega y}
\end{aligned}
\tag{2.97}
$$

one obtains

$$w'_A A = \theta'' \int_0^E \omega_0 \, dA = \theta'' \left(\int_0^E \omega \, dA - \bar{\eta}_0 \bar{\xi}_A A + \bar{\xi}_0 \bar{\eta}_A A \right)$$

$$0 = \theta'' \int_0^E \omega_0 \bar{\eta} \, dA = \theta'' (I_{\omega y} - \bar{\xi}_0 I_x) \tag{2.98}$$

$$0 = \theta'' \int_0^E \omega_0 \bar{\xi} \, dA = \theta'' (I_{\omega x} + \bar{\eta}_0 I_y)$$

From the last two equations of Eq. (2.98), the location of the torsion center $S_0(\bar{\xi}_0, \bar{\eta}_0)$ is known as

$$\bar{\xi}_0 = \frac{I_{\omega y}}{I_x}, \quad \bar{\eta}_0 = -\frac{I_{\omega x}}{I_y} \tag{2.99}$$

Compared with Eq. (2.38), it is known that the torsion center $S_0(\bar{\xi}_0, \bar{\eta}_0)$ is identical to the shear center $S(\xi_s, \eta_s)$. We denote the shear center by $S(\xi_0, \eta_0)$ hereafter. From the first equation of Eq. (2.98),

$$w'_A = \frac{\theta''}{A} \int_0^E \omega_0 \, dA \tag{2.100}$$

Thus the warping stress Eq. (2.95) is obtained

$$\sigma_w = E\theta'' \left(\frac{1}{A} \int_0^E \omega_0 \, dA - \omega_0 \right) = E\omega_n \theta'' \tag{2.101}$$

in which ω_n is the *normalized unit warping* defined by

$$\omega_n = \frac{1}{A} \int_0^E \omega_0 \, dA - \omega_0 \tag{2.102}$$

The shear stress due to warping τ_w can be obtained from the equilibrium condition, Eq. (2.15)

$$\begin{aligned} \tau_w t &= \tau_A t_A - \int_0^s \frac{\partial \sigma_w t}{\partial \zeta} \, ds \\ &= \tau_A t_A - E\theta''' \int_0^s \omega_n \, dA \\ &= \tau_A t_A - E S_\omega \theta''' \end{aligned} \tag{2.103}$$

in which S_ω is the warping statical moment defined by

$$S_\omega = \int_0^s \omega_n \, dA \tag{2.104}$$

If the reference point A is selected at the edge, $\tau_A = 0$, then Eq. (2.103) gives

$$\tau_w t = -E S_\omega \theta''' \tag{2.105}$$

EXAMPLE 2-4 *Normalized Warping and its Static Moment* Calculate the distribution of normalized warping ω_n and its static moment S_ω for the section used in Example 2-1.

SOLUTION See Fig. 2.16.

From Example 2-3, the shear center is $S(1.30, -2.73)$

The double sectorial area $\omega_0 = \int_0^s \rho_0 \, ds$

Start s at point A, $\omega_{0A} = 0$

A-B: $\omega_0 = \omega_{0A} + 1.645 s_1$ $\omega_{0B} = 6.58 \text{ in}^2$

B-C: $\omega_0 = \omega_{0B} - 0.675 s_2$ $\omega_{0C} = -0.17 \text{ in}^2$

C-D: $\omega_0 = \omega_{0C} + 8.355 s_3$ $\omega_{0D} = 16.54 \text{ in}^2$

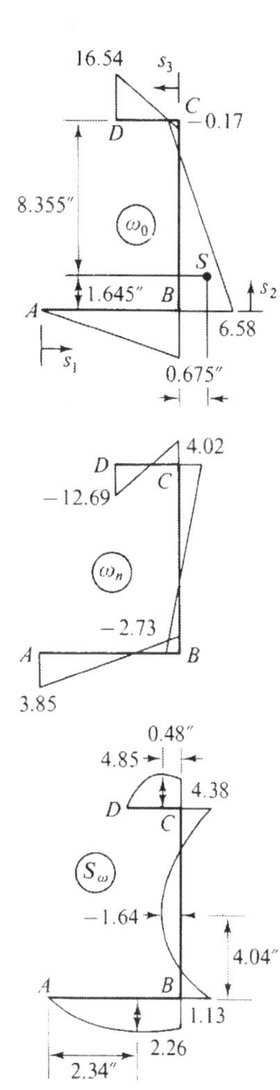

FIGURE 2.16
Normalized warping and its static moment (Example 2.4)

The normalized warping: $\omega_n = \frac{1}{A}\int_A^D \omega_0 t\, ds - \omega_0$

$$A = 0.5(4 + 10 + 2) = 8 \text{ in}^2$$

$$\int_A^D \omega_0 t\, ds = 0.5\left[\int_0^4 1.645 s_1\, ds_1 + \int_0^{10} (6.58 - 0.675 s_2)\, ds_2 + \int_0^2 (-0.17 + 8.355 s_3)\, ds_3\right] = 30.79$$

$$\omega_n = \frac{30.79}{8} - \omega_0 = 3.85 - \omega_0 \qquad \text{shown in Fig. 2.16.}$$

The warping static moment $S_\omega = \int_0^s \omega_n t\, ds$

Start s at point A

$$\text{A-B:}\quad S_\omega = -0.412 s_1^2 + 1.93 s_1 \text{ in}^4$$

$$\text{B-C:}\quad S_\omega = 0.169 s_2^2 - 1.365 s_2 + 1.128 \text{ in}^4$$

$$\text{C-D:}\quad S_\omega = -2.089 s_3^2 + 2.010 s_3 + 4.378 \text{ in}^4$$

Note $\qquad S_\omega \approx 0$ at point $D(s_3 = 2)$

A part of the torsional moment T_w is resisted by the warping shear stresses, from Eq. (2.105),

$$\begin{aligned} T_w &= \int \tau_w t \rho_0\, ds \\ &= -E\theta''' \int \rho_0 \left(\int_0^s \omega_n t\, ds\right) ds \\ &= -E\theta''' \left[\left(\int \rho_0\, ds\right)\int \omega_n t\, ds - \int \omega_n t \left(\int \rho_0\, ds\right) ds\right] \\ &= -E\theta''' \left[\omega_0 \Big|_0^E \left(\int \omega_n t\, ds\right) - \int \omega_n \omega_0 t\, ds\right] \end{aligned} \tag{2.106}$$

From Eq. (2.102),

$$\begin{aligned} \int \omega_n t\, ds &= 0 \\ \int \omega_0 \omega_n t\, ds &= -\int \omega_n^2 t\, ds \end{aligned} \tag{2.107}$$

Thus, Eq. (2.106) is reduced to

$$T_w = -E\theta''' \int \omega_n^2 t \, ds = -EI_\omega \theta''' \tag{2.108}$$

In which I_ω is the *warping moment of inertia* defined by

$$I_\omega = \int_0^E \omega_n^2 t \, ds \tag{2.109}$$

2.5 DIFFERENTIAL EQUATION OF TORSION

The total twisting moment M_z is the sum of the St. Venant torsion T_{sv} in Eq. (2.50) and the warping torsion T_w in Eq. (2.108). Thus

$$M_z = T_{sv} + T_w = GK_T\theta' - EI_\omega\theta''' \tag{2.110}$$

In the case that a distributed torque

$$m_z = -\frac{dM_z}{dz} \tag{2.111}$$

is applied, the torsion equation, Eq. (2.110), becomes the differential equation of the fourth order

$$EI_\omega\theta^{IV} - GK_T\theta'' = m_z \tag{2.112}$$

The general solutions of Eqs. (2.110) and (2.112) are

$$\begin{aligned} \theta &= C_1 + C_2 \cosh \lambda z + C_3 \sinh \lambda z + \frac{M_z}{\lambda^2 EI_\omega} z \\ \theta &= C_4 + C_5 z + C_6 \cosh \lambda z + C_7 \sinh \lambda z - \frac{m_z}{2\lambda^2 EI_\omega} z^2 \end{aligned} \tag{2.113}$$

in which

$$\lambda = \sqrt{\frac{GK_T}{EI_\omega}} \tag{2.114}$$

and C_1 to C_7 are the integration constants to be determined from boundary conditions.

The boundary conditions for torsion are a combination of the following:

$$\begin{aligned} &1.\ \text{twisting angle restrained } \theta = 0 \\ &2.\ \text{twisting free } \theta''' - \lambda^2\theta' = 0 \\ &3.\ \text{warping restrained } \theta' = 0 \\ &4.\ \text{warping free } \theta'' = 0 \end{aligned} \tag{2.115}$$

Table 2.1 presents solutions of torsion for most commonly used boundary conditions.

Table 2.1 SOLUTIONS OF TORSION

$$EI_\omega\theta''' - GK_T\theta' = -M_z \qquad\qquad EI_\omega\theta^{IV} - GK_T\theta'' = m_z$$

$$\lambda^2 = \frac{GK_T}{EI_\omega}$$

$$\theta''' - \lambda^2\theta' = -\frac{M_z}{EI_\omega} \qquad\qquad \theta^{IV} - \lambda^2\theta'' = \frac{m_z}{EI_\omega}$$

$$\theta = \frac{M_z}{\lambda^3 EI_\omega}(C_1 + C_2\cosh\lambda z + C_3\sinh\lambda z + \lambda z)$$

$$\theta = \frac{m_z}{\lambda^2 GK_T}(C_4 + C_5\lambda z + C_6\cosh\lambda z + C_7\sinh\lambda z - \tfrac{1}{2}\lambda^2 z^2)$$

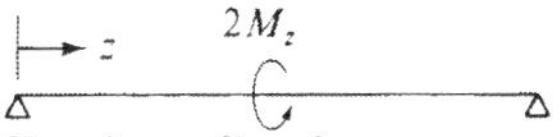

z, $2M_z$

$\theta = \theta'' = 0$ $\quad\theta' = 0$

$L/2$

$$C_1 = C_2 = 0$$

$$C_3 = -\frac{1}{2\cosh\frac{\lambda L}{2}}$$

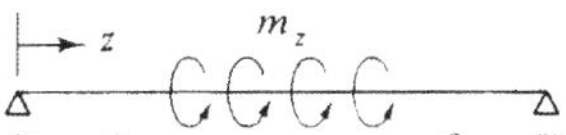

z, m_z

$\theta = \theta'' = 0$ $\quad\theta = \theta'' = 0$

$$C_4 = -C_6 = -1$$

$$C_5 = \frac{\lambda L}{2} - \frac{(1-\sinh\lambda L)(1-\cosh\lambda L)}{\lambda L\sinh\lambda L}$$

$$C_7 = \frac{1-\cosh\lambda L}{\sin\lambda L}$$

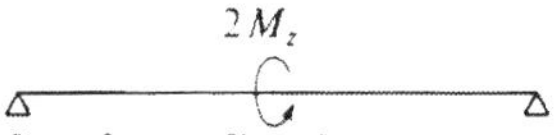

$2M_z$

$\theta = \theta' = 0$ $\quad\theta' = 0$

$L/2$

$$C_1 = -C_2 = \frac{1-\cosh\frac{\lambda L}{2}}{\sinh\frac{\lambda L}{2}}$$

$$C_3 = -\frac{1}{2}$$

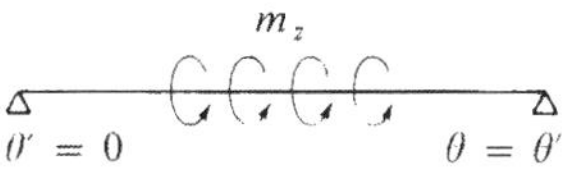

m_z

$\theta = \theta' = 0$ $\quad\theta = \theta' = 0$

$$C_4 = -C_6 = \frac{-\lambda L}{2}\,\frac{\lambda L - 2\sinh\lambda L + \lambda L\cosh\lambda L}{2 - 2\cosh\lambda L + \lambda L\sinh\lambda L}$$

$$C_5 = -C_7 = \frac{1}{2}\lambda L$$

M_z

$\theta = \theta'' = 0$ $\quad\theta'' = 0$

L

$$C_1 = C_2 = C_3 = 0$$

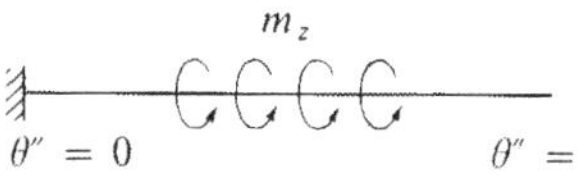

m_z

$\theta = \theta'' = 0$ $\quad\theta'' = 0$, $\quad\theta''' - \lambda^2\theta' = 0$

$$C_4 = -C_6 = -1$$

$$C_5 = \lambda L$$

$$C_7 = \frac{1-\cosh\lambda L}{\sinh\lambda L}$$

M_z

$\theta = \theta' = 0$ $\quad\theta'' = 0$

L

$$C_1 = -C_2 = -\tanh\lambda L$$

$$C_3 = -1$$

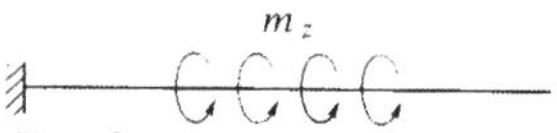

m_z

$\theta = \theta' = 0$ $\quad\theta'' = 0$, $\quad\theta''' - \lambda^2\theta' = 0$

$$C_4 = -C_6 = -\frac{1+\lambda L\sinh\lambda L}{\cosh\lambda L}$$

$$C_5 = -C_7 = \lambda L$$

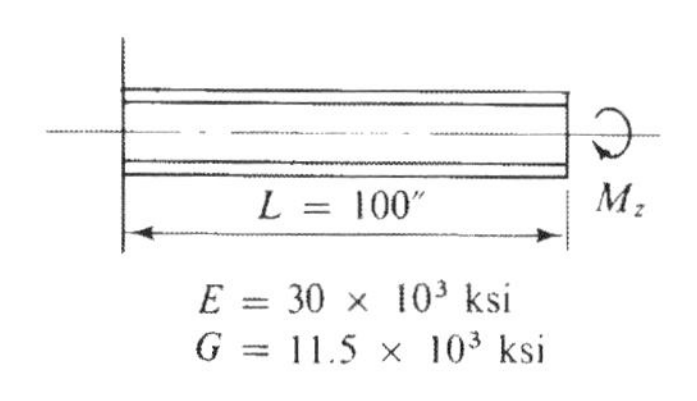

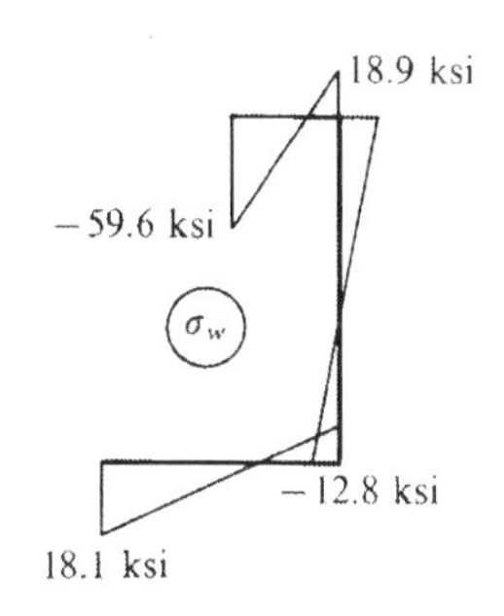

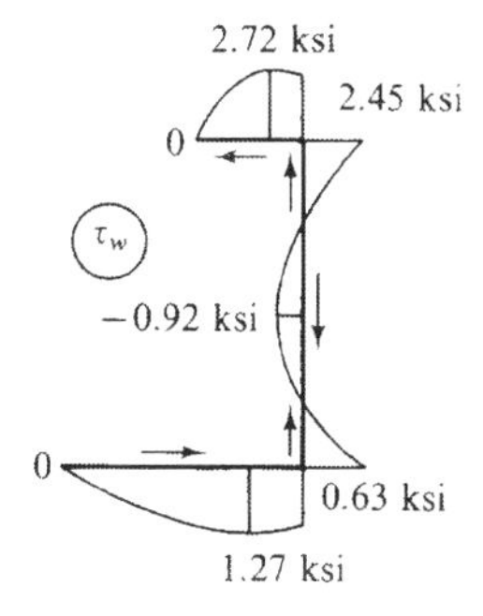

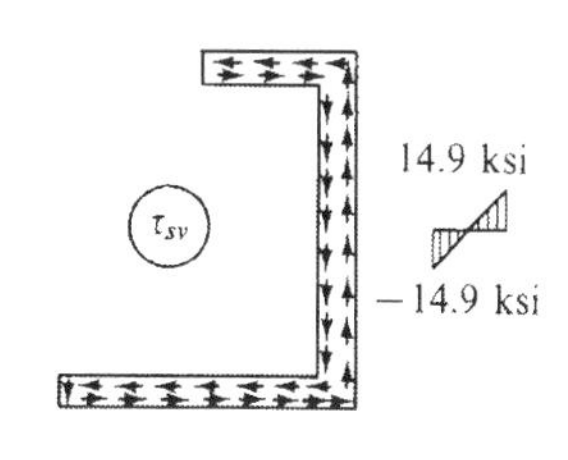

FIGURE 2.17
Stresses in cantilever (Example 2.5)

EXAMPLE 2-5 For the cantilever beam shown in Fig. 2.17, determine

1. warping normal stress at the fixed end
2. warping shear stress at the fixed end
3. St. Venant shear stress
4. twisting angle at the free end

when a twisting moment $M_z = 20k$-in is applied at the free end, $L = 100$ in. The section of the beam is same as Example 2-1.

SOLUTION Warping moment of inertia, from Example 2-4

$$I_\omega = \int_A^D \omega_n^2 t\, ds$$

$$= 0.5 \int_0^4 (3.85 - 1.645 s_1)^2\, ds_1 + 0.5 \int_0^{10} (-2.73 + 0.675 s_2)^2\, ds_2$$

$$+ 0.5 \int_0^2 (4.02 - 8.355 s_3)^2\, ds_3$$

$$= 7.8 + 21.1 + 42.1 = 71.0 \text{ in}^6$$

St. Venant torsion constant

$$K_T = \frac{1}{3}\sum t^3 b = \frac{1}{3} \times 0.5^3 \times (4 + 10 + 2) = 0.667 \text{ in}^4$$

The governing differential equation of the problem is, from Eq. (2.110)

$$\theta''' - \lambda^2\theta' = -\frac{M_z}{EI_\omega}$$

in which

$$\lambda^2 = \frac{GK_T}{EI_\omega} = \frac{11.5 \times 0.667}{30 \times 71} = 3.60 \times 10^{-3} \text{ in}^{-2}, \quad \lambda = 0.06 \text{ in}^{-1}$$

The general solution is from Eq. (2.113)

$$\theta = C_1 + C_2 \cosh \lambda z + C_3 \sinh \lambda z + \frac{M_z}{\lambda^2 EI_\omega} z$$

The boundary conditions are

at $z = 0$, rotation angle restrained, $\theta = 0$: $C_1 + C_2 = 0$

warping restrained, $\theta' = 0$: $C_3\lambda + \dfrac{M_z}{\lambda^2 EI_\omega} = 0$

at $z = L$, warping free, $\theta'' = 0$: $C_2\lambda^2 \cosh \lambda L + C_3\lambda^2 \sinh \lambda L = 0$

Solving for C_1, C_2 and C_3, we obtain

$$C_1 = -\frac{M_z}{\lambda^3 EI_\omega}\tanh \lambda L, \quad C_2 = \frac{M_z}{\lambda^3 EI_\omega}\tanh \lambda L, \quad C_3 = -\frac{M_z}{\lambda^3 EI_\omega}$$

which result the following solutions

$$\theta = \frac{M_z}{\lambda^3 EI_\omega}(-\tanh \lambda L + \tanh \lambda L \cosh \lambda z - \sinh \lambda z + \lambda z)$$

$$\theta' = \frac{M_z}{\lambda^2 EI_\omega}(\tanh \lambda L \sinh \lambda z - \cosh \lambda z + 1)$$

$$\theta'' = \frac{M_z}{\lambda EI_\omega}(\tanh \lambda L \cosh \lambda z - \sinh \lambda z)$$

$$\theta''' = \frac{M_z}{EI_\omega}(\tanh \lambda L \sinh \lambda z - \cosh \lambda z)$$

At the fixed end ($z = 0$)

$$\theta_o = 0$$

At the free end ($z = L$)

$$\theta_L = \frac{M_z}{\lambda^3 EI_\omega}(\lambda L - \tanh \lambda L)$$

$$\theta'_o = 0 \qquad \theta'_L = \frac{M_z}{\lambda^2 EI_\omega}\left(1 - \frac{1}{\cosh \lambda L}\right)$$

$$\theta''_o = \frac{M_z}{\lambda EI_\omega}\tanh \lambda L \qquad \theta''_L = 0$$

$$\theta'''_o = -\frac{M_z}{EI_\omega} \qquad \theta'''_L = -\frac{M_z}{EI_\omega}\frac{1}{\cos \lambda L}$$

1. Warping normal stress at the fixed end, from Eq. (2.101),

$$\sigma_w = E\omega_n\theta''_o = \frac{M_z\omega_n}{\lambda I_\omega}\tanh \lambda L = \frac{20\omega_n}{0.06 \times 71}\tanh(0.06 \times 100) = 4.7\omega_n$$

Using the ω_n distribution calculated in Example 2-4 (Fig. 2.16), the warping normal stress is obtained as shown in Fig. 2.17.

2. Warping shear stress at the fixed end, from Eq. (2.105)

$$\tau_w = -\frac{ES_\omega}{t}\theta'''_o = \frac{M_z}{tI_\omega}S_\omega = \frac{20S_\omega}{0.5 \times 71} = 0.56S_\omega$$

Using the S_ω distribution calculated in Example 2-4 (Fig. 2.16), the warping shear stress is obtained as shown in Fig. 2.17.

3. Shear stress due to St. Venant torsion, from Eq. (2.86)

$$(\tau_{sv})_{\max} = \pm tG\theta'$$

at the fixed end

$$(\tau_{sv})_{\max} = \pm tG\theta'_o = 0$$

at the free end

$$(\tau_{sv})_{\max} = \pm tG\theta'_L = \pm\frac{tGM_z}{\lambda^2 EI_\omega}\left(1 - \frac{1}{\cosh \lambda L}\right) = \pm\frac{0.5 \times 11.5 \times 20\left(1 - \frac{1}{202}\right)}{3.60 \times 10^{-3} \times 30 \times 71}$$

$$= \pm 14.9 \text{ ksi}$$

The St. Venant shear stress distribution is shown in Fig. 2.17.

4. Angle of twist at the free end

$$\theta_L = \frac{M_z}{\lambda^3 EI_\omega}(\lambda L\text{-}\tanh \lambda L) = \frac{20(6-1)}{0.06^3 \times 30 \times 10^3 \times 71} = 0.217 \text{ rad} = 12.5 \text{ degree}$$

2.6 DIFFERENTIAL EQUATIONS OF BEAM-COLUMNS

The formulation of beam-column equations requires two basic steps. One of these is the development of generalized stress-strain relationship, which is the relation between moments and curvatures of a beam-column segment of unit length. It is established with the assumption of linear strain distribution over the cross section and the use of a stress-strain relation of the material. The second step is the equilibrium consideration, which establishes the equilibrium condition between internal and external forces under the deflected configuration of the beam-column. This is usually achieved through the rotation of coordinates. Combining these two steps results in the basic differential equations of the beam-column problem.

2.6.1 Generalized Stress-Strain Relationship

Assuming arbitrary coordinates ξ, η and ω on the cross section, the normal strain ε is related to the generalized strains: axial strain, ε_0, biaxial curvatures Φ_ξ and Φ_η and warping curvature, $\theta'' = \theta_\zeta''$, by (Prob. 2.4)

$$\varepsilon = \varepsilon_0 + \eta\Phi_\xi - \xi\Phi_\eta - \omega\theta_\zeta'' = \varepsilon_0 + \eta\theta_\xi' - \xi\theta_\eta' - \omega\theta_\zeta'' \tag{2.116}$$

in which θ_ξ, θ_η and θ_ζ are rotation angles of the cross section about the coordinate axes ξ, η and ζ, respectively. These are related to the curvatures and axial displacement by

$$\theta_\xi' = \Phi_\xi, \quad \theta_\eta' = \Phi_\eta \qquad \text{and} \qquad w' = \varepsilon_0 \tag{2.117}$$

While the generalized stresses are related to the stress by

$$\begin{aligned}
&\text{axial force} && N = \int \sigma\, dA \\
&\text{bending moment} && M_\xi = \int \sigma\eta\, dA \\
&\text{bending moment} && M_\eta = -\int \sigma\xi\, dA \\
&\text{warping moment} && M_\omega = \int \sigma\omega\, dA
\end{aligned} \tag{2.118}$$

Since the stress-strain relation of elastic material is

$$\sigma = E\varepsilon \tag{2.119}$$

the generalized stress and generalized strain relations are obtained by combining Eqs. (2.116), (2.118) and (2.119) as

$$\begin{Bmatrix} N \\ M_\xi \\ M_\eta \\ M_\omega \end{Bmatrix} = E \begin{bmatrix} A & S_y & -S_x & -S_\omega \\ S_y & I_x & -I_{xy} & -I_{\omega y} \\ -S_x & -I_{xy} & I_y & I_{\omega x} \\ S_\omega & I_{\omega y} & -I_{\omega x} & -I_\omega \end{bmatrix} \begin{Bmatrix} \varepsilon_0 \\ \theta_\xi' \\ \theta_\eta' \\ \theta_\zeta'' \end{Bmatrix} \tag{2.120}$$

in which

$$
\begin{aligned}
A &= \int dA, \quad S_x = \int \xi \, dA, \quad S_y = \int \eta \, dA, \quad S_\omega = \int \omega \, dA \\
I_x &= \int \eta^2 \, dA, \quad I_y = \int \xi^2 \, dA, \quad I_\omega = \int \omega^2 \, dA \\
I_{xy} &= \int \xi\eta \, dA, \quad I_{\omega x} = \int \omega\xi \, dA, \quad I_{\omega y} = \int \omega\eta \, dA
\end{aligned}
\tag{2.121}
$$

where the integration is carried out over the entire cross section. If the axes ξ and η are principal coordinates and ω is the normalized warping about the shear center,

$$S_x = S_y = S_\omega = I_{xy} = I_{\omega x} = I_{\omega y} = 0 \tag{2.122}$$

then Eq. (2.120) is reduced to

$$
\begin{aligned}
N &= EA\varepsilon_0 \\
M_\xi &= EI_x \theta'_\xi \\
M_\eta &= EI_y \theta'_\eta \\
M_\omega &= -EI_\omega \theta''_\zeta
\end{aligned}
\tag{2.123}
$$

The first four conditions in Eq. (2.122) are obvious from the definitions of principal axes and normalized warping. The last two conditions, $I_{\omega x} = I_{\omega y} = 0$ can be shown by using Eq. (2.102), Eq. (2.94), and Eq. (2.38) as

$$
\begin{aligned}
\int_0^E \omega_n \xi t \, ds &= \int_0^E \left[\frac{\xi}{A} \int_0^E \omega_0 \, dA - \omega_0 \xi \right] t \, ds \\
&= -\int_0^E \omega_0 \xi t \, ds \\
&= -\int_0^E (\omega + \eta_0 \xi - \eta_0 \xi_A - \xi_0 \eta + \xi_0 \eta_A) \xi t \, ds \\
&= -I_{\omega x} - \eta_0 I_y = 0
\end{aligned}
\tag{2.124}
$$

The *warping moment* M_ω is often called the *bimoment*. This is because, for an I-shaped cross section, the normalized warping is $\omega_n = \xi\eta$ and the warping moment becomes

$$
\begin{aligned}
M_\omega &= \int \sigma \omega_n \, dA = \int \sigma \xi \eta \, dA \\
&= \int_{UF} \sigma \xi \frac{D-t}{2} \, dA + \int_{WEB} \sigma \eta 0 \, dA + \int_{LF} \sigma \xi \left(-\frac{D-t}{2} \right) dA \\
&= \frac{D-t}{2} \left(\int_{UF} \sigma \xi \, dA - \int_{LF} \sigma \xi \, dA \right)
\end{aligned}
\tag{2.125}
$$

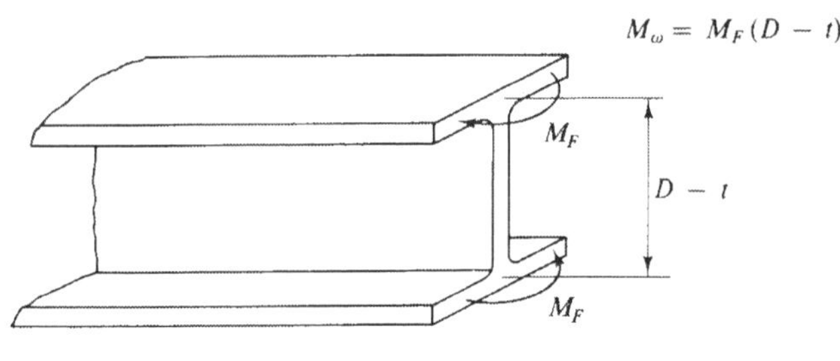

FIGURE 2.18
Bimoment of I-section

in which D is the depth of the cross section and t is the thickness of the flanges and $\int_{UF} \sigma\xi\, dA$ and $\int_{LF} \sigma\xi\, dA$ indicate the integration over the upper flange and the lower flange, respectively. These are equal to the in-plane bending moment of each flange $M_F = \int_A \sigma\xi\, dA$

$$\int_{UF} \sigma\xi\, dA = -\int_{LF} \sigma\xi\, dA = M_F \tag{2.126}$$

Thus, the warping moment, M_ω, is known to be the in-plane flange moment M_F times the distance $(D - t)$ between the two flanges as shown in Fig. 2.18

$$M_\omega = (D - t)M_F \tag{2.127}$$

This is the bimoment of an I-shaped section which is in self-equilibrium.

The twisting moment T_w due to warping shear stress is related to the warping moment M_ω in Eq. (2.120) by (Prob. 2.5).

$$\begin{aligned} T_w &= \int_0^E \rho\tau t\, ds \\ &= ES_\omega\varepsilon_0' + EI_{\omega y}\theta_\xi'' - EI_{\omega x}\theta_\eta'' - EI_\omega\theta_\zeta''' \\ &\quad - \omega_E[EA\varepsilon_0' + ES_y\theta_\xi'' - ES_x\theta_\eta'' - ES_\omega\theta_\zeta''' - \tau_0 t_0] \\ &= M_\omega' - \omega_E[EA\varepsilon_0' + ES_y\theta_\xi'' - ES_x\theta_\eta'' - ES_\omega\theta_\zeta''' - \tau_0 t_0] \end{aligned} \tag{2.128a}$$

This is the twisting moment about the origin of the coordinates. If we select the pole for the unit warping such that the terms in the square brackets in the right hand side of Eq. (2.128*a*) may be zero, then

$$T_w = M_\omega' \tag{2.128b}$$

This condition is satisfied when ξ and η are principal axes and the normalized warping ω_n is taken for ω. In this case, Eq. (2.128*b*) is further simplified as

$$T_w = -EI_\omega\theta_\zeta''' \tag{2.129}$$

This twisting moment is acting about the shear center. The twisting moment T_{sv} due to St. Venant shear stress is given in Eq. (2.50) by

$$T_{sv} = GK_T\theta'_\zeta \tag{2.130}$$

This is a (uniform) twisting and is independent of coordinate axis.

The total internal twisting moment due to the shear stress is thus given by

$$M_\zeta = T_w + T_{sv} = -EI_\omega\theta'''_\zeta + GK_T\theta'_\zeta \tag{2.131}$$

It is to be noted that this twisting moment is acting about the shear center.

As a second order effect, the normal stress σ may produce a torsional moment if a warping deformation exists in the cross section. As shown in Fig. 2.19, the direction of the normal stress σ is inclined by the angle

$$a\theta'_\zeta = \sqrt{(\xi - \xi_0)^2 + (\eta - \eta_0)^2}\,\theta'_\zeta \tag{2.132}$$

Thus its contribution to the torsional moment about the shear center is

$$T_\sigma = \int \sigma a^2\theta'_\zeta\, dA = \bar{K}\theta'_\zeta \tag{2.133}$$

This is known as the *Wagner effect*. In Eq. (2.133), a is the distance from a point on the cross section to the shear center and $\bar{K}$ is called *Wagner coefficient* which is defined by

$$\bar{K} = \int a^2\sigma\, dA = \int \sigma[(\xi - \xi_0)^2 + (\eta - \eta_0)^2]\, dA \tag{2.134}$$

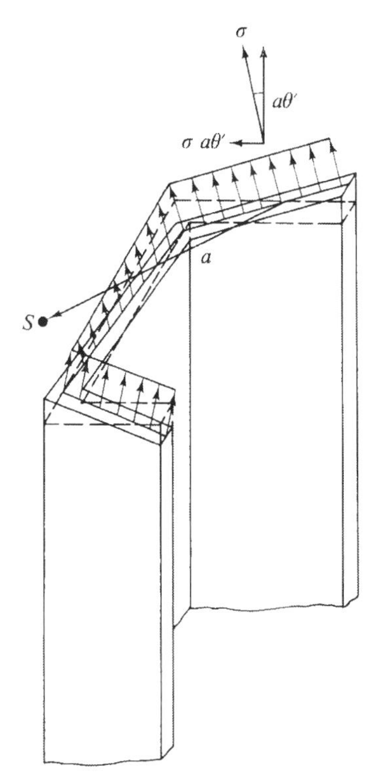

FIGURE 2.19
Torsional moment due to normal stress (Wagner effect)

where the normal stress σ is obtained by combining Eq. (2.116), (2.119) and (2.120) as

$$\sigma = E(1 \quad \eta \quad -\xi \quad -\omega)\begin{Bmatrix} \varepsilon_0 \\ \theta'_\xi \\ \theta'_\eta \\ \theta''_\zeta \end{Bmatrix}$$

$$= (1 \quad \eta \quad -\xi \quad -\omega)\begin{bmatrix} A & S_y & -S_x & -S_\omega \\ S_y & I_x & -I_{xy} & -I_{\omega y} \\ -S_x & -I_{xy} & I_y & I_{\omega x} \\ S_\omega & I_{\omega y} & -I_{\omega x} & -I_\omega \end{bmatrix}^{-1}\begin{Bmatrix} N \\ M_\xi \\ M_\eta \\ M_\omega \end{Bmatrix} \tag{2.135}$$

If the principal axes are taken for ξ and η and the normalized warping about the shear center for ω, then

$$\sigma = \frac{N}{A} + \frac{M_\xi}{I_x}\eta - \frac{M_\eta}{I_y}\xi + \frac{M_\omega}{I_\omega}\omega \tag{2.136}$$

then the Wagner coefficient becomes

$$\begin{aligned} \bar{K} = {} & N\left[\frac{1}{A}(I_x + I_y) + \xi_0^2 + \eta_0^2\right] \\ & + M_\xi\left[\frac{1}{I_x}\left(\int \eta\xi^2\, dA + \int \eta^3\, dA\right) - 2\eta_0\right] \\ & - M_\eta\left[\frac{1}{I_y}\left(\int \xi^3\, dA + \int \xi\eta^2\, dA\right) - 2\xi_0\right] \\ & + \frac{M_\omega}{I_\omega}\left(\int \omega\xi^2\, dA + \int \omega\eta^2\, dA\right) \end{aligned} \tag{2.137}$$

Further, if the cross section is doubly symmetric and the shear center coincides with the centroid, $\xi_0 = 0$ and $\eta_0 = 0$, Eq. (2.137) reduces to

$$\bar{K} = \frac{N}{A}(I_x + I_y) + \frac{M_\omega}{I_\omega}\left(\int \omega\xi^2\, dA + \int \omega\eta^2\, dA\right) \tag{2.138}$$

From this it is known that the biaxial bending normal stresses do not contribute to the twisting moment due to Wagner effect for the case of an elastic doubly symmetric cross section. When the axial stress due to normal force N is of prime importance, then, Eq. (2.134) may be written in the simple form

$$\bar{K} \approx \frac{N}{A}\int [(\xi - \xi_0)^2 + (\eta - \eta_0)^2]\, dA = N\frac{I_0}{A} = Nr_0^2 \tag{2.139}$$

in which I_0 is the polar moment of inertia and r_0 is the polar radius of gyration given by

$$r_0 = \sqrt{\frac{I_0}{A}} = \sqrt{\frac{1}{A}\int [(\xi - \xi_0)^2 + (\eta - \eta_0)^2]\, dA} \tag{2.140}$$

The total internal twisting moment are the summation of T_w and T_{sv} in Eqs. (2.128b) and (2.130) and T_σ in Eq. (2.133), i.e.,

$$M_\zeta = T_w + T_{sv} + T_\sigma = M'_\omega + (GK_T + \bar{K})\theta'_\zeta \tag{2.141}$$

Taking the principal axes for ξ and η and the normalized warping about the shear center for ω, the total internal twisting moment is

$$M_\zeta = T_w + T_{sv} + T_\sigma = -EI_\omega\theta'''_\zeta + (GK_T + \bar{K})\theta'_\zeta \tag{2.142}$$

This twisting moment is acting about the shear center.

Summing up, the generalized stress-strain relations are

$$\begin{Bmatrix} N \\ M_\xi \\ M_\eta \\ M_\omega \end{Bmatrix} = E \begin{bmatrix} A & S_y & -S_x & -S_\omega \\ S_y & I_x & -I_{xy} & -I_{\omega y} \\ -S_x & -I_{xy} & I_y & I_{\omega x} \\ S_\omega & I_{\omega y} & -I_{\omega x} & -I_\omega \end{bmatrix} \begin{Bmatrix} \varepsilon_0 \\ \theta'_\xi \\ \theta'_\eta \\ \theta''_\zeta \end{Bmatrix} \tag{2.143}$$

and the associated twisting moment about the shear center is

$$\{M_\zeta\} = (-EI_\omega \quad GK_T + \bar{K}) \begin{Bmatrix} \theta'''_\zeta \\ \theta'_\zeta \end{Bmatrix} \tag{2.144}$$

The generalized stresses $(N, M_\xi, M_\eta, M_\omega)$ are associated with the normal stress σ and the corresponding generalized strains $(\varepsilon_0, \theta'_\xi, \theta'_\eta, \theta''_\zeta)$ or $(\varepsilon_0, \Phi_\xi, \Phi_\eta, \theta''_\zeta)$ are associated with the axial strains ε. The twisting moment M_ζ is associated with shear stresses. Since shear deformation is neglected in the analysis, the twisting moment M_ζ along with the shear forces V_ξ and V_η are considered *reactions*. They are needed only for equilibrium considerations.

2.6.2 External Forces

In order to examine the external forces, consider a beam-column in a deflected configuration shown in Fig. 2.20. At the bottom, the forces $\{F_o\}$ and the moment $[M_o]$ are applied:

$$\{F_o\} = \begin{Bmatrix} F_{xo} \\ F_{yo} \\ F_{zo} \end{Bmatrix} \quad \text{and} \quad \{M_o\} = \begin{Bmatrix} M_{xo} \\ M_{yo} \\ M_{zo} \end{Bmatrix} \tag{2.145}$$

At an arbitrary section, $z = z_c$, the deflections of the origin C of the local coordinate ξ and η are u and v. Thus, the location of C is defined by

$$C: \ (u, v, z) \tag{2.146}$$

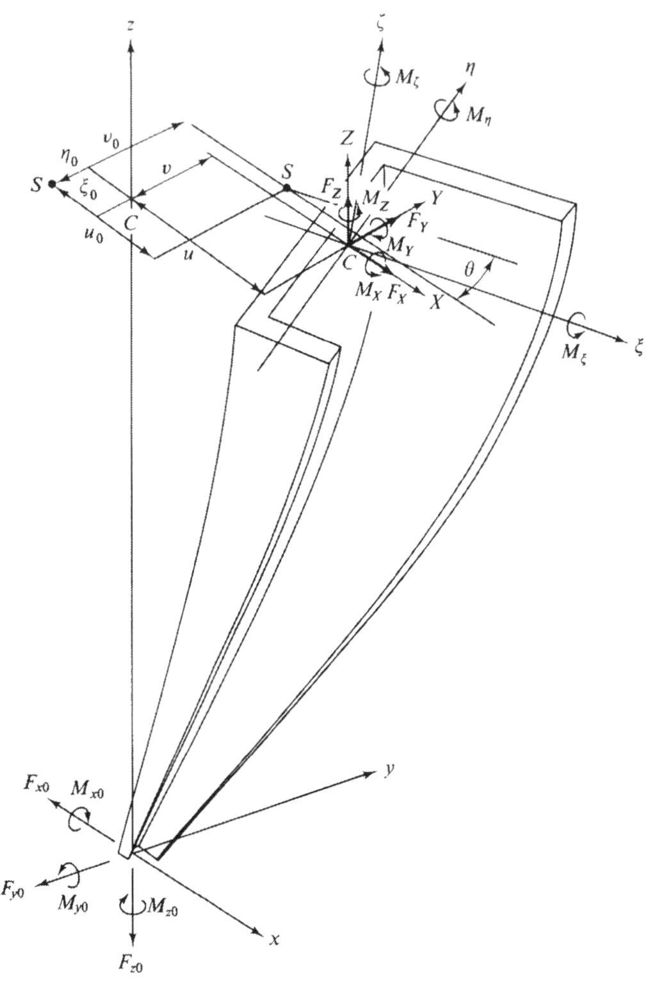

FIGURE 2.20
Equilibrium of beam-column in space

If we take another local coordinate (X, Y, Z) which is parallel to the global coordinates (x, y, z), the forces $\{F_x\}$ and the moments $\{M_x\}$ acting on the cross section are obtained from the equilibrium conditions of forces, $\{F_x\} = \{F_o\}$

$$\begin{Bmatrix} F_X \\ F_Y \\ F_Z \end{Bmatrix} = \begin{Bmatrix} F_{xo} \\ F_{yo} \\ F_{zo} \end{Bmatrix} \tag{2.147}$$

and moments $\{M_X\} = \{M_o\} - [L_c]\{F_o\}$ or

$$\begin{Bmatrix} M_X \\ M_Y \\ M_Z \end{Bmatrix} = \begin{Bmatrix} M_{xo} \\ M_{yo} \\ M_{zo} \end{Bmatrix} - \begin{bmatrix} 0 & -z & v \\ z & 0 & -u \\ -v & u & 0 \end{bmatrix} \begin{Bmatrix} F_{xo} \\ F_{yo} \\ F_{zo} \end{Bmatrix} \tag{2.148}$$

2.6.3 Rotation of Coordinates

In order to equate the internal forces $\{M_\xi\}$ in Eq. (2.143) and $\{M_\zeta\}$ in Eq. (2.144) to the external forces $\{M_x\}$ in Eq. (2.148), a relation between the two coordinate axes (ξ, η, ζ) and (X, Y, Z) is needed. They are related through the rotation matrix $[R]$, i.e., $\{\xi\} = [R]\{X\}$, or

$$\begin{Bmatrix} \xi \\ \eta \\ \zeta \end{Bmatrix} = \begin{bmatrix} \cos(\xi X) & \cos(\xi Y) & \cos(\xi Z) \\ \cos(\eta X) & \cos(\eta Y) & \cos(\eta Z) \\ \cos(\zeta X) & \cos(\zeta Y) & \cos(\zeta Z) \end{bmatrix} \begin{Bmatrix} X \\ Y \\ Z \end{Bmatrix} \tag{2.149}$$

in which cos ( ) indicates the directional cosine of the two axes. If we denote the small rotation angles of ξ, η, ζ axes about the X, Y and Z axes by θ_x, θ_y and θ_z, respectively, then the rotation matrix $[R]$ can be approximately written in an anti-symmetric form as

$$[R] = \begin{bmatrix} \cos\theta_y \cos\theta_z & \sin\theta_z & -\sin\theta_y \\ -\sin\theta_z & \cos\theta_z \cos\theta_x & \sin\theta_x \\ \sin\theta_y & -\sin\theta_x & \cos\theta_x \cos\theta_y \end{bmatrix} \tag{2.150}$$

Noting that the ξ and η axes are perpendicular to the centroidal deflection curve and taking the effect of axial elongation into consideration, we have

$$\tan\theta_x = -\frac{v'}{1+\varepsilon_0}, \quad \tan\theta_y = \frac{u'}{1+\varepsilon_0} \quad \text{and} \quad \theta_z = \theta \tag{2.151}$$

in which prime indicates differentiation with respect to z. Making use of the relations,

$$\cos\theta = \frac{1}{\sqrt{1+\tan^2\theta}} \quad \text{and} \quad \sin\theta = \frac{\tan\theta}{\sqrt{1+\tan^2\theta}} \tag{2.152}$$

the rotation matrix is expressed in the form

$$[R] = \begin{bmatrix} \dfrac{1+\varepsilon_0}{\sqrt{(1+\varepsilon_0)^2+(u')^2}}\cos\theta & \sin\theta & -\dfrac{1}{\sqrt{(1+\varepsilon_0)^2+(u')^2}}u' \\ -\sin\theta & \dfrac{1+\varepsilon_0}{\sqrt{(1+\varepsilon_0)^2+(v')^2}}\cos\theta & -\dfrac{1}{\sqrt{(1+\varepsilon_0)^2+(v')^2}}v' \\ \dfrac{1}{\sqrt{(1+\varepsilon_0)^2+(u')^2}}u' & \dfrac{1}{\sqrt{(1+\varepsilon_0)^2+(v')^2}}v' & \dfrac{(1+\varepsilon_0)^2}{\sqrt{(1+\varepsilon_0)^2+(v')^2}\sqrt{(1+\varepsilon_0)^2+(u')^2}} \end{bmatrix} \tag{2.153}$$

If we assume that the deformations are small, then, Eq. (2.153) reduces to the simple form

$$[R] = \begin{bmatrix} 1 & \theta & -u' \\ -\theta & 1 & -v' \\ u' & v' & 1 \end{bmatrix} \tag{2.154}$$

from which rotation angles to be substituted into Eq. (2.143) and Eq. (2.144) are

$$\begin{Bmatrix} \theta'_\xi \\ \theta'_\eta \\ \theta'_\zeta \end{Bmatrix} = \begin{bmatrix} 1 & \theta & -u' \\ -\theta & 1 & -v' \\ u' & v' & 1 \end{bmatrix} \begin{Bmatrix} -v'' \\ u'' \\ \theta' \end{Bmatrix} = \begin{Bmatrix} -v'' + \theta u'' - u'\theta' \\ u'' + \theta v'' - v'\theta' \\ \theta' - u'v'' + v'u'' \end{Bmatrix} \tag{2.155}$$

$$\begin{Bmatrix} \theta''_\xi \\ \theta''_\eta \\ \theta''_\zeta \end{Bmatrix} = \begin{Bmatrix} -v''' + \theta u''' - u'\theta'' \\ u''' + \theta v''' - v'\theta'' \\ \theta'' - u'v''' + v'u''' \end{Bmatrix} \tag{2.156}$$

$$\theta'''_\zeta \approx \theta''' - u''v''' + v''u''' \tag{2.157}$$

in which the fourth order derivative terms are neglected in the approximation. The axial strain can be written as (Prob. 2.6)

$$\varepsilon_0 = w' + \frac{1}{2}(u')^2 + \frac{1}{2}(v')^2 \tag{2.158}$$

and

$$\varepsilon'_0 = w'' + u'u'' + v'v'' \tag{2.159}$$

2.6.4 Equilibrium Equations (Note: Sections 2.6.4 and 2.7 have been updated by the author. Please download the correct text at www.jrosspub.com/wav).

Since forces $\{F\}$ do not alter along the beam-column, equilibrium conditions are considered only for moments. The internal moments are obtained by substituting Eqs. (2.155) to (2.157) into Eqs. (2.143) and (2.144)

$$\begin{Bmatrix} N \\ M_\xi \\ M_\eta \\ M_\omega \end{Bmatrix}_{\mathrm{INT}} = E \begin{bmatrix} A & S_y & -S_x & -S_\omega \\ S_y & I_x & -I_{xy} & -I_{\omega y} \\ -S_x & -I_{xy} & I_y & I_{\omega x} \\ S_\omega & I_{\omega y} & -I_{\omega x} & -I_\omega \end{bmatrix} \begin{Bmatrix} w' + \frac{1}{2}(u')^2 + \frac{1}{2}(v')^2 \\ -v'' + \theta u'' - u'\theta' \\ u'' + \theta v'' - v'\theta' \\ \theta'' - u'v''' + v'u''' \end{Bmatrix} \tag{2.160}$$

and

$$\{M_\zeta\}_{\mathrm{INT}} = (-EI_\omega \quad GK_T + \bar{K}) \begin{Bmatrix} \theta''' - u''v''' + v''u''' \\ \theta' - u'v'' + v'u'' \end{Bmatrix} \tag{2.161}$$

The external moments in the (ξ, η, ζ) coordinates are from Eq. (2.148), $\{M_\xi\} = [R]\{M_x\} = [R](\{M_o\} - [L_c]\{F_o\})$ or

$$\begin{Bmatrix} M_\xi \\ M_\eta \\ M_\zeta \end{Bmatrix}_{\mathrm{EXT}} = \begin{bmatrix} 1 & \theta & -u' \\ -\theta & 1 & -v' \\ u' & v' & 1 \end{bmatrix} \left(\begin{Bmatrix} M_{xo} \\ M_{yo} \\ M_{zo} \end{Bmatrix} - \begin{bmatrix} 0 & -z & v \\ z & 0 & -u \\ -v & u & 0 \end{bmatrix} \begin{Bmatrix} F_{xo} \\ F_{yo} \\ F_{zo} \end{Bmatrix} \right)$$
$$= \begin{bmatrix} 1 & \theta & -u' \\ -\theta & 1 & -v' \\ u' & v' & 1 \end{bmatrix} \begin{Bmatrix} M_x - vF_{zo} \\ M_y + uF_{zo} \\ M_{zo} + vF_{xo} - uF_{yo} \end{Bmatrix} \tag{2.162}$$

in which M_x and M_y are bending moments of an undeflected member

$$M_x = M_{xo} + zF_{yo}$$
$$M_y = M_{yo} - zF_{xo} \tag{2.163}$$

(Note: Sections 2.6.4 and 2.7 have been updated by the author. Please download the correct text at www.jrosspub.com/wav).

Equating the internal moments in Eq. (2.160) and Eq. (2.161) to the external moments in Eq. (2.162), we obtain

$$
E\begin{bmatrix} A & S_y & -S_x & -S_\omega \\ S_y & I_x & -I_{xy} & -I_{\omega y} \\ -S_x & -I_{xy} & I_y & I_{\omega x} \\ S_\omega & I_{\omega y} & -I_{\omega x} & -I_\omega \end{bmatrix}
\begin{Bmatrix} w' + \frac{1}{2}(u')^2 + \frac{1}{2}(v')^2 \\ -v'' + \theta u'' - u'\theta' \\ u'' + \theta v'' - v'\theta' \\ \theta'' - u'v''' + v'u''' \end{Bmatrix}
= \begin{bmatrix} 1 & 0 & 0 & 0 & 0 \\ 0 & 1 & \theta & -u' & 0 \\ 0 & -\theta & 1 & -v' & 0 \\ 0 & 0 & 0 & 0 & 1 \end{bmatrix}
\begin{Bmatrix} F_{zo} \\ M_x - vF_{zo} \\ M_y + uF_{zo} \\ M_{zo} + vF_{xo} - uF_{yo} \\ M_\omega \end{Bmatrix} \tag{2.164}
$$

and

$$
(-EI_\omega \quad GK_T + \bar{K}) \begin{Bmatrix} \theta''' - u''v''' + v''u''' \\ \theta' - u'v'' + v'u'' \end{Bmatrix}
= (u' \quad v' \quad 1) \begin{Bmatrix} M_x - (v + \eta_0)F_{zo} \\ M_y + (u + \xi_0)F_{zo} \\ M_{zo} + (v + \eta_0)F_{xo} - (u + \xi_0)F_{yo} \end{Bmatrix} \tag{2.165}
$$

Equation (2.164) is the equilibrium equation of axial force, biaxial bending moments and warping moment with respect to the origin of the arbitrary coordinates (ξ, η) on the cross section, and Eq. (2.165) is the equilibrium equation of twisting moment about the *pole* or torsion center (ξ_0, η_0) of the sectorial area. Thus, the external twisting moment in the right hand side of Eq. (2.165) has been corrected from Eq. (2.162) by substituting

$$
u + \xi_0 \quad \text{for } u \qquad \text{and} \qquad v + \eta_0 \quad \text{for } v \tag{2.166}
$$

Since these differential equilibrium equations are highly nonlinear, some simplifying assumptions must be adopted before an actual solution procedure is attempted.

As the first simplification, neglect the higher order terms containing the product of derivatives of displacements, then

$$
E\begin{bmatrix} A & S_y & -S_x & -S_\omega \\ S_y & I_x & -I_{xy} & -I_{\omega y} \\ -S_x & -I_{xy} & I_y & I_{\omega x} \end{bmatrix}
\begin{Bmatrix} w' \\ -v'' + \theta u'' \\ u'' + \theta v'' \\ \theta'' \end{Bmatrix}
= \begin{bmatrix} 1 & 0 & 0 & 0 \\ 0 & 1 & \theta & -u' \\ 0 & -\theta & 1 & -v' \end{bmatrix}
\begin{Bmatrix} F_{zo} \\ M_x - vF_{zo} \\ M_y + uF_{zo} \\ M_{zo} + vF_{xo} - uF_{yo} \end{Bmatrix} \tag{2.167}
$$

$$(-EI_\omega \quad GK_T + \bar{K})\begin{Bmatrix}\theta''' \\ \theta'\end{Bmatrix} = (u' \quad v' \quad 1)\begin{Bmatrix}M_x - (v + \eta_0)F_{zo} \\ M_y + (u + \xi_0)F_{zo} \\ M_{zo} + (v + \eta_0)F_{xo} - (u + \xi_0)F_{yo}\end{Bmatrix} \tag{2.168}$$

If the principal axes are taken for the cross section coordinates (ξ, η) and the shear center S (ξ_0, η_0) is taken as the pole of the normalized warping, Eq. (2.167) and Eq. (2.168) will be much simplified. This simplification will be presented later in Eq. (2.172).

This simplification is only applicable in elastic range. However, for a doubly symmetric section such as a wide flange shape, the shifting of the shear center from the centroid may be small even in the plastic range, thus, we may assume

$$S_\omega = I_{\omega x} = I_{\omega y} = 0$$

Equation of equilibrium, Eq. (2.167), is now simplified,

$$E\begin{bmatrix} A & S_y & -S_x \\ S_y & I_x & -I_{xy} \\ -S_x & -I_{xy} & I_y \end{bmatrix}\begin{Bmatrix} w' \\ -v'' + \theta u'' \\ u'' + \theta v'' \end{Bmatrix} = \begin{bmatrix} 1 & 0 & 0 & 0 \\ 0 & 1 & \theta & -u' \\ 0 & -\theta & 1 & -v' \end{bmatrix}\begin{Bmatrix} F_{zo} \\ M_x - vF_{zo} \\ M_y + uF_{zo} \\ M_{zo} + vF_{xo} - uF_{yo} \end{Bmatrix} \tag{2.169}$$

Now let us rewrite Eqs. (2.169) and (2.168) using the following relations [see Eq. (2.163)]

$$\begin{aligned} F_{xo} &= -M'_y \\ F_{yo} &= M'_x \\ F_{zo} &= N = -P, \quad w' = \varepsilon_0 \end{aligned} \tag{2.170}$$

then

$$\begin{aligned} & E(S_y + \theta S_x)v'' + E(S_x - \theta S_y)u'' - EA\varepsilon_0 = P \\ & -E(I_x + \theta I_{xy})v'' - E(I_{xy} - \theta I_x)u'' + ES_y\varepsilon_0 \\ & = [M_x + vP] + \theta[M_y - uP] - u'(M_{zo} - vM'_y - uM'_x) \\ & E(I_y - \theta I_{xy})u'' + E(I_{xy} + \theta I_y)v'' - ES_x\varepsilon_0 \\ & = [M_y - uP] - \theta[M_x + vP] - v'(M_{zo} - vM'_y - uM'_x) \\ & -EI_\omega\theta''' + (GK_T + \bar{K})\theta' = u'[M_x + (v + \eta_0)P] + v'[M_y - (u + \xi_0)P] \\ & \qquad + M_{zo} - (v + \eta_0)M'_y - (u + \xi_0)M'_x \end{aligned} \tag{2.171}$$

These are the basic equations to be used in Chap. 7 for plastic analysis of beam-columns [see Eq. (7.1)] where the sectional properties S_x, I_x, etc., are evaluated for the elastic, unyielded portion of the cross section.

(Note: Sections 2.6.4 and 2.7 have been updated by the author. Please download the correct text at www.jrosspub.com/wav).

In an elastic analysis, if ξ and η are the principal axes of the cross section and ω is the normalized unit warping with respect to the shear center $S(\xi_0, \eta_0)$,

$$S_x = S_y = S_\omega = I_{xy} = I_{\omega x} = I_{\omega y} = 0$$

then, Eqs. (2.171) are reduced to

$$\begin{aligned}
&-EA\varepsilon_0 = P \\
&-EI_x(v'' - \theta u'') = (v - \theta u)P + M_x + \theta M_y - u'(M_{zo} - vM'_y - uM'_x) \\
&EI_y(u'' + \theta v'') = -(u + \theta v)P + M_y - \theta M_x - v'(M_{zo} - vM'_y - uM'_x) \\
&-EI_\omega \theta''' + (GK_T + \bar{K})\theta' = u'[M_x + (v + \eta_0)P] + v'[M_y - (u + \xi_0)P] \\
&\qquad + M_{zo} - (v + \eta_0)M'_y - (u + \xi_0)M'_x
\end{aligned} \tag{2.172}$$

In the case of elastic section, the shear center $S(\xi_0, \eta_0)$ can be clearly defined. Further, the rotation of the section θ takes place with respect to the shear center; thus Eq. (2.172) can be rewritten in terms of displacements of the shear center,

$$u_0 = u - \theta\eta_0 \qquad \text{and} \qquad v_0 = v + \theta\xi_0$$

For their derivatives, however, the following approximations are used

$$u'_0 = u', \quad v'_0 = v' \qquad \text{and} \qquad u''_0 = u'', \quad v''_0 = v''$$

Thus the last three equations of Eq. (2.172) become

$$\begin{aligned}
-EI_x v''_0 &= (v_0 - \theta\xi_0 - \theta u_0 - \theta^2\eta_0)P + M_x + \theta M_y \\
&\quad - u'_0[M_{zo} - (v_0 - \theta\xi_0)M'_y - (u_0 + \theta\eta_0)M'_x] \\
EI_y u''_0 &= -(u_0 + \theta\eta_0 + \theta v_0 - \theta^2\xi_0)P + M_y - \theta M_x \\
&\quad - v'_0[M_{zo} - (v_0 - \theta\xi_0)M'_y - (u_0 + \theta\eta_0)M'_x] \\
-EI_\omega\theta''' + (GK_T + \bar{K})\theta' &= u'_0[M_x + (v_0 - \theta\xi_0 + \eta_0)P] \\
&\quad + v'_0[M_y - (u_0 + \theta\eta_0 + \xi_0)P] \\
&\quad + M_{zo} - (v_o - \theta\xi_0 + \eta_0)M'_y - (u_0 + \theta\eta_0 + \xi_0)M'_x
\end{aligned} \tag{2.173}$$

Further, if we neglect all the nonlinear terms of displacements, we have

$$\begin{aligned}
&EI_x v''_0 + (v_0 - \theta\xi_0)P + M_x + \theta M_y - u'_0 M_{zo} = 0 \\
&EI_y u''_0 + (u_0 + \theta\eta_0)P - M_y + \theta M_x + v'_0 M_{zo} = 0 \\
&EI_\omega\theta''' - (GK_T + \bar{K})\theta' + (u'_0\eta_0 - v'_0\xi_0)P + u'_0 M_x + v'_0 M_y \\
&\qquad + M_{zo} - (v_0 - \theta\xi_0 + \eta_0)M'_y - (u_0 + \theta\eta_0 + \xi_0)M'_x = 0
\end{aligned} \tag{2.174}$$

These are the basic equations of elastic beam-columns.

In the case that the cross section is doubly symmetric so that the shear center coincides with the centroid,

(Note: Sections 2.6.4 and 2.7 have been updated by the author. Please download the correct text at www.jrosspub.com/wav).

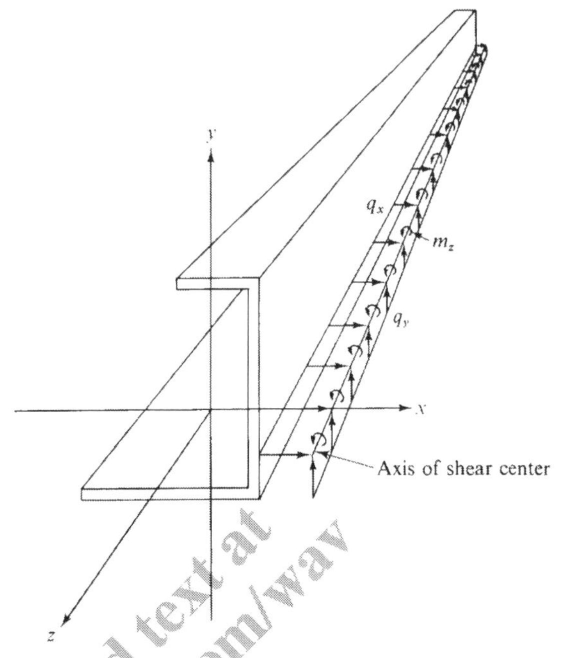

FIGURE 2.21
Beam-column with distributed load

$$\xi_0 = \eta_0 = 0 \tag{2.175}$$

then Eqs. (2.174) can be further reduced to

$$\begin{aligned} EI_x v'' + vP + M_x + \theta M_y - u'M_{zo} &= 0 \\ EI_y u'' + uP - M_y + \theta M_x + v'M_{zo} &= 0 \\ EI_\omega \theta''' - (GK_T + \bar{K})\theta' + u'M_x + v'M_y + M_{zo} - vM'_y - uM'_x &= 0 \end{aligned} \tag{2.176}$$

Further, if the load is applied symmetrically at both ends,

$$M_x = M_{xo}, \quad M_y = M_{yo}, \quad M_{zo} = 0, \quad M'_x = M'_y = 0 \tag{2.177}$$

Thus

$$\begin{aligned} EI_x v'' + vP + M_{xo} + \theta M_{yo} &= 0 \\ EI_y u'' + uP - M_{yo} + \theta M_{xo} &= 0 \\ EI_\omega \theta''' - (GK_T + \bar{K})\theta' + u'M_{xo} + v'M_{yo} &= 0 \end{aligned} \tag{2.178}$$

These are the most fundamental differential equations of beam-columns in space.

If a set of uniformly distributed loads q_x, q_y and m_z are applied along the shear center as shown in Fig. 2.21, the following fourth order differential equations instead of Eqs. (2.174) are to be used, i.e.,

$$
\begin{aligned}
&EI_x v_0^{\mathrm{IV}} + (v_0'' - \theta''\xi_0)P + \theta'' M_y + 2\theta' M_y' + \theta M_y'' - u_0''' M_{zo} - 2u_0'' M_{zo}' + M_x'' = 0 \\
&EI_y u_0^{\mathrm{IV}} + (u_0'' + \theta''\eta_0)P + \theta'' M_x + 2\theta' M_x' + \theta M_x'' + v_0''' M_{zo} + 2v_0'' M_{zo}' - M_y'' = 0 \\
&EI_\omega \theta^{\mathrm{IV}} - (GK_T + \bar{K})\theta'' - \bar{K}'\theta' + u_0''(\eta_0 P + M_x) - v_0''(\xi_0 P - M_y) \qquad (2.179a) \\
&\qquad + \theta'\xi_0 M_y' - (v_0 - \theta\xi_0 + \eta_0)M_y'' \\
&\qquad - \theta'\eta_0 M_x' - (u_0 + \theta\eta_0 + \xi_0)M_x'' + M_z' = 0
\end{aligned}
$$

in which

$$
\begin{aligned}
M_x &= M_{xo} + F_{yo}z - \frac{1}{2}q_y z^2 \\
M_y &= M_{yo} - F_{xo}z + \frac{1}{2}q_x z^2 \qquad (2.179b) \\
M_z &= M_{zo} - m_z z
\end{aligned}
$$

are to be substituted instead of Eq. (2.163). Equations (2.179*a*) are the general differential equations used in Chap. 4. Boundary conditions for these differential equations are

$$
\begin{array}{lll}
\text{displacement} & u, v & (= 0 \text{ if supported}) \\
\text{slope} & u', v' & (= 0 \text{ if fixed}) \\
\text{bending moment} & EI_x v'' = M_{xo} + \theta M_{yo} & (= 0 \text{ if simply supported}) \\
 & EI_y u'' = M_{yo} - \theta M_{xo} & \\
\text{shear force} & EI_x v''' = -V_y - \theta V_x & (= 0 \text{ for free end}) \\
 & EI_y u''' = -V_x + \theta V_y & \qquad (2.180) \\
\text{rotation angle} & \theta & (= 0 \text{ if restrained}) \\
\text{twist} & \theta' & (= 0 \text{ if warping restrained}) \\
\text{warping moment} & EI_\omega \theta'' = M_\omega & (= 0 \text{ if warping free}) \\
\text{twisting moment} & EI_\omega \theta''' - (GK_T + \bar{K})\theta' = M_z & (= 0 \text{ for free end})
\end{array}
$$

2.7 SUMMARY (Note: Sections 2.6.4 and 2.7 have been updated by the author. Please download the correct text at www.jrosspub.com/wav).

In the analysis of torsion, the coordinate of double sectorial area or unit warping plays an important role (Fig. 2.13)

$$
\begin{aligned}
&\text{centroidal} && \omega = \int_0^s \rho \, ds \\
&\text{shear center} && \omega_0 = \int_0^s \rho_0 \, ds = \omega + \eta_0(\xi - \xi_A) - \xi_0(\eta - \eta_A) \qquad (2.181) \\
&\text{normalized} && \omega_n = \frac{1}{A}\int_0^E \omega_0 t \, ds - \omega_0
\end{aligned}
$$

(Note: Sections 2.6.4 and 2.7 have been updated by the author. Please download the correct text at www.jrosspub.com/wav).

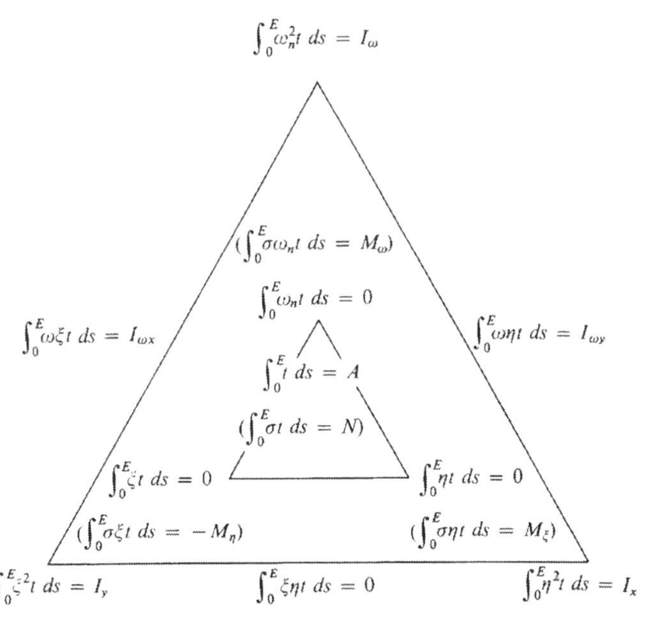

FIGURE 2.22
Relations among section properties

Using these unit warpings together with the principal coordinate axes ξ and η, the sectional properties are

$$\begin{gathered}\int_0^E t\, ds = A \\ \int_0^E \xi t\, ds = \int_0^E \eta t\, ds = \int_0^E \omega_n t\, ds = 0 \\ \int_0^E \xi\eta t\, ds = 0, \quad \int_0^E \xi\omega t\, ds = I_{\omega x}, \quad \int_0^E \eta\omega t\, ds = I_{\omega y} \\ \int_0^E \xi^2 t\, ds = I_y, \quad \int_0^E \eta^2 t\, ds = I_x, \quad \int_0^E \omega_n^2 t\, ds = I_\omega\end{gathered} \tag{2.182}$$

Mutual relationships among these sectional properties are shown schematically in Fig. 2.22. Using these, the characteristics of torsion are analyzed.

Normal stress due to bending moments M_ξ and M_η is

$$\sigma = \frac{1}{I_x I_y - I_{xy}^2}\left[(I_y M_\xi + I_{xy} M_\eta)\eta - (I_{xy} M_\xi + I_x M_\eta)\xi\right] \tag{2.183}$$

Shear stress due to shear forces V_ξ and V_η is

$$\tau t = \frac{-1}{I_x I_y - I_{xy}^2}\left[(I_x V_\xi - I_{xy} V_\eta)\int_0^s \xi t\, ds + (I_y V_\eta - I_{xy} V_\xi)\int_0^s \eta t\, ds\right] + \tau_A t_A \tag{2.184}$$

with

$$\tau_A t_A = \begin{cases} 0 & \text{(for open section)} \\ \dfrac{(I_y V_\eta - I_{xy} V_\xi) \oint \dfrac{\int_0^s \eta t\, ds}{t} ds + (I_x V_\xi - I_{xy} V_\eta) \oint \dfrac{\int_0^s \xi t\, ds}{t} ds}{(I_x I_y - I_{xy}^2) \oint \dfrac{ds}{t}} & \text{(for closed section)} \end{cases} \tag{2.185}$$

The location of the shear center is

$$\begin{aligned} \xi_0 &= \frac{I_y I_{\omega y} - I_{xy} I_{\omega x}}{I_x I_y - I_{xy}^2} \\ \eta_0 &= -\frac{I_x I_{\omega x} - I_{xy} I_{\omega y}}{I_x I_y - I_{xy}^2} \end{aligned} \tag{2.186}$$

The St. Venant torsion equation is

$$T_{sv} = G K_T \theta' \tag{2.187}$$

with

$$K_T = \begin{cases} \dfrac{1}{3} \int t^3\, ds & \text{(for open section)} \\ 4A_0^2 \Big/ \oint \dfrac{1}{t} ds & \text{(for closed section)} \end{cases} \tag{2.188}$$

Shear stress due to the uniform torsion T_{sv} is

$$\tau = \begin{cases} \dfrac{3t T_{sv}}{\int t^3\, ds} & \text{on surface} \quad \text{(open section)} \\ \dfrac{T_{sv}}{2A_0 t} & \text{uniform} \quad \text{(closed section)} \end{cases} \tag{2.189}$$

The normal stress σ_w and the shear stress τ_w due to warping are

$$\begin{aligned} \sigma_w &= E\omega_n \theta'' \\ \tau_w &= \tau_A \left(\frac{t_A}{t}\right) - \frac{1}{t} E S_\omega \theta''' \end{aligned} \tag{2.190}$$

The torsional moment due to the warping shear stress is

$$T_w = -E I_\omega \theta''' \tag{2.191}$$

The differential equations of torsion are

$$\begin{aligned} E I_\omega \theta''' - G K_T \theta' &= -M_z && \text{for end torque} \\ E I_\omega \theta^{\text{IV}} - G K_T \theta'' &= m_z && \text{for distributed torque} \end{aligned} \tag{2.192}$$

(Note: Sections 2.6.4 and 2.7 have been updated by the author. Please download the correct text at www.jrosspub.com/wav).

The general solutions are, respectively

$$\theta = C_1 + C_2 \cosh \lambda z + C_3 \sinh \lambda z + \frac{M_z}{\lambda^2 EI_\omega} z$$
$$\theta = C_4 + C_5 z + C_6 \cosh \lambda z + C_7 \sinh \lambda z - \frac{m_z}{2\lambda^2 EI_\omega} z^2 \tag{2.193}$$

with

$$\lambda = \sqrt{\frac{GK_T}{EI_\omega}} \tag{2.194}$$

The constants C_1 to C_7 are determined from boundary conditions: $\theta = 0$, $\theta' = 0$ $\theta'' = 0$ or $\theta''' - \lambda^2\theta' = 0$. The normal stress is

$$\sigma = E(\varepsilon_0 - \eta v'' - \xi u'' - \omega_n \theta'') \tag{2.195}$$

When ξ and η are principal axes and ω_n is normalized warping, the generalized stress-strain relations are obtained as

$$N = EA\varepsilon_0$$
$$M_\xi = -EI_x v''$$
$$M_\eta = EI_y u'' \tag{2.196}$$
$$M_\omega = -EI_\omega \theta''$$

and the other forces related to shear stresses such as

$$V_\xi = -EI_y u'''$$
$$V_\eta = -EI_x v'''$$
$$T_{sv} = GK_T\theta' \tag{2.197}$$
$$T_w = -EI_\omega \theta'''$$

are considered to be reactions which are needed only for equilibrium considerations. They are directly related to the generalized stresses by:

$$V_\eta = M'_\xi, \quad V_\xi = -M'_\eta \qquad \text{and} \qquad T_w = M'_\omega \tag{2.198}$$

For the second order analysis we add the torsional moment due to the normal stress (Wagner effect),

$$T_\sigma = \bar{K}\theta' \tag{2.199}$$

with

$$\bar{K} = \int a^2 \sigma t \, ds = \int [(\xi - \xi_0)^2 + (\eta - \eta_0)^2]\sigma \, dA \tag{2.200}$$

The differential equations are obtained by equating the internal moments to the external moments under deformed configuration of the beam-column. The general

(Note: Sections 2.6.4 and 2.7 have been updated by the author. Please download the correct text at www.jrosspub.com/wav).

equations have highly nonlinear terms as was seen in Eq. (2.164) and Eq. (2.165). These equations are derived using the rotation matrix, Eq. (2.154) instead of Eq. (2.153). If all the nonlinear terms of displacements are neglected, we have the approximate equations

$$\begin{aligned} &EI_x v_0'' + (v_0 - \theta\xi_0)P + M_x + \theta M_y - u_0' M_{zo} = 0 \\ &EI_y u_0'' + (u_0 + \theta\eta_0)P - M_y + \theta M_x + v_0' M_{zo} = 0 \\ &EI_\omega \theta''' - (GK_T + \bar{K})\theta' + (u_0'\eta_0 - v_0'\xi_0)P + u_0' M_x + v_0' M_y + M_{zo} \\ &\qquad - (v_0 - \theta\xi_0 + \eta_0)M_y' - (u_0 + \theta\eta_0 + \xi_0)M_x' = 0 \end{aligned} \tag{2.201}$$

These simplified equations are often used as the basis for analyses. It is to be noted that the first two equations are equilibrium of bending moment about the principal axes on the cross section while the last equation is equilibrium of twisting moment about the shear center $S(\xi_0, \eta_0)$. All equations are written in terms of displacements of the shear center: u_0, v_0 and θ.

2.8 PROBLEMS

2.1 A beam of a π-shaped cross section is subjected to a concentrated lateral load Q at its mid-span but off-centered as shown in Fig. 2.23. Determine the maximum normal stress σ_{max} and the maximum shear stress τ_{max}. Also determine the vertical deflection of point A shown.

FIGURE 2.23
Centrally loaded beam for Prob. 2.1

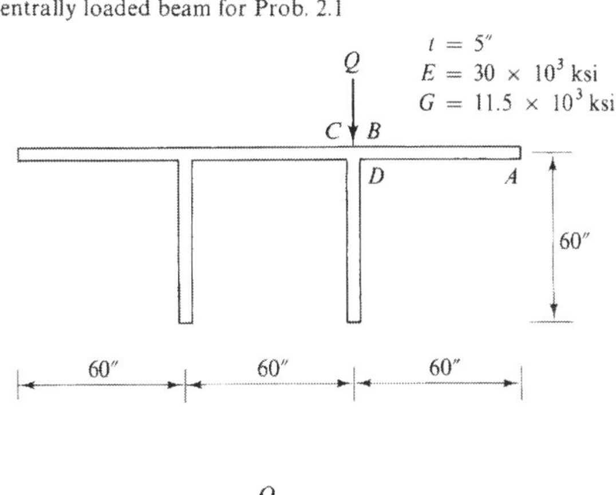

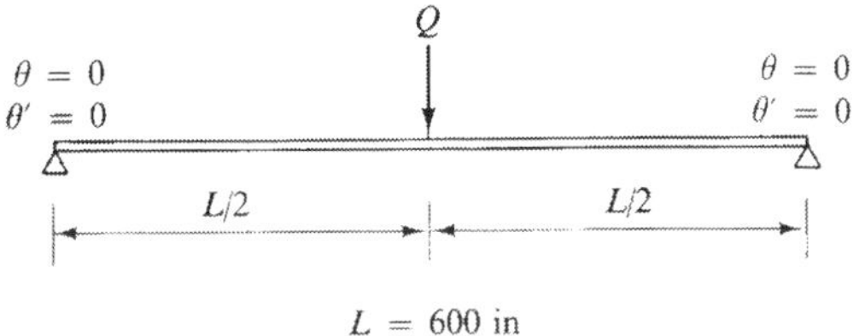

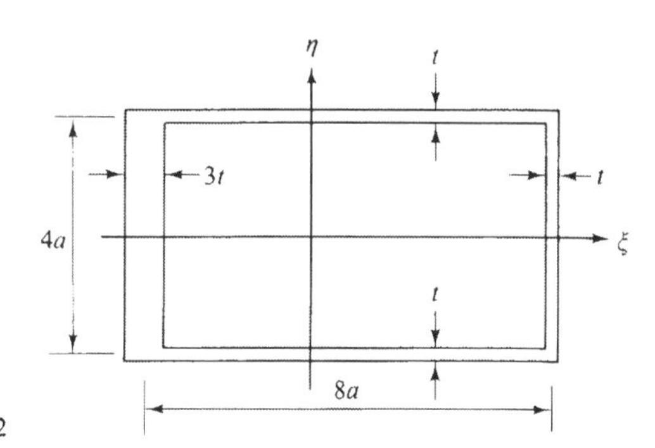

FIGURE 2.24
Box section for Prob. 2.2

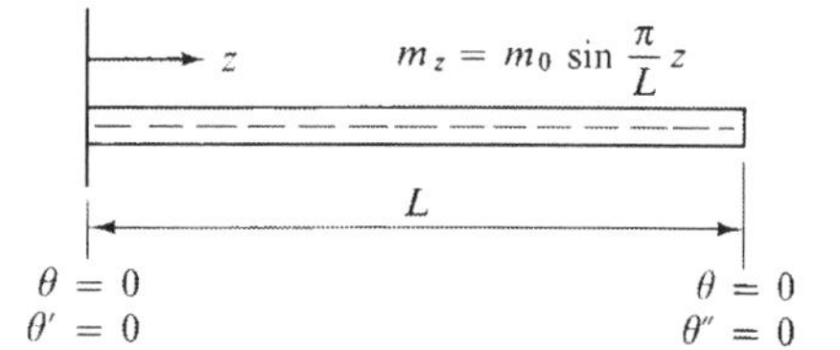

FIGURE 2.25
Beam under distributed torque for Prob. 2.3

$\theta = 0$
$\theta' = 0$

$\theta = 0$
$\theta'' = 0$

2.2 For the box section shown in Fig. 2.24, determine St. Venant torsion constant K_T, location of shear center $S(\xi_0, \eta_0)$ and shear stresses due to shear force V_η applied through the shear center.

2.3 Find the twisting angle θ for the cantilever beam which is subjected to a distributed torque as shown in Fig. 2.25.

2.4 Show that normal strain in general coordinates ξ, η and ω can be expressed by Eq. (2.116), or

$$\varepsilon = \varepsilon_0 + \eta\Phi_\xi - \xi\Phi_\eta - \omega\theta_\zeta''$$

Derive equations for ε_0, Φ_ξ and Φ_η.

2.5 Derive Eq. (2.128*a*) or

$$T_w = M_\omega' - \omega_E\left[EA\varepsilon_0' + ES_y\theta_\xi'' - ES_x\theta_\eta'' - ES_\omega\theta_\zeta''' - \tau_0 t_0\right]$$

Discuss in what condition the relation $T_\omega = M_\omega'$ holds true.

2.6 Show Eq. (2.158) or

$$\varepsilon_0 = w' + \frac{1}{2}(u')^2 + \frac{1}{2}(v')^2$$

2.9 REFERENCES

Galambos, T. V., *Structural Members and Frames*, Prentice-Hall, Inc. (1968).

Kollbrunner, C. F. and Basler, K., *Torsion*, Springer-Verlag, (1966).

Timoshenko, S. P. and Goodier, J. N., *Theory of Elasticity*, 2nd edn, McGraw-Hill Book Company, (1951).

Vlasov, V. Z., *Thin-Walled Elastic Beams*, 2nd edn, National Science Foundation, Washington, D. C. and Department of Commerce, U.S.A., by the Israel Program for Scientific Translations, Jerusalem, p. 292, (1961).

3

LATERAL BUCKLING OF BEAMS AND BEAM-COLUMNS

N. S. Trahair

3.1 INTRODUCTION

When a beam is bent in its stiffer principal plane, it usually deflects only in that plane. However, if the beam does not have sufficient lateral stiffness or lateral support to ensure that this is so, then the beam may buckle out of the plane of loading, as shown in Fig. 3.1. The load at which this buckling occurs may be substantially less than the beam's in-plane load carrying capacity, as indicated in Fig. 3.2, which shows the variations of the moment capacity with slenderness. While short beams can reach the full plastic moment M_p, more slender beams may buckle at moments M_c which are significantly less than M_p.

For an idealized perfectly straight elastic beam, there are no out-of-plane displacements until the applied moment reaches the critical value M_e, when the beam buckles by deflecting laterally and twisting, as shown in Fig. 3.1. Thus lateral buckling involves lateral bending and axial torsion. These two actions are interdependent, and when the beam deflects laterally, the applied moment exerts a component torque about the deflected longitudinal axis which causes the beam to twist, while twisting of the beam causes the applied moment to exert a component lateral bending moment which causes the beam to deflect. This behavior, which is

FIGURE 3.1
Lateral buckling of a cantilever

important for long unrestrained I-beams whose resistances to lateral bending and torsion are low, is called *elastic flexural-torsional buckling.*

The failure of a perfectly straight slender beam is initiated when the additional stresses induced by elastic lateral buckling cause first yield. However, a perfectly straight beam of intermediate slenderness may yield before the buckling load is reached, because of the combined effects of in-plane bending stresses and residual stresses, and may subsequently buckle plastically, as indicated in Fig. 3.2. For very short beams, the plastic buckling load may be higher than the in-plane plastic collapse load, in which case the design load-carrying capacity of the beam is not affected by lateral buckling.

A beam-column which is bent in its stiffer principal plane and compressed [see Fig. 3.3(b)] may also fail prematurely by buckling out of the plane of loading. This behavior is closely related to the flexural-torsional buckling of beams and columns. A two-dimensional rigid frame is composed of beam-columns, and may also buckle laterally under the action of in-plane loading.

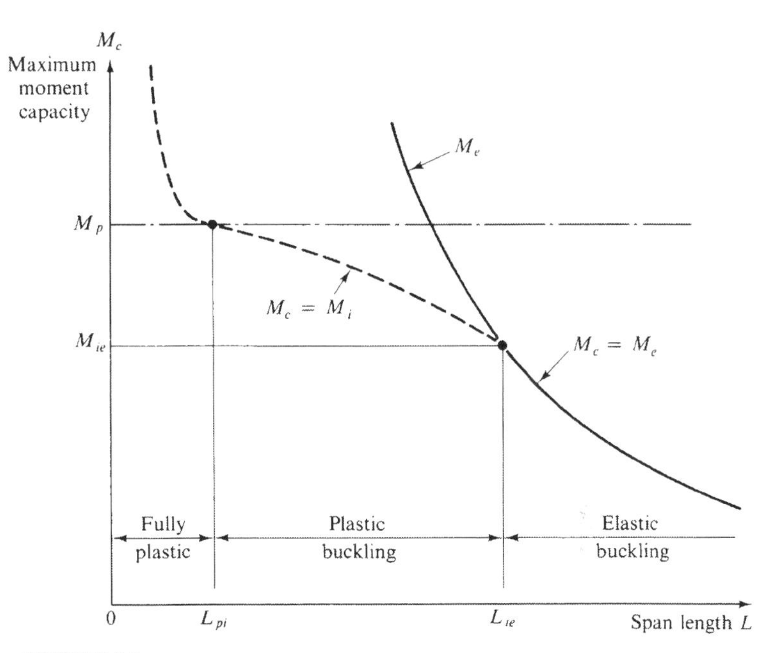

FIGURE 3.2
Maximum moment capacities of beams

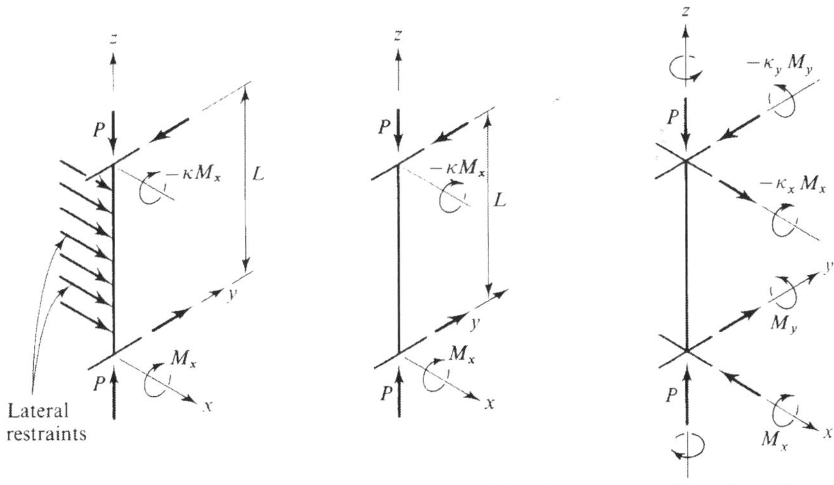

(a) *In-Plane Behavior*
Column deflects v in yz plane only

(b) *Flexural-Torsional Buckling*
Column deflects v in yz plane plane then buckles by deflecting u in xz plane and twisting θ

(c) *Biaxial Bending*
Column deflects uv and twists θ

FIGURE 3.3
Beam-column behavior

In this chapter, the behavior and design of beams, beam-columns, and frames which fail by yielding and lateral buckling are discussed. It is assumed that local buckling of the compression flanges or of the webs does not occur. The general behavior and design of beams and beam-columns bent about both principal axes [see Fig. 3.3(c)] are discussed in later chapters.

3.2 ELASTIC SIMPLY SUPPORTED BEAMS

3.2.1 Equal End Moments

A simply supported beam which is bent in its stiffer principal plane by equal and opposite end moments is shown in Fig. 3.4(a). The beam is elastic and of uniform doubly symmetric I-section. The beam supports prevent both lateral deflection and twist, but the flange ends are free to warp.

The beam will buckle at an elastic critical moment M_e when a deflected and twisted equilibrium position, such as that shown in Fig. 3.4, is possible. The differential equilibrium equations of minor axis bending and torsion of the beam in this position are

$$-EI_y u'' = M_e \theta \tag{3.1}$$

and

$$GK_T \theta' - EI_\omega \theta''' = M_e u' \tag{3.2}$$

in which each prime (′) indicates one differentiation with respect to z. In these equations, $M_e\theta$ is the lateral bending moment induced by the twisting θ of the beam and $M_e u'$ is the torque induced by the lateral deflection u, while the right-hand rule is used to determine the signs of the twist and of the external moment and torque. These equations can also be obtained from the more general Eq. (2.178)

FIGURE 3.4
Buckling of a simply supported I-beam

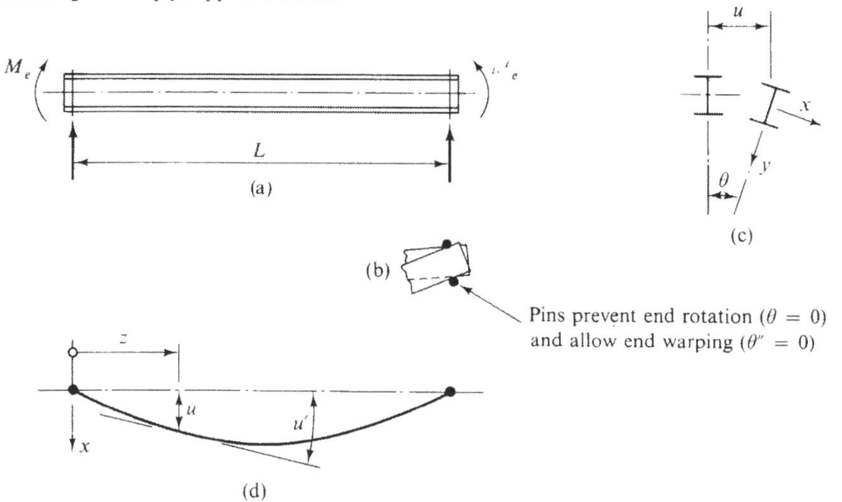

by substituting $M_{xo} = M_e$ and $M'_x = 0$ (equal and opposite major axis end moments), $P = 0$ (no axial load), $M_{yo} = M'_y = 0$ (no minor axis end moments), and $\bar{K} = 0$ (doubly symmetric cross-section with $P = 0$). When these equations are both satisfied at all points along the beam, then the deflected and twisted position is one of equilibrium. This position can be found by differentiating Eq. (3.2) and substituting Eq. (3.1), whence

$$GK_T\theta'' - EI_\omega\theta^{IV} = M_e u'' = M_e\left(-\frac{M_e\theta}{EI_y}\right) \tag{3.3}$$

or

$$\theta^{IV} - \frac{GK_T}{EI_\omega}\theta'' - \frac{M_e^2}{EI_yEI_\omega}\theta = 0 \tag{3.4}$$

The general solution of this equation is

$$\theta = A_1 \sinh \alpha_1 z + A_2 \cosh \alpha_1 z + A_3 \sin \alpha_2 z + A_4 \cos \alpha_2 z \tag{3.5}$$

in which A_1, A_2, A_3 and A_4 are constants of integration and

$$\alpha_1 = \sqrt{\frac{GK_T}{2EI_\omega} + \sqrt{\left(\frac{GK_T}{2EI_\omega}\right)^2 + \frac{M_e^2}{EI_yEI_\omega}}} \tag{3.6}$$

and

$$\alpha_2 = \sqrt{\frac{-GK_T}{2EI_\omega} + \sqrt{\left(\frac{GK_T}{2EI_\omega}\right)^2 + \frac{M_e^2}{EI_yEI_\omega}}} \tag{3.7}$$

The support conditions of end twisting prevented and end warping free are represented by

$$(\theta)_o = (\theta'')_o = (\theta)_l = (\theta'')_l = 0 \tag{3.8}$$

If Eq. (3.5) is substituted into the first two of these, then

$$A_2 + A_4 = 0 \tag{3.9}$$

and

$$\alpha_1^2 A_2 - \alpha_2^2 A_4 = 0 \tag{3.10}$$

or

$$A_2 = A_4 = 0 \tag{3.11}$$

If Eqs. (3.5) and (3.11) are substituted into the last two of Eq. (3.8), then

$$A_1 \sinh \alpha_1 L + A_3 \sin \alpha_2 L = 0 \tag{3.12}$$

and

$$\alpha_1^2 A_1 \sinh \alpha_1 L - \alpha_2^2 A_3 \sin \alpha_2 L = 0 \tag{3.13}$$

which can be solved as

$$A_1(\alpha_1^2 + \alpha_2^2) \sinh \alpha_1 L = 0 \tag{3.14}$$

and

$$A_3(\alpha_1^2 + \alpha_2^2) \sin \alpha_2 L = 0 \tag{3.15}$$

Because α_1 is always positive, Eq. (3.14) can only be satisfied when

$$A_1 = 0 \tag{3.16}$$

Eq. (3.15) is satisfied when

$$A_3 = 0 \tag{3.17}$$

in which case $\theta = 0$ everywhere, and the beam is in the stable pre-buckled position. Eq. (3.15) is also satisfied when

$$\sin \alpha_2 L = 0 \tag{3.18}$$

or

$$\alpha_2 L = n\pi \tag{3.19}$$

in which n is an integer. If this is substituted into Eq. (3.7), then after some rearrangement,

$$M_e = \frac{n\pi}{L}\sqrt{EI_y GK_T}\sqrt{1 + \frac{n^2\pi^2 EI_\omega}{L^2 GK_T}} \tag{3.20}$$

The lowest critical moment corresponds to the lowest value of the integer n, and so this is given by

$$M_e = \frac{\pi}{L}\sqrt{EI_y GK_T}\sqrt{1 + \frac{\pi^2 EI_\omega}{L^2 GK_T}} \tag{3.21}$$

Beams of narrow rectangular section can be considered as the special case of I-beams with zero warping rigidity EI_ω, and for which Eq. (3.21) reduces to the simpler form

$$M_e = \frac{\pi}{L}\sqrt{EI_y GK_T} \tag{3.22}$$

On the other hand, very thin-walled I-beams have comparatively low torsional rigidities GK_T, and if these are ignored, then Eq. (3.21) reduces to

$$M_e = \frac{\pi^2}{L^2}\sqrt{EI_y EI_\omega} \tag{3.23}$$

The twisted shape at buckling can be obtained by substituting Eqs. (3.11), (3.16) and (3.19) (with $n = 1$) into Eq. (3.5), whence

$$\theta = A_3 \sin \pi z/L \tag{3.24}$$

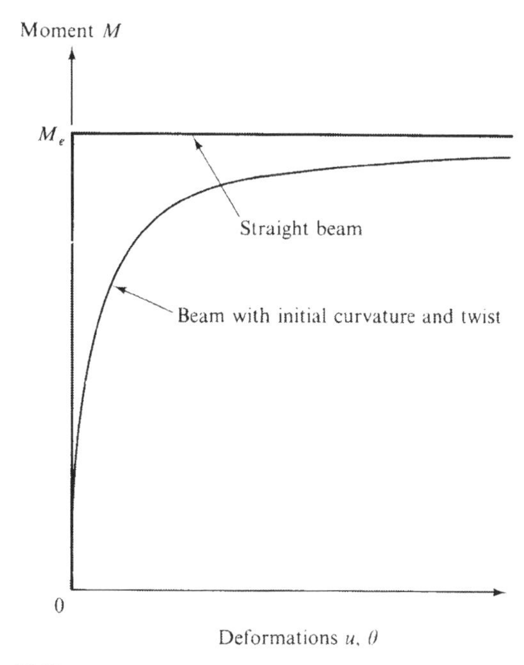

FIGURE 3.5
Lateral deflection and twist of a beam with equal end moments

in which A_3 represents the central twist of the beam and is of indeterminate magnitude, as shown in Fig. 3.5. The deflected shape at buckling can be obtained by substituting Eq. (3.24) into Eq. (3.1), by integrating twice, and by substituting the boundary conditions $(u)_o = (u)_l = 0$, whence

$$u = \frac{M_e A_3}{\pi^2 EI_y/L^2} \sin \pi z/L \tag{3.25}$$

in which $M_e A_3/(\pi^2 EI_y/L^2)$ represents the indeterminate central deflection.

Equation (3.21) is independent of the major axis flexural rigidity EI_x. This is the result of the assumption that the vertical deflections are small enough to be neglected, which is justifiable when EI_x is very much greater than the minor axis flexural rigidity EI_y. This assumption produces conservative estimates of the elastic critical moment, as the true critical moments are equal to the values obtained from Eq. (3.21) divided by the correction factor

$$\sqrt{(1 - EI_y/EI_x)[1 - GK_T(1 + \pi^2 EI_\omega/L^2 GK_T)/EI_x]}$$

(Baker, Horne and Heyman, 1956, Trahair and Woolcock, 1973). This correction factor, which is just less than unity for most beam sections but may be significantly less than unity for column sections, is usually neglected in design. Nevertheless, its value approaches zero as EI_y approaches EI_x, so that the true elastic critical moment approaches infinity. Thus a beam which is bent about its weak axis does not buckle, which is intuitively obvious. Other studies (Clark and Knoll, 1958, Vacharajittiphan *et al.*, 1974) have shown that, in some cases of beams with minor axis end restraints, the correction factor may be close to unity, and that it is generally prudent to ignore the effect of major axis curvature.

3.2.2 Unequal End Moments

A simply supported beam with unequal major axis end moments M_e and κM_e is shown in Fig. 3.6(a). The differential equilibrium equations of minor axis bending and torsion of the beam when in a buckled position are

$$-EI_y u'' = M_x \theta \tag{3.26}$$

and

$$GK_T \theta' - EI_\omega \theta''' = M_x u' - V_y u \tag{3.27}$$

in which the major axis bending moment M_x and shear V_y are given by

$$M_x = M_e - (1 - \kappa) M_e z / L \tag{3.28}$$

and

$$V_y = M'_x = -(1 - \kappa) M_e / L \tag{3.29}$$

These equations are equivalent to those which can be obtained from the more general Eqs. (2.176) and (2.170).

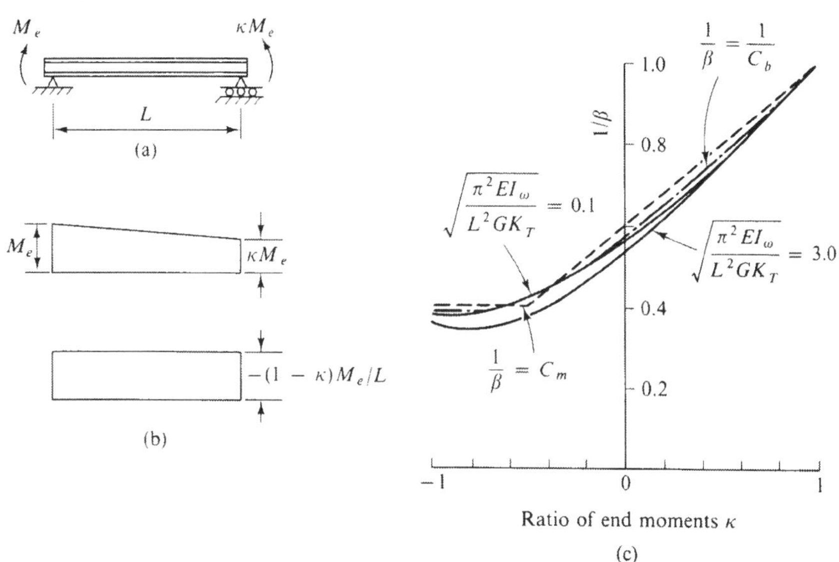

Buckling of beams with unequal end moments. (a) Beam. (b) Moment and shear diagrams. (c) Reciprocals of moment modification factors.

Closed form solutions of these equations are not available. In the past, hand methods of calculation were used to solve differential equations such as these. Details of many of these are given in standard texts (such as Timoshenko and Gere, 1961, Bleich, 1952, and the Structural Stability Research Council Guide edited by Johnston, 1976). While some of these hand methods have been successfully adapted for use with computers, new methods of calculating buckling loads have been developed which were specifically designed for use with computers, and which allow very much larger and more complicated problems to be solved than previously. Some of these computer methods have been extensively used for lateral buckling problems, and these include the finite element method (Barsoum and Gallagher, 1970, Nethercot and Rockey 1971b), the finite integral method (Brown and Trahair, 1968, Vacharajittiphan and Trahair, 1975), the direct stiffness method (Vacharajittiphan and Trahair, 1974), and the transfer matrix method (Yoshida and Imoto, 1973). The direct stiffness method is presented in Section 3.11.

Numerical solutions have been obtained for the beam with unequal end moments shown in Fig. 3.6(a) (Horne, 1954). These can be conveniently expressed in the form of

$$M_e = \beta \frac{\pi\sqrt{EI_y GK_T}}{L}\sqrt{1 + \frac{\pi^2 EI_\omega}{L^2 GK_T}} \tag{3.30}$$

in which the dimensionless factor β accounts for the effect of the non-uniform distribution of the bending moment M_x on elastic flexural-torsional buckling. The variation of $1/\beta$ with the end moment ratio is shown in Fig. 3.6(c) for the two extreme values of the beam parameter $\sqrt{\pi^2 EI_\omega / L^2 GK_T}$ of 0.1 and 3. Also shown in Fig. 3.6(c) is the approximation

$$\beta \simeq C_b = 1.75 - 1.05\kappa + 0.3\kappa^2 \leq 2.56 \tag{3.31}$$

and the even simpler approximation

$$\frac{1}{\beta} \simeq C_m = 0.6 + 0.4\kappa \geq 0.4 \tag{3.32}$$

For beams in uniform bending ($\kappa = 1$), Eqs. (3.31) and (3.32) give $\beta = 1$, in which case Eq. (3.30) reduces to Eq. (3.21). For beams under moment gradient ($\kappa < 1$), the factor β is greater than 1, and the resistance to lateral buckling is higher. The maximum resistance is achieved in beams which are almost bent in double curvature ($\kappa \approx -0.8$), and is approximately 2.5 times the value obtained from Eq. (3.21) for uniform bending.

3.2.3 Central Concentrated Load

A simply supported beam with a central concentrated load P acting at a distance $\bar{a}$ above the shear center axis of the beam is shown in Fig. 3.7(a). When the beam buckles by deflecting laterally and twisting, the line of action of the load

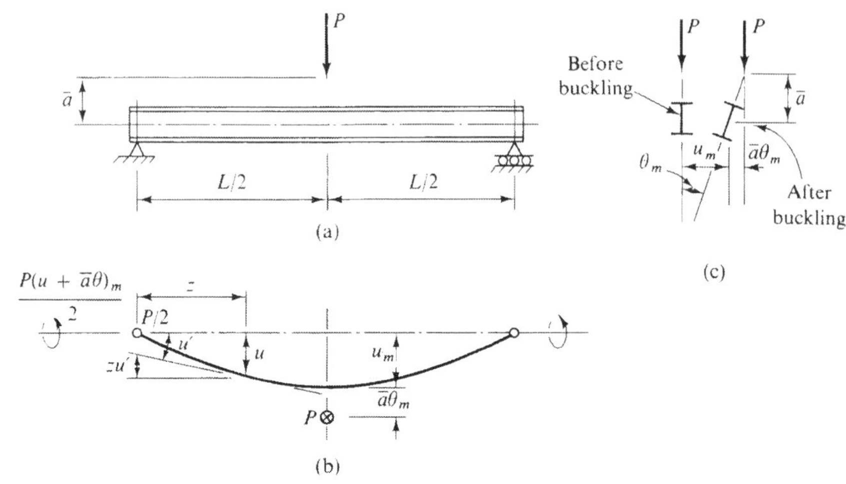

FIGURE 3.7
I-beam with central concentrated load

moves with the central cross section, but remains vertical, as shown in Fig. 3.7(c). The case when the load acts above the shear centre is more dangerous than that for shear center loading because of the additional torque $P\bar{a}\theta_m$ which increases the twisting of the beam and decreases its resistance to buckling.

The differential equilibrium equations of minor axis bending and torsion of the beam when in a buckled position are

$$-EI_y u'' = M_x \theta \tag{3.33}$$

and

$$GK_T\theta' - EI_\omega\theta''' = \frac{P}{2}(u + \bar{a}\theta)_m(1 - 2\langle z - L/2\rangle^0) + M_x u' - V_y u \tag{3.34}$$

in which $(P/2)(u + \bar{a}\theta)_m$ is the end torque and M_x and V_y are given by

$$M_x = Pz/2 - P\langle z - L/2\rangle \tag{3.35}$$

$$V_y = P/2 - P\langle z - L/2\rangle^0 \tag{3.36}$$

in which the values of the second terms are taken as zero when the values inside the Macaulay brackets $\langle\ \rangle$ are negative.

$$\langle S\rangle = \begin{cases} 0 & (S < 0) \\ S & (S \geq 0) \end{cases} \qquad \langle S\rangle^0 = \begin{cases} 0 & (S < 0) \\ 1 & (S > 0) \end{cases}$$

Numerical solutions of these equations for the dimensionless critical load $PL^2/\sqrt{EI_yGK_T}$ are available (Vlasov, 1961, Timoshenko and Gere, 1961, Column Research Committee of Japan, 1971, Anderson and Trahair, 1972), and some of these are shown in Fig. 3.8, in which the dimensionless height δ of the point of application of the load is generally given by

$$\delta = \frac{\bar{a}}{L}\sqrt{\frac{EI_y}{GK_T}} \tag{3.37}$$

For shear center loading ($\delta = 0$), the elastic critical moment at midspan $M_e = PL/4$ increases with the beam parameter $\sqrt{\pi^2 EI_\omega / L^2 GK_T}$ in much the same way as does the critical moment of beams with equal and opposite end moments [see Eq. (3.21)]. Thus the critical moment $PL/4$ can be written in the form

$$M_e = \frac{PL}{4} = \beta \frac{\pi\sqrt{EI_y GK_T}}{L}\sqrt{1 + \pi^2 EI_\omega / L^2 GK_T} \tag{3.38}$$

in which the moment modification factor β which accounts for the effect of the non-uniform distribution of major axis bending moment is now approximately equal to 1.35. Equation (3.38) only differs from Eqs. (3.21) and (3.30) by the different value of β.

The elastic critical load also varies with the load height parameter δ, and although the resistance to buckling is high when the load acts below the shear center axis ($\delta < 0$), it decreases significantly as the point of application rises, as shown by the dashed curves in Fig. 3.8. For equal flanged I-beams, the parameter δ can be transformed into

$$\frac{2\bar{a}}{h} = \frac{\pi(\bar{a}/L)\sqrt{EI_y/GK_T}}{\sqrt{\pi^2 EI_y h^2/4L^2 GK_T}} = \pi\frac{\delta}{\sqrt{\pi^2 EI_\omega / L^2 GK_T}} \tag{3.39}$$

in which h is the distance between flange centroids and $I_\omega = I_y h^2/4$. The variation of the critical load with $2\bar{a}/h$ is shown by the solid lines in Fig. 3.8, and it can be seen that the differences between top ($2\bar{a}/h = 1$) and bottom ($2\bar{a}/h = -1$) flange loading increase

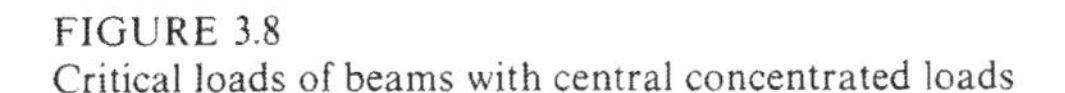

FIGURE 3.8
Critical loads of beams with central concentrated loads

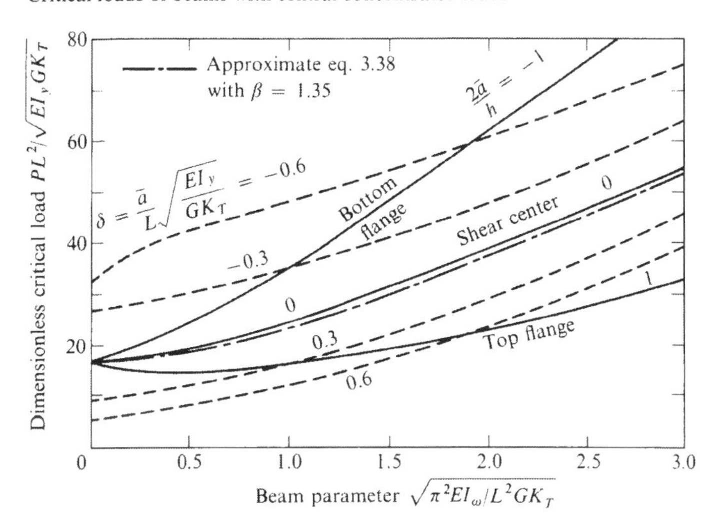

with the geometrical parameter $\sqrt{\pi^2 EI_\omega / L^2 GK_T}$. Thus, this effect is more important for deep beam-type sections of short span than for shallow column-type sections of long span. Approximate expressions for the variations of the modification factor β with the beam parameter $\sqrt{\pi^2 EI_\omega / L^2 GK_T}$ which account for the dimensionless load height $2\bar{a}/h$ for equal flanged I-beams are given by Nethercot and Rockey (1971a).

3.2.4 Other Loading Conditions

The effects of the distribution of the applied load along the length of a simply supported beam on its elastic buckling strength have been investigated numerically in many different studies, and tabulations of critical loads are available (Vlasov, 1961, Timoshenko and Gere, 1961, Clark and Hill, 1960, Structural Stability Research Council Guide edited by Johnston, 1976, Galambos, 1968, Column Research Committee of Japan, 1971, Nethercot and Rockey, 1971a, Anderson and Trahair, 1972, Nethercot, 1972c). Some approximate solutions for the maximum moments $M_{\max} = M_e$ at elastic buckling of simply supported beams which are loaded along their shear center axes are given in Fig. 3.9 by the moment modification factors β in the equation

FIGURE 3.9
Moment modification factors for simply supported beams with shear center loading

Loading	Bending moment	$M_{\max}$	β
 M M		M	1.00
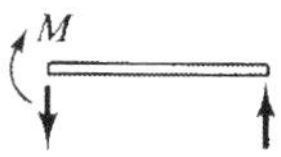 M		M	1.75
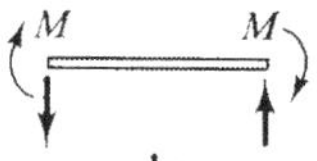 M M		M	2.56
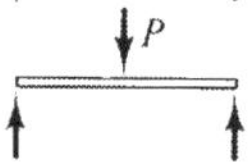 P	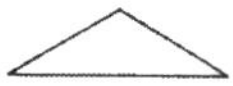	$PL/4$	1.35
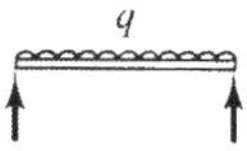 q		$qL^2/8$	1.13
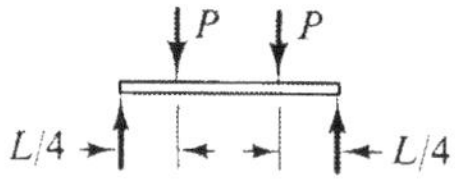 P P, L/4, L/4		$PL/4$	1.04
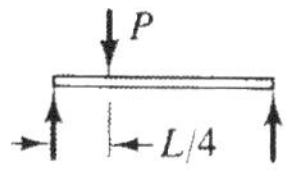 P, L/4	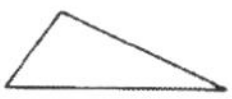	$3PL/16$	1.44

$$M_{\max} = M_e = \beta \frac{\pi\sqrt{EI_y GK_T}}{L} \sqrt{1 + \pi^2 EI_\omega / L^2 GK_T} \tag{3.40}$$

The more dangerous loadings are those for which the values of β are lower. It can be seen from Fig. 3.9 that these loadings are those which produce more nearly constant distributions of major axis bending moment, and that the worst case is that of equal and opposite end moments ($\kappa = 1$).

3.2.5 Monosymmetric Beams

The elastic buckling behavior of simply supported beams of uniform doubly symmetric cross section has been discussed in the previous sections. The buckling behavior of a *monosymmetric* I-beam (i.e., an I-beam such as that shown in Fig. 3.10 whose flanges are different and whose only section axis of symmetry is that through the midline of the web) is a little different, however, because its torsional behavior is different. In the case of simply supported beams with equal and opposite end moments, the differential equation for torsional equilibrium is

$$(GK_T + M_e\beta_x)\theta' - EI_\omega \theta''' = M_e u' \tag{3.41}$$

which differs from the corresponding Eq. (3.2) for doubly symmetric I-beams by the inclusion of the term $M_e\beta_x\theta'$. This is equal to the torque exerted by transverse components of the longitudinal bending stresses which arise when the beam is twisted, and is equivalent to the Wagner effect in Eq. (2.133) of

$$\bar{K}\theta' = \theta' \int_A \sigma a^2 \, dA \tag{3.42}$$

in which σ is the longitudinal stress and a is the distance from a point (x, y) in the cross section to the shear center (x_o, y_o).

In the case of monosymmetric I-beams, $x_o = 0$, while the moments M_e induce stresses $\sigma = M_e y / I_x$. Thus

$$\bar{K} = M_e \left\{ \frac{\int_A x^2 y \, dA + \int_A y^3 \, dA}{I_x} - 2y_0 \right\} = M_e \beta_x \tag{3.43}$$

in which β_x is a monosymmetry parameter of the cross section. An explicit expression for β_x for monosymmetric I-sections is given in Fig. 3.10, and this can also be used for tee-sections by putting the flange thickness t_1 or t_2 equal to zero. Also given in Fig. 3.10 is an explicit expression for the warping section constant I_ω of a monosymmetric I-section. For a tee-section, I_ω is zero.

The action of the torque $M_e\beta_x\theta'$ can be thought of as changing the effective torsional rigidity of the section from GK_T to $(GK_T + M_e\beta_x)$, and is related to the effect which causes some short concentrically loaded compression members to buckle torsionally. In that case the compressive stresses exert a disturbing torque so that

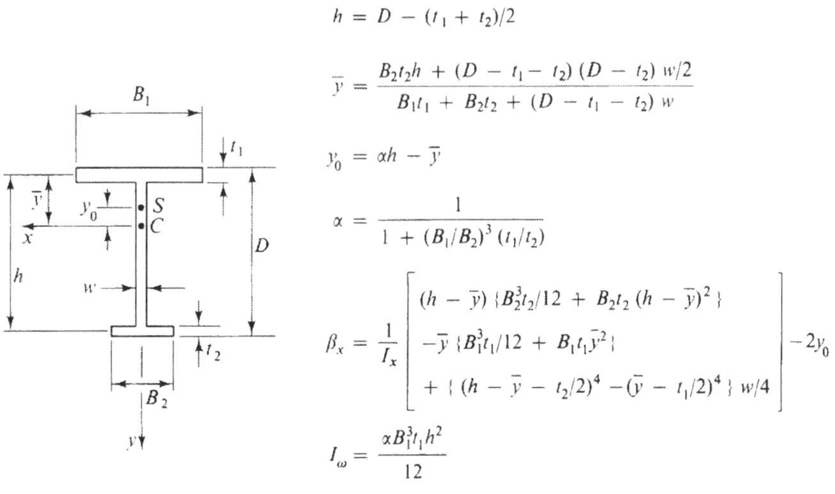

FIGURE 3.10
Properties of monosymmetric I-sections

there is a reduction in the effective torsional rigidity. In doubly symmetric beams the disturbing torque exerted by the compressive bending stresses is exactly balanced by the restoring torque due to the tensile stresses, and β_x is zero. In monosymmetric beams, however, there is an imbalance which is dominated by the stresses in the small flange which is the further from the shear center. Thus, when the small flange is in compression there is a reduction in the effective torsional rigidity ($M_e\beta_x$ is negative), while the reverse is true ($M_e\beta_x$ is positive) when the small flange is in tension. Consequently, the resistance to buckling is increased when the large flange is in compression, and decreased when the small flange is in compression.

The critical moment M_e of simply supported monosymmetric beams with equal and opposite end moments can be obtained by substituting the effective torsional rigidity $(GK_T + M_e\beta_x)$ for GK_T in Eq. (3.21) for doubly symmetric beams, and rearranging, whence

$$M_e = \frac{\pi}{L}\sqrt{EI_y GK_T}\left\{\sqrt{1 + \frac{\pi^2 EI_\omega}{L^2 GK_T} + \left(\frac{\pi\gamma_m}{2}\right)^2} + \frac{\pi\gamma_m}{2}\right\} \tag{3.44}$$

in which the monosymmetry parameter γ_m (which has the same sign as $M_e\beta_x$) is

$$\gamma_m = \frac{\beta_x}{L}\sqrt{\frac{EI_y}{GK_T}} \tag{3.45}$$

and the warping section constant I_ω is as given in Fig. 3.10. The elastic critical stress

$$\sigma_e = \frac{M_e}{S_{x\min}} \tag{3.46}$$

in which $S_{x\min}$ is the lesser major axis elastic section modulus, is the nominal maximum stress in the beam at elastic buckling. It always occurs in the smaller flange, and may be tensile or compressive.

The evaluation of the monosymmetry parameter γ_m is not straight forward, and it has been suggested that a more easily calculated measure of the monosymmetry of the cross section should be used, such as

$$\rho = I_{yc}/I_y \tag{3.47}$$

where I_{yc} is the section minor axis second moment of area of the compression flange. The values of ρ range from 0 (tee-beam with the flange in tension) to 1 (tee-beam with the flange in compression). The variations of the dimensionless elastic critical moment $M_e L/\sqrt{EI_y GK_T}$ of a particular type of monosymmetric beam (Nethercot, 1975d) with the values of ρ are shown in Fig. 3.11. This beam was derived from a doubly symmetric I-beam by reducing one of the flange widths. The seemingly anomalous position of the curve for a tee-beam with its flange in compression ($\rho = 1$) is a consequence of the very rapid changes in the value of I_ω used in Eq. (3.44) which occur at high values of ρ.

It has also been suggested that the very simple approximation

$$\sigma_e = \frac{\pi^2 EI_{yc}}{L^2}\frac{D}{S_{x\min}} \tag{3.48}$$

should be used for light gauge sections for which the torsional rigidity GK_T is comparatively small [see Eq. (3.23)]. This suggestion is equivalent to assuming that

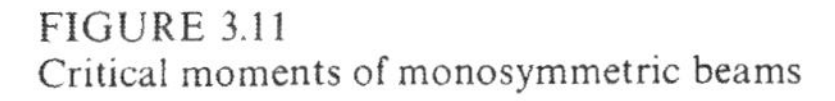

FIGURE 3.11
Critical moments of monosymmetric beams

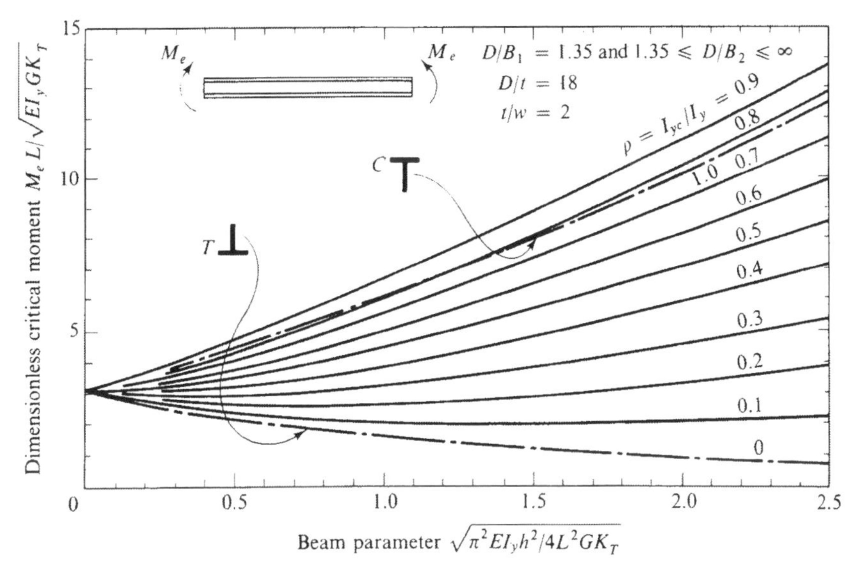

beam buckling can be approximated by the buckling of the compression flange alone when it acts as an axially loaded column.

The elastic flexural-torsional buckling of simply supported monosymmetric beams with other loading conditions has been investigated numerically, and tabulated solutions are available for beams with central concentrated loads or uniformly distributed loads (Column Research Committee of Japan, 1971, Anderson and Trahair, 1972).

3.2.6 Non-Uniform Beams

Non-uniform beams are often more efficient than beams of constant section, and are frequently used in situations where the major axis bending moment varies along the length of the beam. Non-uniform beams of narrow rectangular section are usually tapered in their depth. Non-uniform I-beams may be tapered in their depth, or less commonly in their flange width, and rarely in their flange thickness, while steps in flange width or thickness are common.

Depth reductions in narrow rectangular beams produce significant reductions in their minor axis flexural rigidities EI_y and torsional rigidities GK_T. Because of this, there are also significant reductions in their resistances to flexural-torsional buckling. Closed form solutions for the elastic critical loads of many tapered

FIGURE 3.12
Elastic buckling loads of tapered beams

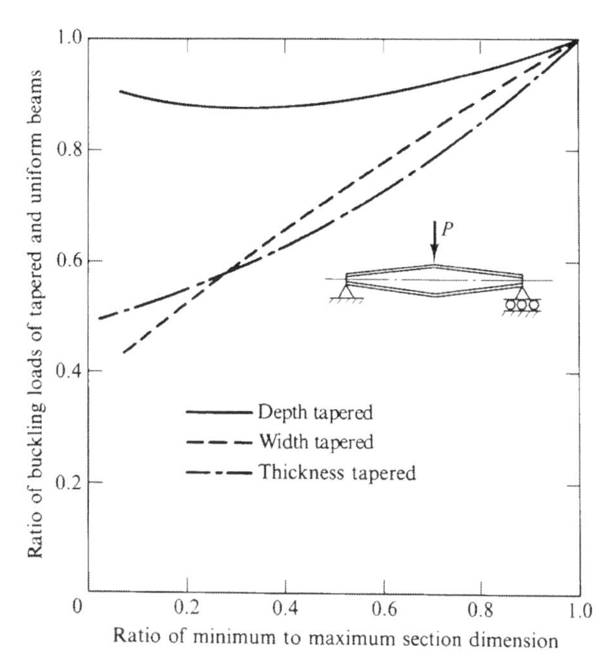

beams and cantilevers are given in the papers cited by the Column Research Committee of Japan (1971), Kitipornchai and Trahair (1972) and Nethercot (1973a).

Depth reductions in I-beams have no effect on the minor axis flexural rigidity EI_y, and little effect on the torsional rigidity GK_T, although they produce significant reductions in the warping rigidity EI_ω. It follows that the resistance to buckling of a beam which does not depend primarily on its warping rigidity is comparatively insensitive to depth tapering, as shown in Fig. 3.12 for beams with central concentrated loads acting at the top flange. Design approximations for web-tapered I-beams have been proposed by Morrell and Lee (1974).

While depth reductions may have comparatively small effects, reductions in the flange width cause significant reductions in GK_T and even greater reductions in EI_y and EI_ω, while reductions in flange thickness cause corresponding reductions in EI_y and EI_ω and in GK_T. Thus the resistance to buckling varies significantly with changes in the flange geometry, as shown in Fig. 3.12. Numerical methods of calculating the elastic critical loads of tapered I-beams have been developed by Kitipornchai and Trahair (1972) and Nethercot (1973a). Design approximations for flange-tapered I-beams have been proposed by Nethercot (1973a) and Taylor, Dwight and Nethercot (1974).

The elastic buckling of I-beams with stepped flanges has been investigated by Trahair and Kitipornchai (1971), and many solutions have been tabulated. The buckling of monosymmetric tapered I-beams has also been studied (Kitipornchai and Trahair 1975c).

3.3 ELASTIC CANTILEVERS

The support conditions of cantilevers differ from those of simply supported beams in that a cantilever is usually completely fixed at one end and completely free at the other. The elastic buckling solution for a cantilever in uniform bending caused by an end moment M can be obtained from the solution given by Eq. (3.21) for simply supported beams by replacing the beam length L by twice the cantilever length $2L$, whence

$$M_e = \frac{\pi\sqrt{EI_y GK_T}}{2L}\sqrt{1 + \frac{\pi^2 EI_\omega}{4L^2 GK_T}} \tag{3.49}$$

The elastic buckling of a cantilever with a concentrated end load P applied at a distance $\bar{a}$ above the shear center can be predicted from the solutions of the differential equations of minor axis bending and torsion

$$-EI_y u'' = M_x \theta \tag{3.50}$$

and

$$GK_T\theta' - EI_\omega\theta''' = P(u + \bar{a}\theta)_l + M_x u' - V_y u \tag{3.51}$$

in which

$$M_x = -P(L - z) \tag{3.52}$$

and

$$V_y = P \tag{3.53}$$

These solutions must satisfy the fixed end ($z = 0$) boundary conditions of

$$(u)_0 = (\theta)_0 = (u')_0 = (\theta')_0 = 0 \tag{3.54}$$

and the condition that the end $z = L$ is free to warp, i.e.

$$(\theta'')_l = 0 \tag{3.55}$$

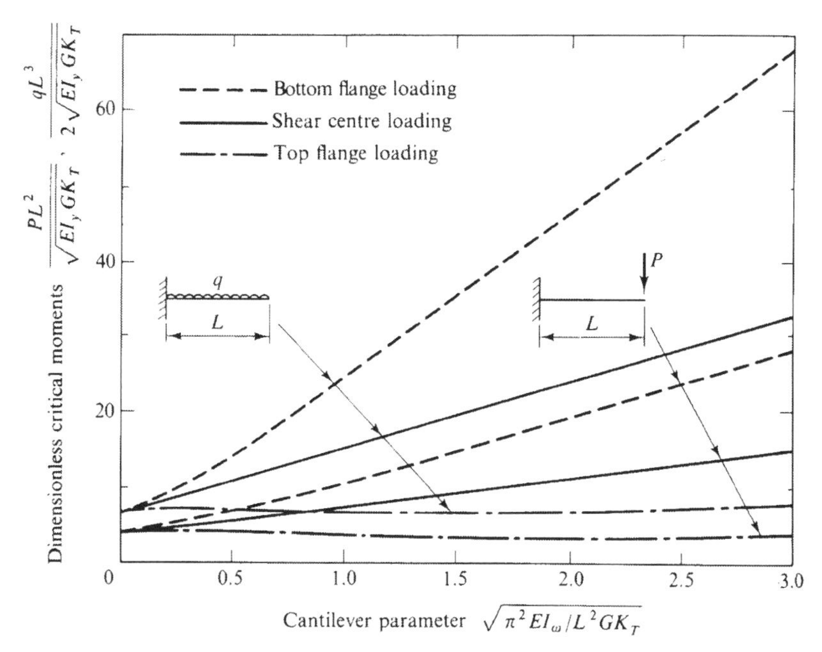

FIGURE 3.13
Elastic buckling loads of cantilevers

Numerical solutions of these equations are available (Anderson and Trahair, 1972, Nethercot, 1973b), and some of these are shown in Fig. 3.13 as plots of the dimensionless critical moments $PL^2/\sqrt{EI_y GK_T}$ for bottom flange, shear center and top flange loading. Also shown in Fig. 3.13 are plots of the dimensionless critical moments $qL^3/2\sqrt{EI_y GK_T}$ of cantilevers with uniformly distributed loads q.

The elastic buckling of monosymmetric cantilevers has been investigated, and many numerical solutions have been tabulated (Anderson and Trahair, 1972). Studies have also been made of the buckling of stepped cantilevers (Massey and McGuire, 1971) and tapered cantilevers (Nethercot, 1973a).

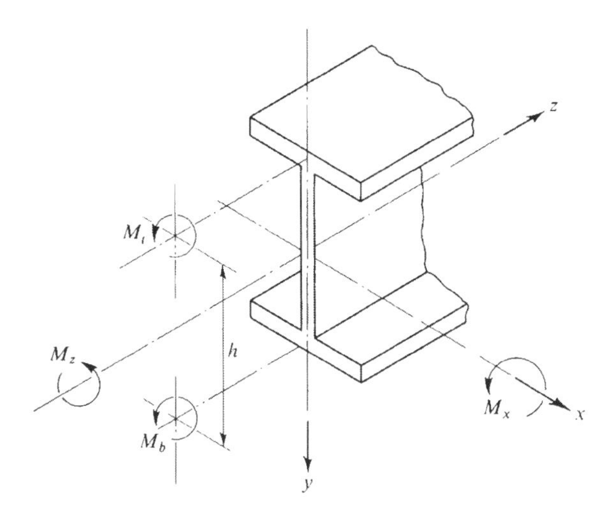

FIGURE 3.14
End restraining moments

3.4 ELASTIC RESTRAINED BEAMS

3.4.1 End Restraints

When a beam forms part of a rigid-jointed structure, the adjacent members elastically restrain the ends of the beam (i.e., they induce restraining moments which are proportional to the end rotations). These restraining actions significantly modify the elastic critical load which causes the beam to buckle. Four different types of restraining moment may act at each end of a beam, as shown in Fig. 3.14. They are:

(a) the major axis end moment M_x which provides restraint about the major axis,

(b) the bottom flange end moment M_b and

(c) the top flange end moment M_t, which provide restraints about the minor axis and against end warping, and

(d) the axial torque M_z which provides restraint against end twisting.

The major axis end restraining moments M_x vary directly with the applied loads, and can be determined by a conventional in-plane bending analysis. The degree of restraint experienced at one end of the beam depends on the major axis stiffness α_x of the adjacent member (which is defined as the ratio of the end moment to the end rotation). This is most conveniently expressed by the ratio R_1 of the actual restraining moment to the maximum moment required to prevent major axis end rotation. Thus R_1 varies from 0 when there is no restraining moment to 1 when there is no end rotation. For beams which are symmetrically loaded and restrained, the restraint parameter R_1 is related to the stiffness α_x of the adjacent members by

$$\alpha_x = \frac{EI_x}{L}\frac{2R_1}{1 - R_1} \tag{3.56}$$

Although the major axis end moments are independent of the buckling deformations, they do affect the critical load of the beam because of their effect on the in-plane bending moment distribution (see Sec. 3.2.4). Many particular cases have been studied, and tabulations of critical loads are available (Austin, Yegian and Tung, 1955, Trahair, 1965, 1966b, 1968a, Column Research Committee of Japan, 1971).

On the other hand, the flange end moments M_b and M_t remain zero until the critical load of the beam is reached, and then increase in proportion to the flange end rotations. Again, the degree of end restraint can be expressed by the ratio of the actual restraining moment to the maximum value required to prevent end rotation. Thus, the minor axis end restraint parameter R_2 (which describes the relative magnitude of the restraining moment $M_b + M_t$) varies between 0 and 1, and the end warping restraint parameter R_4 [which describes the relative magnitude of the differential flange end moments $(M_t - M_b)/2$] varies from 0 when the ends are free to warp to 1 when end warping is prevented.

The particular case of a symmetrically restrained beam with equal and opposite end moments is shown in Fig. 3.15. It is assumed that the minor axis and end warping restraints take the form of equal rotational restraints which act at each flange end and whose stiffnesses are such that

$$\frac{\text{Flange End Moment}}{\text{Flange End Rotation}} = \frac{-EI_y}{L}\left(\frac{R}{1 - R}\right) \tag{3.57}$$

FIGURE 3.15
Elastic buckling of end restrained beams

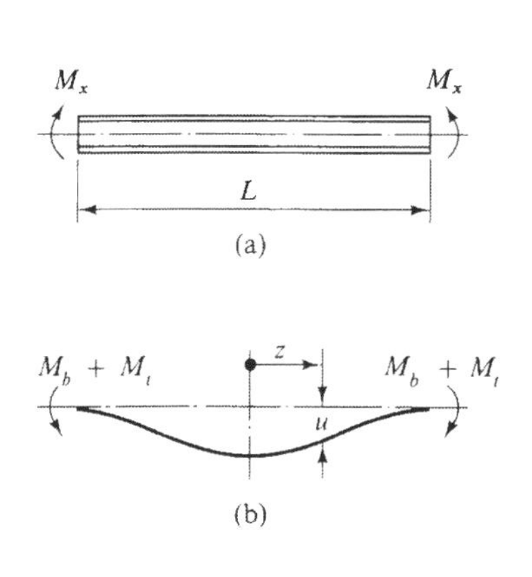

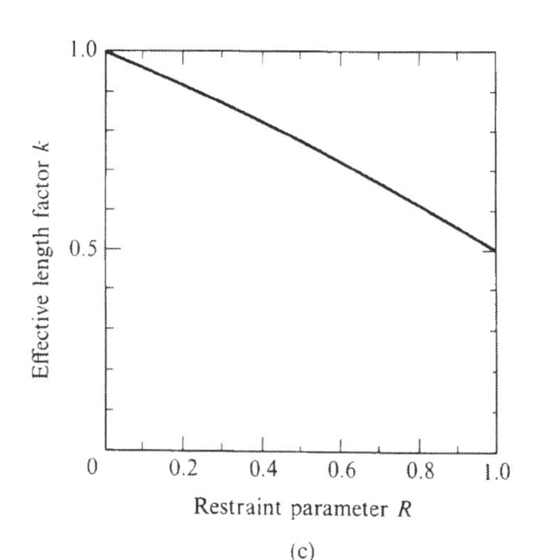

in which case

$$R_2 = R_4 = R \tag{3.58}$$

It can be shown (Trahair, 1970) that the elastic critical moment M_e at which the restrained beam buckles elastically is given by

$$M_e = \frac{\pi}{L_e}\sqrt{EI_yGK_T}\sqrt{1 + \frac{\pi^2 EI_\omega}{L_e^2 GK_T}} \tag{3.59}$$

in which the *effective length* L_e is related to the span L by

$$L_e = kL \tag{3.60}$$

and the *effective length factor* k is given by the solution of

$$\frac{R}{1-R} = \frac{-\pi}{2k}\cot\frac{\pi}{2k} \tag{3.61}$$

It can be seen from the solutions of this equation shown in Fig. 3.15(c) that the effective length factor k decreases from 1 to 0.5 as the restraint parameter R increases from 0 to 1. The corresponding elastic critical moments obtained from Eqs. (3.59) and (3.60) increase from the unrestrained values (corresponding to $k = 1.0$) to values for fixed beams ($k = 0.5$) which range from two times the unrestrained value in the case of beams with negligible warping rigidity ($EI_\omega \rightarrow 0$) to four times for beams with negligible torsional rigidity ($GK_T \rightarrow 0$).

The solutions for the effective length factor k obtained from Eq. (3.61) are exactly the same as those obtained from the effective length chart of Fig. 3.16 for braced compression members (which is equivalent to the corresponding nomogram of the Structural Stability Research Council Guide edited by Johnston, 1976) when the stiffness ratios G_A and G_B (i.e., the ratios of the beam stiffness to the stiffnesses of the restraints at the ends A and B of the beam) are taken as

$$G_A = G_B = \frac{1-R}{R} \tag{3.62}$$

This suggests that the effective length factors for beams with unequal end restraints ($G_A \neq G_B$) can be approximated by using the values given by Fig. 3.16.

The elastic buckling of symmetrically restrained beams with unequal end moments has also been analyzed (Nethercot and Trahair, 1976a) while solutions have been obtained for many other minor axis and end warping restraint conditions (Austin, Yegian, and Tung, 1955, Trahair, 1965, 1966b, 1968a, 1970, Nethercot, 1972d, Nethercot and Rockey, 1973).

The end torques M_z which resist end twisting also remain zero until the critical load is reached, and then increase with the end twists. It has been assumed that the ends of all the beams discussed so far are rigidly restrained against end twisting. When the end restraints are elastic instead of rigid, some end twisting occurs during buckling and the critical load is reduced. Analytical studies (Trahair, 1965)

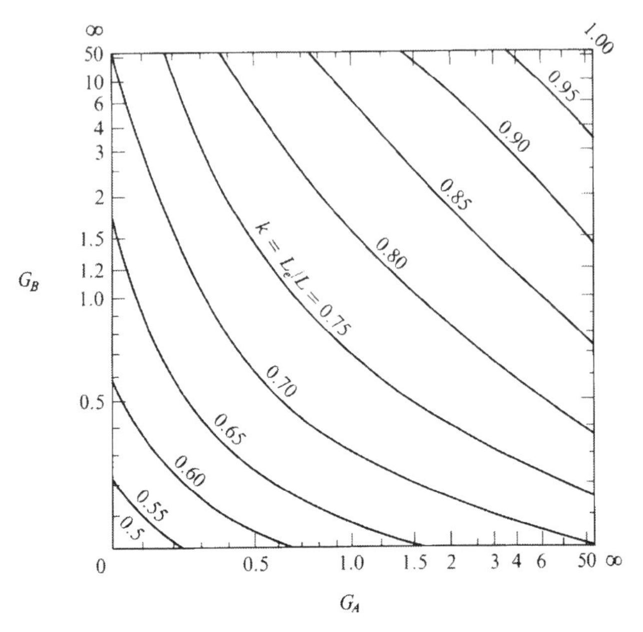

FIGURE 3.16
Effective length factors for braced compression members

of beams with elastic torsional end restraints have shown that the ratio of the reduced critical load P_{er} to the load P_e for a rigidly restrained beam can be approximated by

$$\frac{P_{er}}{P_e} = 1 - R_3^{0.95}\left[A_1 + A_2\left(\frac{\pi^2 EI_\omega}{L^2 GK_T}\right)^{0.95}\right] \tag{3.63}$$

in the range $0.9 \leq P_{er}/P_e \leq 1.0$, where R_3 is the torsional end restraint parameter defined by

$$\alpha_R = \frac{GK_T}{L}\frac{1}{R_3} \tag{3.64}$$

in which α_R is the rotational stiffness of the restraining member. The constants A_1 and A_2 are

$$A_1 = A_2 = 1.1 \tag{3.65}$$

approximately for a beam with a central concentrated load at the shear center Values for other loading conditions can be determined from other critical load solutions (Trahair, 1965).

Another situation for which end twisting is not prevented is illustrated in Fig. 3.17, where the bottom flange of a beam is simply supported at its end and prevented from twisting but the top flange is unrestrained. In this case, beam buckling may be accentuated by distortion of the cross section which results in the

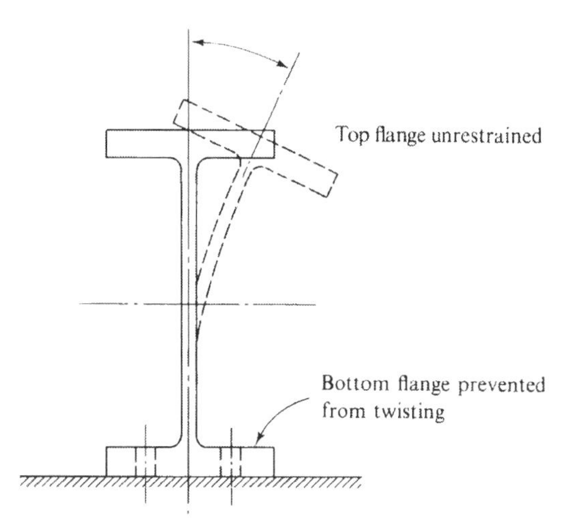

FIGURE 3.17
End distortion of a beam with an unrestrained top flange

web bending shown in Fig. 3.17. Comparatively little is known of the effects of web flexibility on lateral buckling, although it has been suggested (Bartels and Bos, 1973) that these are less important for long shallow beams.

The preceding discussion has dealt with the effects of each type of end restraint, but most beams in rigid-jointed structures have all types of elastic restraint acting simultaneously. Many such cases of combined restraints have been analyzed, and tabular and graphical solutions are available (Austin, Yegian, and Tung, 1955, Trahair, 1966a, b, 1968a).

3.4.2 Intermediate Restraints

The critical moment of a beam may be substantially increased by the provision of an intermediate restraint which prevents the beam from deflecting laterally and twisting at the restraint point. When a simply supported beam with equal and opposite end moments has a rigid restraint at mid-span, then its buckled shape changes from that given by Eqs. (3.24) and (3.25) to (Timoshenko and Gere, 1961)

$$\frac{\theta}{(\theta)_{l/4}} = \frac{u}{(u)_{l/4}} = \sin\frac{\pi z}{L/2} \tag{3.66}$$

and its elastic critical moment is given by

$$\frac{M_e(L/2)}{\sqrt{EI_y GK_T}} = \pi\sqrt{1 + \frac{\pi^2 EI_\omega}{(L/2)^2 GK_T}} \tag{3.67}$$

The intermediate restraint need not be completely rigid, but may be elastic, provided its translational and rotational stiffnesses exceed certain minimum values.

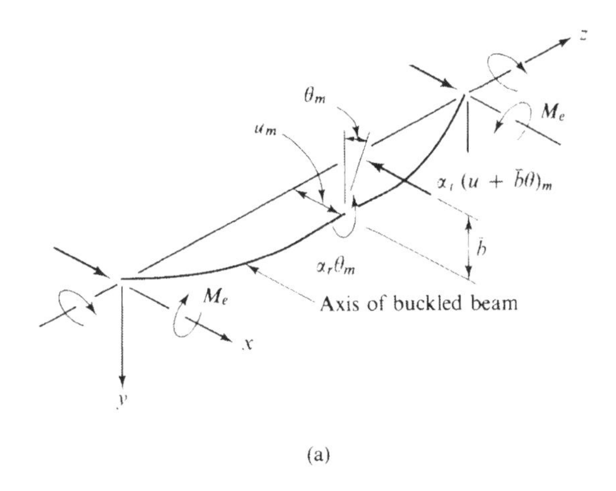

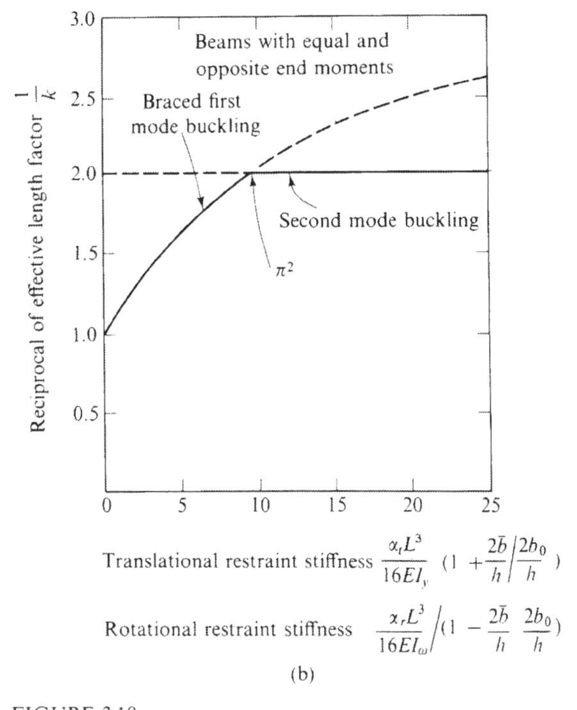

FIGURE 3.18
Beam with elastic intermediate restraints. (a) Beam. (b) Effective length factors

In Fig. 3.18 is shown a beam with equal end moments which has a central translational restraint of stiffness α_t (where α_t is the ratio of lateral force exerted by the restraint to the lateral deflection of the beam measured at the height $\bar{b}$ of the restraint), and a central rotational restraint of stiffness α_r (where α_r is the ratio of the torque exerted by the restraint to the twist of the beam). The elastic critical moment M_e can be expressed in the standard form of Eqs. (3.59) and (3.60) (Mutton and Trahair, 1973) when the stiffnesses α_t, α_r are related to the effective length factor k by

$$\frac{\alpha_t L^3}{16EI_y}\left(1+\frac{2\bar{b}/h}{2b_0/h}\right)=\frac{\left(\frac{\pi}{2k}\right)^3\cot\frac{\pi}{2k}}{\frac{\pi}{2k}\cot\frac{\pi}{2k}-1} \tag{3.68}$$

$$\frac{\alpha_r L^3}{16EI_\omega}\bigg/\left(1-\frac{2\bar{b}}{h}\frac{2b_0}{h}\right)=\frac{\left(\frac{\pi}{2k}\right)^3\cot\frac{\pi}{2k}}{\frac{\pi}{2k}\cot\frac{\pi}{2k}-1} \tag{3.69}$$

in which

$$b_0=\frac{M_e}{\pi^2EI_y/(kL)^2}=\frac{h}{2}\sqrt{1+\pi^2EI_\omega/k^2L^2GK_T} \tag{3.70}$$

These relationships are shown graphically in Fig. 3.18(b).

It can be seen that the effective length factor k varies from 1 when the restraints are of zero stiffness to 0.5 when

$$\frac{\alpha_t L^3}{16EI_y}\left(1+\frac{2\bar{b}/h}{2b_0/h}\right)=\frac{\alpha_r L^3}{16EI_\omega}\bigg/\left(1-\frac{2\bar{b}}{h}\frac{2b_0}{h}\right)=\pi^2 \tag{3.71}$$

If the restraint stiffnesses exceed these values, the beam buckles in the second mode with zero central deflection and twist at a critical moment which corresponds to $k = 0.5$.

When the height $\bar{b}$ of the translational restraint above the shear center is equal to $h^2/4b_0$, then the required rotational stiffness α_r given by Eq. (3.71) is zero. Since b_0 is never less than $h/2$ [see Eq. (3.70)], it follows that a top flange translational restraint of stiffness

$$\alpha_m=\frac{16\pi^2EI_y/L^3}{1+h/2b_0} \tag{3.72}$$

is always sufficient to brace the beam into the second mode. By using Eq. (3.70) and $k = 0.5$, this minimum stiffness can be expressed as

$$\alpha_m=\frac{8M_e}{hL}\frac{1}{1+\sqrt{1+4\pi^2EI_\omega/L^2GK_T}} \tag{3.73}$$

and the greatest value of this is

$$\alpha_m=\frac{4M_e}{Lh} \tag{3.74}$$

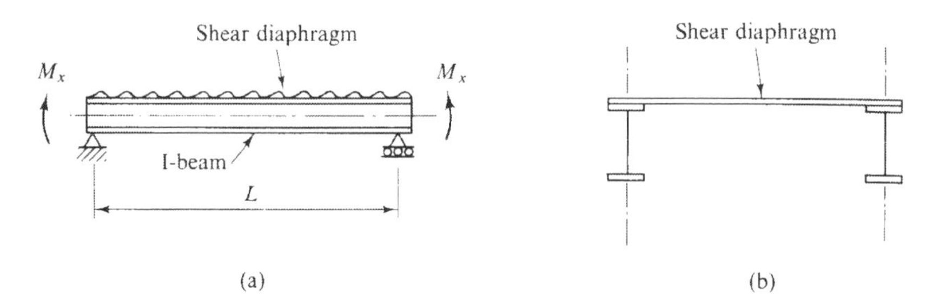

FIGURE 3.19
Diaphragm braced I-beam. (a) Elevation. (b) Section

The flange force P_f at elastic buckling can be approximated by

$$P_f = M_e/h \tag{3.75}$$

and so the minimum top flange translational stiffness can be approximated by

$$\alpha_m = 4P_f/L \tag{3.76}$$

The influence of intermediate restraints on beams with central concentrated and uniformly distributed loads has also been studied, and many values of the minimum restraint stiffnesses required to cause the beams to buckle as if rigidly braced have been determined (Nethercot, 1972e, 1973c, Nethercot and Rockey, 1972, Mutton and Trahair, 1973). The effects of diaphragm bracing on the lateral buckling of simply supported beams with equal and opposite end moments (see Fig. 3.19) have also been investigated (Nethercot and Trahair, 1975), and a simple method of determining whether a diaphragm is capable of providing full bracing has been developed.

3.5 PLASTIC STEEL BEAMS

3.5.1 Equal End Moments

The solution M_e for the critical moment M_c of a perfectly straight simply supported I-beam with equal end moments given by Eq. (3.21) is only valid when the elastic critical stress $\sigma_e = M_e/S_x$ is less than the stress required to cause first yield. In a short span steel beam, yielding occurs before the critical moment is reached, and significant portions of the beam are plastic or strain-hardened when buckling commences. The effective rigidities of these plastic and strain-hardened portions are reduced, and consequently, the critical moment is also reduced. There have been many investigations of plastic buckling, and most of these have been cited in recent studies by Trahair and Kitipornchai (1972), Nethercot (1972b, 1973e, 1974a, b, 1975a), and Fukumoto and Kubo (1972).

For beams with equal and opposite end moments ($\kappa = 1$), the distribution of

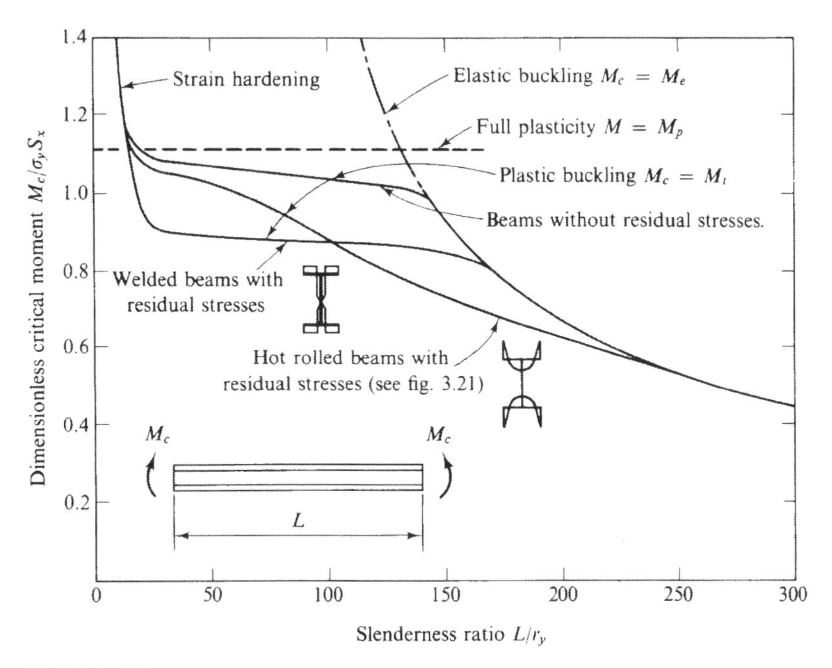

FIGURE 3.20
Lateral buckling strengths of simply supported I-beams

yield across the section does not vary along the beam, and when there are no residual stresses, the plastic critical moment M_i can be calculated from a modified form of Eq. (3.21) as

$$M_i = \frac{\pi}{L}\sqrt{(EI_y)_r(GK_T)_r}\sqrt{1 + \frac{\pi^2(EI_\omega)_r}{L^2(GK_T)_r}} \tag{3.77}$$

in which the subscripted quantities $(\ )_r$ are the reduced rigidities which are effective at buckling. Estimates of these rigidities can be obtained by using the tangent moduli of elasticity which are appropriate to the varying stress levels throughout the section. Thus the values of E and G are used in the elastic areas, while the strain-hardening moduli E_{st} and G_{st} are used in the plastic and strain-hardened areas. When the effective rigidities calculated in this way are used in Eq. (3.77), a *lower bound* estimate of the critical moment is determined (this estimate is a lower bound because it is obtained by assuming strain-hardening unloading of the plastic and strain-hardened regions, rather than elastic unloading). The variation of the dimensionless critical moment $M_c/\sigma_y S_x$ (in which $\sigma_y S_x = M_{yx}$ is the nominal first yield moment) with the slenderness ratio L/r_y of a typical stress-relieved rolled steel section is shown in Fig. 3.20. In the plastic buckling range, the critical moment increases almost linearly with decreasing slenderness from the first yield moment M_{yx} to the full plastic moment $M_p(= \sigma_y Z_x)$ which is reached soon after the

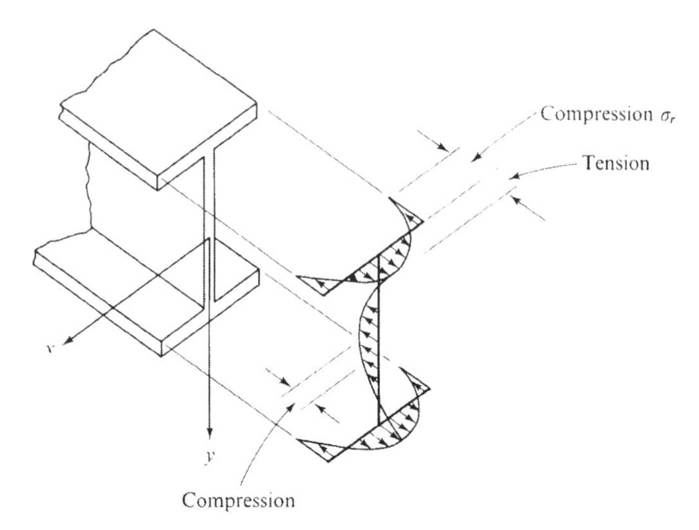

FIGURE 3.21
Idealized residual stress pattern

flanges are fully yielded, when buckling is controlled by the strain-hardening moduli E_{st}, G_{st}.

The plastic critical moment of a beam with residual stresses can be obtained in a similar manner, except that the pattern of yielding is not symmetrical about the section major axis, so that a modified form of Eq. (3.44) for monosymmetric I-beams must be used instead of Eq. (3.77). The plastic critical moment varies markedly with both the magnitude and the distribution of the residual stresses. The moment at which plastic buckling initiates depends on the magnitude of the residual compressive stresses at the flange tips, where yielding causes significant reductions in the effective rigidities $(EI_y)_r$ and $(EI_\omega)_r$. The flange tip stresses are comparatively high in hot-rolled beams, especially those with high ratios of flange to web area (Young, 1975), and so plastic buckling is initiated comparatively early in these beams, as shown in Fig. 3.20. The residual stresses in hot-rolled beams decrease away from the flange tips (see Fig. 3.21), and so the extent of yielding increases and the effective rigidities decrease steadily as the applied moment increases. Because of this, the plastic critical moment decreases in an approximately linear fashion as the slenderness increases, as shown in Fig. 3.20.

In beams fabricated by welding flange plates to web plates, the compressive residual stresses at the flange tips, which increase with the welding heat input, are usually somewhat smaller than those in hot-rolled beams, and so the initiation of plastic buckling is delayed, as shown in Fig. 3.20. However, the variations of the residual stresses across the flanges are smaller in welded beams, and so once flange yielding is initiated, it spreads quickly through the flange with little increase in moment. This causes large reductions in the plastic critical moments of short beams, as indicated in Fig. 3.20.

3.5.2 Other Loading Conditions

When a beam has a more general loading than that of equal and opposite end moments, the in-plane bending moment varies along the beam, and so when yielding occurs its distribution also varies. Because of this, the beam acts as if non-uniform, and the torsion equilibrium equation becomes more complicated. There have been a number of investigations of the effects of different loading arrangements on plastic buckling, and most of these have been cited in recent studies by Nethercot (1973d, 1975a), Yoshida and Imoto (1973), Kitipornchai and Trahair (1975a, b), and Nethercot and Trahair (1976b).

Some numerical solutions for hot-rolled steel beams with a number of different loading arrangements are shown in Fig. 3.22. The most severe loading case is that of equal and opposite end moments ($\kappa = 1$), for which yielding is constant along the beam so that the resistance to lateral buckling is everywhere reduced. Less severe is the case of an unbraced beam with central concentrated load (not shown), for which yielding is confined to a small central portion of the beam, so that any reductions in the section properties are limited to this region. Even less severe cases are those of beams with unequal end moments M and κM, where yielding is confined to small portions near the supports, for which the reductions in the section properties are comparatively unimportant. The least critical case is that of equal end moments that bend the beam in double curvature ($\kappa = -1$), for which the moment gradient is steepest and the regions of yielding are most limited.

The limiting slenderness of a beam which can reach the full plastic moment M_p depends very much on the loading arrangement. It has been suggested (Nethercot and Trahair, 1976b) that for beams with end moments M and κM, this limit could be defined by particular values of a modified beam slenderness $\sqrt{M_p/M_e}$ (which

FIGURE 3.22
Plastic buckling of beams with unequal end moments

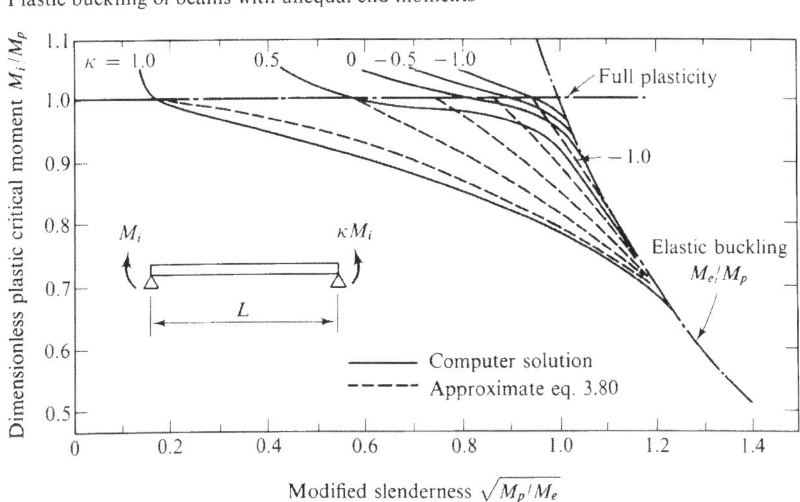

is the beam equivalent of the modified column slenderness $\sqrt{\sigma_y/\sigma_e} = (L/r_y)\sqrt{\sigma_y/\pi^2 E}$ in which σ_e is the elastic buckling stress of the column). For many *short* span beams of universal or wide flange section, the modified slenderness can be closely approximated by

$$\sqrt{\frac{M_p}{M_e}} = 0.01\,\frac{L}{r_y}\sqrt{\frac{\sigma_y}{36}} \tag{3.78}$$

in which the yield stress σ_y is in units of ksi (1 ksi = 6.9 MPa). The limiting values suggested for beams with $M_i = M_p$ are

$$\left(\sqrt{\frac{M_p}{M_e}}\right)_{pi} = \sqrt{\frac{0.39 - 0.30\kappa - 0.07\kappa^2}{0.70}} \tag{3.79}$$

in which M_e is the elastic buckling moment given by Eqs. (3.30) and (3.31) for beams with end moments M and κM.

It has also been suggested that the plastic buckling moments M_i shown in Fig. 3.22 may be approximated by parabolic relationships given by

$$\frac{M_i}{M_p} = 0.70 + 0.30\left(\frac{1 - 0.70M_p/M_e}{0.61 + 0.30\kappa + 0.07\kappa^2}\right) \tag{3.80}$$

for $0.70 \le M_i/M_p \le 1.00$. The limiting values of $(\sqrt{M_p/M_e})_{pi}$ given by Eq. (3.79) can be obtained from Eq. (3.80) by setting $M_i = M_p$.

Approximations have also been developed for beams with transverse loads (Nethercot and Trahair, 1976b).

3.6 DESIGN CRITERIA FOR STEEL BEAMS

3.6.1 Behavior of Real Beams

Real beams differ from the idealized perfectly straight beams whose buckling behavior was discussed in the preceding sections. Real beams have small imperfections such as initial curvature, twist, eccentricity of load, or horizontal load components. These imperfections cause real beams to behave differently to the idealized beams which do not deflect laterally or twist until the critical buckling load is reached, as shown by the elastic buckling curve in Fig. 3.23. Real beams deflect and twist as soon as loading is commenced, and these deformations increase rapidly as the critical buckling load is approached, as shown by the elastic bending and twisting curve in Fig. 3.23. In the case of beams without residual stresses, elastic bending and twisting continue until a limiting stress σ_l is reached at which first yield occurs. The maximum strength of the beam is attained soon after this. The effect of residual stresses is to depress the limiting stress at which first yield occurs, and also the nominal maximum stress σ_m at failure. For a short beam, the maximum stress σ_m

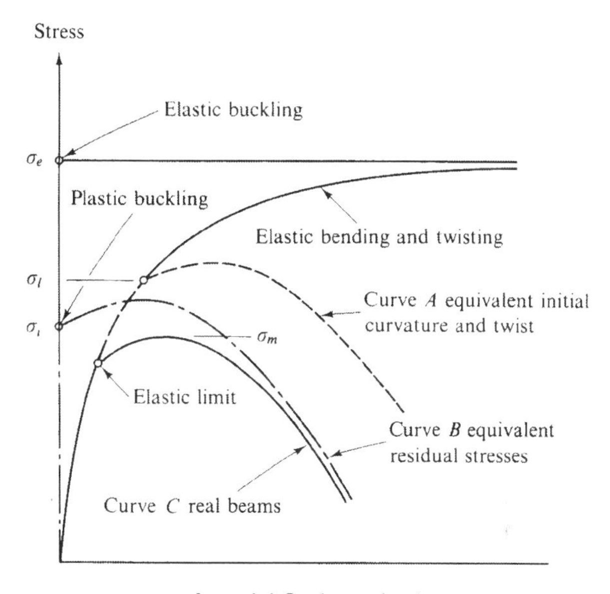

FIGURE 3.23
Behavior of real beams

is often reasonably close to the nominal plastic buckling stress $\sigma_i = M_i/S_x$, while the limiting stress σ_l often provides a close approximation to the ultimate strength of a slender beam.

It is possible to make an accurate analysis of the behavior of a particular real beam which includes the effects of known imperfections and residual stresses by treating the beam as a special case of a biaxially loaded beam-column and by using the methods to be presented in Chaps. 7–12. However, the use of such a sophisticated analysis is not warranted because the magnitudes of the imperfections and residual stresses are uncertain. Instead, design rules are usually based on a simple analysis for one type of equivalent imperfection, or for an equivalent residual stress, the magnitudes of these being chosen so as to make some allowance for the neglected residual stresses or imperfections.

3.6.2 U.S. Specification

The maximum permissible compressive bending stresses σ_{bc} of the AISC 1969 Specification have their origin in the elastic buckling stresses

$$\sigma_e = M_e/S_x \tag{3.81}$$

However, this and Eq. (3.21) are simplified by making the approximations (see Fig. 3.24)

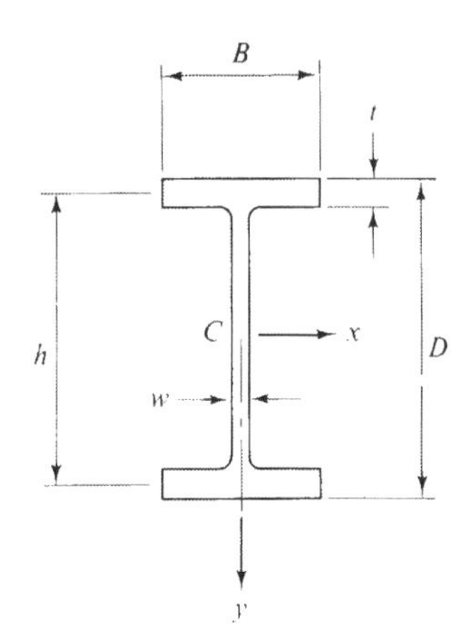

FIGURE 3.24
I-section

$$\left.\begin{aligned} S_x &= BtD\{1 + (D - 2t)w/6Bt\} \\ I_y &= B^3t/6 \\ K_T &= 2Bt^3/3 + Dw^3/3 \\ I_\omega &= I_y D^2/4 \\ \frac{1 + (D - 2t)w^3/2Bt^3}{\{1 + (D - 2t)w/6Bt\}^2} &= 1 \\ E = 2.59G &= 29\,000 \text{ ksi} \\ &\quad (200\,000 \text{ MPa}) \end{aligned}\right\} \tag{3.82}$$

so that the elastic critical stress σ_e can be approximated by

$$\sigma_e \simeq \sqrt{\left(\frac{18\,900}{LD/Bt}\right)^2 + \left(\frac{286\,200}{(L/r_t)^2}\right)^2} \text{ ksi} \tag{3.83}$$

in which r_t is the radius of gyration about the y axis of the compression flange plus one sixth of the web, and is given by

$$r_t = \sqrt{\frac{B^2/12}{[1 + (D - 2t)w/6Bt]}} \tag{3.84}$$

If the second term of Eq. (3.83) is ignored, then a conservative approximation for σ_e is obtained which corresponds to that of Eq. (3.22), and an estimate of the maximum permissible stress σ_{bc} can be obtained by dividing this approximation for σ_e by a factor of safety of 1.57 approximately, whence

$$\sigma_{bc} = \frac{12\,000}{LD/Bt} \text{ ksi} \tag{3.85}$$

Alternatively, if the first term of Eq. (3.83) is ignored, then another conservative approximation for σ_e is obtained [which corresponds to that of Eq. (3.23)], and another estimate for σ_{bc} can be obtained by dividing by 1.69 approximately, whence

$$\sigma_{bc} = \frac{170\,000}{(L/r_t)^2} \text{ ksi} \tag{3.86}$$

However, these results are only valid for elastic beams, and must be reduced for short beams which buckle plastically. The AISC 1969 Specification provides this reduction by adopting a parabolic reduction formula

$$\sigma_i = \sigma_y \left[1 - \frac{1}{2} \left(\frac{3}{4} \right)^2 \frac{\sigma_y}{\sigma_e} \right] \tag{3.87}$$

for the nominal plastic buckling stress σ_i, which is similar in form to that used for the plastic buckling of compression members. For short beams, elastic buckling can be closely approximated by ignoring the first term of Eq. (3.83), and if a factor of safety of 3/2 is used, then the reduced maximum permissible stress σ_{bc} can be obtained as

$$\sigma_{bc} = \frac{2\sigma_y}{3} \left[1 - \frac{1}{2} \left(\frac{3}{4} \right)^2 \frac{\sigma_y (L/r_t)^2}{286\,200} \right] \tag{3.88}$$

This equation is used when $\sqrt{\sigma_y (L/r_t)^2/286\,200} < 4/3$, and Eq. (3.86) when $\sqrt{\sigma_y (L/r_t)^2/286\,200} > 4/3$.

FIGURE 3.25
Maximum permissible stresses of the AISC 1969 specification

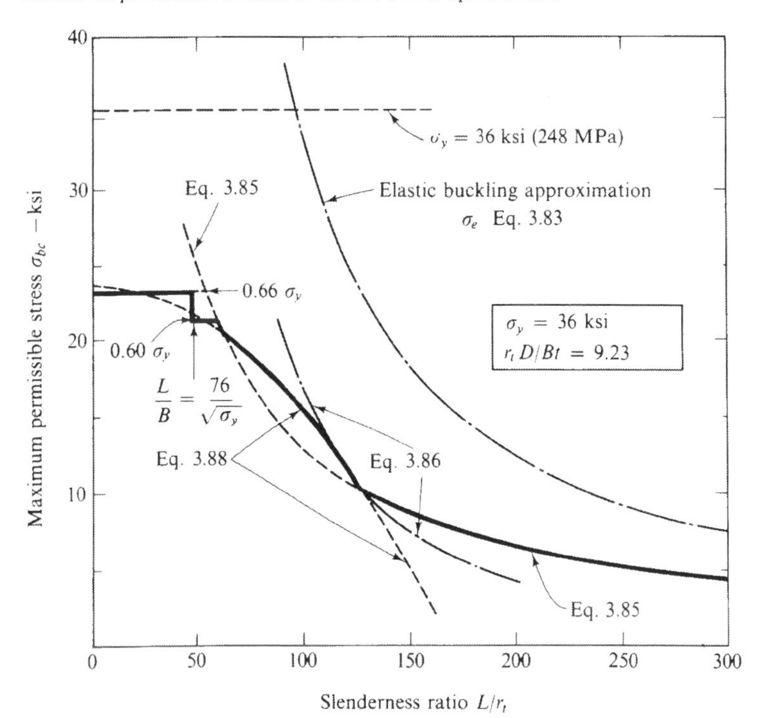

These maximum permissible bending stresses σ_{bc} are shown graphically in Fig. 3.25 for a beam for which $\sigma_y = 36$ ksi (248 MPa) and $r_t D/Bt = 9.23$. For slender beams $\{\sqrt{\sigma_y}(L/r_t)^2/286\,200 > 4/3\}$, the maximum permissible stress σ_{bc} is the higher of the values given by Eqs. (3.85) and (3.86) [the higher value is used because both Eqs. (3.85) and (3.86) are derived from conservative approximations to Eq. (3.83) for the elastic buckling stress σ_e]. For less slender beams $\{\sqrt{\sigma_y}(L/r_t)^2/286\,200 < 4/3\}$, the maximum permissible stress σ_{bc} is the higher of the values given by Eqs. (3.85) and (3.88), with a maximum value of

$$\sigma_{bc} = 0.60\sigma_y \tag{3.89}$$

which corresponds to a nominal factor of safety of $1/0.6 \simeq 1.67$ against first yield. For short beams ($L/B < 76/\sqrt{\sigma_y}$ and $L/B < 20\,000t/D\sigma_y$), the maximum permissible stress is

$$\sigma_{bc} = 0.66\sigma_y \tag{3.90}$$

which corresponds to a factor of safety of $1/0.6 \simeq 1.67$ against full plasticity if the shape factor is discounted to 1.1.

It can be seen from Fig. 3.25 that several calculations must be made before σ_{bc} can be determined, except for very short or very long beams. This is the penalty which must be paid for the simplicity of the approximations made for the elastic critical stress σ_e. Fortunately, graphs of the maximum permissible moments for beams with $\sigma_y = 36$ ksi (248 MPa) are available (American Institute of Steel Construction, 1970), which allow the designer to avoid these calculations. It should also be noted that the AISC 1969 Specification allows more precise analyses to be made, which could in fact be simpler than the several calculations.

The AISC 1969 Specification allows the moment modification factors C_b given by Eq. (3.31) (but with a maximum value of 2.3) to be used for beams with unequal end moments, and these are included in the numerators of Eqs. (3.85) and (3.86), and in the denominator of the second term of Eq. (3.88). However, no allowance is made for the effects of end restraints, and the distance L is conservatively taken as the unbraced length between points of effective lateral and torsional restraint. On the other hand, the potentially dangerous condition of top flange loading is ignored.

3.6.3 British Standard

The British Standard BS449 : 1969 uses a set of design rules which is based on the limiting stress σ_l of an elastic beam with an equivalent initial curvature which is the same as that used in the design of compression members. The limiting stress is modified to account for the real behavior of beams, and is divided by an appropriate safety factor to obtain the maximum permissible stress. For a plate girder the maximum permissible compressive stress σ_{bc} at the working load is obtained from the elastic critical stress σ_e by using a tabular conversion. This is

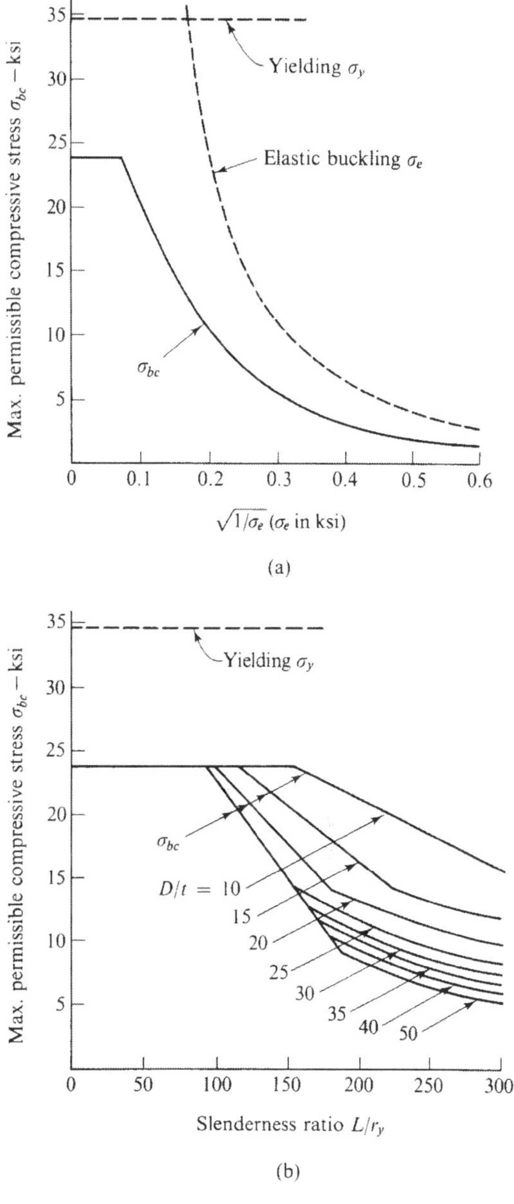

FIGURE 3.26
Maximum permissible compressive stresses according to BS 449: 1969 for beams of grade 43 steel. (a) Plate girders. (b) Hot-rolled I-beams

shown graphically in Fig. 3.26(a) [for a Grade 43 steel which has a nominal yield stress σ_y of 35 ksi (240 MPa)], and compared with the yield and elastic buckling stresses σ_y and σ_e to demonstrate the margins of safety incorporated. The elastic critical stress is calculated from Eqs. (3.21) and (3.81) by using the approximations (see Fig. 3.24),

$$\begin{aligned} S_x &= 1.1BtD \\ I_y &= B^3t/6 \\ K_T &= 0.9Bt^3 \\ I_\omega &= I_yD^2/4 \\ B &= 4.2r_y \\ E &= 2.5G = 30\,000 \text{ ksi} \\ & \qquad (207\,000 \text{ MPa}) \end{aligned} \tag{3.91}$$

whence

$$\sigma_e \simeq \left(\frac{638}{L/r_y}\right)^2 \sqrt{1 + \frac{1}{20}\left(\frac{L}{r_y}\frac{t}{D}\right)^2} \text{ ksi} \tag{3.92}$$

In some cases this value of σ_e is increased by 20 percent. The formulation for simply supported beams is also used for cantilevers built-in at one end and free at the other, while no allowance is made for the effects of the loading arrangement on elastic buckling, except that some account is taken of the increased danger of top flange loading by requiring the value of σ_e to be calculated for a 20 percent increase in the member length L.

A slightly different method of conversion is used for hot-rolled I-beams, in which the maximum permissible stresses σ_{bc} [see Fig. 3.26(b)] are tabulated directly for given values of the slenderness ratio L/r_y and the cross section parameter D/t. This method is somewhat more convenient, as the maximum permissible stress σ_{bc} is determined directly, and the intermediate calculation of the elastic critical stress σ_e from the values of L/r_y and D/t is not required.

The BS449 : 1969 also includes design rules for beams of monosymmetric cross section, for non-uniform beams, and for end restrained beams. An allowance is made for the effects of monosymmetry by increasing or decreasing the elastic critical stress σ_e calculated from Eq. (3.92) in accordance with the ratio of the minor axis stiffness of the compression flange to that of the complete beam. The effects of non-uniformity are allowed for by making a decrease to the flange thickness t to be used in Eq. (3.92), the value of which depends on the ratio of the total flange areas at the points of minimum and maximum bending moment. The increases in the resistance to lateral buckling caused by end restraints are included by allowing the member length L in Eq. (3.92) to be replaced by the effective length L_e, the magnitude of which varies with the degree and type of end restraint.

Recently, modifications to these rules have been proposed by Taylor, Dwight, and Nethercot (1974). Included in these is a different conversion from elastic buckling, in which the design ultimate moment M_u of a hot-rolled I-beam is calculated from the elastic critical moment M_e and the full plastic moment M_p by using the relationship

$$\frac{M_u}{M_p} = \left(\frac{1 + (1 + \eta)M_e/M_p}{2}\right) - \sqrt{\left(\frac{1 + (1 + \eta)M_e/M_p}{2}\right)^2 - \frac{M_e}{M_p}} \not> 1 \tag{3.93}$$

in which

$$\eta = 0.007 \sqrt{\frac{\pi^2 E}{\sigma_y}} \left(\sqrt{\frac{M_p}{M_e}} - 0.4\right) \tag{3.94}$$

Equations (3.93) and (3.94) are similar to the limiting stress relationship for the first yield in beams with initial curvature and twist, except that the full plastic moment M_p is introduced in place of the yield stress. The form of the imperfection parameter η given by Eq. (3.94) was adopted so as to produce reasonable agreement between the values of M_u and the ultimate moment capacities obtained from tests (Dibley, 1969). In particular, Eq. (3.93) predicts that the full plastic moment M_p is reached by beams for which $\sqrt{M_p/M_e}$ is less than 0.4. The values of M_u/M_p given by Eqs. (3.93) and (3.94) are plotted in Fig. 3.27 for beams with $E \simeq 30\,000$ ksi (207 000 MPa) and $\sigma_y = 36$ ksi (248 MPa) for which Eq. (3.94) becomes

$$\eta \simeq 0.64 \sqrt{\frac{36}{\sigma_y}} \left(\sqrt{\frac{M_p}{M_e}} - 0.40\right) \tag{3.95}$$

The above conversion for hot-rolled I-beams is also recommended for welded beams, except that the yield stress σ_y is reduced by approximately 3 ksi (20 MPa)

FIGURE 3.27
Proposed design ultimate moments for a new British code

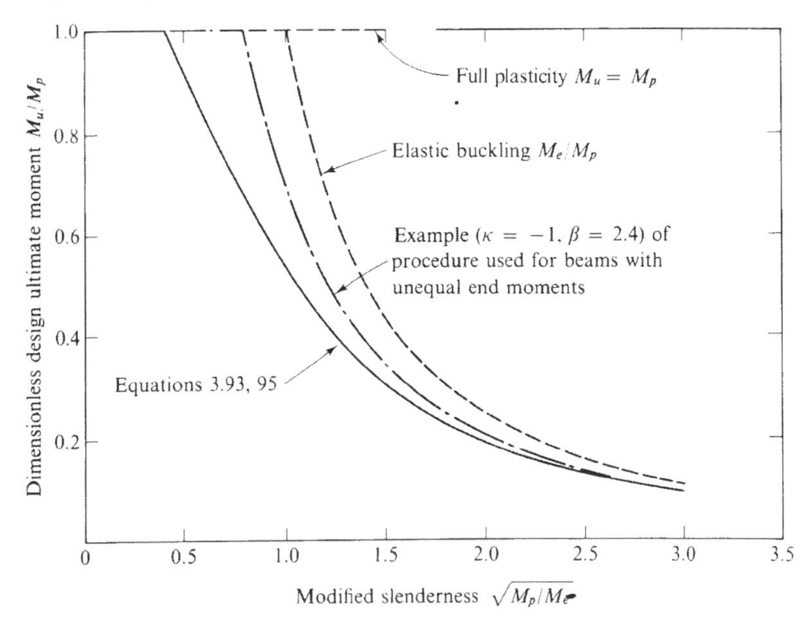

in order to make some allowance for the decreases in the plastic buckling resistance caused by the welding residual stresses.

In the proposed modifications, the method of determining the elastic critical stress σ_e or moment M_e is also a little different. For beams in uniform bending, a return to the precision of Eqs. (3.21) and (3.81) is recommended, and it is suggested that values of the relevant section properties of hot-rolled I-beams should be tabulated, while improved versions of the simple approximation of Eq. (3.92) are given. The effects of non-uniform bending are accounted for, and values are given of the moment modification factors β (see Fig. 3.9) to be used in Eq. (3.40) to calculate the elastic critical moment M_e. However, a different procedure is proposed for beams with unequal end moments, in which the elastic critical moment M_e is determined for the uniform moment condition ($\kappa = 1$, $\beta = 1$), and the value of M_u obtained by substituting this into Eq. (3.93) is then multiplied by the moment modification factor β to obtain the design ultimate moment (which must not exceed M_p). The difference between the two procedures is illustrated in Fig. 3.27 for an extreme case where the moment modification factor β is 2.4. It can be seen that the second procedure (for beams with unequal end moments) leads to significantly higher estimates of the moment capacity, especially for beams of intermediate slenderness which fail by plastic buckling. This reflects, to some extent, the increased resistance to plastic buckling of beams under moment gradient.

3.6.4 Australian Standard

The Australian Standard AS 1250-1975 uses a set of simple semi-empirical equations to relate the maximum permissible stress σ_b to the yield stress σ_y and the elastic critical stress σ_e, and these are shown in Fig. 3.28. For slender beams for which the elastic critical stress σ_e is less than the yield stress σ_y, the maximum permissible stress σ_b given by

$$\frac{\sigma_b}{\sigma_e} = 0.55 - 0.10\frac{\sigma_e}{\sigma_y} \tag{3.96}$$

depends mainly on the elastic critical stress, and has a factor of safety against elastic buckling which varies between $1/0.45 \simeq 2.22$ and $1/0.55 \simeq 1.82$. For beams of intermediate length for which σ_e is greater than σ_y, yielding is more important, and the maximum permissible stress σ_b is given by

$$\frac{\sigma_b}{\sigma_y} = 0.95 - 0.50\sqrt{\frac{\sigma_y}{\sigma_e}} \tag{3.97}$$

This equation provides a transition whose shape approximates that of the typical plastic buckling curve shown in Fig. 3.2. The strength of a short length beam is governed by yielding, and by the resistance of the compression flange to local buckling. Thus the maximum permissible compressive stress σ_{bc} varies with the width-thickness ratio of the compression flange outstand in the range from $0.66\sigma_y$

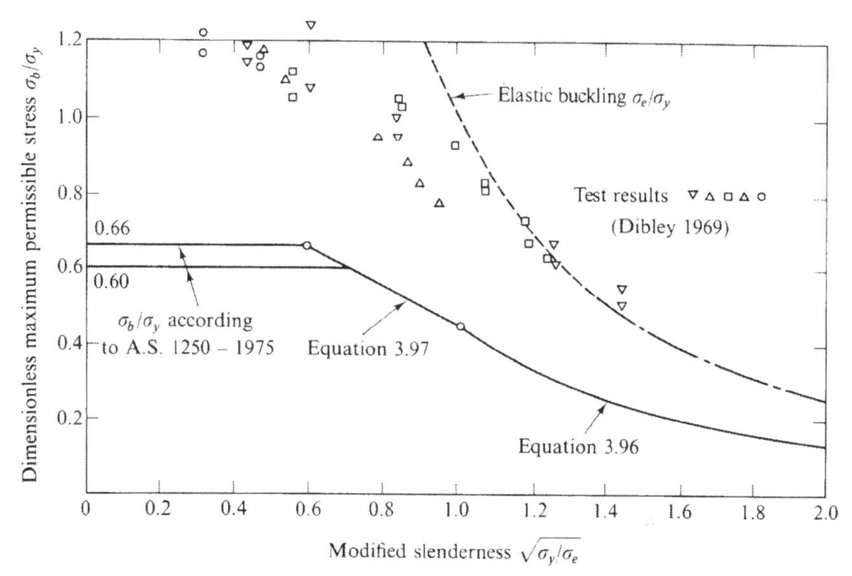

FIGURE 3.28
Maximum permissible stresses of the A.S. 1250, 1975

to $0.60\sigma_y$. Equations (3.96) and (3.97) are shown in Fig. 3.28. The test results also shown in Fig. 3.28 indicate that these equations provide a reasonably consistent load factor which varies between $1/0.60 \simeq 1.67$ for short beams and $1/0.55 \simeq 1.82$ for long beams. However, the load factors for intermediate length beams under uniform moment may be somewhat lower.

In determining the maximum permissible stress σ_b from Eq. (3.96) or (3.97), it is first necessary to calculate the elastic critical stress σ_e, which is the maximum stress in the beam at elastic buckling. The AS 1250-1975 allows two different methods of doing this. In the first, an elastic flexural-torsional buckling analysis is made, or published results may be used to determine the maximum moment M_e at elastic buckling. The elastic critical stress σ_e is calculated from this, and then the maximum permissible stress σ_b is determined from Eq. (3.96) or (3.97). This method is likely to lead to economical structures.

In the second method, the elastic critical stress σ_e is calculated approximately using

$$\sigma_e \simeq \left(\frac{620}{kL/r_y}\right)^2 \sqrt{1 + \frac{1}{20}\left(\frac{kL\,t}{r_y\,D}\right)^2} \text{ ksi} \tag{3.98}$$

(in some cases 1.2 times this result is used). Where appropriate, allowances are made for the effects of top flange loading, monosymmetry, non-uniformity, and end restraints which are similar to those of the BS449:1969 (see Sec. 3.6.3). For convenience, the AS 1250-1975 tabulates values of σ_b which are based on Eqs. (3.98) and (3.96) or (3.97). In using these, the designer calculates the values of

kL/r_y and D/t for the beam to be checked, and selects the maximum permissible stress σ_b from the appropriate table. Designers usually prefer this simpler method, even though it ignores the effects of the in-plane moment distribution, and often results in less economical structures. Design charts are also available (Trahair, 1974) for any Australian universal section of Grade 250 steel, which has a nominal yield stress of approximately 36 ksi (250 MPa). These charts are based on the more accurate Eqs. (3.21) and (3.81), and allow the working load design moment M_d to be determined directly. These charts also allow the direct selection of a suitable beam section.

3.6.5 New Design Criteria

There have been a number of suggestions made for improving existing design criteria. Two of these are based on the maximum moment capacity envelopes shown in Fig. 3.2 which are defined by the full plastic moments M_p of short span beams; by the plastic buckling moments M_i of beams of intermediate span; and by the elastic buckling moments M_e of long span beams. The nominal design moment capacities M_n are derived by reducing these envelopes so as to make allowances for the effects of geometrical imperfections and for the variability of the actual capacities, and also to provide margins of safety against failure.

Methods of calculating the full plastic moment M_p and the elastic buckling moment M_e are well established, and it is only in the representation of the plastic buckling moment M_i and the limiting span lengths L_{pi} (at which $M_i = M_p$) and L_{ie} (at which $M_i = M_e$) that the proposed maximum moment capacity envelopes differ. Galambos and Ravindra (1974) proposed the simple linear approximation

$$M_i = M_p - (M_p - M_{ie})\frac{(L - L_{pi})}{(L_{ie} - L_{pi})} \tag{3.99}$$

for hot-rolled I-beams, in which M_{ie} is the elastic buckling moment of a beam of length L_{ie}. They also proposed that M_{ie} should be approximated by the moment at which the tips of the compression flange first yield, so that

$$M_{ie} \simeq (\sigma_y - \sigma_r)S_x \tag{3.100}$$

in which σ_r is the maximum compressive residual stress at the flange tips. With this approximation, the corresponding length L_{ie} can be calculated from available information on elastic buckling. Galambos and Ravindra reviewed past proposals for determining the limiting span length L_{pi}, and decided to use

$$L_{pi} = 240 r_y/\sqrt{\sigma_y} \tag{3.101}$$

(in which σ_y is in ksi) for beams under nearly uniform moment ($0.5 \leq \kappa \leq 1.0$), and

$$L_{pi} = 390 r_y/\sqrt{\sigma_y} \tag{3.102}$$

for beams under moment gradient ($-1.0 \leq \kappa < 0.5$).

However, recent studies have shown that Eq. (3.101) is too optimistic for unrestrained beams [the original derivation (Lay and Galambos, 1965) was for beams with very high end restraints], while it is known that Eq. (3.102) is unnecessarily conservative for beams under high moment gradients ($\kappa \to -1$). A more accurate approximation for the lower limiting span length L_{pi} is provided by the modified slenderness $(\sqrt{M_p/M_e})_{pi}$ [Eq. (3.79) and Nethercot and Trahair, 1976b] of

$$\left(\sqrt{\frac{M_p}{M_e}}\right)_{pi} = \sqrt{\frac{0.39 - 0.30\kappa - 0.07\kappa^2}{0.70}} \tag{3.103}$$

For many short span hot-rolled I-beams, this lower slenderness limit can be closely approximated by

$$\left(\sqrt{\frac{M_p}{M_e}}\right)_{pi} = 0.01\,\frac{L_{pi}}{r_y}\sqrt{\frac{\sigma_y}{36}} \tag{3.104}$$

when the yield stress σ_y is in ksi, and so

$$L_{pi} \simeq 100 r_y \sqrt{36/\sigma_y}\sqrt{\frac{0.39 - 0.30\kappa - 0.07\kappa^2}{0.70}} \tag{3.105}$$

Nethercot and Trahair (1976b) have suggested that an accurate approximation for the plastic buckling moment M_i can be obtained from Eq. (3.80). This equation fits an approximating parabola between the limit defined by Eq. (3.103), and that of

$$M_{ie} = 0.70 M_p \tag{3.106}$$

When a more accurate approximation than M_p is needed for very short span beams, then the numerical results obtained by Nethercot and Trahair (1976b) suggest

$$M_i = M_p\left\{1.2 - 0.2\sqrt{\frac{M_p}{M_e}}\sqrt{\frac{0.70}{0.39 - 0.30\kappa - 0.07\kappa^2}}\right\} \tag{3.107}$$

in the range $1 \le M_i/M_p \le 1.1$. Equations (3.80) and (3.106) are valid for the usual hot-rolled I-beams with $\sigma_y \simeq 36$ ksi (248 MPa), but may need to be modified for beams with other yield stresses and residual stresses.

Galambos and Ravindra (1974) have proposed a similar basis for welded I-beams to that of Eq. (3.99) for hot-rolled I-beams, but with higher residual stresses σ_r being used to calculate M_{ie} from Eq. (3.100). However, other studies (Fukumoto, Fujiwara, and Watanabe, 1971, Fukumoto and Kubo, 1972, and Nethercot, 1974b) have shown that this is not entirely satisfactory, and Nethercot (1974b) has suggested that

$$M_i = M_{ie} \tag{3.108}$$

should be used in the range $L_{pi} \le L \le L_{ie}$.

Account may be taken of the effects of geometrical imperfections and of the variability of the actual moment capacities of real beams by making reductions from the maximum moment capacity envelope defined by M_p, M_i, and M_e. Reductions have been proposed by Nethercot and Trahair (1976b) which vary with a modified slenderness $\sqrt{M_p/M_c}$, in which M_c is the appropriate buckling moment. These proposals are that the ultimate moment capacity M_u for stocky beams $(0 \le \sqrt{M_p/M_c} \le 1)$ should be

$$M_u = M_p \tag{3.109}$$

that M_u for beams of intermediate slenderness $(1.0 \le \sqrt{M_p/M_c} \le 1.1)$ should be

$$M_u = (1.57M_c/M_p - 0.57)M_c \tag{3.110}$$

and that M_u for slender beams $(1.1 \le \sqrt{M_p/M_c})$ should be

$$M_u = (0.95 - 0.27M_c/M_p)M_c \tag{3.111}$$

These simple equations were chosen as a lower bound to the experimental results (Dibley, 1969, Kitipornchai and Trahair, 1975b) shown in Fig. 3.29. The reductions they provide are greatest near $\sqrt{M_p/M_c} = 1.1$, and produce design ultimate moments M_u which approach the full plastic moments M_p of stocky beams as M_p/M_c decreases towards 1.0, and which approach 95 percent of the buckling moments M_c of slender beams as M_p/M_c increases.

A different set of reductions was suggested by Galambos and Ravindra (1974)

FIGURE 3.29
Proposed design basis for ultimate moment capacity

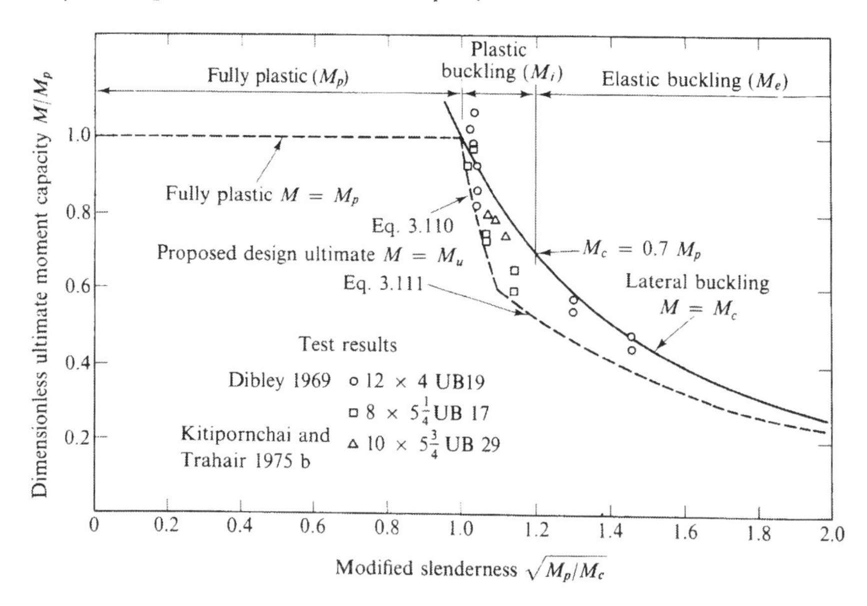

in their proposals for a *load and resistance factor design method.* In this method, the ultimate strength of a design is satisfactory when

$$\phi M_n \geq \Sigma \gamma_k Q_{nk} \tag{3.112}$$

in which Q_{nk} and γ_k are the load effects (moments) and load factors, respectively, and M_n is the nominal moment resistance given by the design specification. The resistance factor ϕ is related directly to the variability of the resistance and the degree of safety required, and indirectly to the variability of the loads. In particular, the variability of the resistance is influenced by the variabilities of the material properties σ_y, E, and E_{st}, of the section dimensions, of the residual stresses and geometrical imperfections, and of the theoretical predictions for and experimental measurements of the resistance.

Galambos and Ravindra (1974) proposed that the appropriate values of M_p, M_i [Eq. (3.99)], and M_e should be used for the nominal moment resistance M_n, and determined ϕ factors for a hot-rolled I-beam and a welded I-beam. For the hot-rolled beam, ϕ was 0.89 for low slenderness ($M_n = M_p$), varied from 0.88 to 0.78 for intermediate slenderness ($M_n = M_i$), and was 0.84 for high slenderness ($M_n = M_e$). Thus the variation of ϕM_n with slenderness is similar to that of the reduced ultimate moment capacities M_u predicted by Eqs. (3.109) to (3.111), in that the greatest reductions are for intermediate slendernesses, while the ϕ factor ratio of $0.84/0.89 \simeq 0.94$ for very slender and stocky beams is very close to the limiting M_u ratio of 0.95/1.00 given by Eqs. (3.109) and (3.111). This suggests that if the nominal moment resistance M_n was redefined as

$$M_n = M_u \tag{3.113}$$

instead of M_p, M_i, and M_e, then a constant resistance factor of

$$\phi = 0.89 \tag{3.114}$$

could be adopted. However, Galambos and Ravindra (1974) proposed that the near average constant value of

$$\phi = 0.86 \tag{3.115}$$

should be used with the nominal moment resistance M_n defined by M_p, M_i, and M_e. This appears to penalize short beams, and to be overoptimistic for some beams of intermediate slenderness.

Galambos and Ravindra (1974) determined ϕ factors for the welded beam which were similar to those for the hot-rolled beam, except for intermediate slendernesses, where ϕ varied from 0.88 to 0.51, reflecting to some extent the more serious effects of the residual stress distributions in welded beams. Because of these greater variations, they proposed that variable ϕ factors should be used for welded beams. However, the use of Eq. (3.99) is likely to lead to inaccurate estimates of M_c for welded beams, and it is anticipated that a more satisfactory design method could be developed by using Eq. (3.108) to estimate M_c.

3.7 CONTINUOUS BEAMS

3.7.1 Elastic Beams

A continuous beam is perhaps the simplest type of rigid-jointed plane frame structure, and can be regarded as a series of beam spans which are rigidly connected together at supports. In general, there are interactions between adjacent spans during buckling which depend on the geometry and loading of the complete beam.

The effects of these interactions on the elastic buckling modes of a three span continuous beam are shown in Fig. 3.30. When only the end spans are loaded ($P_2 = 0$), they are restrained during buckling by the center span, and the buckled shape has inflection points in the end spans, as shown in Fig. 3.30(b). When only the center span is loaded ($P_1 = 0$), it is restrained by the end spans, and the buckled shape has inflection points in the center span, as shown in Fig. 3.30(c). Between these two extremes exists the zero interaction load combination for which the buckled shape has inflection points at the internal supports, as shown in Fig. 3.30(d), and each span buckles as if unrestrained in the buckling plane.

The corresponding effects of the interactions on the elastic buckling load combinations are shown in Fig. 3.31. It can be seen that as the loads P_1 on the end spans increase from zero, increases the critical load P_2 of the center span until a maximum value is reached, and that a similar effect occurs as the center span load P_2 increases from zero. These increases are caused by the changes in the

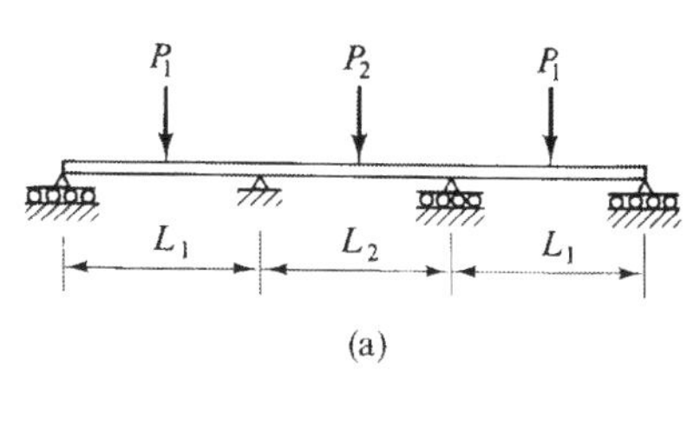

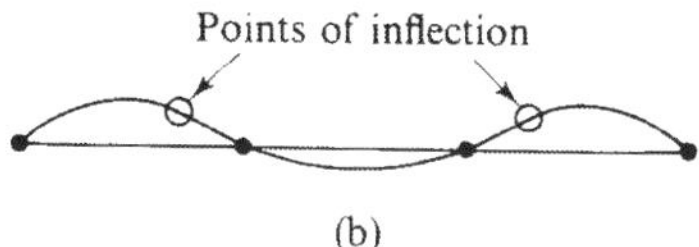

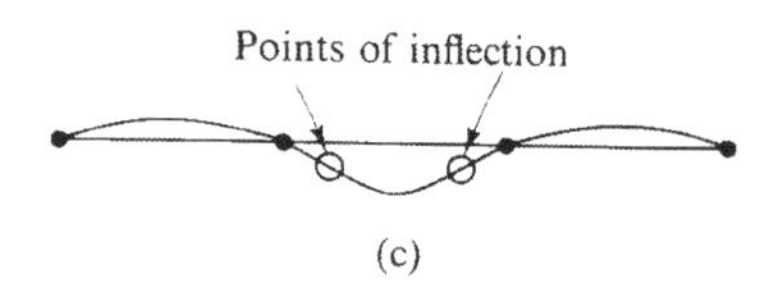

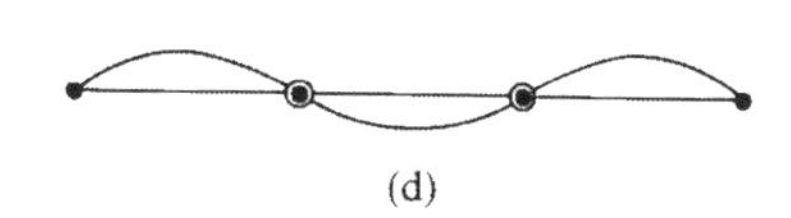

FIGURE 3.30
Buckling modes for symmetrical three-span continuous beam. (a) Elevation. (b) Mode 1, $P_2 = 0$. (c) Mode 2, $P_1 = 0$. (d) Mode 3, zero interaction.

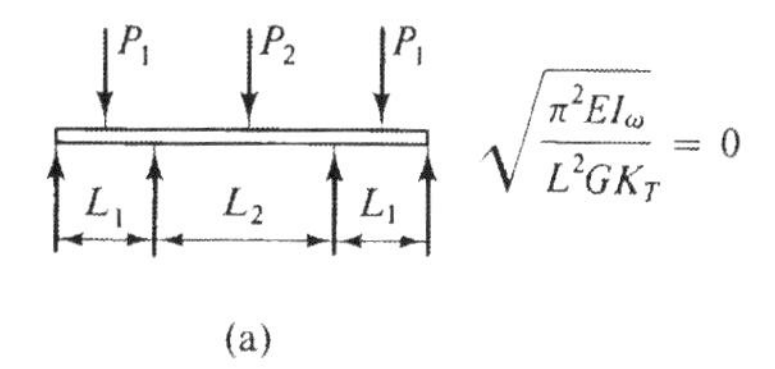

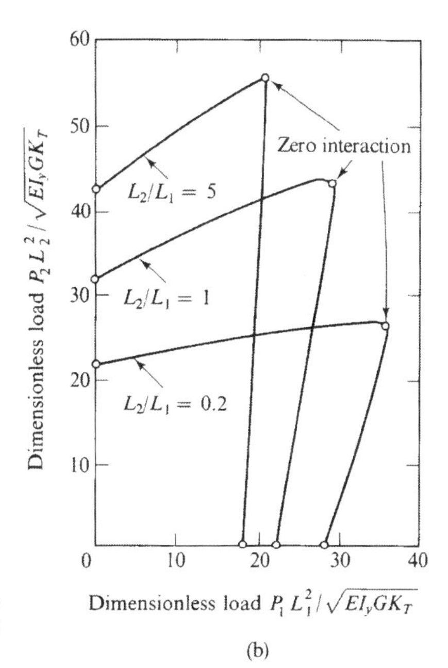

FIGURE 3.31
Elastic critical loads of symmetrical three-span beams. (a) Three-span beam. (b) Critical load combinations

in-plane moment distributions of the restrained spans, which are more important than the decreases caused by the reductions in the restraint offered by the restraining span. Because of this, there may be two different critical values of the load P_2, in which case the beam is only stable when the load P_2 lies between these two values. The lower of these corresponds to the situation where the buckling of the end spans is restrained by the center span. If P_2 is decreased below this lower value, then the central bending moments in the restrained end spans are increased, and they buckle. On the other hand, if P_2 is increased above the upper critical value (at which buckling of the center span is restrained by the end spans), then the central bending moment in the center span is increased, and it buckles.

The elastic buckling of continuous beams has been investigated (Hartmann, 1967, Trahair, 1968b, 1969), and accurate solutions have been determined for a number of beams by using simple extensions of some of the computer methods used for single span beams (see Sect. 3.2.2). More recently, a number of general purpose computer programs have been developed for calculating the elastic buckling loads of continuous beams (Powell and Klingner, 1970, Nethercot, 1972a) and rigid-jointed plane frames (Vacharajittiphan and Trahair, 1974b, 1975). These programs are based on either the finite element, the finite integral, or the direct stiffness methods of analysis, but it is possible to develop similar programs using other methods, such as the finite difference method presented in Chapter 10, or the finite segment method presented in Chapter 11. The programs contain series of analyses of the buckling of the individual elements of the beam or frame which are assembled together in accordance with the equilibrium and compatibility requirements which must be satisfied at the junctions between elements. Some of the programs only require simple data which describe the geometry of the beam or frame, its supports and restraints, and its loading system (the direct stiffness method is described in Sect. 3.11). Unfortunately, these programs are not yet generally available to designers. An alternative which would permit the calculation of the elastic buckling loads of a continuous beam is to provide comprehensive tables of accurate numerical solutions. However, this is not practicable because of the large number of parameters required to describe a beam and its loading system. Instead, comparatively simple methods have been developed which enable the designer to calculate elastic buckling loads. In one such method, which can be used for continuous beams which are prevented from twisting and deflecting at their supports, reasonable accuracy is achieved by allowing for the buckling interactions between adjacent spans and by making use of available accurate tabulations of the effects of elastic end restraints (see Sect. 3.4.1) on lateral buckling. Unfortunately, the available tabulations are too detailed for them to be easily used, while results are only available for a few loading and restraint cases. Even a gross simplification of this method, in which the interaction diagram for the buckling loads of two adjacent segments [see Fig. 3.31(b)] is approximated by two straight lines, has proved too complex for use in routine design, possibly for the same reasons as those cited above, and also because of the indirect calculation method which must be used to locate the approximating straight lines.

An even simpler but less accurate method for beams which consist of a series of segments whose ends are prevented from deflecting laterally and twisting by supports or braces, was proposed by Salvadori in 1951. In this method, the effects of lateral continuity between adjacent segments are ignored, and each segment is regarded as being simply supported laterally. Thus the elastic buckling of each segment is analysed for its in-plane moment distribution [and here the moment modification factors in Eq. (3.31) or Fig. 3.9 may be used] and an effective length L_e equal to the segment length L. The so-determined elastic critical moment of each segment is then used to evaluate a corresponding beam load set, and the lowest of these is taken as the elastic critical load set. This method produces a lower bound estimate which is sometimes remarkably close to the true critical load set.

However, this is not always the case, and so a much more accurate but still reasonably simple method has been developed (Nethercot and Trahair, 1976a). In this method, the accuracy of the lower bound estimate (obtained as described

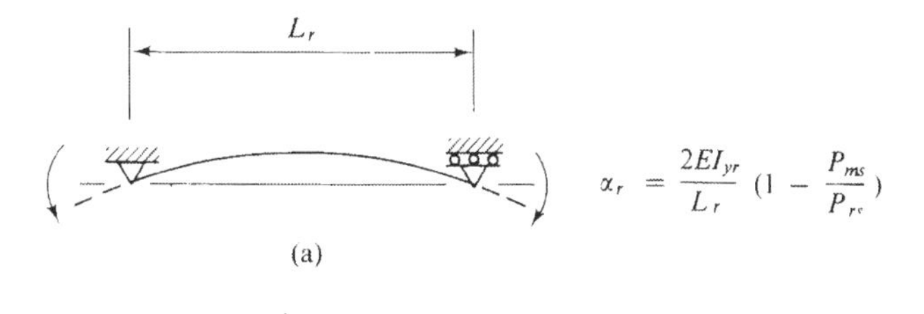

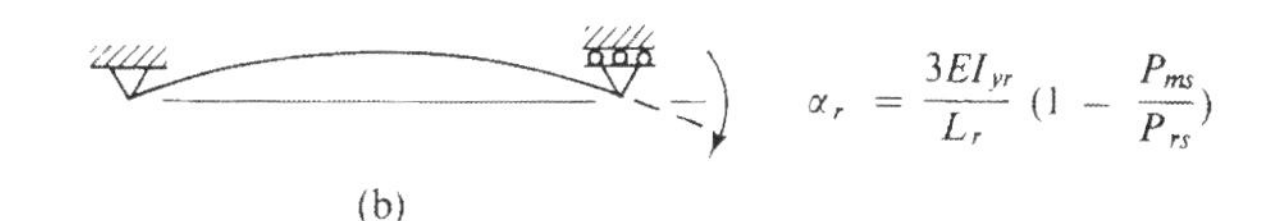

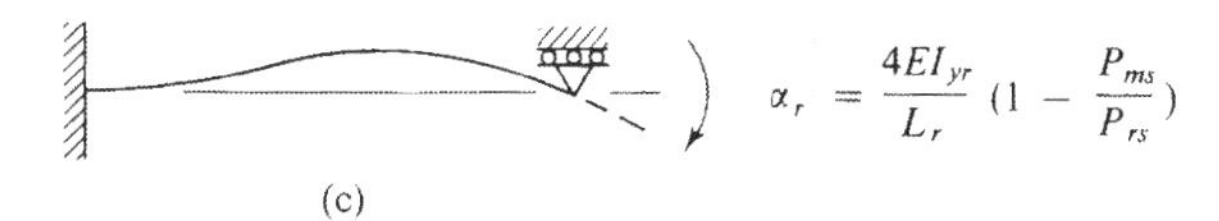

FIGURE 3.32
Stiffness approximation for restraining segments. (a) Segment providing equal end restraints. (b) Segment hinged at far end. (c) Segment fixed at far end

above) is improved by allowing for the interactions between the critical segment and the adjacent segments at buckling. This is done by using the simple approximations shown in Fig. 3.32 for the destabilizing effects of the in-plane bending moments on the stiffnesses of the adjacent segments, and by approximating the restraining effects of these segments on the critical segment by using the effective length chart of Fig. 3.16 for braced compression members to estimate the effective length of the critical beam segment, as suggested in Sect. 3.4.1. A step by step summary of the method is as follows.

1. Determine the properties EI_y, GK_T, EI_ω, of each segment.
2. Analyze the in-plane bending moment distribution through the beam, and determine the moment modification factors β for each segment from Eq. (3.31) or Fig. 3.9.
3. Assume all effective length factors $k(= L_e/L)$ to be equal to unity.
4. Calculate the critical moment M_e of each segment from

$$\frac{M_e L_e}{\sqrt{EI_y GK_T}} = \beta\pi\sqrt{1 + \frac{\pi^2 EI_\omega}{L_e^2 GK_T}} \tag{3.116}$$

   and the corresponding beam buckling loads P_s.

5. Determine a lower bound estimate of the beam buckling load as the lowest value P_{ms} of the loads P_s, and identify the segment associated with this as the critical segment AB. [This is the approximate method (Salvadori, 1951) discussed above, in which the effects of lateral continuity are ignored].

6. If a more accurate estimate of the beam buckling load is required, use the values P_{ms} and P_{rsA}, P_{rsB} calculated in Step 4 together with Fig. 3.32 to approximate the stiffnesses α_{rA}, α_{rB} of the segments adjacent to the critical segment AB.

7. Calculate the stiffness α_m of the critical segment AB from

$$\alpha_m = 2EI_y/L_m \tag{3.117}$$

8. Calculate the stiffness ratios G_A and G_B from

$$G_A = \alpha_m/\alpha_{rA}, \quad G_B = \alpha_m/\alpha_{rB} \tag{3.118}$$

9. Determine the effective length factor k for the critical segment AB from Fig. 3.16, and the effective length $L_e = kL_m$.

10. Calculate the elastic critical moment M_e of the critical segment AB using Eq. (3.116), and from this the corresponding improved approximation of the elastic critical load P_e of the beam.

The application of both the simple and the more accurate methods is demonstrated in the following calculations for the laterally continuous beam shown in Fig. 3.33.

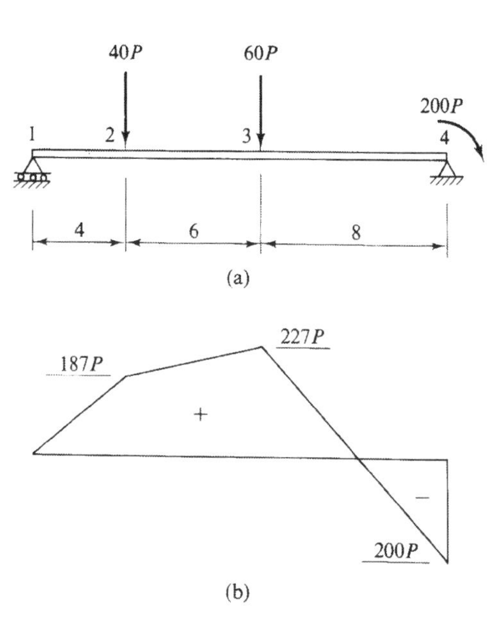

FIGURE 3.33
Laterally continuous beam example. (a) Beam. (Lateral deflection and twist prevented at supports and load.) (b) In-plane moment distribution

1. $EI_y = 1.5 \times 10^6, GK_T = 2.0 \times 10^4,$
   $EI_\omega = 2.5 \times 10^4$
   $L_{12} = 4, L_{23} = 6, L_{34} = 8.$

2. $M_1 = 0, M_2 = 187P, M_3 = 227P, M_4 = -200P$
   $\kappa_{12} = 0/187 = 0.000, \beta_{12} = 1.75$
   $\kappa_{23} = 187/227 = 0.824, \beta_{23} = 1.09$
   $\kappa_{34} = -200/227 = -0.882, \beta_{34} = 2.56$

3. $k_{12} = 1.00, k_{23} = 1.00, k_{34} = 1.00$

4. $$M_{e12} = \frac{1.75 \times \pi \times \sqrt{1.5 \times 10^6 \times 2.0 \times 10^4}\sqrt{1 + \pi^2 \times 2.5 \times 10^4/4^2 \times 2.0 \times 10^4}}{4}$$

$$= \frac{1.75 \times \pi \times 1.73 \times 10^5 \times 1.33}{4} = 317 \times 10^3$$

$$P_{s12} = 317 \times 10^3/187 = 1.70 \times 10^3$$

$$M_{e23} = \frac{1.09 \times \pi \times 1.73 \times 10^5 \times 1.16}{6} = 114 \times 10^3$$

$$P_{s23} = 114 \times 10^3/227 = 0.50 \times 10^3$$

$$M_{e34} = \frac{2.56 \times \pi \times 1.73 \times 10^5 \times 1.09}{8} = 190 \times 10^3$$

$$P_{s34} = 190 \times 10^3/227 = 0.84 \times 10^3$$

5. $P_{ms} = 0.50$ and critical segment is 23.

6. $$\alpha_{12} = \frac{3 \times 1.5 \times 10^6}{4}\left(1 - \frac{0.50}{1.70}\right) = 0.794 \times 10^6$$

$$\alpha_{34} = \frac{3 \times 1.5 \times 10^6}{8}\left(1 - \frac{0.50}{0.84}\right) = 0.228 \times 10^6$$

7. $$\alpha_{23} = \frac{2 \times 1.5 \times 10^6}{6} = 0.500 \times 10^6$$

8. $$G_2 = \frac{0.500 \times 10^6}{0.794 \times 10^6} = 0.63$$

$$G_3 = \frac{0.500 \times 10^6}{0.228 \times 10^6} = 2.20$$

9. $k_{23} = 0.78$

10. $$M_{e23} = \frac{1.09 \times \pi \times 1.73 \times 10^5 \times \sqrt{1 + \pi^2 \times 2.5 \times 10^4/0.78^2 \times 6^2 \times 2.0 \times 10^4}}{0.78 \times 6}$$

$$= 157 \times 10^3$$

$$P_e = 157 \times 10^3/227 = 0.69 \times 10^3$$

which is a 38 percent increase over the Salvadori lower bound of 0.50.

It should be noted that while the calculations for the lower bound (the first five steps) are made for all segments, those for the improved estimate are only made for the critical segment, and so comparatively little extra effort is involved.

3.7.2 Plastic Beams

The lateral buckling of partially yielded continuous beams is influenced by two separate effects. The first of these is concerned with the interactions between laterally continuous segments during plastic buckling. These interactions are substantially affected by the significant reductions of the lateral stiffnesses of the segments caused by yielding. The plastic buckling of beams which are continuous laterally but statically determinate with respect to in-plane bending has been analyzed by Nethercot (1975b), while careful experiments have been made by Dibley (1969).

A simple hand method of analyzing approximately the plastic buckling of beams with braces and supports which prevent lateral deflection and twist has been developed by Nethercot and Trahair (1976b), who modified their approximate method described in Sect. 3.7.1 for elastic beams. In this modified method, the plastic buckling moments M_i of the segments between brace and support points are first calculated by assuming that the segments are unrestrained laterally. This is done by using the method explained in Sect. 3.5.2, in which the elastic buckling moment M_e is first calculated [using Eqs. (3.30) and (3.31) for example], and then modified if necessary [as in Eq. (3.80)] so as to include the effect of the in-plane moment distribution on the plastic buckling moment M_i.

The interaction between plastic segments during buckling is analyzed approximately by an extension of the procedure described in Sect. 3.7.1 for elastic continuous beams. For this extension, modifications are made so that the effective plastic lateral stiffnesses $\alpha_m, \alpha_{rA}, \alpha_{rB}$ of the critical segment AB and its adjacent segments can be calculated approximately. In the absence of a more precise method and with the purpose of developing as simple a method as possible, it was proposed that the elastic lateral stiffness of the critical segment given by Eq. (3.117) should be multiplied by the ratio M_i/M_e of the approximate plastic and elastic buckling moments already calculated for the critical segment. The previous approximation from Fig. 3.32 for the elastic stiffness of an adjacent restraining segment is similarly multiplied by the ratio M_i/M_e calculated for the restraining segment. (It may be noted, however, that this latter approximation is likely to be extremely conservative, as the calculated value of M_e is likely to be much greater than the actual moment in the segment at buckling.) These modifications allow the stiffness ratios to be calculated, the effective length factor k to be determined from Fig. 3.16, and the new elastic buckling moment M_e for the critical segment to be calculated from Eq. (3.116). The plastic buckling moment M_i can then be estimated from Eq. (3.80). A comparison with accurate solutions has shown that the approximate method produces reasonably accurate and consistently safe predictions.

The second influence on the plastic lateral buckling of continuous beams is that of the plastic redistribution of in-plane bending moment which takes place in statically indeterminate beams. This redistribution changes the in-plane moment distributions in the beam segments from the initial elastic distributions, and consequently changes the lateral buckling behavior of the segments. The plastic buckling of indeterminate continuous beams has been analyzed by Yoshida and Imoto (1973) and by Yoshida, Nethercot, and Trahair (1975), while experiments have been conducted on stocky plastic beams by Bansal (1971) and on beams of intermediate slenderness by Poowannachaikul and Trahair (1974).

The studies of Poowannachaikul and Trahair (1974) and Yoshida, Nethercot and Trahair (1975) have shown that the in-plane moment redistributions are less severe than predicted by rigid-plastic theory because of the effects of residual stresses and strain-hardening, and that the changes in the plastic buckling loads are gradual. It is suggested that a reasonable approximation to the plastic buckling loads of a statically indeterminate continuous beam can be obtained from the lower of the two estimates calculated for the elastic and the plastic collapse moment distributions by the modified method described earlier in this section.

3.8 ELASTIC BEAM-COLUMNS

It is convenient to discuss the behavior of beam-columns under the three separate headings of *in-plane behavior*, *flexural-torsional buckling*, and *biaxial bending*. When a beam-column is bent about its weaker principal axis, or when it is prevented from deflecting laterally while being bent about its stronger principal axis [as shown in Fig. 3.3(a)], then its action is confined to the plane of bending. This in-plane behavior is discussed in Volume 1. When a beam-column which is bent about its stronger principal axis [as shown in Fig. 3.3(b)] is not restrained laterally, then it may buckle out of the plane of bending by deflecting laterally and twisting. This flexural-torsional buckling is discussed in this and the following sections. More generally, a beam-column may be bent about both principal axes and twisted, as shown in Fig. 3.3(c). This biaxial bending is discussed in the other chapters of this volume.

3.8.1 Equal End Moments

A simply supported beam-column which is bent about its major axis by end moments M_x and κM_x and compressed by an axial force P is shown in Fig. 3.3(b). The beam-column supports prevent both lateral deflection and twist, but the flange ends are free to warp. It is assumed that the beam-column is elastic and of uniform doubly symmetric I-section, and the particular case is considered for which the end moments are equal and opposite ($\kappa = 1$).

When the applied load and moments reach the critical buckling values, a deflected and twisted equilibrium position is possible. The differential equilibrium equations for such a position are, from Eq. (2.178)

$$-EI_y u'' = M_x\theta + Pu \tag{3.119}$$

for minor axis bending, and

$$GK_T\theta' - EI_\omega\theta''' = M_x u' + Pr_0^2\theta' \tag{3.120}$$

for torsion, in which $r_0 = \sqrt{(I_x + I_y)/A}$ is the polar radius of gyration.

These equations only differ from the corresponding Eqs. (3.1) and (3.2) for beams by the additional lateral bending moment Pu induced by the deflection of the beam and the additional torque $Pr_0^2\theta'$ induced by the twisting. This additional torque is exerted by transverse components of the longitudinal stresses σ, and is equivalent to the corresponding torque in Eq. (3.42) of

$$-\bar{K}\theta' = -\theta'\int_A \sigma a^2\, dA \tag{3.121}$$

in which a is the distance to the shear center. The moments and forces in the beam-column induce stresses

$$\sigma = \frac{yM_x}{I_x} - \frac{P}{A} \tag{3.122}$$

and so for a doubly symmetric cross section,

$$\bar{K} = M_x\left\{\frac{\int_A x^2 y\, dA + \int_A y^3\, dA}{I_x}\right\} - \frac{P}{A}\int_A (x^2 + y^2)\, dA \tag{3.123}$$

which reduces to

$$\bar{K} = -Pr_0^2 \tag{3.124}$$

The support conditions of end deflection and twisting prevented and end warping free are represented by

$$\left.\begin{aligned}(u)_o = (u)_l &= 0\\ (\theta)_o = (\theta)_l &= 0\\ (\theta'')_o = (\theta'')_l &= 0\end{aligned}\right\} \tag{3.125}$$

These conditions are automatically satisfied if we assume the sinusoidal deflected and twisted shapes

$$\left.\begin{aligned}u &= B_3 \sin \pi z/L\\ \theta &= A_3 \sin \pi z/L\end{aligned}\right\} \tag{3.126}$$

in which B_3 and A_3 are the central deflection and twist.

If these shapes are substituted, then Eqs. (3.119) and (3.120) become

$$\left.\begin{aligned}\left(\frac{\pi^2 EI_y}{L^2} - P\right) B_3 \sin\frac{\pi z}{L} - M_x A_3 \sin\frac{\pi z}{L} = 0\\ -M_x B_3 \frac{\pi}{L}\cos\frac{\pi z}{L} + \left(GK_T - Pr_0^2 + \frac{\pi^2 EI_\omega}{L^2}\right) A_3 \frac{\pi}{L}\cos\frac{\pi z}{L} = 0\end{aligned}\right\} \tag{3.127}$$

These equations can be rearranged as

$$\left.\begin{aligned}\frac{B_3}{A_3} &= \frac{M_x}{\pi^2 EI_y/L^2 - P} \\ \frac{B_3}{A_3} &= \frac{r_0^2}{M_x}\left\{\frac{GK_T}{r_0^2}\left(1 + \frac{\pi^2 EI_\omega}{L^2 GK_T}\right) - P\right\}\end{aligned}\right\} \tag{3.128}$$

and the ratio B_3/A_3 can be eliminated, whence

$$\frac{M_x^2}{M_e^2} = \left[\left(1 - \frac{P}{P_{ey}}\right)\left(1 - \frac{P}{P_z}\right)\right] \tag{3.129}$$

in which

$$P_{ey} = \pi^2 EI_y/L^2 \tag{3.130}$$

is the minor axis flexural buckling load of an axially loaded column, and

$$P_z = \frac{GK_T}{r_0^2}\left(1 + \frac{\pi^2 EI_\omega}{L^2 GK_T}\right) \tag{3.131}$$

is the corresponding torsional buckling load, and

$$M_e = r_0\sqrt{P_{ey}P_z} = \frac{\pi\sqrt{EI_y GK_T}}{L}\sqrt{1 + \frac{\pi^2 EI_\omega}{L^2 GK_T}} \tag{3.132}$$

is the flexural-torsional buckling moment of a beam in uniform bending [see Sect. 3.2.1].

Thus the deflected and twisted shapes of Eqs. (3.126) satisfy Eqs. (3.119) and (3.120) at every point when the moment M_x and the load P reach the values which satisfy Eq. (3.129). This latter equation therefore defines the combinations of moment and load which cause the beam-column to buckle elastically. It can be seen that for the limiting case of $M_x = 0$, the beam-column buckles as an axially loaded column at the lower of P_{ey} and P_z, and that when $P = 0$, it buckles as a beam at the critical moment M_e. Some more general solutions of Eq. (3.129) are shown in Fig. 3.34.

This derivation of Eq. (3.129) neglects the amplification of the in-plane bending moment by the axial load, and also the effects of in-plane curvature on lateral buckling. When these are accounted for, a very close approximation to the buckling combination of M_x and P can be obtained (Vacharajittiphan, Woolcock, and Trahair, 1974) from

$$\frac{M_x^2}{M_e^2} = \left(1 - \frac{P}{P_{ex}}\right)\left(1 - \frac{P}{P_{ey}}\right)\left(1 - \frac{P}{P_z}\right) \tag{3.133}$$

in which

$$P_{ex} = \pi^2 EI_x/L^2 \tag{3.134}$$

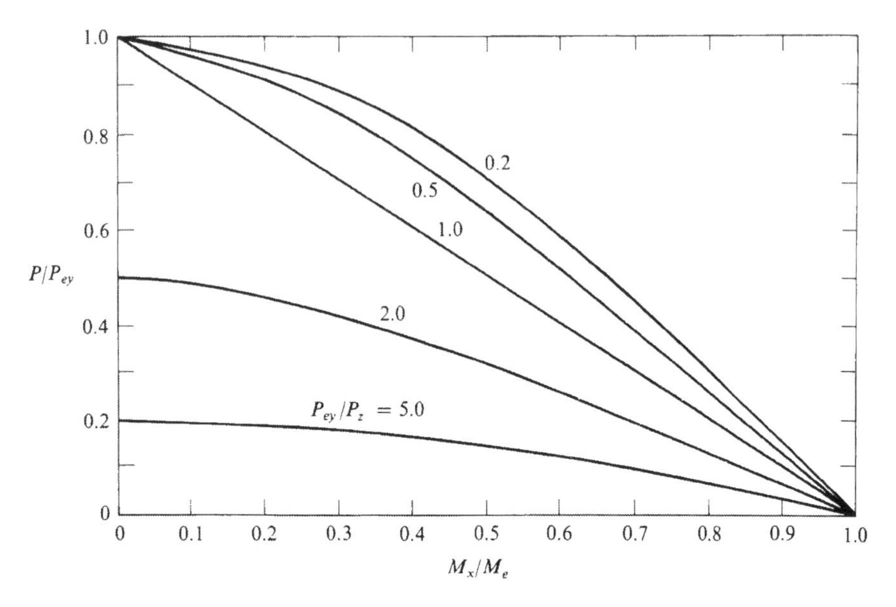

FIGURE 3.34
Elastic critical load combinations for beam-columns with equal end moments

is the major axis flexural buckling load of an axially loaded column, and the value of M_e is obtained by including the in-plane curvature correction discussed in Sect. (3.2.1). A conservative and slightly less accurate approximation is obtained if this latter correction is omitted.

The maximum possible value of P is the least of P_{ex}, P_{ey}, and P_z. For most sections this maximum value is much less than P_{ex}, and so Eq. (3.133) is very close to Eq. (3.129). For most hot-rolled steel members, P_{ey} is less than P_z, and so $(1 - P/P_z) > (1 - P/P_{ey})(1 - P/P_{ex})$, in which case Eq. (3.133) can be safely approximated by the interaction equation

$$\frac{P}{P_{ey}} + \frac{1}{(1 - P/P_{ex})} \frac{M_x}{M_e} = 1 \tag{3.135}$$

3.8.2 Unequal End Moments

The elastic flexural-torsional buckling of simply supported beam-columns with unequal major axis end moments M_x and κM_x has been investigated numerically, and tabulated solutions are available (Horne, 1954, Salvadori, 1955, 1956, Column Research Committee of Japan, 1971).

Horne (1954) tabulated values of a factor $\sqrt{F}$ which allow the unequal end moments to be treated as equivalent equal and opposite end moments $M_x/\sqrt{F}$. Thus the elastic buckling of beam-columns with unequal end moments can be approximated by modifying Eq. (3.133) to

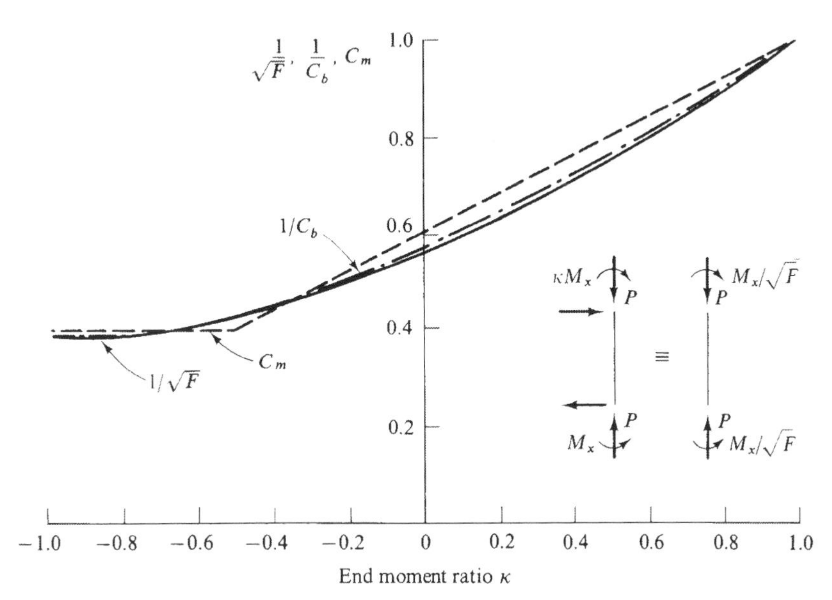

FIGURE 3.35
Equivalent equal end moments for beam-columns

$$\frac{(M_x/\sqrt{F})^2}{M_e^2} = \left(1 - \frac{P}{P_{ex}}\right)\left(1 - \frac{P}{P_{ey}}\right)\left(1 - \frac{P}{P_z}\right) \tag{3.136}$$

or by a similarly modified Eq. (3.135).

The variation of $1/\sqrt{F}$ with the end moment ratio κ is shown in Fig. 3.35. The factor $\sqrt{F}$ is the same as the moment modification factor β used in Eq. (3.30) for beams with unequal end moments, and shown in Fig. 3.6. It can therefore be closely approximated by the factors C_b and $1/C_m$ given in Eqs. (3.31) and (3.32). Thus the modified Eq. (3.135) can be written as

$$\frac{P}{P_{ey}} + \frac{1}{(1 - P/P_{ex})}\frac{(M_x/\beta)}{M_e} = 1 \tag{3.137}$$

in which M_e is given by Eq. (3.132). Alternatively, the original Eq. (3.135) can be used provided the value used for M_e is that for the unequal end moments, and is calculated from Eq. (3.30) by using the approximations of Eqs. (3.31) or (3.32) for the moment modification factor β.

3.8.3 Elastic End Restraints

The effects of elastic end restraints on the lateral buckling of beam-columns are similar to those on beams. A solution can be found for the case of a beam-column with equal and opposite end moments ($\kappa = 1$) which has equal minor axis rotational restraints acting at each flange end whose stiffnesses can be expressed in the form

$$\frac{\text{Flange End Moment}}{\text{Flange End Rotation}} = -\frac{EI_y}{L}\frac{R}{1-R} \tag{3.138}$$

in which R is a restraint parameter. The flange moments at the end $z = L$ are

$$M_{t,b} = \tfrac{1}{2}EI_y[u'' \pm (h/2)\theta'']_l \tag{3.139}$$

while the corresponding flange end rotations are $[u' \pm (h/2)\theta']_l$. Thus the restraint conditions of Eq. (3.138) can be expressed as

$$\frac{L[u'' \pm (h/2)\theta'']_l}{[u' \pm (h/2)\theta']_l} = -\frac{2R}{1-R} \tag{3.140}$$

or

$$\frac{L(u'')_l}{(u')_l} = \frac{L(\theta'')_l}{(\theta')_l} = -\frac{2R}{1-R} \tag{3.141}$$

The other boundary conditions for the end $z = L$ are those of end deflection and twisting prevented, i.e.,

$$(u)_l = (\theta)_l = 0 \tag{3.142}$$

Similar conditions apply at the end $z = 0$.

The differential equilibrium equations for a buckled position of the restrained beam-column are

$$\left.\begin{aligned} -EI_y u'' &= M_x\theta + Pu - (M_t + M_b) \\ \text{and} \qquad GK_T\theta' - EI_\omega\theta''' &= M_x u' + Pr_0^2\theta' \end{aligned}\right\} \tag{3.143}$$

It can be shown that these differential equations and the boundary conditions of Eqs. (3.140), (3.141), and (3.142) are satisfied by the buckled shape

$$u = \frac{M_x}{P_{ey} - P}\theta = A\{\cos \pi(z - L/2)/kL - \cos \pi/2k\} \tag{3.144}$$

in which

$$P_{ey} = \pi^2 EI_y/k^2L^2 \tag{3.145}$$

is now the flexural buckling load of a restrained column, and the effective length factor k is the solution of

$$\frac{R}{1-R} = \frac{-\pi}{2k}\cot\frac{\pi}{2k} \tag{3.146}$$

provided

$$\frac{M_x^2}{M_e^2} = \left(1 - \frac{P}{P_{ey}}\right)\left(1 - \frac{P}{P_z}\right) \tag{3.147}$$

is satisfied, in which

$$P_z = \frac{GK_T}{r_0^2}\left(1 + \frac{\pi^2 EI_\omega}{k^2L^2GK_T}\right) \tag{3.148}$$

is now the torsional buckling load of a restrained column, and

$$M_e = \frac{\pi\sqrt{EI_yGK_T}}{kL}\sqrt{1 + \frac{\pi^2 EI_\omega}{k^2L^2GK_T}} = r_0\sqrt{P_{ey}P_z} \tag{3.149}$$

is the lateral buckling moment of a restrained beam [see Eqs. (3.59) and (3.60)]. The combination of moments and loads which cause buckling are therefore given by Eq. (3.147), which is the same as the corresponding Eq. (3.129) for unrestrained beam-columns, except for the familiar use of the effective length kL instead of the actual length L in the definitions of P_{ey}, P_z, and M_e. Thus the solutions shown in Fig. 3.34 can also be applied to end-restrained beam-columns provided the effective length factor k obtained from Eq. (3.146) is used to evaluate P_{ey}, P_z, and M_e. Equation (3.146) is the same as Eq. (3.61) for the effective lengths of end-restrained beams, and so the solution shown in Fig. 3.15 can also be used for the effective length factors of restrained beam-columns.

The solutions for the effective length factor k obtained from Eq. (3.146) are the same as those obtained from the effective length chart of Fig. 3.16 for braced compression members (see also Sect. 3.4.1) when the stiffness ratios G_A, G_B (i.e., the ratios of the beam-column stiffness to the stiffnesses of the restraints at the ends A, B of the member) are taken as

$$G_A = G_B = \frac{1 - R}{R} \tag{3.150}$$

This suggests that the effective length factors for beam-columns with unequal end restraints ($G_A \neq G_B$) can be approximated in the same way as those for beams by using the values given by Fig. 3.16.

A further approximation may be suggested for restrained beam-columns with unequal end moments ($\kappa \neq 1$), in which Eq. (3.136) or (3.137) is used, together with the appropriate moment modification factors $\sqrt{F}$ or β obtained from Fig. 3.35 and the effective length factor k obtained from Fig. 3.16.

3.9 PLASTIC STEEL BEAM-COLUMNS

The solutions obtained in Sect. 3.8 for the flexural-torsional buckling of beam-columns are only valid while they remain elastic. When the in-plane loading on a beam-column causes prior yielding, the effective rigidities of some sections of the member are reduced, and buckling may occur at a load which is significantly less than the elastic buckling load, as indicated in Fig. 3.36. An example of this is given in Fig. 3.37 (Galambos and Fukumoto, 1966), in which are shown the variations with the major axis slenderness L/r_x of the elastic critical moment (the dotted line)

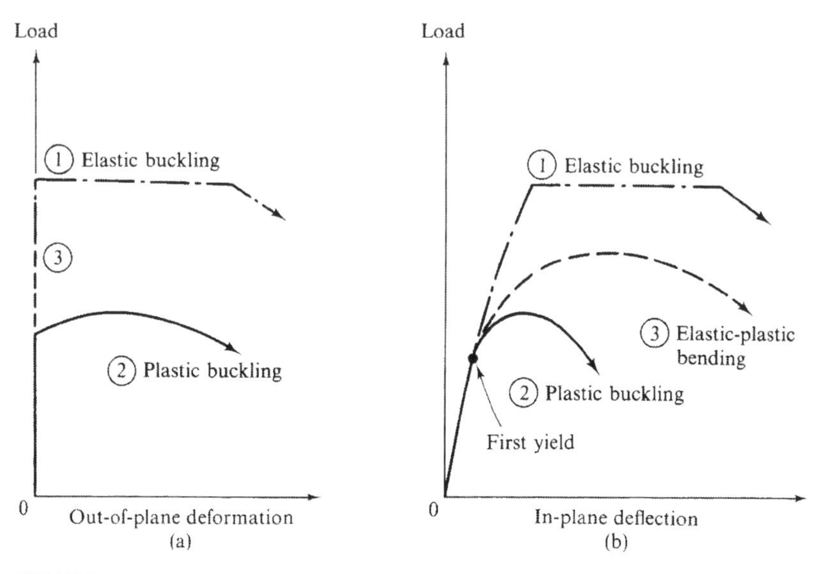

FIGURE 3.36
Flexural-torsional buckling of beam-columns. (a) Out-of-plane behavior. (b) In-plane Behavior

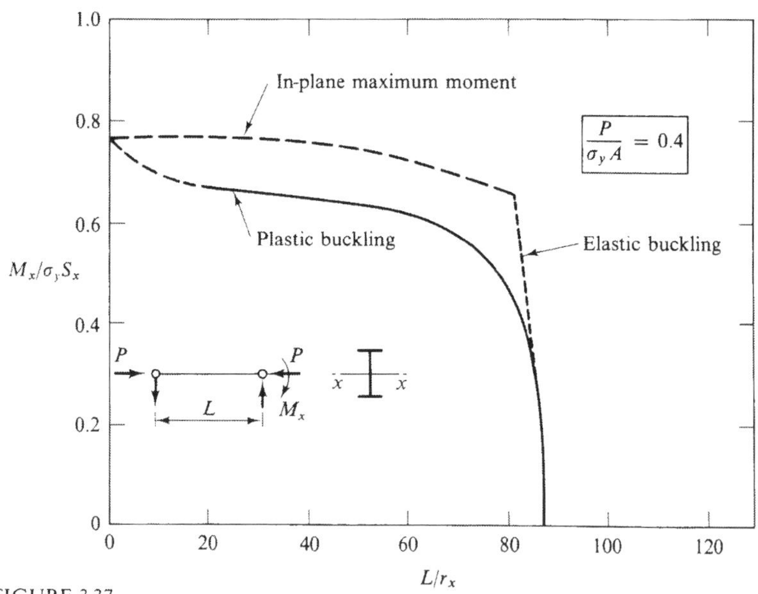

FIGURE 3.37
Plastic flexural-torsional buckling of a beam column

and the plastic critical moment (the full line). Also shown is the variation of the maximum in-plane moment (the dashed line).

A detailed study of the flexural-torsional buckling of plastic beam-columns was made by Lim and Lu (1970). They first reviewed previous research and then conducted a systematic theoretical investigation of the plastic buckling of a range of restrained isolated beam-columns under various combinations of axial load and unequal end moments. They concluded that the effects of the prebuckling deformations and of minor axis flexural restraints were important for long beam-columns, and that the effects of warping restraints were only important in short beam-columns.

Lim and Lu's theoretical work was based on lower bound tangent modulus solutions obtained by assuming that the modulus of elasticity is equal to zero in any yielded regions (instead of the strain-hardening value, as discussed in Sect. 3.5.1). This assumption is unnecessarily conservative, as it leads to the conclusion that every beam-column buckles laterally, no matter how close its brace spacing or how low its slenderness. Because of this, Lim and Lu also obtained upper bound reduced modulus solutions (see Chap. 6 in Vol. 1). They also studied the effects of geometrical imperfections, and used their results to extrapolate to a "maximum strength" of a perfect beam-column which lies between the upper and lower bound buckling solutions.

The plastic buckling of three-story continuous beam-columns with simple supports at top and bottom was also studied by Lim and Lu (1970). A range of different geometrical and loading arrangements were considered, and the interactions between the story-heights during buckling were investigated. It was concluded that these interactions were favorable in respect of the most critical story-height, in that its buckling load was increased above that calculated by ignoring the interactions. On the other hand, the interactions were detrimental to the least critical story-height, and the buckling load calculated for this story-height by ignoring the interactions erred on the unsafe side. These conclusions conform with those reached by Salvadori (1951) in his study of the elastic buckling of laterally continuous beams, and which form the basis for his lower bound procedure (see Sect. 3.7.1) for calculating elastic critical loads. It seems likely, therefore, that the methods discussed in Sect. 3.7.2 for allowing for the buckling interactions between the adjacent segments of plastic beams could be extended to beam-columns.

3.10 DESIGN CRITERIA FOR STEEL BEAM-COLUMNS

The ultimate strength of a beam-column whose failure is influenced by its susceptibility to flexural-torsional buckling can be estimated very simply by adapting the linear interaction equation for in-plane bending about the major axis [Eq. (14.30) of Volume 1]

$$\frac{P}{P_u} + \frac{C_m}{(1 - P/P_{ex})} \frac{M_x}{M_{ux}} = 1 \tag{3.151}$$

in which P_u is the ultimate strength of an axially loaded column ($M_x = 0$) which buckles in-plane, and M_{ux} is the ultimate in-plane strength (which is equal to the full plastic moment M_{px}) of a beam ($P = 0$).

This equation is simply modified for flexural-torsional buckling by redefining P_u as the ultimate minor axis strength of an axially loaded column, and M_{ux} as the ultimate strength of a beam with equal and opposite end moments ($\kappa = 1$) which fails either by in-plane plasticity at M_{px} or at a moment which is reduced by the influence of flexural-torsional buckling. The basis for this modification is the remarkable similarity between Eq. (3.151) and the simple linear approximation of Eq. (3.137) for the elastic flexural-torsional buckling of beam-columns, and the close correspondence shown in Fig. 3.35 between the unequal end moment conversion factors C_m for in-plane failure and β (or $1/\sqrt{F}$) for flexural-torsional buckling.

Galambos (1968) found that Eq. (3.151) often led to conservative predictions. He suggested that a more accurate estimate of the plastic critical buckling load of a beam-column could be obtained by modifying its elastic critical load in a similar way to that used in Eq. (3.87) (Sect. 3.6.2) to obtain the plastic critical load of a beam. Thus the plastic critical load P_i of a beam-column can be expressed in terms of the nominal in-plane first yield load P_{fy} given by

$$\frac{P_{fy}}{P_y} = 1 - \frac{M_x}{M_{yx}} \tag{3.152}$$

and of the elastic critical load P given by Eq. (3.137), as

$$\frac{P_i}{P_{fy}} = 1 - \frac{1}{4}\frac{P_{fy}}{P} \tag{3.153}$$

while $P > 0.5P_{fy}$. In these equations, $P_y = A\sigma_y$ is the squash load, and $M_{yx} = \sigma_y S_x$ is the first yield moment when P is zero.

The theoretical investigations of Lim and Lu (1970) showed that while the loads calculated from the modified form of Eq. (3.151) were sometimes greater than their predicted tangent modulus buckling loads, they more generally provided conservative estimates of their "maximum strengths" (see Sect. 3.9), especially those for beam-columns of low slenderness. Following this finding, they proposed that Eq. (3.151) should be modified to

$$\frac{P}{P_u} + \frac{C_m}{(1 - P/P_{ex})}\frac{M_x\{1 - 100/(L/r_x)^2\}}{M_{ux}} = 1 \tag{3.154}$$

when L/r_x is greater than 10, provided that M_x does not exceed M_{ux}.

In the AISC 1969 Specification, the working load method of designing beam-columns against flexural-torsional buckling is governed either by

$$\frac{f_a}{0.60\sigma_y} + \frac{f_{bc}}{\sigma_{bc}} \leq 1 \tag{3.155}$$

and

$$\frac{f_a}{\sigma_a} + \frac{C_m}{(1 - 23f_a/12\sigma_{ex})}\frac{f_{bc}}{\sigma_{bc}} \leq 1 \tag{3.156}$$

or, if f_a/σ_a is less than 0.15, by

$$\frac{f_a}{\sigma_a} + \frac{f_{bc}}{\sigma_{bc}} \leq 1 \tag{3.157}$$

In these equations, f_a and f_{bc} are the axial and compressive bending stresses caused by the working loads, σ_{ex} is the stress at which an axially loaded column ($f_{bc} = 0$) buckles elastically about the major axis, σ_a is the maximum permissible stress in an axially loaded column [Eqs. (14.7) and (14.8) of Volume 1], and σ_{bc} is the maximum permissible stress in a beam ($f_a = 0$) with equal and opposite end moments, reduced if necessary to allow for the effects of flexural-torsional buckling (see Sec. 3.6.2). Eqs. (3.155), (3.156), and (3.157) were obtained from the corresponding equations [Eqs. (14.35), (14.33) and (14.34) of Volume 1] for the in-plane design of beam-columns by making similar modifications to allow for the effects of flexural-torsional buckling to those made to Eq. (3.151) to change it from an in-plane ultimate strength interaction equation.

The effects of flexural-torsional buckling on the plastic design of beam-columns are allowed for approximately in the AISC 1969 Specification by reducing the moment M_{ux} used in the modified form of Eq. (3.151) from M_{px} to

$$M_m = M_{px}\left\{1.07 - \frac{(L/r_y)\sqrt{\sigma_y}}{3160}\right\} \leq M_{px} \tag{3.158}$$

In the British Standard BS 449:1969, the working load method of designing beam-columns against flexural-torsional buckling is governed by

$$\frac{f_a}{\sigma_a} + \frac{f_{bc}}{\sigma_{bc}} \leq 1 \tag{3.159}$$

in which σ_{bc} is also reduced if necessary to allow for the effects of flexural-torsional buckling. However, this equation takes no account of the effects of unequal end moments, while the omission of the factor $1/(1 - 23f_a/12\sigma_{ex})$ means that the increases in the in-plane bending moments caused by the axial load are neglected.

In the Australian Standard AS1250-1975, the working load method of designing beam-columns against flexural-torsional buckling is governed by equations similar to Eqs. (3.155), (3.156), and (3.157), except that a factor of 0.60 is used to multiply σ_{ex} instead of 12/23.

3.11 DIRECT STIFFNESS ANALYSIS OF ELASTIC LATERAL BUCKLING

The instability of a beam or plane frame structure may be classified in accordance with its load-deformation characteristic (see Fig. 3.38) as being of the bifurcation or non-bifurcation type. For the bifurcation type of instability, the structure deforms in a stable configuration as the load is increased from zero until a critical load (at the point of bifurcation on the load-deformation path) is reached, and the system has two states of equilibrium, the original configuration which is now unstable,

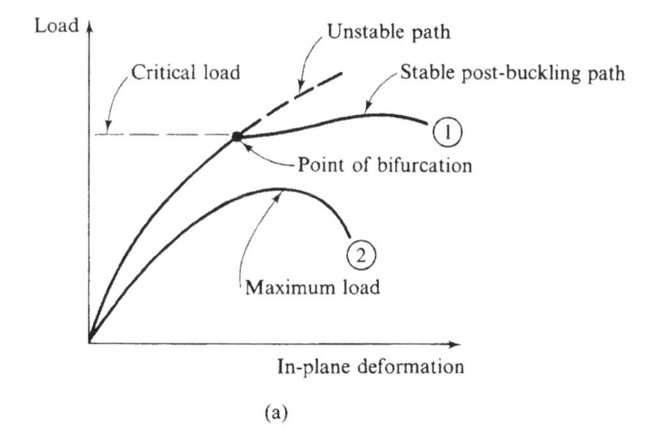

(a)

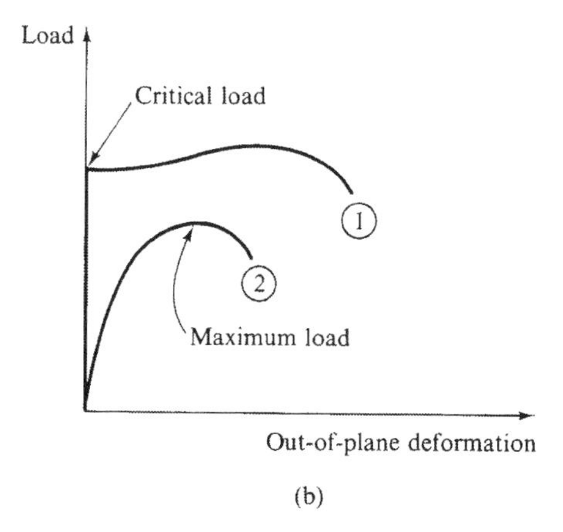

(b)

(1) In-plane loading (bifurcation)
(2) Three-dimensional loading (non-bifurcation)

FIGURE 3.38
Load-deformation characteristics of plane frames. (a) In-plane characteristics. (b) Out-of-plane characteristics

and a new stable configuration. The structure prefers the stable configuration and changes its deformations suddenly from the unstable to the stable state. Structures with three-dimensional loading exhibit the non-bifurcation type of instability, in which the deflections increase until a maximum load is reached, beyond which static equilibrium can only be sustained by decreasing the load.

Many methods of calculating the critical or maximum loads have been developed. These methods may be classified as energy methods, as differential equation solutions, or as stiffness approaches, although they are all closely related. The energy methods and the methods of solution of the governing differential equilibrium equations have been employed successfully for isolated members or small structures. In particular, much of the basic theory of elastic lateral buckling has been developed from rigorous solutions of the differential equations (Secs. 3.2.1, 3.8.1, for example), while many of the numerical solutions presented in this chapter were obtained by energy methods or from approximate solutions of the differential equations obtained by numerical integration (see Chap. 9), by the finite difference method (see Chap. 10), or by other methods.

However, these methods become very complicated when applied to larger elastic structures or to partially plastic structures. In the past decade, the matrix methods of structural analysis have revolutionized solution techniques for large elastic structures because of their simplicity and systematic method of solution. One of the better known and more widely used of these is the *direct stiffness* (or *displacement*) *method* (Harrison, 1973). This can be used in two different ways, depending on the load-deformation characteristic of the structure, and these can be referred to as the determinantal and the deteriorated (or tangent) stiffness methods. The former method makes use of the fact that the determinant of the complete stiffness matrix of the structure is zero at the critical loading condition, and so the use of this method is associated with the bifurcation type of instability (buckling or eigenvalue approach, see Chap. 1, Vol. 1). The tangent stiffness approach is a technique applied in the non-linear structural analysis of the non-bifurcation type of instability. In this case the stiffness matrix is constantly modified along the load-deformation path, and the maximum load is obtained from the full load-deformation characteristic (Fig. 3.38) (load-deflection approach, see Chap. 1, Vol. 1). This latter method is used in Chaps. 6, 7, 11 and 12.

The flexural-torsional buckling of a plane frame subjected to loading in its plane is of the bifurcation type, and the out-of-plane deformations remain zero until the critical loading condition is reached [Fig. 3.38(b)]. Thus the in-plane behavior of the frame up to the critical load can be analyzed independently of the out-of-plane buckling behavior (using a non-linear in-plane analysis when necessary). The solutions from this in-plane analysis can then be substituted in the flexural-torsional equations which govern the out-of-plane buckling of the frame. These equations can be solved numerically to determine the member stiffness matrices, and these can be transformed and assembled to form the frame stiffness matrix. The critical load which causes the frame to buckle can then be found from the condition that the determinant of this matrix should be zero.

A number of different numerical methods may be used to determine the member stiffness matrices for flexural-torsional buckling, including the finite difference method, numerical integration, the finite element method, and the finite integral method. In later sub-sections, the use of the finite integral method is discussed, and a computer program based on this method is described. The program can carry out a first or second order in-plane small displacement analysis, and the effects of the in-plane deformations on the flexural-torsional buckling loads may be included.

3.11.1 Differential Equilibrium Equations

The usual assumptions for the flexural-torsional buckling analysis can be summarized as follows:

1. The structure remains elastic.
2. The in-plane deformations of the structure prior to buckling are small but finite. However, the out-of-plane deformations and the changes in the in-plane deformations are infinitesimally small when the structure buckles laterally out of its plane.
3. Before buckling, each applied load acts in the plane of the structure.
4. During buckling, each applied load remains constant in magnitude and direction but its line of action may move out of the original plane of the structure, as shown in Fig. 3.39.

In order to define the position and attitude of a member cross section, three right-hand sets of orthogonal axes are used. The frame axis system consists of the X, Y, Z axes which are fixed in position and direction. The X axis is perpendicular to the initial plane of the frame, and the Y and Z axes lie in this plane. The member axes x, y, z have their origin at one end A of the member [Fig. 3.40(a)]. The x axis is parallel to the X axis, and is positive in the same sense, while the z axis coincides with the centroidal axis of the undeformed member. Finally, the third set of axes consists of the local major and minor principal axes ξ and η of a cross-

FIGURE 3.39
Forces and moments acting on portion of a typical member buckling

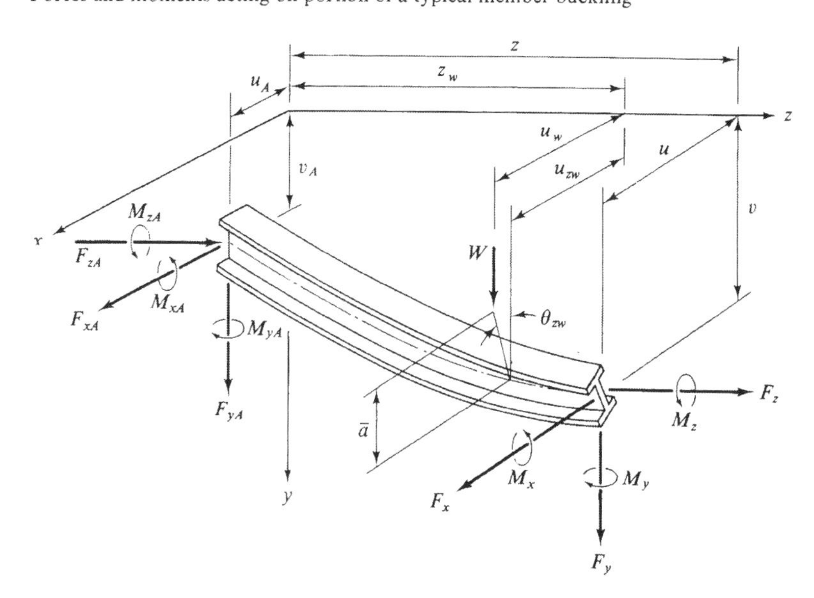

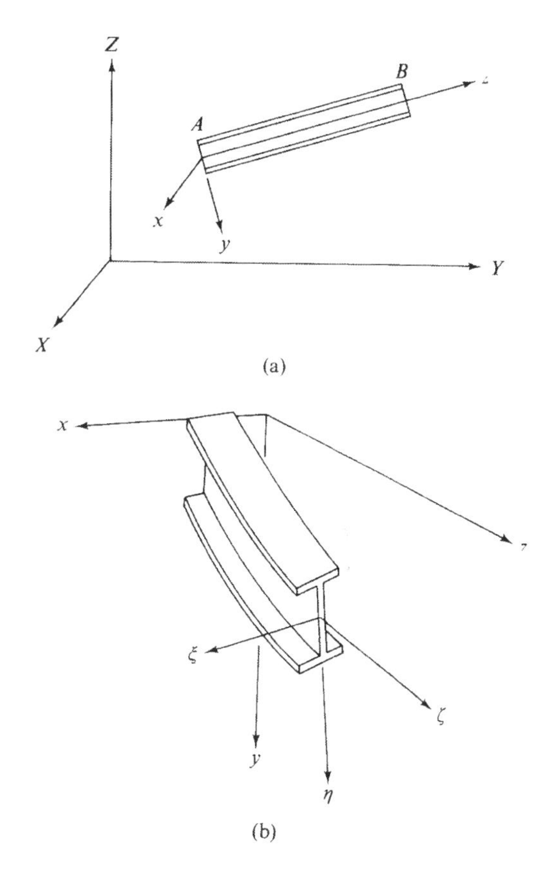

FIGURE 3.40
Orthogonal axis systems. (a) Frame and member co-ordinate axes before loading. (b) Member and local co-ordinate axes at buckling

section of the deformed member [Fig. 3.40(b)] and the local ζ axis which is tangential to the deformed centroidal axis.

When the structure buckles, a typical member takes up the equilibrium position shown in Fig. 3.39. The cross section at a distance z from A deflects u and v parallel to the x and y axes respectively, and rotates θ about the centroidal axis ζ of the deformed member.

The differential equations which govern the lateral buckling of beams and beam-columns can be obtained by using the results of Chap. 2. If the member is elastic and of doubly symmetric cross section, then the internal moments of resistance are obtained from Eqs. (2.123) and (2.142) as

$$\left.\begin{aligned} M_\xi &= EI_x\theta'_\xi \\ M_\eta &= EI_y\theta'_\eta \\ M_\zeta &= -EI_\omega\tau'' + (GK_T + \bar{K})\tau \end{aligned}\right\} \tag{3.160}$$

in which τ is the twist of the member. Relationships between the curvatures θ'_ξ, θ'_η and twist τ and the deformations u, v, and θ can be obtained by considering the rates of change along the length of the member of the direction cosines (Love, 1944) contained in the rotation matrix (Vacharajittiphan, Woolcock, and Trahair, 1974) alternatively (Vacharajittiphan *et al.*, 1974)

$$[T_R] = \begin{bmatrix} 1 & \theta - u'v' & -u' - v'\theta \\ -\theta & 1 & -v' \\ u' & v' & 1 \end{bmatrix} \tag{3-161}$$

[which is equivalent to Eq. (2.154) when the v' terms are neglected]. Thus Eq. (3.160) becomes

$$\left.\begin{aligned} M_\xi &= -EI_x v'' \\ M_\eta &= EI_y(u'' + \theta v'') \\ M_\zeta &= -EI_\omega(\theta' - u'v'')'' + (GK_T + \bar{K})(\theta' - u'v'') \end{aligned}\right\} \tag{3.162}$$

The external moments acting about axes parallel to the x, y, z axes can be obtained from Eq. (2.148) by using the relative displacements $(u - u_A)$ and $(v - v_A)$ as

$$\begin{Bmatrix} M_x \\ M_y \\ M_z \end{Bmatrix} = -\begin{Bmatrix} M_{xA} \\ M_{yA} \\ M_{zA} \end{Bmatrix} + [T_T]\begin{Bmatrix} F_{xA} \\ F_{yA} \\ F_{zA} \end{Bmatrix} + \begin{Bmatrix} -(z - z_w)\langle W\rangle \\ 0 \\ (u - u_w)\langle W\rangle \end{Bmatrix} \tag{3.163}$$

in which

$$[T_T] = \begin{bmatrix} 0 & -z & (v - v_A) \\ z & 0 & -(u - u_A) \\ -(v - v_A) & (u - u_A) & 0 \end{bmatrix} \tag{3.164}$$

It should be noted that the directions of the forces and moments at point A have been reversed from those of Chap. 2 for convenience in the direct stiffness method. The last term in Eq. (3.163) allows for the effects of any transverse load W. The brackets $\langle\ \rangle$ indicate that the quantity inside them is zero when z is less than z_w, the distance of the load W from the end A. The lateral displacement u_w of the load is given by

$$u_w = u_{zw} + \bar{a}\theta_{zw} \tag{3.165}$$

in which u_{zw} and θ_{zw} are the lateral displacement of the centroid and rotation of the cross section at $z = z_w$, and $\bar{a}$ is the height of the load W above the member centroidal axis (see Fig. 3.39).

The x, y, z external moments of Eq. (3.163) can be transformed to moments acting about the ξ, η, and ζ axes by multiplying them by the rotation matrix T_R, so that

$$\begin{Bmatrix} M_\xi \\ M_\eta \\ M_\zeta \end{Bmatrix} = [T_R]\begin{Bmatrix} M_x \\ M_y \\ M_z \end{Bmatrix} \tag{3.166}$$

The differential equations for lateral buckling can then be obtained by equating these external moments to the internal moments of resistance of Eq. (3.162), whence

$$\begin{Bmatrix} -EI_x v'' \\ EI_y(u'' + v''\theta) \\ -EI_\omega \tau'' + (GK_T + \bar{K})\tau \end{Bmatrix} = [T_R][T_T]\begin{Bmatrix} F_{xA} \\ F_{yA} \\ F_{zA} \end{Bmatrix} - [T_R]\begin{Bmatrix} M_{xA} \\ M_{yA} \\ M_{zA} \end{Bmatrix} + [T_R]\begin{Bmatrix} -(z - z_w)\langle W \rangle \\ 0 \\ (u - u_w)\langle W \rangle \end{Bmatrix} \tag{3.167}$$

The term $\bar{K}$ in the torsion equation is the increase in the torsional rigidity due to the axial tension force F_ζ which occurs when the member has a twist τ, and is equal to $F_\zeta r_0^2$.

For flexural-torsional buckling it is assumed that all the buckling deformations u, θ and their derivatives and the buckling actions F_{xA}, M_{yA}, M_{zA} are infinitesimally small. The first of Eq. (3.167) can therefore be separated as

$$-EI_x v'' = -zF_{yA} + (v - v_A)F_{zA} - M_{xA} - (z - z_w)\langle W \rangle \tag{3.168}$$

which describes the in-plane deformed equilibrium state before buckling occurs. The remaining two equations of Eq. (3.167) become linear when the solutions of this in-plane equation are substituted, and govern the flexural-torsional buckling of the structure.

3.11.2 Direct Stiffness Method

The stiffness or displacement method has been widely used in structural analysis because it provides a simple and systematic method of computation. Details of the method can be found in many standard textbooks (Harrison, 1973), and only a brief description (with particular reference to frame structures) will be given here.

In the direct stiffness method, the relationships between the actions and deformations at the ends of a member are obtained directly from the governing differential equations when these can be solved exactly, as in the first or second order elastic in-plane analysis. In other cases, these equations can be solved numerically, or if they are too complicated, then the relationships can be obtained from the energy principle by using an assumed set of displacement functions, as in the finite element method. In all cases, the relationships can be expressed as

$$\{SR_0\} = [k_0]\{\delta_0\} + \{SR_{F0}\} \tag{3.169}$$

in which $\{SR_0\}$ and $\{\delta_0\}$ are the vectors of the actions (or generalized stresses) and the corresponding deformations (or generalized strains) at the ends of a member. The square matrix $[k_0]$ is the member stiffness matrix, and the vector $\{SR_{F0}\}$ contains the fixed-end reactions which are required to balance any loads acting on the member (when there are no end deformations).

The actions and deformations of an ith member are transformed from the orthogonal xyz member axis system to the orthogonal XYZ structure axis system, and vice versa, by a rotation matrix $[T]_i$ and its inverse $[T]_i^{-1} = [T]_i^T$. Thus

$$\{SR\}_i = [T]_i^T \{SR_0\}_i \tag{3.170}$$

which describes the static equilibrium relationships between the two sets of actions, and

$$\{\delta_0\}_i = [T]_i \{\delta\}_i \tag{3.171}$$

which describes the compatibility relationships between the member end deformations in the two axis systems. By combining Eqs. (3.169) to (3.171), the action-displacement relationships in the structure axis system can be expressed as

$$\{SR\}_i = [k]_i \{\delta\}_i + \{SR_F\}_i \tag{3.172}$$

where

$$[k]_i = [T]_i^T [k_0]_i [T]_i \tag{3.173}$$

Similar operations can be performed for each member of the structure.

These transformed member relationships can then be assembled to form

$$\{P\} = [K]\{\delta\} + \{P_F\} \tag{3.174}$$

in which $\{P\}$ and $\{P_F\}$ are the vectors of the external joint loads of the frame and of the fixed-end loads, and $\{\delta\}$ is the vector of the displacements of the frame joints. The identity between some of the individual member displacements $\{\delta\}_i$ is a reflection of the compatibility conditions at the joints, and permits the reduction of these to the frame displacements $\{\delta\}$. The related correspondences between some of the individual member actions $\{SR\}_i$ and $\{SR_F\}_i$ allows the addition of these in accordance with the joint equilibrium equations, and the consequent reduction to the frame loads $\{P\}$ and $\{P_F\}$. These additions and reductions are duplicated in forming the frame stiffness matrix $[K]$ from the member stiffness matrices $[k]_i$.

Before Eq. (3.174) can be solved for the joint displacements $\{\delta\}$, the effects of the boundary conditions must be included by modifying the appropriate rows and columns in the stiffness matrix $[K]$. When a joint is elastically restrained against one deformation, then the joint restraining force R is related to the joint deformation δ by

$$-R = K\delta \tag{3.175}$$

in which K is the stiffness of the elastic restraint. The effect of the restraint force R on the equilibrium relationship between the corresponding external joint load P and the member actions $\{SR\}_i$ can be included in the stiffness matrix $[K]$ by adding the restraint stiffness to the corresponding diagonal term. When a joint is fixed against one deformation, the restraint stiffness becomes infinite and the diagonal term in the matrix $[K]$ is replaced by a very large number. Alternatively, the appropriate row and column in the matrix $[K]$ and the terms in the load vectors $\{P\}$ and $\{P_F\}$ are made zero, while the diagonal term is replaced by unity.

For a given set of applied loads $\{P\}$ (and corresponding $\{P_F\}$), Eq. (3.174) can be solved immediately for the joint displacements $\{\delta\}$ provided $[K]$ is known. It is then possible to calculate the member end deformations $\{\delta_0\}_i$ from Eq. (3.171), and subsequently the end actions $\{SR_0\}_i$ can be determined from Eq. (3.169). However, the stiffness matrices $[k]_i$ and $[K]$ may depend on the applied loads or on the joint displacements, as well as on the geometry of the frame. In this case Eqs. (3.174) are non-linear and iterative methods must be used to solve them.

In the special case of instability of the bifurcation type, the stiffness matrix $[K]$ is a direct function of the load factor, while Eqs. (3.174) can be arranged so that the vector $\{\delta\}$ only contains joint deformations associated with buckling, and the vectors $\{P\}$ and $\{P_F\}$ only contain the corresponding zero forces, i.e.,

$$[K]\{\delta\} = \{0\} \tag{3.176}$$

One solution of this set of equations corresponds to the stable prebuckling state for which all the joint deformations $\{\delta\}$ associated with buckling are zero. The equations are also satisfied when the determinant of the complete matrix $[K]$ is zero, i.e.,

$$|K| = 0 \tag{3.177}$$

FIGURE 3.41
In-plane member end actions and deformations. (a) Member end actions. (b) Member end deformations

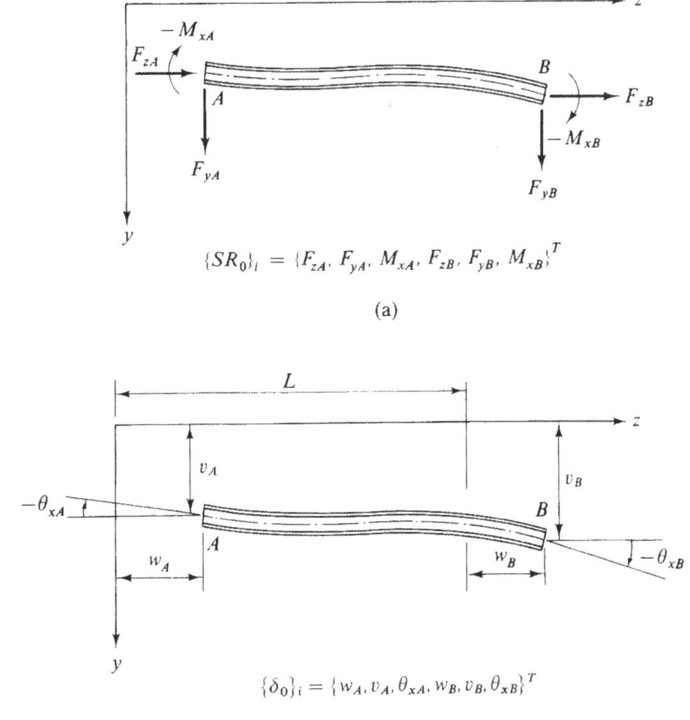

and the lowest load factor for which this is true defines the lowest critical value associated with a particular loading pattern.

3.11.3 In-Plane Analysis

The differential equations governing the non-linear in-plane behavior of a member are given by Eq. (3.168) and the axial load-deformation relationship. For the small displacement elastic analysis, the non-linearity arises from the fact that the presence of the axial force in each member changes its bending stiffness, and also because the member stiffness relationships must be formulated for the displaced positions of the members.

When the axial force in a member is known, the differential equation can be solved, and the required member in-plane stiffness matrix $[k_0]$ and the fixed-end actions $\{SR_{Fo}\}$ can be obtained. For a uniform member subjected to end actions only (Fig. 3.41), the member stiffness matrix which relates the end actions $\{SR_0\}_i$ and deformations $\{\delta_0\}_i$ specified in Fig. 3.41 is given by (Livesley, 1964)

$$[k_0]_i = \begin{bmatrix} k_{11} & k_{12} \\ k_{21} & k_{22} \end{bmatrix} \tag{3.178}$$

in which

$$\left.\begin{aligned}
[k_{11}] &= \begin{bmatrix} EA/L & 0 & 0 \\ 0 & 2s(1+c)EI_x/L^3 - F_z/L & s(1+c)EI_x/L^2 \\ 0 & s(1+c)EI_x/L^2 & sEI_x/L \end{bmatrix} \\
[k_{12}] = [k_{21}]^T &= \begin{bmatrix} -EA/L & 0 & 0 \\ 0 & -2s(1+c)EI_x/L^3 + F_z/L & s(1+c)EI_x/L^2 \\ 0 & -s(1+c)EI_x/L^2 & scEI_x/L \end{bmatrix} \\
[k_{22}] &= \begin{bmatrix} EA/L & 0 & 0 \\ 0 & 2s(1+c)EI_x/L^3 - F_z/L & -s(1+c)EI_x/L^2 \\ 0 & -s(1+c)EI_x/L^2 & sEI_x/L \end{bmatrix}
\end{aligned}\right\} \tag{3.179}$$

and

$$\left.\begin{aligned}
s &= \frac{\omega(\omega + \cot\omega - \omega\cot^2\omega)}{1 - \omega\cot\omega} \\
sc &= \frac{\omega(\omega - \cot\omega + \omega\cot^2\omega)}{1 - \omega\cot\omega}
\end{aligned}\right\} \tag{3.180}$$

in which

$$\omega = \frac{\pi}{2}\sqrt{\frac{-F_z}{\pi^2 EI_x/L^2}} \tag{3.181}$$

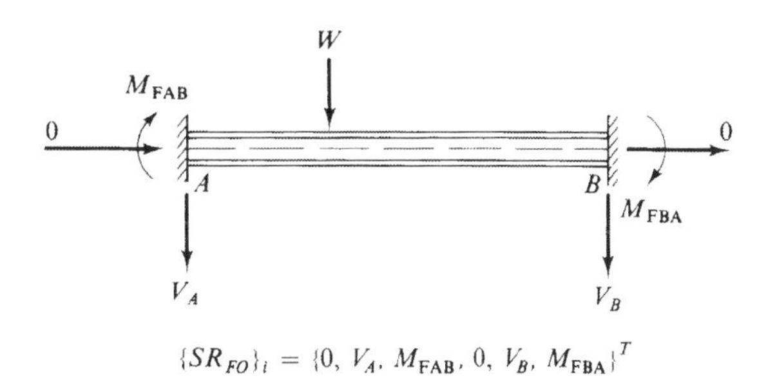

FIGURE 3.42
In-plane fixed-end actions

It should be noted that when F_z is a tensile force, then Eq. (3.181) will cause the circular function cot ω in Eq. (3.180) to change to a hyperbolic function.

In general, the frame is subjected to both bending and axial actions, and the distribution of the member axial forces in the frame is initially unknown. In this case, the solutions have to be obtained iteratively. For the first cycle of the iteration, the first order linear member stiffness matrix and the corresponding fixed-end actions (see Fig. 3.42) for each member are used in assembling the frame stiffness matrix $[K]$ and the load vector $\{P_F\}$. Eqs. (3.174) are then solved for the first order joint deformations, and the member actions are determined, including the member axial forces. These are used to revise the member stiffness matrices and the fixed-end actions for the next cycle. This process is repeated until there are no significant differences between the solutions of two successive cycles. The flow chart of the in-plane analysis section of a computer program is shown in Fig. 3.43.

3.11.4 Flexural-Torsional Buckling Analysis

3.11.4.1 *Finite integral formulation of the differential equations*

The critical load factor at which a rigid plane frame buckles elastically under a particular loading pattern can be determined by solving the differential equations [Eqs. (3.167)] which govern flexural-torsional buckling. For any estimate of the critical load factor, the in-plane equation [Eq. (3.168)] can be solved separately as described in the previous section. The in-plane solutions required to linearize the flexural-torsional buckling equations are the end resultants M_{xA}, F_{yA}, F_{zA} and the deformations v, v', v'' for each member of the frame. The buckling differential equations can then be used to formulate the member stiffness matrix $[k_O]$, and the solution for buckling of the whole structure can be found by employing the direct stiffness approach.

The member stiffness matrix $[k_O]$ can be formulated from solutions of the flexural-torsional buckling equations obtained by using the method of *finite integrals* (Brown and Trahair, 1968, 1975). In this method, indefinite integration of a function

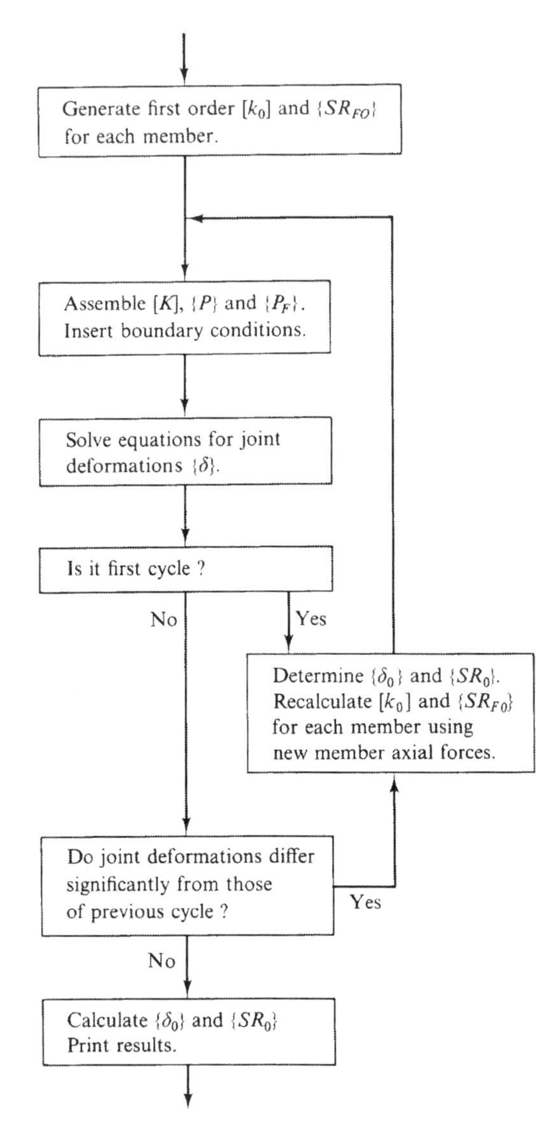

FIGURE 3.43
Flow chart for in-plane second order analysis

with respect to the independent variable is replaced by a series of definite integrals at selected values of the independent variable. Thus the continuous identity

$$u' = \int_A^z u''\, dz + u'_A \tag{3.182}$$

is replaced by the series of approximate identities

$$\{u'\} = [N]\{u''\} + u'_A\{1\} \tag{3.183}$$

in which $\{u''\}$ and $\{u'\}$ are vectors of the values of the function u'' and its integral u' at the selected values of z, $[N]$ is a square matrix of coefficients, and $\{1\}$ is defined as the vector $\{1, 1, 1, \dots 1\}^T$ This replacement may be repeated if a second integration is required, and so

$$\{u\} = [N]\{u'\} + u_A\{1\} \tag{3.184}$$

or

$$\{u\} = [N]^2\{u''\} + u'_A\{z\} + u_A\{1\} \tag{3.185}$$

The approximating coefficients in the matrix $[N]$ may be obtained by fitting polynomial curves to the values of the function and integrating. In the case where the intervals H between the selected values of z are equal and quadratics are fitted over pairs of adjacent intervals, then it can be shown (Brown and Trahair, 1968) that

$$[N] = \frac{H}{12}\begin{bmatrix} 0 & 0 & 0 & & & & \\ 5 & 8 & -1 & & & & \\ 4 & 16 & 4 & & & & \\ 4 & 16 & 9 & 8 & -1 & & \\ 4 & 16 & 8 & 16 & 4 & & \\ \vdots & \vdots & \vdots & \vdots & \vdots & \vdots & \vdots \end{bmatrix} \tag{3.186}$$

The method of finite integrals can be used to obtain numerical solutions of differential equations. This is done by replacing the continuous differential equation which must be satisfied everywhere by a series of simultaneous equations which represent the differential equation at a series of discrete points. All but the highest differential coefficients in these equations are eliminated by replacing them by linear combinations of the highest differential coefficients and of the constants of integration, these combinations being determined by the method of finite integrals. The resulting simultaneous equations may be combined with the boundary conditions, and solutions for the highest differential coefficients and the constants of integration may be found by the conventional methods of solving simultaneous equations.

In the case of stability problems of the bifurcation type, the right hand sides of the simultaneous equations are all zero. For example, the differential equation governing the flexural buckling of a pin-ended column is

$$EI_y u'' + Pu = 0 \tag{3.187}$$

and this can be replaced by the finite integral approximation

$$(EI_y[I] + P[N]^2)\{u''\} + P(u'_A\{z\} + u_A\{1\}) = \{0\} \tag{3.188}$$

in which $[I]$ is a unit matrix. When the boundary conditions

$$u_A = u_B = 0 \tag{3.189}$$

are substituted into Eq. (3.185), then u'_A can be expressed in the form

$$u'_A = -\frac{1}{L}\{A_i\}^T\{u''\} \tag{3.190}$$

in which the elements of the row vector $\{A_i\}^T$ are the last row of the matrix product $[N]^2$. If this is substituted, then Eq. (3.188) becomes

$$\left(EI_y[I] + P[N]^2 - \frac{Pz}{L}[A]\right)\{u''\} = \{0\} \tag{3.191}$$

in which $[A]$ is a square matrix, each row of which is identically the same as $\{A_i\}^T$. The critical values of the load P can then be found from the condition that the determinant of the matrix $(EI_y[I] + P[N]^2 - (Pz/L)[A])$ is zero at buckling.

In using the finite integral method to formulate the member stiffness matrix for flexural-torsional buckling, the lower differential coefficients in the buckling equations [Eqs. (3.167)] are eliminated by using Eqs. (3.183) and (3.185), together with

$$\left.\begin{aligned}\{\tau'\} &= [N]\{\tau''\} + \tau'_A\{1\} \\ \{\tau\} &= [N]^2\{\tau''\} + \tau'_A\{z\} + \tau_A\{1\}\end{aligned}\right\} \tag{3.192}$$

and

$$\{\theta\} = [N][v''][N]\{u''\} + u'_A\{v' - v'_A\} + [N]^3\{\tau''\} + \tau'_A\left\{\frac{z^2}{2}\right\} + \tau_A\{z\} + \theta_A\{1\} \tag{3.193}$$

which is obtained from the twist relationship used to transform the last of Eqs. (3.160) to the last of Eqs. (3.162), and in which $[v'']$ is a diagonal matrix with the values of the diagonal terms equal to $\{v''\}$. When the effects of small in-plane curvatures v'' are neglected, then Eq. (3.193) simplifies to

$$\{\theta\} = [N]^3\{\tau''\} + \tau'_A\left\{\frac{z^2}{2}\right\} + \tau_A\{z\} + \theta_A\{1\} \tag{3.194}$$

Using Eqs. (3.183), (3.185), (3.192) and (3.193), the second and third differential equations of Eq. (3.167) for each member can now be rearranged and expressed as a system of linear equations in which the unknowns are the values of u'', τ'' at each member node, the end deformations $\{\delta_{oA}\} = \{u_A, u'_A, \theta_A\}^T, \tau_A, \tau'_A$, and the buckling actions $\{SR_{oA}\} = \{F_{xA}, M_{yA}, M_{zA}\}^T$. Thus these nodal equations can be represented as

$$[Q_u]\begin{Bmatrix}u'' \\ \tau''\end{Bmatrix} + [Q_\tau]\begin{Bmatrix}\tau'_A \\ \tau_A\end{Bmatrix} + [Q_\delta]\{\delta_{oA}\} + [Q_S]\{SR_{oA}\} = \{0\} \tag{3.195}$$

in which $\{u'', \tau''\}^T$ is a vector containing the unknown values of u'' and τ'', and $[Q_u], [Q_\tau], [Q_\delta], [Q_S]$ are matrices with elements obtained from combinations of various terms in Eqs. (3.167), (3.183), (3.185), (3.192), and (3.193). There are two of these equations for each member node, and so the total number of these equations is equal to the total number of the unknown values of u'' and τ'' for the member.

In order to obtain the stiffness relationships between the buckling deformations $\{\delta_o\}$ and the actions $\{SR_o\}$ at the ends of a member, eight additional equations are required, and these can be obtained from the warping boundary and member equilibrium equations. The unknowns τ_A and τ'_A in Eq. (3.194) which are related to the differential flange rotations and moments associated with warping could be included in the vectors of the deformations $\{\delta_{oA}\}$ and actions $\{SR_{oA}\}$ respectively. This would imply that there should be four buckling deformations and four corresponding actions at each end of a member. However, it is assumed in this section that each member is elastically restrained against warping at both ends, and that two warping boundary equations can be formulated. These warping equations are satisfied internally by the buckling deformations of each member, and thus the vectors $\{\delta_{oA}\}$ and $\{SR_{oA}\}$ remain as the three end deformations and the three actions as used in Eq. (3.195).

The warping boundary conditions at joints have been studied by Vacharajittiphan and Trahair (1974a). When a member is elastically restrained against warping at both ends, the boundary equations can be expressed as

$$\left.\begin{aligned} \tau'_A &= (K_w)_A \tau_A \\ \tau'_B &= -(K_w)_B \tau_B \end{aligned}\right\} \tag{3.196}$$

in which K_w is the warping restraint stiffness. This has a zero value when the end is free to warp, and a very large value when it is prevented from warping, while the values for intermediate restraint conditions can be obtained from the results presented by Vacharajittiphan and Trahair (1974a). By using the finite integral equations [Eqs. (3.192)] these two warping equations can be expressed as

$$[Q_{uw}] \begin{Bmatrix} u'' \\ \tau'' \end{Bmatrix} + [Q_{\tau w}] \begin{Bmatrix} \tau'_A \\ \tau_A \end{Bmatrix} = \{0\} \tag{3.197}$$

Six more equations are required to relate the deformations and actions at the ends A and B. Three of these can be obtained directly from Eqs. (3.183), (3.185), and (3.193), which require that the deformations given by combinations of the highest derivatives must be compatible with the deformations $\{\delta_{oB}\} = \{u_B, u'_B, \theta_B\}^T$ at the end B. The three equations can be represented as

$$\{\delta_{oB}\} = [C_u] \begin{Bmatrix} u'' \\ \tau'' \end{Bmatrix} + [C_\tau] \begin{Bmatrix} \tau'_A \\ \tau_A \end{Bmatrix} + [C_\delta]\{\delta_{oA}\} + [C_S]\{SR_{oA}\} \tag{3.198}$$

in which $[C_S]$ is a zero matrix. Three member equilibrium equations are required for the actions $\{SR_{oB}\} = \{F_{xB}, M_{yB}, M_{zB}\}^T$, and these are obtained from the equilibrium equation

$$F_{xB} = -F_{xA} \tag{3.199}$$

and the last two of Eq. (3.163). These three linear equations can be expressed in the form

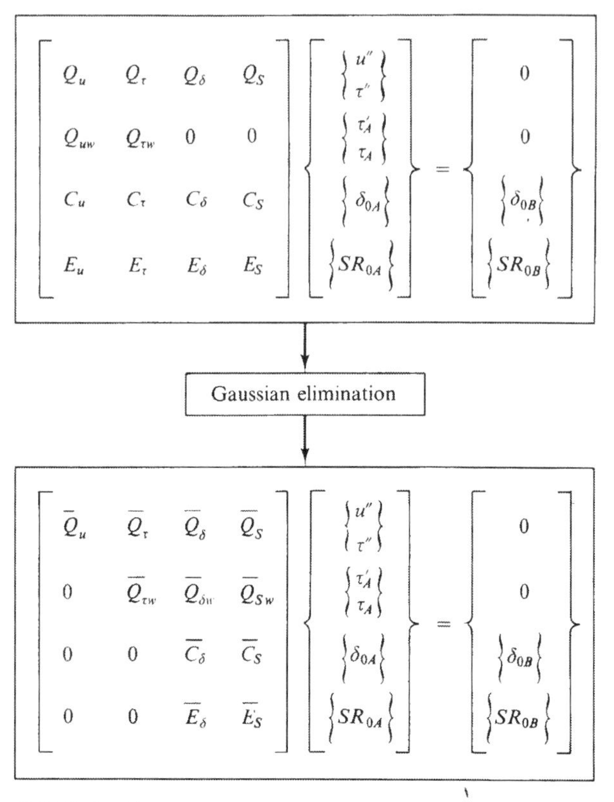

FIGURE 3.44
Operations on the member buckling equations

$$\{SR_{oB}\} = [E_u]\begin{Bmatrix} u'' \\ \tau'' \end{Bmatrix} + [E_\tau]\begin{Bmatrix} \tau'_A \\ \tau_A \end{Bmatrix} + [E_\delta]\{\delta_{oA}\} + [E_S]\{SR_{oA}\} \tag{3.200}$$

3.11.4.2 *Member and frame stiffness matrices*

The member stiffness matrix $[k_o]$ can be obtained from Eqs. (3.195), (3.197), (3.198) and (3.200) by using Gaussian elimination as shown in Fig. 3.44. Thus the terms under the diagonals of the matrices $[Q_u]$ and $[Q_{\tau w}]$ and all the terms of $[Q_{uw}]$, $[C_u]$, $[C_\tau]$, $[E_u]$, $[E_\tau]$ are reduced to zero, while the matrices $[C_\delta]$, $[C_S]$, $[E_\delta]$, $[E_S]$ are modified to $[C_\delta]$, $[C_S]$, $[E_\delta]$, $[E_S]$ respectively. This process modifies the compatibility [Eq. (3.198)] and member equilibrium equations [Eq. (3.200)], so that after rearrangement the relationships between the end deformations and actions can be expressed as

$$\begin{bmatrix} C_S & 0 \\ E_S & -I \end{bmatrix}\begin{Bmatrix} SR_{oA} \\ SR_{oB} \end{Bmatrix} = \begin{bmatrix} -C_\delta & I \\ -E_\delta & 0 \end{bmatrix}\begin{Bmatrix} \delta_{oA} \\ \delta_{oB} \end{Bmatrix} \tag{3.201}$$

in which $[I]$ is a unit matrix. By solving Eq. (3.201) using Gaussian elimination, the actions $\{SR_o\} = \{SR_{oA}^T, SR_{oB}^T\}^T$ can be expressed in terms of the end deformations $\{\delta_o\} = \{\delta_{oA}^T, \delta_{oB}^T\}^T$ as

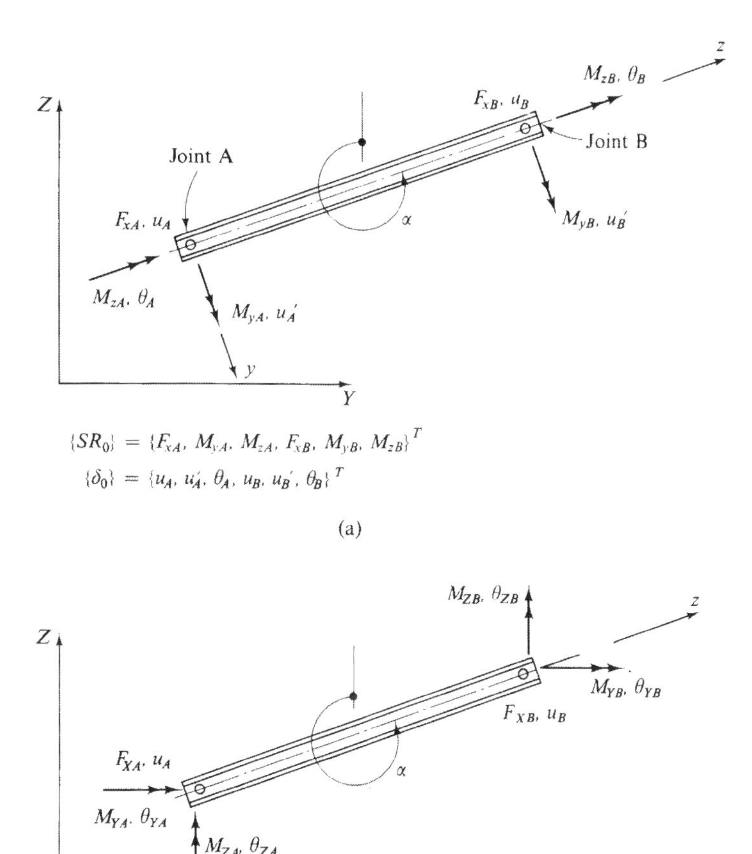

FIGURE 3.45
Transformation of buckling actions and deformations. (a) Actions and deformations in x, y, z system. (b) Actions and deformations in X, Y, Z system

$$\{SR_o\} = [k_o]\{\delta_o\} \tag{3.202}$$

in which $[k_o]$ is the member stiffness matrix for flexural-torsional buckling.

The member stiffness matrix $[k_o]$ in the member x, y, z axis system can be transformed to the stiffness matrix $[k]$ in the frame X, Y, Z axes as in Eq. (3.173) by using the transformation matrix $[T]$. The matrix $[T]$ can be obtained by using either the equilibrium transformation of the buckling actions or the compatibility transformation of the deformations indicated in Fig. 3.45, whence

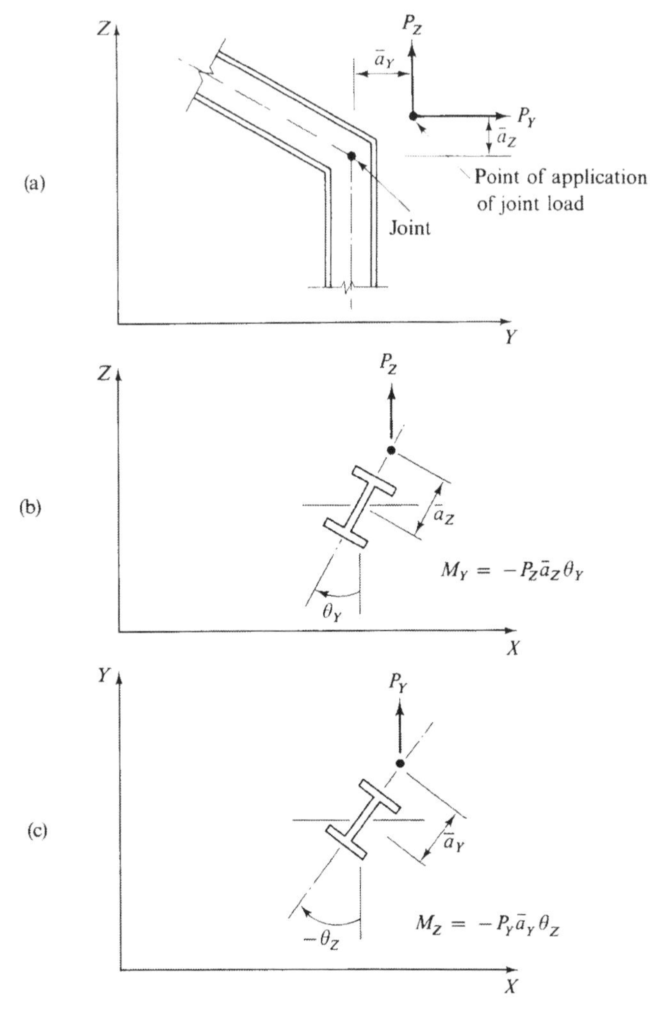

FIGURE 3.46
Effects of loading away from joint centroid at buckling

$$[T] = \begin{bmatrix} 1 & 0 & 0 \\ 0 & \cos\alpha & \sin\alpha \\ 0 & -\sin\alpha & \cos\alpha \end{bmatrix} \tag{3.203}$$

in which α is the angle between the member z axis and the frame Z axis, measured in the positive sense of the X axis.

The frame stiffness matrix $[K]$ is then assembled from the member stiffness matrices $[k]$, and any boundary conditions associated with the out-of-plane joint deformations are applied by modifying the appropriate rows and columns of the matrix $[K]$, as described in Sec. 3.11.2. The matrix $[K]$ must also be modified when any of the applied joint loads act at a small distance away from the centroid of the joint (Fig. 3.46). These loads produce additional bending moments about the Y and Z axes when the joint rotates during buckling, which are given by

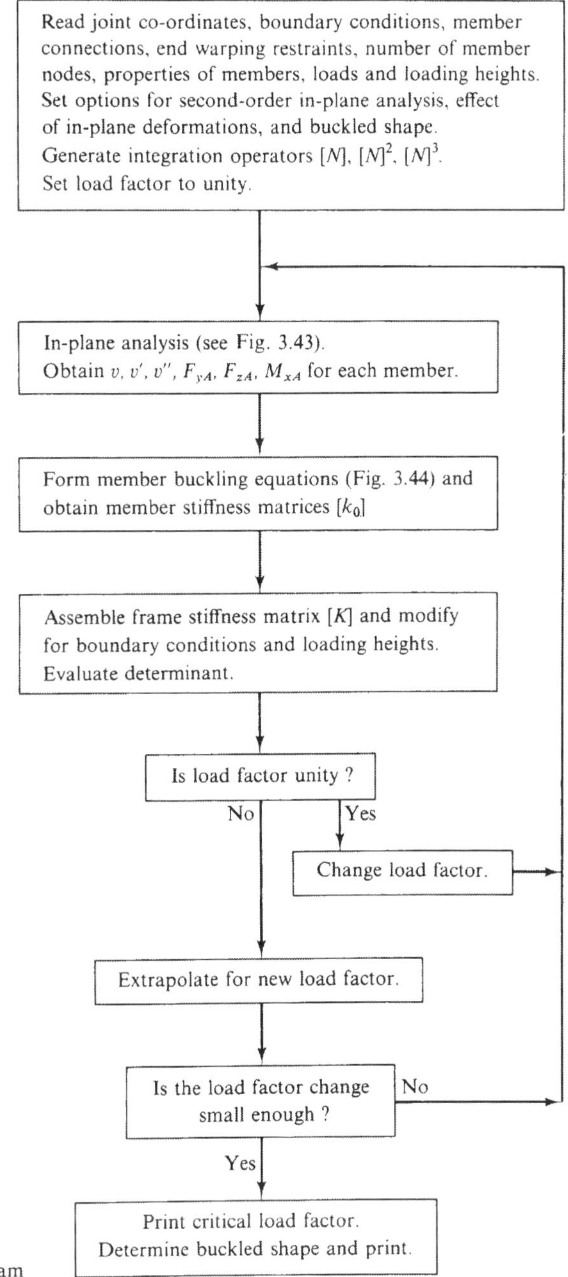

FIGURE 3.47
Flow chart of complete computer program

$$\left.\begin{aligned} M_Y &= -P_Z\bar{a}_Z\theta_Y \\ M_Z &= -P_Y\bar{a}_Y\theta_Z \end{aligned}\right\} \tag{3.204}$$

in which P_Y and P_Z are the loads acting at the joint in the Y and Z directions, $\bar{a}_Y$ and $\bar{a}_Z$ are the heights of the loads above the centroid of the joint, and θ_Y and θ_Z are the angles of rotation of the joint about axes parallel to the Y and Z axes respectively. Eq. (3.204) is used to modify the frame stiffness matrix $[K]$ by adding appropriate terms to its diagonal terms which correspond to these joint rotations.

3.11.4.3 *Buckling solution*

A flow chart of a computer method of finding the buckling solution is given in Fig. 3.47. The first step involves the reading in of all the data which define the structure, its loading system, and the types of in-plane and buckling analysis to be performed. The integration operators are also generated. The in-plane analysis is then performed, as discussed in Sec. 3.11.3 and Fig. 3.43.

Following this, the member buckling stiffness matrices $[k_o]$ are then formed, and the frame stiffness matrix $[K]$ is assembled. Because there are no applied loads acting on the structure which directly cause out-of-plane behavior, the out-of-plane load vectors $\{P\}$ and $\{P_F\}$ in Eq. (3.174) are zero. Thus the solution for flexural-torsional buckling is obtained when the magnitudes of the in-plane loads are such that the determinant of the out-of-plane frame stiffness matrix $[K]$ is zero. An iterative procedure is used, in which an initial estimate is made of the critical load factor and the determinant $|K|$ is evaluated. The initial estimate of the load factor is then modified, and a new value of the determinant is calculated. An extrapolation procedure is then used to approach the value of the critical load factor, for which the determinant is zero. The iteration is terminated when successive estimates of the critical load factor differ by less than a prescribed value.

The corresponding buckled shape can then be determined, if required, by arbitrarily assigning a unit value to the last non-zero unknown of the joint deformations $\{\delta\}$. The remaining equations of the frame stiffness matrix $[K]$ are then solved, and a complete set of joint deformations $\{\delta\}$ is obtained. The member end deformations $\{\delta_o\}$ are then calculated by using Eq. (3.171). The member buckling equations [Eqs. (3.195) and (3.197–200)] are regenerated, and the member stiffness matrix $[k_o]$ is obtained to determine the end actions $\{SR_o\}$ from Eq. (3.202). The highest derivatives $\{u'', \tau''\}^T$ and $\{\tau'_A, \tau_A\}^T$ for the member are then calculated by solving Eq. (3.195) after substituting for the known vectors $\{\delta_{oA}\}$ and $\{SR_{oA}\}$. Finally, the deformations u, θ at each node of the member are obtained by integration using Eqs. (3.183), (3.185), (3.192) and (3.193).

3.11.5 Application to Rigid-Jointed Frames

The systematic method of analysis described in the previous sections can be applied to the flexural-torsional buckling of single members and of rigid-jointed plane frames

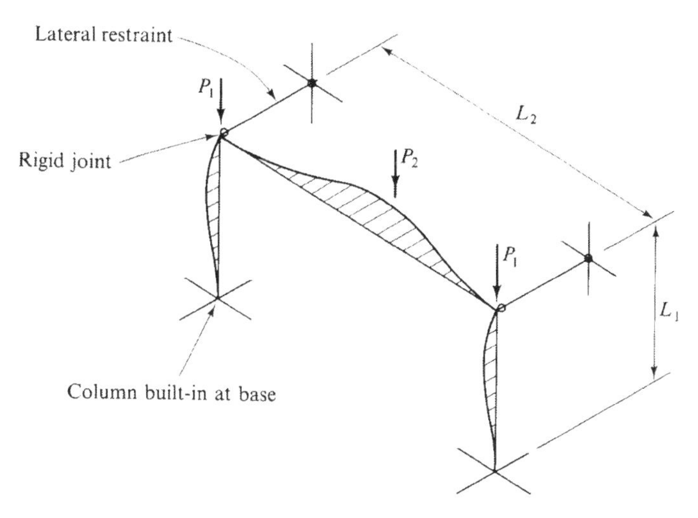

FIGURE 3.48
Flexural-torsional buckling of a portal frame

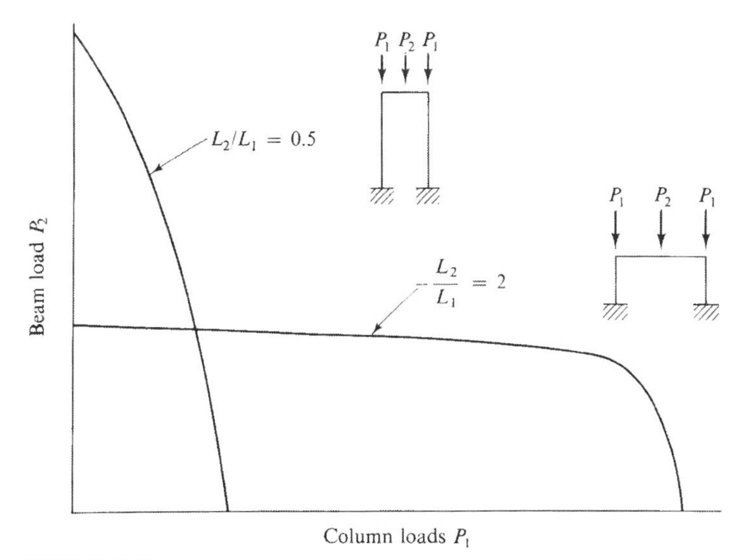

FIGURE 3.49
Elastic buckling loads of portal frames

with any in-plane loading pattern. A computer program has been prepared (Vacharajittiphan and Trahair, 1974b) for which the only input data required are those which specify the geometry of the frame and its members, any degrees of freedom or elastic restraints at the joints, the warping conditions at both ends of each member, the number of nodes for each member, the cross-sectional properties, and the loading pattern. Data are also read in to specify whether the second-order in-plane analysis is to be carried out, whether the effect of in-plane deformations prior to flexural-torsional buckling is to be included in the analysis, and whether the buckled shape is to be determined (see Fig. 3.47).

Interaction diagrams for the critical loads of a number of elastic rigid-jointed frames have been determined (Vacharajittiphan, 1974, Vacharajittiphan and Trahair, 1975), and two of these (for the portal frame shown in Fig. 3.48) are given in Fig. 3.49. These diagrams show that the region of stability is convex, as it also is for flexural-torsional buckling of continuous beams (Fig. 3.31). Because of this convexity, linear interpolations made between known critical load sets are conservative.

3.12 PROBLEMS

3.1 An elastic simply supported I-beam has an initial lack of straightness $u_o = u_{mo} \sin \pi z/L$, and an initial twist $\theta_o = \theta_{mo} \sin \pi z/L$, where u_{mo} and θ_{mo} are related by

$$\frac{u_{mo}}{\theta_{mo}} = \frac{M_e}{\pi^2 EI_y/L^2} \tag{3.205}$$

and M_e is the elastic critical moment given by Eq. (3.21). Show that equal and opposite end moments M_x cause the beam to deflect and twist

$$\begin{aligned} u &= u_m \sin \pi z/L \\ \theta &= \theta_m \sin \pi z/L \end{aligned} \tag{3.206}$$

and that u_m and θ_m are given by

$$\frac{u_m}{u_{mo}} = \frac{\theta_m}{\theta_{mo}} = \frac{M_x/M_e}{1 - M_x/M_e} \tag{3.207}$$

Comment on the significance of these results.

3.2 A simply supported steel I-beam whose properties are given in Fig. 3.50 has a central concentrated load applied in such a way that lateral deflection and twist are prevented at midspan ($u_m = \theta_m = 0$). Determine the elastic buckling load if the beam span is 35 ft (10.67 m).

3.3 A simply supported narrow rectangular steel beam has a central concentrated load acting at its top surface. The beam is 10 in (254 mm) deep, 1 in (25.4 mm) wide and its span is 5 ft

FIGURE 3.50
Properties of an I-section beam

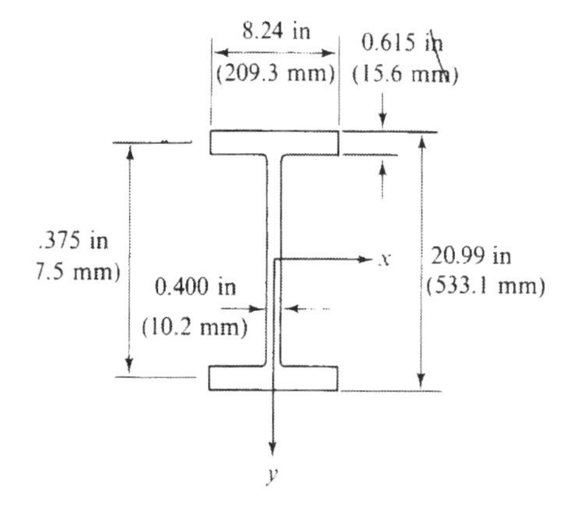

	Imperial		S.I.	
Symbol	Value	Units	Value	Units
A	18.35	in^2	118.4×10^2	mm^2
I_x	1330	in^4	55359×10^4	mm^4
r_x	8.54	in	217	mm
S_x	126.4	in^3	2071×10^3	mm^3
Z_x	144.1	in^3	2362×10^3	mm
I_y	57.5	in^4	2393×10^4	mm
r_y	1.77	in	45.0	mm
r_t	1.93	in	49.0	mm
I_ω	5968	in^6	1.603×10^{12}	mm^6
K_T	1.97	in^4	82.0×10^4	mm
E	30000	ksi	207000	MPa
G	12000	ksi	83000	MPa
σ_y	45.0	ksi	310	MPa

(1.52 m). Determine the elastic buckling load if the load is free to move laterally with the beam during buckling.

3.4 A monosymmetric tee-beam is fabricated by cutting off one flange of a steel beam whose cross section is shown in Fig. 3.50. If the tee-beam is simply supported over a span of 12 ft (3.66 m) and has equal and opposite end moments, determine the elastic buckling moment

(a) when the flange is in compression, and

(b) when the flange is in tension.

3.5 A 12 ft (3.66 m) long steel cantilever has the cross section shown in Fig. 3.50. Determine the elastic buckling value of the load which acts at the free end of the top flange.

3.6 The flanges of an elastic I-beam with equal and opposite end moments ($\kappa = 1$) have equal end rotational restraints whose stiffnesses are such that Eq. (3.57) holds. Prove that the buckled shape of the beam is given by

$$u = \frac{M_e}{\pi^2 EI_y/k^2L^2}\theta = A\left\{\cos\frac{\pi(z - L/2)}{kL} - \cos\frac{\pi}{2k}\right\} \tag{3.208}$$

in which $z = 0$ at the left hand end of the beam and M_e and k are as given by Eqs. (3.59), (3.60) and (3.61).

3.7 A simply supported I-beam with equal and opposite end moments ($\kappa = 1$) has midspan elastic translational and torsional restraints (Fig. 3.18) whose stiffnesses α_t and α_r satisfy Eqs. (3.68), (3.69) and (3.70). Prove that the buckled shape of the beam is given by

$$u = \frac{M_e\theta}{\pi^2 EI_y/k^2L^2} = A\left(\frac{z}{L} - \frac{\sin \pi z/kL}{(\pi/k)\cos \pi/2k}\right) \tag{3.209}$$

in which M_e is as given by Eqs. (3.59) and (3.60). Hence, prove that the limiting values of α_t and α_r which cause the beam to act as if rigidly braced are given by Eq. (3.71).

3.8 Determine the plastic buckling load of the beam of Prob. 3.2.

FIGURE 3.51
Laterally continuous beam. (a) Beam. (b) Moment diagram

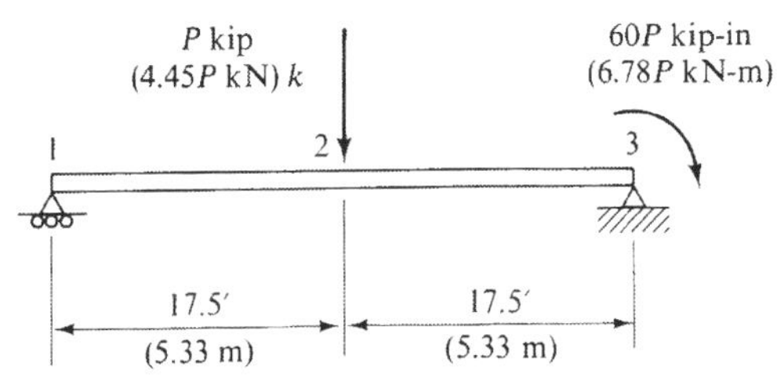

Lateral deflection and twist prevented at points 1, 2, and 3
Beam properties as in Fig. 3.50.

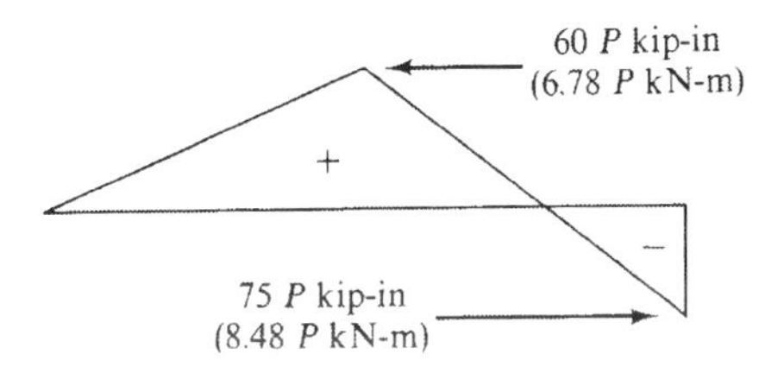

3.9 Determine the design working load capacity of the beam of Prob. 3.2 by using the AISC 1969 Specification.

3.10 Determine the design ultimate load capacity of the beam of Prob. 3.2 by using both

(a) the method of Taylor, Dwight and Nethercot (1974) described in Sec. 3.6.3, and

(b) the method of Nethercot and Trahair (1976b) described in Sec. 3.6.5.

3.11 Determine the elastic buckling load parameter of the braced beam shown in Fig. 3.51.

3.12 A simply supported steel beam-column has an axial load P kip (4.45 P kN) and unequal end moments of 0 and 20 P kip-in (2.26 P kN-m). The length of the beam-column is 17.5 ft (5.33 m) and its properties are given in Fig. 3.50. Determine the value of P at elastic buckling.

3.13 For the beam-column of Prob. 3.12, determine

(a) the value of P which satisfies the modified form of Eq. (3.151) when P_u and M_{ux}/C_m are approximated by 1.7 times the corresponding working load values of the AISC 1969 Specification, and

(b) the value of P which satisfies Eq. (3.154) proposed by Lim and Lu (1970), and

(c) the value of P at plastic buckling by using the method of Galambos (1968) described in Sec. 3.10.

3.13 REFERENCES

AISC, Specification for the Design, Fabrication and Erection of Structural Steel for Buildings, American Institute of Steel Construction, (1969).

AISC, Manual of Steel Construction, 7th edn, American Institute of Steel Construction, (1970).

Anderson, J. M. and Trahair, N. S., "Stability of Monosymmetric Beams and Cantilevers," *Journal of the Structural Division, ASCE,* Vol. 98, No. ST1, Proc. Paper 8646, Jan., pp. 269–286, (1972).

Austin, W. J., Yegian, S. and Tung, T. P., "Lateral Buckling of Elastically End-Restrained Beams," *Proceedings of the ASCE,* Vol. 81, Sep. No. 673, April, pp. 1–25, (1955).

Baker, J. F., Horne, M. R. and Heyman, J., *The Steel Skeleton,* Vol. II, Cambridge University Press, (1956).

Bansal, J. P., "The Lateral Instability of Continuous Steel Beams," *AISI Project No. 157,* Beams in Rigid Building Frames, Report No. 3, Department of Civil Engineering, The University of Texas, Austin, August, (1971).

Barsoum, R. S. and Gallagher, R. H. "Finite Element Analysis of Torsional and Torsional-Flexural Stability Problems," *International Journal for Numerical Methods in Engineering,* Vol. 2, pp. 335–352, (1970).

Bartels, D. and Bos, C. A. M., "Investigation of the Effect of the Boundary Conditions on the Lateral Buckling Phenomenon, Taking Account of Cross Sectional Deformation," *Heron,* Vol. 19, No. 1, pp. 3–26, (1973).

Bleich, F., *Buckling Strength of Metal Structures,* McGraw-Hill Book Company, Inc., New York, (1952).

British Standards Institution, *BS449: 1969 Specification for the Use of Structural Steel in Building,* BSI, London, (1969).

Brown, P. T. and Trahair, N. S., "Finite Integral Solution of Differential Equations," *Civil Engineering Transactions,* Institution of Engineers, Australia, Vol. CE10, No. 2, Oct., pp. 193–196, (1968).

Brown, P. T. and Trahair, N. S., "Finite Integral Solution of Torsion and Plate Problems," *Civil Engineering Transactions,* Institution of Engineers, Australia, Vol. CE17, No. 2, pp. 59–63, (1975).

Clark, J. W. and Hill, H. N., "Lateral Buckling of Beams," *Journal of the Structural Division, ASCE,* Vol. 86, No. ST7, Proc. Paper 2559, July, pp. 175–196, (1960).

Clark, J. W. and Knoll, A. H., "Effect of Deflection on Lateral Buckling Strength," *Journal of the Engineering Mechanics Division, ASCE,* Vol. 84, No. EM2, Paper No. 1596, April, pp. 1–18, (1958).

Column Research Committee of Japan, *Handbook of Structural Stability,* Corona Publishing Company, Ltd., Tokyo, (1971).

Dibley, J. E., "Lateral-Torsional Buckling of I-Sections in Grade 55 Steel," *Proceedings, I.C.E.,* Vol. 43, Aug., pp. 559–627, (1969).

Fukumoto, Y., Fujiwara, M. and Watanabe, N., "Inelastic Lateral Buckling Tests on Welded Beams and Girders," *Proceedings, Jap. Soc. Civ. Engineers*, May, pp. 39–51, (1971).

Fukomoto, Y. and Kubo, M., "Lateral Buckling Strength of Girders with Bracing Systems," *Prelim Report, IABSE*, Ninth Congress, Amsterdam, May, pp. 299–304, (1972).

Galambos, T. V., *Structural Members and Frames*, Prentice-Hall, Inc., Englewood Cliffs, New Jersey, (1968).

Galambos, T. V. and Fukomoto, Y., "Inelastic Lateral-Torsional Buckling of Beam-Columns," *Journal of the Structural Division, ASCE*, Vol. 92, No. ST2, Proc. Paper 4770, April, pp. 41–61, (1966).

Galambos, T. V. and Ravindra, M. K., "Load and Resistance Factor Design Criteria for Steel Beams," *Research Report No. 27*, Department of Civil and Environmental Engineering, Washington University, St. Louis, (1974).

Harrison, H. B., *Computer Methods in Structural Analysis*, Prentice-Hall, Inc., Englewood Cliffs, New Jersey, (1973).

Hartmann, A. J., "Elastic Lateral Buckling of Continuous Beams," *Journal of the Structural Division, ASCE*, Vol. 93, No. ST4, Proc. Paper 5363, Aug., pp. 11–26, (1967).

Horne, M. R., "The Flexural-Torsional Buckling of Members of Symmetrical I-Section Under Combined Thrust and Unequal Terminal Moments," *Quarterly Journal of Mechanics and Applied Mathematics*, Vol. 7, Part 4, pp. 410–426, (1954).

Johnston, B. G. (ed.), *The Structural Stability Research Council Guide to Stability Design Criteria for Metal Structures*, John Wiley, 3rd edn, (1976).

Kitipornchai, S. and Trahair, N. S., "Elastic Stability of Tapered I-Beams," *Journal of the Structural Division, ASCE*, Vol. 98, No. ST3, Proc. Paper 8775, March, pp. 713–728, (1972).

Kitipornchai, S. and Trahair, N. S., "Buckling of Inelastic I-Beams Under Moment Gradient," *Journal of the Structural Division, ASCE*, Vol. 101, No. ST5, Proc. Paper 11 295, May, pp. 991–1004, (1975a).

Kitipornchai, S. and Trahair, N. S., "Inelastic Buckling of Simply Supported Steel I-Beams," *Journal of the Structural Division, ASCE*, Vol. 101, No. ST7, Proc. Paper 11 419, July, pp. 1333–1347, (1975b).

Kitipornchai, S. and Trahair, N. S., "Elastic Behavior of Tapered Monosymmetric I-Beams," *Journal of the Structural Division, ASCE*, Vol. 101, No. ST8, Proc. Paper 11 479, Aug., pp. 1661–1678, (1975c).

Lay, M. G. and Galambos, T. V., "Inelastic Steel Beams Under Uniform Moment," *Journal of the Structural Division, ASCE*, Vol. 91, No. ST6, Proc. Paper 4566, December, pp. 67–93, (1965).

Lim, L. C. and Lu, L.-W., "The Strength and Behavior of Laterally Unsupported Columns," *Fritz Engineering Laboratory Report No. 329.5*, Lehigh University, Bethlehem, Pa., June, pp. 1–234, (1970).

Livesley, R. K., *Matrix Methods of Structural Analysis*, Pergamon Press, Oxford, (1964).

Love, A. E. H., *A Treatise on the Mathematical Theory of Elasticity*, 4th edn, Dover Publications, New York, (1944).

Massey, C. and McGuire, P. J., "Lateral Stability of Nonuniform Cantilevers," *Journal of the Engineering Mechanics Division, ASCE*, Vol. 97, No. EM3, Proc. Paper 8172, June, pp. 673–686, (1971).

Morrell, M. L. and Lee, G. C., "Allowable Stress for Web-Tapered Beams," *Welding Research Council Bulletin*, No. 192, February, pp. 1–12, (1974).

Mutton, B. R. and Trahair, N. S., "Stiffness Requirements for Lateral Bracing," *Journal of the Structural Division, ASCE*, Vol. 99, No. ST10, Proc. Paper 10 086, Oct., pp. 2167–2182, (1973).

Nethercot, D. A., "Recent Progress in the Application of the Finite Element Method to Problems of the Lateral Buckling of Beams," *Proceedings, Conference on Finite Element Methods in Civil Engineering*, Engineering Institute of Canada, Montreal, June, p. 367, (1972a).

Nethercot, D. A., "Factors Affecting the Buckling Stability of Partially Plastic Beams," *Proceedings, Institution of Civil Engineers*, Vol. 53, Sept., pp. 285–304, (1972b).

Nethercot, D. A., "The Effect of Load Position on the Lateral Stability of Beams," *Proceedings, Conference on Finite Element Methods in Civil Engineering*, Engineering Institute of Canada, Montreal, June, p. 347, (1972c).

Nethercot, D. A., "Influence of End Support Conditions on the Stability of Transversely Loaded Beams," *Building Science*, Vol. 7, pp. 87–94, (1972d).

Nethercot, D. A., "Bracing of Slender Members," *Civil Engng. and Pub. Wks. Rev.*, Vol. 67, No. 796, Nov., p. 1158, (1972e).

Nethercot, D. A., "Lateral Buckling of Tapered Beams," *Publications, IABSE*, Vol. 33-II, pp. 173–192, (1973a).

Nethercot, D. A., "The Effective Lengths of Cantilevers as Governed by Lateral Buckling," *The Structural Engineer*, Vol. 51, No. 5, May, pp. 161–168, (1973b).

Nethercot, D. A., "Buckling of Laterally or Torsionally Restrained Beams," *Journal of the Engineering Mechanics Division, ASCE*, Vol. 99, No. EM4, Proc. Paper 9903, Aug., pp. 773–791, (1973c).

Nethercot, D. A., "The Solution of Inelastic Lateral Stability Problems by the Finite Element Method," *Proceedings, 4th Australasian Conference on Mechanics of Structures and Materials*, Brisbane, pp. 183–190, (1973d).

Nethercot, D. A., "Inelastic Buckling of Monosymmetric I-Beams," *Journal of the Structural Division, ASCE*, Vol. 99, No. ST7, Technical Note, July, pp. 1696–1701, (1973e).
Nethercot, D. A., "Residual Stresses and their Influence Upon the Lateral Buckling of Rolled Steel Beams," *The Structural Engineer*, Vol. 52, No. 3, March, pp. 89–96, (1974a).
Nethercot, D. A., "Buckling of Welded Beams and Girders," *Publications, IABSE*, Vol. 34-I, pp. 107–121, (1974b).
Nethercot, D. A., "Inelastic Buckling of Steel Beams Under Non-Uniform Moment," *The Structural Engineer*, Vol. 53, No. 2, Feb., pp. 73–78, (1975a).
Nethercot, D. A., "Effective Lengths of Partially Plastic Steel Beams," *Journal of the Structural Division, ASCE*, Vol. 101, No. ST5, Technical Note, May, pp. 1163–1166, (1975b).
Nethercot, D. A., "Bending and Buckling of Welded Hybrid Beams," *Research Report R41*, Dept. of Civil and Structural Engineering, University of Sheffield, March, (1975c).
Nethercot, D. A., Private Communication to N. S. Trahair, (1975d).
Nethercot, D. A. and Rockey, K. C., "A Unified Approach to the Elastic Lateral Buckling of Beams," *The Structural Engineer*, Vol. 49, No. 7, July, pp. 321–330, (1971a).
Nethercot, D. A. and Rockey, K. C., "Finite Element Solutions for the Buckling of Columns and Beams," *Int. J. Mech. Sci.*, Vol. 13, pp. 945–949, (1971b).
Nethercot, D. A. and Rockey, K. C., "The Lateral Buckling of Beams Having Discrete Intermediate Restraints," *The Structural Engineer*, Vol. 50, No. 10, Oct., pp. 391–403, (1972).
Nethercot, D. A. and Rockey, K. C., "Lateral Buckling of Beams with Mixed End Conditions," *The Structural Engineer*, Vol. 51, No. 4, April, pp. 133–138, (1973).
Nethercot, D. A. and Trahair, N. S., "Design of Diaphragm Braced I-Beams," *Journal of the Structural Division, ASCE*, Vol. 101, No. ST10, Proc. Paper 11 615, Oct., pp. 2045–2061, (1975).
Nethercot, D. A. and Trahair, N. S., "Lateral Buckling Approximations for Elastic Beams," *The Structural Engineer*, Vol. 54, No. 6, June, pp. 197–204, (1976a).
Nethercot, D. A. and Trahair, N. S., "Inelastic Lateral Buckling of Determinate Beams," *Journal of the Structural Division, ASCE*, Vol. 102, No. ST4, Proc. Paper 12020, April, 701–717, (1976b).
Poowannachaikul, T. and Trahair, N. S., "Inelastic Buckling of Continuous Steel I-Beams," *Research Report R51*, Department of Civil and Structural Engineering, University of Sheffield, November, (1974).
Powell, G. and Klingner, R., "Elastic Lateral Buckling of Steel Beams," *Journal of the Structural Division, ASCE*, Vol. 96, No. ST9, Proc. Paper 7555, September, pp. 1919–1932, (1970).
Salvadori, M. G., "Lateral Buckling of Beams of Rectangular Cross-Section under Bending and Shear," *Proceedings, 1st U.S. National Congress of Applied Mechanics*, p. 403, (1951).
Salvadori, M. G., "Lateral Buckling of I-Beams," *Transactions, ASCE*, Vol. 120, Paper 2773, pp. 1165–1177, (1955).
Salvadori, M. G., "Lateral Buckling of Eccentrically Loaded I-Columns," *Transactions, ASCE*, Vol. 121, Paper 2836, pp. 1163–1178, (1956).
Standards Association of Australia, *AS 1250-1975 SAA Steel Structures Code*, SAA, Sydney, (1975).
Taylor, J. C., Dwight, J. B. and Nethercot, D. A., "Buckling of Beams and Struts: Proposals for a New British Code," *Proceedings, Conference on Metal Structures and the Practising Engineer*, Inst. Engrs. Aust., Melbourne, Nov., pp. 27–31, (1974).
Timoshenko, S. P. and Gere, J. M., *Theory of Elastic Stability*, 2nd edn, McGraw-Hill Book Company, (1961).
Trahair, N. S., "Stability of I-Beams with Elastic End Restraints," *Journal of the Institution of Engineers, Australia*, Vol. 37, No. 6, June, pp. 157–168, (1965).
Trahair, N. S., "The Bending Stress Rules of the Draft ASCAl," *Journal of the Institution of Engineers, Australia*, Vol. 38, No. 6, June, pp. 131–141, (1966a).
Trahair, N. S., "Elastic Stability of I-Beam Elements in Rigid-Jointed Structures," *Journal of the Institution of Engineers, Australia*, Vol. 38, No. 7–8, July–Aug., pp. 171–180, (1966b).
Trahair, N. S., "Elastic Stability of Propped Cantilevers," *Civil Engineering Transactions*, Institution of Engineers, Australia, Vol. CE10, No. 1, April, pp. 94–100, (1968a).
Trahair, N. S., "Interaction Buckling of Narrow Rectangular Continuous Beams," *Civil Engineering Transactions*, Institution of Engineers, Australia, Vol. CE10, No. 2, Oct., pp. 167–172, (1968b).
Trahair, N. S., "Elastic Stability of Continuous Beams," *Journal of the Structural Division, ASCE*, Vol. 95, No. ST6, Proc. Paper 6632, June, pp. 1295–1312, (1969).
Trahair, N. S., Discussion of "Lateral Torsional Buckling of I-Sections in Grade 55 Steel" by J. E. Dibley, *Proceedings, Institution of Civil Engineers*, Vol. 46, May, pp. 97–101, (1970).
Trahair, N. S., *Design of Steel Beams*, 2nd edn, Australian Institute of Steel Construction, Sydney, (1974).
Trahair, N. S. and Kitipornchai, S., "Elastic Lateral Buckling of Stepped I-Beams," *Journal of the Structural Division, ASCE*, Vol. 97, No. ST10, Proc. Paper 8445, Oct., pp. 2535–2548, (1971).

Trahair, N. S. and Kitipornchai, S., "Buckling of Inelastic I-Beams Under Uniform Moment," *Journal of the Structural Division, ASCE,* Vol. 98, No. ST11, Proc. Paper 9339, Nov., pp. 2551–2566, (1972).

Trahair, N. S. and Woolcock, S. T., "Effect of Major Axis Curvature on I-Beam Stability," *Journal of the Engineering Mechanics Division, ASCE,* Vol. 99, No. EM1, Proc. Paper 9548, Feb., pp. 85–98, (1973).

Vacharajittiphan, P., "Stability of Frame Structures," Ph.D. Thesis, Department of Civil Engineering, University of Sydney, Sydney, (1974).

Vacharajittiphan, P. and Trahair, N. S., "Warping and Distortion at I-Section Joints," *Journal of the Structural Division, ASCE,* Vol. 100, No. ST3, Proc. Paper 10 390, March, pp. 547–564, (1974a).

Vacharajittiphan, P. and Trahair, N. S., "Direct Stiffness Analysis of Lateral Buckling," *Journal of Structural Mechanics,* Vol. 3, No. 2, pp. 107–137, (1974b).

Vacharajittiphan, P. and Trahair, N. S., "Analysis of Lateral Buckling in Plane Frames," *Journal of the Structural Division, ASCE,* Vol. 101, No. ST7, Proc. Paper 11 430, July, pp. 1497–1516, (1975).

Vacharajittiphan, P., Woolcock, S. T. and Trahair, N. S., "Effect of In-Plane Deformation on Lateral Buckling," *Journal of Structural Mechanics,* Vol. 3, No. 1, pp. 29–60, (1974).

Vlasov, V. Z., *Thin-Walled Elastic Beams,* Israel Program for Scientific Translations, Jerusalem, (1961).

Yoshida, H. and Imoto, Y., "Inelastic Lateral Buckling of Restrained Beams," *Journal of the Engineering Mechanics Division, ASCE,* Vol. 99, No. EM2, Proc. Paper 9666, April, pp. 343–366, (1973).

Yoshida, H., Nethercot, D. A. and Trahair, N. S., "Analysis of Inelastic Buckling of Continuous Beams," *Research Report R65,* Department of Civil and Structural Engineering, University of Sheffield, August, (1975).

Young, B. W., "Residual Stresses in Hot-Rolled Sections," *Proceedings, International Colloquium on Column Strength, IABSE,* Vol. 23, pp. 25–38, (1975).

4

ANALYSIS OF ELASTIC BEAM-COLUMNS

4.1 INTRODUCTION

The theory of bending and torsion of thin-walled elastic beams presented in Chap. 2 is now applied to solve the behavior of biaxially loaded beam-columns. This is a generalization of beam-column problems to consider their space behavior under general three-dimensional loading conditions. Here, as in Chaps. 3 and 4 of Vol. 1 of this book, the analyses are limited within the elastic range of the material so that geometric nonlinearity is the only possible source of complication of the analysis. The elastic theory presented here can serve as a theoretical basis for plastic analysis of biaxially loaded beam-columns to be presented in Chaps. 7 to 12. Chapters 5 and 6 present the strength and behavior of beam-column segments where the complication of the analysis is caused almost entirely by the non-linear properties of the material.

All assumptions made previously in Chap. 2 are again reviewed here so that all equations derived in Chap. 2 may be used to solve the general equations derived herein. The basic assumptions required in this chapter are:

1. No distortion of the shape of cross-section—this together with elasticity assumption of material guarantees invariability of sectional properties.

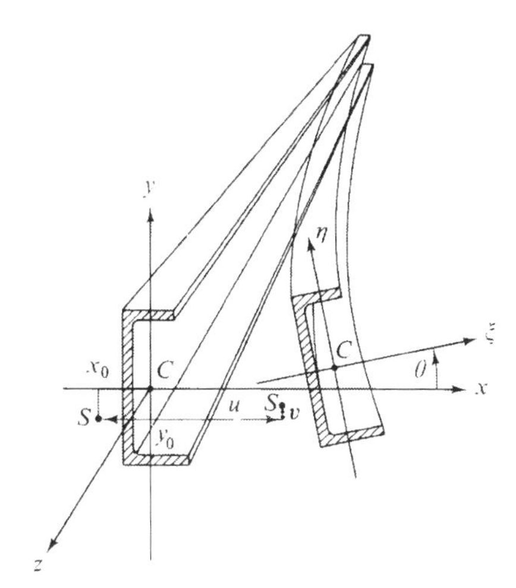

FIGURE 4.1
Deformation of a beam-column

2. Thin-walled open cross-section—only stresses on a cross section are normal stress, and shear stress in the direction of contour of a cross-section.

3. Shear deformation ignored—it follows that the warping deformation of a cross-section can be calculated by the rule of sectorial area.

4. Initially straight beam-column—the effect of initial lateral deflections and twist on the behavior of a beam-column can be neglected when compared with the end eccentricities of the applied thrust. Effect of an initially imperfect beam-column is discussed briefly in Sect. 4.5.

5. Small deformation—the lateral deflections u, v and the twisting angle θ are small so that the governing differential equations remain linear. Effect of finite deformation is discussed briefly in Sect. 4.6.

6. No lateral loads—the beam-columns are loaded only at their ends.

In addition to these assumptions, the coordinate system (x, y, z) are defined in accordance with the principal axes of the cross section having the origin at the *centroid*, C, but the displacements (u, v, θ) of a cross section are taken with respect to its *shear center*, S, at location (x_0, y_0) as shown in Fig. 4.1. We start here with the basic equations for a beam-column element dz as shown in Fig. 4.2. This element is subjected to axial forces $N, N + N'\,dz$, and bending moments M_x, $M_x + M_x'\,dz$ and M_y, $M_y + M_y'\,dz$ in which all derivatives are with respect to z. In Fig. 4.2, positive M_x, M_y and N, represented as vectors, point in the positive x-, y- and z-direction respectively, using the right hand screw rule for moment on the positive section. The values of M_x, M_y, and N at a section z are produced by the loads applied at both ends with respect to the undeformed bar. The equilibrium equation of the element after deformation of u, v and θ are obtained from

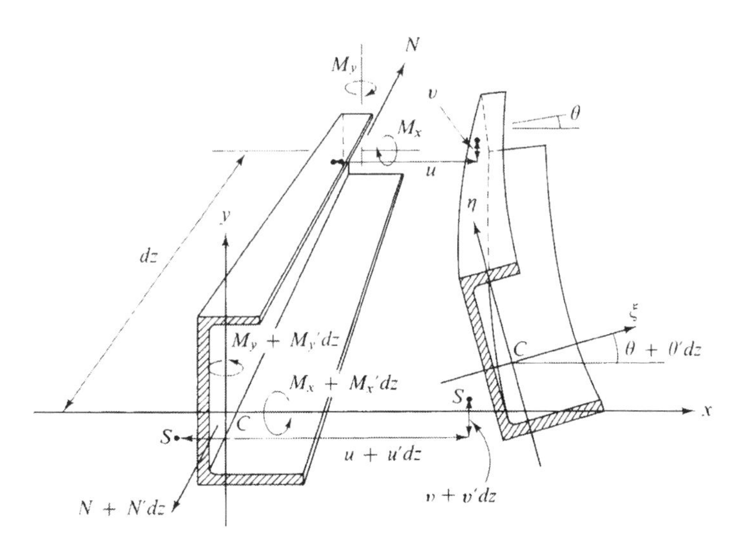

FIGURE 4.2
Equilibrium of beam-column element

Eq. (2.179a) and Eq. (2.179b), neglecting the end twisting moment and distributed load,

$$
\begin{aligned}
&EI_y u^{IV} - [N(u' + y_0\theta')]' + (M_x\theta)'' = 0 \\
&EI_x v^{IV} - [N(v' - x_0\theta')]' + (M_y\theta)'' = 0 \qquad (4.1)\\
&EI_\omega \theta^{IV} - GK_T\theta'' - (\bar{K}\theta')' - y_0(Nu')' + x_0(Nv')' + M_x u'' + M_y v'' = 0
\end{aligned}
$$

in which $\bar{K}$ is the Wagner effect coefficient [Eq. (2.137)],

$$\bar{K} = \int \sigma[(x - x_0)^2 + (y - y_0)^2]\, dA = \frac{I_0}{A} N + \beta_x M_x - \beta_y M_y + \beta_\omega M_\omega \qquad (4.2)$$

I_0 is the polar moment of inertia of the section about the shear center,

$$I_0 = \int [(x - x_0)^2 + (y - y_0)^2]\, dA = I_x + I_y + A(x_0^2 + y_0^2) \qquad (4.3)$$

and

$$
\begin{aligned}
\beta_x &= \frac{1}{I_x}\left(\int y^3\, dA + \int x^2 y\, dA\right) - 2y_0 \\
\beta_y &= \frac{1}{I_y}\left(\int x^3\, dA + \int x y^2\, dA\right) - 2x_0 \qquad (4.4)\\
\beta_\omega &= \frac{1}{I_\omega}\left(\int \omega x^2\, dA + \int \omega y^2\, dA\right)
\end{aligned}
$$

Equations (4.1) are the general differential equations for elastic beam-columns in three dimensional space (Timoshenko, 1945), which are to be simplified according to special loading cases discussed in what follows.

4.2 BEAM-COLUMN WITH EQUAL END ECCENTRICITIES—EXACT APPROACH

The beam-column is subjected to an axial compression P with equal end eccentricities e_x and e_y at both ends as shown in Fig. 4.3. The applied loads are constant along the beam-column,

$$N = -P, \quad M_x = -Pe_y \quad \text{and} \quad M_y = Pe_x \tag{4.5}$$

Inserting this in Eq. (4.1), we find

$$\begin{aligned} &EI_y u^{IV} + Pu'' - P(e_y - y_0)\theta'' = 0 \\ &EI_x v^{IV} + Pv'' + P(e_x - x_0)\theta'' = 0 \\ &EI_\omega \theta^{IV} - (GK_T + \bar{K})\theta'' + P[(e_x - x_0)v'' - (e_y - y_0)u''] = 0 \end{aligned} \tag{4.6}$$

Integrating Eq. (4.6) twice, we have

$$\begin{aligned} &EI_y u'' + Pu - P(e_y - y_0)\theta = C_1 z + C_2 \\ &EI_x v'' + Pv + P(e_x - x_0)\theta = C_3 z + C_4 \\ &EI_\omega \theta'' - (GK_T + \bar{K})\theta + P(e_x - x_0)v - P(e_y - y_0)u = C_5 z + C_6 \end{aligned} \tag{4.7}$$

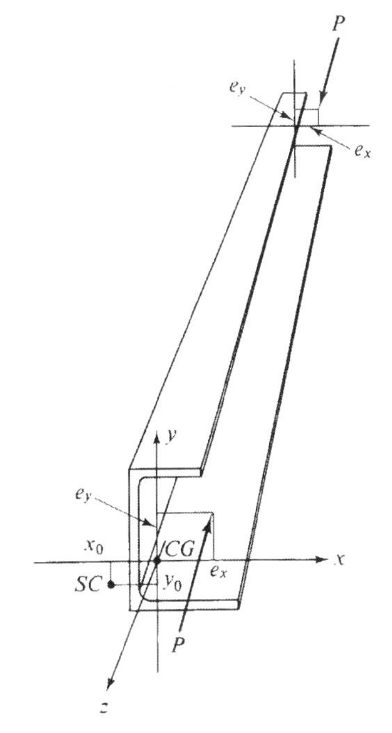

FIGURE 4.3
Beam-column with equal end eccentricity

This is a system of three linear non-homogeneous differential equations with the three unknowns u, v, and θ. An exact solution is possible and has been reported by Culver (1966a). The complete solution consists of two parts: a particular solution and a complementary solution. Since all terms on the right side of Eq. (4.7) are linear functions of z, it follows that the particular solutions u_p, v_p and θ_p of Eq. (4.7) must satisfy the relations

$$\begin{aligned} Pu - P(e_y - y_0)\theta &= C_1 z + C_2 \\ Pv + P(e_x - x_0)\theta &= C_3 z + C_4 \\ -P(e_y - y_0)u + P(e_x - x_0)v - (GK_T + \bar{K})\theta &= C_5 z + C_6 \end{aligned} \tag{4.8}$$

from which we can solve for u, v and θ as

$$\begin{aligned} u_p &= -\frac{1}{\Delta}\left\{\left[\frac{GK_T + \bar{K}}{P} + (e_x - x_0)^2\right](C_1 z + C_2) + (e_x - x_0)(e_y - y_0)(C_3 z + C_4) \right. \\ &\quad \left. - (e_y - y_0)(C_5 z + C_6)\right\} \\ v_p &= -\frac{1}{\Delta}\left\{\left[\frac{GK_T + \bar{K}}{P} + (e_y - y_0)^2\right](C_3 z + C_4) + (e_x - x_0)(e_y - y_0)(C_1 z + C_2) \right. \\ &\quad \left. + (e_x - x_0)(C_5 z + C_6)\right\} \\ \theta_p &= \frac{1}{\Delta}\{(e_y - y_0)(C_1 z + C_2) - (e_x - x_0)(C_3 z + C_4) + (C_5 z + C_6)\} \end{aligned} \tag{4.9}$$

in which

$$\Delta = P\begin{vmatrix} 1 & 0 & -(e_y - y_0) \\ 0 & 1 & e_x - x_0 \\ -(e_y - y_0) & e_x - x_0 & -\dfrac{GK_T + \bar{K}}{P} \end{vmatrix} \tag{4.10}$$

The complementary solutions u_c, v_c and θ_c of Eq. (4.7) are obtained by assuming

$$\begin{aligned} u &= Ae^{\lambda z} \\ v &= Be^{\lambda z} \\ \theta &= Ce^{\lambda z} \end{aligned} \tag{4.11}$$

Substituting the complementary functions (4.11) into the left hand side of Eq. (4.7), and factoring the common terms $e^{\lambda z}$, and equating the coefficients of $e^{\lambda z}$ to zero gives

$$\begin{bmatrix} (EI_y\lambda^2 + P) & 0 & -P(e_y - y_0) \\ 0 & (EI_x\lambda^2 + P) & P(e_x - x_0) \\ -P(e_y - y_0) & P(e_x - x_0) & (EI_\omega\lambda^2 - GK_T - \bar{K}) \end{bmatrix}\begin{Bmatrix} A \\ B \\ C \end{Bmatrix} = \begin{Bmatrix} 0 \\ 0 \\ 0 \end{Bmatrix} \tag{4.12}$$

In order to have a non-trivial solution for u, v, and θ, the determinant of the coefficient matrix must be zero which leads to the auxiliary equation,

$$\begin{aligned}&(EI_y\lambda^2 + P)(EI_x\lambda^2 + P)(EI_\omega\lambda^2 - GK_T - \bar{K}) - P^2(EI_x\lambda^2 + P)(e_y - y_0)^2\\&- P^2(EI_y\lambda^2 + P)(e_x - x_0)^2 = 0\end{aligned} \tag{4.13}$$

The solutions for λ^2 of Eq. (4.13) have two negative roots $-\lambda_1^2$, $-\lambda_2^2$ and one positive root λ_3^2, hence, the corresponding six solutions for λ are

$$\pm i\lambda_1, \quad \pm i\lambda_2 \quad \text{and} \quad \pm\lambda_3 \tag{4.14}$$

which are to be evaluated numerically. The complementary solutions for u, v and θ have the forms

$$\begin{aligned}u_c &= \frac{P(e_y - y_0)}{EI_y}\left\{\sum_{j=1}^{2}\frac{1}{\lambda_y^2 + \lambda_j^2}(J_j \sin\lambda_j z + K_j\cos\lambda_j z)\right.\\&\quad\left.+ \frac{1}{\lambda_y^2 + \lambda_3^2}(J_3\sinh\lambda_3 z + K_3\cosh\lambda_3 z)\right\}\\v_c &= \frac{P(e_x - x_0)}{EI_x}\left\{\sum_{j=1}^{2}\frac{1}{\lambda_x^2 + \lambda_j^2}(J_j \sin\lambda_j z + K_j\cos\lambda_j z)\right.\\&\quad\left.+ \frac{1}{\lambda_x^2 + \lambda_3^2}(J_3\sinh\lambda_3 z + K_3\cosh\lambda_3 z)\right\}\\\theta_c &= \sum_{j=1}^{2}(J_j\sin\lambda_j z + K_j\cos\lambda_j z) + J_3\sinh\lambda_3 z + K_3\cosh\lambda_3 z\end{aligned} \tag{4.15}$$

in which

$$\lambda_x^2 = \frac{P}{EI_x} \quad \text{and} \quad \lambda_y^2 = \frac{P}{EI_y} \tag{4.16}$$

The complete solution to Eq. (4.7) is given by summation of the complementary solution Eq. (4.15) and the particular solution Eq. (4.9),

$$\begin{aligned}u &= u_c + u_p\\v &= v_c + v_p\\\theta &= \theta_c + \theta_p\end{aligned} \tag{4.17}$$

The twelve integration constants

$$J_1, J_2, J_3, K_1, K_2, K_3, C_1, C_2, C_3, C_4, C_5, C_6$$

can be determined by the following twelve boundary conditions,

$$\begin{aligned}&u = 0, v = 0, \theta = 0: \quad \text{geometric boundary conditions}\\&u'' = \frac{Pe_x}{EI_y}, v'' = \frac{Pe_y}{EI_x}: \quad \text{static boundary conditions}\end{aligned} \tag{4.18}$$

and

$$\theta'' = 0: \quad \text{warping free}$$

or

$$\theta' = 0: \quad \text{warping restrained} \tag{4.19}$$

at both ends

$$z = 0 \qquad \text{and} \qquad z = L$$

Since an algebraic evaluation of the integration constants is not feasible,

FIGURE 4.4
Midspan displacements of beam-columns (Culver, 1966a) (1 kip = 4.45 kN, 1 in = 2.54 cm)

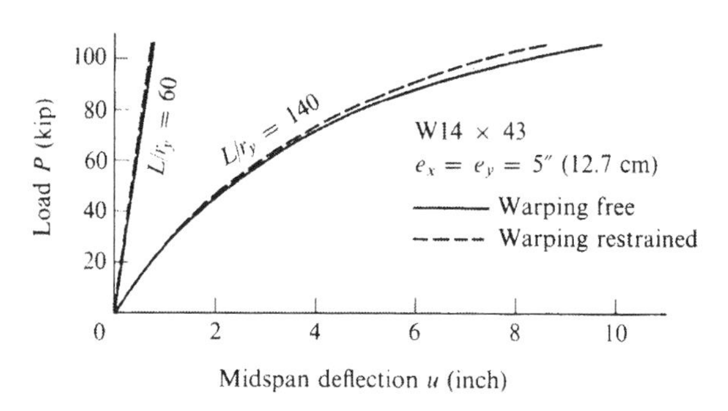

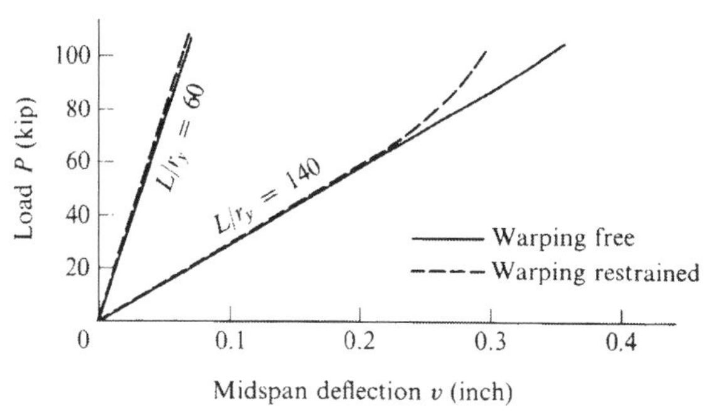

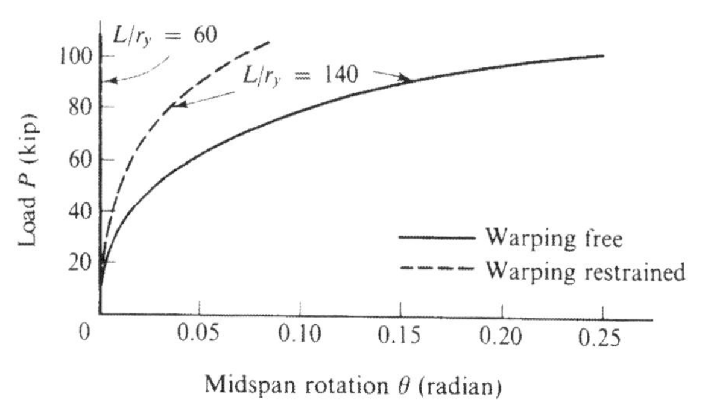

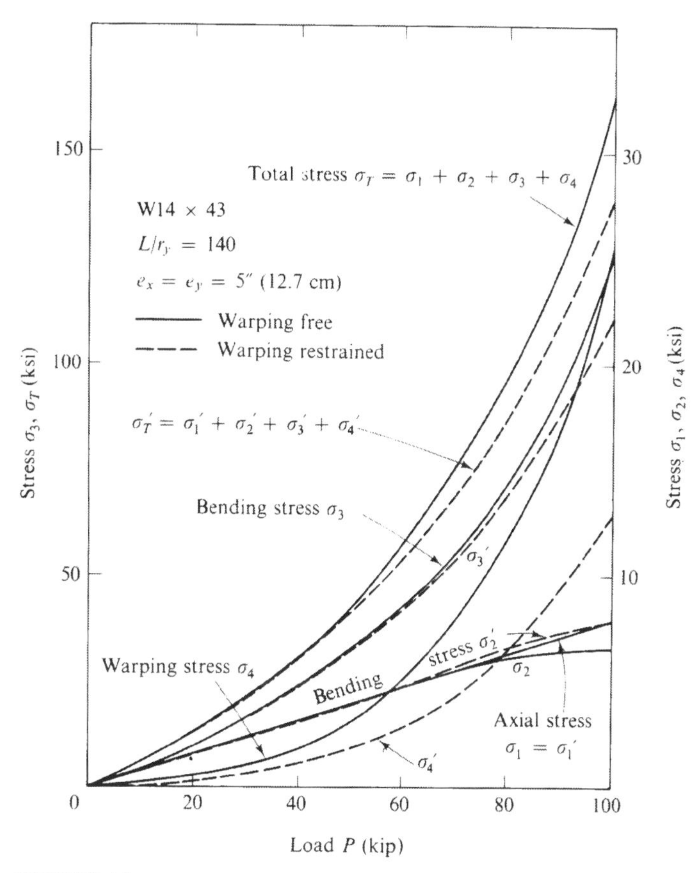

FIGURE 4.5
Comparison in stresses due to warping restraint (Culver, 1966a) (1 kip = 4.45 kN, 1 in = 2.54 cm, 1 ksi = 6.9 MN/m^2)

Culver (1966a) solved the problem numerically for a beam-column of wide-flange cross section W14 × 43. Two slenderness ratios were selected,

$$L/r_y = 60 \qquad \text{and} \qquad 140 \tag{4.20a}$$

The end eccentricities at both ends are taken to be,

$$e_x = 5 \text{ in (12.7 cm)} \qquad \text{and} \qquad e_y = 5 \text{ in (12.7 cm)} \tag{4.20b}$$

The boundary conditions for warping are either free ($\theta'' = 0$) or restrained ($\theta' = 0$). His numerical results are summarized in Figs. 4.4 and 4.5. Figure 4.4 shows the load-midspan deflections u, v and rotation θ, respectively. It is seen that warping restraint has little effect on the lateral deflections u, v, especially for less slender members. However, warping restraint has a significant strengthening effect on the rotational stiffness of a slender beam-column. Figure 4.5 shows normal stresses σ in the direction of the fibers of the beam-column as a function of load P. The total normal stresses σ_T are produced by three effects, axial compression σ_1, bending σ_2 and σ_3, and warping σ_4 of the middle line of the cross section. The total stresses σ_T for combined axial force and bending and warping moments are given by superposition of σ_1, σ_2, σ_3 and σ_4.

4.3 BEAM-COLUMN WITH EQUAL END ECCENTRICITIES—APPROXIMATE APPROACH

In the exact approach described above, evaluation for the integration constants was made by a somewhat lengthy numerical calculation. In order to find a simple solution for Eq. (4.6), lateral deflections u and v, and rotation θ can be approximated by fixed functions in such a way as to match their boundary conditions at both ends. Peköz and Celebi (1969) presented such a solution using the Galerkin method for which the functions for deflections and rotation are assumed to have the forms

$$\begin{aligned} u &= A_0 u_0(z) - \frac{Pe_x}{2EI_y} z(L - z) \\ v &= B_0 v_0(z) - \frac{Pe_y}{2EI_x} z(L - z) \\ \theta &= C_0 \theta_0(z) \end{aligned} \tag{4.21}$$

The functions $u_0(z)$, $v_0(z)$ and $\theta_0(z)$ are assumed to satisfy the geometric boundary conditions by the following functions:

$$\begin{aligned} &\text{For simple support:} \quad f(z) = \sin \frac{\pi z}{L} \\ &\text{For fixed support:} \quad f(z) = \sin \frac{\lambda z}{L} - \sinh \frac{\lambda z}{L} - \alpha\left(\cos \frac{\lambda z}{L} - \cosh \frac{\lambda z}{L}\right) \end{aligned} \tag{4.22}$$

in which

$$\alpha = \frac{\sinh \lambda - \sin \lambda}{\cosh \lambda - \cos \lambda} \quad \text{and} \quad \lambda = \frac{3}{2}\pi \tag{4.23}$$

Thus, the lateral deflection functions u and v as given by Eq. (4.21) satisfy the static boundary conditions at both ends, $z = 0$ and $z = L$,

$$u'' = \frac{Pe_x}{EI_y} \quad \text{and} \quad v'' = \frac{Pe_y}{EI_x} \tag{4.24}$$

Applying the Galerkin method to the differential equations (4.6), using the assumed deflection functions, Eq. (4.21), the following algebraic equations are obtained (see Sec. 3.7, Chap. 3, Vol. 1),

$$A_0\left(EI_y\int_0^L u_0^{IV}u_0\,dz + P\int_0^L u_0''u_0\,dz\right) - P(e_y - y_0)C_0\int_0^L \theta_0''u_0\,dz = -\frac{P^2e_x}{EI_y}\int_0^L u_0\,dz$$

$$B_0\left(EI_x\int_0^L v_0^{IV}v_0\,dz + P\int_0^L v_0''v_0\,dz\right) + P(e_x - x_0)C_0\int_0^L \theta_0''v_0\,dz = -\frac{P^2e_y}{EI_x}\int_0^L v_0\,dz \tag{4.25}$$

$$C_0\left[EI_\omega\int_0^L \theta_0^{IV}\theta_0\,dz - (GK_T + \bar{K})\int_0^L \theta_0''\theta_0\,dz\right] - P(e_y - y_0)A_0\int_0^L u_0''\theta_0\,dz$$
$$+ P(e_x - x_0)B_0\int_0^L v_0''\theta_0\,dz = P^2\left[\frac{(e_y - y_0)e_x}{EI_y} - \frac{(e_x - x_0)e_y}{EI_x}\right]\int_0^L \theta_0\,dz$$

Hence the system of three differential equations is reduced to a system of three linear equations with the unknowns A_0, B_0 and C_0 the solution of which does not present any difficulties. Using Eqs. (4.22), Eqs. (4.25) can be rearranged in the matrix form:

$$\begin{bmatrix} P_y - P & 0 & P(e_y - y_0)K'_{13} \\ 0 & P_x - P & -P(e_x - x_0)K'_{23} \\ P(e_y - y_0)K'_{31} & -P(e_x - x_0)K'_{32} & r_0^2(P_z - P) \end{bmatrix}\begin{Bmatrix} A_0 \\ B_0 \\ C_0 \end{Bmatrix} = \begin{Bmatrix} -\dfrac{P^2}{P_y}e_xK_1 \\ -\dfrac{P^2}{P_x}e_yK_2 \\ P^2\left[\dfrac{(e_y - y_0)e_x}{P_y} - \dfrac{(e_x - x_0)e_y}{P_x}\right]K_3 \end{Bmatrix} \tag{4.26}$$

in which the following notations are used:

$$P_x = K_{11}\frac{\pi^2EI_x}{L^2} \qquad \text{flexural buckling load about } x\text{-axis}$$

$$P_y = K_{22}\frac{\pi^2EI_y}{L^2} \qquad \text{flexural buckling load about } y\text{-axis} \tag{4.27}$$

$$P_z = \frac{1}{r_0^2}\left(K_{33}\frac{\pi^2EI_\omega}{L^2} + GK_T\right) \qquad \text{torsional buckling load}$$

in which

$$\bar{K} = \int \sigma[(x - x_0)^2 + (y - y_0)^2]\,dA = -Pr_0^2 \tag{4.28a}$$

Table 4.1 COEFFICIENTS K [PEKÖZ AND CELEBI (1969)]

Boundary Condition at $z = 0, L$	Coefficients										
	K_{11}	K_{22}	K_{33}	K_1	K_2	K_3	K'_{13}	K'_{31}	K'_{23}	K'_{32}	K_{23}
$u'' = v'' = \theta'' = 0$	1.0000	1.0000	1.0000	1.2732	1.2732	1.2732	1.0000	1.0000	1.0000	1.0000	1.0000
$u'' = v' = \theta'' = 0$	1.0000	4.1223	1.0000	1.2732	—	1.2732	1.0000	1.0000	0.5507	1.4171	0.8834
$u' = v' = \theta'' = 0$	4.1223	4.1223	1.0000	—	—	1.2732	0.5507	1.4171	0.5507	1.4171	0.8834
$u'' = v' = \theta' = 0$	1.0000	4.1223	4.1223	1.2732	—	0.6597	1.4171	0.5507	1.0000	0.8834	1.0000
$u' = v' = \theta' = 0$	4.1223	4.1223	4.1223	—	—	0.6597	1.0000	1.0000	1.0000	1.0000	1.0000
$u'' = v'' = \theta' = 0$	1.0000	1.0000	4.1223	1.2732	1.2732	0.6597	1.4171	0.5507	1.4171	0.5507	0.8834

If the normal stress due to warping is negligibly small compared with those due to normal force and bending, then the Wagner effect coefficient is simplified by substituting Eq. (4.5) into Eq. (4.3), and

$$r_0^2 = x_0^2 + y_0^2 + \frac{1}{A}(I_x + I_y) + \beta_x e_y + \beta_y e_x \tag{4.28b}$$

Note that in general r_0^2 is not a purely cross sectional property but depends on the eccentricities e_x and e_y also. For the special case of a doubly-symmetric cross section x_0, y_0, β_x, and β_y are zero and r_0 is equal to the polar radius of gyration. The ten coefficients denoted by K_i or K_{ij} are constants. The values of these constants are obtained by integration of the basic deflections u_0, v_0, θ_0 and their derivatives [Eqs. (4.22)]. The numerical values of these constants are listed in Table 4.1 for several boundary conditions. Solving for A_0, B_0 and C_0 from Eq. (4.26), deflections of the beam-column are obtained by Eq. (4.21). Since Eqs. (4.26) are non-homogeneous in general except for the special case of an axially loaded column, $e_x = e_y = 0$, deflections increase as load is increased. In the special case, however, buckling takes place. If a singly symmetric section ($y_0 = 0$) is subjected to an axial compression force P in the plane of symmetry ($e_y = 0$), Eqs. (4.26) turn out to be an independent equation of in-plane bending about x-axis and two homogeneous simultaneous equations of flexural torsional buckling,

$$(P_y - P)A_0 = -\frac{P^2}{P_y} e_x K_1 \tag{4.29a}$$

$$\begin{bmatrix} P_x - P & -P(e_x - x_0)K'_{23} \\ -P(e_x - x_0)K'_{32} & r_0^2(P_z - P) \end{bmatrix} \begin{Bmatrix} B_0 \\ C_0 \end{Bmatrix} = \begin{Bmatrix} 0 \\ 0 \end{Bmatrix} \tag{4.29b}$$

from which three critical loads are obtained

$$P_{cr} = P_y \tag{4.29c}$$

$$P_{cr} = \frac{(P_x + P_z) \pm \sqrt{(P_x - P_z)^2 + 4P_x P_z \left(\dfrac{e_x - x_0}{r_0}\right)^2 K_{23}^2}}{2\left[1 - \left(\dfrac{e_x - x_0}{r_0}\right)^2 K_{23}^2\right]} \tag{4.29d}$$

in which

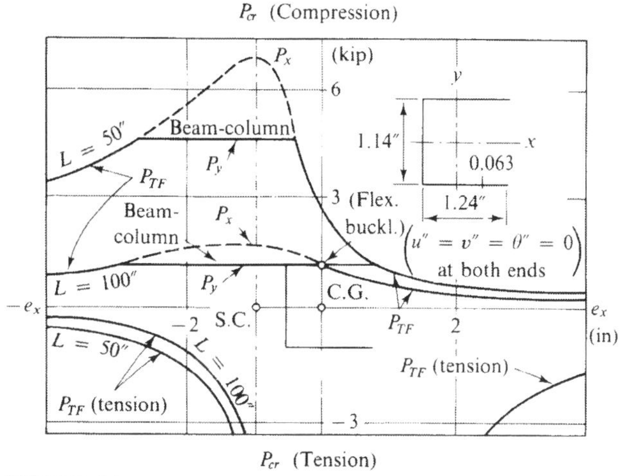

FIGURE 4.6
Typical buckling load plots for two channel members (Peköz and Celebi, 1969) (1 kip = 4.45 kN, 1 in = 2.54 cm)

$$K_{23} = \sqrt{K'_{23}K'_{32}}$$

The value of K_{23} is also listed in Table 4.1.

Peköz and Celebi (1969) plotted these buckling loads for members of channel section under symmetric loading conditions for two cases $L = 50$ in (127 cm), and 100 in (254 cm). For a given beam-column (P_y, P_z, r_0 known) and a given end eccentricity e_x, Fig. 4.6 shows the critical load P_{cr} which will cause lateral torsional buckling of the member. Two solutions from Eq. (4.29d) are possible: one for compression buckling and the other for tension buckling. The peak points of the two compression curves are the flexural buckling loads about the strong axis bending P_x. This occurs only when the load P is applied on the shear center. The curves are cut off by the weak axis flexural buckling load P_y (horizontal lines marked P_y). The portions of the curves controlled by torsional flexural buckling mode are marked P_{TF} in Fig. 4.6.

Approximate solutions to the differential equations have also been reported by others. Dabrowski (1961) considered the problem of biaxial bending from an energy standpoint and utilized the Rayleigh-Ritz method to obtain a solution (see Sec. 3.7, Chap. 3, Vol. 1). Thurlimann (1953), on the other hand, dealt directly with the equations of biaxial bending and obtained a solution using results from the problem of in-plane bending. Prawel and Lee (1964) solved the problem using an electric analog computer. The approximate solutions of Thurlimann and Dabrowski lead to essentially the same result given here. In each case, the simultaneous differential equations are reduced to a set of simultaneous algebraic equations. It should be noted that their approximate solutions fail to satisfy the statical boundary conditions [Eq. (4.24)], yet their numerical results are found within 1 percent of the exact values given in the preceding section.

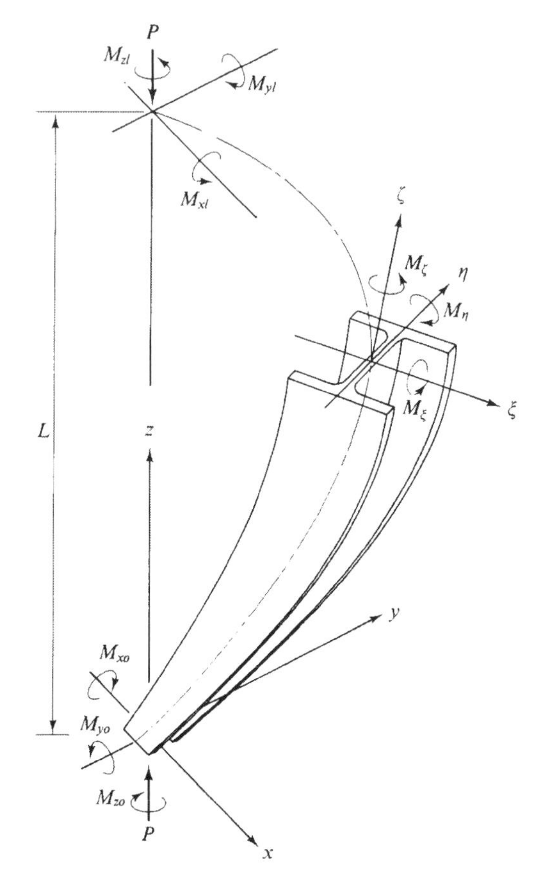

FIGURE 4.7
Beam-column with unequal end moments

4.4 BEAM-COLUMNS WITH UNEQUAL END ECCENTRICITIES

The H-beam-column shown in Fig. 4.7 is subjected to an axial thrust P and unequal end moments due to biaxial eccentricities e_x and e_y of the thrust P applied at both ends,

$$\begin{aligned} M_{xo} &= -Pe_{yo}, \quad M_{yo} = Pe_{xo} \quad \text{at} \quad z = 0 \\ M_{xl} &= -Pe_{yl}, \quad M_{yl} = Pe_{xl} \quad \text{at} \quad z = L \end{aligned} \tag{4.30}$$

The bending moments M_x and M_y at a section z produced by the end moments, excluding the bending moment produced by the thrust P are:

$$\begin{aligned} M_x &= M_{xo} + \frac{z}{L}(M_{xl} - M_{xo}) \\ M_y &= M_{yo} + \frac{z}{L}(M_{yl} - M_{yo}) \end{aligned} \tag{4.31}$$

Here, positive M_x, M_y and M_z represented as vectors, point in the positive x-, y- and z-direction, respectively. The twisting moments M_{zo} and M_{zl} at both ends are always equal from the equilibrium condition, they are called *reactions* since they are associated with the rotation restraints at the ends, $\theta = 0$. It follows that the twisting moment M_z about the axis of undeflected beam-column is zero,

$$M_z = M_{zo} = M_{zl} = 0 \tag{4.32}$$

Equating these external moments at a section z, or using Eq. (2.162) with $M_{zo} = 0$ and taking linear terms only,

$$\begin{aligned} M_{\xi(\text{ext})} &= M_x + Pv + M_y\theta \\ M_{\eta(\text{ext})} &= M_y - Pu - M_x\theta \\ M_{\zeta(\text{ext})} &= M_x u' + M_y v' - \frac{v}{L}(M_{yl} - M_{yo}) - \frac{u}{L}(M_{xl} - M_{xo}) \end{aligned} \tag{4.33}$$

to the internal moments,

$$\begin{aligned} M_{\xi(\text{int})} &= -EI_x v'' \\ M_{\eta(\text{int})} &= EI_y u'' \\ M_{\zeta(\text{int})} &= -EI_\omega \theta''' + (GK_T + \bar{K})\theta' \end{aligned} \tag{4.34}$$

the governing differential equations are obtained, [Eq. (2.176)],

$$\begin{aligned} &EI_y u'' + Pu + M_x\theta = M_y \\ &EI_x v'' + Pv + M_y\theta = -M_x \\ &EI_\omega \theta''' - (GK_T + \bar{K})\theta' + M_x u' + M_y v' - \frac{v}{L}(M_{yl} - M_{yo}) - \frac{u}{L}(M_{xl} - M_{xo}) = 0 \end{aligned} \tag{4.35}$$

These equations can also be derived by integrating Eq. (4.1) and utilizing the boundary conditions, Eq. (4.30), with the doubly symmetric condition for the cross section, $x_0 = y_0 = 0$. This is a system of three linear non-homogeneous differential equations with the three unknowns u, v, and θ. Chen and Atsuta (1973) have solved these equations in closed form assuming the deflected shapes of the beam-column as

$$\begin{aligned} u &= A_0 \sin\frac{\pi z}{L} + \frac{z}{L}\left(1 - \frac{z}{L}\right)\left(A_1 + A_2\frac{z}{L}\right) \\ v &= B_0 \sin\frac{\pi z}{L} + \frac{z}{L}\left(1 - \frac{z}{L}\right)\left(B_1 + B_2\frac{z}{L}\right) \\ \theta &= C_0 \sin\frac{\pi z}{L} + \frac{z}{L}\left(1 - \frac{z}{L}\right)\left(C_1 + C_2\frac{z}{L}\right) \end{aligned} \tag{4.36}$$

in which A_i, B_i, C_i $(i = 0, 1, 2)$ are constants to be determined from the governing equations and boundary conditions. These deflection functions satisfy the geometric (simply supported) boundary conditions except for the conditions of warping deformations,

$$u = 0, \quad v = 0, \quad \theta = 0 \qquad \text{at} \qquad z = 0 \qquad \text{and} \qquad L \tag{4.37}$$

The warping restraint factors, $\gamma_{\omega o}$ and $\gamma_{\omega l}$, at the ends of the beam-column are introduced by

$$\gamma_{\omega o} = \frac{C_2 - C_1}{C_2 + \pi C_0}$$
$$\gamma_{\omega l} = \frac{2C_2 + C_1}{C_2 - \pi C_0} \tag{4.38}$$

so that the warping boundary conditions may be satisfied at both ends:

$$\gamma_\omega = 0 \qquad \text{implies} \qquad \theta'' = 0 \qquad \text{(warping free)}$$
$$\gamma_\omega = 1 \qquad \text{implies} \qquad \theta' = 0 \qquad \text{(warping restrained)} \tag{4.39}$$

From Eq. (4.38), the constants, C_1 and C_2 are solved in terms of C_0,

$$C_1 = -\pi\gamma_1 C_0$$
$$C_2 = \pi\gamma_2 C_0 \tag{4.40}$$

in which

$$\gamma_1 = \frac{2\gamma_{\omega o} + \gamma_{\omega l} - 2\gamma_{\omega o}\gamma_{\omega l}}{3 - \gamma_{\omega o} - \gamma_{\omega l}}$$
$$\gamma_2 = \frac{\gamma_{\omega o} - \gamma_{\omega l}}{3 - \gamma_{\omega o} - \gamma_{\omega l}} \tag{4.41}$$

Substituting the deflection functions, Eqs. (4.36), into the governing differential equations, Eq. (4.35), results in

$$\frac{EI_y}{L^2}\left[\pi^2 A_0 \sin\frac{\pi z}{L} - 2(A_2 - A_1) + 6A_2\frac{z}{L}\right]$$
$$- P\left[A_0 \sin\frac{\pi z}{L} + \frac{z}{L}\left(1 - \frac{z}{L}\right)\left(A_1 + A_2\frac{z}{L}\right)\right]$$
$$- M_x\left[C_0 \sin\frac{\pi z}{L} + \frac{z}{L}\left(1 - \frac{z}{L}\right)\left(C_1 + C_2\frac{z}{L}\right)\right] = -M_y$$

$$\frac{EI_x}{L^2}\left[\pi^2 B_0 \sin\frac{\pi z}{L} - 2(B_2 - B_1) + 6B_2\frac{z}{L}\right]$$
$$- P\left[B_0 \sin\frac{\pi z}{L} + \frac{z}{L}\left(1 - \frac{z}{L}\right)\left(B_1 + B_2\frac{z}{L}\right)\right]$$
$$- M_y\left[C_0 \sin\frac{\pi z}{L} + \frac{z}{L}\left(1 - \frac{z}{L}\right)\left(C_1 + C_2\frac{z}{L}\right)\right] = M_x$$

$$
\begin{aligned}
&\frac{EI_\omega}{L^2}\left[\pi^3 C_0 \cos\frac{\pi z}{L} + 6C_2\right] \\
&\quad + (GK_T + \bar{K})\left[\pi C_0 \cos\frac{\pi z}{L} + C_1 + 2(C_2 - C_1)\frac{z}{L} - 3C_2\left(\frac{z}{L}\right)^2\right] \\
&\quad - M_x\left[\pi A_0 \cos\frac{\pi z}{L} + A_1 + 2(A_2 - A_1)\frac{z}{L} - 3A_2\left(\frac{z}{L}\right)^2\right] \\
&\quad - M_y\left[\pi B_0 \cos\frac{\pi z}{L} + B_1 + 2(B_2 - B_1)\frac{z}{L} - 3B_2\left(\frac{z}{L}\right)^2\right] \\
&\quad + (M_{yl} - M_{yo})\left[B_0 \sin\frac{\pi z}{L} + \frac{z}{L}\left(1 - \frac{z}{L}\right)\left(B_1 + B_2\frac{z}{L}\right)\right] \\
&\quad + (M_{xl} - M_{xo})\left[A_0 \sin\frac{\pi z}{L} + \frac{z}{L}\left(1 - \frac{z}{L}\right)\left(A_1 + A_2\frac{z}{L}\right)\right] = 0
\end{aligned}
\tag{4.42}
$$

Applying the static boundary conditions

$$
\begin{aligned}
M_x &= M_{xo}, \quad M_y = M_{yo} \quad \text{at} \quad z = 0 \\
M_x &= M_{xl}, \quad M_y = M_{yl} \quad \text{at} \quad z = L
\end{aligned}
\tag{4.43}
$$

to the first two equations in Eqs. (4.42), the four constants are determined:

$$
\begin{aligned}
A_1 &= \frac{-L^2}{6EI_y}(M_{yl} + 2M_{yo}) \\
A_2 &= \frac{-L^2}{6EI_y}(M_{yl} - M_{yo}) \\
B_1 &= \frac{L^2}{6EI_x}(M_{xl} + 2M_{xo}) \\
B_2 &= \frac{L^2}{6EI_x}(M_{xl} - M_{xo})
\end{aligned}
\tag{4.44}
$$

Evaluation of the last equation in Eqs. (4.42) at the ends, $z = 0$ and $z = L$, gives the following relation considering the condition (4.32),

$$
\begin{aligned}
&(\pi A_0 + A_1)M_{xo} + (\pi A_0 + A_1 + A_2)M_{xl} + (\pi B_0 + B_1)M_{yo} \\
&\quad + (\pi B_0 + B_1 + B_2)M_{yl} - 2\pi^3 C_0 \frac{EI_\omega}{L^2} \\
&\quad - (2\pi C_0 + 2C_1 + C_2)(GK_T + \bar{K}) = 0
\end{aligned}
\tag{4.45}
$$

In order to determine A_0, B_0, and C_0, evaluate the first two equations of (4.42) at the midspan of the beam-column ($z = L/2$) and get

$$
\begin{aligned}
&\frac{\pi^2 EI_y}{L^2}\left(A_0 + \frac{2}{\pi^2}A_1 + \frac{1}{\pi^2}A_2\right) - P\left(A_0 + \frac{1}{4}A_1 + \frac{1}{8}A_2\right) \\
&\qquad - \bar{M}_x\left(C_0 + \frac{1}{4}C_1 + \frac{1}{8}C_2\right) = -\bar{M}_y \\
&\frac{\pi^2 EI_x}{L^2}\left(B_0 + \frac{2}{\pi^2}B_1 + \frac{1}{\pi^2}B_2\right) - P\left(B_0 + \frac{1}{4}B_1 + \frac{1}{8}B_2\right) \\
&\qquad - \bar{M}_y\left(C_0 + \frac{1}{4}C_1 + \frac{1}{8}C_2\right) = \bar{M}_x
\end{aligned}
\tag{4.46}
$$

in which

$$
\begin{aligned}
\bar{M}_x &= \frac{1}{2}(M_{xo} + M_{xl}) \\
\bar{M}_y &= \frac{1}{2}(M_{yo} + M_{yl})
\end{aligned}
\tag{4.47}
$$

Using Eqs. (4.40) and (4.44), (4.46) together with (4.45) gives the three simultaneous equations in terms of the three unknowns, A_0, B_0, and C_0 which can be used to evaluate these constants. The simultaneous equations for A_0, B_0, and C_0 can be expressed in the matrix form as

$$
[K]\{U\} = \{F\} \tag{4.48}
$$

in which

$$
[K] = \begin{bmatrix}
\dfrac{\pi^2 EI_y}{L^2} - P & 0 & -\bar{M}_x \\
0 & \dfrac{\pi^2 EI_x}{L^2} - P & -\bar{M}_y \\
-\bar{M}_x & -\bar{M}_y & \dfrac{\pi^2 EI_\omega}{L^2\gamma_c} - [Pr_0^2 - GK_T]\gamma
\end{bmatrix}
\tag{4.49}
$$

$$
\{U\} = \begin{Bmatrix} A_0 \\ B_0 \\ C_0\gamma_c \end{Bmatrix} \tag{4.50}
$$

$$
\{F\} = \begin{Bmatrix}
-\dfrac{PL^2}{8EI_y}\bar{M}_y \\
\dfrac{PL^2}{8EI_x}\bar{M}_x \\
\dfrac{L^2}{12\pi E}\left(\dfrac{1}{I_x} - \dfrac{1}{I_y}\right)(4\bar{M}_x\bar{M}_y + M_{xo}M_{yo} + M_{xl}M_{yl})
\end{Bmatrix}
\tag{4.51}
$$

and

$$\gamma_c = 1 + \frac{\pi}{4}\gamma_1 + \frac{\pi}{8}\gamma_2$$
$$\gamma = \left(1 + \gamma_1 + \frac{1}{2}\gamma_2\right) \Big/ \gamma_c \tag{4.52}$$

From Eqs. (4.49) to (4.51), all information about elastic stability of a beam-column is obtained. First, when the beam-column is a centrally loaded column, $\{F\} = 0$, the simultaneous equations, Eq. (4.48), become homogeneous as

$$\begin{bmatrix} P_y - P & 0 & 0 \\ 0 & P_x - P & 0 \\ 0 & 0 & r_0^2(P_z - P)\gamma \end{bmatrix} \begin{Bmatrix} A_0 \\ B_0 \\ C_o\gamma_c \end{Bmatrix} = \begin{Bmatrix} 0 \\ 0 \\ 0 \end{Bmatrix} \tag{4.53}$$

In order that a nontrivial solution for A_0, B_0, and C_0 exists, det $[K] = 0$, from which three stability limit loads are solved:

$$P_y = \frac{\pi^2 EI_y}{L^2}$$
$$P_x = \frac{\pi^2 EI_x}{L^2} \tag{4.54}$$
$$P_z = \frac{1}{r_0^2}\left(\frac{\pi^2 EI_\omega}{L^2\gamma_c r} + GK_T\right)$$

The first two equations in Eq. (4.54) give flexural buckling loads about the weak axis and the strong axis of the cross section, respectively. The last equation gives torsional buckling load. If end warpings are not restrained ($\gamma_1 = \gamma_2 = 0$), then this torsion buckling load coincides with the well-known expression such as Timoshenko's and Gere's solution (1961). Lateral torsional buckling can also be examined. Assuming in-plane bending moment about x-axis only (M_{xo}, M_{xl}), Eq. (4.48) yields

$$\begin{bmatrix} P_y - P & 0 & -\bar{M}_x \\ 0 & P_x - P & 0 \\ -\bar{M}_x & 0 & r_0^2(P_z - P)\gamma \end{bmatrix} \begin{Bmatrix} A_0 \\ B_0 \\ C_0\gamma_c \end{Bmatrix} = \begin{Bmatrix} 0 \\ \dfrac{\pi^2}{8}\dfrac{P}{P_x}\bar{M}_x \\ 0 \end{Bmatrix} \tag{4.55a}$$

or

$$B_0 = \frac{\pi^2}{8}\frac{P\bar{M}_x}{P_x(P_x - P)} \tag{4.55b}$$

and

$$\begin{bmatrix} P_y - P & -\bar{M}_x \\ -\bar{M}_x & r_0^2(P_z - P)\gamma \end{bmatrix} \begin{Bmatrix} A_0 \\ C_0\gamma_c \end{Bmatrix} = \begin{Bmatrix} 0 \\ 0 \end{Bmatrix} \tag{4.56}$$

Equating the determinant of the coefficient matrix [Eq. (4.56)] to zero, the lateral torsional buckling condition is obtained as

$$(P_y - P)(P_z - P) = \left(\frac{\bar{M}_x}{r_0}\right)^2 \frac{1}{\gamma} \tag{4.57}$$

Solving for P, the lateral torsional buckling load is obtained

$$P_{cr} = \frac{1}{2}\left\{(P_y + P_z) \pm \sqrt{(P_y - P_z)^2 + 4\frac{1}{\gamma r_0^2}\bar{M}_x^2}\right\} \tag{4.58}$$

This is a similar expression to that of Eq. (4.29d) if the axes x and y are interchanged, because previously y axis was considered to be the axis of symmetry.

When the beam-column is subjected to biaxial bending as well as axial thrust, Eqs. (4.48) are not homogeneous; the beam-column must be treated as a load-deflection problem. The solutions for this problem exist only when

$$\det[K] \geq 0 \tag{4.59}$$

which leads to the condition

$$r_0^2\gamma(P_z - P) \geq \frac{\bar{M}_y^2}{P_x - P} + \frac{\bar{M}_x^2}{P_y - P} \tag{4.60}$$

This criterion can be checked conveniently by defining the stability function

$$F_{st} = \frac{\dfrac{\bar{M}_y^2}{P_x - P} + \dfrac{\bar{M}_x^2}{P_y - P}}{r_0^2\gamma(P_z - P)} \tag{4.61}$$

in which

$$\begin{array}{llll} F_{st} < 1 & \text{implies} & \det[K] > 0: & \text{stable} \\ F_{st} = 1 & \text{implies} & \det[K] = 0: & \text{stability limit} \\ F_{st} > 1 & \text{implies} & \det[K] < 0: & \text{unstable} \end{array} \tag{4.62}$$

when the beam-column is stable ($F_{st} < 1$), Eq. (4.48) can be solved for A_0, B_0 and C_0 as

$$\begin{aligned} A_0 &= -\frac{1}{P_y - P}\left(\frac{\pi^2}{8}\frac{P}{P_y}\bar{M}_y - C_0\gamma_c\bar{M}_x\right) \\ B_0 &= \frac{1}{P_x - P}\left(\frac{\pi^2}{8}\frac{P}{P_x}\bar{M}_x + C_0\gamma_c\bar{M}_y\right) \\ C_0 &= \frac{\pi}{12\gamma_c}\left(\frac{F_{st}}{1 - F_{st}}\right) \\ &\times \frac{\left(\dfrac{1}{P_x} - \dfrac{1}{P_y}\right)(4\bar{M}_x\bar{M}_y + M_{xo}M_{yo} + M_{xl}M_{yl}) - \dfrac{3\pi}{2}\left(\dfrac{P/P_y}{P_y - P} - \dfrac{P/P_x}{P_x - P}\right)\bar{M}_x\bar{M}_y}{\dfrac{\bar{M}_x^2}{P_y - P} + \dfrac{\bar{M}_y^2}{P_x - P}} \end{aligned} \tag{4.63}$$

Since all the deflection functions have been determined, the elastic column has been solved under biaxial loading. The moments at an arbitrary section can now be computed from Eqs. (4.34) and (4.36).

$$M_\xi = P_x\left[B_0 \sin\frac{\pi z}{L} + \frac{2}{\pi^2}(B_1 - B_2) + \frac{6}{\pi^2}B_2\frac{z}{L}\right]$$

$$M_\eta = -P_y\left[A_0 \sin\frac{\pi z}{L} + \frac{2}{\pi^2}(A_1 - A_2) + \frac{6}{\pi^2}A_2\frac{z}{L}\right]$$

$$M_\zeta = \frac{\pi}{L}\left(\frac{\pi^2 EI_\omega}{L^2} + GK_T + \bar{K}\right)C_0 \cos\frac{\pi z}{L} \qquad (4.64)$$

$$+ \frac{1}{L}(GK_T + \bar{K})C_1\left(1 - 2\frac{z}{L}\right)$$

$$+ \frac{1}{L}\left[\frac{6EI_\omega}{L^2} + (GK_T + \bar{K})\left(2\frac{z}{L} - 3\frac{z^2}{L^2}\right)\right]C_2$$

These expressions for moments are useful for determining the ultimate strength of a beam-column and for finding the location of the critical section as will be discussed later in Chap. 7.

4.5 BEAM-COLUMNS WITH INITIAL IMPERFECTIONS

In the preceding sections, the differential equations governing the elastic response of biaxially loaded beam-columns are formulated on the assumption that the beam-column is perfectly straight and free from residual stress prior to load application. The influence of initial deflections, twist and residual stress on the biaxial bending behavior of beam-column is examined here. The discussion is limited to beam-columns under symmetric loading conditions. The differential equations for a biaxially loaded beam-column of wide-flange cross section will first be presented. These equations will then be extended to the general thin-walled cross-sectional shape for which the shear center and the centroid are not coincident.

Figure 4.8 shows the cross section of the wide-flange shape which is initially deflected and twisted by the amount u_0, v_0 and θ_0. The quantities u_0, v_0 and θ_0 vary along the length of the beam-column. The principal axes of the initially displaced cross sections are denoted by the x' and y' axes. The z' axis is tangent to the displaced center line of the beam-column. A residual stress distribution over the cross section is also shown in Fig. 4.8, which is assumed to be in the direction of the z' axis.

Under the application of the load P, the beam-column will deflect and twist. The additional deflections and twist are denoted by u, v and θ; u and v are the displacement in the x' and y' directions and θ is the twist about z' axis. The ξ, η and ζ axes are used to denote the displaced x'-, y'- and z'-axes. These displacements are shown in Fig. 4.8. All these displacements are assumed to be small.

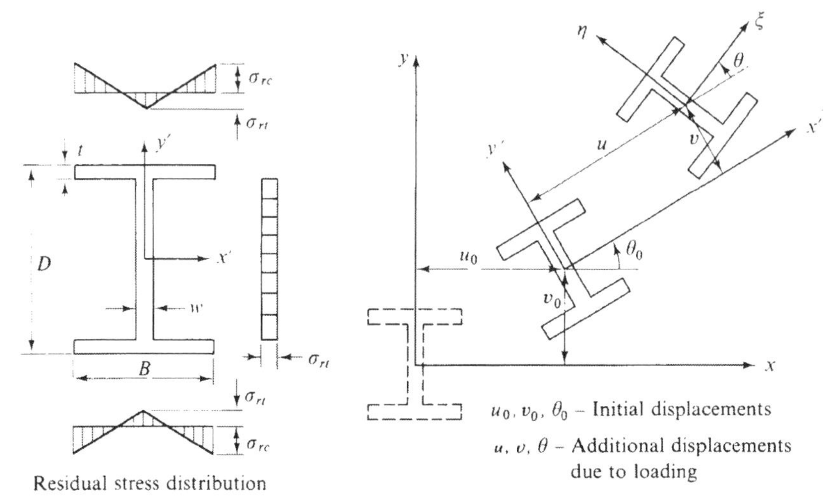

FIGURE 4.8
Displacements in plane of cross section and assumed distribution of residual stress

Equating the external moments at a section z produced by the end loads [see Eqs. (4.33)]

$$
\begin{aligned}
M_{\xi(\text{ext})} &= M_x + P(v_0 + v) + M_y(\theta_0 + \theta) \\
M_{\eta(\text{ext})} &= M_y - P(u_0 + u) - M_x(\theta_0 + \theta) \\
M_{\zeta(\text{ext})} &= M_x(u_0' + u') + M_y(v_0' + v) - \bar{K}(\theta_0' + \theta')
\end{aligned}
\tag{4.65}
$$

to the internal moments [see Eqs. (4.34)], the governing differential equations are obtained,

$$
\begin{aligned}
EI_y u'' + P(u_0 + u) + M_x(\theta_0 + \theta) &= M_y \\
EI_x v'' + P(v_0 + v) + M_y(\theta_0 + \theta) &= -M_x \\
EI_\omega \theta''' - (GK_T + \bar{\bar{K}})\theta' + M_x(u_0' + u') + M_y(v_0' + v') &= \bar{K}\theta_0'
\end{aligned}
\tag{4.66}
$$

in which $\bar{\bar{K}}$ is defined as

$$
\begin{aligned}
\bar{\bar{K}} &= \int (\sigma + \sigma_r)(x^2 + y^2)\, dA \\
&= \int \sigma(x^2 + y^2)\, dA + \int \sigma_r(x^2 + y^2)\, dA \\
&= \bar{K} + \bar{B}
\end{aligned}
\tag{4.67}
$$

where the normal stress σ is produced by the applied load P and the residual stress existed in the cross section is denoted by σ_r. The first and second integrals in the right hand side of Eq. (4.67) are denoted by $\bar{K}$ and $\bar{B}$, respectively. The term $\bar{B}$ in this equation represents the contribution of the residual stresses to the internal twist moment. The value of $\bar{K}$ for a general thin-walled cross section is defined in

Eq. (4.28). For the special case of a wide-flange shape of cross section $\bar{K}$ is equal to $-P(I_x + I_y)/A$ where $(I_x + I_y)$ is the polar moment of inertia and A is the area of the cross section. Using for σ_r the distribution shown in Fig. 4.8, the integral defined for $\bar{B}$ becomes

$$\bar{B} = \int \sigma_r(x^2 + y^2)\, dA$$
$$= \frac{w(D-2t)^3}{12}|\sigma_{rt}| - \frac{tB^3}{8}\left(|\sigma_{rc}| - \frac{|\sigma_{rt}|}{3}\right) - \frac{(D-t)^2 tB}{4}(|\sigma_{rc}| - |\sigma_{rt}|) \tag{4.68}$$

in which $|\sigma_{rt}|$ and $|\sigma_{rc}|$ denote the absolute values of the maximum tensile and compressive residual stresses, respectively. Expression (4.68) was first obtained by Fukumoto and Galambos (1966).

It should be noted that the displacements u and v in Eq. (4.66) are measured in the directions of x' and y' axes, respectively (Fig. 4.8). M_x and M_y at a section z are the bending moments produced by a compressive load P with end eccentricities e_x and e_y with respect to the undeformed straight bar, i.e., $M_x = -Pe_y$ and $M_y = Pe_x$.

Equation (4.66) may be extended to the case of the biaxially loaded beam-column of general cross sectional shape under symmetric loading conditions. Proceeding as before and noting that the beam-column will twist about an axis through the shear center we find [Eq. (2.174)]

$$\begin{aligned} &EI_y u'' + P[u_0 + u + y_0(\theta_0 + \theta)] + M_x(\theta_0 + \theta) = M_y \\ &EI_x v'' + P[v_0 + v - x_0(\theta_0 + \theta)] + M_y(\theta_0 + \theta) = -M_x \\ &EI_\omega \theta''' - (GK_T + \bar{K} + \bar{B})\theta' + M_x(u_0' + u') + M_y(v_0' + v') \\ &\qquad + Py_0(u_0' + u') - Px_0(v_0' + v') = \bar{K}\theta_0' \end{aligned} \tag{4.69}$$

The last two terms in the left hand side of the third equation represent the contribution of axial load P which acts in the centroid of the cross section. The load P has components $P(u_0' + u')$ and $P(v_0' + v')$ acting perpendicular to the ζ-axis which produce the twisting moments $P(u_0' + u')y_0$ and $-P(v_0' + v')x_0$.

For a biaxially loaded beam-column, the bending moments M_x and M_y at a section z have the values $-Pe_y$ and Pe_x, respectively, it follows that Eqs. (4.69) may be rewritten in the form

$$\begin{aligned} &EI_y u'' + P[u - (e_y - y_0)(\theta_0 + \theta)] = P(e_x - u_0) \\ &EI_x v'' + P[v + (e_x - x_0)(\theta_0 + \theta)] = P(e_y - v_0) \\ &EI_\omega \theta''' - (GK_T + \bar{K} + \bar{B})\theta' + P[(e_x - x_0)(v_0' + v') \\ &\qquad - (e_y - y_0)(u_0' + u')] = \bar{K}\theta_0' \end{aligned} \tag{4.70}$$

If the residual stresses $\bar{B}$, and the initial deflections u_0 and v_0, and twist θ_0 are assumed to be zero, Eqs. (4.70) are exactly the same as those derived by Goodier

(1942) for a perfect straight beam-column. Goodier's equations were solved exactly by Culver (1966a) and approximately by Thürlimann (1953), Dabrowski (1961), and Prawel and Lee (1964) for symmetrically loaded beam-columns with simply supported boundary conditions at both ends. More general boundary conditions were treated by Horne (1954), Trahair (1969), Syal and Sharma (1970, 1971), Tebedge and Tall (1973), Vinnakota and Aysto (1974) and Vinnakota and Aoshima (1974). For the special case of only initially twisted beam-columns, Eqs. (4.70) are exactly the same as those derived by Thurlimann (1953). The differential equations for an initially twisted beam-column of *finite* magnitude were given by Zickel (1956). Zickel's equations are nonlinear and no general solution has been obtained. Thürlimann (1953) modified Zickel's equations and obtained an approximate solution for the case of small initial twist. The present form of Eqs. (4.70) was derived by Culver (1966b). An exact solution of Eqs. (4.70) was also obtained by Culver (1966b).

Before proceeding with the solution of Eqs. (4.70), it is necessary to know the initial deflected shape of the beam-column (u_0, v_0, θ_0). In the following the initial deflections of the member are assumed to be

$$\begin{aligned} u_0 &= B_0 \sin \frac{\pi z}{L} \\ v_0 &= C_0 \sin \frac{\pi z}{L} \\ \theta_0 &= D_0 \sin \frac{\pi z}{L} \end{aligned} \tag{4.71}$$

which satisfy the simply supported boundary conditions at both ends $z = 0, L$

$$u_0 = v_0 = \theta_0 = u_0'' = v_0'' = \theta_0'' = 0 \tag{4.72}$$

The boundary conditions to be considered in the following problem are

$$\begin{aligned} &u = v = \theta = \theta'' = 0 \\ &u'' = \frac{Pe_x}{EI_y}, \quad v'' = \frac{Pe_y}{EI_x} \end{aligned} \tag{4.73}$$

for both ends $z = 0$ and L. $\theta'' = 0$ implies that warping is unrestrained at both ends.

Proceeding as in Sec. 4.2, differentiating the first two equations of (4.70) twice and the third one once and substituting the initial deflections (4.71) to these fourth order differential equations, the following equations are obtained

$$\begin{aligned} &EI_y u^{\text{IV}} + P[u'' - (e_y - y_0)\theta''] = P[B_0 - (e_y - y_0)D_0]\frac{\pi^2}{L^2} \sin \frac{\pi z}{L} \\ &EI_x v^{\text{IV}} + P[v'' + (e_x - x_0)\theta''] = P[C_0 + (e_x - x_0)D_0]\frac{\pi^2}{L^2} \sin \frac{\pi z}{L} \\ &EI_\omega \theta^{\text{IV}} - [GK_T + \bar{K} + \bar{B}]\theta'' + P[(e_x - x_0)v'' - (e_y - y_0)u''] \\ &\quad = P[(e_x - x_0)C_0 - (e_y - y_0)B_0 + r_0^2 D_0]\frac{\pi^2}{L^2} \sin \frac{\pi z}{L} \end{aligned} \tag{4.74}$$

This is a system of three linear inhomogeneous differential equations with the three unknowns u, v, and θ. The general solution of Eqs. (4.74) consists of a particular solution and a complementary solution. The complementary solution is obtained from the homogeneous part of Eqs. (4.74), which is exactly the same as Eqs. (4.6) with the exception of the added term $\bar{B}$. These homogeneous equations comprise the mathematical model for a biaxially loaded beam-column in which the initial imperfections, u_0, v_0 and θ_0 are zero which have been solved exactly in Sec. 4.2. Proceeding as previously, the complementary solution for Eqs. (4.74) are given by the previous Eqs. (4.9) and Eqs. (4.15) [or Eqs. (4.17)].

Using the method of undetermined coefficients, the particular solution for Eqs. (4.74) is obtained by assuming the deflected shapes of the member to be the form

$$\begin{aligned} u_p &= C_7 \sin\frac{\pi z}{L} + C_8 \cos\frac{\pi z}{L} \\ v_p &= C_9 \sin\frac{\pi z}{L} + C_{10} \cos\frac{\pi z}{L} \\ \theta_p &= C_{11} \sin\frac{\pi z}{L} + C_{12} \cos\frac{\pi z}{L} \end{aligned} \tag{4.75}$$

Inserting Eq. (4.75) in Eqs. (4.74) and equating the coefficients of similar terms in these equations gives the following values for the constants $C_7 \ldots C_{12}$:

$$C_8 = C_{10} = C_{12} = 0 \tag{4.76}$$

$$C_7 = \frac{1}{\Delta}\begin{vmatrix} B_0 - e_y D_0 & 0 & e_y \\ C_0 + e_x D_0 & \dfrac{P_x}{P} - 1 & -e_x \\ r_0^2 D_0 + e_x C_0 - e_y B_0 & -e_x & r_0^2\left(\dfrac{P_T}{P} + \dfrac{\bar{B}}{\bar{K}} - 1\right) \end{vmatrix} \tag{4.77}$$

$$C_9 = \frac{1}{\Delta}\begin{vmatrix} \dfrac{P_y}{P} - 1 & B_0 - e_y D_0 & e_y \\ 0 & C_0 + e_x D_0 & -e_x \\ e_y & r_0^2 D_0 + e_x C_0 - e_y B_0 & r_0^2\left(\dfrac{P_T}{P} + \dfrac{\bar{B}}{\bar{K}} - 1\right) \end{vmatrix} \tag{4.78}$$

$$C_{11} = \frac{1}{\Delta}\begin{vmatrix} \dfrac{P_y}{P} - 1 & 0 & B_0 - e_y D_0 \\ 0 & \dfrac{P_x}{P} - 1 & C_0 + e_x D_0 \\ e_y & -e_x & r_0^2 D_0 + e_x C_0 - e_y B_0 \end{vmatrix} \tag{4.79}$$

in which

$$\Delta = \begin{vmatrix} \dfrac{P_y}{P} - 1 & 0 & e_y \\ 0 & \dfrac{P_x}{P} - 1 & -e_x \\ e_y & -e_x & r_0^2\left(\dfrac{P_T}{P} + \dfrac{\bar{B}}{\bar{K}} - 1\right) \end{vmatrix} \tag{4.80}$$

and

$$P_T = \frac{1}{r_0^2}\left(\frac{\pi^2 EI_\omega}{L^2} + GK_T\right) \tag{4.81}$$

which corresponds to a special case of P_z defined in Eqs. (4.54) when warping is unrestrained, $\gamma_1 = \gamma_2 = 1$. P_x and P_y are the same as defined by Eqs. (4.54).

The complete solution of Eqs. (4.74) is given by Eqs. (4.9), (4.15) of Sect. 4.2, together with Eqs. (4.75). The twelve integration constants

$$J_1, J_2, J_3, K_1, K_2, K_3, C_1, C_2, C_3, C_4, C_5, C_6$$

may be determined from the boundary conditions in Eqs. (4.73).

Since the constants C_8, C_{10}, C_{12} are zero [see Eq. (4.76)], the complete solution of Eqs. (4.74) may be written in the form

$$\begin{aligned} u &= u_R + C_7 \sin\frac{\pi z}{L} \\ v &= v_R + C_9 \sin\frac{\pi z}{L} \\ \theta &= \theta_R + C_{11} \sin\frac{\pi z}{L} \end{aligned} \tag{4.82}$$

From Eqs. (4.77), (4.78) and (4.79), we know that C_7, C_9 and C_{11} are dependent only on the initial deflections B_0, C_0, and D_0 [Eq. (4.71)]. The second term in each of Eqs. (4.82), therefore, represents the effect of initial deflections on the behavior of biaxially loaded beam-columns. The deflections u_R, v_R and θ_R in the first term in each of Eqs. (4.82) represent the beam-column deflections with the term $\bar{B}$ included which reflects the effect of residual stress on the behavior of biaxially loaded beam-columns in which the initial imperfections, u_0, v_0 and θ_0 are zero.

The solution described above was evaluated numerically by Culver (1966b) for the exactly same problems as given in Sect. 4.2 [Eq. (4.20)] with the only exception of the term $\bar{B}$. The term $\bar{B}$ for the present problems was calculated from Eq. (4.68) with

$$|\sigma_{rc}| = 0.39\sigma_y \qquad \text{and} \qquad \sigma_y = 33 \text{ ksi } (228 \text{ MN/m}^2)$$

which results in $\bar{B} = 99.6$ kip-in (11.3 kN-m).

It is found that the increase in the deflections u_R, v_R and θ_R as a result of the residual stress was less than 1 percent for both cases. The effect of residual stress on the *elastic* behavior of biaxially loaded beam-column is therefore negligible. This fact may also be observed from the coefficient of θ'' in the third equation of (4.74) where, for all practical beam-columns, the quantity $\bar{B}$ is much smaller than that of $GK_T + \bar{K}$. The residual stress will, of course, have a significant effect on decreasing the elastic limit of the beam-column beyond which the elastic analysis becomes invalid.

Numerical results for initially *curved* and *twisted* beam-columns with no residual stress $\bar{B} = 0$ were also solved in order to compare the numerical results with those obtained previously in Sect. 4.2 for a perfectly straight beam-column. The values used here for the initial deflections B_0 and C_0 are the maximum values that can occur in wide-flange shapes rolled to the tolerance of existing ASTM Specifications (Standard A6-64, 1965), which result in

$$B_0 = C_0 = -0.118 \text{ in } (-0.3 \text{ cm}) \qquad \text{for} \qquad L/r_y = 60$$

and

$$B_0 = C_0 = -0.276 \text{ in } (-0.7 \text{ cm}) \qquad \text{for} \qquad L/r_y = 140$$

Because ASTM Standards (1965) do not specify tolerance for initial twist, values of

$$D_0 = \frac{1}{500}\left(\frac{L}{D}\right) = \begin{cases} +0.0173 \text{ radians for } L/r_y = 60 \\ +0.0403 \text{ radians for } L/r_y = 140 \end{cases}$$

were used.

The numerical results obtained indicate that the initial imperfections result in a significant increase in member deflections and rotations when compared with those of a perfect beam-column solution. The influence of these imperfections on the percentage increase of the total normal stress σ_T in the beam-column is found to be essentially constant throughout the entire range of loading. At the center line of the member, the percentage increases of σ_T of the imperfect member over that of the perfect member are found to be 4.3 percent for $L/r_y = 60$ and 8.7 percent for $L/r_y = 140$.

4.6 FINITE DEFORMATION OF BIAXIALLY LOADED BEAM-COLUMNS

As so far discussed, if the deformations of a beam-column are small, *linear* ordinary differential equations can be formulated (*second-order analysis*). If the displacements become large, however, or if post-buckling behavior is important, higher order terms must be included in the derivation resulting in *nonlinear* differential equations (*third-order analysis*).

The beam-column is shown in Fig. 4.7 which is subjected to unequal end moments M_{xo}, M_{yo}, M_{xl}, M_{yl} about the principal centroidal axes and an axial thrust P

at the centroid. A general thin-walled open cross section with the shear center (x_0, y_0) is considered here.

At an arbitrary section z, the bending moments M_ξ, M_η and the twisting moment M_ζ about the principal axes ξ and η of the section produced by the external applied loads are obtained from Eq. (2.162) with the rotation matrix of Eq. (2.153),

$$
\begin{aligned}
M_{\xi(\text{ext})} &= \frac{1+\varepsilon_0}{\sqrt{1+(u_c')^2}}\left[M_{xo} + \frac{z}{L}(M_{xl} - M_{xo}) + Pv_c\right]\cos\theta \\
&\quad + \left[M_{yo} + \frac{z}{L}(M_{yl} - M_{yo}) - Pu_c\right]\sin\theta \\
&\quad - \frac{1}{\sqrt{1+(u_c')^2}}\left[M_z - \frac{v_c}{L}(M_{yl} - M_{yo}) - \frac{u_c}{L}(M_{xl} - M_{xo})\right]u_c' \\
M_{\eta(\text{ext})} &= \frac{1+\varepsilon_0}{\sqrt{1+(v_c')^2}}\left[M_{yo} + \frac{z}{L}(M_{yl} - M_{yo}) - Pu_c\right]\cos\theta \\
&\quad - \left[M_{xo} + \frac{z}{L}(M_{xl} - M_{xo}) + Pv_c\right]\sin\theta \\
&\quad - \frac{1}{\sqrt{1+(v_c')^2}}\left[M_z - \frac{v_c}{L}(M_{yl} - M_{yo}) - \frac{u_c}{L}(M_{xl} - M_{xo})\right]v_c' \\
M_{\zeta(\text{ext})} &= \frac{1}{\sqrt{1+(u_c')^2}}\left[M_{xo} + \frac{z}{L}(M_{xl} - M_{xo}) + Pv_c\right]u_c' \\
&\quad + \frac{1}{\sqrt{1+(v_c')^2}}\left[M_{yo} + \frac{z}{L}(M_{yl} - M_{yo}) - Pu_c\right]v_c' \\
&\quad + \frac{(1+\varepsilon_0)^2}{\sqrt{1+(u_c')^2}\sqrt{1+(v_c')^2}}\left[M_z - \frac{v_c}{L}(M_{yl} - M_{yo}) - \frac{u_c}{L}(M_{xl} - M_{xo})\right]
\end{aligned}
\tag{4.83}
$$

These are the external moments about the centroidal axes ξ, η and ζ expressed in terms of the displacements of the centroid, u_c, v_c and θ.

The internal moments are obtained from Eq. (2.143) and Eq. (2.144) using the curvatures about ξ, η and ζ axes,

$$
\begin{aligned}
M_{\xi(\text{int})} &= EI_x\theta_\xi' \\
M_{\eta(\text{int})} &= EI_y\theta_\eta' \\
M_{\zeta(\text{int})} &= -EI_\omega\theta_\zeta''' + (GK_T + \overline{K})\theta_\zeta'
\end{aligned}
\tag{4.84}
$$

These curvatures are transformed again, making use of the rotation matrix in general expression, Eq. (2.153),

$$\begin{Bmatrix} \theta'_\xi \\ \theta'_\eta \\ \theta'_\zeta \end{Bmatrix} = \begin{bmatrix} \dfrac{1+\varepsilon_0}{\sqrt{1+(u'_c)^2}}\cos\theta & \sin\theta & -\dfrac{1}{\sqrt{1+(u'_c)^2}}u'_c \\ -\sin\theta & \dfrac{1+\varepsilon_0}{\sqrt{1+(v'_c)^2}}\cos\theta & -\dfrac{1}{\sqrt{1+(v'_c)^2}}v'_c \\ \dfrac{1}{\sqrt{1+(u'_c)^2}}u'_c & \dfrac{1}{\sqrt{1+(v'_c)^2}}v'_c & \dfrac{(1+\varepsilon_0)^2}{\sqrt{1+(v'_c)^2}\sqrt{1+(u'_c)^2}} \end{bmatrix} \begin{Bmatrix} \theta'_x \\ \theta'_y \\ \theta'_z \end{Bmatrix} \tag{4.85}$$

in which the curvatures in the global coordinates are given using the deflections,

$$\begin{aligned} \theta'_x &= \Phi_x = \frac{-v''_c}{\left[\sqrt{1+(v'_c)^2}\right]^3} \\ \theta'_y &= \Phi_y = \frac{u''_c}{\left[\sqrt{1+(u'_c)^2}\right]^3} \\ \theta'_z &= \theta' \end{aligned} \tag{4.86}$$

Combining Eq. (4.84), Eq. (4.85), and Eq. (4.86) together, the internal moments are obtained

$$\begin{aligned} M_{\xi(\text{int})} &= EI_x\left[-\frac{1+\varepsilon_0}{\sqrt{1+(u'_c)^2}}\frac{v''_c}{\left[\sqrt{1+(v'_c)^2}\right]^3}\cos\theta \right. \\ &\quad \left. + \frac{u''_c}{\left[\sqrt{1+(u'_c)^2}\right]^3}\sin\theta - \frac{1}{\sqrt{1+(u'_c)^2}}u'_c\theta'\right] \\ M_{\eta(\text{int})} &= EI_y\left[\frac{1+\varepsilon_0}{\sqrt{1+(v'_c)^2}}\frac{u''_c}{\left[\sqrt{1+(u'_c)^2}\right]^3}\cos\theta \right. \\ &\quad \left. + \frac{v''_c}{\left[\sqrt{1+(v'_c)^2}\right]^3}\sin\theta - \frac{1}{\sqrt{1+(v'_c)^2}}v'_c\theta'\right] \\ M_{\zeta(\text{int})} &= -EI_\omega\left[\frac{(1+\varepsilon_0)^2}{\sqrt{1+(u'_c)^2}\sqrt{1+(v'_c)^2}}\theta''' - \frac{1}{\sqrt{1+(u'_c)^2}\left[\sqrt{1+(v'_c)^2}\right]^3}(u'_c v''_c)'' \right. \\ &\quad \left. + \frac{1}{\sqrt{1+(v'_c)^2}\left[\sqrt{1+(u'_c)^2}\right]^3}(u''_c v'_c)''\right] \\ &\quad + (GK_T + \bar{K})\left[\frac{(1+\varepsilon_0)^2}{\sqrt{1+(u'_c)^2}\sqrt{1+(v'_c)^2}}\theta' - \frac{1}{\sqrt{1+(u'_c)^2}\left[\sqrt{1+(v'_c)^2}\right]^3}u'_c v''_c \right. \\ &\quad \left. + \frac{1}{\sqrt{1+(v'_c)^2}\left[\sqrt{1+(u'_c)^2}\right]^3}u''_c v'_c\right] \end{aligned} \tag{4.87}$$

In the derivation of $M_{\zeta(\text{int})}$ above, derivatives of $\sqrt{1+(u'_c)^2}$ and $\sqrt{1+(v'_c)^2}$ are small and neglected. Since the external twisting moment $M_{\zeta(\text{ext})}$ is written about the centroid while the internal twisting moment $M_{\zeta(\text{int})}$ is about the shear center, then

$M_{\zeta(\text{ext})}$ is rewritten to be an external twisting moment about the shear center by substituting

$$u_c + x_0 \quad \text{for} \quad u_c \quad \text{and} \quad v_c + y_0 \quad \text{for} \quad v_c$$

Thus, one obtains

$$\begin{aligned} M_{\zeta(\text{ext})} = {} & \frac{1}{\sqrt{1+(u_c')^2}}\left[M_{xo} + \frac{z}{L}(M_{xl} - M_{xo}) + P(v_c + y_0)\right]u_c' \\ & + \frac{1}{\sqrt{1+(v_c')^2}}\left[M_{yo} + \frac{z}{L}(M_{yl} - M_{yo}) - P(u_c + x_0)\right]v_c' \\ & + \frac{(1+\varepsilon_0)^2}{\sqrt{1+(u_c')^2}\sqrt{1+(v_c')^2}}\left[M_z - \frac{M_{yl} - M_{yo}}{L}(v_c + y_0) - \frac{M_{xl} - M_{xo}}{L}(u_c + x_0)\right] \end{aligned} \tag{4.88}$$

Before we equate the external moments to the internal moments, the displacements of the centroid u_c and v_c are transformed into the displacements of the shear center u and v using,

$$\begin{aligned} u_c &= u + x_0\left(1 - \frac{\cos\theta}{\sqrt{1+(u_c')^2}}\right) + y_0\frac{\sin\theta}{\sqrt{1+(v_c')^2}} \approx u + y_0\theta \\ v_c &= v + y_0\left(1 - \frac{\cos\theta}{\sqrt{1+(v_c')^2}}\right) - x_0\frac{\sin\theta}{\sqrt{1+(u_c')^2}} \approx v - x_0\theta \end{aligned} \tag{4.89}$$

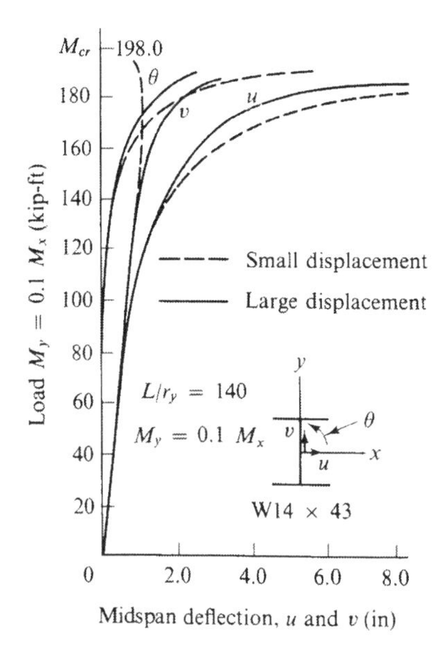

FIGURE 4.9
Deformation of beam due to biaxial bending (Soltis and Christiano, 1972) (1 in = 2.54 cm, 1 kip = 4.45 kN)

and using the following approximations

$$\sqrt{1+(u_c')^2} \approx \sqrt{1+(u')^2} \qquad \text{and} \qquad \sqrt{1+(v_c')^2} \approx \sqrt{1+(v')^2}$$

we obtain the equilibrium equations of a space beam-column in the finite deformation analysis as

$$\begin{aligned}
&EI_x\left[-\frac{\cos\theta(1+\varepsilon_0)}{\left[\sqrt{1+(v')^2}\right]^3}(v''-\theta''x_0)+\frac{\sin\theta}{[1+(u')^2]}(u''+\theta''y_0)-(u'+\theta'y_0)\theta'\right]\\
&\quad=\left[M_{xo}+\frac{z}{L}(M_{xl}-M_{xo})+P(v-\theta x_0)\right]\cos\theta(1+\varepsilon_0)\\
&\qquad+\sqrt{1+(u')^2}\left[M_{yo}+\frac{z}{L}(M_{yl}-M_{yo})-P(u+\theta y_0)\right]\sin\theta\\
&\qquad-\left[M_z-\frac{1}{L}(v-\theta x_0)(M_{yl}-M_{yo})-\frac{1}{L}(u+\theta y_0)(M_{xl}-M_{xo})\right](u'+\theta y_0)\\
&EI_y\left[\frac{\cos\theta(1+\varepsilon_0)}{\left[\sqrt{1+(u')^2}\right]^3}(u''+\theta''y_0)+\frac{\sin\theta}{[1+(v')^2]}(v''-\theta''x_0)-(v'-\theta'x_0)\theta'\right]\\
&\quad=\left[M_{yo}+\frac{z}{L}(M_{yl}-M_{yo})-P(u+\theta y_0)\right]\cos\theta(1+\varepsilon_0)\\
&\qquad-\sqrt{1+(v')^2}\left[M_{xo}+\frac{z}{L}(M_{xl}-M_{xo})+P(v-\theta x_0)\right]\sin\theta\\
&\qquad-\left[M_z-\frac{1}{L}(v-\theta x_0)(M_{yl}-M_{yo})-\frac{1}{L}(u+\theta y_0)(M_{xl}-M_{xo})\right](v'-\theta x_0)\\
&-EI_\omega\left[(1+\varepsilon_0)^2\theta'''-\frac{1}{1+(v')^2}[(u'+\theta'y_0)(v''-\theta''x_0)]''\right.\\
&\qquad\left.+\frac{1}{1+(u')^2}[(u''+\theta''y_0)(v'-\theta'x_0)]''\right]+(GK_T+K)\left[(1+\varepsilon_0)^2\theta'\right.\\
&\qquad\left.-\frac{1}{1+(v')^2}(u'+\theta'y_0)(v''-\theta''x_0)+\frac{1}{1+(u')^2}(u''+\theta''y_0)(v'-\theta'x_0)\right]\\
&\quad=\sqrt{1+(v')^2}\left[M_{xo}+\frac{z}{L}(M_{xl}-M_{xo})+P(v-\theta x_0+y_0)\right](u'+\theta'y_0)\\
&\qquad+\sqrt{1+(u')^2}\left[M_{yo}+\frac{z}{L}(M_{yl}-M_{yo})-P(u+\theta y_0+x_0)\right](v'-\theta'x_0)\\
&\qquad+(1+\varepsilon_0)^2\left[M_z-\frac{1}{L}(v-\theta x_0+y_0)(M_{yl}-M_{yo})\right.\\
&\qquad\left.-\frac{1}{L}(u+\theta y_0+x_0)(M_{xl}-M_{xo})\right]
\end{aligned} \tag{4.90}$$

The equations for second-order analysis (4.6) can be obtained as a special case of Eq. (4.90) by setting all the nonlinear terms equal to zero, such as

$$(u')^2 = (v')^2 = u''\theta = v''\theta = 0$$

and by assuming

$$\tan\theta = \theta, \quad \cos\theta = 1, \quad \varepsilon_o \ll 1 \tag{4.91}$$

This linearization process is equivalent to the assumption that the deformations of the beam-column are small, so that their products and products of their derivatives are negligible.

Soltis and Christiano (1972) solved the highly nonlinear differential equations similar to that of Eqs. (4.90) by replacing them with a set of finite-difference equations and using Newton-Raphson iteration method. Near the bifurcation point, a functional iteration method is used, because it is more reliable than the Newton-Raphson method when deformations become large.

Numerical results are reported for a W14 × 43 beam-column having slenderness ratio $L/r_y = 140$. Simply supported, free to warp boundary conditions are adopted for the cases considered, i.e.,

$$u = v = \theta = \theta'' = 0 \quad \text{at} \quad z = 0 \quad \text{and} \quad z = L \tag{4.92}$$

A comparison of second- and third-order theories is made in Figs. 4.9 and 4.10. Figure 4.9 shows the end moment mid-span displacement curves for the beam-column subjected to equal end moments $M_y = 0.1M_x$ with $P = 0$. Differences between the two theories are seen virtually indistinguishable when the load is less than 0.7 M_{cr}.

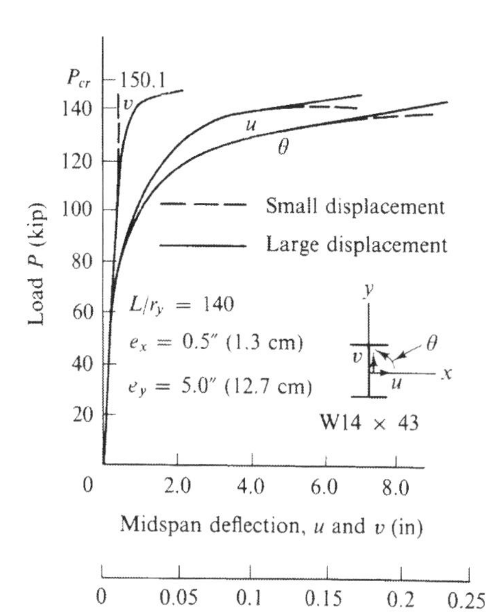

FIGURE 4.10
Deformation of beam-column due to eccentric loading (Soltis and Christiano, 1972) (1 in = 2.54 cm, 1 kip = 4.45 kN)

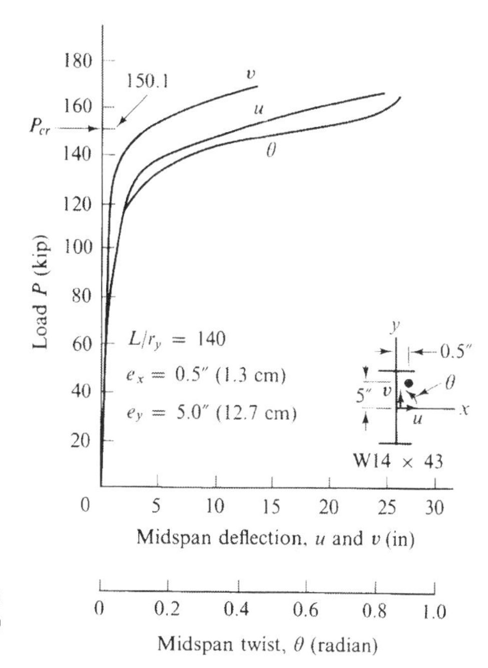

FIGURE 4.11
Deformation of beam-column beyond critical load (Soltis and Christiano, 1972) (1 in = 2.54 cm, 1 kip = 4.45 kN)

As the load exceeds approximately 0.8 M_{cr}, however, the difference becomes significant. It is seen that the second-order theory tends to overestimate the displacement in the weak axis direction u, and twist θ, while the displacement in the strong axis direction v is greatly underestimated. In Fig. 4.10, an axial thrust P is applied with biaxial eccentricity of $e_x = 0.5$ in and $e_y = 5.0$ in (12.7 cm). A similar trend is apparent for the case of the eccentrically loaded beam-column (Fig. 4.10).

In order to examine the post-buckling behavior of the elastic beam-column, the load is increased up to 1.1 P_{cr} and the results of Fig. 4.10 are replotted in Fig. 4.11. It is seen that the beam-column still has some stiffness after buckling provided it remains within the elastic range of the material, as assumed in the theory.

4.7 SUMMARY

The differential equations of biaxially loaded beam-columns considering the effects of large deflection and including the effects of higher order terms due to axial loads and torsional resistance (third-order analysis) are presented in Eq. (4.90) in which the exact expressions for curvature, terms related to axial shortening caused by torsion, the direction-cosines between deformed and undeformed coordinate systems, and products of derivatives are all included in the formulation. For small deflection analysis, we consider only second-order terms such that $\varepsilon_0 \ll 1$ and all products of deformations and products of their derivatives are negligible, and furthermore, the twist angle θ is considered small such that $\sin\theta = \theta$, $\cos\theta = 1$, and $\tan\theta = \theta$. Equations (4.90) then reduce to the form [Eq. (2.174)]

$$\begin{gathered}
EI_y u'' + P(u + y_0\theta) - \left[M_{yo} + \frac{z}{L}(M_{yl} - M_{yo})\right] \\
+ \left[M_{xo} + \frac{z}{L}(M_{xl} - M_{xo})\right]\theta + v'M_z = 0 \\
EI_x v'' + P(v - x_0\theta) + \left[M_{xo} + \frac{z}{L}(M_{xl} - M_{xo})\right] \\
+ \left[M_{yo} + \frac{z}{L}(M_{yl} - M_{yo})\right]\theta - u'M_z = 0 \\
EI_\omega\theta''' - (GK_T + \bar{K})\theta' + Py_0u' - Px_0v' + u'\left[M_{xo} + \frac{z}{L}(M_{xl} - M_{xo})\right] \\
+ v'\left[M_{yo} + \frac{z}{L}(M_{yl} - M_{yo})\right] - \frac{v}{L}(M_{yl} - M_{yo}) - \frac{u}{L}(M_{xl} - M_{xo}) + M_z = 0
\end{gathered} \tag{4.93}$$

In which it was assumed that $(u + \theta y_0)' = u'$ and $(v - \theta x_0)' = v'$. If all moments are the same at both ends,

$$\begin{aligned}
M_{xo} &= M_{xl} = \bar{M}_x \\
M_{yo} &= M_{yl} = \bar{M}_y \\
M_z &= \bar{M}_z
\end{aligned} \tag{4.94}$$

then Eqs. (4.93) can be further reduced to the form

$$\begin{aligned}
&EI_y u'' + P(u + y_0\theta) - \bar{M}_y + \bar{M}_x\theta + v'\bar{M}_z = 0 \\
&EI_x v'' + P(v - x_0\theta) + \bar{M}_x + \bar{M}_y\theta - u'\bar{M}_z = 0 \\
&EI_\omega\theta''' - (GK_T + \bar{K})\theta' + u'(\bar{M}_x + Py_0) + v'(\bar{M}_y - Px_0) + \bar{M}_z = 0
\end{aligned} \tag{4.95}$$

Finally, if the cross section is of a doubly symmetric shape, the shear center coincides with the centroid,

$$x_0 = y_0 = 0 \tag{4.96}$$

then, Eqs. (4.95) reduce to the simple form

$$\begin{aligned}
&EI_y u'' + Pu + \bar{M}_x\theta + v'\bar{M}_z = \bar{M}_y \\
&EI_x v'' + Pv + \bar{M}_y\theta - u'\bar{M}_z = -\bar{M}_x \\
&EI_\omega\theta''' - (GK_T + \bar{K})\theta' + u'\bar{M}_x + v'\bar{M}_y = -\bar{M}_z
\end{aligned} \tag{4.97}$$

The exact solutions for these differential equations are possible only for very simple cases and more general cases require a numerical integration. An approximate solution for the simplest case, Eq. (4.97), with no twisting moment and warping free end conditions can be obtained by assuming the displacement functions of the form

$$u = A \sin\frac{\pi z}{L} - \frac{\bar{M}_y}{2EI_y} z(L - z)$$
$$v = B \sin\frac{\pi z}{L} + \frac{\bar{M}_x}{2EI_x} z(L - z) \tag{4.98}$$
$$\theta = C \sin\frac{\pi z}{L}$$

which satisfy all the simply supported boundary conditions at both ends $z = 0$ and $z = L$,

$$\begin{aligned} &u = 0, \qquad v = 0, \qquad \theta = 0: \quad \text{geometric} \\ &EI_y u'' = \bar{M}_y, \quad EI_x v'' = -\bar{M}_x, \quad \theta'' = 0: \quad \text{static} \end{aligned} \tag{4.99}$$

Inserting the displacement functions (4.98) in Eqs. (4.97) and evaluating the appropriate quantities at the mid-span ($z = L/2$) of the member result in

$$\begin{bmatrix} P_y - P & 0 & -\bar{M}_x \\ 0 & P_x - P & -\bar{M}_y \\ -\bar{M}_x & -\bar{M}_y & r_0^2(P_z - P) \end{bmatrix} \begin{Bmatrix} A \\ B \\ C \end{Bmatrix} = \begin{Bmatrix} -\dfrac{\pi^2}{8}\dfrac{P}{P_y}\bar{M}_y \\ \dfrac{\pi^2}{8}\dfrac{P}{P_x}\bar{M}_x \\ \dfrac{\pi}{2}\left(\dfrac{1}{P_x} - \dfrac{1}{P_y}\right)\bar{M}_x\bar{M}_y \end{Bmatrix} \tag{4.100}$$

in which the characteristic values of the coefficient matrix, the elastic buckling loads, are defined by

$$P_x = \frac{\pi^2 EI_x}{L^2}: \quad \text{strong-axis flexural buckling load}$$
$$P_y = \frac{\pi^2 EI_y}{L^2}: \quad \text{weak-axis flexural buckling load} \tag{4.101}$$
$$P_z = \left(\frac{\pi^2 EI_\omega}{L^2} + GK_T\right)\Big/ r_0^2: \quad \text{torsional buckling load}$$

and the following notation has been used,

$$\bar{K} = \int \sigma(x^2 + y^2)\, dA = -\frac{P}{A}(I_x + I_y) = -Pr_0^2 \tag{4.102}$$

The coefficients of the displacement functions, Eq. (4.98), are solved from Eq. (4.100),

$$A = -\left(\frac{\pi^2}{8}\frac{P}{P_y}\bar{M}_y - \bar{M}_x C\right) \Big/ (P_y - P)$$

$$B = \left(\frac{\pi^2}{8}\frac{P}{P_x}\bar{M}_x + \bar{M}_y C\right) \Big/ (P_x - P)$$

$$C = \frac{\pi^2}{8}\frac{\bar{M}_x \bar{M}_y}{P_x P_y} P \frac{\dfrac{P_x}{P_y - P} - \dfrac{P_y}{P_x - P} - \dfrac{4}{\pi}\dfrac{P_y - P_x}{P}}{\dfrac{\bar{M}_x^2}{P_y - P} + \dfrac{\bar{M}_y^2}{P_x - P} - r_0^2(P_z - P)} \tag{4.103}$$

Thus, the maximum displacements at the mid-span of the member are obtained as

$$u_{max} = A - \frac{L^2}{8EI_y}\bar{M}_y = -\frac{1}{P_y - P}\left(\frac{\pi^2}{8}\bar{M}_y - C\bar{M}_x\right)$$

$$v_{max} = B + \frac{L^2}{8EI_x}\bar{M}_x = \frac{1}{P_x - P}\left(\frac{\pi^2}{8}\bar{M}_x + C\bar{M}_y\right) \tag{4.104}$$

$$\theta_{max} = C$$

The differential equations of an initially bent (u_0, v_0) and twisted (θ_0) beam-column with residual stress σ_r (or $\bar{B}$) subjected to a symmetric compressive force P with end eccentricities e_x and e_y are

$$EI_y u'' + P[u - (e_y - y_0)(\theta_0 + \theta)] = P(e_x - u_0)$$

$$EI_x v'' + P[v + (e_x - x_0)(\theta_0 + \theta)] = P(e_y - v_0) \tag{4.105}$$

$$EI_\omega \theta''' - (GK_T + \bar{K} + \bar{B})\theta' + P[(e_x - x_0)(v_0' + v') - (e_y - y_0)(u_0' + u')] = -Pr_0^2\theta_0'$$

in which

$$\bar{K} = -Pr_0^2$$

$$r_0^2 = x_0^2 + y_0^2 + \frac{1}{A}(I_x + I_y) + \beta_x e_x + \beta_y e_y$$

$$\beta_x = \frac{1}{I_x}\left(\int y^3\, dA + \int x^2 y\, dA\right) - 2y_0 \tag{4.106}$$

$$\beta_y = \frac{1}{I_y}\left(\int x^3\, dA + \int y^2 x\, dA\right) - 2x_0$$

$$\bar{B} = \int \sigma_r (x^2 + y^2)\, dA$$

If the residual stresses, σ_r and the initial deflections, u_0 and v_0, and the initial twist θ_0, are all assumed to be zero, Eqs. (4.105) are exactly the same as those, Eqs. (4.95), for a perfectly straight beam-column which is free of residual stress.

Assuming the initial imperfections of the beam-column axis as the half wave of a sine curve, the exact solution of Eqs. (4.105) for a pin-ended member can be obtained in closed form. The general solution consists of a particular solution and a complementary solution. The particular solution represents the effect of initial deflections on the behavior of biaxially loaded beam-columns. The complementary solution, resulting from the homogeneous part of Eqs. (4.105), represents the deflections of a perfect straight beam-column which includes the effect of residual stress. For a wide-flange shape of cross section, the effect of this residual stress on the elastic behavior of a biaxially loaded beam-column is found negligible. The initial imperfections, however, may substantially reduce the limit of elastic behavior of the beam-column.

4.8 PROBLEMS

4.1 Show that symmetric loading of a cylindrical tubular beam-column does not produce torsional deformation.

4.2 Examine that in-plane bending moments can produce out-of-plane deformation, for which torsional deformation is always accompanied.

4.3 Design a column of W14 × 43 section whose bending deformation about the weak axis is restrained and whose flexural buckling load about the strong axis coincides with its torsional buckling load. Determine the length and buckling load of the column.

4.9 REFERENCES

ASTM Standard A6-64, "General Requirements for Delivery of Rolled Steel Plates, Shapes, Sheet Piling and Bars for Structural Use," American Society for Testing and Materials, Philadelphia, Pa., January, (1965).

Chen, W. F. and Atsuta, T., "Ultimate Strength of Biaxially Loaded Steel H-Columns," *Journal of the Structural Division, ASCE*, Vol. 99, No. ST3, Proc. Paper 9613, March, pp. 469–489, (1973).

Culver, C. G., "Exact Solution of the Biaxial Bending Equations," *Journal of the Structural Division, ASCE*, Vol. 92, No. ST2 Proc. Paper 4772, April, pp. 63–83, (1966a).

Culver, C. G., "Initial Imperfections in Biaxial Bending," *Journal of the Structural Division, ASCE*, Vol. 92, No. ST3, Proc. Paper 4846, June, pp. 119–135, (1966b).

Dabrowski, R., "Dunnwanding Stäbe unter Zweiachsig Aussermittigen Druck," *Der Stahlbau*, Wilhelm Ernst & Sohn, Berlin-Wilmersdorf, Germany, December, (1961).

Fukumoto, Y. and Galambos, T. V., "Inelastic Lateral-Torsional Buckling of Beam-Columns," *Journal of the Structural Division, ASCE*, Vol. 92, No. ST2, Proc. Paper 4770, April, pp. 41–61, (1966).

Goodier, J. N., "Torsional and Flexural Buckling of Bars of Thin Walled Open Section Under Compressive and Bending Loads," *Journal of Applied Mechanics, Transactions, ASME*, Vol. 64, September, p. A-103, (1942).

Horne, M. R., "The Flexural-Torsional Buckling of Members of Symmetrical I-Section Under Combined Thrust and Unequal Terminal Moments," *Quarterly Journal of Mechanics and Applied Mathematics*, Vol. 7, Part 4, p. 410, (1954).

Peköz, T. B. and Celebi, N., "Torsional Flexural Buckling of Thin-Walled Sections Under Eccentric Load," *Cornell Engineering Research Bulletin* 69-1, Cornell University, Ithaca, New York, September, (1969).

Prawel, S. P. and Lee, G. C., "Biaxial Flexure of Columns by Analog Computers," *Journal of the Engineering Mechanics Division, ASCE*, Vol. 90, No. EM1, Proc. Paper 3805, February, pp. 83–111, (1964).

Soltis, L. A. and Christiano, P., "Finite Deformation of Biaxially Loaded Columns," *Journal of the Structural Division, ASCE*, Vol. 98, No. ST12, December, Proc. Paper 9407, pp. 2647–2662, (1972).

Syal, I. C. and Sharma, S. S., "Elastic Behavior of Biaxially Loaded Steel Columns," *Journal of the Structural Division, ASCE*, Vol. 96, No. ST3, Proc. Paper 7143, March, pp. 469–486, (1970).

Syal, I. C. and Sharma, S. S., "Biaxially Loaded Beam-Column Analysis," *Journal of the Structural Division, ASCE*, Vol. 97, No. ST9, Proc. Paper 8384, September, pp. 2245–2259, (1971).

Tebedge, N. and Tall, L., "Linear Stability Analysis of Beam-Columns," *Journal of the Structural Division, ASCE,* Vol. 99, No. ST12, Proc. Paper 10232, December, pp. 2439–2457, (1973).

Thürlimann, B., "Deformation of and Stresses in Initially Twisted Open Cross Section," *Report No. 3, Brown University,* Providence, Rhode Island, June, (1953).

Timoshenko, S. P., "Theory of Bending, Torsion and Buckling of Thin-Walled Members of Open Cross Section," *Journal of Franklin Institute,* Philadelphia, Pa., Vol. 239, No. 3, March, p. 201, No. 4, April, p. 249; No. 5, May, p. 343, (1945).

Timoshenko, S. P. and Gere, J. M., "Theory of Elastic Stability," McGraw-Hill Book Co., New York, (1961).

Trahair, N. S., "Restrained Elastic Beam-Columns," *Journal of the Structural Division, ASCE,* Vol. 95, No. ST12, Proc. Paper 6941, December, pp. 2641–2664, (1969).

Vinnakota, S. and Aoshima, Y., "Spatial Behavior of Rotationally and Directionally Restrained Beam-Columns," *Publications, International Association for Bridge and Structural Engineering,* Zurich, Vol. 34-II, pp. 169–194, (1974).

Vinnakota, S. and Aysto, P., "Inelastic Spatial Stability of Restrained Beam-Columns," *Journal of the Structural Division, ASCE,* Vol. 100, No. ST11, Proc. Paper 10 919, November, pp. 2235–2254, (1974).

Zickel, J., "Pretwisted Beams and Columns," *Journal of Applied Mechanics,* Vol. 23, (1956).

5

STRENGTH OF BEAM-COLUMN SEGMENTS

5.1 INTRODUCTION

In the following two chapters the *strength* and *behavior* of beam-column segments subjected to compression combined with biaxial bending are presented. As the name implies, here we are only concerned with *short* beam-columns for which the effect of lateral deflections on the magnitudes of bending moments is negligible. As a result, the maximum strength occurs when the entire cross section is fully plastic or yielded in the case of metal compression members or when the maximum strain (or stress) attains some prescribed value in the case of reinforced concrete members. This is known as *stress problem of first order*. Material yielding or failure is the primary cause of the strength limit of the member. This chapter presents methods of analysis in predicting this limiting strength, and simple formulas to enable designers to assess this strength. The elastic-plastic behavior of beam-column segments subjected to biaxial loading is the subject of the following chapter.

In determining maximum strength of beam-column segments subjected to axial load and biaxial bending, ideal nonlinear stress-strain relations for steel and/or concrete, conservation of entire plane sections or each of the thin-walled flat plates of which the section is composed are usually assumed. In the case of reinforced concrete

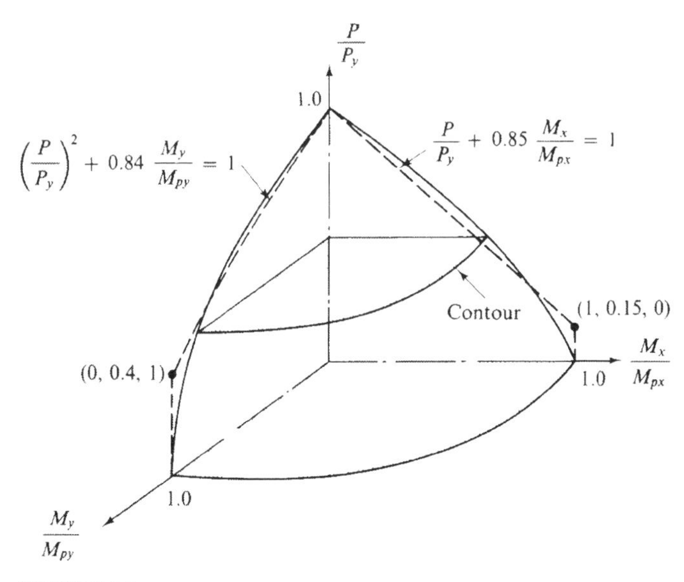

FIGURE 5.1
Interaction surface for steel H-section

or composite members, no slip and no tension resistance by concrete are further assumed.

When the position of the neutral axis is known or assumed, the magnitude of the resultant axial force P and bending moments M_x and M_y, which result in the fully plastification of the cross section or produce the prescribed limit strain of the material, can be determined directly using equations of equilibrium. When the position of the neutral axis is not known, that is, when the generalized stresses P, M_x and M_y are prescribed, the equations of equilibrium can be solved only by the method of successive approximations. All such procedures generally involve more or less tedious cycles of numerical calculations. These calculations result in a function which defines an interaction surface relating P, M_x and M_y as shown in Fig. 5.1.

Limit or interaction surface

This basic surface is an important concept in plastic theory, that of *limit yield surface* or simply *limit surface*. Characteristics of limit surfaces have been discussed in some detail in Chap. 2, Vol. 1, of the book. Figure 5.1 shows a typical maximum strength interaction limit surface for steel H-beam-column segments. A point in this three-dimensional generalized-stress space represents, for a given H-section having known values of yield load P_y and fully plastic moments M_{px} and M_{py} about strong and weak axes of bending of the cross section, respectively, a certain combination of biaxial bending moment and axial load. If the point lies within the *convex* surface, then the combination is one that can be carried by the segment. A point on the limit surface represents a combination of biaxial bending moment and axial load that just cause the section to become fully plastic or just reach the prescribed limit strain. A point outside the convex limit surface represents an impossible state. This chapter aims to establish interaction or limit surfaces for various types of beam-column cross section.

The shaded area in Fig. 5.1 attempts to represent, in two dimensions, what is essentially a three-dimensional surface describing the maximum strength of a beam-column segment subjected to axial load and biaxial bending moment. Here, the interaction surface is pictured graphically as a surface formed by a series of interaction curves drawn radially from the P-axis. For a given axial load P/P_y, the maximum biaxial bending resistance of a segment assumes the typical shape of contour of the surface shown in Fig. 5.1. The advantage of expressing the strength in terms of these contours of the surface is that design charts can readily be prepared. Furthermore, these interaction curves can be used as a basis for developing simple approximate formulas.

In Fig. 5.1, the solid lines on the mutually perpendicular planes represent the actual limit or failure curves under the relevant restricted loading conditions. The dotted lines represent previously established approximations. In particular the dotted lines on plane $P - M_x$ or $P - M_y$ are fully plastic moment acting at the so-called *plastic hinge* section of a beam considering the effect of axial load. These in-plane strength curves have been previously established. From these end points, expressions for biaxial moment interaction curves can be developed. This is described in the latter part of this chapter.

The two theorems of limit analysis, the upper-bound theorem and the lower-bound theorem are applicable here for generating interaction relations of beam-column segments. These limit theorems, as summarized briefly in Chap. 2, Vol. 1 of this book, permit lower and upper bounds on the limit or failure surface to be found conveniently when the materials are idealized as elastic perfectly plastic. An exact solution is obtained if the applications of both lower-bound method and upper-bound method result in an identical solution. The methods of limit analysis are powerful in that they provide an effective means in examining the accuracy of a solution. A comprehensive treatment of the subject of limit analysis of perfect plasticity as applied to metal, soil and concrete is given in a recent book by Chen (1975).

A brief historical sketch

Studies on the interaction relations of steel beam-column cross sections under biaxial loading have been made by several researchers. Ringo, McDonough and Baseheart (1973) treated wide-flange sections as composed of thin plate elements. In the analysis, the lower-bound technique of limit analysis is applied using a straight neutral axis passing through both upper and lower flanges. Design nomographs for many commonly used wide-flange shapes along with approximate interaction equations for some selected base sections are developed.

Morris and Fenves (1969) have developed interaction equations of rectangular section, wide-flange section and circular section considering most of the possible cases for the locations of neutral axis. The solution is essentially based on the lower-bound approach. They also consider the effect of St. Venant torsion shear stress on the interaction surfaces.

Santathadaporn and Chen (1970) have developed interaction equations of rectangular section and wide-flange sections applying the lower- and upper-bound techniques of limit analysis. In the lower-bound analysis, they started with the general assumption that the neutral axis is a curved line. In order to determine the shape of this neutral curved line, the variational calculus technique was utilized. In the upper-bound solution of wide-flange section, warping deformation was taken into consideration. Upper- and lower-bound methods were found in Sect. 5.2 to result in an almost identical solution.

All above mentioned solutions are given in somewhat lengthy analytical expressions in terms of different cases depending on the locations of the neutral axis. These solutions are all approximate in nature. Further, it is almost always impossible or at least impractical to cover all possible locations of neutral axis even for a particular shape of wide-flange section.

Treating a general thin-walled section to be a combination of rectangular elements, Chen and Atsuta (1972, 1974) have succeeded in demonstrating that the problem could be solved by *superposition*. By deriving a simple expression for rectangular area in terms of the neutral axis location, any commonly used shape could then be solved by adding and subtracting rectangular areas. The method is valid for both symmetric and unsymmetric sections. It covers all possible locations of neutral axis and results in an exact solution. Detail of the method is presented in Sect. 5.3 for symmetric cases and in Sect. 5.4 for unsymmetric cases. Interaction curves for circular, double-web, wide-flange, angle, tee and channel sections are presented in Sect. 5.5, and approximate interaction formulas for wide-flange sections are developed in Sect. 5.6.

Studies on the ultimate strength of reinforced concrete sections subjected to compression combined with biaxial bending have received considerable attention. The concept of using failure or limit surfaces has been presented by Bresler (1960). Some typical interaction curves for square sections along with a simple design formula are presented in Sect. 5.7.

5.2 APPROXIMATE METHOD OF ANALYSIS

In the following three sections, lower and upper bound theorems of limit analysis are applied to obtain interaction relationships relating axial force P and biaxial bending moments M_x and M_y acting in two perpendicular directions on a steel cross section under the condition that the entire section is fully plastic. Herein, the maximum safe domain and the minimum collapse domain are obtained approximately by use of the *variational calculus* and the minimum procedure, respectively. The technique is first applied to a rectangular section, and then extended for a wide-flange section. Exact methods of analysis using the concept of *superposition* are given in the following sections.

5.2.1 Interaction Equation of Rectangular Section

Consider a rectangular section of $b \times d$ which is fully yielded by tension $(+\sigma_y)$ or compression $(-\sigma_y)$ bordered by a neutral line $y = f(x)$ as shown in Fig. 5.2. The interaction relations among axial thrust P and bending moments M_x and M_y are uniquely determined, once the neutral line is correctly selected. The variational calculus method and the lower- and upper-bound theorems of limit analysis are applied herein for the determination of the shape of the neutral line $y = f(x)$ and the subsequent interaction relations (Santathadaporn and Chen, 1970).

Lower bound solution

The lower bound theorem of limit analysis states that the load computed on the basis of an assumed equilibrium state of stress distribution which does not violate the yield condition will be less than or at best equal to the true limit or ultimate load.

A system of stress distribution shown in Fig. 5.2 satisfies the yield condition and equilibrium for a rectangular section. This has simple tension above and simple compression below the neutral plane $y = f(x)$. The axial force P and the bending moments M_x and M_y are positive when the axial force causes tension and the bending moments produce tensile stress in the first quadrant of the coordinate system shown in Fig. 5.2. The equilibrium consideration then gives

$$
\begin{aligned}
P &= -\int_{-b/2}^{+b/2} 2\sigma_y f(x)\,dx \\
M_x &= \int_{-b/2}^{+b/2} \sigma_y\left[(d^2/4) - f^2(x)\right]dx \\
M_y &= -\int_{-b/2}^{+b/2} 2\sigma_y x f(x)\,dx
\end{aligned}
\tag{5.1}
$$

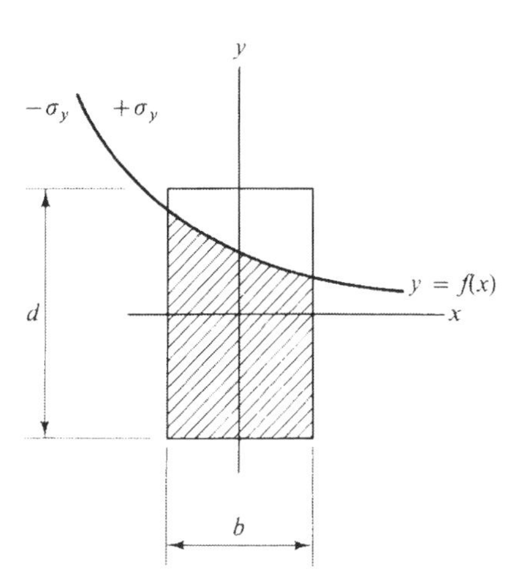

FIGURE 5.2
Rectangular section with neutral line

where $f(x)$ is as yet an unknown function. The problem of finding the interaction curve may now be stated as follows: Given a value of P and M_y, determine a function $f(x)$ so as to maximize M_x subjected to the constraint conditions P and M_y. With Lagrange's multiplier denoted by λ_1 and λ_2, one can write the functional to be optimized as

$$H = \sigma_y[(d^2/4) - f^2(x)] - \lambda_1[2\sigma_y x f(x)] - \lambda_2[2\sigma_y f(x)] \tag{5.2}$$

Since all integrands in P, M_x and M_y involve only $f(x)$ and none of its derivatives, the Euler differential equation will be finite rather than differential. With the equation

$$\frac{\partial H}{\partial f(x)} - \frac{d}{dx}\left[\frac{\partial H}{\partial f'(x)}\right] = 0 \tag{5.3}$$

it follows that

$$f(x) = -\lambda_1 x - \lambda_2 \tag{5.4}$$

The neutral axis therefore must be a straight line and the integrals for P, M_x and M_y can be evaluated

$$\begin{aligned} P &= 2\sigma_y b\lambda_2 \\ M_x &= 2\sigma_y\left[\frac{bd^2}{8} - \frac{\lambda_1^2}{3}\left(\frac{b}{2}\right)^3 - \lambda_2^2\left(\frac{b}{2}\right)\right] \\ M_y &= \frac{1}{6}\sigma_y b^3 \lambda_1 \end{aligned} \tag{5.5}$$

Eliminating λ_1 and λ_2 and introducing the dimensionless quantities

$$\begin{aligned} p &= \frac{P}{P_y} \\ m_x &= \frac{M_x}{M_{px}} \\ m_y &= \frac{M_y}{M_{py}} \end{aligned} \tag{5.6}$$

where

$$\begin{aligned} P_y &= \sigma_y\, bd \\ M_{px} &= \frac{1}{4}\sigma_y\, bd^2 \\ M_{py} &= \frac{1}{4}\sigma_y\, db^2 \end{aligned} \tag{5.7}$$

leads to the particularly simple interaction equation,

$$p^2 + m_x + \frac{3}{4}m_y^2 = 1 \tag{5.8}$$

Table 5.1 INTERACTION EQUATIONS FOR RECTANGULAR SECTION

Case	Location of N.A.	Equation	Valid for
1	y, x, N.A.	$p^2 + m_x + \frac{3}{4} m_y{}^2 = 1$	$m_x \geqslant \frac{2}{3}(1 - p)$ $m_y \leqslant \frac{2}{3}(1 - p)$
2	y, x, N.A.	$p^2 + \frac{3}{4} m_x^2 + m_y = 1$	$m_x \leqslant \frac{2}{3}(1 - p)$ $m_y \geqslant \frac{2}{3}(1 - p)$
3	y, x, N.A.	$p + \frac{9}{4}\left[1 - \frac{m_x}{2(1-p)}\right]\left[1 - \frac{m_y}{2(1-p)}\right] = 1$	$m_x \geqslant \frac{2}{3}(1 - p)$ $m_y \geqslant \frac{2}{3}(1 - p)$

valid for

$$m_y \leq \frac{2}{3}(1 - p) \leq m_x \tag{5.9}$$

Equation (5.8) is derived for the particular case the neutral axis passing through the two vertical sides of the rectangle (case 1 in Table 5.1). There are also two other possible locations of the neutral axis (case 2 and 3 in Table 5.1) depending upon the relative magnitude of p, m_x and m_y. Their interaction equations are summarized in Table 5.1. The interaction curves corresponding to Table 5.1 are shown in Fig. 5.3.

Upper bound solution

The upper bound theorem of limit analysis states that the load computed on the basis of an assumed fully plastic velocity field by equating of the external and internal rate of work for such a field will give an upper bound solution for the collapse or limit load.

In Fig. 5.4, $\dot{\varepsilon}_0$ is the assumed strain rate at the centroid of the section and $\dot{\kappa}_0$ is the assumed curvature rate. The generalized stresses which are associated with these strain and curvature rates are axial force P, and the bending moment M which can be resolved into two components of moment M_x and M_y about the principal axes. Here, it is assumed that the deformation pattern (velocity field) is a pure bending and

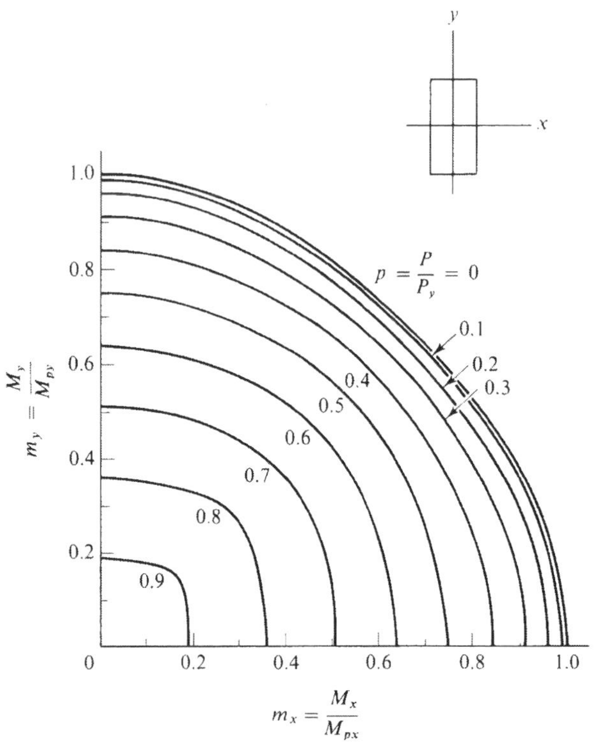

FIGURE 5.3
Interaction curves for rectangular section

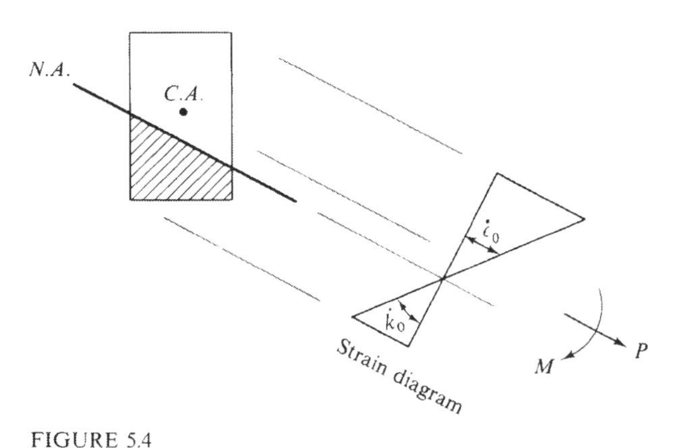

FIGURE 5.4
Strain distribution

hence specified completely by $\dot{\varepsilon}_0$ and $\dot{\kappa}_0$. The stress distribution associated with this pattern of deformation is exactly the same as the lower bound stress distribution. Therefore, the lower bound solution must, in fact, be the correct one to within the limits of the proposed theory.

5.2.2 Interaction Equation of Wide-Flange Section

The most popular structural shape is the wide-flange section, for which the interaction relations are obtained by the similar approach to that for rectangular section.

Lower bound solution

Guided by the results for a rectangular section, the neutral axis is assumed to be a straight line,

$$x = -\lambda_3 y - \lambda_4 \tag{5.10}$$

With the simplification of the stress distribution in the web as shown in Fig. 5.5., the following interaction equations are obtained

$$\begin{aligned} P &= 4\sigma_y t\lambda_3 y_1 + 2\sigma_y w y_1 \\ M_x &= \frac{4}{3}\sigma_y\left[\left(\frac{d}{2}\right)^3 - \left(\frac{d}{2} - t\right)^3\right]\lambda_3 + \sigma_y w\left[\left(\frac{d}{2} - t\right)^2 - y_1^2\right] \\ M_y &= 2\sigma_y\left\{\left(\frac{b}{2}\right)^2 t - \frac{\lambda_3^2}{3}\left[\left(\frac{d}{2}\right)^3 - \left(\frac{d}{2} - t\right)^3\right] - (\lambda_3 y_1)^2 t\right\} \end{aligned} \tag{5.11}$$

where

$$y_1 = \lambda_4/\lambda_3 \tag{5.12}$$

If dimensionless quantities are defined by Eqs. (5.6) with

$$\begin{aligned} P_y &= \sigma_y A = \sigma_y[2bt + w(d - 2t)] \\ M_{px} &= \sigma_y Z_x = \sigma_y\left[bt(d - t) + w\left(\frac{d}{2} - t\right)^2\right] \\ M_{py} &= \sigma_y Z_y = \sigma_y\left[\frac{tb^2}{2} + \frac{1}{4}w^2(d - 2t)\right] \end{aligned} \tag{5.13}$$

the interaction equations may finally be written in the form

$$\begin{aligned} p &= \frac{2}{A}[2t\lambda_3 y_1 + w y_1] \\ m_x &= \frac{1}{Z_x}\left\{\frac{4}{3}\left[\left(\frac{d}{2}\right)^3 - \left(\frac{d}{2} - t\right)^3\right]\lambda_3 + w\left[\left(\frac{d}{2} - t\right)^2 - y_1^2\right]\right\} \\ m_y &= \frac{2}{Z_y}\left\{\left(\frac{b}{2}\right)^2 t - \frac{1}{3}\left[\left(\frac{d}{2}\right)^3 - \left(\frac{d}{2} - t\right)^3\right]\lambda_3^2 - t(\lambda_3 y_1)^2\right\} \end{aligned} \tag{5.14}$$

valid for

$$\begin{aligned} &0 \le y_1 \le (d/2) - t \\ &\frac{w/2}{y_1 + (d/2) - t} \le \lambda_3 \le \frac{b/2}{(d/2) + y_1} \end{aligned} \tag{5.15}$$

Table 5.2 INTERACTION EQUATIONS FOR WIDE-FLANGE SECTIONS

Case	Location of N.A.	Equation	Valid for	
			Lower bound	Upper bound
1	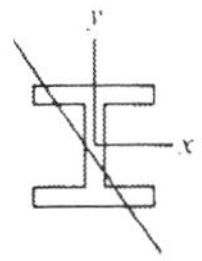	$p = \frac{2}{A}[2t\lambda_3 y_1 + wy_1]$ $m_x = \frac{1}{Z_x}\left\{\frac{4}{3}n_3\lambda_3 + w\left[\left(\frac{d}{2}-t\right)^2 - y_1^2\right]\right\}$ $m_y = \frac{2}{Z_y}\left[\frac{tb^2}{4} - \frac{1}{3}n_3\lambda_3^2 - t\lambda_3^2 y_1^2\right]$	$0 \leqslant y_1 \leqslant \frac{d}{2} - t$ $\frac{\frac{w}{2}}{y_1 + \frac{d}{2} - t} \leqslant \lambda_3 \leqslant \frac{\frac{b}{2}}{\frac{d}{2} + y_1}$	$0 \leqslant y_1 \leqslant \frac{d}{2} - t$ $\frac{\frac{w}{2}}{y_1 + \frac{d}{2} - t} \leqslant \lambda_3 \leqslant \frac{\frac{b}{2}}{\frac{d}{2} - \frac{t}{2} +}$
2	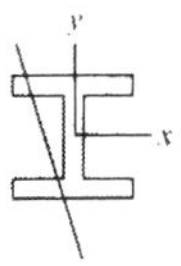	$p = \frac{2}{A}\left[2t\lambda_3 y_1 + w\left(\frac{d}{2} - t\right)\right]$ $m_x = \frac{4}{3}\frac{n_3\lambda_3}{Z_x}$ $m_y = \frac{2}{Z_y}\left[\frac{tb^2}{4} - \frac{1}{3}n_3\lambda_3^2 - t\lambda_3^2 y_1^2\right]$	$\frac{d}{2} - t \leqslant y_1$ $\frac{\frac{w}{2}}{y_1 + \frac{d}{2} - t} \leqslant \lambda_3 \leqslant \frac{\frac{b}{2}}{\frac{d}{2} + y_1}$	$\frac{d}{2} - \frac{t}{2} \leqslant y_1$ $\frac{\frac{w}{2}}{y_1 + \frac{d}{2} - t} \leqslant \lambda_3 \leqslant \frac{\frac{b}{2}}{\frac{d}{2} - \frac{t}{2} + y}$
3	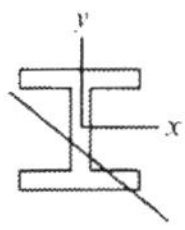	$p = \frac{1}{A}[bt - n_2\lambda_3 + 2t\lambda_3 y_1 + 2wy_1]$ $m_x = \frac{1}{Z_x}\left\{\frac{bt}{2}(d - t) + \frac{2}{3}n_3\lambda_3 - n_2\lambda_3 y_1 + w\left[\left(\frac{d}{2} - t\right)^2 - y_1^2\right]\right\}$ $m_y = \frac{1}{Z_y}\left[\frac{tb^2}{4} - \frac{1}{3}n_3\lambda_3^2 + n_2\lambda_3^2 y_1 - t\lambda_3^2 y_1^2\right]$	$0 \leqslant y_1 \leqslant \frac{d}{2} - t$ $\frac{\frac{b}{2}}{y_1 + \frac{d}{2} - t} \leqslant \lambda_3 \leqslant \frac{\frac{b}{2}}{\frac{d}{2} - y_1}$	$0 \leqslant y_1 \leqslant \frac{d}{2} - t$ $\frac{\frac{b}{2}}{y_1 + \frac{d}{2} - \frac{t}{2}} \leqslant \lambda_3 \leqslant \frac{\frac{b}{2}}{\frac{d}{2} - \frac{t}{2} - y}$
4	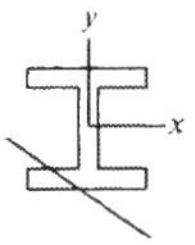	$p = \frac{1}{A}\left[bt - n_2\lambda_3 + 2t\lambda_3 y_1 + 2w\left(\frac{d}{2} - t\right)\right]$ $m_x = \frac{1}{Z_x}\left[\frac{bt}{2}(d - t) + \frac{2}{3}n_3\lambda_3 - n_2\lambda_3 y_1\right]$ $m_y = \frac{1}{Z_y}\left[\frac{tb^2}{4} - \frac{1}{3}n_3\lambda_3^2 + n_2\lambda_3^2 y_1 - t\lambda_3^2 y_1^2\right]$	$\frac{d}{2} - \frac{t}{2} \leqslant y_1$ $\frac{\frac{b}{2}}{y_1 + \frac{d}{2} - t} \leqslant \lambda_3 \leqslant \frac{\frac{b}{2}}{y_1 - \frac{d}{2} + t}$	$\frac{d}{2} - \frac{t}{2} \leqslant y_1$ $\frac{\frac{b}{2}}{y_1 + \frac{d}{2} - \frac{t}{2}} \leqslant \lambda_3 \leqslant \frac{\frac{b}{2}}{y_1 - \frac{d}{2} +}$
5a	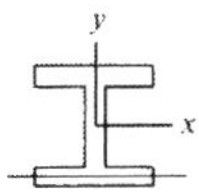	$m_y = 0$ $m_x = \frac{A}{2Z_x}\left[d(1 - p) - \frac{A}{2b}(1 - p)^2\right]$	$\frac{w(d - 2t)}{A} \leqslant p \leqslant 1$	
5b	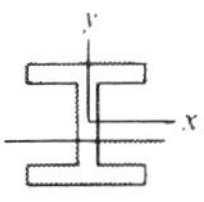	$m_y = 0$ $m_x = 1 \quad \frac{A^2p^2}{4wZ_x}$	$0 \leqslant p \leqslant \frac{w(d - 2t)}{A}$	
6a	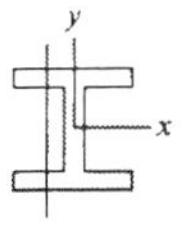	$m_x = 0$ $m_y = \frac{A}{2Z_y}\left[b(1 - p) - \frac{A}{4t}(1 - p)^2\right]$	$\frac{wd}{A} \leqslant p \leqslant 1$	
6b	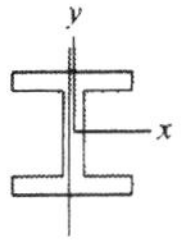	$m_x = 0$ $m_y = 1 - \frac{A^2p^2}{4dZ_y}$	$0 \leqslant p \leqslant \frac{wd}{A}$	

Remarks:

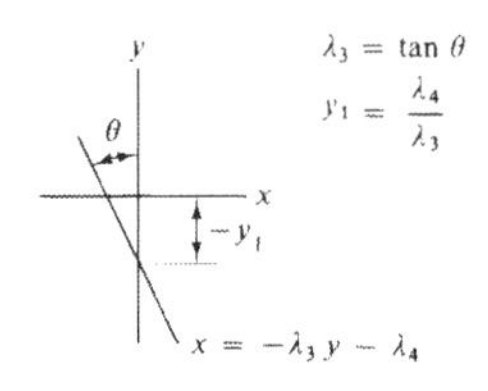

$$A = 2bt + w(d - 2t)$$

$$Z_x = bt(d - t) + w\left(\frac{d}{2} - t\right)^2$$

$$Z_y = \frac{1}{2}tb^2 + \frac{1}{4}w^2(d - 2t)$$

$$n_2 = \left(\frac{d}{2}\right)^2 - \left(\frac{d}{2} - t\right)^2$$

$$= t(d - t)$$

Lower bound

$$n_1 = \left(\frac{d}{2}\right)^3 - \left(\frac{d}{2} - t\right)^3$$

$$= \frac{3}{4}t(d - t)^2 + \frac{t^3}{4}$$

Upper bound

$$n_1 = \frac{3}{4}t(d - t)^2$$

Once the values of p and m_x are given, the values of λ_3 and y_1 which define the location of the neutral axis can be determined by the first two equations in Eq. (5.14). The value of m_y is then computed from the last equation in Eq. (5.14). The solution is valid for the case in which the neutral axis passes through the top flange, the web, and the bottom flange (Fig. 5.5). Different locations of neutral axis and their corresponding interaction equations are summarized in Table 5.2.

Upper bound solution

It is assumed that the deformation pattern is a pure bending for each of the three-thin-wall flat plates, that is, the plane cross sections remain plane after deformation for the two flanges as well as for the web (Fig. 5.6). They are assumed to behave independently and exert no restraint upon one another except the compatibility conditions at the junction between the flange and the web. Then if $\dot{\kappa}$ is the rate of curvature and $\dot{\varepsilon}$ the strain rate at the centroid for each of the three thin-wall plates and if the subscripts t, b and w denote the top flange, bottom flange and web respectively, the assumed deformation field can be specified completely by six variables with an $\dot{\varepsilon}$ and

FIGURE 5.5
Simplified stress distribution

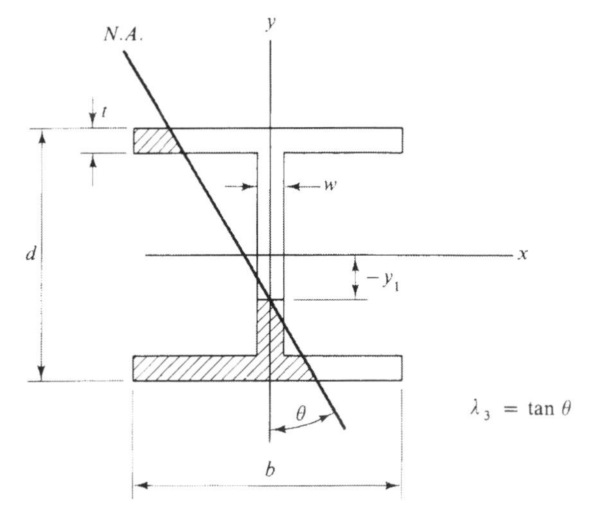

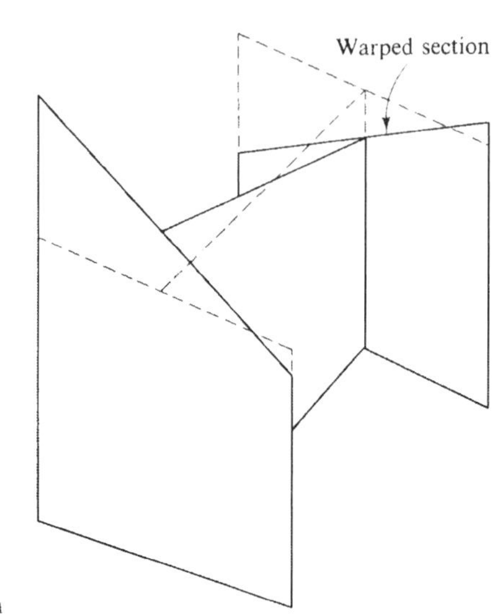

FIGURE 5.6
A failure mechanism with a warped section

FIGURE 5.7
Strain distribution

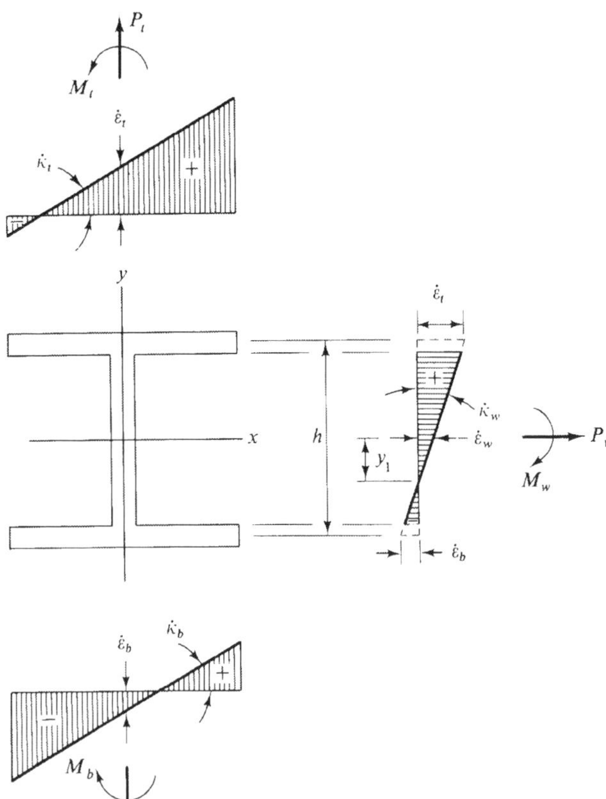

a $\dot{\kappa}$ for each of the three plates. The six variables are not completely independent and must satisfy the compatibility conditions (Fig. 5.7)

$$\begin{aligned} \dot{\varepsilon}_t &= \dot{\varepsilon}_w + h\dot{\kappa}_w/2 \\ \dot{\varepsilon}_b &= \dot{\varepsilon}_w - h\dot{\kappa}_w/2 \end{aligned} \tag{5.16}$$

where h is the distance from center of the top flange to the center of the bottom flange.

The rate of energy dissipation across the section is

$$\dot{W}_i = \int \sigma_y |\dot{\varepsilon}| \, dA \tag{5.17}$$

where $\dot{\varepsilon}$ is the rate of strain at any point.

Considering the web first, the strain rate $\dot{\varepsilon}$ is given by

$$\dot{\varepsilon} = (y + y_1)\dot{\kappa}_w \tag{5.18}$$

where

$$y_1 = \dot{\varepsilon}_w/\dot{\kappa}_w \tag{5.19}$$

which is the distance from the point of zero strain to the center of the web (see Fig. 5.7).

Thus, the rate of dissipation of energy in the web is

$$\sigma_y w \left[\frac{(h-t)^2}{4} \dot{\kappa}_w + \frac{\dot{\varepsilon}_w^2}{\dot{\kappa}_w} \right] \tag{5.20}$$

Similarly, the rate of energy dissipation for the top and the bottom flanges are

$$\begin{aligned} &\sigma_y t \left[\frac{b^2}{4} \dot{\kappa}_t + \left(\frac{h}{2} + \frac{\dot{\varepsilon}_w}{\dot{\kappa}_w} \right)^2 \frac{\dot{\kappa}_w^2}{\dot{\kappa}_t} \right] \\ &\sigma_y t \left[\frac{b^2}{4} \dot{\kappa}_b + \left(\frac{h}{2} - \frac{\dot{\varepsilon}_w}{\dot{\kappa}_w} \right)^2 \frac{\dot{\kappa}_w^2}{\dot{\kappa}_b} \right] \end{aligned} \tag{5.21}$$

The rates of external work for the web and the flanges are

$$\begin{aligned} &P_w \dot{\varepsilon}_w + M_w \dot{\kappa}_w \\ &P_t \left(\dot{\varepsilon}_w + \frac{h}{2} \dot{\kappa}_w \right) + M_t \dot{\kappa}_t \\ &P_b \left(\dot{\varepsilon}_w - \frac{h}{2} \dot{\kappa}_w \right) + M_b \dot{\kappa}_b \end{aligned} \tag{5.22}$$

The upper bound solution is obtained by equating the total rate of work done by the external forces to the total rate of dissipation.

$$\dot{W}_e = \dot{W}_i \tag{5.23}$$

or

$$\begin{aligned} P\dot{\varepsilon}_w + M_x\dot{\kappa}_w + M_t\dot{\kappa}_t + M_b\dot{\kappa}_b \\ = \sigma_y t\left[\frac{b^2}{4}\dot{\kappa}_t + \left(\frac{h}{2} + \frac{\dot{\varepsilon}_w}{\dot{\kappa}_w}\right)^2 \frac{\dot{\kappa}_w^2}{\dot{\kappa}_t}\right] \\ + \sigma_y t\left[\frac{b^2}{4}\dot{\kappa}_b + \left(\frac{h}{2} - \frac{\dot{\varepsilon}_w}{\dot{\kappa}_w}\right)^2 \frac{\dot{\kappa}_w^2}{\dot{\kappa}_b}\right] \\ + \sigma_y w\left[\frac{(h-t)^2}{4}\dot{\kappa}_w + \frac{\dot{\varepsilon}_w^2}{\dot{\kappa}_w}\right] \end{aligned} \tag{5.24}$$

where

$$\begin{aligned} P &= P_w + P_t + P_b \\ M_x &= M_w + P_t h/2 - P_b h/2 \\ M_y &= M_t + M_b \end{aligned} \tag{5.25}$$

In order to derive an upper bound solution on the interaction curve, arbitrary values of strain rate and curvature rate can be assumed. However, each different deformation pattern assumed may result in a different set of stress resultants over the wide-flange section and hence will correspond to a different stress boundary value problem solved. For example, if the entire wide-flange section is assumed to remain plane (Fig. 5.8), the associated stress field corresponding to the deformation pattern is the one used for the lower bound calculations. Hence the stress resultants are P, M_x and M_y. On the other hand, if the more general deformation pattern of Fig. 5.6 is used, additional warping moments acting in the planes of each of the two flanges, equal in magnitude but opposite in sense, will result from the associated stress field. Therefore, a different stress boundary value problem is solved.

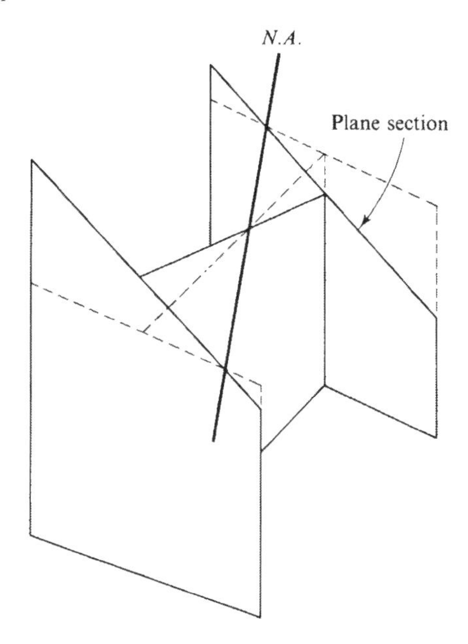

FIGURE 5.8
A failure mechanism with a plane section

Here, we consider, for simplicity, only the particular deformation pattern of Fig. 5.7 as an illustrative example of evaluating these integrals. With

$$\dot{\kappa}_t = \dot{\kappa}_b = \dot{\kappa}_f$$

Eq. (5.24) reduces to

$$\begin{aligned} &P\dot{\varepsilon}_w + M_x\dot{\kappa}_w + M_y\dot{\kappa}_f \\ &\quad = 2\sigma_y t\left[\frac{b^2}{4}\dot{\kappa}_f + \frac{h^2}{4}\frac{\dot{\kappa}_w^2}{\dot{\kappa}_f} + \frac{\dot{\varepsilon}_w^2}{\dot{\kappa}_f}\right] \\ &\quad + \sigma_y w\left[\frac{(h-t)^2}{4}\dot{\kappa}_w + \frac{\dot{\varepsilon}_w^2}{\dot{\kappa}_w}\right] \end{aligned} \tag{5.26}$$

According to the upper bound theorem of limit analysis, the best choice of the deformation pattern corresponds to the minimum value of the loads P, M_x and M_y. Equation (5.26) can be rewritten in the form

$$P = \frac{1}{\dot{\varepsilon}_w}[\dot{W}_i - M_x\dot{\kappa}_w - M_y\dot{\kappa}_f] \tag{5.27}$$

The function P has a minimum value with respect to $\dot{\varepsilon}_w$ when $\partial P/\partial\dot{\varepsilon}_w = 0$, that is

$$\frac{\partial \dot{W}_i}{\partial \dot{\varepsilon}_w} = \frac{1}{\dot{\varepsilon}_w}[\dot{W}_i - M_x\dot{\kappa}_w - M_y\dot{\kappa}_f] = P \tag{5.28}$$

In general it can be shown that these conditions are

$$\begin{aligned} P &= \partial\dot{W}_i/\partial\dot{\varepsilon}_w \\ M_x &= \partial\dot{W}_i/\partial\dot{\kappa}_w \\ M_y &= \partial\dot{W}_i/\partial\dot{\kappa}_f \end{aligned} \tag{5.29}$$

which gives us the interaction equation in terms of the parameters y_1 and $\lambda_3 (= \dot{\kappa}_w/\dot{\kappa}_f)$

$$\begin{aligned} p &= \frac{2}{A}[2t\lambda_3 y_1 + wy_1] \\ m_x &= \frac{1}{Z_x}\left\{th^2\lambda_3 + w\left[\frac{(h-t)^2}{4} - y_1^2\right]\right\} \\ m_y &= \frac{2}{Z_y}t\left[\frac{b^2}{4} - \frac{h^2}{4}\lambda_3^2 - \lambda_3^2 y_1^2\right] \end{aligned} \tag{5.30}$$

valid for

$$0 \le y_1 \le \frac{d}{2} - t$$

and

$$\frac{w/2}{y_1 + d/2 - t} \le \lambda_3 \le \frac{b/2}{d/2 - t/2 + y_1} \tag{5.31}$$

Table 5.3 COMPARISON OF LOWER AND UPPER BOUND SOLUTIONS

Section	p	m_x	m_y Lower Bound	m_y Upper bound
W12 × 31	0.2	0.929	0	0
	0.2	0.836	0.303	0.303
	0.2	0.743	0.473	0.472
	0.2	0.651	0.617	0.617
	0.2	0.558	0.737	0.737
	0.2	0.465	0.832	0.832
	0.2	0.372	0.903	0.903
	0.2	0.279	0.950	0.950
	0.2	0.186	0.975	0.975
	0.2	0	0.993	0.993

	p	m_x	m_y Lower bound	m_y Upper bound
W14 × 426	0.2	0.904	0	0
	0.2	0.814	0.315	0.313
	0.2	0.678	0.530	0.524
	0.2	0.588	0.646	0.642
	0.2	0.497	0.743	0.741
	0.2	0.407	0.823	0.821
	0.2	0.316	0.885	0.884
	0.2	0.226	0.930	0.929
	0.2	0.136	0.958	0.958
	0.2	0	0.981	0.981

The results of upper bound solutions for other cases corresponding to the lower bound solutions are also given in Table 5.2. As can be seen, the difference between the lower and the upper bound solutions is the value of n_3 which differs by $t^3/4$ only. If the thickness of the flange is small, this value is negligible.

Table 5.3 is the comparison of the numerical results obtained from the lower and the upper bound solutions for W12 × 31 and W14 × 426. Small difference is observed from W14 × 426. It can be noticed that the values of the lower bound solution are slightly higher than those of the upper bound solution. This is due to the assumption made in the upper bound solution that the section is thin and the strain rate was computed by the average value at the middle plane of the thickness. Equations (5.30) may also be derived directly from lower bound approach by assuming the stress distribution shown in Fig. 5.9.

5.2.3 Effect of Torsion

Morris and Fenves (1969) derived lower bound interaction equations for rectangular, wide flange, box and circular tube sections in which shear stress due to St. Venant torsion is taken into consideration. The distribution of the shear stress is assumed to

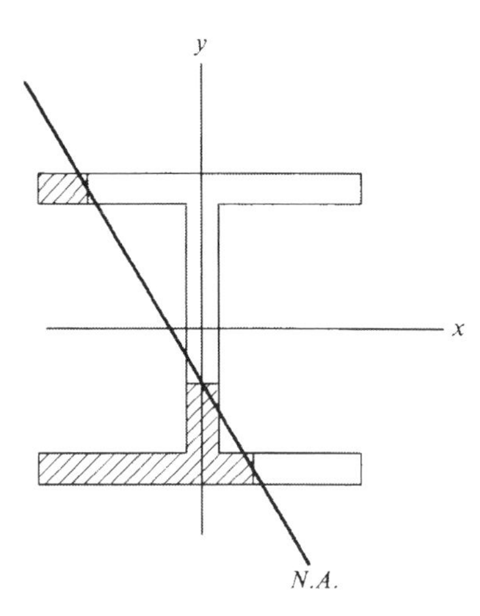

FIGURE 5.9
An assumed stress distribution

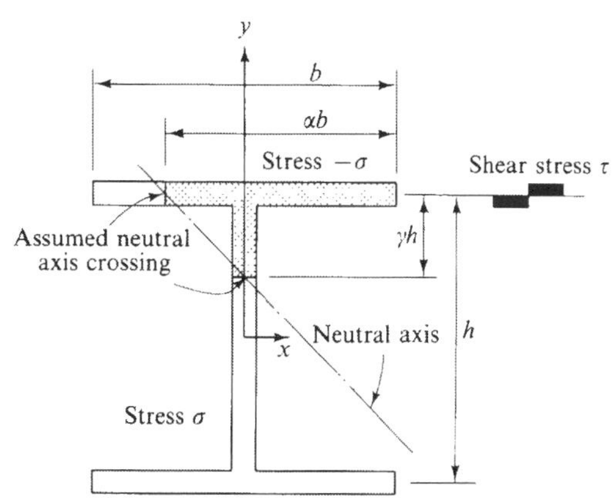

FIGURE 5.10
Assumed stress distribution

be constant over the entire cross section but in the opposite direction for each side of every plate element as shown in Fig. 5.10. This assumption simplifies the calculation of torsional moment or torque T in relation to the shear stress τ,

$$T = \sum_{i=1}^{N} \left(\frac{b_i t_i^2}{2} \right) \tau \tag{5.32}$$

in which summation is to be carried out for all plate elements of $b_i \times t_i$. Thus the fully plastic torsional moment T_p is expressed by

$$T_p = \sum_{i=1}^{N} \frac{b_i t_i^2}{2} \tau_y \tag{5.33}$$

in which τ_y is the value of shearing yield stress and the magnitude of torsional moment can be measured by a parameter t,

$$t = \frac{T}{T_p} = \frac{\tau}{\tau_y} \tag{5.34}$$

The yield condition at every point on the section is assumed to follow the von Mises yield criterion,

$$\sigma^2 + 3\tau^2 = \sigma_y^2 (= 3\tau_y^2) \tag{5.35}$$

from which it is known that the yield value due to normal stress σ is reduced by the factor

$$\frac{\sigma}{\sigma_y} = \sqrt{1 - t^2} \tag{5.36}$$

These results imply that effect of torsional moment t can be taken into consideration by simply substituting

$$\begin{aligned} p &\text{ by } \sqrt{1 - t^2}\, p \\ m_x &\text{ by } \sqrt{1 - t^2}\, m_x \\ m_y &\text{ by } \sqrt{1 - t^2}\, m_y \end{aligned} \tag{5.37}$$

in the interaction equations relating p, m_x and m_y. The interaction surface thus obtained can be interpreted geometrically as reducing all coordinates by the factor $\sqrt{1 - t^2}$ as shown in Fig. 5.11. The outer surface in Fig. 5.11 corresponds to zero torsional moment and intersects each of the coordinate axes at a value of 1. The inner surface corresponds to a normalized torsional moment, t, and has the shape as the outer surfaces; all coordinates are reduced by the factor $\sqrt{1 - t^2}$ The surface element marked 1-a-c in Fig. 5.11 corresponds to the particular neutral axis position shown in Fig. 5.10 for the H-shaped section. Equations defining various surface elements for rectangular, circular tube, wide-flange, and box sections are given in Sect. 5.9, Problems 5.4 to 5.7.

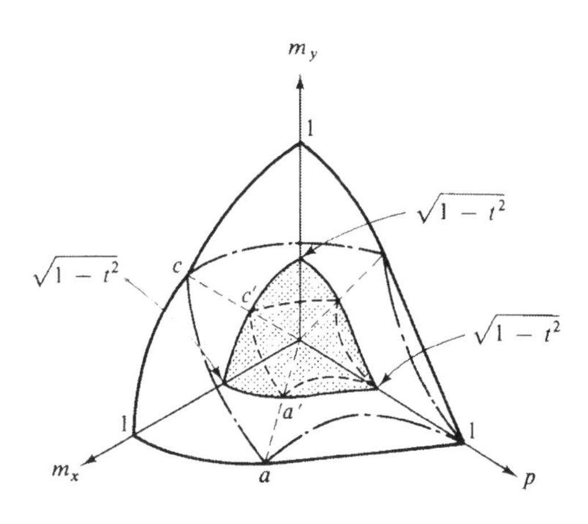

FIGURE 5.11
Family of lower bound yield surfaces

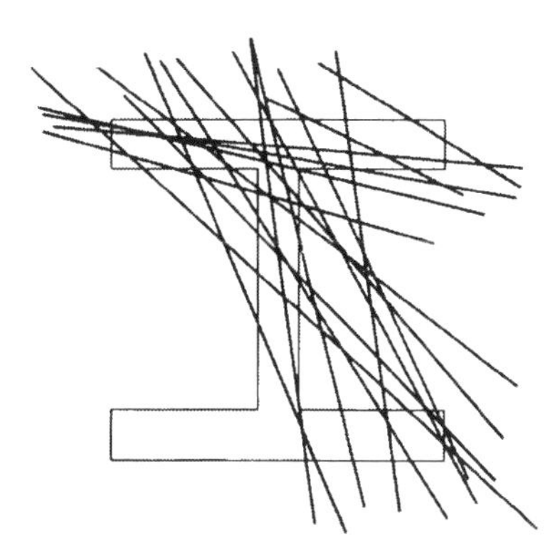

FIGURE 5.12
Cases of neutral axis location for wide flange section

5.3 EXACT METHOD FOR SYMMETRIC SECTION

The interaction equations described in the previous section are not exact except for rectangular section. This is clear from the fact that these solutions obtained by the lower- and upper-bound techniques of limit analysis do not coincide. This approximation appears to be inevitable as long as interaction equations are searched from each possible case of neutral axis location. There are too many cases of neutral axis locations as demonstrated in Fig. 5.12 for the case of a wide-flange section.

For this reason, most researchers neglect some special cases. For example, the special cases where the neutral axis cuts through an edge of a flange plate are generally not considered. Further, stress gradients through a plate thickness are often neglected as can be seen in Figs. 5.9 and 5.10.

Later, we shall see that these approximate solutions result only in small errors when compared with exact solutions. Thus, they are considered sufficiently accurate for practical design purposes. However, in actual design, it is desirable to have simple expressions which are applicable to general cross sections and cover all possible cases of neutral axis location. If the expressions can further be shown to be *exact*, then the method will be extremely powerful. This was done by Chen and Atsuta (1972). In the following, we shall derive all interaction equations of doubly symmetric cross sections using only the exact solution of rectangular section, along with the concept of superposition.

5.3.1 Exact Interaction Relations for Rectangular Section

Consider a rectangular section of width 2b and depth 2d, as shown in Fig. 5.13. If the section is fully plastified, the normal stress can be assumed to be $-\sigma_y$ above the neutral axis (NA) and $+\sigma_y$ below the NA without loss of generality. Only axial stress is considered and the effect of shear stress on the yielding is neglected.

The location of the NA can be determined by two independent parameters, vertical distance e and angle with horizontal axis θ. The resultant forces of the section are obtained uniquely as functions of e and θ:

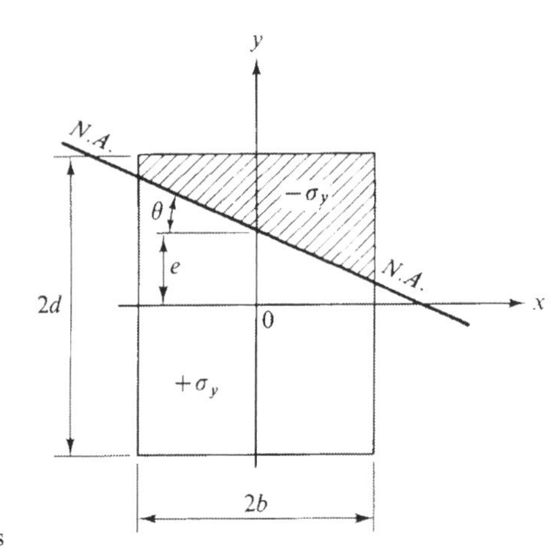

FIGURE 5.13
Rectangular section and neutral axis

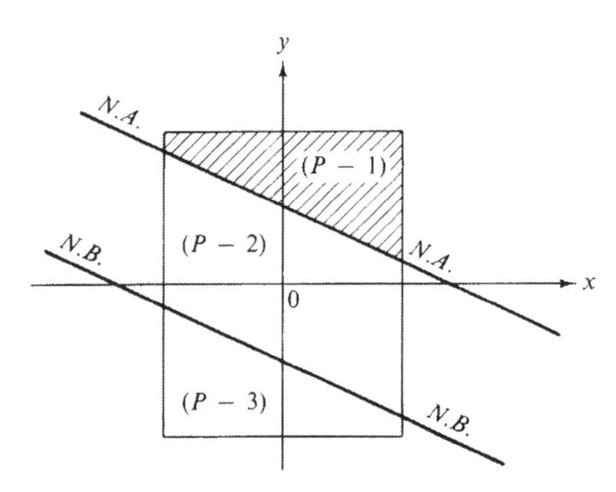

FIGURE 5.14
Partitioning of rectangular section

Axial force.

$$P = f_p(e, \theta) = \sigma_y A_p(e, \theta)$$

Moment about horizontal axis.

$$M_x = f_x(e, \theta) = \sigma_y Q_x(e, \theta) \tag{5.38}$$

Moment about vertical axis.

$$M_y = f_y(e, \theta) = \sigma_y Q_y(e, \theta)$$

These functions are to be determined by the two limit analyses, the lower bound analysis and the upper bound analysis. The resulting functions, derived in the succeeding sections, are the same and thus exact.

Lower bound solution

As the state of stress given in Fig. 5.13 satisfies the yield condition and the equilibrium conditions, integration of stress over the section gives a lower bound solution for the biaxial bending forces.

The rectangular section can be divided into three parts: P-1, P-2, and P-3 by the neutral axis (NA) and a straight line NB, as shown in Fig. 5.14. Lines NB and NA are symmetric with respect to the origin. From the symmetry with respect to both the x-axis and y-axis, part P-2 contributes only to axial force P and parts P-1 and P-3 contribute only to bending moments M_x and M_y. Axial force P and bending moments M_x and M_y are considered positive when the axial force causes tension and the bending moments produce compressive stress in the first quadrant of the coordinate system shown. This definition of sign convention differs from that of Sect. 5.2 (page 198). Then the three resultant forces can be expressed as

$$\begin{aligned} P &= \sigma_y A_2 \\ M_x &= \sigma_y A_1 e_{y1} + \sigma_y A_3 e_{y3} \\ M_y &= \sigma_y A_1 e_{x1} + \sigma_y A_3 e_{x3} \end{aligned} \tag{5.39}$$

in which A_1, A_2, and A_3 are areas of parts P-1, P-2, and P-3, respectively, and e_{xj} and e_{yj} are the centroidal coordinates of portion P-j. From symmetry again

$$\begin{aligned} A_3 &= A_1 \\ e_{x3} &= e_{x1} \\ e_{y3} &= e_{y1} \\ e_{x2} &= e_{y2} = 0 \end{aligned} \tag{5.40}$$

Thus

$$A_2 = A - 2A_1 \tag{5.41}$$

in which $A = 4bd$ is total area of the rectangular section. Equations (5.39) become

$$\begin{aligned} P &= \sigma_y A_2 = \sigma_y A_p \\ M_x &= 2\sigma_y A_1 e_{y1} = \sigma_y Q_x \\ M_y &= 2\sigma_y A_1 e_{x1} = \sigma_y Q_y \end{aligned} \tag{5.42}$$

in which

$$\begin{aligned} A_p &= A - 2A_1 \\ Q_x &= 2A_1 e_{y1} \\ Q_y &= 2A_1 e_{x1} \end{aligned} \tag{5.43}$$

In order to express A_1, e_{x1}, and e_{y1} in terms of two parameters e and θ, we now define a pair of parentheses that have the meaning

$$\langle S \rangle = \begin{cases} 0 & \text{when} \quad S \le 0 \\ S & \text{when} \quad S \ge 0 \end{cases} \tag{5.44}$$

or in a single expression

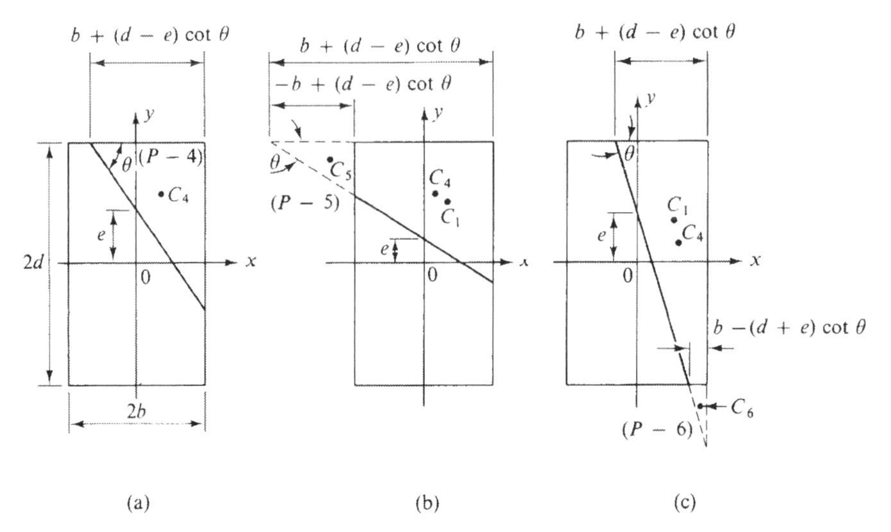

FIGURE 5.15
Possible locations of neutral axis

$$\langle S \rangle = \frac{S + |S|}{2} \tag{5.45}$$

Considering part P-1, there are three different cases possible, as shown in Fig. 5.15. For each case, area and centroid can be obtained using the specially defined parentheses:

Part 4

$$\begin{aligned} A_4 &= \frac{1}{2}\langle b \tan\theta + d - e\rangle^2 \cot\theta \\ e_{y4} &= \frac{1}{3}(-\,b\tan\theta + 2d + e) \\ e_{x4} &= \frac{1}{3}(2b\tan\theta - d + e)\cot\theta \end{aligned} \tag{5.46}$$

Part 5

$$\begin{aligned} A_5 &= \frac{1}{2}\langle -\,b\tan\theta + d - e\rangle^2 \cot\theta \\ e_{y5} &= \frac{1}{3}(b\tan\theta + 2d + e) \\ e_{x5} &= \frac{1}{3}(2b\tan\theta + d - e)\cot\theta \end{aligned} \tag{5.47}$$

Part 6

$$A_6 = \frac{1}{2}\langle b \tan\theta - d - e\rangle^2 \cot\theta$$

$$e_{y6} = \frac{1}{3}(b \tan\theta + 2d - e) \tag{5.48}$$

$$e_{x6} = \frac{1}{3}(2b \tan\theta + d + e)\cot\theta$$

and

$$A_1 = A_4 - A_5 - A_6$$

$$A_1 e_{x1} = A_4 e_{x4} + A_5 e_{x5} - A_6 e_{x6} \tag{5.49}$$

$$A_1 e_{y1} = A_4 e_{y4} - A_5 e_{y5} + A_6 e_{y6}$$

Equations (5.43) become

$$A_p = 4bd - \langle b \tan\theta + d - e\rangle^2 \cot\theta + \langle -b \tan\theta + d - e\rangle^2 \cot\theta + \langle b \tan\theta - d - e\rangle^2 \cot\theta$$

$$Q_x = \frac{1}{3}\langle b \tan\theta + d - e\rangle^2(-b \tan\theta + 2d + e)\cot\theta - \frac{1}{3}\langle -b \tan\theta + d - e\rangle^2(b \tan\theta + 2d + e)\cot\theta + \frac{1}{3}\langle b \tan\theta - d - e\rangle^2(b \tan\theta + 2d - e)\cot\theta$$

and (5.50)

$$Q_y = \frac{1}{3}\langle b \tan\theta + d - e\rangle^2(2b \tan\theta - d + e)\cot^2\theta + \frac{1}{3}\langle -b \tan\theta + d - e\rangle^2(2b \tan\theta + d - e)\cot^2\theta - \frac{1}{3}\langle b \tan\theta - d - e\rangle^2(2b \tan\theta + d + e)\cot^2\theta$$

Define the nondimensional forces as

$$p = \frac{P}{P_y}; \quad m_x = \frac{M_x}{M_{px}}; \quad m_y = \frac{M_y}{M_{py}} \tag{5.51}$$

in which

$$P_y = \sigma_y A; \quad M_{px} = \sigma_y Z_x; \quad M_{py} = \sigma_y Z_y \tag{5.52}$$

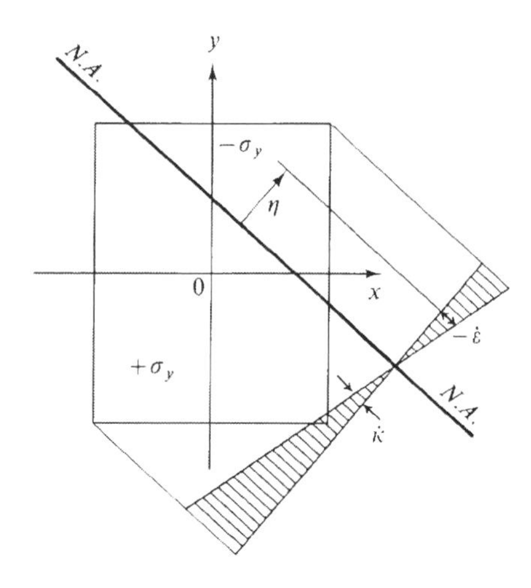

FIGURE 5.16
Distribution of strain rate

Therefore

$$p = \frac{A_p}{A}; \quad m_x = \frac{Q_x}{Z_x}; \quad m_y = \frac{Q_y}{Z_y} \tag{5.53}$$

in which

$$\begin{aligned} A &= 4bd = \text{total area of the section} \\ Z_x &= 2bd^2 = \text{plastic section modulus about } x\text{-axis} \\ Z_y &= 2b^2d = \text{plastic section modulus about } y\text{-axis} \end{aligned} \tag{5.54}$$

The resultant forces, p, m_x, and m_y, have now been expressed as functions of e and θ. The values of θ are valid in the region

$$0° < \theta < 90° \tag{5.55}$$

Upper bound solution

An upper bound solution can be obtained by equating internal rate of energy dissipation $\dot{D}_I$ due to assumed strain field $\dot{\varepsilon}$ to rates of external work $\dot{W}_E$ due to increments of resultant forces $\dot{P}$, $\dot{M}_x$, and $\dot{M}_y$. Assume the rate of strain field as (Fig. 5.16)

$$\dot{\varepsilon} = -\eta\dot{\kappa} \tag{5.56}$$

in which η is the distance from fibre to the NA. The rate of internal dissipation is

$$\begin{aligned} \dot{D}_I &= \int_A \sigma\dot{\varepsilon}\, dA = \sigma_y\dot{\kappa}\left(\int_{A_1} \eta\, dA - \int_{A_2} \eta\, dA - \int_{A_3} \eta\, dA\right) \\ &= \sigma_y\dot{\kappa}(\eta_1 A_1 + \eta_2 A_2 + \eta_3 A_3) \end{aligned} \tag{5.57}$$

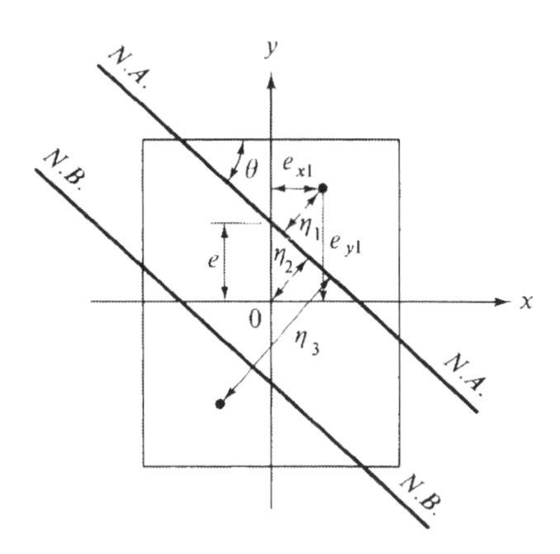

FIGURE 5.17
Geometry of centroidal points

in which η_i is the distance from neutral axis NA to the centroids of part P-i in η coordinate and

$$\eta_3 = 2\eta_2 + \eta_1; \quad A_3 = A_1 \tag{5.58}$$

From Fig. 5.17, η_1 and η_2 are related to e, θ, e_{x1}, and e_{y1} as

$$\begin{aligned} \eta_1 &= e_{x1} \sin\theta + e_{y1} \cos\theta - e\cos\theta \\ \eta_2 &= e\cos\theta \end{aligned} \tag{5.59}$$

and the dissipation becomes

$$\begin{aligned} \dot{D}_I &= \sigma_y \dot{\kappa}[2\eta_1 A_1 + (2A_1 + A_2)\eta_2] \\ &= \sigma_y \dot{\kappa}(2A_1 e_{x1} \sin\theta + 2A_1 e_{y1} \cos\theta + A_2 e \cos\theta) \end{aligned} \tag{5.60}$$

The rate of external work is

$$\dot{W}_E = \dot{\varepsilon}_o P + \dot{\kappa}_x M_x + \dot{\kappa}_y M_y \tag{5.61}$$

in which

$$\begin{aligned} \dot{\varepsilon}_o &= \text{strain rate at the centroid 0} \\ \dot{\kappa}_x &= \text{curvature rate about } x\text{-axis} \\ \dot{\kappa}_y &= \text{curvature rate about } y\text{-axis} \end{aligned} \tag{5.62}$$

and

$$\begin{aligned} \dot{\varepsilon}_o &= \dot{\kappa} e \cos\theta \\ \dot{\kappa}_x &= \dot{\kappa} \cos\theta \\ \dot{\kappa}_y &= \dot{\kappa} \sin\theta \end{aligned} \tag{5.63}$$

Then

$$\dot{W}_E = \dot{\kappa}(Pe\cos\theta + M_x \cos\theta + M_y \sin\theta) \tag{5.64}$$

Equating the rate of internal energy dissipation to the rate of external work, $\dot{W}_E = \dot{D}_I$

$$\begin{aligned} &Pe\cos\theta + M_x\cos\theta + M_y\sin\theta \\ &\quad = \sigma_y(A_2 e\cos\theta + 2A_1 e_{y1}\cos\theta + 2A_1 e_{x1}\sin\theta) \end{aligned} \tag{5.65}$$

is obtained. Equation (5.65) must be valid for all values of e and θ. It follows that

$$\begin{aligned} P &= \sigma_y A_2 \\ M_x &= 2\sigma_y e_{y1} A_1 \\ M_y &= 2\sigma_y e_{x1} A_1 \end{aligned} \tag{5.66}$$

which is identical to Eq. (5.42) obtained from a lower bound analysis. Therefore, it can be concluded that the solution is exact.

5.3.2 Exact Interaction Relations for Double Web Sections

A double web section as shown in Fig. 5.18 is a general case, and Eqs. (5.50) and (5.54) can be extended to the general case. Function A_p for the general case, e.g., can be obtained by first finding functions $A_p(b_0, d_0)$, $A_p(b_0, d_1)$, $A_p(b_1, d_1)$, and $A_p(b_2, d_1)$ corresponding to the rectangular sections $2b_0 \times 2d_0$, $2b_0 \times 2d_1$, $2b_1 \times 2d_1$, and $2b_2 \times 2d_1$, respectively. Function A_p for the double web section is then obtained by the simple algebraic summation

$$A_p = A_p(b_0, d_0) - A_p(b_0, d_1) + A_p(b_1, d_1) - A_p(b_2, d_1) \tag{5.67}$$

Similarly

$$\begin{aligned} Q_x &= Q_x(b_0, d_0) - Q_x(b_0, d_1) + Q_x(b_1, d_1) - Q_x(b_2, d_1) \\ Q_y &= Q_y(b_0, d_0) - Q_y(b_0, d_1) + Q_y(b_1, d_1) - Q_y(b_2, d_1) \\ A &= 4(b_0 d_0 - b_0 d_1 + b_1 d_1 - b_2 d_1) \\ Z_x &= 2(b_0 d_0^2 - b_0 d_1^2 + b_1 d_1^2 - b_2 d_1^2) \\ Z_y &= 2(b_0^2 d_0 - b_0^2 d_1 + b_1^2 d_1 - b_2^2 d_1) \end{aligned} \tag{5.68}$$

Particular cases may then be handled using these equations and the special properties in each case.

Rectangular Section ($B \times D$).

$$\begin{aligned} &b_0 = \frac{1}{2}B, \quad b_1 = \frac{1}{2}B, \quad b_2 = 0 \\ &d_0 = \frac{1}{2}D, \quad d_1 = 0 \end{aligned} \tag{5.69}$$

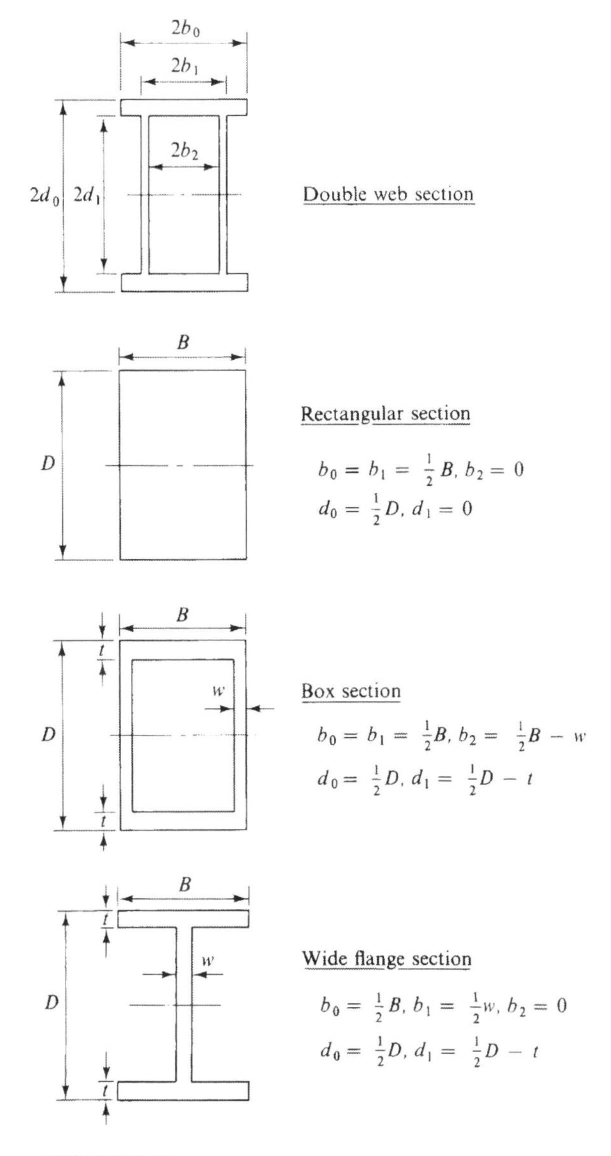

FIGURE 5.18
Particular cases of double web section

Box Section $(B \times D, t, w)$.

$$b_0 = \frac{1}{2}B, \quad b_1 = \frac{1}{2}B, \quad b_2 = \frac{1}{2}B - w$$
$$d_0 = \frac{1}{2}D, \quad d_1 = \frac{1}{2}D - t \tag{5.70}$$

Wide-Flange Section $(B \times D, t, w)$.

$$b_0 = \frac{1}{2}B, \quad b_1 = \frac{1}{2}w, \quad b_2 = 0$$
$$d_0 = \frac{1}{2}D, \quad d_1 = \frac{1}{2}D - t \tag{5.71}$$

5.3.3 Exact Interaction Relations for Circular Section

Consider a solid circular section of radius r as shown in Fig. 5.19. The NA is at distance e from the center and makes angle θ with the x-axis. Part P-1 above the NA is assumed fully stressed by σ_y. Area and centroid of this portion are given by

$$A_1 = r^2\left(\phi - \frac{1}{2}\sin 2\phi\right)$$
$$e_{x1} = \frac{1}{2}r\sin\theta\,\frac{\sin\phi - \frac{1}{3}\sin 3\phi}{\phi - \frac{1}{2}\sin 2\phi} \tag{5.72}$$
$$e_{y1} = \frac{1}{2}r\cos\theta\,\frac{\sin\phi - \frac{1}{3}\sin 3\phi}{\phi - \frac{1}{2}\sin 2\phi}$$

in which

$$\phi = \cos^{-1}\frac{e}{\bar{r}}$$
$$\frac{e}{\bar{r}} = \frac{e}{r} \quad \text{when} \quad \frac{e}{r} \leq 1 \tag{5.73}$$
$$= 1 \quad \text{when} \quad \frac{e}{r} \geq 1$$

or

$$\frac{e}{\bar{r}} = 1 - \langle 1 - \frac{e}{r} \rangle \tag{5.74}$$

and the corresponding quantities are derived as

$$A_p(r) = A - 2A_1 = \pi r^2 - 2r^2\left(\phi - \frac{1}{2}\sin 2\phi\right)$$
$$Q_x(r) = 2A_1 e_{y1} = r^3\cos\theta\left(\sin\phi - \frac{1}{3}\sin 3\phi\right) \tag{5.75}$$
$$Q_y(r) = 2A_1 e_{x1} = r^3\sin\theta\left(\sin\phi - \frac{1}{3}\sin 3\phi\right)$$

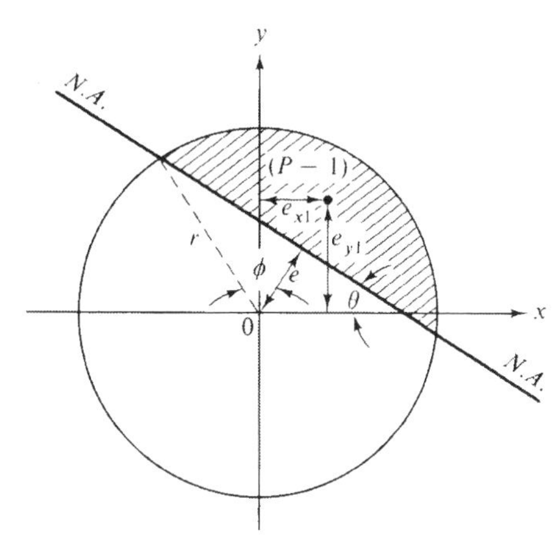

FIGURE 5.19
Solid circular section

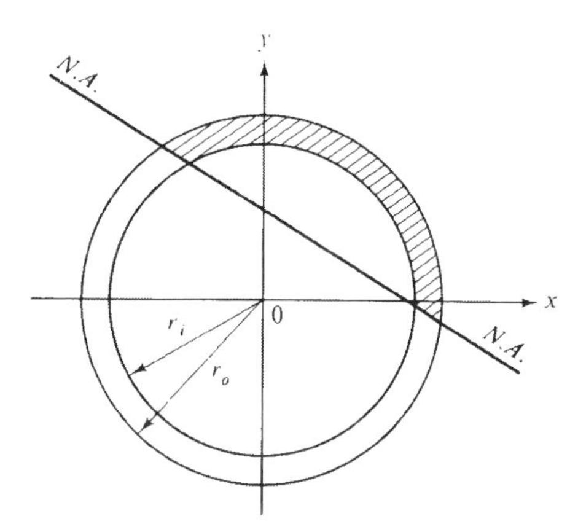

FIGURE 5.20
Circular tube section

and

$$A = \pi r^2$$

$$Z_x = \frac{4}{3} r^3 \tag{5.76}$$

$$Z_y = \frac{4}{3} r^3$$

Using Eqs. (5.75) and (5.76), corresponding expressions for a circular tube section of external radius r_0 and internal radius r_i can be obtained (Fig. 5.20)

$$\begin{aligned} A_p &= A_p(r_0) - A_p(r_i) \\ Q_x &= Q_x(r_0) - Q_x(r_i) \\ Q_y &= Q_y(r_0) - Q_y(r_i) \end{aligned} \tag{5.77}$$

and

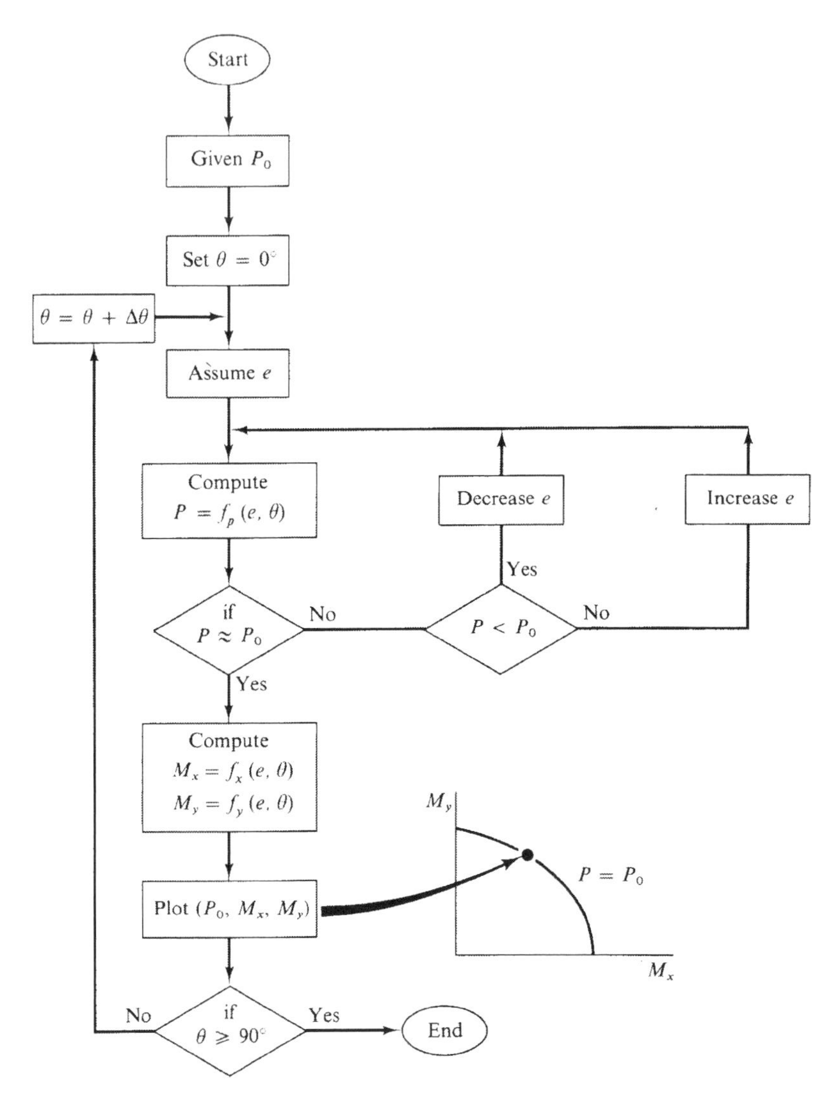

FIGURE 5.21
Flow diagram of interaction curve plotting

$$A = \pi(r_0^2 - r_i^2)$$

$$Z_x = \frac{4}{3}(r_0^3 - r_i^3) \tag{5.78}$$

$$Z_y = \frac{4}{3}(r_0^3 - r_i^3)$$

The procedure in obtaining the interaction surface from Eqs. (5.38) is now stated as follows. Since the resultant forces P, M_x and M_y are given in terms of the

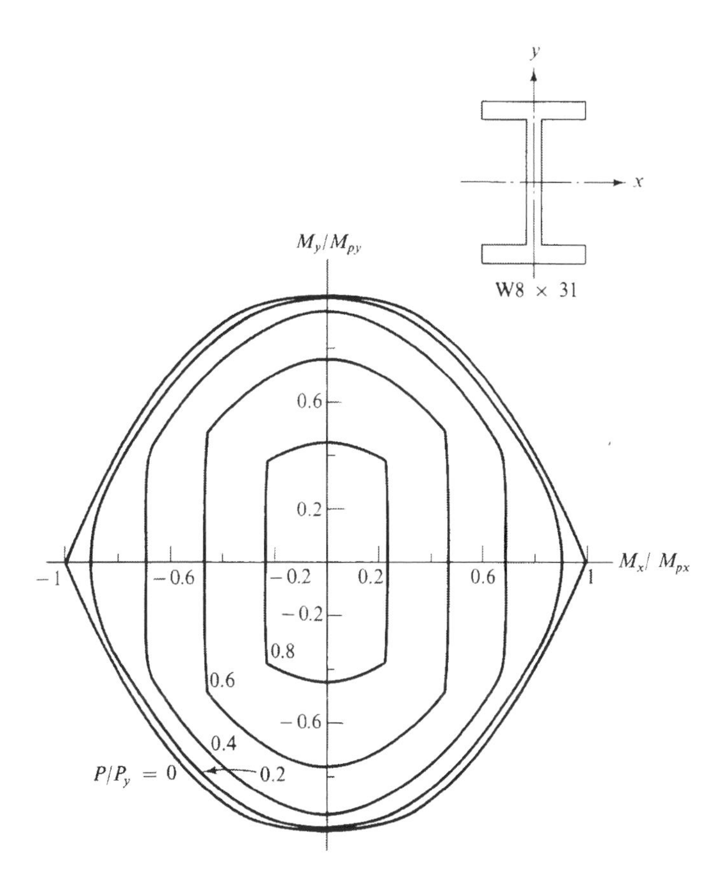

FIGURE 5.22
Interaction curves of a wide-flange section

parameters e and θ, an arbitrary choice of e and θ will result a point on the interaction surface. In other words, all possible combinations of e and θ will generate the entire interaction surface. As previously mentioned, however, the interaction curves of M_x and M_y for a given constant value of P are more desirable for design purposes. This can be done by the iterative procedure explained by the flow diagram shown in Fig. 5.21.

Figure 5.22 shows typical numerical results obtained for a particular wide-flange section, W8 × 31. Since the section is doubly symmetric, the interaction surface is also doubly symmetric. Thus, it is sufficient to illustrate the entire surface by its first quadrant only. This is demonstrated in Fig. 5.23 for W14 × 426. Referring to the interaction curves for W14 × 426 (Fig. 5.23), the solid lines represent the present exact solutions and the dotted lines are the results reported in Table 5.2 by Santathadaporn and Chen (1970). They are practically identical to each other except in a small region. This small difference results from the fact that in Table 5.2 it is assumed that the NA cuts through the web horizontally, as shown in Fig. 5.24(a).

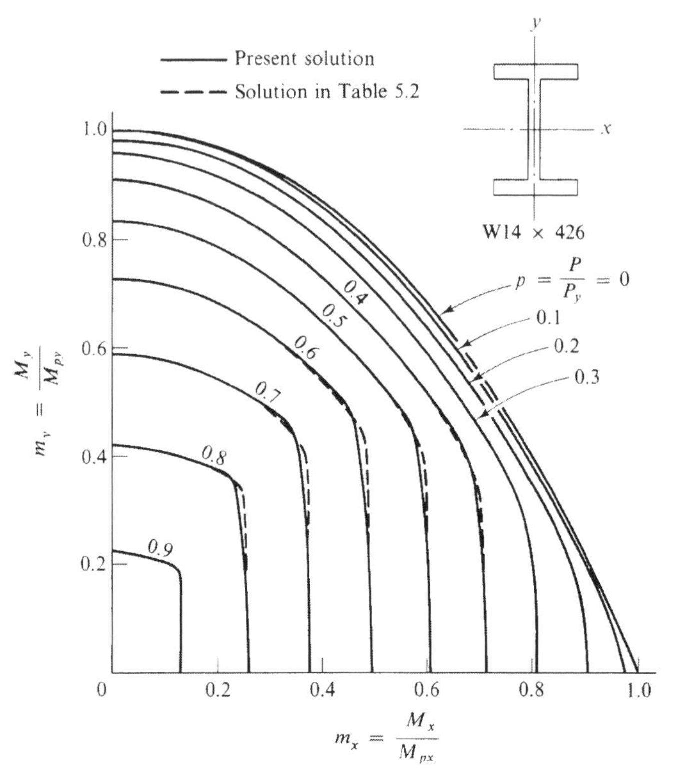

FIGURE 5.23
Interaction curves for a wide flange section

Furthermore, in Table 5.2 the cases where the NA cuts through an edge of the flange plate were omitted [Fig. 5.24(a)]. As a consequence, a sharp corner appears on each of the interaction curves. Such corners do not show up on the exact interaction curves as can be seen in the figure.

Note that the present solution is exact only for the idealized wide-flange shape [Fig. 5.24(b)]. An actual wide-flange shape has, of course, rounded corners and edges [Fig. 5.24(c)], which are not accounted for herein. Thus, emphasis must not be put on its exactness, rather on its simplicity in calculations.

5.4 EXACT METHOD FOR GENERAL SECTION

The above-mentioned technique for symmetric sections is now extended to the case of general section. Consider a general section with area A_0 which is composed of n-elements with areas $A_1, A_2, \ldots A_n$ as shown in Fig. 5.25. The total area A_0 and the static moments of area Q_{xo} and Q_{yo} of the total area are given by the summation of all its component elements, i.e.,

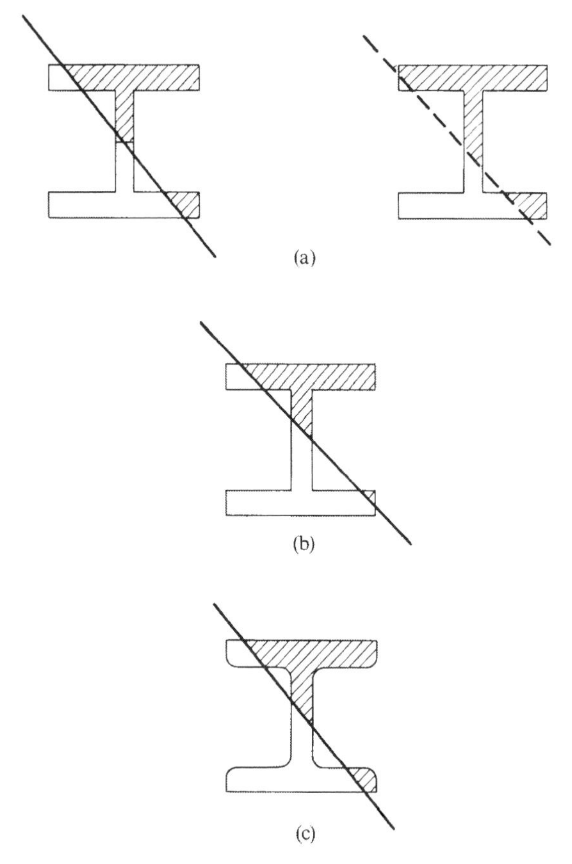

FIGURE 5.24
Wide flange sections and neutral axes (a) Idealized section with approximate NA (Table 5.2). (b) Idealized section with exact NA. (c) Actual section with exact NA

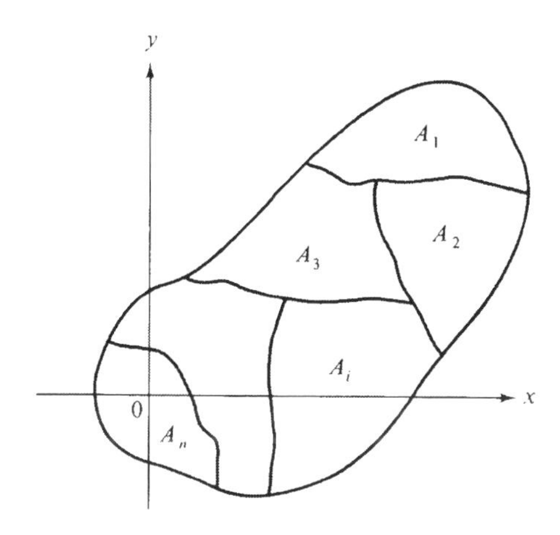

FIGURE 5.25
Partitioning of general section

$$
\begin{aligned}
A_0 &= \int_{A_0} dA = \sum_{i=1}^{n} \int_{A_i} dA = \sum_{i=1}^{n} A_i \\
Q_{xo} &= \int_{A_0} y\,dA = \sum_{i=1}^{n} \int_{A_i} y\,dA = \sum_{i=1}^{n} Q_{xi} \\
Q_{yo} &= \int_{A_0} x\,dA = \sum_{i=1}^{n} \int_{A_i} x\,dA = \sum_{i=1}^{n} Q_{yi}
\end{aligned} \tag{5.79}
$$

in which

$$
A_i = \int_{A_i} dA, \quad Q_{xi} = \int_{A_i} y\,dA, \quad Q_{yi} = \int_{A_i} x\,dA \tag{5.80}
$$

5.4.1 Lower Bound Analysis

A lower bound solution is obtained from an assumed stress field and equilibrium conditions. Figure 5.26 shows the i-th element A_i of the section. The neutral axis NA divides the element A_i into the upper portion A_i^- and the lower portion A_i^+

$$
A_i = A_i^- + A_i^+ \tag{5.81}
$$

Let us assume that the lower portion A_i^+ is yielded by tension $(+\sigma_y)$ and the upper portion A_i^- by compression $(-\sigma_y)$.

The contribution of the i-th element to the resultant forces of the entire section is calculated by,

$$
\begin{aligned}
P_i &= \sigma_y \int_{A_i^+} dA - \sigma_y \int_{A_i^-} dA = \sigma_y(A_i - 2A_i^-) \\
M_{xi} &= -\sigma_y \int_{A_i^+} y\,dA + \sigma_y \int_{A_i^-} y\,dA = -\sigma_y(Q_{xi} - 2Q_{xi}^-) \\
M_{yi} &= -\sigma_y \int_{A_i^+} x\,dA + \sigma_y \int_{A_i^-} x\,dA = -\sigma_y(Q_{yi} - 2Q_{yi}^-)
\end{aligned} \tag{5.82}
$$

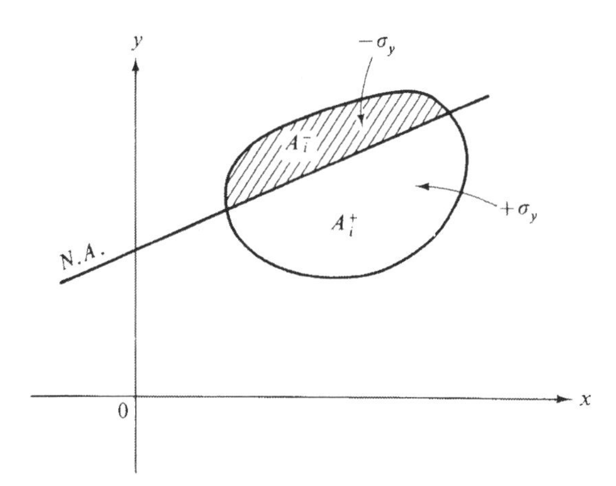

FIGURE 5.26
Neutral axis and an element A_i

in which

$$A_i^- = \int_{A_i^-} dA, \quad Q_{xi}^- = \int_{A_i^-} y\,dA, \quad Q_{yi}^- = \int_{A_i^-} x\,dA \tag{5.83}$$

Thus the resultant forces for the total section are

$$\begin{aligned} P &= \sum_{i=1}^{n} P_i = \sigma_y \sum_{i=1}^{n} (A_i - 2A_i^-) = \sigma_y \left(A_0 - 2 \sum_{i=1}^{n} A_i^- \right) \\ M_x &= \sum_{i=1}^{n} M_{xi} = -\sigma_y \sum_{i=1}^{n} (Q_{xi} - 2Q_{xi}^-) = -\sigma_y \left(Q_{xo} - 2 \sum_{i=1}^{n} Q_{xi}^- \right) \\ M_y &= \sum_{i=1}^{n} M_{yi} = -\sigma_y \sum_{i=1}^{n} (Q_{yi} - 2Q_{yi}^-) = -\sigma_y \left(Q_{yo} - 2 \sum_{i=1}^{n} Q_{yi}^- \right) \end{aligned} \tag{5.84}$$

5.4.2 Upper Bound Analysis

An upper bound solution is obtained from an assumed strain rate field and the use of virtual work principle. Assuming a straight neutral axis, the strain rate field becomes a linear function of coordinates x and y,

$$\dot{\varepsilon} = \dot{\varepsilon}_o - \dot{\kappa}_x y - \dot{\kappa}_y x \tag{5.85}$$

in which

$\dot{\varepsilon}_o$ = strain rate at the origin

$\dot{\kappa}_x$ = curvature rate about x-axis

$\dot{\kappa}_y$ = curvature rate about y-axis

Considering again an element A_i which is divided by neutral axis NA into A_i^+ and A_i^- as shown in Fig. 5.26. The rate of internal energy dissipation in this element is

$$\dot{D}_{Ii} = \int_{A_i} \sigma \dot{\varepsilon}\,dA = \sigma_y \int_{A_i^+} \dot{\varepsilon}\,dA - \sigma_y \int_{A_i^-} \dot{\varepsilon}\,dA \tag{5.86}$$

Using Eq. (5.81) and substituting $\dot{\varepsilon}$ of Eq. (5.85), we have

$$\begin{aligned} \dot{D}_{Ii} &= \sigma_y \int_{A_i} \dot{\varepsilon}\,dA - 2\sigma_y \int_{A_i^-} \dot{\varepsilon}\,dA \\ &= \sigma_y \int_{A_i} (\dot{\varepsilon}_o - \dot{\kappa}_x y - \dot{\kappa}_y x)\,dA - 2\sigma_y \int_{A_i^-} (\dot{\varepsilon}_o - \dot{\kappa}_x y - \dot{\kappa}_y x)\,dA \\ &= \sigma_y \dot{\varepsilon}_o \left(\int_{A_i} dA - 2 \int_{A_i^-} dA \right) - \sigma_y \dot{\kappa}_x \left(\int_{A_i} y\,dA - 2 \int_{A_i^-} y\,dA \right) \\ &\quad - \sigma_y \dot{\kappa}_y \left(\int_{A_i} x\,dA - 2 \int_{A_i^-} x\,dA \right) = \sigma_y \dot{\varepsilon}_o (A_i - 2A_i^-) - \sigma_y \dot{\kappa}_x (Q_{xi} - 2Q_{xi}^-) \\ &\quad - \sigma_y \dot{\kappa}_y (Q_{yi} - 2Q_{yi}^-) \end{aligned} \tag{5.87}$$

Thus the rate of internal energy dissipation in the entire section A_0 is

$$\dot{D}_I = \sum_{i=1}^{n} \dot{D}_{Ii} = \sigma_y \left[\dot{\varepsilon}_o \left(A_0 - 2 \sum_{i=1}^{n} A_i^- \right) - \dot{\kappa}_x \left(Q_{xo} - 2 \sum_{i=1}^{n} Q_{xi}^- \right) - \dot{\kappa}_y \left(Q_{yo} - 2 \sum_{i=1}^{n} Q_{yi}^- \right) \right] \tag{5.88}$$

The rate of external work is given by

$$\dot{W}_E = \dot{\varepsilon}_o P + \dot{\kappa}_x M_x + \dot{\kappa}_y M_y \tag{5.89}$$

Equating the rate of internal energy dissipation $\dot{D}_I$ to the rate of external work $\dot{W}_E$,

$$\dot{\varepsilon}_o P + \dot{\kappa}_x M_x + \dot{\kappa}_y M_y = \sigma_y \left[\dot{\varepsilon}_o \left(A_0 - 2 \sum_{i=1}^{n} A_i^- \right) - \dot{\kappa}_x \left(Q_{xo} - 2 \sum_{i=1}^{n} Q_{xi}^- \right) - \dot{\kappa}_y \left(Q_{yo} - 2 \sum_{i=1}^{n} Q_{yi}^- \right) \right] \tag{5.90}$$

If we assume

$$P = \sigma_y \left(A_0 - 2 \sum_{i=1}^{n} A_i^- \right)$$

$$M_x = -\sigma_y \left(Q_{xo} - 2 \sum_{i=1}^{n} Q_{xi}^- \right) \tag{5.91}$$

$$M_y = -\sigma_y \left(Q_{yo} - 2 \sum_{i=1}^{n} Q_{yi}^- \right)$$

then Eq. (5.90) is always satisfied. Equation (5.91) is therefore an upper bound solution.

The lower-bound solution, Eq. (5.84), and the upper-bound solution, Eq. (5.91), are identical, thus these equations are proved to be exact.

FIGURE 5.27
A line element and the neutral axis

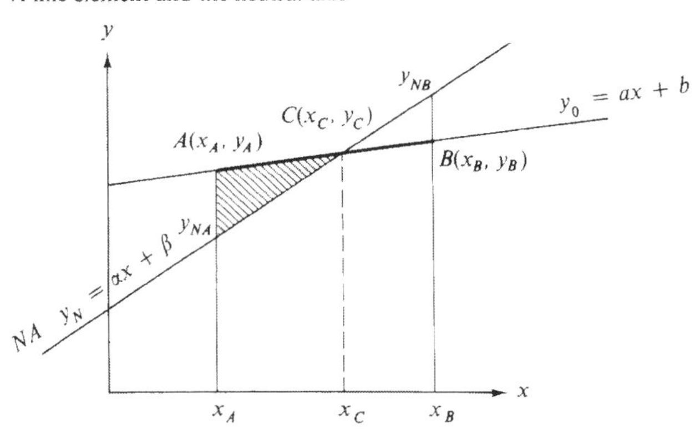

From the structural analysis point of view, it is convenient to take the origin 0 of the coordinate axes x and y at the centroid of the section, thus,

$$Q_{xo} = \int_{A_0} y\,dA = 0$$
$$Q_{yo} = \int_{A_0} x\,dA = 0 \tag{5.92}$$

The x- and y-axes are not necessarily the principal axes of the section.

Now the resultant forces of the section are given by

$$P = \sigma_y\left(A_0 - 2\sum_{i=1}^{n} A_i^-\right)$$
$$M_x = 2\sigma_y \sum_{i=1}^{n} Q_{xi}^- \tag{5.93}$$
$$M_y = 2\sigma_y \sum_{i=1}^{n} Q_{yi}^-$$

It is worth while to note here that the values of P, M_x, M_y in Eq. (5.93) can be obtained by treating only the compressive yield portion A_i^- of each element A_i.

5.4.3 Derivation of Interaction Equations

The equations for A_i^-, Q_{xi}^- and Q_{yi}^- for the i-th element of the general section will be first derived in what follows and the resulting equations for the entire section will then be obtained by the method of superposition.

Consider a straight line element AB as shown in Fig. 5.27, which is defined by the two end points,

$$A(x_A, y_A), \quad B(x_B, y_B) \tag{5.94}$$

The ordinate of the intermediate point is given by

$$y_0 = ax + b \tag{5.95}$$

in which

$$a = \frac{y_B - y_A}{x_B - x_A}$$
$$b = \frac{x_B y_A - x_A y_B}{x_B - x_A} \tag{5.96}$$

As shown in Fig. 5.27, the neutral axis NA defined by

$$y_N = \alpha x + \beta \tag{5.97}$$

intersects the line AB at the point C of the coordinates,

$$x_C = \frac{b - \beta}{\alpha - a} \quad \text{and} \quad y_C = \frac{b - \beta}{\alpha - a}a + b \tag{5.98}$$

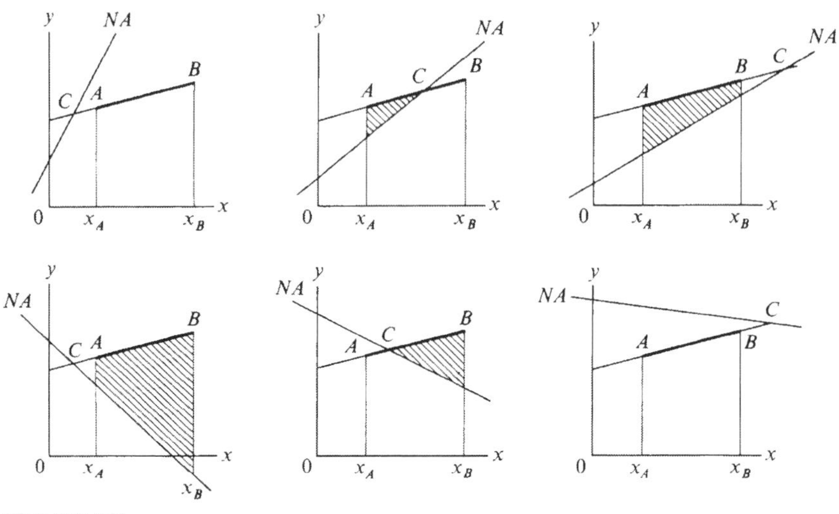

FIGURE 5.28
Possible cases for neutral axis location

The shaded area below the line AB and above the NA is of primary concern in the computations (Fig. 5.27). There are six different cases possible according to the relative locations of the NA and the line AB as shown in Fig. 5.28. It will be convenient herein to introduce the following specially defined integral

$$\oint_A^B S\,dx = \begin{cases} \displaystyle\int_A^B S\,dx & (A \le B) \\ 0 & (B \le A) \end{cases} \tag{5.99}$$

This integral is useful in the derivation of the general interaction equations together with the specially defined parentheses $\langle\ \rangle$ in Eq. (5.44).

Referring to Fig. 5.28, the contributions of the shaded area can be expressed by

$$\begin{aligned} \bar{A} &= \int_A^B \langle y_0 - y_N \rangle\,dx \\ \bar{Q}_x &= \frac{1}{2}\int_A^B \langle y_0 - y_N \rangle (y_0 + y_N)\,dx \\ \bar{Q}_y &= \int_A^B \langle y_0 - y_N \rangle\, x\,dx \end{aligned} \tag{5.100}$$

Taking the point C into consideration, one obtains

$$\bar{A} = \begin{cases} \displaystyle\oint_A^C (y_0 - y_N)dx - \oint_B^C (y_0 - y_N)dx & (\alpha > 0) \\ \displaystyle\oint_C^B (y_0 - y_N)dx - \oint_C^A (y_0 - y_N)dx & (\alpha < 0) \end{cases}$$

$$\bar{Q}_x = \begin{cases} \dfrac{1}{2}\displaystyle\oint_A^C (y_0^2 - y_N^2)\,dx - \dfrac{1}{2}\oint_B^C (y_0^2 - y_N^2)\,dx & (\alpha > 0) \\ \dfrac{1}{2}\displaystyle\oint_C^B (y_0^2 - y_N^2)\,dx - \dfrac{1}{2}\oint_C^A (y_0^2 - y_N^2)\,dx & (\alpha < 0) \end{cases} \tag{5.101}$$

$$\bar{Q}_y = \begin{cases} \displaystyle\oint_A^C (y_0 - y_N)x\,dx - \oint_B^C (y_0 - y_N)x\,dx & (\alpha > 0) \\ \displaystyle\oint_C^B (y_0 - y_N)x\,dx - \oint_C^A (y_0 - y_N)x\,dx & (\alpha < 0) \end{cases}$$

Substituting Eq. (5.97) into Eqs. (5.101) and making use of Eq. (5.98), one obtains

$$\bar{A} = \bar{A}(A, B, NA) = \frac{1}{2(\alpha - a)}[\langle y_A - y_{NA}\rangle^2 - \langle y_B - y_{NB}\rangle^2]$$

$$\begin{aligned} \bar{Q}_x &= \bar{Q}_x(A, B, NA) \\ &= \frac{1}{6(\alpha - a)}[\langle y_A - y_{NA}\rangle^2(y_A + y_{NA} + y_C) - \langle y_B - y_{NB}\rangle^2(y_B + y_{NB} + y_C)] \end{aligned} \tag{5.102}$$

$$\begin{aligned} \bar{Q}_y &= \bar{Q}_y(A, B, NA) \\ &= \frac{1}{6(\alpha - a)}[\langle y_A - y_{NA}\rangle^2(2x_A + x_C) - \langle y_B - y_{NB}\rangle^2(2x_B + x_C)] \end{aligned}$$

in which y_{NA} and y_{NB} are the ordinates of the neutral axis NA at x_A and x_B, respectively (Fig. 5.27),

$$\begin{aligned} y_{NA} &= \alpha x_A + \beta \\ y_{NB} &= \alpha x_B + \beta \end{aligned} \tag{5.103}$$

It is to be noted that Eqs. (5.102) cover all six cases shown in Fig. 5.28.

For the i-th element A_i with area bounded by the points A_1 B_1 B_2 A_2 as shown in Fig. 5.29, Eqs. (5.102) can be used to obtain solutions. The function for A_i^- for the i-th element, for example, can be obtained by first finding the functions $A(A_1, B_1,$

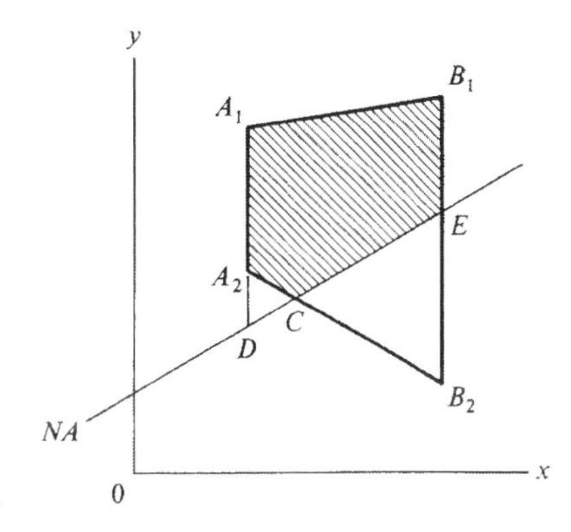

FIGURE 5.29
Element A_i and the NA

Table 5.4 CALCULATION EXAMPLE OF AN INTERACTION POINT, $\alpha = -1$, $\beta = 0.95$, FIG. 5.30

Equations or Figures	Fig. 30(a)	Fig. 30(b)	Fig. 30(c)	Fig. 30(d)
Fig. 5.30	$A_1(x_{A1}, y_{A1})$ $B_1(x_{B1}, y_{B1})$ $A_2(x_{A2}, y_{A2})$ $B_2(x_{B2}, y_{B2})$ $C_1(x_{C1}, y_{C1})$ $C_2(x_{C2}, y_{C2})$	(−0.786, 2.713) (2.214, 2.713) (−0.786, −1.287) (2.214, −1.287) (−1.763, 2.713) (2.237, −1.287)	(−0.411, 2.713) (2.214, 2.713) (−0.411, −0.912) (2.214, −0.912) (−1.763, 2.713) (1.862, −0.912)	$p = 0.6$ $m_x = 0.623$ $m_y = -0.138$
Eq. 5.103	$y_{NA} = \alpha x_A + \beta$ $y_{NB} = \alpha x_B + \beta$	1.736 −1.264	1.361 −1.264	
	$\alpha - a$	−1	−1	
Eq. 5.104	$A_i^- = \bar{A}(A_1, B_1, NA) - \bar{A}(A_2, B_2, NA)$			Eq. 5.107(a)
Eq. 5.102	$= \langle y_{A1} - y_{NA} \rangle^2$ $- \langle y_{B1} - y_{NB} \rangle^2$ $- \langle y_{A2} - y_{NA} \rangle^2$ $+ \langle y_{B2} - y_{NB} \rangle^2$ $= (\ )/2(\alpha - a)$	$\langle 0.977 \rangle^2 = 0.955$ $-\langle 3.977 \rangle^2 = -15.817$ $-\langle -3.023 \rangle^2 = -0$ $+\langle -0.023 \rangle^2 = +0$ $(-14.862)/(-2) = 7.431$	$\langle 1.352 \rangle^2 = 1.828$ $-\langle 3.977 \rangle^2 = -15.817$ $-\langle -2.273 \rangle^2 = -0$ $+\langle 0.352 \rangle^2 = +0.124$ $(-13.865)/(-2) = 6.933$	$p = 1 - 2\Sigma A_i^-/A_0$ $= 1 - 2 \times 0.498/2.484$ $= \boxed{0.600}$ $= 0.498 (= \Sigma A_i^-)$
Eq. 5.104	$Q_{xi}^- = \bar{Q}_x(A_1, B_1, NA) - \bar{Q}_x(A_2, B_2, NA)$			Eq. 5.107(b)
Eq. 5.102	$= \langle\ \rangle^2 (y_{A1} + y_{NA} + y_{C1})$ $- \langle\ \rangle^2 (y_{B1} + y_{NB} + y_{C1})$ $- \langle\ \rangle^2 (y_{A2} + y_{NA} + y_{C2})$ $+ \langle\ \rangle^2 (y_{B2} + y_{NB} + y_{C2})$ $= (\ )/6(\alpha - a)$	$0.955 \times 7.162 = 6.834$ $-15.817 \times 4.162 = -65.830$ -0 $+0$ $(-58.996)/(-6) = 9.840$	$1.828 \times 6.787 = 12.407$ $-15.817 \times 4.162 = -65.830$ -0 $+0.124 \times (-3.088) = -0.383$ $(-53.806)/(-6) = 8.975$	$m_x = 2\Sigma Q_{xi}^-/Z_{xo}$ $= 2 \times 0.865/2.776$ $= \boxed{0.623}$ $= 0.865 (= \Sigma Q_{xi}^-)$
Eq. 5.104	$Q_{yi}^- = \bar{Q}_y(A_1, B_1, NA) - \bar{Q}_y(A_2, B_2, NA)$			Eq. 5.107(c)
Eq. 5.102	$= \langle\ \rangle^2 (2x_{A1} + x_{C1})$ $- \langle\ \rangle^2 (2x_{B1} + x_{C1})$ $- \langle\ \rangle^2 (2x_{A2} + x_{C2})$ $+ \langle\ \rangle^2 (2x_{B2} + x_{C2})$ $= (\ )/6(\alpha - a)$	$0.955 \times (-3.335) = -3.185$ $-15.817 \times 2.665 = -42.152$ -0 $+0$ $(-45.337)/(-6) = 7.556$	$1.828 \times (-2.585) = -4.725$ $-15.817 \times (2.665) = -42.152$ -0 $+0.124 \times 6.290 = +0.780$ $(-46.097)/(-6) = 7.683$	$m_y = 2\Sigma Q_{yi}^-/Z_{yo}$ $= 2 \times (-0.127)/1.848$ $= \boxed{-0.138}$ $= -0.127 (= \Sigma Q_{yi}^-)$

NA) and $\bar{A}(A_2, B_2, NA)$ corresponding to the areas A_1 B_1 ED and A_2 CD respectively. The function A_i^- for the area A_1 B_1 EC A_2 (shaded area in Fig. 5.29) is then obtained by the simple algebraic summation

$$A_i^- = \bar{A}(A_1, B_1, NA) - \bar{A}(A_2, B_2, NA)$$

Similarly,

$$Q_{xi}^- = \bar{Q}_x(A_1, B_1, NA) - \bar{Q}_x(A_2, B_2, NA) \qquad (5.104)$$

$$Q_{yi}^- = \bar{Q}_y(A_1, B_1, NA) - \bar{Q}_y(A_2, B_2, NA)$$

where the four points are located as

$$A_1(x_{A1}, y_{A1}), \quad B_1(x_{B1}, y_{B1}), \quad A_2(x_{A2}, y_{A2}), \quad B_2(x_{B2}, y_{B2}) \qquad (5.105)$$

Summing up these values for all n-elements, the resultant forces are obtained from Eq. (5.93).

Using the following fully plastic quantities of the section

$$P_y = \sigma_y A_o$$

$$M_{px} = \sigma_y Z_{xo} \qquad (5.106)$$

$$M_{py} = \sigma_y Z_{yo}$$

to define the following nondimensional variables from Eq. (5.93)

$$p = \frac{P}{P_y} = 1 - 2\sum_{i=1}^{n} A_i^-/A_o$$

$$m_x = \frac{M_x}{M_{px}} = 2\sum_{i=1}^{n} Q_{xi}^-/Z_{xo} \qquad (5.107)$$

$$m_y = \frac{M_y}{M_{py}} = 2\sum_{i=1}^{n} Q_{yi}^-/Z_{yo}$$

in which

$$A_o = \text{total area of the section}$$

$$Z_{xo} = \text{plastic section modulus about } x\text{-axis} \qquad (5.108)$$

$$Z_{yo} = \text{plastic section modulus about } y\text{-axis}$$

Equations (5.107) give the exact interaction relations of the section in terms of the parameters α and β. The procedure in obtaining the interaction surface or interaction curves is the same as that explained in Sec. 5.3 for doubly symmetric sections.

5.4.4 A Worked Example

As an example of the actual procedure, a point on a yield surface for the angle section L 4 × 3 × 3/8 is calculated in Table 5.4. The section is treated as a subtraction of two rectangular sections 4 × 3″ and 3.625 × 2.625″ as shown in Fig. 5.30.

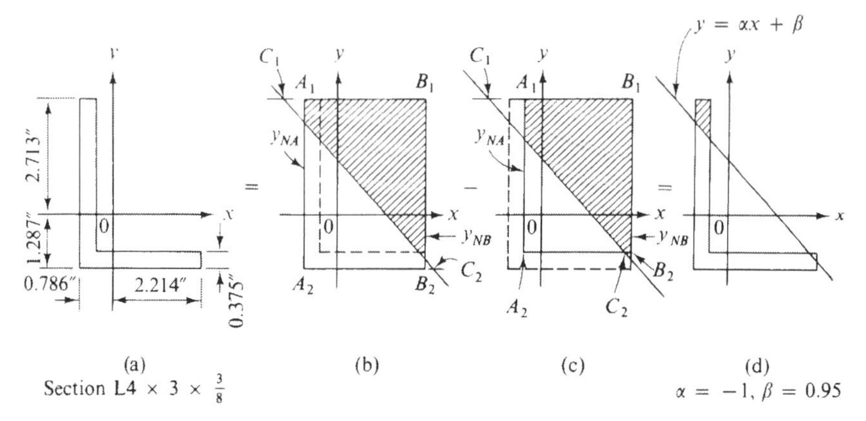

FIGURE 5.30
An illustrative example given in Table 5.4 (a) Section L4 × 3 × $\frac{3}{8}$, (b) Rectangular section 4 × 3, (c) Rectangular section 3.625 × 2.625, (d) $\alpha = -1$, $\beta = 0.95$

The plastic section properties are

$$A_o = 2.484 \text{ in}^2 \qquad Z_{xo} = 2.776 \text{ in}^3 \qquad Z_{yo} = 1.848 \text{ in}^3 \tag{5.109}$$

The neutral axis is assumed to be

$$y = -x + 0.95$$

or

$$\alpha = -1 \quad \text{and} \quad \beta = 0.95 \tag{5.110}$$

The values A_i, Q_{xi} and Q_{yi} to be substituted in Eqs. (5.93) are

$$\begin{aligned} A_1^- &= 7.431, \quad A_2^- = 6.933 \quad \text{and} \quad \Sigma A_i^- = 0.498 \text{ in}^2 \\ Q_{x1}^- &= 9.840, \quad Q_{x2}^- = 8.975 \quad \text{and} \quad \Sigma Q_{xi}^- = 0.865 \text{ in}^3 \\ Q_{y1}^- &= 7.556, \quad Q_{y2}^- = 7.683 \quad \text{and} \quad \Sigma Q_{yi}^- = -0.127 \text{ in}^3 \end{aligned} \tag{5.111}$$

The generalized stresses are obtained from Eqs. (5.93) as

$$\begin{aligned} P &= (2.484 - 2 \times 0.498)\sigma_y = 1.488\,\sigma_y \\ M_x &= 2 \times 0.865\,\sigma_y = 1.730\,\sigma_y \\ M_y &= -2 \times 0.127\,\sigma_y = -0.254\,\sigma_y \end{aligned} \tag{5.112}$$

The corresponding nondimensional interaction point on the interaction surface $(m_x - m_y - p)$ plot is then given by

$$p = 0.6, \quad m_x = 0.623 \quad \text{and} \quad m_y = -0.138 \tag{5.113}$$

This point is shown by the circle in Fig. 5.32(a).

5.5 INTERACTION CURVES FOR STRUCTURAL STEEL SECTIONS

Using the present method, the interaction curves of various shapes of steel cross section are presented in Figs. 5.31 and 5.32. Figure 5.31 shows the interaction curves of doubly symmetric sections of circular-tube, double-web and wide-flange shapes; their interaction curves are also doubly symmetric. Figure 5.32 shows the interaction curves of non-symmetric sections of angle shape and sections with only one axis of symmetry such as channel, tee and double-angle shapes. The largest loop of the curves represents the interaction curve without thrust ($p = 0$). When p increases, the loop becomes closer to a square or a triangle. However, the convexity condition of yield surface of interaction surface is not violated. Since the section of angle shape has no axes of symmetry, its interaction curves are not symmetric with respect to any axis. For channel, tee and double-angle shapes, their interaction curves are symmetric with respect to either M_x or M_y axis. It should be noted, however, that in the three dimensional plot ($m_x - m_y - p$ space), the interaction surface is always symmetric with respect to the origin. This follows directly from the condition that tension and compression yield stress levels are equal.

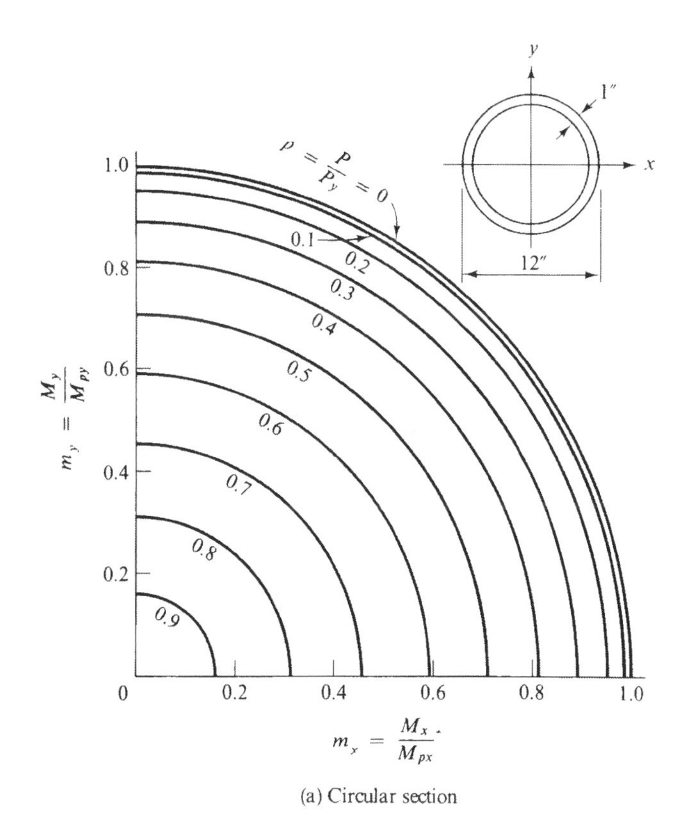

FIGURE 5.31 (a) and (b)
Interaction curves for doubly symmetric sections, (a) Circular section, (b) Double web section

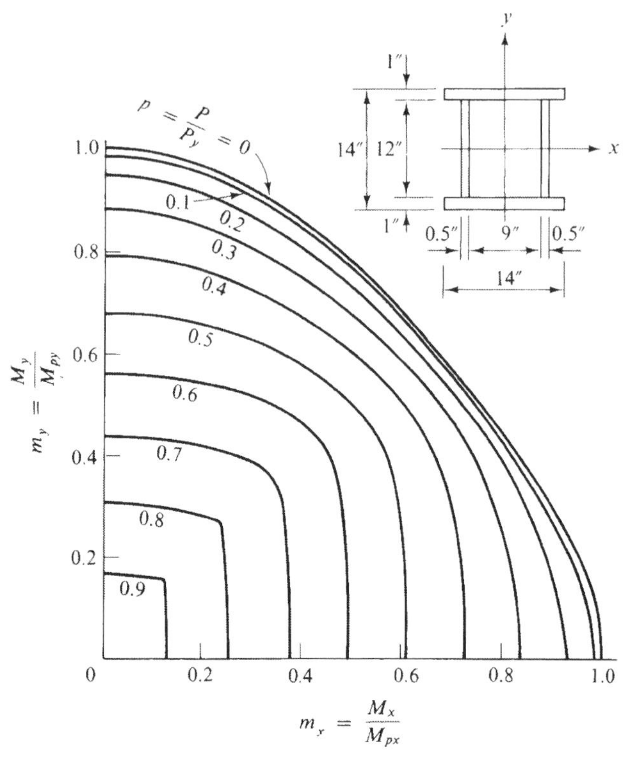

(b) Double web section

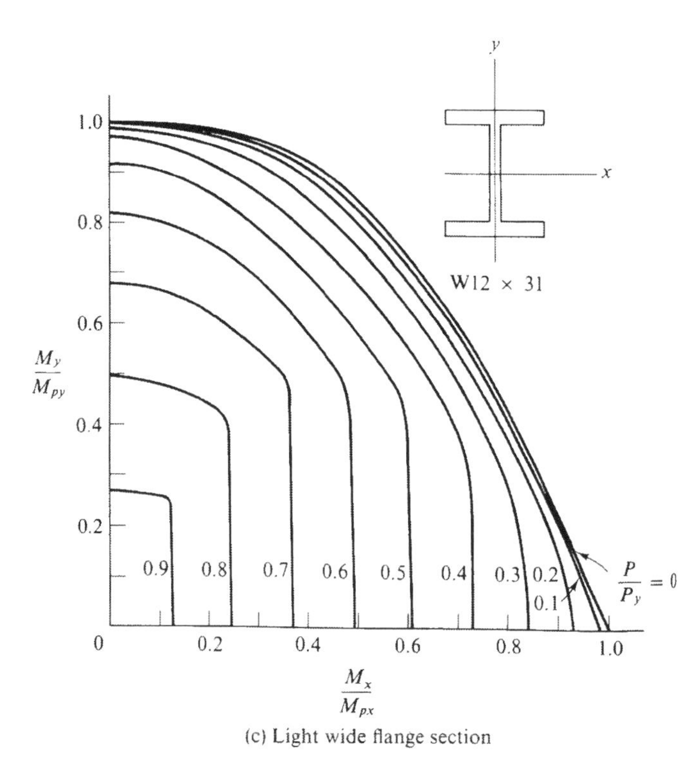

(c) Light wide flange section

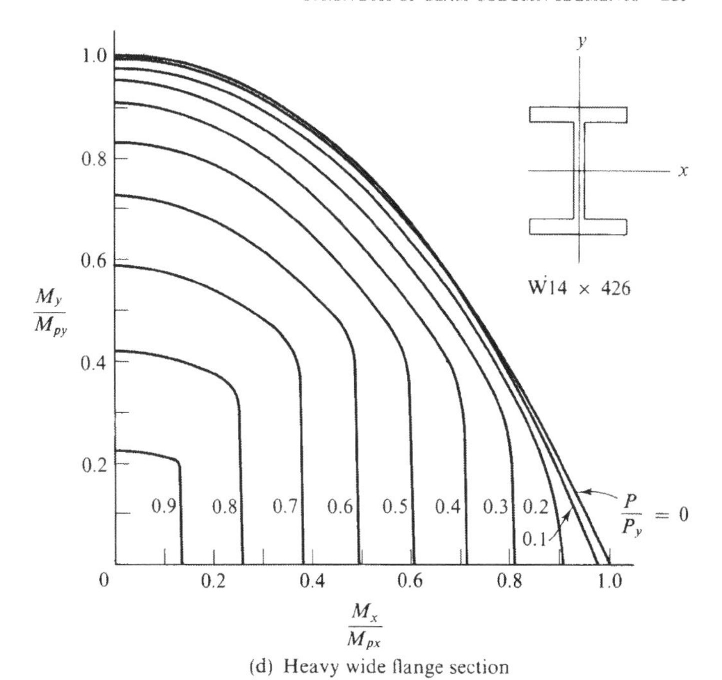

(d) Heavy wide flange section

FIGURE 5.31 (c) and (d)
Interaction curves for doubly symmetric sections. (c) A light wide flange section, (d) A heavy wide flange section

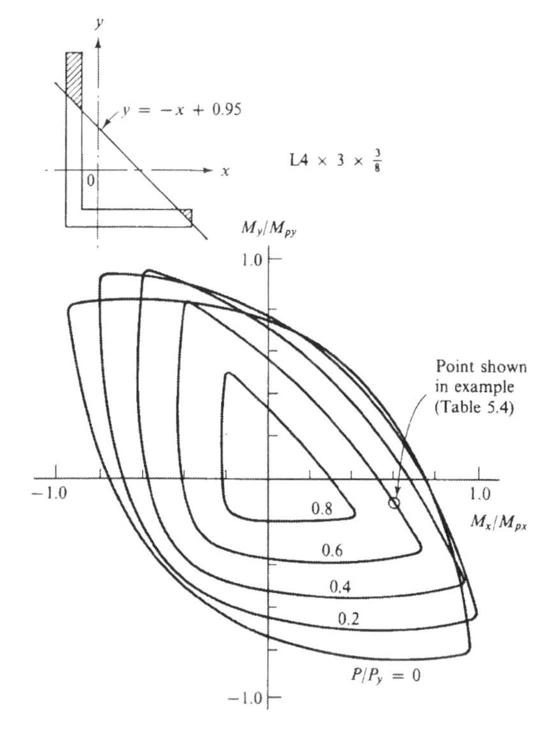

(a) Angle section

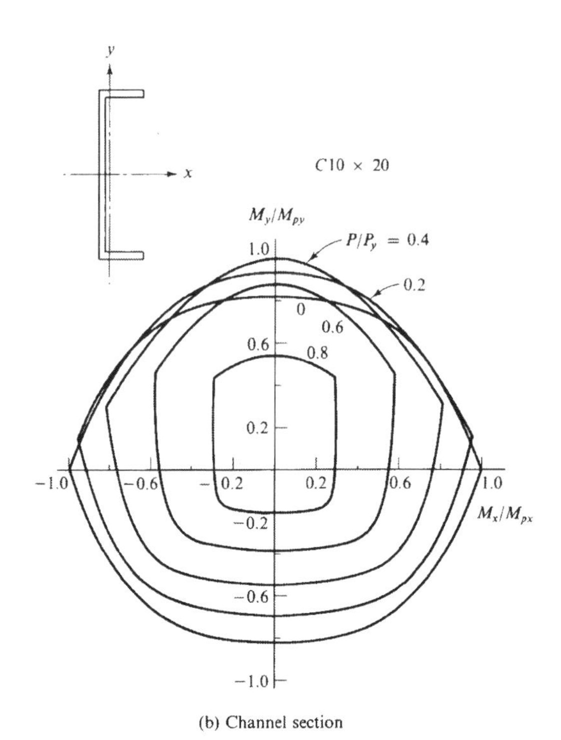

(b) Channel section

FIGURE 5.32 (a) and (b)
Interaction curves for sections with no or one axis of symmetry, (a) Angle section, (b) Channel section.

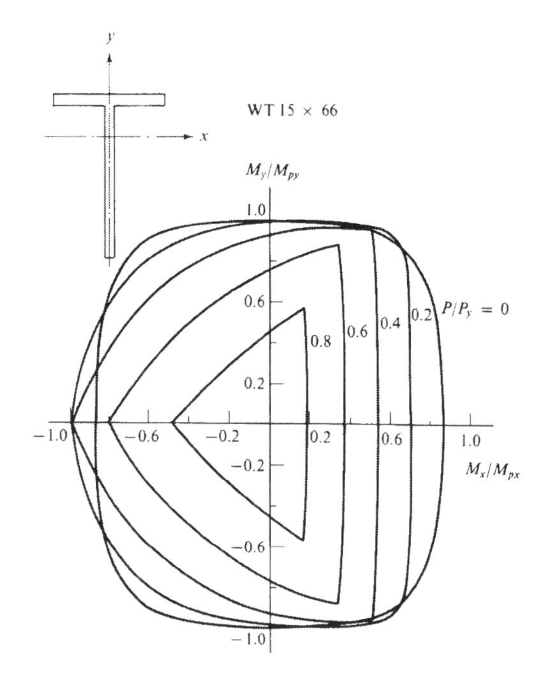

(c) Structural tee section

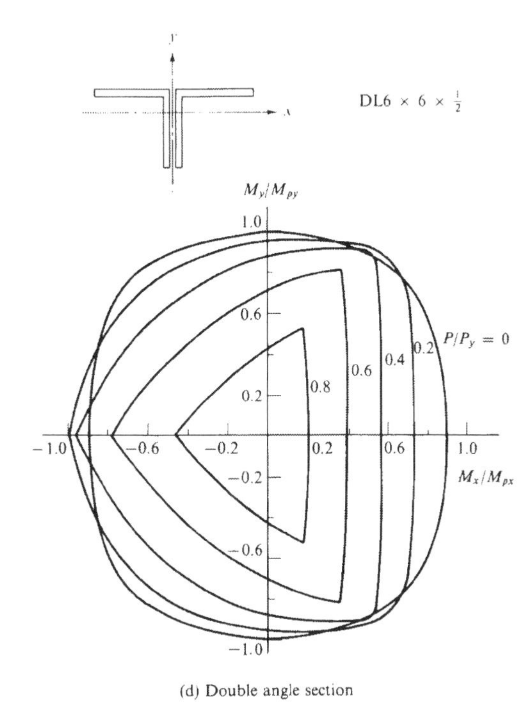

FIGURE 5.32 (c) and (d)
Interaction curves for sections with one axis of symmetry, (c) Structural tee section, (d) Double angle section

In order to investigate the effect of cross section sizes on the interaction curves, two extreme sizes of wide-flange shapes (W12 × 31 and W14 × 426) are selected and plotted in Fig. 5.33, from which it is known that the lighter the section, the stronger the moment carrying capacity. But the difference is not very large. This is especially true for wide-flange shapes with the same height as shown in Fig. 5.34. Similar comparisons are made for other structural shapes in Fig. 5.35. For each shape of cross section, two values of thrust ($p = 0.2$ and 0.8) are shown. The dotted curves in the figures represent a somewhat averaged curve for a particular shape of cross section. It was found that the effect of section weights on the interaction curves is much less than that of section sizes. Two extreme cases of weight were found to give interaction curves almost identical to the average curves. It can therefore be concluded that the interaction curves presented in Figs. 5.31 and 5.32 may be considered as representative curves for those shapes of sections. It is of interest to note that interaction curves of structural tee shape are found to be almost identical to those of double-angle shapes.

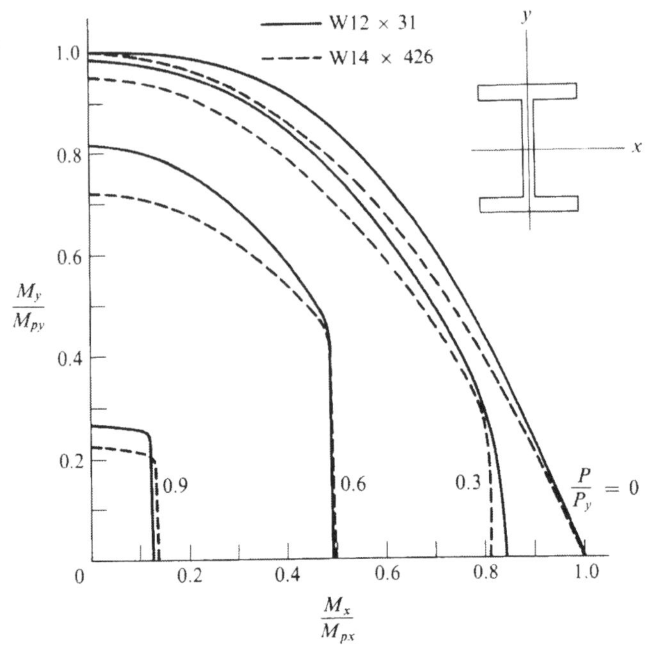

FIGURE 5.33
Comparison of interaction curves between heavy and light sections

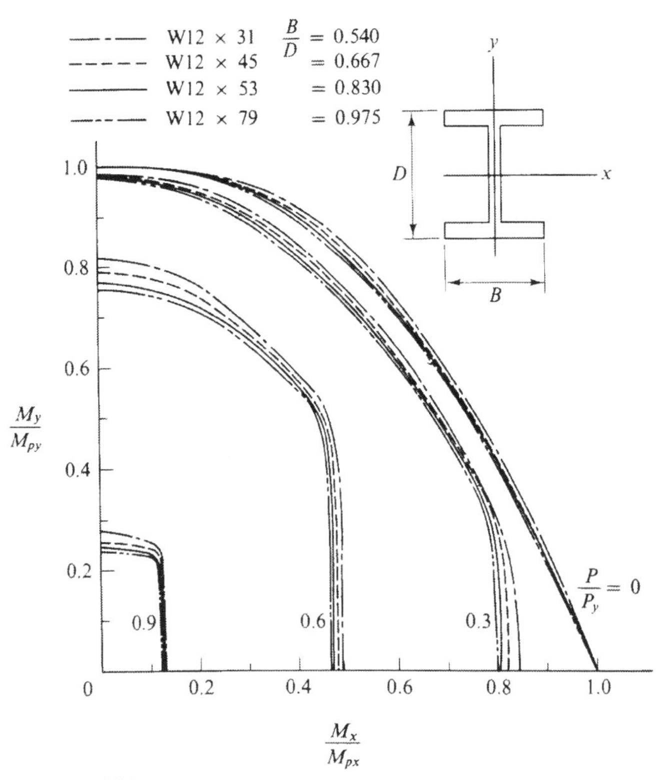

FIGURE 5.34
Comparison of interaction curves of biaxial bending for various ratios of B/D

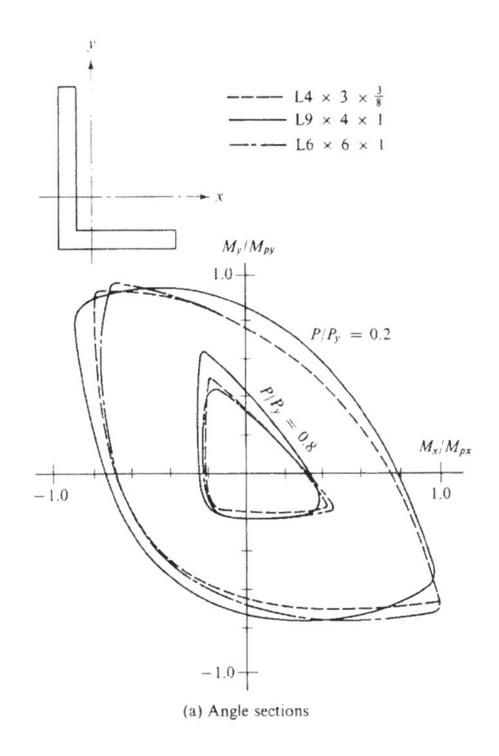

(a) Angle sections

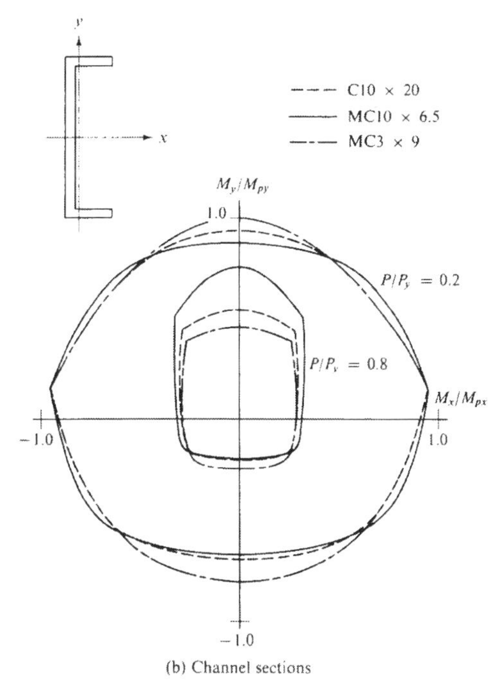

(b) Channel sections

FIGURE 5.35 (a) and (b)
Comparison of interaction curves for angle and channel sections

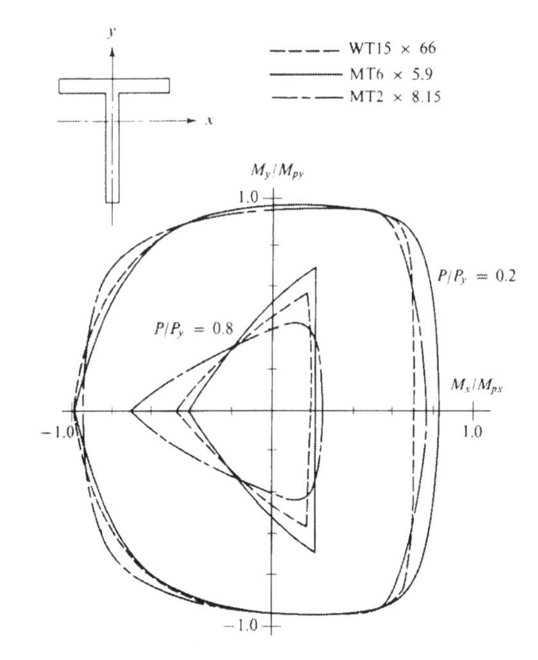

(c) Structural tee sections

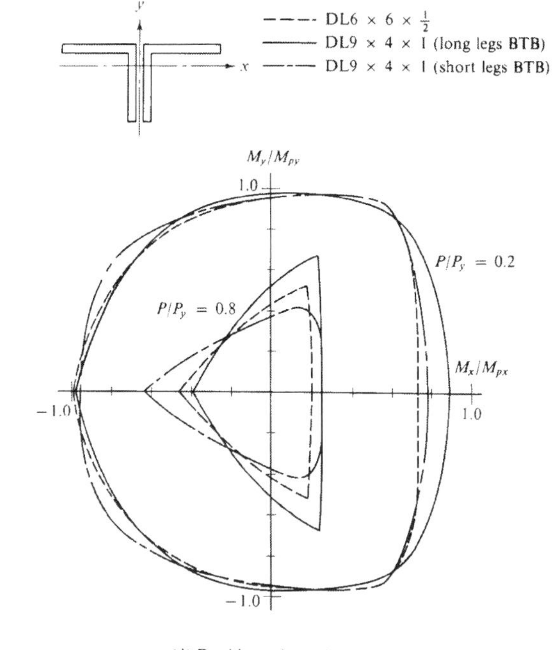

(d) Double angle sections

FIGURE 5.35 (c) and (d)
Comparison of interaction curves for structural tee and double angle sections

Table 5.5 GROUP SELECTION TABLE

W Section	Group	P/P_b	M_x/M_{bx}	M_y/M_{by}
W8 × 17	7	0.305	0.18	0.02
W8 × 20	7	0.375	0.23	0.02
W8 × 24	4	0.449	0.30	0.30
W8 × 28	4	0.526	0.35	0.35
W8 × 31	3	1.000	0.09	1.00
W8 × 35	2	1.000	0.10	0.10
W8 × 40	3	1.301	0.13	1.32
W8 × 58	3	1.918	0.18	2.00
W8 × 67	5	1.000	0.10	0.10
W10 × 33	6	0.536	0.35	0.43
W10 × 39	4	0.732	0.60	0.57
W10 × 45	4	0.861	0.70	0.68
W10 × 54	1	0.570	0.38	0.38
W10 × 60	1	0.632	0.44	0.45
W10 × 66	2	1.913	0.24	0.25
W10 × 72	3	2.372	0.29	3.00
W10 × 77	2	2.236	0.27	0.30
W10 × 89	2	2.591	0.32	0.34
W10 × 100	2	2.923	0.35	0.38
W12 × 40	6	0.642	0.60	0.50
W12 × 45	6	0.720	0.63	0.60
W12 × 50	6	0.720	0.70	0.65
W12 × 53	4	1.000	1.00	0.98
W12 × 58	3	1.853	0.28	2.34
W12 × 65	3	2.099	0.31	3.16
W12 × 72	3	2.332	0.35	3.50
W12 × 79	3	2.567	0.38	3.84
W12 × 85	2	2.435	0.37	0.37
W12 × 92	3	3.004	0.46	4.60
W12 × 99	3	3.241	0.50	4.90
W12 × 106	3	3.481	0.54	5.30
W12 × 120	3	3.934	0.60	6.10
W12 × 133	2	3.862	0.57	0.60
W12 × 161	5	2.405	0.35	0.35
W12 × 190	5	2.845	0.45	0.45
W14 × 43	7	0.796	0.78	0.08
W14 × 48	7	0.898	0.88	0.08
W14 × 53	7	1.000	1.00	0.10
W14 × 61	6	1.000	1.00	1.00
W14 × 68	6	1.121	1.10	1.12
W14 × 74	6	1.226	1.20	1.23
W14 × 78	4	1.496	1.70	7.70
W14 × 84	3	2.700	0.47	4.04
W14 × 87	1	0.917	0.92	0.90
W14 × 95	1	1.000	0.98	1.00
W14 × 103	1	1.089	1.08	1.10
W14 × 119	1	1.263	1.26	1.23
W14 × 127	1	1.348	1.36	1.35
W14 × 136	1	1.440	1.44	1.45
W14 × 142	1	1.514	1.50	1.60
W14 × 150	1	1.609	1.62	1.70
W14 × 176	1	1.887	1.92	2.05
W14 × 184	1	1.983	1.96	2.15
W14 × 193	1	2.076	2.20	2.25
W14 × 202	2	5.982	1.03	1.16
W14 × 211	2	6.241	1.07	1.22
W14 × 219	2	6.491	1.12	1.26
W14 × 228	2	6.768	1.15	1.32
W14 × 237	2	7.028	1.20	1.40
W14 × 246	2	7.306	1.25	1.44
W14 × 264	2	7.848	1.40	1.56
W14 × 287	5	4.366	0.75	0.85
W14 × 314	5	4.794	0.78	0.93
W14 × 342	5	5.224	0.95	1.01
W14 × 398	5	6.100	1.05	1.10
W14 × 426	5	6.549	1.15	1.20

5.6 INTERACTION RELATIONS FOR WIDE-FLANGE SECTIONS

5.6.1 Design Nomographs

At a preliminary design stage, some rough but general information on interaction relations is desired. For this purpose, Ringo, *et al.* (1973) developed a set of design charts for most wide-flange shapes in form of nomographs. Seventy-three wide-flange shapes are classified into seven groups as listed in Table 5.5. For each group, a basic section is specified for which the interaction curves are prepared. In order to allow this interaction graph, with a family of P curves, to be used for the analysis of all cross sections in the same group, geometric relationships between thrust P_b and moments M_{bx}, M_{by} on the base section and thrust P and moments M_x, M_y on all other sections of the group are developed. Typical interaction curves along with their geometrical relationships are shown in Figs. 5.36 and 5.37.

As illustrated in Fig. 5.36 for the "Group 1", interaction relations between M_x and M_y of an actual section are obtained through their constant ratios to M_{bx}

FIGURE 5.36
Nomographs for group 1, 1 kip = 4.45 kN, 1 in-kip = 11.3 cm-kN

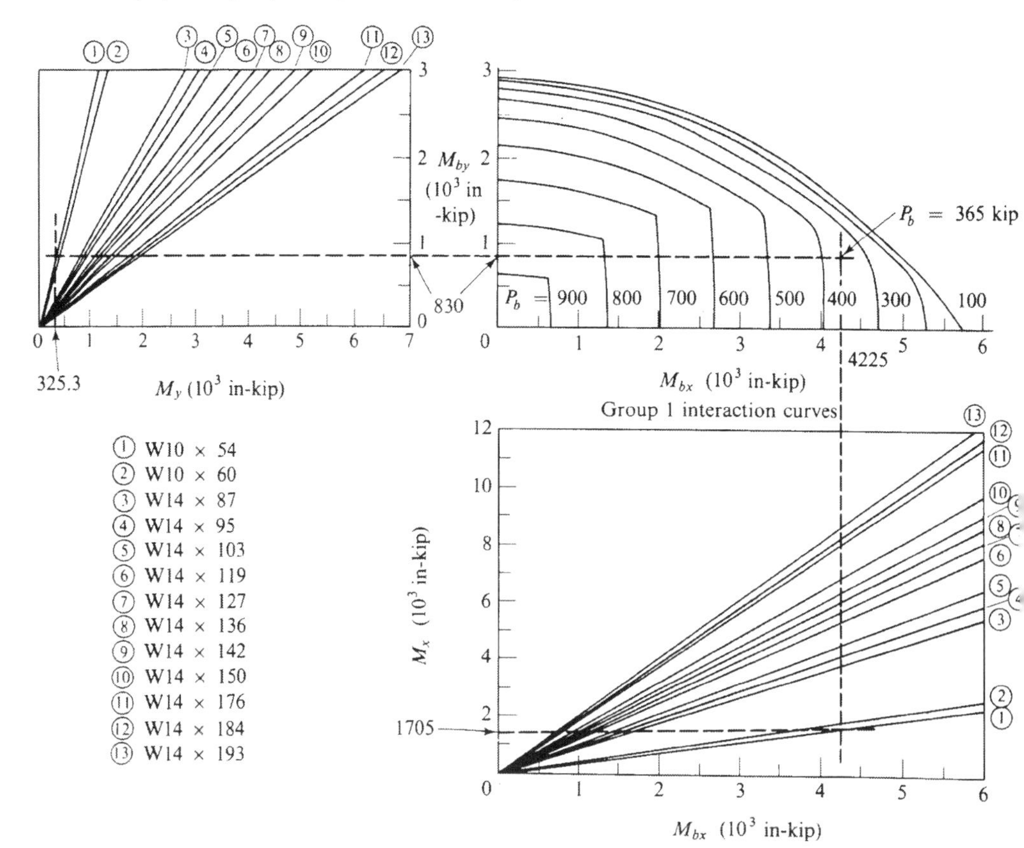

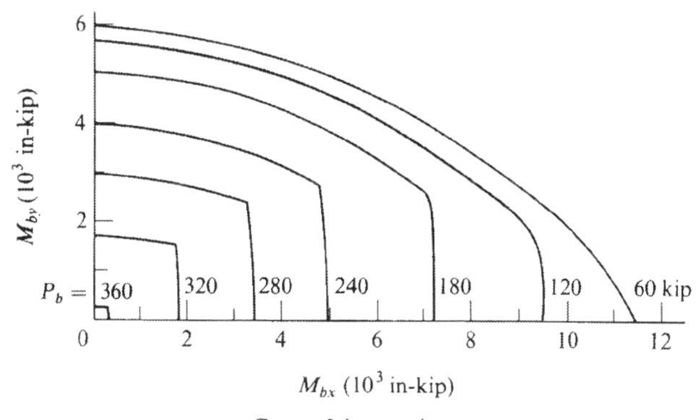

Group 2 interaction curve

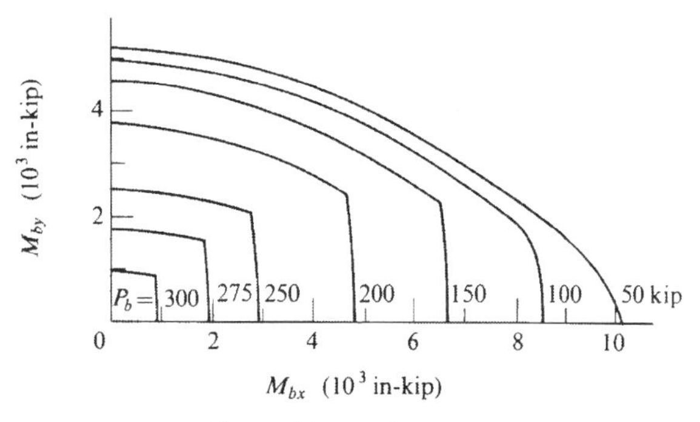

Group 3 interaction curve

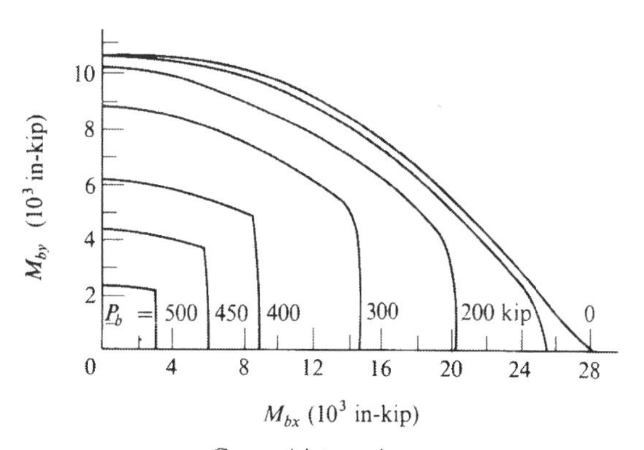

Group 4 interaction curve

FIGURE 5.37 (a)
Interaction curves for basic section (groups 2 to 4) 1 kip = 4.45 kN, 1 in-kip = 11.3 cm-kN

and M_{by}. The ratios for each section are also listed in Table 5.5. In the upper right of Fig. 5.36, the family of interaction curves for the base section (W14 × 95) is shown. In the lower right and upper left of Fig. 5.36, the straight lines relating the various other sections of the group to the base section are shown. An illustrative example

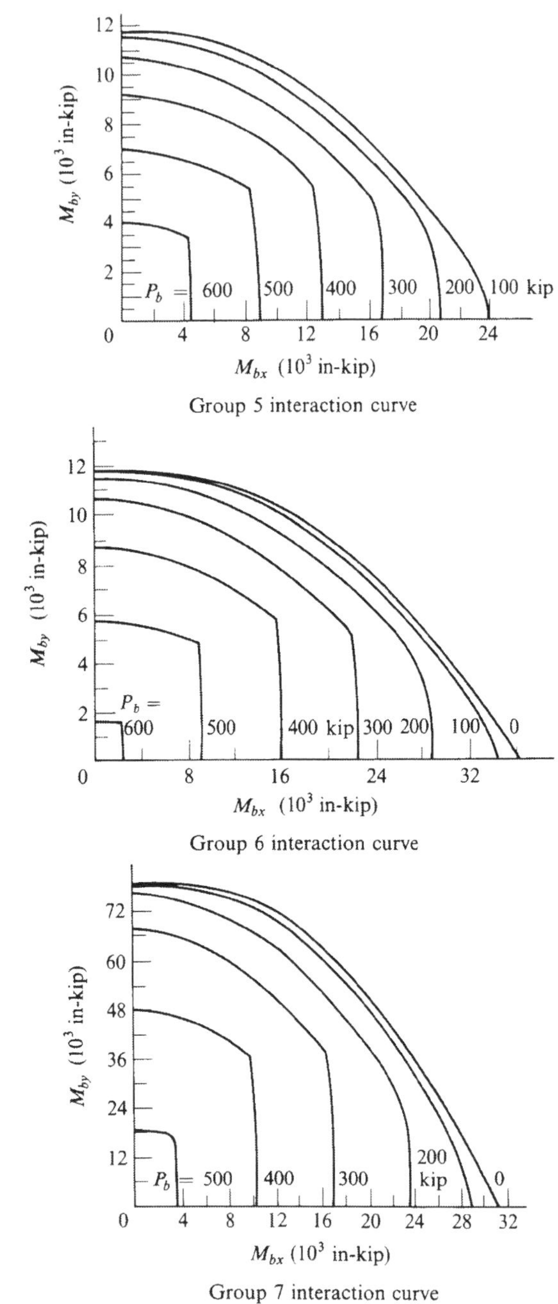

FIGURE 5.37 (b)
Interaction curves for basic sections (Groups 5 to 7) (1 kip = 4.45 kN, 1 in-kip = 113 mm-kN)

shows how one uses the graphs to relate the values of M_x and M_y and the corresponding maximum load P for a given section. The details of this example are given below.

$$\text{applied } M_x = 1705.0 \text{ kip-in.}$$

$$\text{applied } M_y = 325.3 \text{ kip-in.}$$

Selecting a W10 × 54 column section of ASTM A36 steel ($\sigma_y = 36$ ksi), find the maximum compressive thrust, P.

From Table 5.5, the W10 × 54 is a member of Group 1, in which the W14 × 95 shown in Fig. 5.36 is the base section and has a load capacity $P_b = 365$ kip. This is determined from the equivalent M_{bx} and M_{by} values, both of which may be determined graphically as indicated on Fig. 5.36. Also from Table 5.5, one obtains the ratio $P/P_b = 0.570$ for the W10 × 54. Therefore P is computed as $0.570 \times 365 = 208$ kip. This is the maximum thrust which can be carried by the member when the given moments exist on the braced cross section. The interaction curves for the other six basic sections are shown in Fig. 5.37. These are valid for wide flange sections of mild steel ($\sigma_y = 36$ ksi or 248.4 MN/m^2). These nomographs are especially useful for design of biaxially loaded columns in the case that the critical section occurs at the end of the members.

5.6.2 Simple Interaction Equations

To handle the interaction relation analytically, simple interaction equations are required. As a general form of the interaction equation, assume

$$\frac{m_x^\alpha}{1 - p^\beta} + \frac{m_y^\mu}{1 - p^\nu} + p^\gamma = 1 \tag{5.114}$$

in which α, β, γ, μ and ν are constants to be determined from the exact interaction relation. For most wide-flange sections, it is found that good agreement results when $\mu = 1$ and $\nu = \infty$.

When strong axis bending moment M_x is large, weak axis bending moment M_y has little effect on the interaction as shown in Fig. 5.31(c) and (d). Thus, the following two equations are assumed as the interaction equations for wide-flange sections:

$$\begin{aligned} f_1(m_x, m_y, p) &= \frac{m_x^\alpha}{1 - p^\beta} + m_y + p^\gamma - 1 = 0 \\ f_2(m_x, m_y, p) &= m_x + p^\delta - 1 = 0 \end{aligned} \tag{5.115}$$

The actual interaction equation, $f(m_x, m_y, p) = 0$, is given by

$$\begin{aligned} f &= f_1(m_x, m_y, p) \quad \text{when} \quad m_y \geq \bar{m}_y \\ f &= f_2(m_x, m_y, p) \quad \text{when} \quad m_y \leq \bar{m}_y \end{aligned} \tag{5.116}$$

and

$$\bar{m}_y = 1 - p^\gamma - \frac{(1 - p^\delta)^\alpha}{1 - p^\beta} \tag{5.117}$$

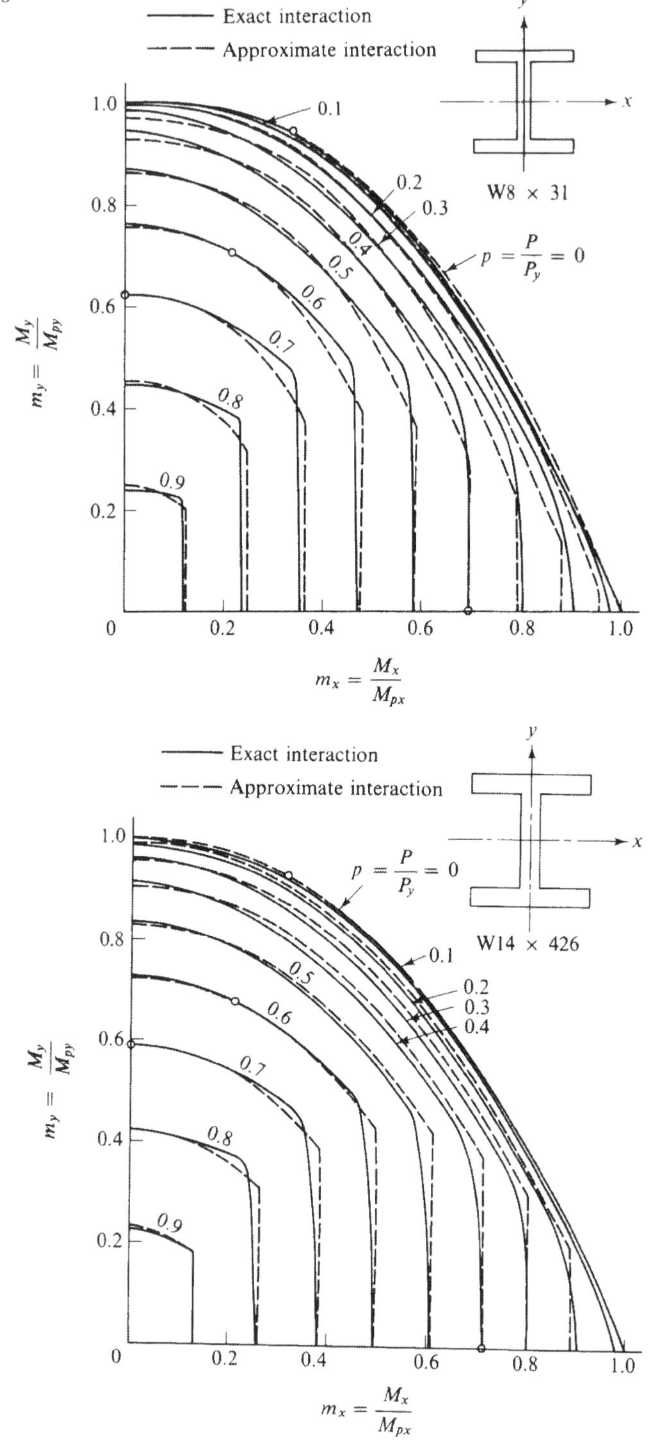

FIGURE 5.38
Approximation of interaction curves for a light-weight wide flange section and a heavy-weight wide flange section

The four constants, α, β, γ and δ, are determined using values of p, m_x and m_y corresponding to four representative points in the exact interaction relation.

As examples, the interaction curves for two wide-flange shapes W8 × 31 and W14 × 426 were calculated and shown in Fig. 5.38. The solid lines show the exact solutions and the dotted lines are results using the simple interaction equations (5.116). The approximation appears to be quite good. In the figures, the four points indicated by the circles are the selected exact points. Their numerical values and the corresponding four constants are as follows:

*W*8 × 31:

$$
\begin{aligned}
p_1 &= 0, & m_{x1} &= 0.34, & m_{y1} &= 0.93 \\
p_2 &= 0.7, & m_{x2} &= 0, & m_{y2} &= 0.62 \\
p_3 &= 0.6, & m_{x3} &= 0.22, & m_{y3} &= 0.70 \\
p_4 &= 0.4, & m_{x4} &= 0.69, & m_{y4} &= 0
\end{aligned}
\tag{5.118}
$$

and

$$\alpha = 2.453, \quad \beta = 1.209, \quad \gamma = 2.714, \quad \delta = 1.987$$

*W*14 × 426:

$$
\begin{aligned}
p_1 &= 0, & m_{x1} &= 0.31, & m_{y1} &= 0.92 \\
p_2 &= 0.7, & m_{x2} &= 0, & m_{y2} &= 0.59 \\
p_3 &= 0.6, & m_{x3} &= 0.23, & m_{y3} &= 0.66 \\
p_4 &= 0.4, & m_{x4} &= 0.71, & m_{y4} &= 0
\end{aligned}
\tag{5.119}
$$

and

$$\alpha = 2.176, \quad \beta = 2.678, \quad \gamma = 2.480, \quad \delta = 1.357$$

5.6.3 Force-Deformation Rate Relations

The analytical description of the interaction relations given in the preceding section may be considered as a suitable basis for a three-dimensional space frame analysis. With this concept, the interaction surface [Eq. (5.115)] is assumed to be the perfectly plastic yield surface, and the force-deformation rate relations can then be derived from the normality condition (flow rule, see Chap. 2, Vol. 1).

Thus, if $f(m_x, m_y, p) = 0$ denotes the yield condition, with $f < 0$ corresponding to stress states below yield, then

$$\dot{\kappa}_x = \frac{\lambda \dfrac{\partial f}{\partial m_x}}{M_{px}}$$

$$\dot{\kappa}_y = \frac{\lambda \dfrac{\partial f}{\partial m_y}}{M_{py}} \tag{5.120}$$

$$\dot{\varepsilon}_o = \frac{\lambda \dfrac{\partial f}{\partial p}}{P_y}$$

If either $f < 0$ or $f = 0$ and $\dot{f} < 0$

$$\dot{\kappa}_x = \dot{\kappa}_y = \dot{\varepsilon}_o = 0 \tag{5.121}$$

in which λ is a positive scalar.

According to the concept of perfect plasticity, the vector representing the deformation rate is normal to the yield surface at a regular point. At a singular point of the yield surface, the deformation rate vector lies within the directions of the normals to the surface at adjacent points. For example, the normals drawn to curve AB and line BC in Fig. 5.39 are the projections on plane $p =$ constant of possible deformation rates for stress points lying on curve AB and line BC. When the stress point lies at corner B in Fig. 5.39, the deformation rate vector lies in the fan bounded by the normals to the sides which meet at the corner. For face AB of the yield surface, $m_y > \bar{m}_y$

FIGURE 5.39
Yield surface and strain vectors

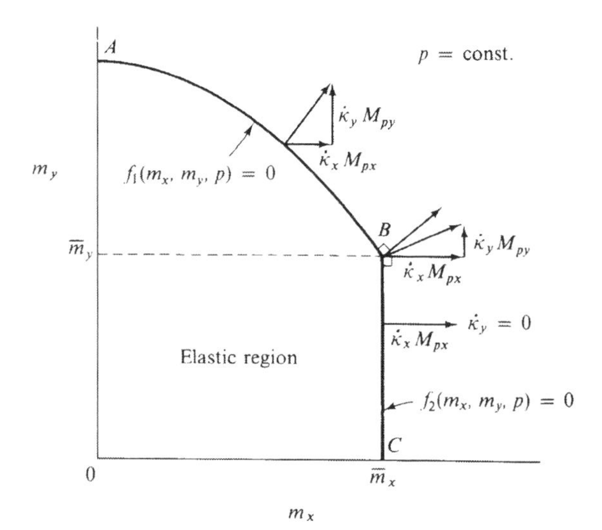

$$\dot{\kappa}_x = \frac{\lambda}{M_{px}} \frac{\alpha}{1 - p^\beta} m_x^{\alpha - 1}$$

$$\dot{\kappa}_y = \frac{\lambda}{M_{py}} \tag{5.122}$$

$$\dot{\varepsilon}_o = \frac{\lambda}{P_y}\left[\frac{\beta p^{\beta - 1}}{(1 - p^\beta)^2} m_x^\alpha + \gamma p^{\gamma - 1}\right]$$

in which $\bar{m}_y$ is the value at the edge of the yield surface as given by Eq. (5.117). For face BC of the yield surface, $m_y < \bar{m}_y$

$$\dot{\kappa}_x = \frac{\lambda}{M_{px}}$$

$$\dot{\kappa}_y = 0 \tag{5.123}$$

$$\dot{\varepsilon}_o = \frac{\lambda}{P_y} \delta p^{\delta - 1}$$

At edge B of the yield surface, $m = \bar{m}_y$, it is convenient to introduce a scalar parameter $\rho (0 \leq \rho \leq 1$ at most) of position and time whose increase corresponds to a transition between the regular faces reckoned in a counterclockwise sense around the yield curve CBA:

$$\dot{\kappa}_x = \frac{\lambda}{M_{px}}\left[1 - \rho + \frac{\alpha}{1 - p^\beta} m_x^{\alpha - 1} \rho\right]$$

$$\dot{\kappa}_y = \frac{\lambda}{M_{py}} \rho \tag{5.124}$$

$$\dot{\varepsilon}_o = \frac{\lambda}{P_y}\left[\frac{\beta p^{\beta - 1}}{(1 - p^\beta)^2} m_x^\alpha \rho + \gamma p^{\gamma - 1} \rho + \delta p^{\delta - 1}(1 - \rho)\right]$$

5.7 INTERACTION RELATIONS FOR REINFORCED CONCRETE SECTIONS

Moment-curvature relationships of reinforced concrete sections will be discussed in Sect. 6.7, Chap. 6. The maximum points on the moment-curvature curves represent the maximum strength of a biaxially loaded beam-column segment. The maximum strength is controlled either by the maximum concrete strain ε_o which occurs at a corner of the section or by the overall stress distribution of the cross section. The ratio of the maximum concrete compressive strain ε_o and the strain at the compression strength ε_c' of concrete lies between the values 2.0 and 3.0 and generally very close to the constant value 3.0 at the maximum strength state. This fact can be used as an ultimate strength criteria for reinforced concrete section. It was also found that the maximum strength of beam-column segments in biaxial bending and compression

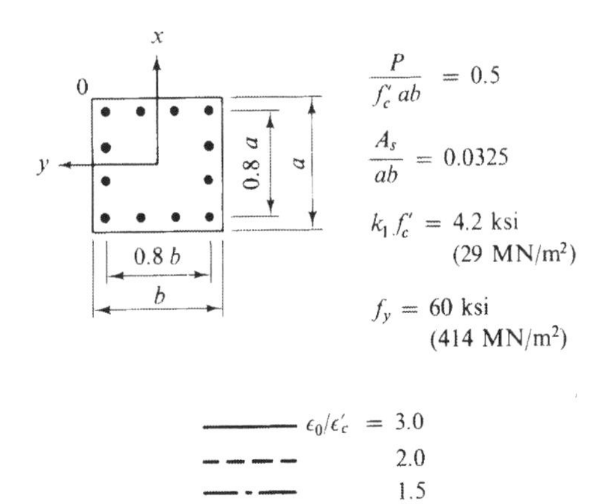

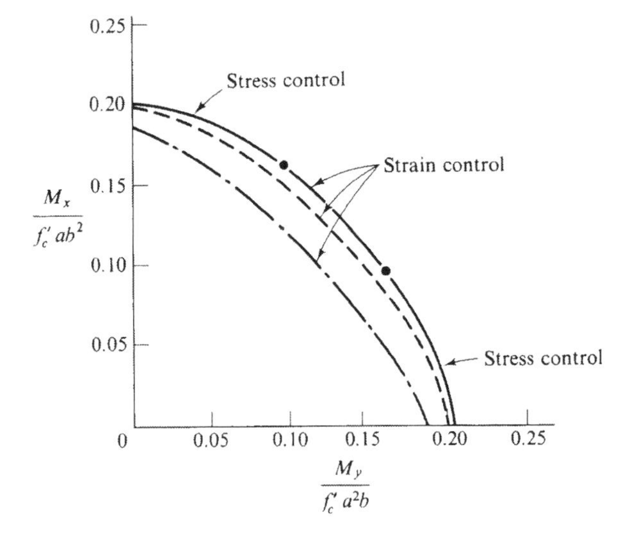

FIGURE 5.40 (a)
Interaction curves for reinforced concrete sections—Standard case

are not unduly sensitive to the variations in the assumed concrete ultimate strain value which is often chosen in the range between 0.003 and 0.004.

In the numerical analysis, the cross section is partitioned by a rectangular grid into a large number of small elemental areas of steel and concrete [see Fig. 6.26(d)]. The moment-thrust-curvature relations to be described in the following chapter are obtained by step-by-step application of an analytically developed linear force-deformation equation (Chen and Shoraka, 1975). The method of analysis is most suitable and efficient for computer solution. Details of the analytical formulation and procedures are given in Chap. 6.

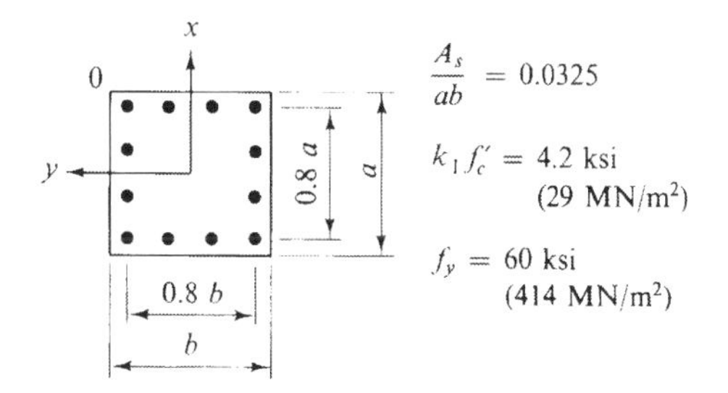

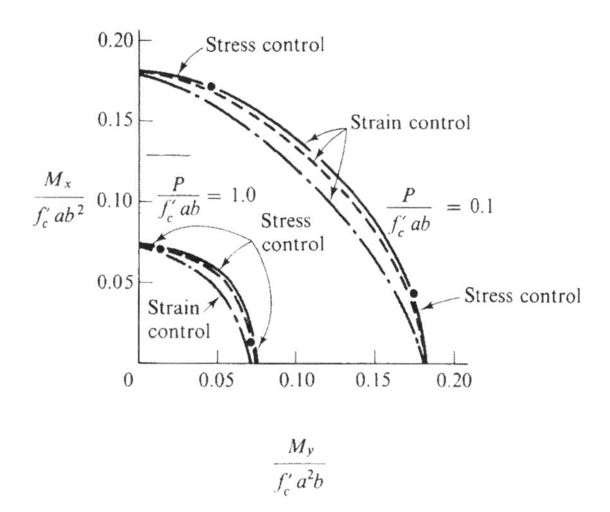

FIGURE 5.40 (b)
Axial compression force effect on the strength of a reinforced concrete section

The maximum loads obtained in this way for the standard cross section [Fig. 5.40(a)] with three values of strain ratio, $\varepsilon_o/\varepsilon_c' = 1.5$, 2.0 and 3.0 are represented by the interaction curves in Figs. 5.40. The small circles in these figures indicate the boundary of the regions where the maximum load is controlled either by the maximum concrete strain or by the overall stress distribution of the cross section. The important factors influencing the maximum load carrying capacity of a biaxially loaded beam-column segment are the axial compression force, P, the concrete quality, $k_1 f_c'$, steel quality, f_y, and percentage of reinforcement A_s/ab, as shown in the insets of Fig. 5.40. Since the interaction curves are nondimensionalized, they can be directly used in analysis and design computations.

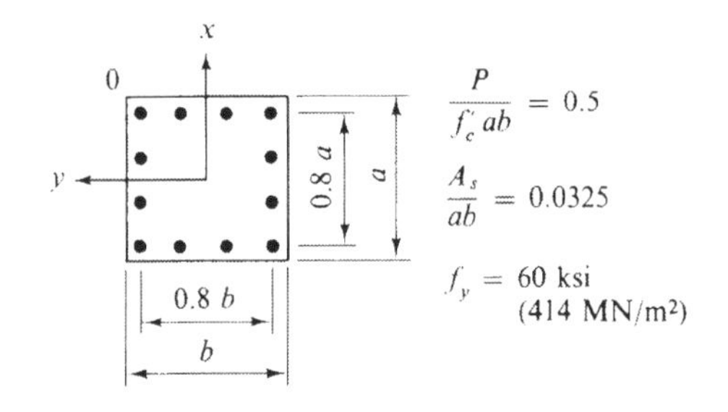

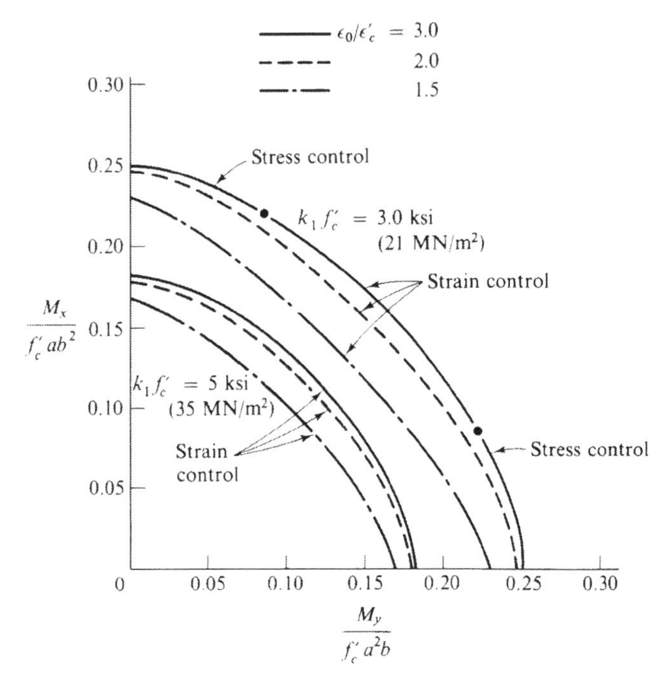

FIGURE 5.40 (c)
Concrete quality effect on the strength of a reinforced concrete section

Simple interaction equations

The general form of the interaction curves shown in Figs. 5.40 may be approximated by a nondimensional interaction equation (Bresler, 1960)

$$\left(\frac{M_x}{M_{xo}}\right)^{\alpha} + \left(\frac{M_y}{M_{yo}}\right)^{\alpha} = 1.0 \tag{5.125}$$

where M_{xo} and M_{yo} represent the load carrying capacities of a particular beam-column segment under compression and uniaxial bending moment about x and y axes, respectively. Thus, for a given compression P, M_{xo} and M_{yo} are the values given on the $M_y = 0$ and $M_x = 0$ axes shown in Fig. 5.40. The value α is the exponent depending

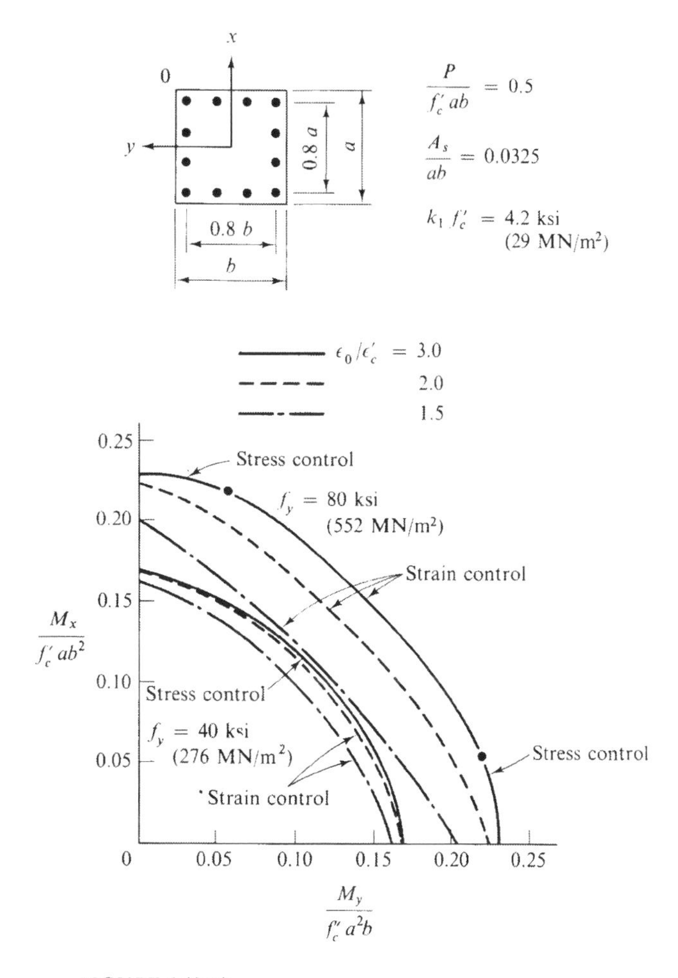

FIGURE 5.40 (d)
Steel quality effect on the strength of a reinforced concrete section

on column dimensions, amount and distribution of steel reinforcement, stress-strain characteristics of steel and concrete, and magnitude of axial compression. For a given compression and a given beam-column characteristic, the value of α is a numerical constant.

The interaction curves given previously in Figs. 5.40(a) and (b) for the particular case of strain ratio $\varepsilon_o/\varepsilon_c' = 1.5$ are now nondimensionalized by the values M_{xo} and M_{yo} and plotted in Fig. 5.41. These curves corresponding to constant values of $P/f_c'ab = 0.1$, 0.5 and 1.0 which may be thought of as "load contours" (inset, Fig. 5.41a). Using Eq. (5.125), values of α are calculated for this beam-column segment. The calculated values of α are found, varying from 1.3 to 1.4 for $P/f_c'\,ab = 0.1$ and 0.5 but jumping to 1.7 for $P/f_c'\,ab = 1.0$. The comparison between the actual curves computed directly on the basis of stress-strain relations and the theoretical curves obtained from

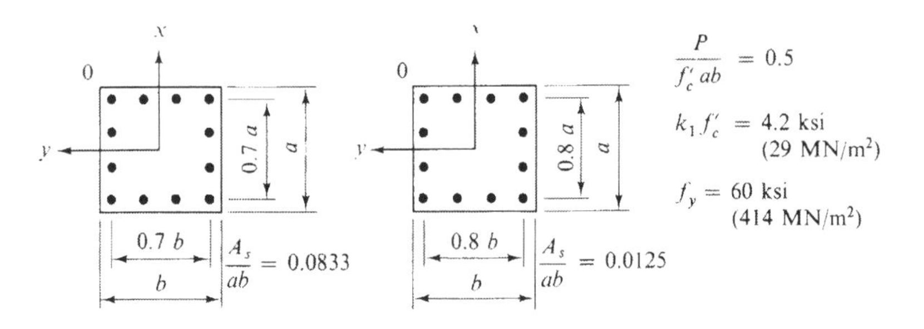

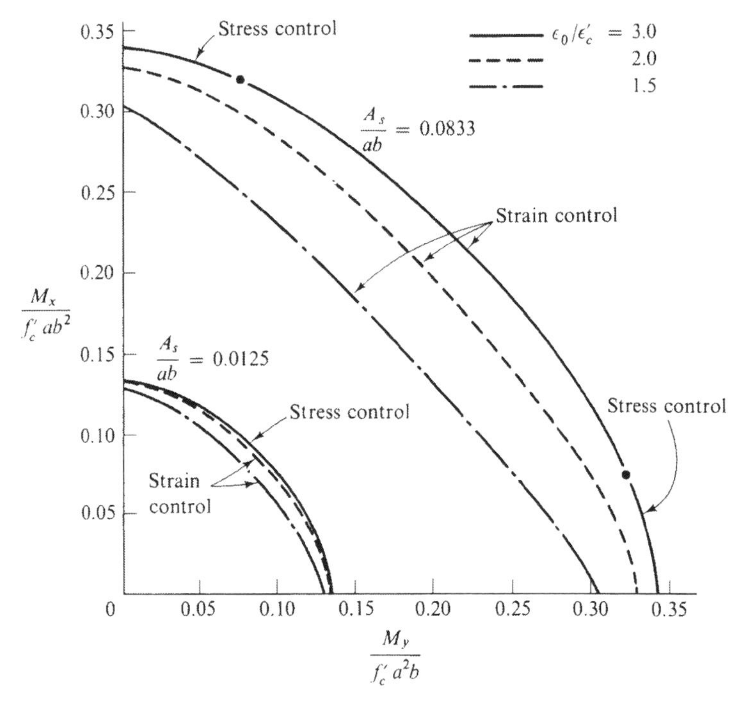

(e) Percentage of reinforcement effect

FIGURE 5.40 (e)
Percentage of steel reinforcement effect on the strength of reinforced concrete sections

Eq. (5.125) is also shown in Fig. 5.41 and good agreement is observed. The values of α for columns with a wide range of variation in values of f'_c, f_y and A_s/ab are tabulated in Table 5.6 for the particular case of strain ratio $\varepsilon_o/\varepsilon'_c = 1.5$ (recommended by ACI). In general, the values of α in the range 1.1 to 1.4 are seen to give a good approximation for all the cases investigated in the low and moderate axial compression range, but large variation in values of α is observed for columns with high axial compression.

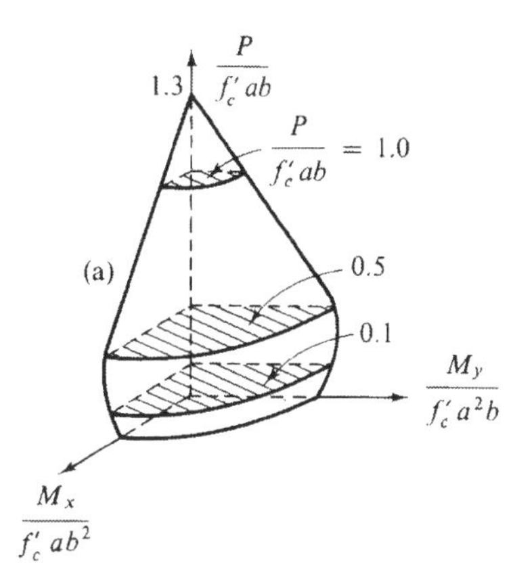

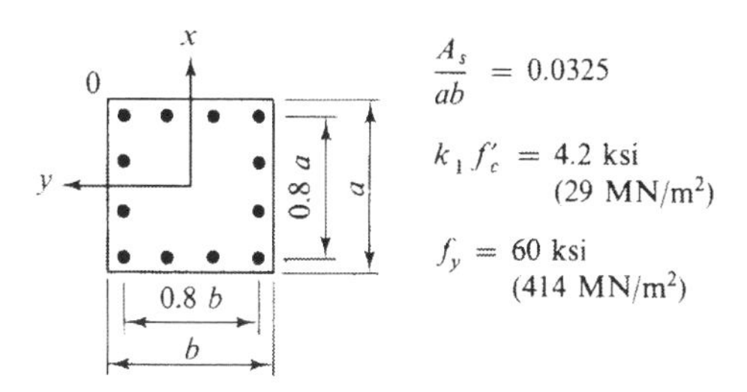

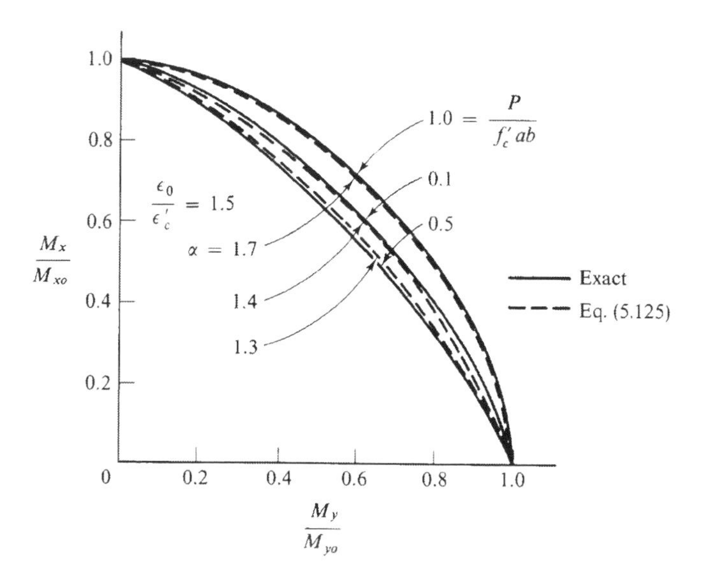

FIGURE 5.41
Comparison of interaction curves

Table 5.6 COMPUTED VALUES OF α IN EQ. 5.125

$\frac{P}{f_c' ab}$	$k_1 f_c'$ (ksi)	f_y (ksi)	$\frac{A_s}{ab}$	α	Note
0.5	4.2	60	0.0325	1.3	
0.1	4.2	60	0.0325	1.4	
1.0	4.2	60	0.0325	1.7	
0.5	3	60	0.0325	1.3	$\frac{\varepsilon_o}{\varepsilon_c'} = 1.5$
0.5	5	60	0.0325	1.4	
0.5	4.2	40	0.0325	1.4	for all cases.
0.5	4.2	80	0.0325	1.2	
0.5	4.2	60	0.0125	1.4	
0.5	4.2	60	0.0833	1.1	

1 ksi = 6.9 MN/m^2.

5.8 SUMMARY

The resultant forces acting on a section under the combined action of thrust and biaxial bending moment are

$$P = \int_A \sigma \, dx \, dy$$
$$M_x = \int_A \sigma y \, dx \, dy \tag{5.126}$$
$$M_y = -\int_A \sigma x \, dx \, dy$$

If the section is fully plastified, the stress σ in a metal section takes either one of the tension yield stress $+\sigma_y$ or the compression yield stress $-\sigma_y$ of the material. The border line between tension and compression is the neutral line NA, which may be expressed by the function,

$$f(x, y) = 0 \tag{5.127}$$

from which the sign of stress at a point (x, y) is known by

$$\text{SIGN}[f(x, y)] = \begin{cases} 1 & \text{for} \quad f(x, y) > 0 \\ -1 & \text{for} \quad f(x, y) < 0 \end{cases} \tag{5.128}$$

Thus, from Eq. (5.126), we have

$$P = \sigma_y \int_A \text{SIGN}[f(x, y)] \, dx \, dy$$
$$M_x = \sigma_y \int_A \text{SIGN}[f(x, y)] \, y \, dx \, dy \tag{5.129}$$
$$M_y = -\sigma_y \int_A \text{SIGN}[f(x, y)] \, x \, dx \, dy$$

Once the neutral axis function $f(x, y)$ is selected, the resultant forces P, M_x, M_y are directly calculated from Eq. (5.129) and a point on the interaction surface can be obtained.

However, an arbitrary choice of the function $f(x, y)$ will result in a smaller interaction surface than the exact one because this is a lower bound approach in the limit analysis. In order to obtain the largest possible interaction surface, which is the exact one, the calculus of variation technique was utilized and found that the neutral line must be a straight line,

$$f(x, y) = \alpha x + \beta y + 1 = 0 \tag{5.130}$$

Thus, Eq. (5.129) has the following form

$$\begin{aligned} P &= \sigma_y \int_A \mathrm{SIGN}(\alpha x + \beta y + 1)\, dx\, dy \\ M_x &= \sigma_y \int_A \mathrm{SIGN}(\alpha x + \beta y + 1)\, y\, dx\, dy \\ M_y &= -\sigma_y \int_A \mathrm{SIGN}(\alpha x + \beta y + 1)\, x\, dx\, dy \end{aligned} \tag{5.131}$$

These are the exact interaction relations in terms of two parameters α and β. Elimination of α and β from these three equations will result in a single interaction equation. Since this is not possible for a general cross section, the following approaches were proposed:

1. Approximate method for specific shapes (Sect. 5.2).
2. Exact method using superposition (Sect. 5.3 and Sect. 5.4).
3. Numerical integration (Sect. 5.7).

For the purpose of practical applications, the exactness of the solution is not very important, but simplicity in expression is preferable, for which some efforts are made here in Sect. 5.6 and Sect. 5.7 to obtain simple interaction equations for various sections including reinforced concrete sections.

5.9 PROBLEMS

5.1 Derive the interaction equation of a regular triangle section.

5.2 Discuss the correlation between the shape of cross section and the shape of interaction surface.

5.3 Discuss the physical meaning of the volume inside the interaction surface.

5.4 The rectangular section shown in Fig. 5.42 is subjected to an axial compressive load P, and to a biaxial bending moment having components M_x and M_y, and a torque T. Three separate cases of full plasticity, cases (1), (2) and (3) in the figure (hatched portions are yielding in compression), must be considered to establish the yield surface in the first octant.

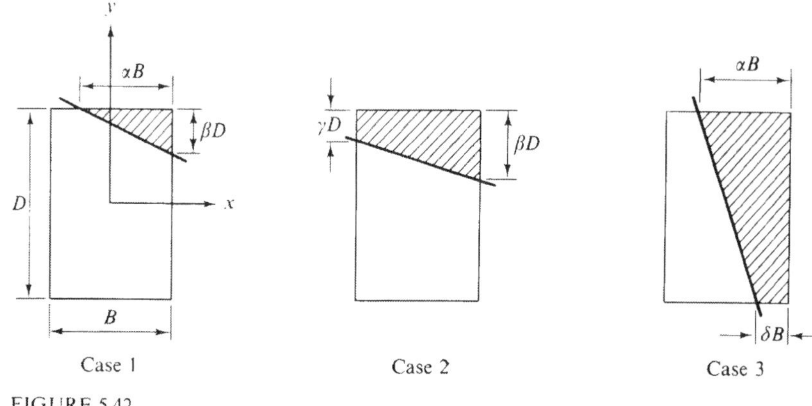

FIGURE 5.42
Figure for Prob. 5.4(a)

Fully plastic stress resultants: $(D \geq B)$

$$P_y = \sigma_y BD, \qquad T_p = \frac{\sigma_y B^2}{2/3}(D - B/3)$$

$$M_{px} = \frac{1}{4}\sigma_y BD^2, \quad M_{py} = \frac{1}{4}\sigma_y B^2 D$$

(a) Prove the following relations

Case 1

$$\left.\begin{aligned} p &= \left(\frac{\sigma}{\sigma_y}\right)(1 - \alpha\beta) \\ m_x &= 4\left(\frac{\sigma}{\sigma_y}\right)\alpha\beta\left(\frac{1}{2} - \frac{\beta}{3}\right) \\ m_y &= 4\left(\frac{\sigma}{\sigma_y}\right)\alpha\beta\left(\frac{1}{2} - \frac{\alpha}{3}\right) \end{aligned}\right\} \tag{5.132}$$

$$1.2p\sqrt{1 - t^2} + 0.6p^2 - \frac{0.8p^3}{\sqrt{1 - t^2}} + 0.9(\sqrt{1 - t^2} - p)(m_x + m_y) - 0.45m_x m_y + t^2 = 1 \tag{5.133}$$

Case 2

$$\left.\begin{aligned} p &= \left(\frac{\sigma}{\sigma_y}\right)(1 - \beta - \gamma) \\ m_x &= \left(\frac{\sigma}{\sigma_y}\right)\left[2\beta + 2\gamma - \frac{4}{3}(\beta + \gamma)^2 + \frac{4}{3}\beta\gamma\right] \\ m_y &= \frac{2}{3}\left(\frac{\sigma}{\sigma_y}\right)(\beta - \gamma) \end{aligned}\right\} \tag{5.134}$$

$$p^2 + \frac{3}{4}m_y^2 + m_x\sqrt{1 - t^2} + t^2 = 1 \tag{5.135}$$

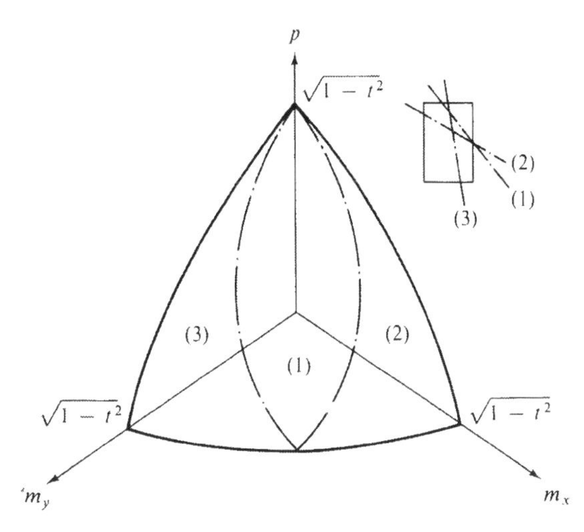

FIGURE 5.43
Figure for Prob. 5.4(b)

Case 3

$$
\left.
\begin{aligned}
p &= \left(\frac{\sigma}{\sigma_y}\right)(1 - \alpha - \delta) \\
m_x &= \frac{2}{3}\left(\frac{\sigma}{\sigma_y}\right)(\alpha - \delta) \\
m_y &= \left(\frac{\sigma}{\sigma_y}\right)\left[2\alpha + 2\delta - \frac{4}{3}(\alpha + \delta)^2 + \frac{4}{3}\alpha\delta\right]
\end{aligned}
\right\} \tag{5.136}
$$

$$
p^2 + m_y\sqrt{1 - t^2} + \frac{3}{4}m_x^2 + t^2 = 1 \tag{5.137}
$$

(b) Show that the interaction eqs. (5.133), (5.135) and (5.137) define the surface elements as shown in Fig. 5.43.

5.5 Repeat Prob. 5.4 for the I-section shown in Fig. 5.44; the thickness of flanges and web is *thin.*

Fully plastic stress resultants:

$$
C = \frac{A_w}{A_f} = \frac{wD}{tB}
$$

$$
P_y = \sigma_y A_f(2 + C), \qquad T_p = \frac{\sigma_y}{\sqrt{3}} A_f\left(t + \frac{Cw}{2}\right)
$$

$$
M_{px} = \sigma_y A_f D(1 + C/4), \qquad M_{py} = \frac{\sigma_y A_f B}{2}
$$

A_f = area of one flange
A_w = area of web
D = depth of section, center to center of flanges
B = width of flange
t = flange thickness = A_f/B, and
w = web thickness $\simeq A_w/D$

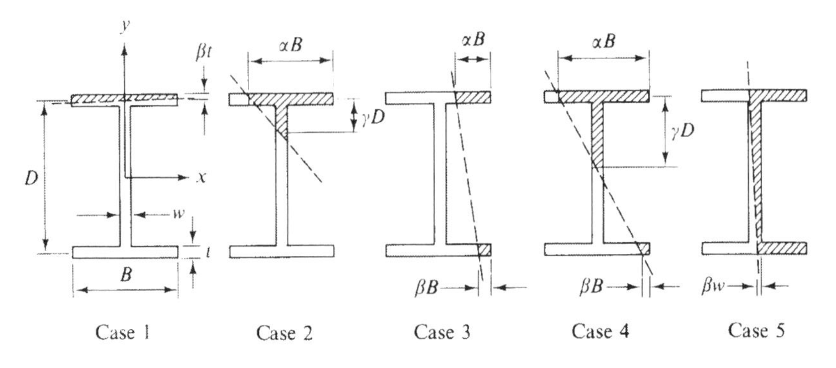

FIGURE 5.44
Figure for Prob. 5.5(a)

(a) Prove the following relations

Case 1

$$\left.\begin{aligned} p &= \left(\frac{\sigma}{\sigma_y}\right)\left(1 - \frac{\beta}{1 + C/2}\right) \\ m_x &= \left(\frac{\sigma}{\sigma_y}\right)\left(\frac{\beta}{1 + C/4}\right) \\ m_y &= \text{indeterminate} \end{aligned}\right\} \tag{5.138}$$

$$p + \frac{4 + C}{4 + 2C} m_x - \sqrt{1 - t^2} = 0 \tag{5.139}$$

Case 2

$$\left.\begin{aligned} p &= \left(\frac{\sigma}{\sigma_y}\right)\left(1 - \frac{\alpha + \gamma C}{1 + C/2}\right) \\ m_x &= \left(\frac{\sigma}{\sigma_y}\right)\left[\frac{\alpha + \gamma C(1 - \gamma)}{1 + C/4}\right] \\ m_y &= 2\left(\frac{\sigma}{\sigma_y}\right)\alpha(1 - \alpha) \end{aligned}\right\} \tag{5.140}$$

$$\frac{(4 + C)m_x}{\sqrt{1 - t^2}} - 2\sqrt{1 - \frac{2m_y}{\sqrt{1 - t^2}}} + \frac{1}{C}\left[\frac{p(2 + C)}{\sqrt{1 - t^2}} + \sqrt{1 - \frac{2m_y}{\sqrt{1 - t^2}}} - 1\right]^2 = 2 + C \tag{5.141}$$

Case 3

$$\left.\begin{aligned} p &= \left(\frac{\sigma}{\sigma_y}\right)\left(1 - \frac{\alpha + \beta}{1 + C/2}\right) \\ m_x &= \left(\frac{\sigma}{\sigma_y}\right)\left(\frac{\alpha - \beta}{1 + C/4}\right) \\ m_y &= 2\left(\frac{\sigma}{\sigma_y}\right)(\alpha - \alpha^2 + \beta - \beta^2) \end{aligned}\right\} \tag{5.142}$$

$$m_y\sqrt{1-t^2}+\left[p(1+C/2)-\frac{C}{2}\sqrt{1-t^2}\right]^2+\left(1+\frac{C}{4}\right)^2 m_x^2+t^2=1 \tag{5.143}$$

Case 4

$$\left.\begin{aligned} p&=\left(\frac{\sigma}{\sigma_y}\right)\left(1-\frac{\alpha+\beta+\gamma C}{1+C/2}\right)\\ m_x&=\left(\frac{\sigma}{\sigma_y}\right)\left[\frac{\alpha-\beta+\gamma C(1-\gamma)}{1+C/4}\right]\\ m_y&=2\left(\frac{\sigma}{\sigma_y}\right)(\alpha-\alpha^2+\beta-\beta^2)\end{aligned}\right\}\quad \gamma=\frac{\alpha-1/2}{\alpha-\beta} \tag{5.144}$$

$$\begin{aligned}&(2+C/2)m_x-C\sqrt{1-t^2}\\ &\quad-\sqrt{(C^2-12)(1-t^2)+[12m_y-(4C+C^2)m_x]\sqrt{1-t^2}+(4+C)^2m_x^2+3p^2(C+2)^2}\\ &\quad+[54C(1-t^2)^{3/2}-54m_y(1-t^2)]/\{(4+C)m_x-2C\sqrt{1-t^2}\\ &\quad+\sqrt{(C^2-12)(1-t^2)+[12m_y-(4C+C^2)m_x]\sqrt{1-t^2}+(4+C)^2m_x^2+3p^2(C+2)^2}\}^2=0\end{aligned} \tag{5.145}$$

Case 5

$$\left.\begin{aligned} p&=\text{indeterminate}\\ m_x&=\text{indeterminate}\\ m_y&=\left(\frac{\sigma}{\sigma_y}\right)\end{aligned}\right\} \tag{5.146}$$

$$m_y^2+t^2=1 \tag{5.147}$$

(b) Show that the interaction eqs. (5.139), (5.141), (5.143), (5.145) and (5.147) define the surface elements as shown in Fig. 5.45.

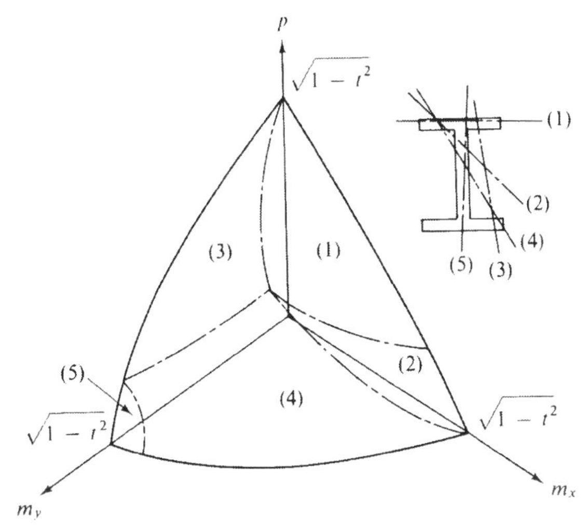

FIGURE 5.45
Figure for Prob. 5.5(b)

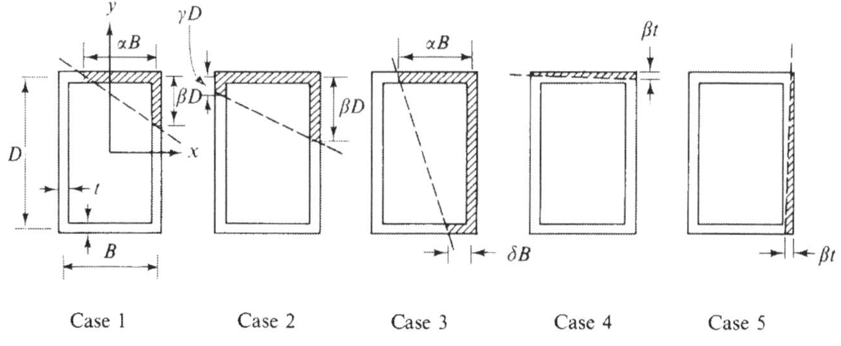

FIGURE 5.46
Figure for Prob. 5.6(a)

5.6 Repeat Prob. 5.5 for the rectangular hollow section shown in Fig. 5.46; the uniform thickness t is *small*.

Fully plastic stress resultants:

$$C = \frac{B}{D}$$

$$P_y = 2\sigma_y Dt(1 + C), \qquad T_p = \frac{2\sigma_y}{\sqrt{3}} tBD$$

$$M_{px} = \sigma_y tD^2\left(C + \frac{1}{2}\right), \quad M_{py} = \sigma_y DBt(1 + C/2)$$

(a) Prove the following relations

Case 1

$$\left.\begin{aligned} p &= \left(\frac{\sigma}{\sigma_y}\right)\left(1 - \frac{\alpha C + \beta}{1 + C}\right) \\ m_x &= \left(\frac{\sigma}{\sigma_y}\right)\left[\frac{\alpha C + \beta(1 - \beta)}{C + 1/2}\right] \\ m_y &= \left(\frac{\sigma}{\sigma_y}\right)\left[\frac{\beta + \alpha(1 - \alpha)C}{1 + C/2}\right] \end{aligned}\right\} \tag{5.148}$$

$$\sqrt{(C + C^2)(\sqrt{1 - t^2} - p) - (C + C^2/2)m_y} + \sqrt{(1 + C)(\sqrt{1 - t^2} - p) - (C + 1/2)m_x} - \frac{(1 + C)(p - \sqrt{1 - t^2})}{\sqrt[4]{1 - t^2}} = 0 \tag{5.149}$$

Case 2

$$\left.\begin{aligned} p &= \left(\frac{\sigma}{\sigma_y}\right)\left(\frac{1 - \beta - \gamma}{1 + C}\right) \\ m_x &= \left(\frac{\sigma}{\sigma_y}\right)\left(\frac{C + \beta(1 - \beta) + \gamma(1 - \gamma)}{C + 1/2}\right) \\ m_y &= \left(\frac{\sigma}{\sigma_y}\right)\left(\frac{\beta - \gamma}{1 + C/2}\right) \end{aligned}\right\} \tag{5.150}$$

$$(1 + C)^2 p^2 + (1 + C/2)^2 m_y^2 + (2C + 1)(m_x\sqrt{1 - t^2} + t^2) = 2C + 1 \tag{5.151}$$

Case 3

$$\left.\begin{aligned} p &= \left(\frac{\sigma}{\sigma_y}\right)\left[\frac{C(1 - \alpha - \delta)}{1 + C}\right] \\ m_x &= \left(\frac{\sigma}{\sigma_y}\right)\left[\frac{C(\alpha - \delta)}{C + 1/2}\right] \\ m_y &= \left(\frac{\sigma}{\sigma_y}\right)\left[\frac{1 + \alpha C(1 - \alpha) + \delta C(1 - \delta)}{1 + C/2}\right] \end{aligned}\right\} \tag{5.152}$$

$$\frac{(1 + C)^2}{C} p^2 + \frac{(C + 1/2)^2}{C} m_x^2 + (C + 2)(m_y\sqrt{1 - t^2} + t^2) = 2 + C \tag{5.153}$$

Case 4

$$\left.\begin{aligned} p &= \left(\frac{\sigma}{\sigma_y}\right)\left(1 - \frac{\beta C}{1 + C}\right) \\ m_x &= \left(\frac{\sigma}{\sigma_y}\right)\left(\frac{\beta C}{C + 1/2}\right) \\ m_y &= \text{indeterminate} \end{aligned}\right\} \tag{5.154}$$

$$p + \left(\frac{C + 1/2}{C + 1}\right) m_x - \sqrt{1 - t^2} = 0 \tag{5.155}$$

Case 5

$$\left.\begin{aligned} p &= \left(\frac{\sigma}{\sigma_y}\right)\left(1 - \frac{\beta}{1 + C}\right) \\ m_x &= \text{indeterminate} \\ m_y &= \left(\frac{\sigma}{\sigma_y}\right)\left(\frac{\beta}{1 + C/2}\right) \end{aligned}\right\} \tag{5.156}$$

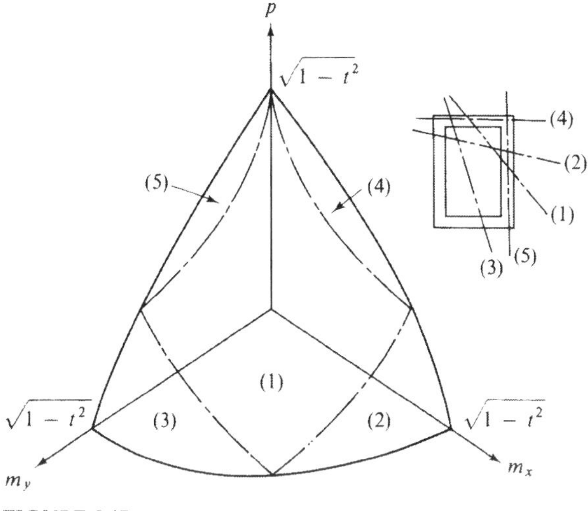

FIGURE 5.47
Figure for Prob. 5.6(b)

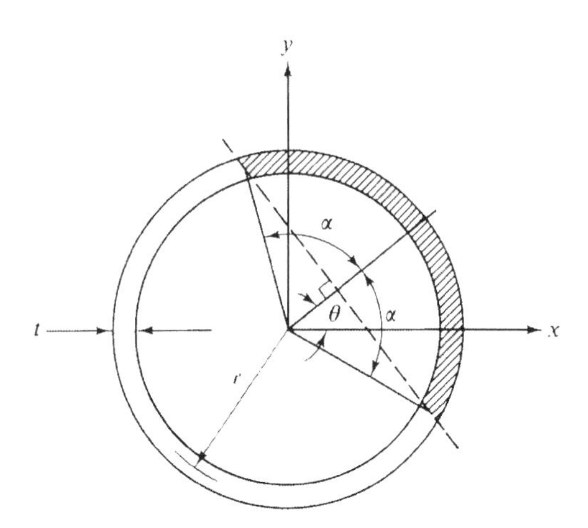

FIGURE 5.48
Figure for Prob. 5.7

$$p + \left(\frac{1 + C/2}{1 + C}\right) m_y - \sqrt{1 - t^2} = 0 \tag{5.157}$$

(b) Show that the interaction equations (5.149), (5.151), (5.153), (5.155) and (5.157) define surface elements as shown in Fig. 5.47.

5.7 Repeat Prob. 5.5 for the thin-walled circular tube shown in Fig. 5.48. Fully plastic stress resultants: A = cross-sectional area

$$P_y = \sigma_y A, \quad T_p = \frac{\sigma_y}{\sqrt{3}} Ar, \quad M_{px} = M_{py} = \frac{2}{\pi}\sigma_y Ar$$

Prove the following relations

$$\left.\begin{aligned} p &= \left(\frac{\sigma}{\sigma_y}\right)\left(1 - \frac{2\alpha}{\pi}\right) \\ m_x &= \left(\frac{\sigma}{\sigma_y}\right)\sin\alpha\sin\theta \\ m_y &= \left(\frac{\sigma}{\sigma_y}\right)\sin\alpha\cos\theta \end{aligned}\right\} \tag{5.158}$$

$$m_x^2 + m_y^2 - (1 - t^2)\sin^2\left[\frac{\pi}{2}\left(1 - \frac{p}{\sqrt{1 - t^2}}\right)\right] = 0 \tag{5.159}$$

5.10 REFERENCES

Bresler, B., "Design Criteria for Reinforced Columns under Axial Load and Biaxial Bending," *Journal of the American Concrete Institute*, Vol. 32, No. 5, November, pp. 481–490, (1960).

Chen, W. F., *Limit Analysis and Soil Plasticity*, Elsevier Scientific Publishing Company, Amsterdam and New York, (1975).

Chen, W. F. and Atsuta, T., "Interaction Equations for Biaxially Loaded Sections," *Journal of the Structural Division, ASCE*, Vol. 98, No. ST5, Proc. paper 8902, May, pp. 1035–1052, (1972).

Chen, W. F. and Atsuta, T., "Interaction Curves for Steel Sections under Axial Load and Biaxial Bending," Canadian Society of Civil Engineering, *The Engineering Journal, EIC*, Vol. 17, No. A-3, March/April, pp. I–VIII, (1974).

Chen, W. F. and Shoraka, M. T., "Tangent Stiffness Method for Biaxial Bending of Reinforced Concrete Columns," International Association for Bridge and Structural Engineering, Vol. 35-I, Zurich, pp. 23–44, (1975).

Morris, G. A. and Fenves, S. J., "Approximate Yield Surface Equations," *Journal of the Engineering Mechanics Division, ASCE*, Vol. 95, No. EM4, Proc. paper 6741, August, pp. 937–954, (1969).

Ringo, B. C., McDonough, J. F. and Baseheart, T. M., "Plastic Design of Biaxially Loaded Beam-Columns," *Engineering Journal, AISC*, Vol. 10, No. 1, 1st Quarter, pp. 6–18, (1973).

Santathadaporn, S. and Chen, W. F., "Interaction Curves for Sections under Combined Biaxial Bending and Axial Force," *Welding Research Council Bulletin*, No. 148, February, (1970).

6
BEHAVIOR OF BEAM-COLUMN SEGMENTS

6.1 INTRODUCTION

Chapter 5 of Volume 1 discussed the in-plane behavior of beam-column segments throughout the entire range of loading up to ultimate load, and developed simple analytical expressions for the moment-curvature-thrust relationship to describe this behavior. This is the basic relationship required in the analysis of any beam-column problems. This chapter attempts to generalize that discussion to the case of beam-column segments subjected to combined axial compression, biaxial bending, and warping deformation. As was seen for in-plane behavior, plastic behavior of beam-column segments is fairly involved and any rigorous analysis which attempts to cover their space behavior is destined to be very complex, even for an elastic-perfectly plastic material. This is due mainly to the fact that it is difficult to determine the elastic-plastic boundary which separates the plastic portion from the elastically remained portion of a cross section. Furthermore, the neutral axis is no longer a straight line. For these reasons, recourse must be had to numerical methods to obtain solutions.

The purpose of this chapter is to describe analytical methods to establish the relations between *generalized strains* (axial strain ε_o, biaxial curvatures Φ_x and Φ_y and warping curvature Φ_ω) and *generalized stresses* (axial thrust P, biaxial bending moments M_x and M_y and warping moment or bi-moment M_ω). Since plastic state of stress and strain depends highly on the history of loading, the generalized stress-strain relation cannot be obtained directly. Step-by-step calculations that follow the history of loading are required. For this reason, it is necessary to establish the relationship between the infinitesimal generalized stress increments $\{\dot{F}\}$ or $\{dF\} = \{dP, dM_x, dM_y, dM_\omega\}$ and the infinitesimal generalized strain increments $\{\dot{\Delta}\}$ or $\{d\Delta\} = \{d\varepsilon_o, d\Phi_x, d\Phi_y, d\Phi_\omega\}$. Once the incremental relationship is established, some iterative procedure must be applied in order to find the path of generalized strains from the path of generalized stresses using the current stress-strain information.

In considering the history of loading, stress-strain characteristics of material must be prescribed. For metal beam-columns, a trilinear curve is assumed herein so that elastic-perfectly plastic and linear strain hardening can be taken into account in addition to elastic unloading and Bauschinger effect. In the analysis only normal stress and strain are treated and transverse shear stresses are not considered. For metal members, actual analysis and formulation are made only for wide-flange shapes although the method of analysis is applicable to general shapes of cross section. Stress-strain relation of the material is evaluated in its rate form as

$$\dot{\sigma} = E_t \dot{\varepsilon} \tag{6.1}$$

in which

$$\dot{\sigma} = \text{stress rate}$$

$$\dot{\varepsilon} = \text{strain rate}$$

$$E_t = \text{tangent modulus at current state}$$

Studies of reinforced concrete members made using this analysis are presented in Sec. 6.8.

6.2 GENERALIZED STRESS-STRAIN RELATIONSHIPS

Since only normal stress and normal strain are considered, deformation quantities related to normal strain are taken as generalized strains of the section:

$$\begin{aligned}
\varepsilon_o &= \frac{1}{A}\int \varepsilon\, dA, && \text{strain at centroid} \\
\Phi_x &= \frac{1}{A}\int \frac{\partial \varepsilon}{\partial Y}\, dA, && \text{bending curvature about } X\text{-axis} \\
\Phi_y &= -\frac{1}{A}\int \frac{\partial \varepsilon}{\partial X}\, dA, && \text{bending curvature about } Y\text{-axis} \\
\Phi_\omega &= \frac{1}{A}\int \frac{\partial^2 \varepsilon}{\partial X\, \partial Y}\, dA, && \text{warping curvature at centroid}
\end{aligned} \tag{6.2}$$

in which A = area of the cross section. Strain distribution is assumed to be a second order polynomial in X and Y coordinate:

$$\varepsilon = \alpha + \beta X + \gamma Y + \delta XY \tag{6.3}$$

Using the doubly symmetric property of the wide-flange shape of cross section

$$\begin{aligned}
\varepsilon_o &= \frac{1}{A}\left(\alpha \int dA + \beta \int X\,dA + \gamma \int Y\,dA + \delta \int XY\,dA\right) = \alpha \\
\Phi_x &= \frac{1}{A}\left(\gamma \int dA + \delta \int X\,dA\right) = \gamma \\
\Phi_y &= -\frac{1}{A}\left(\beta \int dA + \delta \int Y\,dA\right) = -\beta \\
\Phi_\omega &= \frac{1}{A}\left(\delta \int dA\right) = \delta
\end{aligned} \tag{6.4}$$

Eq. (6.3) has the form

$$\varepsilon = \varepsilon_o - \Phi_y X + \Phi_x Y + \Phi_\omega XY \tag{6.5}$$

Eq. (6.5) can be directly obtained from Eq. (2.116) by substituting the normalized unit warping function ω_n for $-\omega$, for a wide flange section, $\omega_n = XY$. If we take the length of a beam-column into consideration, these generalized strains are related with deflections

$$\varepsilon_o = \frac{\partial w}{\partial z}, \quad \Phi_x = -\frac{\partial^2 v}{\partial z^2}, \quad \Phi_y = \frac{\partial^2 u}{\partial z^2}, \quad \Phi_\omega = \frac{\partial^2 \theta}{\partial z^2}$$

The corresponding generalized stresses are determined from the rate of internal energy dissipation

$$\begin{aligned}
\dot{D}_I &= \int \sigma \dot{\varepsilon}\,dA = \dot{\varepsilon}_o \int \sigma\,dA - \dot{\Phi}_y \int \sigma X\,dA \\
&\quad + \dot{\Phi}_x \int \sigma Y\,dA + \dot{\Phi}_\omega \int \sigma XY dA \\
&= \dot{\varepsilon}_o P + \dot{\Phi}_y M_y + \dot{\Phi}_x M_x + \dot{\Phi}_\omega M_\omega
\end{aligned} \tag{6.6}$$

that is,

$$\begin{aligned}
P &= \int \sigma\,dA, && \text{axial thrust} \\
M_x &= \int \sigma Y\,dA, && \text{bending moment about } X\text{-axis} \\
M_y &= -\int \sigma X\,dA, && \text{bending moment about } Y\text{-axis} \\
M_\omega &= \int \sigma XY\,dA, && \text{warping moment or bi-moment about centroid}
\end{aligned} \tag{6.7}$$

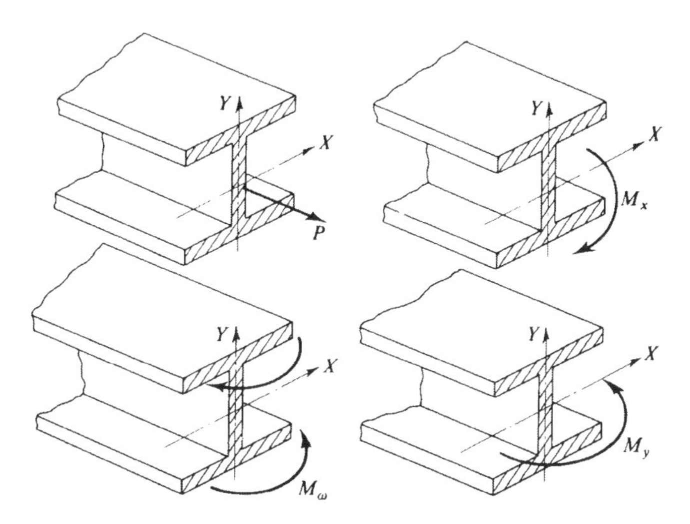

FIGURE 6.1
Generalized stresses

These are the resultant forces or generalized stresses producing normal stress over the entire cross section. The physical meaning of these forces is shown in Fig. 6.1. Details of the explanation of the physical meaning of bimoment were examined in Chap. 2, in which the general definition of *bimoment* for a thin-walled cross section is defined in Eq. (2.118) as

$$\text{bimoment} = M_\omega = \int_A \sigma\omega\, dA \tag{6.8}$$

in which ω represents the double sectorial area swept by the radius from an arbitrary point as the radius moves along the middle line of the cross section from a reference point. For the particular instance of an H-shaped section, the value of $\omega = XY$, thus the expression, M_ω, defined in Eq. (6.7) is obtained [Eq. (2.125)]

$$\begin{aligned} M_\omega &= \int \sigma XY\, dA = \frac{h}{2}\int^{\text{(upper flange)}} \sigma X\, dA - \frac{h}{2}\int^{\text{(lower flange)}} \sigma X\, dA \\ M_\omega &= \frac{h}{2}\left[M_{y(\text{upper flange})} - M_{y(\text{lower flange})}\right] \end{aligned} \tag{6.9}$$

This self-equilibrium bimoment is shown in Fig. 6.1. The warping displacement and bimoment may be demonstrated clearly by considering a simple physical model as shown in Fig. 6.2 [Vlassov (1961)]. The biaxial eccentric load, $4P$, is seen to be decomposed into four components. The first three are statically equivalent to an axial force, $4P$, and two bending moments, M_x, M_y, about two principal axis of the

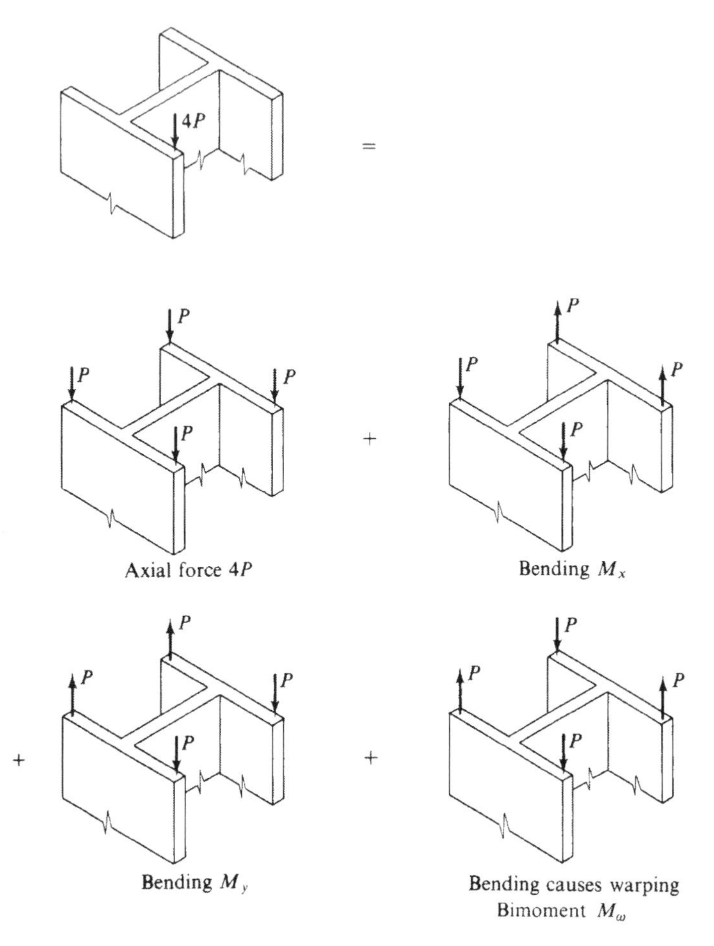

FIGURE 6.2
Decomposition of a biaxial loading

H-section. However, these three equivalent systems do not produce the warping deformation. It is necessary to consider a fourth system which produces zero axial force and zero bending moment resultants on the section. This fourth system termed warping moment (or bimoment) M_ω causes the column to warp or twist. Introducing the nondimensional variables

$$\bar{\sigma} = \frac{\sigma}{\sigma_y}, \quad \bar{\varepsilon} = \frac{\varepsilon}{\varepsilon_y}, \quad x = \frac{X}{\frac{B}{2}}, \quad y = \frac{Y}{\frac{D}{2}} \tag{6.10}$$

in which B and D are width and depth of the cross section, respectively, and initial yielding quantities

$$
\begin{aligned}
P_y &= A\sigma_y\,, \qquad \varepsilon_{oy} = \varepsilon_y \\
M_{yx} &= \frac{I_x\sigma_y}{\frac{D}{2}}\,, \qquad \Phi_{yx} = \frac{\varepsilon_y}{\frac{D}{2}} \\
M_{yy} &= \frac{I_y\sigma_y}{\frac{B}{2}}\,, \qquad \Phi_{yy} = \frac{\varepsilon_y}{\frac{B}{2}} \\
M_{y\omega} &= \frac{I_\omega\sigma_y}{\frac{BD}{4}}\,, \qquad \Phi_{p\omega} = \frac{\varepsilon_y}{\frac{BD}{4}}
\end{aligned}
\tag{6.11}
$$

in which

$$
I_x = \int_A Y^2\,dA\,, \quad I_y = \int_A X^2\,dA\,, \quad I_\omega = \int_A X^2Y^2\,dA
\tag{6.12}
$$

Using Eq. (6.11), the generalized stresses are nondimensionalized as

$$
\begin{aligned}
p &= \frac{P}{P_y} = \frac{1}{A\sigma_y}\int \sigma\,dA = \frac{1}{\bar{A}}\int \bar{\sigma}\,da \\
m_x &= \frac{M_x}{M_{yx}} = \frac{\frac{D}{2}}{I_x\sigma_y}\int \sigma Y\,dA = \frac{1}{\bar{I}_x}\int \bar{\sigma}y\,da \\
m_y &= \frac{M_y}{M_{yy}} = -\frac{\frac{B}{2}}{I_y\sigma_y}\int \sigma X\,dA = -\frac{1}{\bar{I}_y}\int \bar{\sigma}x\,da \\
m_\omega &= \frac{M_\omega}{M_{y\omega}} = \frac{\frac{BD}{4}}{I_\omega\sigma_y}\int \sigma XY\,dA = \frac{1}{\bar{I}_\omega}\int \bar{\sigma}xy\,da
\end{aligned}
\tag{6.13}
$$

in which the nondimensional quantities are

$$
\begin{aligned}
&da = dx\,dy\,, \quad \bar{A} = \iint dx\,dy\,, \quad \bar{I}_x = \iint y^2\,dx\,dy\,, \\
&\bar{I}_y = \iint x^2\,dx\,dy\,, \quad \bar{I}_\omega = \iint x^2y^2\,dx\,dy
\end{aligned}
\tag{6.14}
$$

and the generalized strains are nondimensionalized as

$$
\begin{aligned}
\bar{\varepsilon}_o &= \frac{\varepsilon_o}{\varepsilon_{oy}} = \frac{\varepsilon_o}{\varepsilon_y} = \frac{1}{A}\int \bar{\varepsilon}\, da \\
\bar{\varphi}_x &= \frac{\Phi_x}{\Phi_{yx}} = \frac{\frac{D}{2}}{\varepsilon_y}\,\Phi_x = \frac{1}{A}\int \frac{\partial \bar{\varepsilon}}{\partial y}\, da \\
\bar{\varphi}_y &= \frac{\Phi_y}{\Phi_{yy}} = \frac{\frac{B}{2}}{\varepsilon_y}\,\Phi_y = -\frac{1}{A}\int \frac{\partial \bar{\varepsilon}}{\partial x}\, da \\
\bar{\varphi}_\omega &= \frac{\Phi_\omega}{\Phi_{p\omega}} = \frac{\frac{BD}{4}}{\varepsilon_y}\,\Phi_\omega = \frac{1}{A}\int \frac{\partial^2 \bar{\varepsilon}}{\partial x\, \partial y}\, da
\end{aligned}
\tag{6.15}
$$

Noting that the notation $\bar{\varphi}$ used here for nondimensional curvature is interchangeable with the notation φ customarily used in other chapters. The nondimensional strain distribution including the residual strains ε_r has the form [Eq. (6.5)]

$$
\begin{aligned}
\bar{\varepsilon} &= \frac{\varepsilon}{\varepsilon_y} = \frac{1}{\varepsilon_y}(\varepsilon_o + Y\Phi_x - X\Phi_y + XY\Phi_\omega + \varepsilon_r) \\
&= \bar{\varepsilon}_o + y\bar{\varphi}_x - x\bar{\varphi}_y + xy\bar{\varphi}_\omega + \bar{\varepsilon}_r
\end{aligned}
\tag{6.16}
$$

The relations between the current stress σ and the current strain ε are not known because their relations depend on the previous history of loading. Instead, relations between the stress rate $\dot{\sigma}$ and the strain rate $\dot{\varepsilon}$ are known from Eq. (6.1) using the tangent modulus E_t at the current state. In the nondimensional form, we have,

$$
\dot{\bar{\sigma}} = \frac{\dot{\sigma}}{\sigma_y} = \frac{1}{\sigma_y}E_t\dot{\varepsilon} = \frac{1}{\sigma_y}E_t\varepsilon_y\frac{\dot{\varepsilon}}{\varepsilon_y} = \frac{E_t}{E}\dot{\bar{\varepsilon}} = \bar{E}_t\dot{\bar{\varepsilon}}
\tag{6.17}
$$

in which $\bar{E}_t = E_t/E$. Thus from Eq. (6.16), the stress rate becomes

$$
\dot{\bar{\sigma}} = \bar{E}_t\dot{\bar{\varepsilon}}_o + \bar{E}_t y\dot{\bar{\varphi}}_x \quad \bar{E}_t x\dot{\bar{\varphi}}_y + \bar{E}_t xy\dot{\bar{\varphi}}_\omega
\tag{6.18}
$$

in which it was assumed that the residual strain is independent of time, i.e., $\dot{\varepsilon}_r = 0$.

Generalized stress-strain rate relationships are obtained from Eqs. (6.13), (6.15) and (6.18). Thus,

$$
\begin{aligned}
\dot{p} &= \frac{1}{A}\left(\dot{\bar{\varepsilon}}_o\int \bar{E}_t\, da + \dot{\bar{\varphi}}_x\int \bar{E}_t y\, da - \dot{\bar{\varphi}}_y\int \bar{E}_t x\, da + \dot{\bar{\varphi}}_\omega\int \bar{E}_t xy\, da\right) \\
\dot{m}_x &= \frac{1}{I_x}\left(\dot{\bar{\varepsilon}}_o\int \bar{E}_t y\, da + \dot{\bar{\varphi}}_x\int \bar{E}_t y^2\, da - \dot{\bar{\varphi}}_y\int \bar{E}_t xy\, da + \dot{\bar{\varphi}}_\omega\int \bar{E}_t xy^2\, da\right) \\
\dot{m}_y &= -\frac{1}{I_y}\left(\dot{\bar{\varepsilon}}_o\int \bar{E}_t x\, da + \dot{\bar{\varphi}}_x\int \bar{E}_t xy\, da - \dot{\bar{\varphi}}_y\int \bar{E}_t x^2\, da + \dot{\bar{\varphi}}_\omega\int \bar{E}_t x^2 y\, da\right) \\
\dot{m}_\omega &= \frac{1}{I_\omega}\left(\dot{\bar{\varepsilon}}_o\int \bar{E}_t xy\, da + \dot{\bar{\varphi}}_x\int \bar{E}_t xy^2\, da - \dot{\bar{\varphi}}_y\int \bar{E}_t x^2 y\, da + \dot{\bar{\varphi}}_\omega\int \bar{E}_t x^2 y^2\, da\right)
\end{aligned}
\tag{6.19}
$$

or in a matrix form

$$\{\dot{f}\} = [K]\{\dot{\delta}\} \tag{6.20}$$

in which $\{\dot{f}\}$ = nondimensional generalized stress rate vector; $\{\dot{\delta}\}$ = nondimensional generalized strain rate vector; $[K]$ = stiffness matrix, i.e.,

$$\{\dot{f}\} = \begin{Bmatrix} \dot{p} \\ \dot{m}_x \\ \dot{m}_y \\ \dot{m}_\omega \end{Bmatrix} \qquad \{\dot{\delta}\} = \begin{Bmatrix} \dot{\bar{\varepsilon}}_o \\ \dot{\bar{\varphi}}_x \\ \dot{\bar{\varphi}}_y \\ \dot{\bar{\varphi}}_\omega \end{Bmatrix} \tag{6.21}$$

and

$$[K] = \begin{bmatrix} \int \bar{E}_t \frac{da}{\bar{A}} & \int \bar{E}_t y \frac{da}{\bar{A}} & -\int \bar{E}_t x \frac{da}{\bar{A}} & \int \bar{E}_t xy \frac{da}{\bar{A}} \\ \int \bar{E}_t y \frac{da}{\bar{I}_x} & \int \bar{E}_t y^2 \frac{da}{\bar{I}_x} & -\int \bar{E}_t xy \frac{da}{\bar{I}_x} & \int \bar{E}_t xy^2 \frac{da}{\bar{I}_x} \\ -\int \bar{E}_t x \frac{da}{\bar{I}_y} & -\int \bar{E}_t xy \frac{da}{\bar{I}_y} & \int \bar{E}_t x^2 \frac{da}{\bar{I}_y} & -\int \bar{E}_t x^2 y \frac{da}{\bar{I}_y} \\ \int \bar{E}_t xy \frac{da}{\bar{I}_\omega} & \int \bar{E}_t xy^2 \frac{da}{\bar{I}_\omega} & -\int \bar{E}_t x^2 y \frac{da}{\bar{I}_\omega} & \int \bar{E}_t x^2 y^2 \frac{da}{\bar{I}_\omega} \end{bmatrix} \tag{6.22}$$

In the elastic regime, $\bar{E}_t = 1$, and the stiffness matrix, $[K]$, becomes the unit matrix $[K_0]$, using the doubly symmetric property of the cross section

$$[K_0] = \begin{bmatrix} 1 & 0 & 0 & 0 \\ 0 & 1 & 0 & 0 \\ 0 & 0 & 1 & 0 \\ 0 & 0 & 0 & 1 \end{bmatrix} \tag{6.23}$$

Equation (6.22) for biaxially loaded H-segments was first given by Santathadaporn and Chen (1972) for the special case of axial compression p combined with biaxial bending moment m_x and m_y. The present form of Eqs. (6.21) and (6.22) including warping moment m_ω and warping curvature $\bar{\varphi}_\omega$ was first obtained by Chen and Atsuta (1973).

Residual Stress

The residual stress distribution for a wide-flange section is assumed to be linear in each plate element. The maximum compressive residual stress, σ_{rc}, occurs at the tips of flange plates. Since the residual stress is in self-equilibrium, the maximum tensile residual stress, σ_{rt}, at the junction point of flange and web is

$$\sigma_{rt} = \frac{\sigma_{rc}}{1 + \frac{D}{B}\frac{w}{t} - 2\frac{w}{B}} \tag{6.24}$$

in which

D = depth of wide-flange section

B = width of flange plates

t = thickness of flange plate

w = thickness of web plate.

Using σ_{rc} and σ_{rt}, the distribution of residual stress, σ_r, in the flange can be expressed as

$$\sigma_r = \sigma_{rt} - \frac{\sigma_{rc} + \sigma_{rt}}{\frac{B}{2}} |X| \tag{6.25}$$

or nondimensionally

$$\bar{\sigma}_r = \bar{\sigma}_{rt} - (\bar{\sigma}_{rc} + \bar{\sigma}_{rt}) |x| \tag{6.26}$$

For most shapes of wide-flange cross section the maximum residual stress of

$$\bar{\sigma}_{rc} = \frac{\sigma_{rc}}{\sigma_y} = 0.3 \tag{6.27}$$

is normally assumed. The residual strain is obtained by dividing the residual stress by Young's modulus E. Although, in the above generalized stress-strain relation, Eq. (6.20), the residual stress σ_r or strain ε_r do not appear, they have an effect on the tangent modulus value E_t as a history of loading. Determination of the tangent modulus E_t at the current state, and the value of generalized strains at a subsequent new state are the two main steps of calculations for a solution.

6.3 TANGENT STIFFNESS APPROACH

The generalized stress-strain relation is given in its rate form as seen in Eq. (6.20), which here can be considered as the tangent relation between generalized stress $\{\dot{f}\}$ and generalized strain $\{\dot{\delta}\}$, as illustrated in Fig. 6.3. In this sense, the stiffness matrix $[K]$ as defined in Eq. (6.22) is called *tangent stiffness* at the current state as it represents the tangent of the force-deformation curve as well as the stiffness of the cross section.

The problem to be solved now may be interpreted as follows: Given a state A at which all information about the stresses $\{f_A\}$, strains $\{\delta_A\}$ and the tangent stiffness $[K_A]$ are known, find the generalized strain $\{\delta_B\}$ when the generalized stress $\{f_A\}$ is increased to the value $\{f_B\}$ at the state B as shown in Fig. 6.3. An approximate solution for the generalized strain $\{\delta'_B\}$ can be obtained using Eq. (6.20) as,

$$\{f_B\} - \{f_A\} = [K_A](\{\delta'_B\} - \{\delta_A\}) \tag{6.28}$$

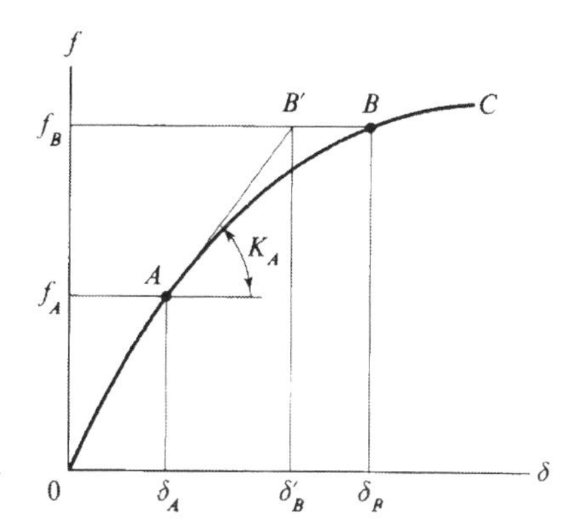

FIGURE 6.3
Tangent stiffness on a generalized stress-strain curve

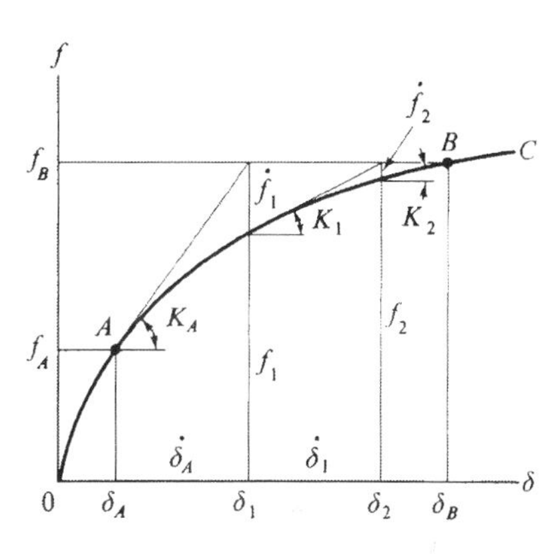

FIGURE 6.4
Procedures in tangent stiffness method

or

$$\{\delta'_B\} = \{\delta_A\} + [K_A]^{-1}(\{f_B\} - \{f_A\}) \tag{6.29}$$

The solution $\{\delta'_B\}$ may deviate considerably from the exact value $\{\delta_B\}$ as illustrated in Fig. 6.3.

Therefore, the generalized strain $\{\delta_B\}$ can only be solved by step-by-step iterative technique. It is easily understood that the error involved in each step cannot be eliminated completely, no matter how small a force increment may be taken. Further, small increments require a considerable amount of computation time. As an efficient way out of these problems, the tangent stiffness method as proposed by Santathadaporn and Chen (1972) is found to be very powerful and efficient for computer solution.

The procedure of the tangent stiffness method is shown graphically in Fig. 6.4. The curve OABC is the true force-deformation curve. The increment of force from A to B is

$$\{\dot{f}_A\} = \{f_B\} - \{f_A\} \tag{6.30}$$

The matrix $[K_A]$ which is equivalent to the slope at point A on the curve is determined from Eqs. (6.22). With the increment of external force $\{\dot{f}_A\}$, the increment of deformations is obtained from

$$\{\dot{\delta}_A\} = [K_A]^{-1}\{\dot{f}_A\} \tag{6.31}$$

in which $[K_A]^{-1}$ is the inverse of the matrix $[K_A]$.

The preceding equation is merely an approximate solution, because $[K_A]$ is calculated before the increment occurs. The tangent stiffness matrix alters slightly as the elastic-plastic boundary moves during the increment of deformation. Nevertheless, the equation gives a good prediction of the increment of deformation, provided that the increment of external force is small. The first estimated deformation is given by the sum of $\{\delta_A\}$ and the incremental deformation $\{\dot{\delta}_A\}$ predicted by Eq. (6.31).

$$\{\delta_1\} = \{\delta_A\} + \{\dot{\delta}_A\} \tag{6.32}$$

The deformation gives rise to internal force $\{f_1\}$ which is not in equilibrium with the external force $\{f_B\}$. The first unbalanced force $\{\dot{f}_1\}$ is computed from

$$\{\dot{f}_1\} = \{f_B\} - \{f_1\} \tag{6.33}$$

The next step is to find a correction vector $\{\dot{\delta}_1\}$ which will be added to $\{\delta_1\}$ in order to eliminate the unbalanced force. Vector $\{\dot{\delta}_1\}$ is obtained from

$$\{\dot{\delta}_1\} = [K_1]^{-1}\{\dot{f}_1\} \tag{6.34}$$

in which $[K_1]^{-1}$ is the inverse of the new tangent stiffness matrix $[K_1]$ corresponding to the state $\{f_1\}$ and $\{\delta_1\}$. The process is repeated until the unbalanced force becomes zero or is within a prescribed tolerant limit.

If the increment of force is small, the first estimate of the increment of deformation from Eq. (6.31) is quite accurate and the subsequent correction will not be necessary. Even with a large incremental force, the solution will generally converge within just a few cycles of iteration. The unbalanced force resulting from each iterative cycle is always smaller than the previous one and diminishes rapidly.

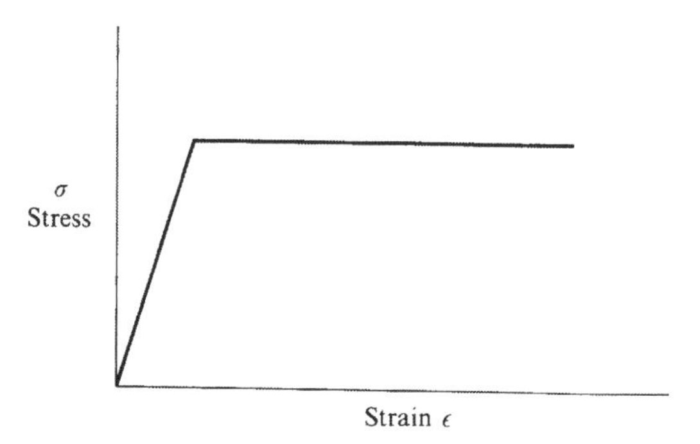

FIGURE 6.5
Idealized stress-strain curve

Some numerical examples of the tangent stiffness method for solutions of biaxially loaded wide-flange section W8 × 31 are reported by Santathadaporn and Chen (1972), in which stress-strain relation of material was assumed to be elastic-

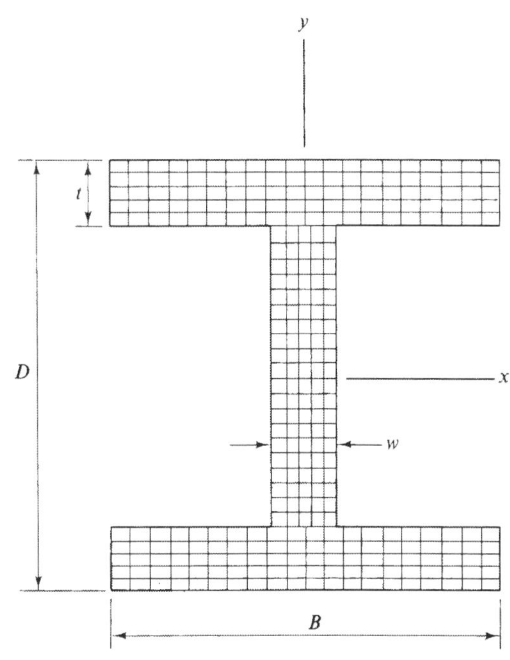

FIGURE 6.6
Wide-flange section divided into finite elements

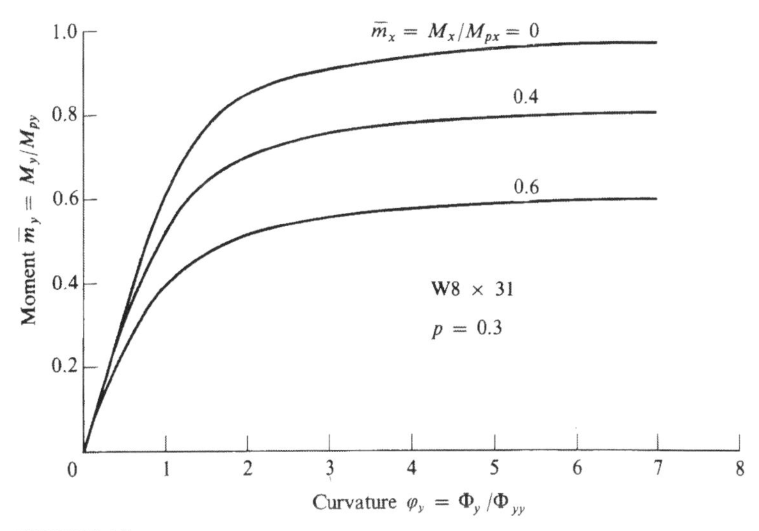

FIGURE 6.7
Moment-curvature curves for biaxial bending

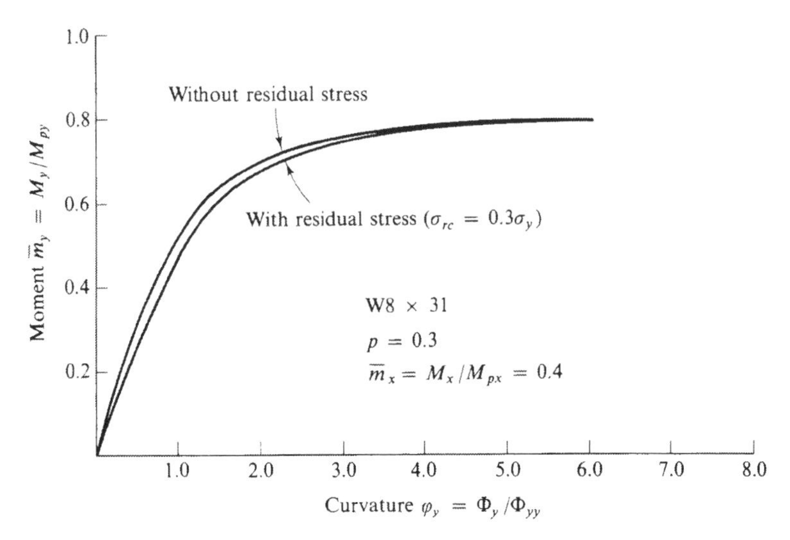

FIGURE 6.8
Effect of residual stress

perfectly plastic as shown in Fig. 6.5 and warping deformation was neglected. The elements of the tangent stiffness matrix were evaluated numerically by dividing the cross section into finite elements as shown in Fig. 6.6. The strain and stress in each element were computed as the average values at its centroid.

Figure 6.7 shows moment-curvature relations for weak axis bending ($\bar{m}_y = M_y/M_{py}$ versus φ_y) for various values of $\bar{m}_x = M_x/M_{px}$ where $\bar{m}$ denotes nondimensional moment with respect to fully plastic state. The axial force, P, was maintained at the constant value $0.3P_y$. The influence of residual stress is shown in Fig. 6.8. As can be seen, the presence of the residual stress has a significant effect on the elastic-plastic behavior of the curve. The plastic limit moment is, of course, not affected by the residual stress.

6.4 RESPONSE OF SEGMENT TO CYCLIC LOADING

The stress-strain relationship is assumed to be trilinear as shown in Fig. 6.9(a). The initial stress-strain curve is composed of three regimes, elastic regime (OA or OA'), plastic regime (AB or $A'B'$), and strain-hardening regime (BC or $B'C'$). The points with a prime indicate the state of stress in a compression regime.

For the monotonically increased loading case, the stress-strain points follow the solid lines, OABC or $OA'B'C'$. In this case the state of stress, σ, is determined uniquely by the state of strain ε.

The plastic unloading behavior is idealized as follows. If the material is compressed after plastic extension up to the point, G, it follows the path, $GG'B'C'$. If unloaded from the point B, the corresponding unloading path is $BD'B'C'$ The

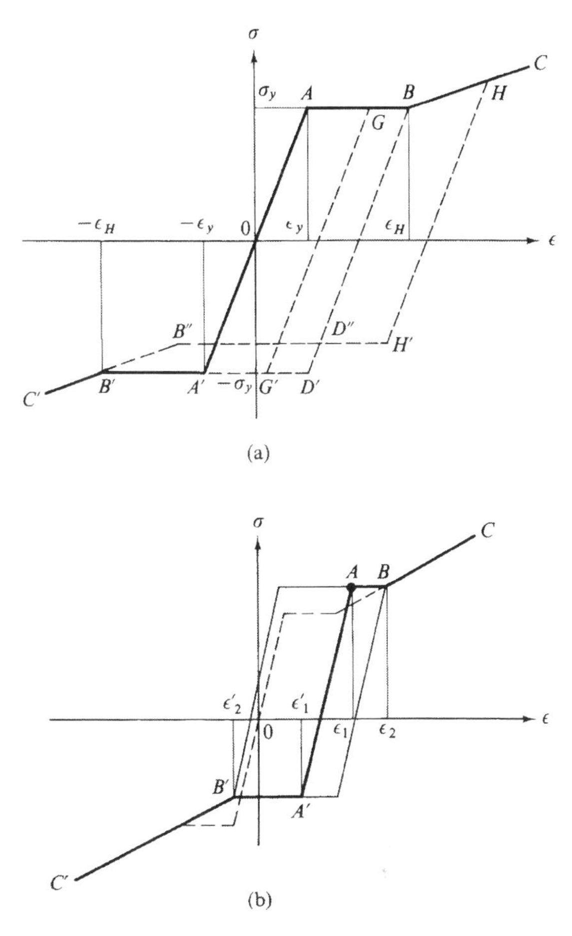

FIGURE 6.9
Idealized stress-strain relationship (a) Initial state of stress strain. (b) Subsequent state of stress strain.

material exhibits no "Bauschinger effect." However, if the material is unloaded and then compressed after some strain hardening up to the point H, the reversed path is $HH'D''B''C'$, and some "Bauschinger effects" are exhibited.

The state of stress, A, as shown in Fig. 6.9(b) is a general case and can be used to establish the limits of each regime. Regimes of stress states are determined by the current strain, ε:

$$
\begin{array}{llll}
\varepsilon'_1 \leq \varepsilon \leq \varepsilon_1, & & & \text{elastic regime} \\
\varepsilon'_2 \leq \varepsilon < \varepsilon'_1 & \text{or} & \varepsilon_1 < \varepsilon \leq \varepsilon_2, & \text{plastic regime} \\
\varepsilon < \varepsilon_2 & \text{or} & \varepsilon_2 < \varepsilon, & \text{strain hardening regime}
\end{array}
\tag{6.35}
$$

in which ε_1, ε'_1, ε_2, and ε'_2 are strain limits for each regime. These strain limits will of course depend on the current strain, ε, as seen in Fig. 6.9(b). The changes of these

Table 6.1 STRESS-STRAIN RELATIONSHIPS INCLUDING HYSTERESIS

Current state		Change in Strain Limits		Tangent Modulus, E_t	
Strain, ε (1)	Stress, σ (2)	$\varepsilon_1, \varepsilon_1'$ (3)	$\varepsilon_2, \varepsilon_2'$ (4)	$\dot{\varepsilon} > 0$ (5)	$\dot{\varepsilon} < 0$ (6)
Tension strain-hardening	$E[(\varepsilon_1 - \varepsilon_1')/2] + E_H\{\varepsilon - [(\varepsilon_2 - \varepsilon_2')/2]\}$	$\varepsilon_1 = \varepsilon$ $\varepsilon_1' = \varepsilon - 2\varepsilon_y$	$\varepsilon_2 = \varepsilon$ $\varepsilon_2' = \varepsilon - 2\varepsilon_H$	E_H	E
ε_2					
plastic	$E[(\varepsilon_1 - \varepsilon_1')/2] + E_H[(\varepsilon_2 + \varepsilon_2')/2]$	$\varepsilon_1 = \varepsilon$ $\varepsilon_1' = \varepsilon - 2\varepsilon_y$	no change	0	E
ε_1					
elastic	$E\{\varepsilon - [(\varepsilon_1 + \varepsilon_1')/2]\} + E_H[(\varepsilon_2 + \varepsilon_2')/2]$	no change	no change	E	E
ε_1'					
plastic	$E[(\varepsilon_1' - \varepsilon_1)/2] + E_H[(\varepsilon_2' + \varepsilon_2)/2]$	$\varepsilon_1' = \varepsilon$ $\varepsilon_1 = \varepsilon + 2\varepsilon_y$	no change	E	0
ε_2'					
Compression strain-hardening	$E[(\varepsilon_1' - \varepsilon_1)/2] + E_H\{\varepsilon - [(\varepsilon_2' - \varepsilon_2)/2]\}$	$\varepsilon_1' = \varepsilon$ $\varepsilon_1 = \varepsilon + 2\varepsilon_y$	$\varepsilon_2' = \varepsilon$ $\varepsilon_2 = \varepsilon + 2\varepsilon_H$	E	E_H

strain limits as a function of the current strain, ε, are given in Table 6.1. Once the strain limits are established, the current state of stress can be determined uniquely by the elastic modulus, E, strain hardening modulus, E_H, the current state of strain, ε, and the new strain limits. These stress-strain relationships are summarized in Table 6.1.

Response to proportional loading

Figure 6.10 is a proportional loading case with the loads $(P/P_y, M_x/M_{px}, M_y/M_{py})$ being increased proportionally and monotonically between 99 percent (point A) and -99 percent (point B) of the full plastic limit state (0.3, 0.6, 0.6). Strain hardening and residual stress are considered but warping deformation is completely restrained ($\Phi_\omega = 0$). Where there is no residual stresses over the cross section, the deformations $(\bar{\varepsilon}_o, \bar{\varphi}_x, \bar{\varphi}_y)$ are found to be exactly the same at point A before and after the cycle of loading. At point B, they have the opposite sign but the magnitude is the same. Since the warping deformation is assumed to be completely restrained, the corresponding *warping moment*, M_ω, required for such a restraint is considered as a reaction. Its magnitude is also found to be the same at points A and B. The presence of residual stress is seen to have some effects on the elastic-plastic behavior of the curves but the effect of material strain hardening on the curves is not found to be significant. The energy dissipation during the first cycle of the loading is also shown in Fig. 6.10.

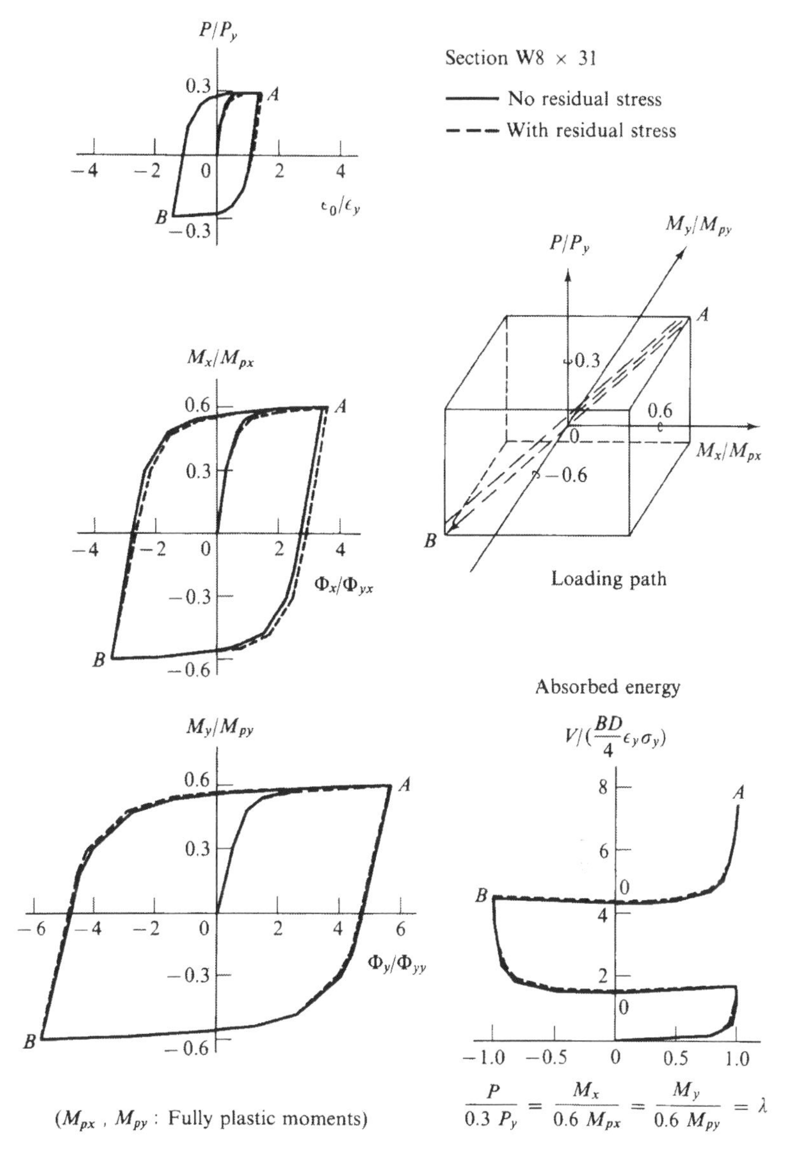

FIGURE 6.10
Proportional loading

Response to nonproportional loading

The influence of *loading path* on the curves is shown in Fig. 6.11. Five different loading paths are chosen in the figure to reach the same loading point A $(P/P_y = 0.3, M_x/M_{px} = 0.6, M_y/M_{py} = 0.6)$.

As can be seen, the same values of deformations are reached for all different loading paths. Although all the loading paths are monotonically increased, a large

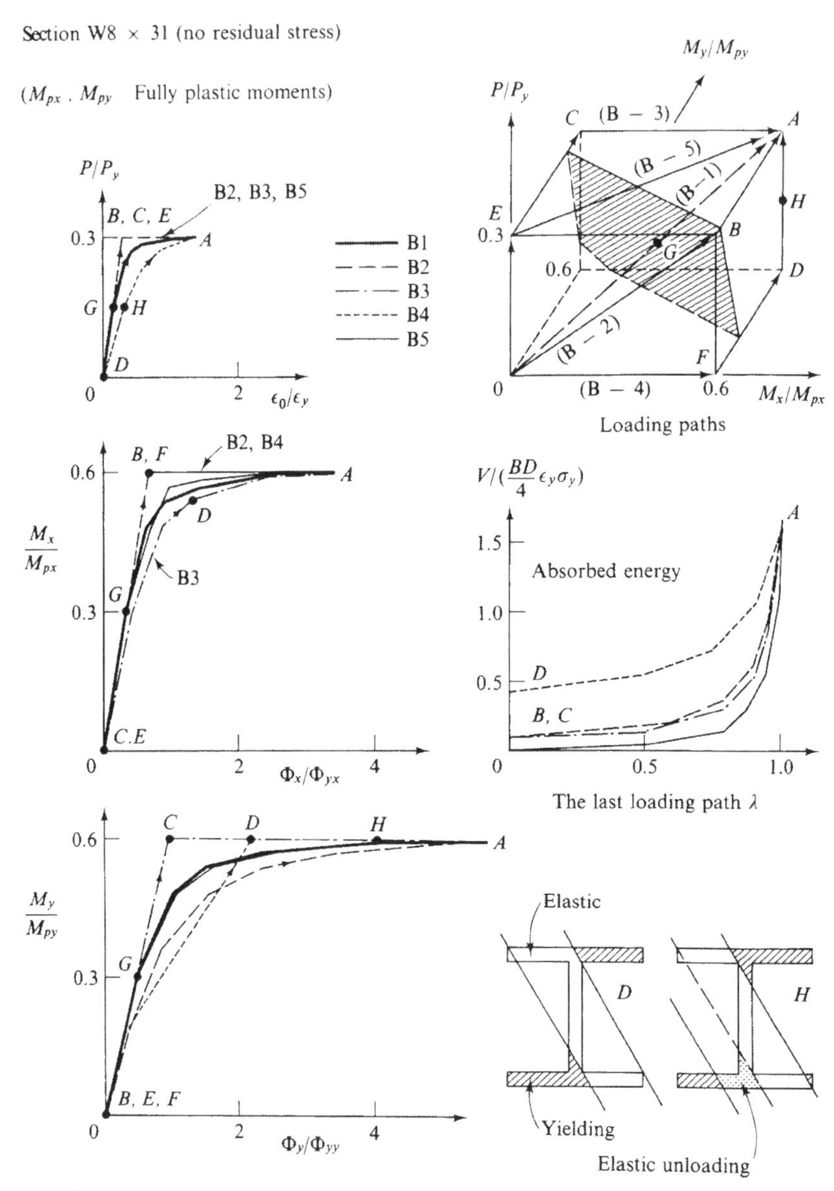

FIGURE 6.11
Effect of loading paths

yielded portion of the section is seen to be unloaded elastically from load point D to H as shown in Fig. 6.11. The energy dissipations corresponding to each loading path are also shown in Fig. 6.11.

Response to cyclic loading

The response of a section under a cyclic loading is of major importance in the

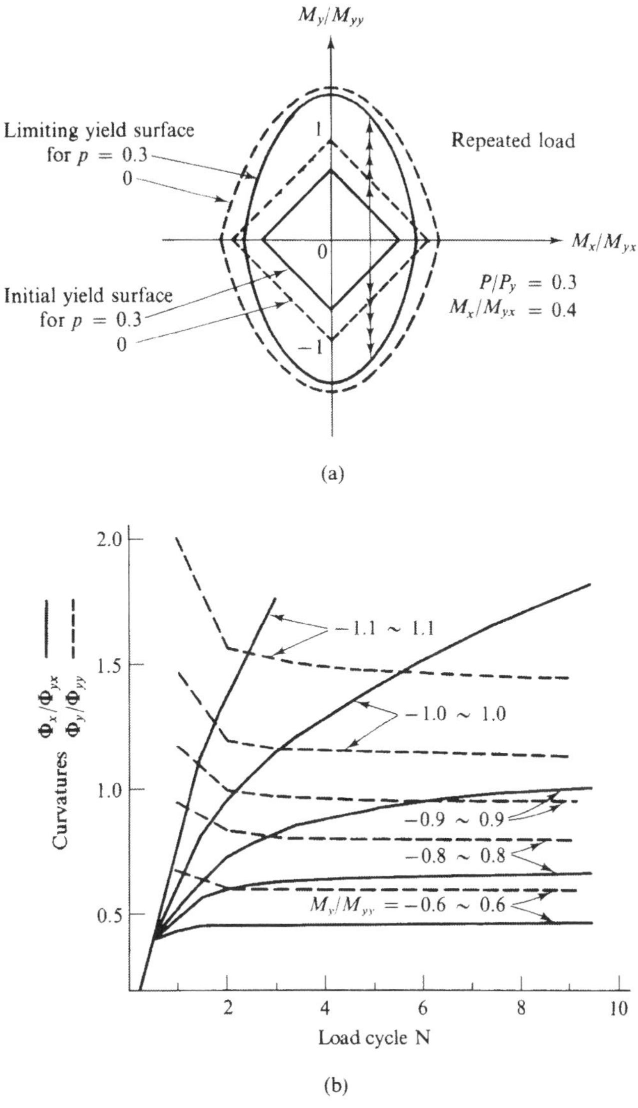

FIGURE 6.12
Repeated loading. (a) Loading condition. (b) Response to cyclic load

low-cycle fatigue and shakedown analysis. In these analyses, estimation of energy dissipation during the loading cycle plays an important role.

Figure 6.12(a) shows the load, $m_y = M_y/M_{yy}$, fluctuating between some limits while the other two loads are kept constant ($p = P/P_y = 0.3$, $m_x = M_x/M_{yx} = 0.4$). The half amplitude of the limits for the moment, m_y, is 0.6, 0.8, 0.9, 1.0, and 1.1. Since the elastic region is bounded by $-0.3 \leq m_y \leq 0.3$, plastic deformations are produced in all cases.

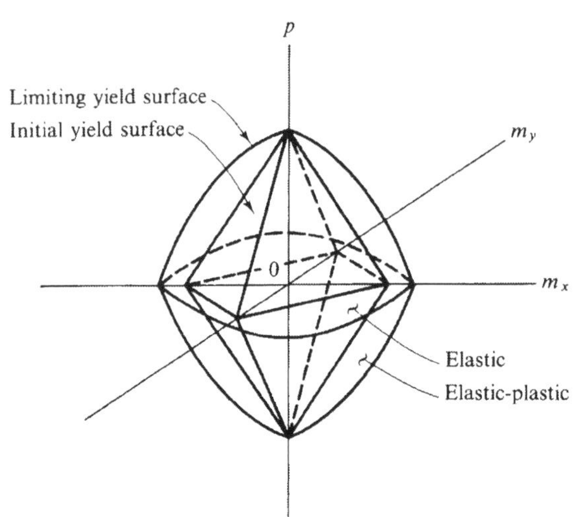

FIGURE 6.13
Yield surfaces in generalized stress space

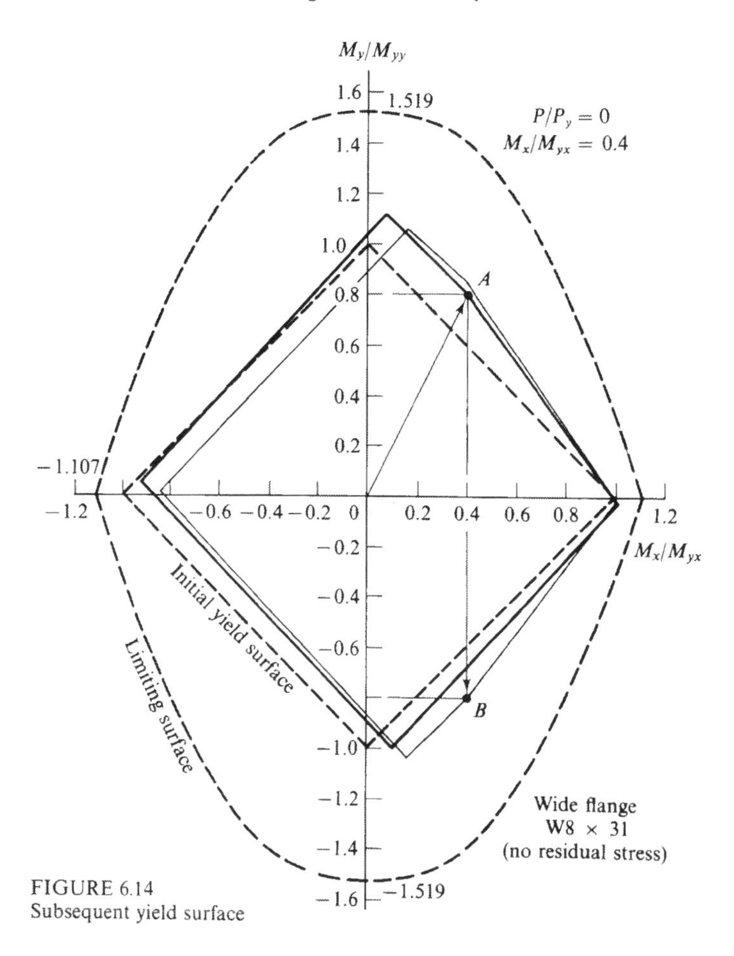

FIGURE 6.14
Subsequent yield surface

Numerical results are shown in Fig. 6.12(b). In all cases, curvature $\bar{\varphi}_y$ decreases with loading cycles and tends to converge to a certain limit value. The deformations, $\bar{\varepsilon}_0$ and $\bar{\varphi}_x$, and the energy dissipation are seen to increase monotonically with loading cycles for large amplitudes but tend to converge for small amplitudes ($m_y < 1.0$).

A theoretical and experimental investigation for the in-plane case of H-segments subjected to combined thrust and cyclic bending moment has been reported by Kato and Akiyama (1969). Some numerical results for plastic response of a symmetric section for an arbitrary history of in-plane loading have been presented in Chap. 5, Vol. 1 [Baron and Venkatesan (1969)].

6.5 SUBSEQUENT YIELD SURFACES

When we plot states of generalized stresses in $m_x - m_y - p$ space (for simplicity of illustration, warping deformation is set to be zero herein), the space is divided into

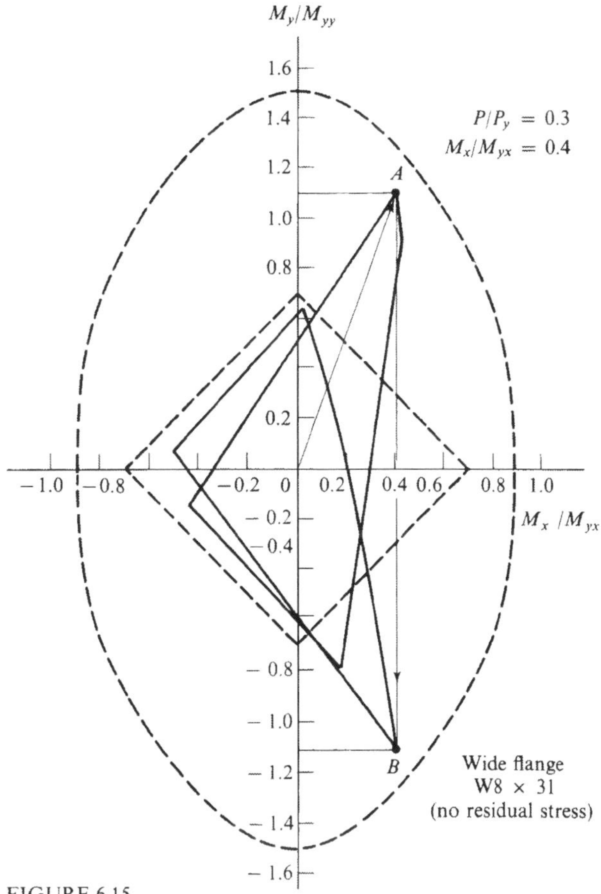

FIGURE 6.15
Transformation of subsequent yield surface

three regimes, i.e., elastic regime, elastic-plastic regime, and unattainable regime as shown in Fig. 6.13. The boundaries of regimes represent the *initial yield surface* and the *limiting yield surface*, respectively. Although the limiting yield surface is a uniquely fixed surface for a given cross section, the initial yield surface alters depending on history of loading. It is interesting to examine the change in shape of the initial yield surface due to cyclic loading as illustrated in Fig. 6.14 and Fig. 6.15.

Figure 6.14 shows *subsequent yield surfaces* due to loadings $p = 0$, $m_x = 0.4$, and m_y between 0.8 and -0.8. The dotted lines are the initial yield surface and the limiting yield surface.

After the first loading (point A), the yield surface translates so that the load point A is now on the subsequent yield surface. There are two interesting points: (1) The opposite side of point A moves towards the origin (Bauschinger Effect); and (2) the subsequent yield surface has a corner at the loading point.

After the second loading (point B), the subsequent yield surface changes again and both points A and B are now inside the surface. Thus the repetition of loading between A and B lies in elastic regime and further plastic deformation has ceased, or the section has *shaken down*.

Figure 6.15 shows changes of the subsequent yield surface due to repeated loading $p = 0.3$, $m_x = 0.4$, and m_y between 1.1 and -1.1. The repeated loading, m_y, is applied between points A and B. When the amplitude of m_y is small ($m_y = 0.8-$

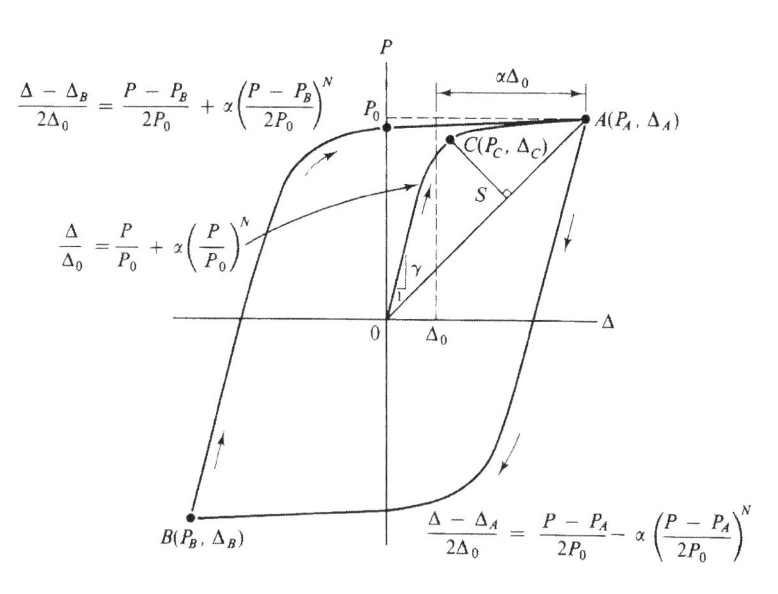

FIGURE 6.16
Curve fitting by Ramberg-Osgood function

Fig. 6.14) the subsequent yield surface tends to change so that the section shakes down. When the amplitude is large ($m_y = 1.1$ Fig. 6.15), the tendency of shakedown is not observed. Thus, plastic strain energy will increase during the repetition of loading, and low cycle fatigue failure will take place.

6.6 FUNCTION FITTING OF GENERALIZED STRESS-STRAIN CURVES

A generalized stress-strain hysteresis curve of a common structural section usually has the shape shown diagrammatically in Fig. 6.16. The curve is composed of three portions: (1) Initial loading portion; (2) unloading portion; and (3) reloading portion. For each portion, a function similar to a Ramberg-Osgood type of equation may be assumed.

Initial Loading Portion $(0 \rightarrow A)$

$$\frac{\Delta}{\Delta_0} = \frac{P}{P_0} + \alpha\left(\frac{P}{P_0}\right)^N \tag{6.36}$$

Unloading Portion $(A \rightarrow B)$

$$\frac{\Delta - \Delta_A}{2\Delta_0} = \frac{P - P_A}{2P_0} - \alpha\left(\frac{P - P_A}{2P_0}\right)^N \tag{6.37}$$

Reloading Portion $(B \rightarrow A)$

$$\frac{\Delta - \Delta_B}{2\Delta_0} = \frac{P - P_B}{2P_0} + \alpha\left(\frac{P - P_B}{2P_0}\right)^N \tag{6.38}$$

Equations (6.37) and (6.38) are directly obtained from Eq. (6.36) by doubling the scale and shifting the origin. The four constants, α, N, Δ_0, and P_0, are to be determined so that a given P-Δ curve can be closely represented by the functions. The following four conditions may be used for the determination of these four constants.

1. P_0 is assumed to be equal to P_A

$$P_0 = P_A \tag{6.39}$$

2. The slope, γ, at the origin must be fitted

$$\Delta_0 = \frac{P_0}{\gamma} \tag{6.40}$$

3. The function satisfies point A

$$\alpha = \frac{\Delta_A}{\Delta_0} - 1 \tag{6.41}$$

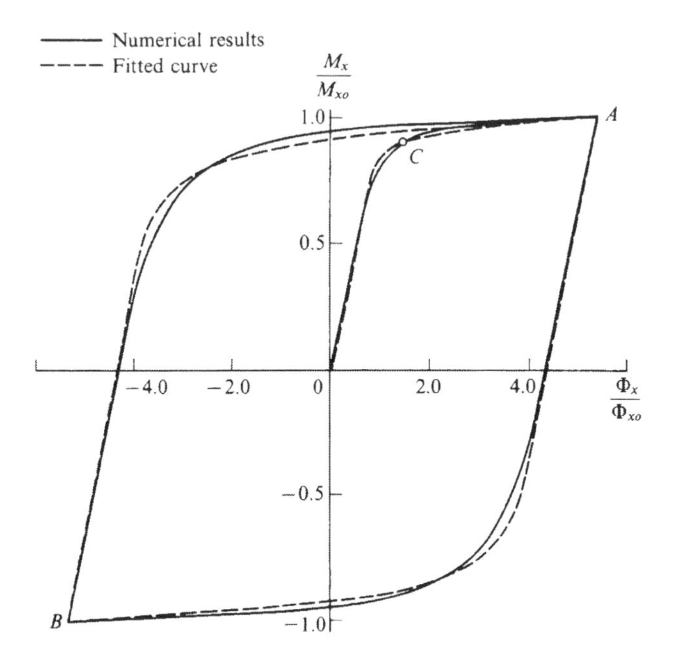

FIGURE 6.17
Moment-curvature curve fitting

4. The farthest point, C, from line OA is selected to satisfy the function

$$N = \frac{\log \dfrac{\gamma\Delta_A - P_A}{\gamma\Delta_C - P_C}}{\log \dfrac{P_A}{P_C}} \tag{6.42}$$

As an example, Fig. 6.17 shows the curve-fitting for the moment-curvature curve ($M_x - \Phi_x$ curve) examined in Fig. 6.10. The solid line is the actual numerical result and the dotted curve represents curve-fitting. The agreement observed is good. In this case, the four constants used in Eqs. (6.36), (6.37) and (6.38) are found to be:

$$\alpha = 4.38; \quad N = 21; \quad \Phi_{xo} = 0.636\Phi_{yx}; \quad M_{xo} = 0.594 M_{px} \tag{6.43}$$

in which M_{px} denotes fully plastic moment about x axis when no axial force or moment about y is acting.

6.7 BEHAVIOR OF FABRICATED TUBULAR STEEL SEGMENTS

Fabricated steel tubular beam-columns as commonly used in off-shore oil-drilling structures are formed from flat steel plate by a cold-rolling into a cylindrical shape, a fully-plastic bending procedure. Because of fabrication limitations such cylindrical formations are usually about ten feet (3 m) long. The rolling of the plate brings together two opposite edges of the plate and this joint is then welded to form a longitudinal seam in a cylindrical "can." A number of such "cans" are then welded circumferentially together end-to-end to form the required column length (Fig. 6.18). Both longitudinal and circumferential welds (between the cans) are full-penetration, multi-pass welds. This process of fabrication results in significant residual stresses in at least two directions in the completed column.

The tangent stiffness equation as derived in Sec. 6.2 for wide-flange shapes of cross section, is extended and applied here to the case of fabricated tubular steel beam-column segments subjected to compression P combined with biaxial bending moments M_x, and M_y. Warping moment and warping deformation do not exist in the case of tubular members.

In the analysis, both longitudinal and circumferential residual stresses are considered using Tresca yield criterion. In particular, allowance has been made for the relieving of circumferential residual stresses due to applied longitudinal loads which cause a "Poisson's ratio effect" on circumferential residual stresses. The effects of interaction between longitudinal and circumferential residual stresses on the elastic-plastic behavior of fabricated tubular steel beam-column segments are presented.

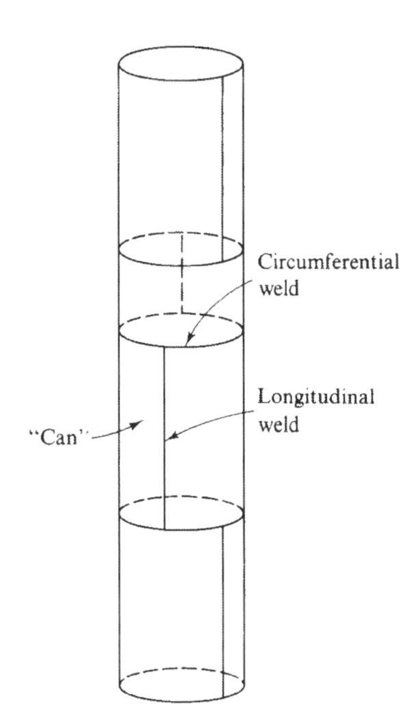

FIGURE 6.18
Fabricated tubular steel columns as used in offshore structures

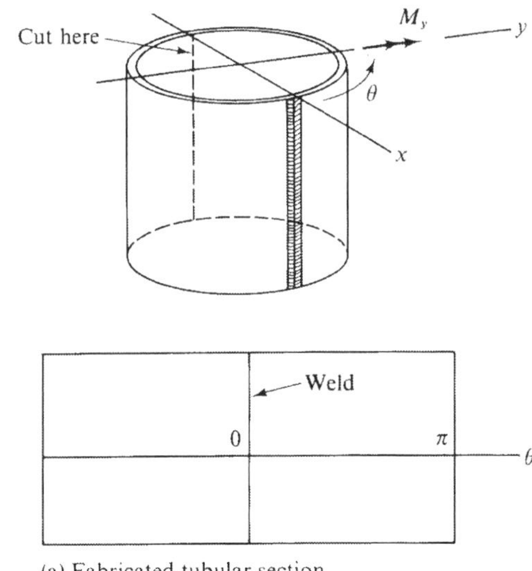

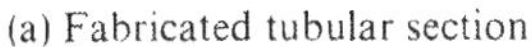
(a) Fabricated tubular section

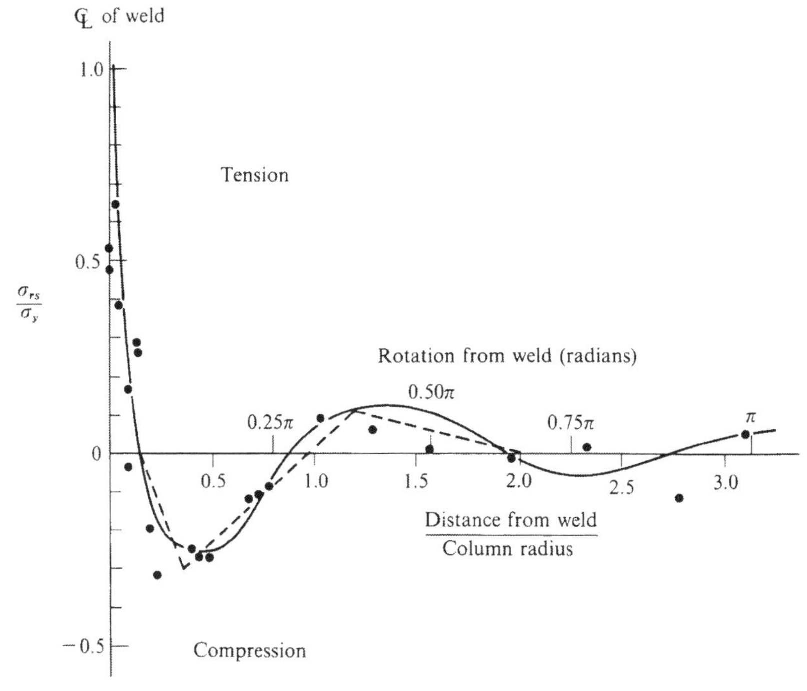

(b) Measured from slicing method

FIGURE 6.19 (a) and (b)
Flat projection of weld-induced longitudinal residual stress distribution

Residual stresses

Longitudinal residual stresses introduced by longitudinal welding of the "cans" can be measured by the method of "sectioning" (Tebedge *et al*, 1973). The residual

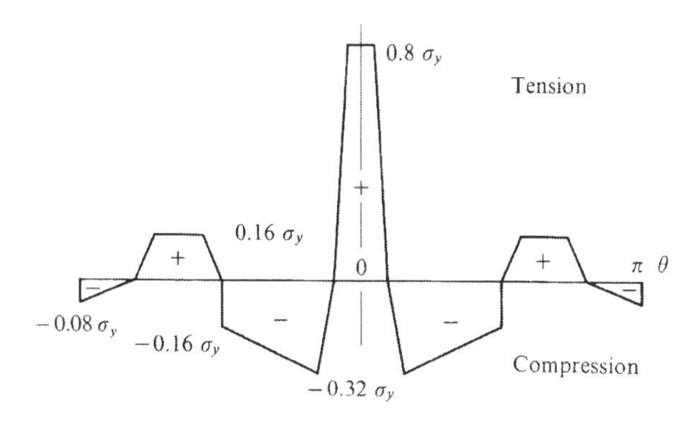

FIGURE 6.19 (c) Multi-linear approximation

stress distribution thus obtained is shown in Fig. 6.19(b). This is a half flat surface projection of a cylindrical shape, cut as shown in Fig. 6.19(a). This distribution represents an average through the wall thickness. The solid curve shows a possible approximation of the test points by a smooth curve reported by Chen and Ross (1976, 1977) and the dotted lines are a straight-line approximation suggested as a simplified alternative. It is noted that near the weld the material has effectively yielded in tension. It is particularly noteworthy that, except near the weld, longitudinal residual stresses as approximated by straight lines in Fig. 6.19(b) go through zero points at distances from the weld equal to multiples of the column radius. The longitudinal residual stress distribution due to the longitudinal welding procedure may be approximated closely by multi-linear lines as shown in Fig. 6.19(c). This multi-linear approximation will be used later in the study of moment-curvature behavior of a fabricated tubular segment.

Circumferential residual stresses can be measured by a hole-drilling technique (Redner, 1974) in which surface strain measurements can be taken of the strain release due to drilling at the base of a small diameter hole in the tubular column wall. For a given location, the experiment can be conducted both from inside and outside surfaces of the tubular column [Fig. 6.20a(i)]. Chen and Ross (1976, 1977) found that the variation in the distribution of circumferential residual stresses through the thickness of the wall at different locations on the same cross section is not significant [Fig. 6.20a(i)]. Figure 6.20a(ii) shows a typical experimental result. The hole-drilling technique is known to have a limited range of validity such that the results near the surface, as well as those taken near the center line of the wall, may contain possible inaccuracies. Thus, the straight-line approximation is dotted in these areas. Figure 6.20a(iii) shows the average circumferential residual stress pattern obtained.

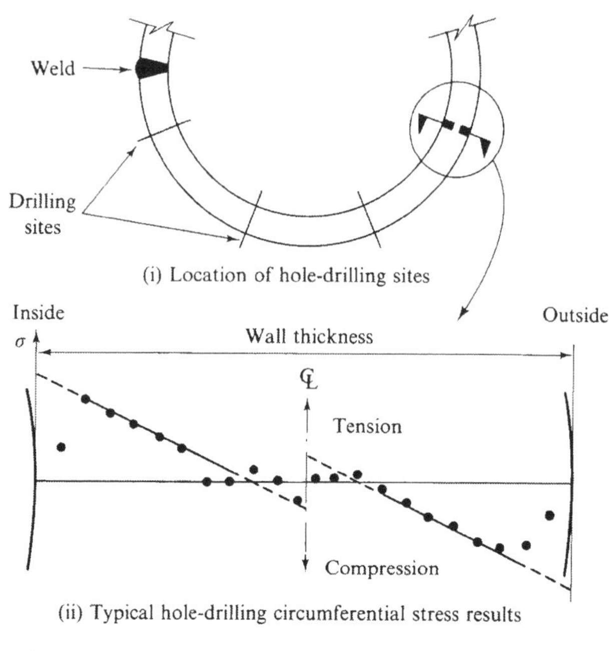

(i) Location of hole-drilling sites

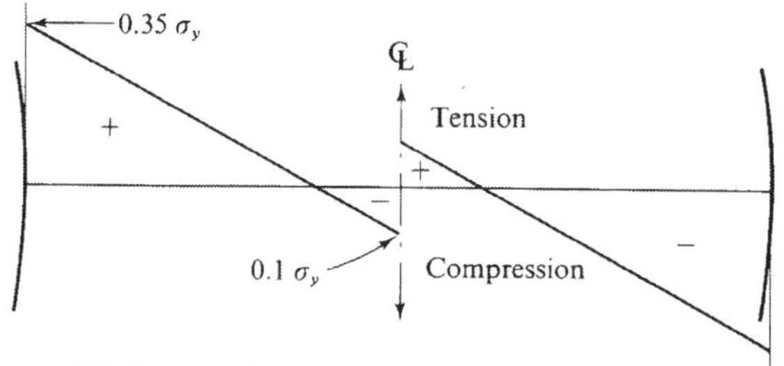

(ii) Typical hole-drilling circumferential stress results

(iii) Average circumferential residual stress pattern

(a) Measured from hole-drilling method

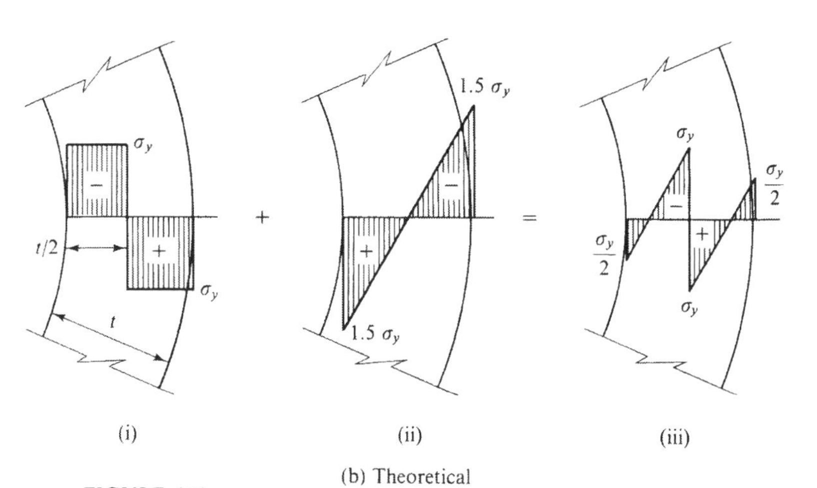

(i) (ii) (iii)

(b) Theoretical

FIGURE 6.20
Circumferential residual stress distribution through wall thickness

If the forming of a flat plate into a cylindrical shape by several cycles of repeated cold-rolling can be idealized as the process of pure bending of a beam, a crude estimate of the pattern of circumferential residual stresses distribution is then possible. This is illustrated in Fig. 6.20(b). In the cold-rolling process, fully plastic yielding of the plate must occur [Fig. 6.20b(i)]. Before welding, the "can" is allowed to "spring back" [Fig. 6.20b(ii)] and it results in the circumferential residual stress distribution shown in Fig. 6.20b(iii). Figure 6.20(b) shows that this estimate differs markedly in magnitude from the measured distribution [Fig. 6.20(a)], but it does show the correct pattern of distribution. Repeated cold-rolling, incomplete spring back and subsequent cooling of the welds, and initial residual stresses existing in the flat plate before forming, can all explain these discrepancies.

Biaxial stress interaction

An important facet of the analysis is that elemental stresses in two directions (i.e., longitudinal and circumferential) are linked by a Poisson's ratio effect. As the longitudinal stresses are applied (whether by application of axial load or of bending moment) the circumferential stresses will be effected, i.e., an increase in compression longitudinal stress produces a circumferential stress (approximately equal to longitudinal stress multiplied by Poisson's ratio) which increases circumferential compression stress and decreases circumferential tension stress. This interaction causes problems in the analysis when either the longitudinal stress is tensile and the circumferential stress is compressive, or vice versa, because in these regimes the Tresca yield curves are at 45° to the biaxial principal stress axes (Fig. 6.21). Not only will this interaction affect the stress state at which the element yields, but it will obviously affect the subsequent loading history of each particular element.

In the elastic range, elastic Poisson's ratio (0.29 for steel) is assumed to be adequate. However, in the plastic range, the case is somewhat different. When both circumferential and longitudinal stresses are either in tension or compression, then the Tresca yield condition indicates that the limiting stress of an element is the uniaxial yield stress σ_y, beyond which the element can assume no more load, but merely deforms plastically. However, in the tension-compression regimes of the Tresca yield diagram, the element stress condition is such that the elemental stress

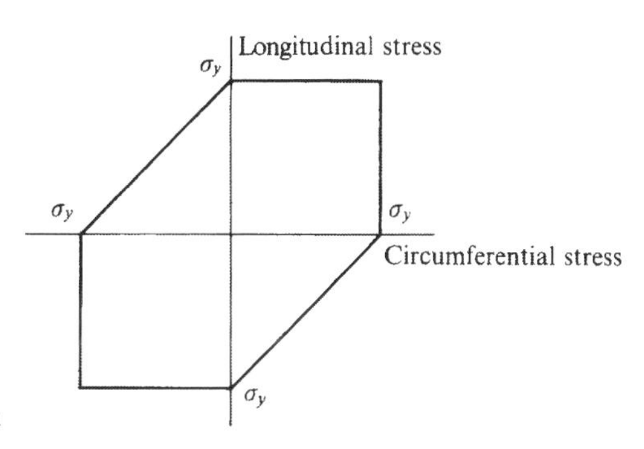

FIGURE 6.21
Tresca yield criterion

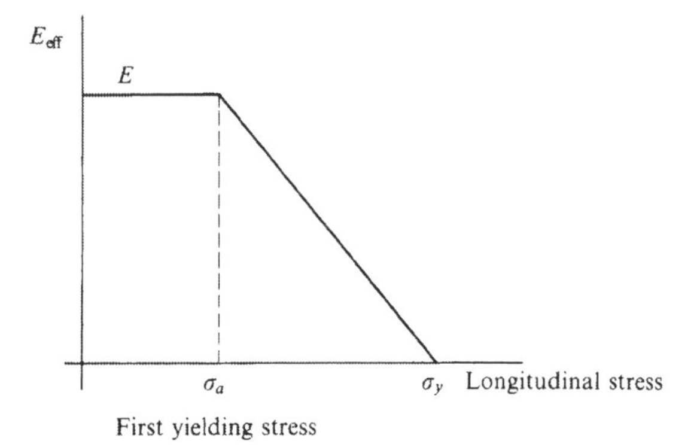

FIGURE 6.22
Effective Young's modulus idealization for biaxial stress condition

state can change while the element still remains on the yielded curve, but merely shifts its position on the sloping lines of the Tresca yield diagram. For this to be so, a Poisson's ratio of 0.5 is necessary, and the limiting longitudinal stress reaches, eventually, the uniaxial yield stress σ_y of the material. In other words, an element which has yielded in either of the tension-compression zones of the Tresca diagram does have some resistance to an increased segment curvature.

A rigorous plasticity solution requires the strain vector to be normal to the yield surface (flow rule) (see Chap. 2, Vol. 1). However, a solution of this type is complicated, and does not become readily accommodated in the expressions of Eq. (6.22). In order to find a relatively simple, yet reasonably accurate solution to this impasse, the concept of effective modulus E_{eff} for biaxial stress state is adopted herein. For an elastic-perfectly plastic material, an element has an elastic modulus, E, which can be assumed constant until the element yields at a longitudinal stress, σ_a. We also know that at the uniaxial yield stress, σ_y, the effective E value is zero. These two conditions were linked with a straight line as shown in Fig. 6.22, to provide a reasonable estimate of decreasing elemental resistance to an increased applied longitudinal loading in the biaxial tension-compression regimes of the Tresca yield diagram.

Moment-curvature curves

With inclusion of residual stresses in two directions, and the stress relieving through "Poisson's ratio effect" considered on an elemental basis, a modified M-P-Φ curve is expected on a tubular section. Figure 6.23 shows a comparison of M-P-Φ curves for a "perfect" tubular section with no residual stresses and that of the same section with due consideration given to both longitudinal and circumferential residual stresses. The curves compared are for the case of constant axial loads $P = 0$, 0.4, 0.6, $0.8P_y$ and uniaxial bending moment M_y (Fig. 6.19a). As expected, the stress relieving phenomena result in the same fully plastic moment capacity, M_{pc} being reached for all the cases considered.

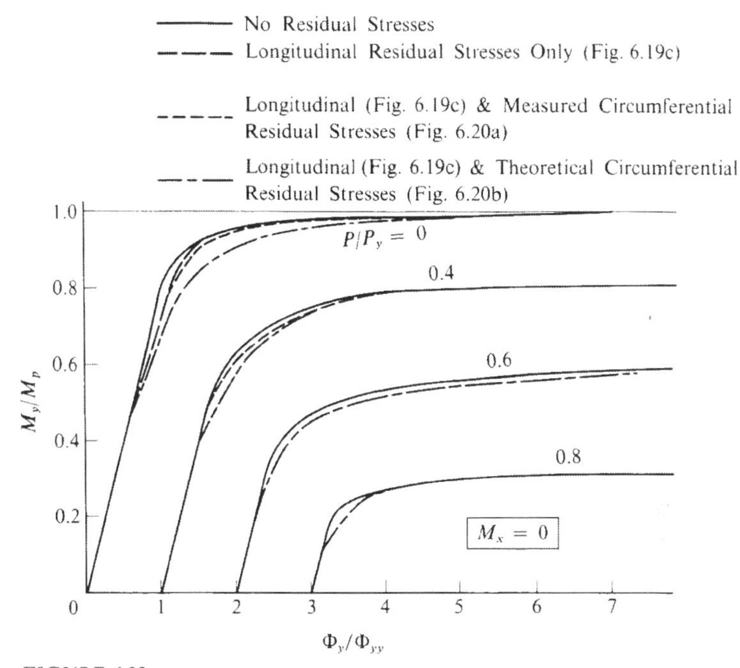

FIGURE 6.23
Effect of inclusion of residual stresses on a moment-curvature curve

Two other sets of M-P-Φ curves for a column section with residual stresses included are also presented. Figures 6.24 show the effects of biaxial bending moment M_x being applied to the segment for two cases ($P = 0$ and $P = 0.4P_y$). As expected, the inclusion of such an effect reduces the fully plastic moment capacity of the segment. Figure 6.25 show the effects of axial loads on the segment for the case of constant biaxial moment $M_x = 0.4M_p$.

6.8 BEHAVIOR OF REINFORCED CONCRETE SECTIONS

The tangent stiffness method developed previously for biaxially loaded steel sections is extended here to the case of reinforced concrete sections as shown in Fig. 6.26. The formulation is based on the following assumptions,

1. Concrete has no tensile strength, and in the usual notation

$$\bar{f}_c = \frac{f_c}{k_1 f_c'} = 0 \qquad \text{when} \qquad \bar{\varepsilon}_c = \frac{\varepsilon_c}{\varepsilon_c'} \leq 0 \tag{6.44}$$

where k_1 is the ratio of strength of concrete in member to specified cylinder compression strength and ε_c' is the concrete strain when concrete stress is $k_1 f_c'$. Here, compressive stress and strain are considered positive.

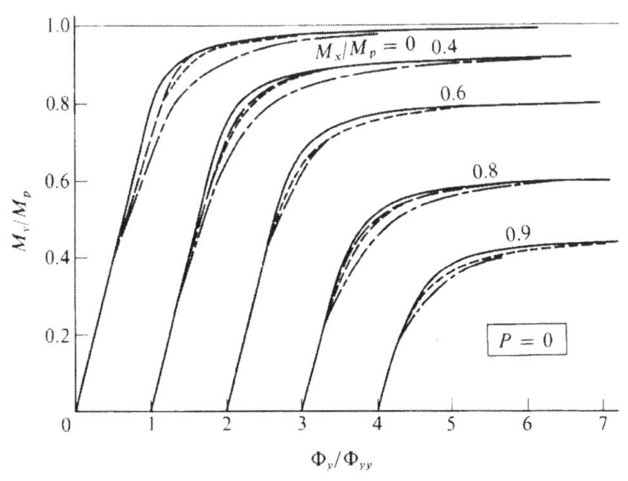

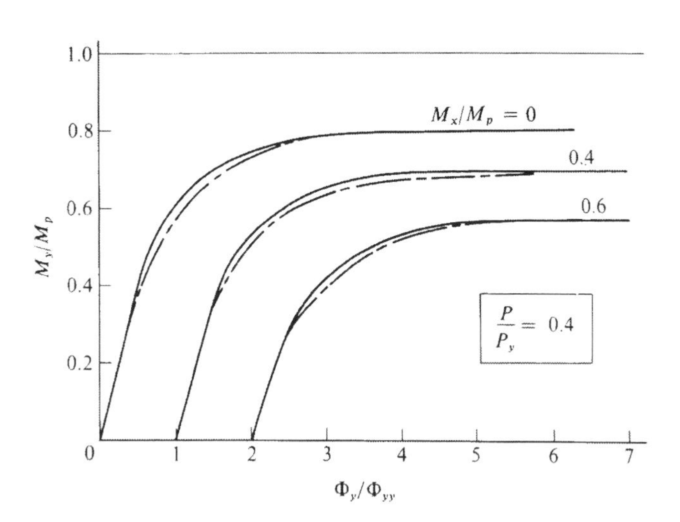

FIGURE 6.24
Moment curvature curves for biaxial bending

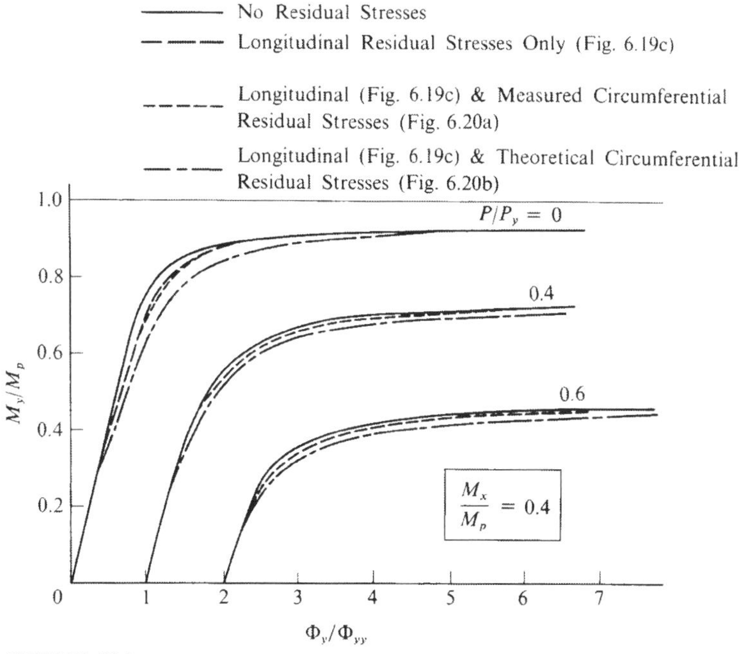

FIGURE 6.25
M-P-Φ curves for axial load and uniaxial bending

2. The stress-strain relationship for concrete in compression is nonlinear and is of the form as shown in Fig. 6.27(a) which can be expressed mathematically as

$$
\begin{aligned}
\bar{f}_c &= \gamma_1 \bar{\varepsilon}_c + (3 - 2\gamma_1)\bar{\varepsilon}_c^2 + (\gamma_1 - 2)\bar{\varepsilon}_c^3 && \text{when} \quad 0 \le \bar{\varepsilon}_c \le 1 \\
\bar{f}_c &= 1 - \frac{1 - 2\bar{\varepsilon}_c + \bar{\varepsilon}_c^2}{1 - 2\gamma_2 + \gamma_2^2} && \text{when} \quad 1 \le \bar{\varepsilon}_c \le \gamma_2 \\
\bar{f}_c &= 0 && \text{when} \quad \bar{\varepsilon}_c \ge \gamma_2
\end{aligned}
\tag{6.45}
$$

where

$$\gamma_1 = \frac{E_c \varepsilon_c'}{k_1 f_c'} \tag{6.46}$$

and γ_2 represents the point of intersection of the stress-strain curve with the strain axis.

3. The stress-strain relationship for steel is elastic perfectly plastic in both tension and compression [Fig. 6.27(b)] and in the usual notation

$$
\begin{aligned}
\bar{f}_s &= \frac{f_s}{f_y} = -1 && \text{when} \quad \bar{\varepsilon}_s = \frac{\varepsilon_s}{\varepsilon_y} < -1 \\
\bar{f}_s &= \bar{\varepsilon}_s && \text{when} \quad -1 \le \bar{\varepsilon}_s \le 1 \\
\bar{f}_s &= 1 && \text{when} \quad \bar{\varepsilon}_s > 1
\end{aligned}
\tag{6.47}
$$

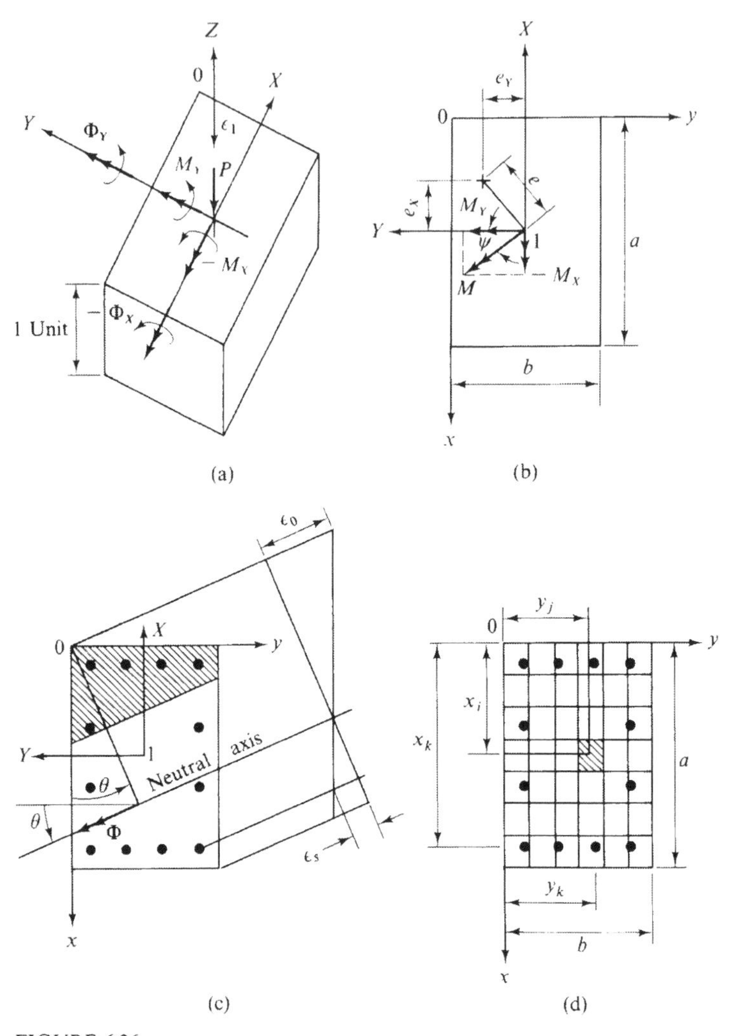

FIGURE 6.26
Moment, curvature and strain in cross section and partitioning of cross section

4. The effects of creep and shrinkage of the concrete are disregarded.
5. Plane sections remain plane before and after bending, implying no warping deformation.

Tangent stiffness matrix

In order to evaluate the internal actions, the concrete area is divided by horizontal and vertical lines into a total of N_c small rectangular elements, ΔA_c [Fig. 6.26(d)]. The total steel area is assumed to be distributed in N_s elements, all of equal area ΔA_s. The relation between ΔA_s and ΔA_c is

$$\Delta A_s = \rho' \Delta A_c \tag{6.48}$$

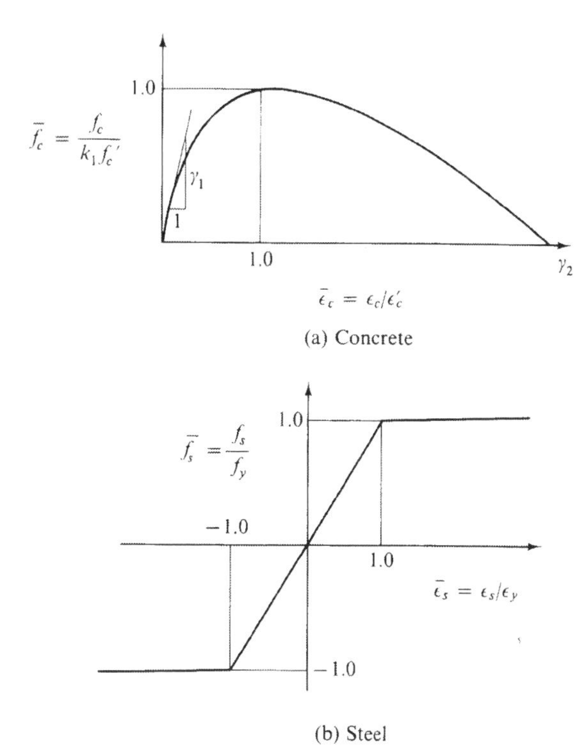

FIGURE 6.27
Stress-strain relations

where

$$\rho' = \frac{N_c}{N_s}\rho, \quad \rho = \frac{A_s}{ab} \tag{6.49}$$

The generalized stress-strain relations in their rate forms are given by Eq. (6.19); in the present case, they become, using the right hand screw rule for moment,

$$\begin{aligned}
\delta M_x &= -\left\{\sum_{i=1}^{N_a}\sum_{j=1}^{N_b} y_j(\delta f_c)_{ij} + \rho' \sum_{k=1}^{N_s} y_k(\delta f_s)_k - \rho' \sum_{k=1}^{N_s} y_k(\delta f_c)_k\right\}\Delta A_c \\
\delta M_y &= \left\{\sum_{i=1}^{N_a}\sum_{j=1}^{N_b} x_i(\delta f_c)_{ij} + \rho' \sum_{k=1}^{N_s} x_k(\delta f_s)_k - \rho' \sum_{k=1}^{N_s} x_k(\delta f_c)_k\right\}\Delta A_c \\
\delta P &= \left\{\sum_{i=1}^{N_a}\sum_{j=1}^{N_b} (\delta f_c)_{ij} + \rho' \sum_{k=1}^{N_s} (\delta f_s)_k - \rho' \sum_{k=1}^{N_s} (\delta f_c)_k\right\}\Delta A_c
\end{aligned} \tag{6.50}$$

where N_a and N_b are the numbers of rows and columns of elemental concrete areas respectively, and N_s is the number of bars.

The incremental changes of stress and strain in concrete are related by

$$\delta f_c = G_c \delta \varepsilon_c \tag{6.51}$$

where

$$G_c = 0 \qquad \text{when} \qquad \varepsilon_c \leq 0$$

$$G_c = \frac{k_1 f_c'}{\varepsilon_c'}\gamma_1 + 2\frac{k_1 f_c'}{\varepsilon_c'^2}(3 - 2\gamma_1)\varepsilon_c + 3\frac{k_1 f_c'}{\varepsilon_c'^3}(\gamma_1 - 2)\varepsilon_c^2 \quad \text{when} \quad 0 < \varepsilon_c \leq \varepsilon_c' \tag{6.52}$$

$$G_c = 2\frac{k_1 f_c'}{\varepsilon_c'(1 - 2\gamma_2 + \gamma_2^2)} - 2\frac{k_1 f_c'}{\varepsilon_c'^2(1 - 2\gamma_2 + \gamma_2^2)}\varepsilon_c \quad \text{when} \quad \varepsilon_c' < \varepsilon_c \leq \varepsilon_c'\gamma_2$$

$$G_c = 0 \qquad \text{when} \qquad \varepsilon_c > \varepsilon_c'\gamma_2$$

The incremental changes in stress and strain of steel are related by

$$\delta f_s = G_s \delta\varepsilon_s \tag{6.53}$$

where

$$G_s = 0 \qquad \text{when} \qquad \varepsilon_s < -\varepsilon_y$$

and

$$G_s = \frac{f_y}{\varepsilon_y} = E_s \qquad \text{when} \qquad -\varepsilon_y \leq \varepsilon_s \leq \varepsilon_y \tag{6.54}$$

and

$$G_s = 0 \qquad \text{when} \qquad \varepsilon_s > \varepsilon_y$$

Substituting δf_c and δf_s from Eqs. (6.51) and (6.53) into Eq. (6.50), we have,

$$\delta M_x = -\left\{\sum_{i=1}^{N_a}\sum_{j=1}^{N_b} y_j (G_c)_{ij}(\delta\varepsilon_c)_{ij} + \rho' \sum_{k=1}^{N_s} y_k (G_s)_k (\delta\varepsilon_s)_k - \rho' \sum_{k=1}^{N_s} y_k (G_c)_k (\delta\varepsilon_c)_k\right\}\Delta A_c$$

$$\delta M_y = \left\{\sum_{i=1}^{N_a}\sum_{j=1}^{N_b} x_i (G_c)_{ij}(\delta\varepsilon_c)_{ij} + \rho' \sum_{k=1}^{N_s} x_k (G_s)_k (\delta\varepsilon_s)_k - \rho' \sum_{k=1}^{N_s} x_k (G_c)_k (\delta\varepsilon_c)_k\right\}\Delta A_c$$

$$\delta P = \left\{\sum_{i=1}^{N_a}\sum_{j=1}^{N_b} (G_c)_{ij}(\delta\varepsilon_c)_{ij} + \rho' \sum_{k=1}^{N_s} (G_s)_k (\delta\varepsilon_s)_k - \rho' \sum_{k=1}^{N_s} (G_c)_k (\delta\varepsilon_c)_k\right\}\Delta A_c \tag{6.55}$$

The strain ε at any point in the cross section with respect to xy coordinate can be expressed in a linear form as

$$\varepsilon = -y\Phi_x + x\Phi_y + \varepsilon_o \tag{6.56}$$

where ε_o is the strain at the corner O (Fig. 6.26). The incremental change of the strain is

$$\delta\varepsilon = -y\delta\Phi_x + x\delta\Phi_y + \delta\varepsilon_o \tag{6.57}$$

or

$$(\delta\varepsilon_c)_{ij} = -y_j\delta\Phi_x + x_i\delta\Phi_y + \delta\varepsilon_o$$

$$(\delta\varepsilon_s)_k = (\delta\varepsilon_c)_k = -y_k\delta\Phi_x + x_k\delta\Phi_y + \delta\varepsilon_o \tag{6.58}$$

where $\delta\varepsilon_o$ is the strain increment at the corner O of the cross section. Combinations of Eqs. (6.55) and (6.58) gives a set of simultaneous linear equations which can be written in the matrix form as

$$\begin{Bmatrix} \delta M_x \\ \delta M_y \\ \delta P \end{Bmatrix} = \begin{vmatrix} Q_{11} & Q_{12} & Q_{13} \\ Q_{21} & Q_{22} & Q_{23} \\ Q_{31} & Q_{32} & Q_{33} \end{vmatrix} \begin{Bmatrix} \delta\Phi_x \\ \delta\Phi_y \\ \delta\varepsilon_o \end{Bmatrix} \tag{6.59}$$

where Q_{ij} is defined as

$$\begin{aligned}
Q_{11} &= \left\{ \sum_{i=1}^{N_a} \sum_{j=1}^{N_b} (y_j)^2 (G_c)_{ij} + \rho' \sum_{k=1}^{N_s} (y_k)^2 (G_s)_k - \rho' \sum_{k=1}^{N_s} (y_k)^2 (G_c)_k \right\} \Delta A_c \\
Q_{22} &= \left\{ \sum_{i=1}^{N_a} \sum_{j=1}^{N_b} (x_i)^2 (G_c)_{ij} + \rho' \sum_{k=1}^{N_s} (x_k)^2 (G_s)_k - \rho' \sum_{k=1}^{N_s} (x_k)^2 (G_c)_k \right\} \Delta A_c \\
Q_{33} &= \left\{ \sum_{i=1}^{N_a} \sum_{j=1}^{N_b} (G_c)_{ij} + \rho' \sum_{k=1}^{N_s} (G_s)_k - \rho' \sum_{k=1}^{N_s} (G_c)_k \right\} \Delta A_c \\
Q_{12} = Q_{21} &= -\left\{ \sum_{i=1}^{N_a} \sum_{j=1}^{N_b} x_i y_j (G_c)_{ij} + \rho' \sum_{k=1}^{N_s} x_k y_k (G_s)_k - \rho' \sum_{k=1}^{N_s} x_k y_k (G_c)_k \right\} \Delta A_c \\
Q_{13} = Q_{31} &= -\left\{ \sum_{i=1}^{N_a} \sum_{j=1}^{N_b} y_j (G_c)_{ij} + \rho' \sum_{k=1}^{N_s} y_k (G_s)_k - \rho' \sum_{k=1}^{N_s} y_k (G_c)_k \right\} \Delta A_c \\
Q_{23} = Q_{32} &= \left\{ \sum_{i=1}^{N_a} \sum_{j=1}^{N_b} x_i (G_c)_{ij} + \rho' \sum_{k=1}^{N_s} x_k (G_s)_k - \rho' \sum_{k=1}^{N_s} x_k (G_c)_k \right\} \Delta A_c
\end{aligned} \tag{6.60}$$

Equation (6.59) can be written as

$$\{dF\} = [Q]\{d\Delta\} \tag{6.61}$$

The symmetric matrix $[Q]$, whose elements are given by Eq. (6.60) is the tangent stiffness matrix.

Numerical Examples (Chen and Shoraka, 1975)

Based upon the equations formulated, the numerical work was performed on the specific case of a square section with the following values as a standard concrete column cross section:

$$\begin{gathered}
a = 24 \text{ in (61 cm)}, \quad b = 24 \text{ in (61 cm)}, \quad N_a = 10, \quad N_b = 10 \\
N_s = 12, \quad \rho = \frac{A_s}{ab} = 0.0325, \quad k_1 f_c' = 4.2 \text{ ksi (29 MN/m}^2) \\
k_1 = 0.85, \quad f_s = 60.0 \text{ ksi (414 MN/m}^2), \quad \varepsilon_c' = 0.002 \\
E_s = 29\,000\,000 \text{ psi (200 000 000 kN/m}^2), \\
E_c = 57\,600 \sqrt{f_c'} \text{ in psi} \quad \text{(for normal weight concrete)} \\
\gamma_2 = 4, \quad \gamma_1 = \text{computed from Eq. (6.46)}.
\end{gathered} \tag{6.62}$$

The elements of the tangent stiffness matrix of the cross section were evaluated numerically by dividing the cross section into finite elements N_c varied from 100 (10 × 10) to 400 (20 × 20) for the square section. The increase in accuracy obtained by using the finer grids was only 0.1 percent. A partitioning of the concrete cross section into 100 elements and the steel areas into 12 elements distributed uniformly around the sides of the section are used herein.

The strain and stress in each element were computed as the average value at its centroid. All force and deformation vectors are nondimensionalized as,

Force vector (Fig. 6.26a)

$$\left\{ \frac{P}{f_c'ab} \quad \frac{M_X}{f_c'ab^2} \quad \frac{M_Y}{f_c'a^2b} \right\} \tag{6.63}$$

Deformation vector (Fig. 6.26a)

$$\left\{ \frac{\Phi_X}{\left(\frac{\varepsilon_c'}{a}\right)} \quad \frac{\Phi_Y}{\left(\frac{\varepsilon_c'}{b}\right)} \quad \frac{\varepsilon_o}{\varepsilon_c'} \right\} \tag{6.64}$$

FIGURE 6.28
Moment-curvature relations: standard case

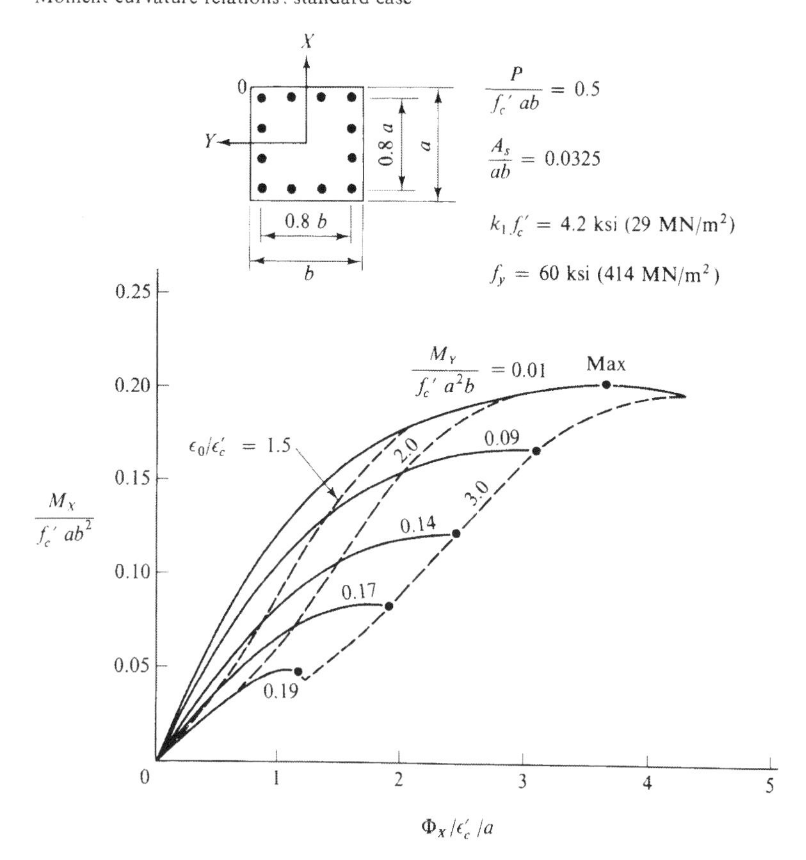

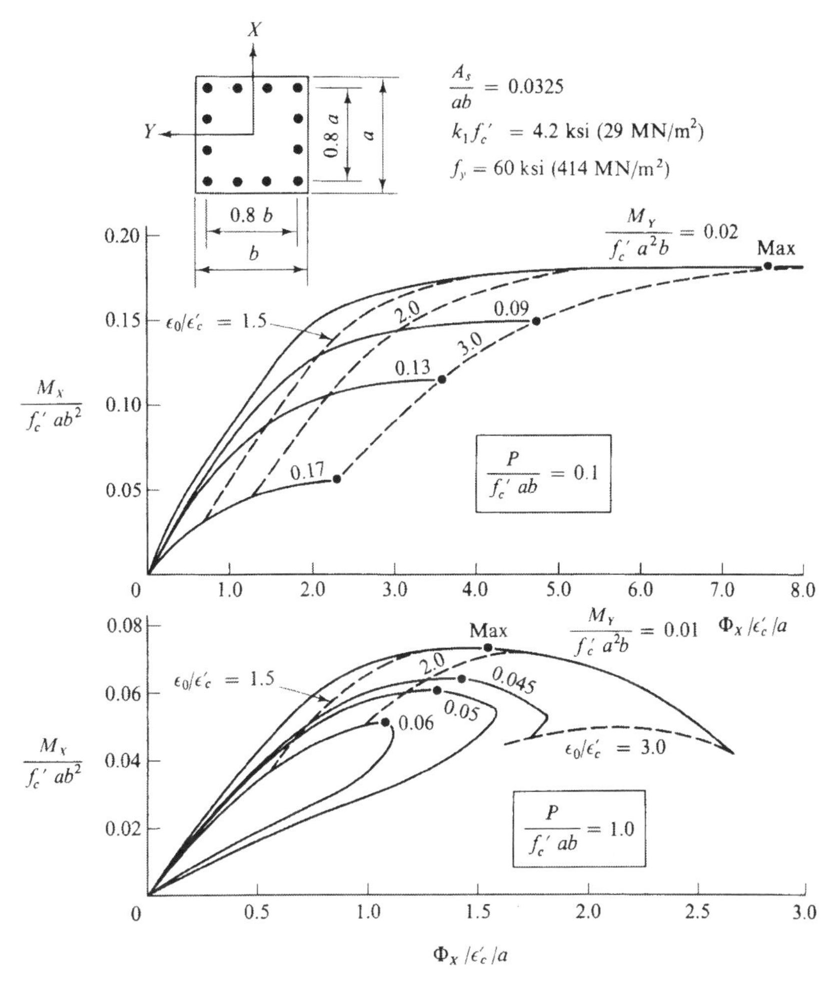

FIGURE 6.29
Moment-curvature relations: axial compression force effect

Example—given path of loading

The moment-curvature curves plotted in Figs. 6.28 to 6.33 are for M_X vs. Φ_X for various values of M_Y. The column section is first loaded axially up to some value and then bent by M_Y to some other value while keeping P constant and finally bent by M_X to failure while keeping P and M_Y constant. The curves have been terminated when the strain ratio $\varepsilon_0/\varepsilon_c'$ reaches the value 3.0. To indicate the magnitude of the strains in the cross section, two other lines of constant $\varepsilon_0/\varepsilon_c' = 1.5$ recommended by ACI (318-71) and 2.0 have been plotted across the main curves (dotted lines in the figures).

The maximum values of the moment are indicated by the small circles in Figs. 6.28 to 6.33 which represent the maximum strength interaction relation of the cross

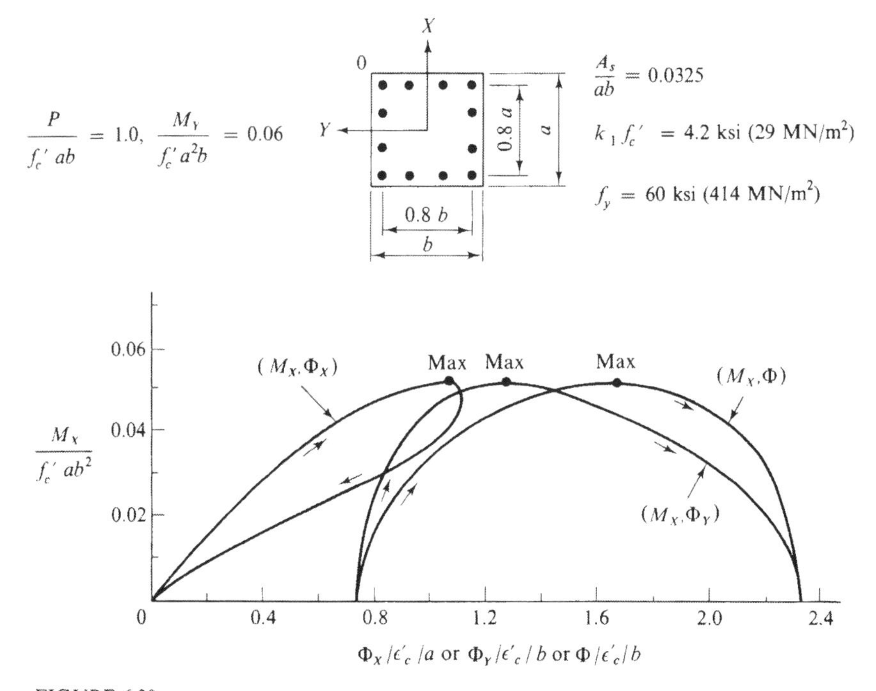

FIGURE 6.30
Moment-curvature relations: complete unloading

section as was discussed in Chap. 5. These moment curvature curves indicate that the maximum strength of short columns in biaxial bending and compression, are not unduly sensitive to the variations in the assumed concrete ultimate strain which is often chosen in the range between 0.003 and 0.004.

The moment curvature curves shown in Fig. 6.28 are considered to be the standard cases. The important factors influencing the behavior of the curves are the magnitude of compression force P, concrete quality $k_1 f_c'$, steel quality f_y, and percentage of reinforcement A_s/ab. The variations of these factors with respect to the standard case are given in Figs. 6.29 to 6.33.

The influence of axial compression force on the moment curvature curves is shown in Fig. 6.29. The unloading of the moment, M_X, with respect to an increase in curvature Φ_X is not seen for the curves $P = 0.1 f_c'ab$ within the range $\varepsilon_o/\varepsilon_c' = 3.0$ but is rather rapid for the curves with $P = 1.0 f_c'ab$. It is also observed, that when $P = 1.0 f_c'ab$ and the bending moment $M_Y \geq 0.05 f_c'a^2b$, there is a very rapid unloading for both moment M_X and curvature Φ_X. The curvature Φ_Y or the resultant curvature Φ is, of course, not unloaded with respect to a decrease in moment M_X, as shown in Fig. 6.30.

The influence of concrete quality $k_1 f_c'$ and steel quality f_y on the moment curvature curves is shown in Figs. 6.31 and 6.32. The results are calculated for concrete with $k_1 f_c' = 3.0$ ksi (21 MN/m^2) and 5.0 ksi (35 MN/m^2) (Fig. 6.31) and for

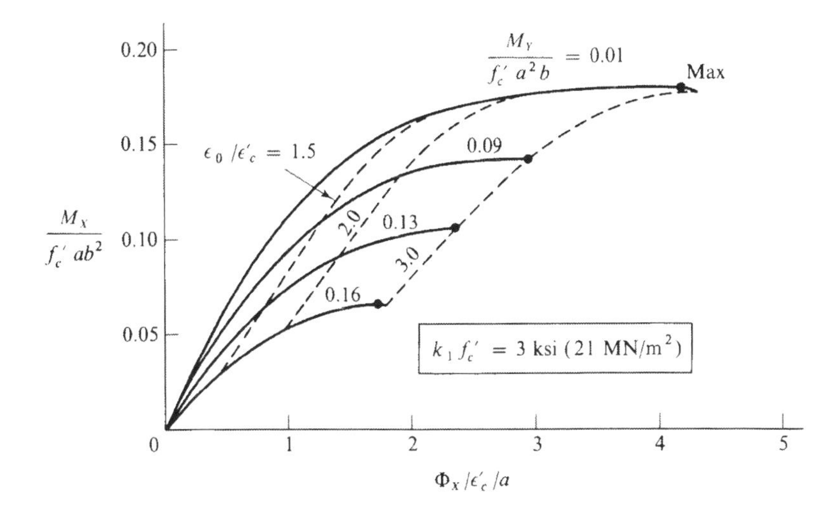

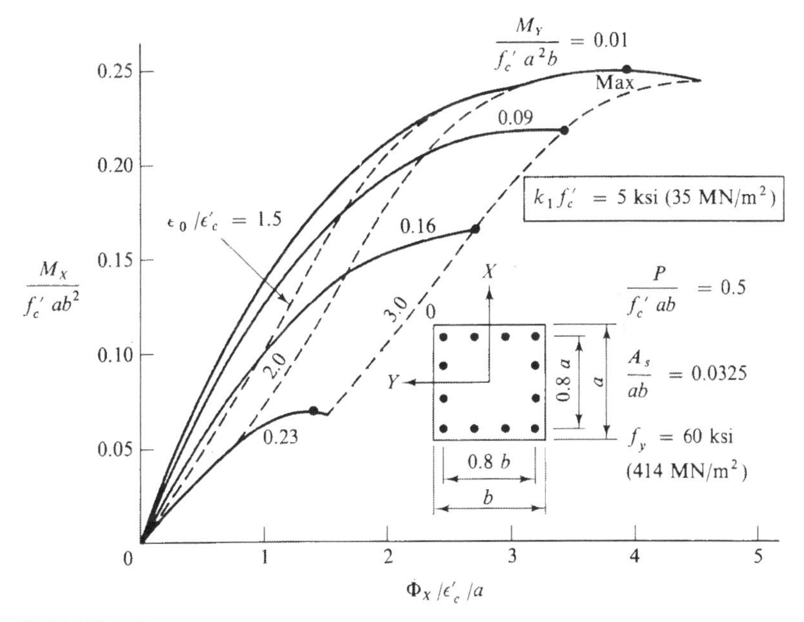

FIGURE 6.31
Moment-curvature relations: concrete quality effect

steel with $f_y = 40$ ksi (276 MN/m^2) and 80 ksi (552 MN/m^2) (Fig. 6.32) respectively. As can be seen, an increase in material qualities significantly increases the stiffness and strength of a biaxially loaded cross section.

Figure 6.33 shows the influence of the percentage reinforcement A_s/ab on the moment curvature relationships. It is evident from the figure that the percentage

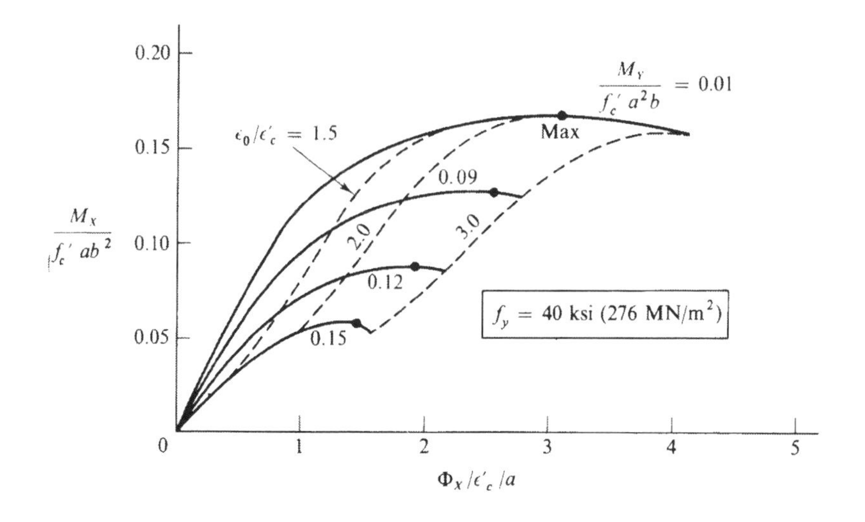

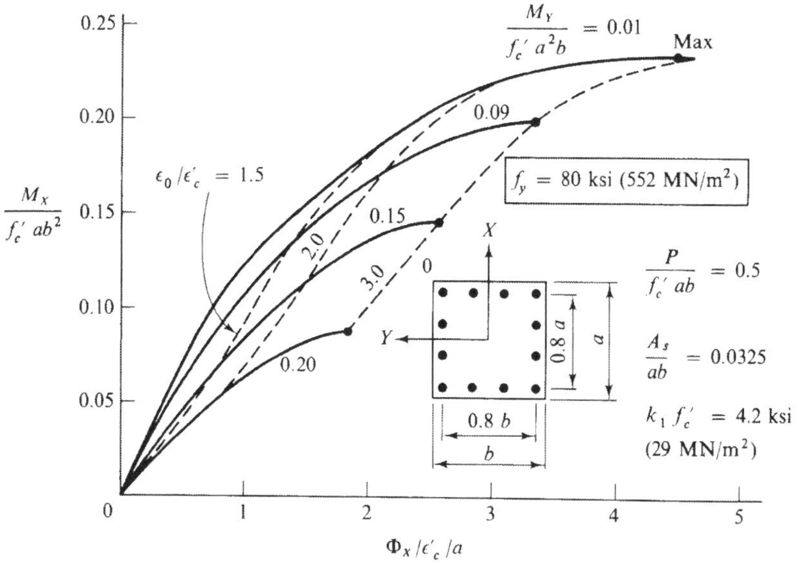

FIGURE 6.32
Moment-curvature relations: steel quality effect

steel reinforcement has an appreciable effect on the behavior of a biaxially loaded cross section.

Example—given mixed path of loading and deformation

The moment-curvature curves plotted in Fig. 6.34 are for e/b vs. Φ for a given set of values of $\theta = \tan^{-1}(\Phi_X/\Phi_Y) = 15°$ and $P/P_u = 0.2$, 0.4, 0.6 and 0.8 where P_u is the failure load of the section for zero eccentricity. In the figure, the column

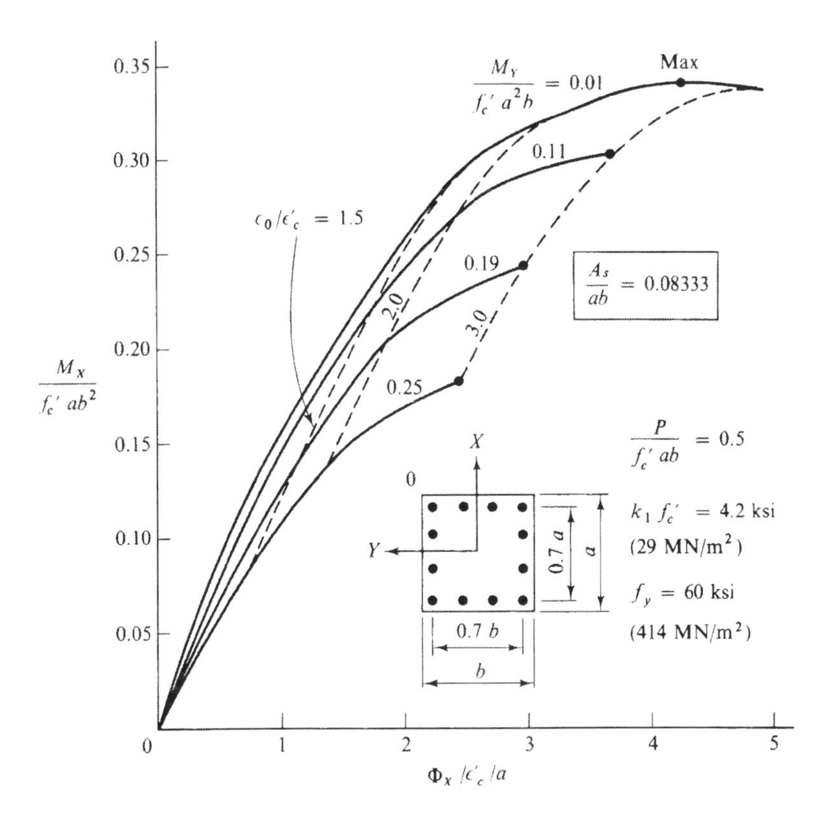

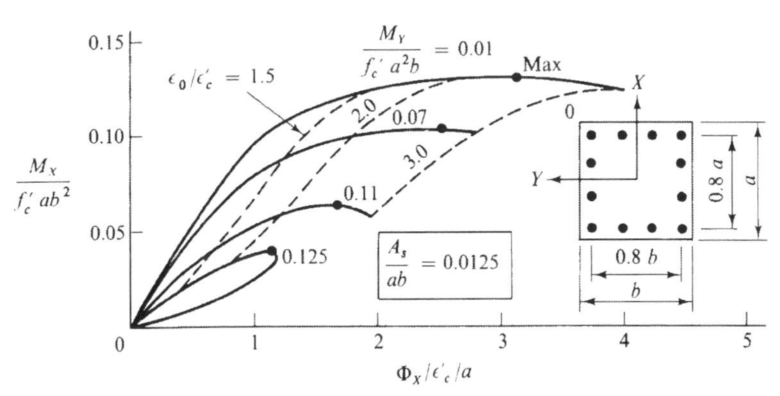

FIGURE 6.33
Moment-curvature relations: percentage of reinforcement effect

section is first loaded axially to some value; and then the axial force P is held constant while the bending curvatures Φ_X and Φ_Y (or $\Phi = \sqrt{\Phi_X^2 + \Phi_Y^2}$) are increased proportionally in magnitude from zero. The corresponding bending moments M_X and M_Y (or $e = M/P = \sqrt{M_X^2 + M_Y^2}/P$) and axial strain ε_o at the corner O can be

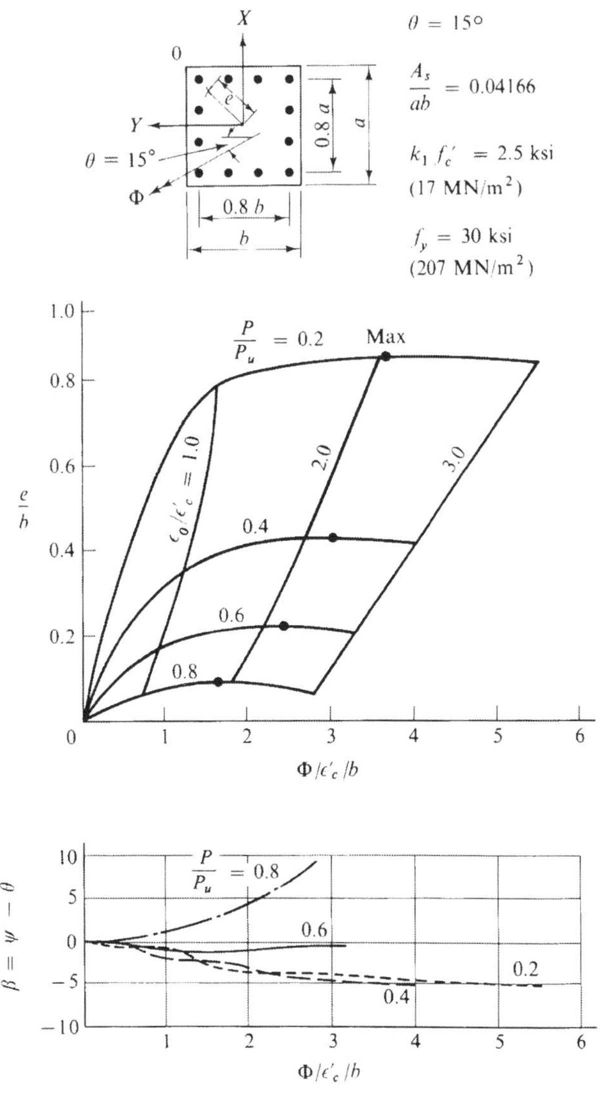

FIGURE 6.34
Moment curvature relations: $\theta = 15°$

obtained by the tangent stiffness method. These moment curvature curves were compared with those obtained previously by Warner (1969) and an excellent agreement was found in all cases.

The maximum difference between the angles θ and ψ, i.e., between the directions of the resultant curvature Φ and resultant moment vectors, $\beta = \psi - \theta$, is also shown in Fig. 6.34. It can be seen that the moment and curvature vectors nearly coincide in direction throughout the entire range of loading. The maximum difference between the two vectors is of the order of ten degrees.

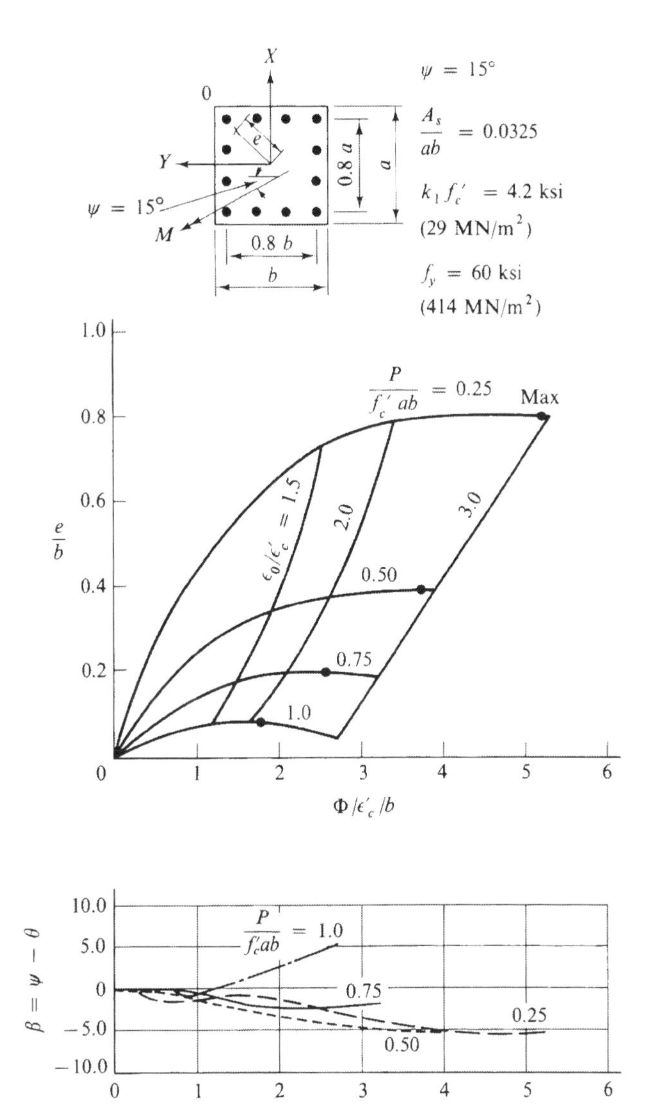

FIGURE 6.35
Moment-curvature relations: $\psi = 15°$

It is also of interest to note that a similar conclusion is also true for the case of other loading paths. For example, in Fig. 6.35; the section is first loaded axially to some constant value and then the axial force P is held constant while the bending moments M_X and M_Y are increased proportionally in magnitude; i.e., $\psi = \tan^{-1}(M_X/M_Y)$. The corresponding bending curvatures Φ_X and Φ_Y and axial strain ε_o can be obtained by the tangent stiffness method. The maximum difference between the angles θ and ψ is again only of the order of ten degrees.

6.9 SUMMARY

The generalized stress $\{F\}$ of an I-shape beam-column segment are defined as the stress resultants over the section

$$
\begin{aligned}
P &= \int \sigma \, dA \\
M_x &= \int \sigma Y \, dA \\
M_y &= -\int \sigma X \, dA \\
M_\omega &= \int \sigma X Y \, dA
\end{aligned}
\tag{6.65}
$$

and the corresponding generalized strains $\{\Delta\}$ are defined by the average strains over the section,

$$
\begin{aligned}
\varepsilon_o &= \frac{1}{A}\int \varepsilon \, dA \\
\Phi_x &= \frac{1}{A}\int \frac{\partial \varepsilon}{\partial Y} \, dA \\
\Phi_y &= -\frac{1}{A}\int \frac{\partial \varepsilon}{\partial X} \, dA \\
\Phi_\omega &= \frac{1}{A}\int \frac{\partial^2 \varepsilon}{\partial X \, \partial Y} \, dA
\end{aligned}
\tag{6.66}
$$

such that the rate of internal energy dissipation in the section is the dot product of the generalized stress vector $\{F\}$ and the generalized strain rate vector $\{\dot{\Delta}\}$, i.e.,

$$
\dot{D}_I = \{\dot{\Delta}\}^T\{F\} = \dot{\varepsilon}_o P + \dot{\Phi}_x M_x + \dot{\Phi}_y M_y + \dot{\Phi}_\omega M_\omega \tag{6.67}
$$

The generalized stresses and generalized strains are related to each other in their rate form $\{\dot{F}\}$ and $\{\dot{\Delta}\}$ by the tangent stiffness matrix $[K]$,

$$
\{\dot{F}\} = [K]\{\dot{\Delta}\}
$$

or

$$
\begin{Bmatrix} \dot{P} \\ \dot{M}_x \\ \dot{M}_y \\ \dot{M}_\omega \end{Bmatrix}
=
\begin{bmatrix}
\int E_t \, dA & \int E_t Y \, dA & -\int E_t X \, dA & \int E_t X Y \, dA \\
\int E_t Y \, dA & \int E_t Y^2 \, dA & -\int E_t X Y \, dA & \int E_t X Y^2 \, dA \\
-\int E_t X \, dA & -\int E_t X Y \, dA & \int E_t X^2 \, dA & -\int E_t X^2 Y \, dA \\
\int E_t X Y \, dA & \int E_t X Y^2 \, dA & -\int E_t X^2 Y \, dA & \int E_t X^2 Y^2 \, dA
\end{bmatrix}
\begin{Bmatrix} \dot{\varepsilon}_o \\ \dot{\Phi}_x \\ \dot{\Phi}_y \\ \dot{\Phi}_\omega \end{Bmatrix}
\tag{6.68}
$$

The relations between the generalized stresses $\{F\}$ and generalized strain $\{\Delta\}$ in the plastic range are non-linear and must be solved by an iterative procedure such as the tangent stiffness method, the procedure of which is summarized in what follows.

When the current strains $\{\Delta_A\}$ at state A are given, the corresponding current stresses $\{F_A\}$ and the tangent stiffness matrix $[K_A]$ are known as functions of the current strains,

$$\begin{aligned} \{F_A\} &= \{F(\Delta_A)\} \\ [K_A] &= [K(\Delta_A)] \end{aligned} \tag{6.69}$$

The problem is to find the values of strains $\{\Delta_B\}$ at state B when the stresses are increased from $\{F_A\}$ to $\{F_B\}$. This can only be achieved by iteration starting from the known state A. The iteration procedures are:

$$\begin{aligned} \{\Delta_i\} &= \{\Delta_{i-1}\} + [K_{i-1}]^{-1}\{dF_{i-1}\} \\ \{F_i\} &= \{F(\Delta_i)\} \\ \{dF_i\} &= \{F_B\} - \{F_i\} \end{aligned} \tag{6.70}$$

for which the starting values are

$$\begin{aligned} \{\Delta_o\} &= \{\Delta_A\} \\ \{F_o\} &= \{F_A\} \\ \{dF_o\} &= \{F_B\} - \{F_A\} \end{aligned} \tag{6.71}$$

and the iteration stops at cycle N for which the unbalanced force $\{dF_N\}$ falls within a prescribed tolerance limit.

Response of steel beam-column segments subject to cyclic biaxial loading is examined. The stress-strain material model used in the analysis considers elastic unloading, strain hardening, and Bauschinger effect. Numerical results are shown for a wide flange section. The effects of loading path and the movement of subsequent yield surfaces are also discussed.

For fabricated tubular steel cross sections as used in offshore structures, the yield of a material element may be caused by both circumferential and longitudinal stresses. The value of E_t in Eq. (6.68) must therefore be replaced by an "effective" modulus E_{eff} in the regions of biaxial tension-compression yielding. The effects of the interaction between longitudinal and circumferential residual stresses on the behavior of such beam-column segments are presented.

Moment curvature relations of reinforced concrete segments are also treated. The important factors influencing the behavior of these curves are discussed.

6.10 PROBLEMS

6.1 Examine that the tangent stiffness matrix $[K]$ in Eq. (6.68) is the Jacobian matrix of the generalized stresses (P, M_x, M_y, M_ω) with respect to the generalized strains $(\varepsilon_o, \Phi_x, \Phi_y, \Phi_\omega)$ i.e.,

$$[K] = \begin{bmatrix} \dfrac{\partial P}{\partial \varepsilon_0} & \dfrac{\partial P}{\partial \Phi_x} & \dfrac{\partial P}{\partial \Phi_y} & \dfrac{\partial P}{\partial \Phi_\omega} \\ \dfrac{\partial M_x}{\partial \varepsilon_0} & \dfrac{\partial M_x}{\partial \Phi_x} & \dfrac{\partial M_x}{\partial \Phi_y} & \dfrac{\partial M_x}{\partial \Phi_\omega} \\ \dfrac{\partial M_y}{\partial \varepsilon_0} & \dfrac{\partial M_y}{\partial \Phi_x} & \dfrac{\partial M_y}{\partial \Phi_y} & \dfrac{\partial M_y}{\partial \Phi_\omega} \\ \dfrac{\partial M_\omega}{\partial \varepsilon_0} & \dfrac{\partial M_\omega}{\partial \Phi_x} & \dfrac{\partial M_\omega}{\partial \Phi_y} & \dfrac{\partial M_\omega}{\partial \Phi_\omega} \end{bmatrix} \tag{6.72}$$

The technique has been utilized by Gurfinkel (1970), and Santathadaporn and Chen (1972).

6.2 The warping curvature Φ_ω in Eq. (6.66) and the bimoment M_ω in Eq. (6.65) can be taken as a generalized strain and a generalized stress only for doubly symmetric sections. Discuss what parameters can be taken as the generalized stresses and generalized strains for unsymmetric sections, for example, L-shaped angle section.

6.3 From the movement of the subsequent yield surface (Fig. 6.14 and Fig. 6.15), realize that the generalized strain hardening rule is neither isotropic nor kinematic hardening. Discuss what type of hardening rule could be applicable.

6.11 REFERENCES

ACI Committee 318-71: Building Code Requirements for Reinforced Concrete (ACI 318-71), American Concrete Institute, Detroit, (1971).

Atsuta, T., "Analyses of Inelastic Beam-Columns," Ph.D. Dissertation, Dept. of Civil Engineering, Lehigh University, Bethlehem, Pa., (1972).

Baron, F. and Venkatesan, M. S., "Inelastic Response for Arbitrary Histories of Loads," *Journal of the Engineering Mechanics Division, ASCE*, Vol. 95, No. EM3, Proc. Paper 6634, June, pp. 763–786, (1969).

Chen, W. F. and Atsuta, T., "Inelastic Response of Column Segments under Biaxial Loads," *Journal of the Engineering Mechanics Division, ASCE*, Vol. 99, No. EM4, Proc. Paper 9957, Aug., pp. 685–701, (1973).

Chen, W. F. and Ross, D. A., "The Axial Strength and Behavior of Cylindrical Columns," OTC paper No. 2683, Eighth Annual Offshore Technology Conference, Houston, Texas, May 3–6, pp. 741–754, (1976). See also *Journal of Petroleum Technology*, Society of Petroleum Engineers of AIME, March, pp. 239–241, (1977).

Chen, W. F. and Ross, D. A., "Tests of Fabricated Tubular Columns," *Journal of the Structural Division, ASCE*, Vol. 103, No. ST3, Proc. Paper 12809, pp. 619–634, March, (1977).

Chen, W. F. and Shoraka, M. T., "Tangent Stiffness Method for Biaxial Bending of Reinforced Concrete Columns," *IABSE Publications*, Vol. 35-I, Zurich, pp. 23–44, (1975).

Gurfinkel, G., "Analysis of Footings Subjected to Biaxial Bending," *Journal of the Structural Division, ASCE*, Vol. 96, No. ST6, Proc. Paper 7329, June, pp. 1049–1059, (1970).

Kato, B. and Akiyama, H., "Inelastic Bar Subjected to Thrust and Cyclic Bending," *Journal of the Structural Division, ASCE*, Vol. 95, No. ST1, Proc. Paper 6335, January, pp. 33–56, (1969).

Redner, S., "Measurement of Residual Stresses by the Blind Hole Drilling Method," Photolastic, Inc., Bulletin TDG-5, Malvern, Pa., (1974).

Santathadaporn, S. and Chen, W. F., "Tangent Stiffness Method for Biaxial Bending," *Journal of the Structural Division, ASCE*, Vol. 98, No. ST1, Proc. Paper 8637, June, pp. 153–163, (1972).

Tebedge, N., Alpstein, G. and Tall, L., "Residual Stress Measurement by the Sectioning Method," *Experimental Mechanics*, Vol. 13, No. 2, February, pp. 88–96, (1973).

Vlassov, V. Z., "Thin-Walled Elastic Beams," National Science Foundation, Washington, D.C., and the Department of Commerce, U.S.A., by the *Israel Program for Scientific Translations*, Jerusalem, 2nd edn, (1961).

Warner, R. F., "Biaxial Moment Thrust Curvature Relations," *Journal of the Structural Division, ASCE*, Vol. 95, No. ST5, Proc. Paper 6564, May, pp. 923–940, (1969).

7

APPROXIMATE DEFLECTION METHOD FOR PLASTIC BEAM-COLUMNS

7.1 INTRODUCTION

In Chap. 3, the special case of flexural-torsional buckling strength of a plane beam-column subjected to loading in its plane is approached and solved from the standpoint of *bifurcation*. There, the out-of-plane deformations of a beam-column are assumed to remain zero until the critical loading condition is reached. Thus the in-plane behavior of the beam-column up to the critical load can be analyzed independently of the out-of-plane buckling behavior. The solutions from this in-plane analysis can then be substituted in the flexural-torsional equations which govern the out-of-plane buckling of the beam-column. However, this cannot be done for the more general loading case in which the beam-columns are subjected to biaxial bending moment as well as axial force.

In the particular case when the material remains in the elastic range, the three coupled differential equations of equilibrium for an *elastic* beam-column can be solved exactly or approximately by the use of formal mathematics. This has been presented in Chap. 4. In the plastic or non-linear range, however, these fourth-order coupled differential equations are intractable and recourse must be had to approximate methods or numerical methods to obtain solutions.

The numerical methods which have been used by various investigators include

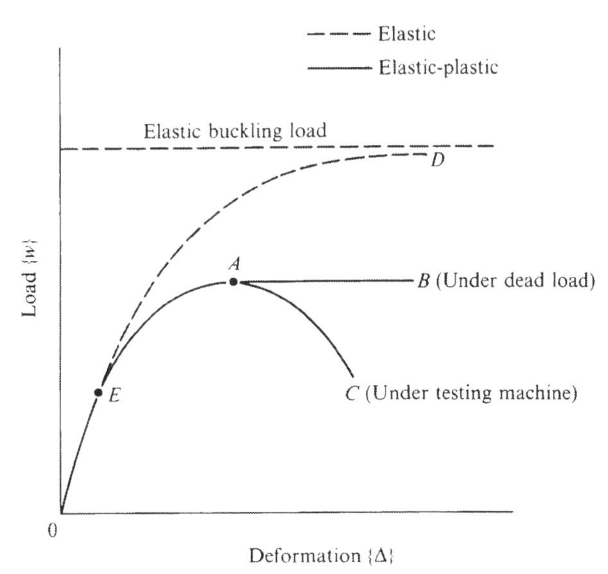

FIGURE 7.1
Load-deformation curves of beam-columns

the numerical integration (Chap. 9), finite difference method (Chap. 10), and the finite element method (Chaps. 11 and 12). In this chapter, approximate deflection methods for obtaining the solutions of elastic-plastic beam-column problems are presented in which all the boundary conditions may not be satisfied. An improved and more rigorous numerical procedure for determining the behavior and strength of a biaxially loaded plastic beam-column is presented in Chap. 8.

Before the actual presentation and discussion of these methods, it is useful to note that for an elastic bending problem, the deformation always increases with the applied load, and a unique equilibrium position of the beam-column is defined for each value of load as shown by the dashed curve *OED* in Fig. 7.1. However, for an elastic-plastic beam-column, the deformations increase initially with the load until a maximum load is reached, beyond which static equilibrium can only be sustained by decreasing the load as illustrated by the curve *OEAC* in Fig. 7.1. The beam-column now has two states of equilibrium corresponding to each value of load. There exists no unique one-to-one correspondence between applied load and deformations. If a beam-column is tested under dead weight loading, i.e., loading in which the load is not relieved at the maximum load carrying capacity, the deformations of the beam-column near this maximum load increase rapidly and become infinitely large at the maximum load (curve *OEAB* in Fig. 7.1). The loading and unloading behavior described above for an elastic-plastic beam-column results, when loading is controlled, as in a testing machine, such that static equilibrium positions are obtained at successively lower values of load but with increasing deformations. The deformational response of a beam-column, from zero to the maximum load and then unloaded with increasing deformations, can be determined by either the approximate methods described in this and following chapters, or by the numerical solution techniques of differential equations to be presented in subsequent chapters, utilizing either a direct or an indirect incremental-load or -deformation approach.

In the approximate deflection methods, the deflected shape of a beam-column is assumed to be defined by a simple function such as orthogonal functions (Sharma and Gaylord, 1969; Lindner, 1972), power series (Syal and Sharma, 1971), or their combinations (Santathadaporn and Chen, 1973), depending upon applied loading and boundary conditions. In most cases, the solution is further simplified by establishing equilibrium of the beam-column only at some selected cross sections where plastic yielding of material is anticipated to be most extensive. For example, the mid-span section is selected for a beam-column loaded symmetrically (Santathadaporn and Chen, 1973). In the analysis, the material is assumed to be ideally elastic-perfectly plastic. Strain hardening and in some cases strain reversal are neglected.

7.2 BASIC EQUATIONS

Consider a beam-column in the fixed coordinate axes x, y, and z (Fig. 4.7). The lateral displacements of the centroid of the deflected profile at any point are u and v, positive in the positive directions of x and y. The angle of twist is θ, taken positive by the right-hand screw rule. The principal axes of the cross-section are ξ and η. The ζ is in the direction of tangent to the deflected axis of the beam-column. The bending moments M_ξ and M_η and the twisting moment M_ζ are considered positive by the right-hand screw rule on the front face. The beam-column is subjected to loads applied only at the ends, i.e., biaxial bending moments M_{xo} and M_{yo}, twisting moment M_{zo}, and axial compression P, at the end $z = 0$, and also M_{xl}, M_{yl}, M_{zl} and P at the other end $z = L$.

Since the beam-column is elastic-plastic, the sectional properties along the length of the member are not constant. Thus, we have to start the analysis with equilibrium equations of the form given in Eq. (2.171), i.e.,

$$
\begin{aligned}
&-[\int E\eta^2\,dA + \theta \int E\xi\eta\,dA]v'' + [\theta \int E\eta^2\,dA - \int E\xi\eta\,dA]u'' + [\int E\eta\,dA]\varepsilon_o \\
&\qquad = M_x + Pv + (M_y - Pu)\theta - u'(M_z - vM'_y - uM'_x) \\
&[\int E\xi\eta\,dA + \theta \int E\xi^2\,dA]v'' - [\theta \int E\xi\eta\,dA - \int E\xi^2\,dA]u'' - [\int E\xi\,dA]\varepsilon_o \\
&\qquad = M_y - Pu - (M_x + Pv)\theta - v'(M_z - vM'_y - uM'_x) \qquad (7.1) \\
&[\int E\eta\,dA + \theta \int E\xi\,dA]v'' - [\theta \int E\eta\,dA - \int E\xi\,dA]u'' - [\int E\,dA]\varepsilon_o = P \\
&EI_\omega\theta''' - (GK_T + \bar{K})\theta' = -M_z - (M_x + Pv + Py_o)u' - (M_y - Pu - Px_0)v' \\
&\qquad + (v + x_0\theta + y_0)M'_y + (u - y_0\theta + x_0)M'_x
\end{aligned}
$$

in which prime (′) indicates one-time derivative with respect to z. In Eq. (7.1), x_0 and y_0 represent the location of shear center, integration is carried out over the entire cross section, and M_x, M_y and M_z are external moments at a section z induced by the applied end moments, i.e.,

$$
\begin{aligned}
M_x &= M_{xo} + \frac{z}{L}(M_{xl} - M_{xo}) \\
M_y &= M_{yo} + \frac{z}{L}(M_{yl} - M_{yo}) \\
M_z &= M_{zo} = M_{zl}
\end{aligned}
\tag{7.2}
$$

It is to be noted that Eqs. (7.1) are derived under the assumption that the warping-torsional rigidity EI_ω and the St. Venant or pure torsional rigidity GK_T do not alter during the yielding of a cross section. Thus, the twisting moment is referred to the shear center which is determined when the entire cross section is elastic.

For an elastic beam-column of symmetric cross section with ξ and η as the centroidal coordinate axes, we have

$$
\begin{aligned}
&\int E\eta^2\, dA = EI_\xi, \quad \int E\xi^2\, dA = EI_\eta, \quad \int E\, dA = EA \\
&\int E\eta\, dA = \int E\xi\, dA = \int E\xi\eta\, dA = 0
\end{aligned}
\tag{7.3}
$$

Equations (7.1) then reduce to the form

$$
\begin{aligned}
&EI_\xi(v'' - u''\theta) + M_x + Pv + (M_y - Pu)\theta - (M_z - vM'_y - uM'_x)u' = 0 \\
&EI_\eta(u'' + v''\theta) - M_y + Pu + (M_x + Pv)\theta + (M_z - vM'_y - uM'_x)v' = 0 \\
&EI_\omega\theta''' - (GK_T + \bar{K})\theta' + (M_x + Pv + Py_0)u' + (M_y - Pu - Px_0)v' \\
&\qquad + M_z - (v + x_0\theta + y_0)M'_y - (u - y_0\theta + x_0)M'_x = 0 \\
&-EA\varepsilon_o = P
\end{aligned}
\tag{7.4}
$$

If the displacements of shear center $u_0 = u - \theta y_0$ and $v_0 = v + \theta x_0$ are used, then Eqs. (7.4) become, after neglecting all the nonlinear terms,

$$
\begin{aligned}
&EI_\xi v''_0 + P(v_0 - x_0\theta) + M_x + M_y\theta - u'_0 M_z = 0 \\
&EI_\eta u''_0 + P(u_0 + y_0\theta) - M_y + M_x\theta + v'_0 M_z = 0 \\
&EI_\omega\theta''' - (GK_T + \bar{K})\theta' + (M_x + Py_0)u'_0 + (M_y - Px_0)v'_0 + M_z \\
&\qquad - (v_0 + y_0)M'_y - (u_0 + x_0)M'_x = 0
\end{aligned}
\tag{7.5}
$$

which are identical to the ones derived previously in Eqs. (2.174) of Chap. 2 for an elastic beam-column of doubly symmetric section. Goodier's equations (4.93) are obtained from Eq. (7.5) assuming shifting of shear center (x_0, y_0) is small.

7.3 TANGENT STIFFNESS APPROACH FOR SYMMETRIC LOADING

Since the system of four differential equations is highly non-linear as caused by the terms which must be integrated over the elastic-plastic cross section, some iterative procedures are needed in order to obtain solutions. In this section, the tangent stiffness approach similar to that described previously in Chap. 6 is applied to obtain solutions for beam-columns loaded symmetrically at both ends, i.e.,

$$M_x = M_{xo} = M_{xl}, \quad M_y = M_{yo} = M_{yl}, \quad M_z = M_{zo} = M_{zl} = 0 \tag{7.6}$$

and

$$M'_x = M'_y = 0$$

In this method the stiffness matrix of an elastic-plastic beam-column is constantly modified along the load-deformation path, and the maximum load is obtained from the full load-deformation characteristic. Two other approximate methods of solutions are presented in subsequent sections.

7.3.1 Tangent Stiffness Matrix

Santathadaporn and Chen (1973) derived the tangent stiffness matrix which relates the rate of change of external force vector $\{\dot{W}\}$ to the rate of change of deformation vector $\{\dot{\Delta}\}$. This is described in what follows.

Since the deformations are symmetric with respect to the mid-span and having the maximum values at the mid-span, the deformation vector $\{\Delta\}$ of the beam-column is therefore defined here by these maximum values of displacements and rotation at the mid-span v_m, u_m, ε_m, and θ_m, in which the subscript, m, denotes the mid-span of the beam-column. In order to determine these four unknowns, the four equilibrium equations of (7.1) must be used. Evaluating Eq. (7.1) at mid-span and making use of Eq. (7.6), we obtain

$$\begin{aligned}
-[\textstyle\int E\eta^2\,dA + \theta_m \int E\xi\eta\,dA]v''_m + [\theta_m \int E\eta^2\,dA - \int E\xi\eta\,dA]u''_m + [\int E\eta\,dA]\varepsilon_m& \\
= M_{xo} + Pv_m + (M_{yo} - Pu_m)\theta_m& \\
[\textstyle\int E\xi\eta\,dA + \theta_m \int E\xi^2\,dA]v''_m - [\theta_m \int E\xi\eta\,dA - \int E\xi^2\,dA]u''_m - [\int E\xi\,dA]\varepsilon_m& \\
= M_{yo} - Pu_m - (M_{xo} + Pv_m)\theta_m& \\
[\textstyle\int E\eta\,dA + \theta_m \int E\xi\,dA]v''_m - [\theta_m \int E\eta\,dA - \int E\xi\,dA]u''_m - [\int E\,dA]\varepsilon_m = P& \\
EI_\omega\theta''''_m - (GK_T + \bar{K})\theta''_m = -(M_{xo} + Pv_m)u''_m - (M_{yo} - Pu_m)v''_m&
\end{aligned} \tag{7.7}$$

In the derivation of Eqs. (7.7), the shear center is assumed to locate at the centroid of the cross section, i.e., $x_0 = 0$ and $y_0 = 0$. The last equation of (7.7) was obtained by differentiating the last equation of (7.1) once with respect to z. This is because the last equation of (7.1) is satisfied identically at mid-span since the twisting moment is always zero at the mid-span of a beam-column under a symmetric loading condition. It is therefore necessary to replace this equation by the rate of change of twisting moment (with respect to z) at mid-span.

The equilibrium equations (7.7) are ones derived directly by equating the external forces $\{F\}_{\text{ext}}$ to the internal forces $\{F\}_{\text{int}}$ at mid-span of the beam-column, in Eqs. (2.160) and (2.162), that is,

$$\{F\}_{\text{ext}} = \begin{Bmatrix} -M_\xi \\ M_\eta \\ -P \\ M_\zeta \end{Bmatrix}_{\text{ext}} = \begin{Bmatrix} -M_{xo} - Pv_m - (M_{yo} - Pu_m)\theta_m \\ M_{yo} - Pu_m - (M_{xo} + Pv_m)\theta_m \\ -P \\ (M_{xo} + Pv_m)u'_m + (M_{yo} - Pu_m)v'_m \end{Bmatrix} \tag{7.8}$$

$$\{F\}_{\text{int}} = \begin{Bmatrix} -M_\xi \\ M_\eta \\ -P \\ M_\zeta \end{Bmatrix}_{\text{int}}$$

$$= \begin{Bmatrix} [\int E\eta^2\,dA + \theta_m \int E\xi\eta\,dA]v''_m + [\int E\xi\eta\,dA - \theta_m \int E\eta^2\,dA]u''_m - [\int E\eta\,dA]\varepsilon_m \\ [\int E\xi\eta\,dA + \theta_m \int E\xi^2\,dA]v''_m + [\int E\xi^2\,dA - \theta_m \int E\xi\eta\,dA]u''_m - [\int E\xi\,dA]\varepsilon_m \\ -[\int E\eta\,dA + \theta_m \int E\xi\,dA]v''_m - [\int E\xi\,dA - \theta_m \int E\eta\,dA]u''_m + [\int E\,dA]\varepsilon_m \\ -EI_\omega\theta'''_m + (GK_T + \bar{K})\theta'_m \end{Bmatrix} \tag{7.9}$$

The rate or incremental equilibrium equations are now obtained by differentiating Eqs. (7.7) with respect to a time-like variable denoted by a dot on a quantity

$$\begin{aligned}
&-[\textstyle\int E\eta^2\,dA + \theta_m \int E\xi\eta\,dA]\dot{v}''_m - P\dot{v}_m + [\theta_m \int E\eta^2\,dA - \int E\xi\eta\,dA]\dot{u}''_m + P\theta_m\dot{u}_m \\
&\qquad + [\textstyle\int E\eta\,dA]\dot{\varepsilon}_m - [(\int E\xi\eta\,dA)v''_m - (\int E\eta^2\,dA)u''_m + M_{yo} - Pu_m]\dot{\theta}_m \\
&\qquad = \dot{M}_{xo} + \dot{P}v_m + (\dot{M}_{yo} - \dot{P}u_m)\theta_m \\
&[\textstyle\int E\xi\eta\,dA + \theta_m \int E\xi^2\,dA]\dot{v}''_m + P\dot{u}_m - [\theta_m \int E\xi\eta\,dA - \int E\xi^2\,dA]\dot{u}''_m + P\theta_m\dot{v}_m \\
&\qquad - [\textstyle\int E\xi\,dA]\dot{\varepsilon}_m + [(\int E\xi^2\,dA)v''_m - (\int E\xi\eta\,dA)u''_m + M_{xo} + Pv_m]\dot{\theta}_m \\
&\qquad = \dot{M}_{yo} - \dot{P}u_m - (\dot{M}_{xo} + \dot{P}v_m)\theta_m \\
&[\textstyle\int E\eta\,dA + \theta_m \int E\xi\,dA]\dot{v}''_m - [\theta_m \int E\eta\,dA - \int E\xi\,dA]\dot{u}''_m \\
&\qquad - [\textstyle\int E\,dA]\dot{\varepsilon}_m + [(\int E\xi\,dA)v''_m - (\int E\eta\,dA)u''_m]\dot{\theta}_m = \dot{P} \\
&(M_{yo} - Pu_m)\dot{v}''_m + Pu''_m\dot{v}_m + (M_{xo} + Pv_m)\dot{u}''_m - Pv''_m\dot{u}_m + EI_\omega\dot{\theta}''''_m - (GK_T + \bar{K})\dot{\theta}''_m \\
&\qquad = -(\dot{M}_{xo} + \dot{P}v_m)u''_m - (\dot{M}_{yo} - \dot{P}u_m)v''_m
\end{aligned} \tag{7.10}$$

These rate equilibrium equations are also applicable to an initially *imperfect* beam-column because the initial deflections and residual stresses are rate-independent.

It is now necessary to express $\dot{u}_m$, $\dot{v}_m$, $\dot{\theta}_m$ and $\dot{\theta}''''_m$ in Eq. (7.10) in terms of $\dot{u}''_m$, $\dot{v}''_m$ and $\dot{\theta}''_m$ so that the resulting equations will be linear. This can be done by assuming the deflected shape functions of the beam-column in the following forms:

$$v = A_1 \sin \frac{\pi z}{L} + 4A_2 \frac{z}{L}\left(1 - \frac{z}{L}\right)$$

$$u = B_1 \sin \frac{\pi z}{L} + 4B_2 \frac{z}{L}\left(1 - \frac{z}{L}\right) \tag{7.11}$$

$$\theta = \begin{cases} \theta_m \sin \dfrac{\pi z}{L} & \text{for warping permitted ends} \\ \dfrac{1}{2}\theta_m\left(1 - \cos \dfrac{2\pi z}{L}\right) & \text{for warping fully restrained ends} \end{cases}$$

from which the following relations among the rates of deformations at the mid-span can be established

$$\dot{v}_m = -\frac{L^2}{\gamma_1}\dot{v}''_m$$

$$\dot{u}_m = -\frac{L^2}{\gamma_2}\dot{u}''_m \tag{7.12}$$

$$\dot{\theta}_m = -\frac{1}{\gamma_3}\left(\frac{L}{\pi}\right)^2 \dot{\theta}''_m \quad \text{and} \quad \dot{\theta}''''_m = -\gamma_3\left(\frac{\pi}{L}\right)^2 \dot{\theta}''_m$$

in which γ_1, γ_2 and γ_3 are the factors depending on the end conditions of the beam-column (Prob. 7.4):

$$\gamma_1 = \begin{cases} \pi^2 & \text{for } M_{xo} = M_{xl} = 0 \\ 8 & \text{for } M_{xo} = M_{xl} \neq 0 \end{cases}$$

$$\gamma_2 = \begin{cases} \pi^2 & \text{for } M_{yo} = M_{yl} = 0 \\ 8 & \text{for } M_{yo} = M_{yl} \neq 0 \end{cases} \tag{7.13}$$

$$\gamma_3 = \begin{cases} 1 & \text{for warping free} \\ 2 & \text{for warping fully restrained} \end{cases}$$

In fact the values for γ_1 and γ_2 for the cases $M_{xo} = M_{xl} \neq 0$ and $M_{yo} = M_{yl} \neq 0$ respectively should lie somewhere between 8 and π^2 depending upon the magnitude of the applied end moments. However, it has been found that the maximum strength of a beam-column is insensitive to these factors, and the values of γ_1 and γ_2 adopted in Eqs. (7.13) are considered to be sufficiently accurate for practical use (Santathadaporn and Chen, 1973). A similar conclusion has also been reached for the in-plane bending of beam-columns (Chen, 1971). Making use of Eq. (7.12), Eqs. (7.10), after some manipulation, can be expressed in matrix form as:

$$\{\dot{W}(4)\} = [R(4 \times 4)]\{\dot{\Delta}(4)\} \tag{7.14}$$

$\{\dot{W}(4)\}$ = the rate of change in external force vector of size 4

$\{\dot{\Delta}(4)\}$ = the rate of change in deformation vector of size 4

$[R(4 \times 4)]$ = the tangent stiffness matrix of size 4×4 which includes both the effect of instability and the effect of yielding or plastification of the cross section.

and

$$\{\dot{W}\} = \begin{Bmatrix} -\dot{M}_{xo} - \dot{P}v_m - (\dot{M}_{yo} - \dot{P}u_m)\theta_m \\ \dot{M}_{yo} - \dot{P}u_m - (\dot{M}_{xo} + \dot{P}v_m)\theta_m \\ -\dot{P} \\ (\dot{M}_{xo} + \dot{P}v_m)u''_m + (\dot{M}_{yo} - \dot{P}u_m)v''_m \end{Bmatrix} \tag{7.15}$$

$$\{\dot{\Delta}\} = \begin{Bmatrix} \dot{v}''_m \\ \dot{u}''_m \\ \dot{\varepsilon}_m \\ \dot{\theta}''_m \end{Bmatrix} \tag{7.16}$$

$$[R] = \begin{bmatrix} \int E\eta^2\,dA + \theta_m \int E\xi\eta\,dA - \dfrac{L^2}{\gamma_1}P & \int E\xi\eta\,dA - \theta_m \int E\eta^2\,dA + \dfrac{L^2}{\gamma_2}P\theta_m & -\int E\eta\,dA & -\left(\dfrac{1}{\gamma_3}\dfrac{L^2}{\pi^2}\right)[(\int E\xi\eta\,dA)v''_m - (\int E\eta^2\,dA)u''_m + M_{yo} - Pu_m] \\ \int E\xi\eta\,dA + \theta_m \int E\xi^2\,dA - \dfrac{L^2}{\gamma_1}P\theta_m & \int E\xi^2\,dA - \theta_m \int E\xi\eta\,dA - \dfrac{L^2}{\gamma_2}P & -\int E\xi\,dA & -\left(\dfrac{1}{\gamma_3}\dfrac{L^2}{\pi^2}\right)[(\int E\xi^2\,dA)v''_m - (\int E\xi\eta\,dA)u''_m + M_{xo} + Pv_m] \\ -\int E\eta\,dA - \theta_m \int E\xi\,dA & -\int E\xi\,dA + \theta_m \int E\eta\,dA & \int E\,dA & -\left(\dfrac{1}{\gamma_3}\dfrac{L^2}{\pi^2}\right)[-(\int E\xi\,dA)v''_m + (\int E\eta\,dA)u''_m] \\ -(M_{yo} - Pu_m) + \dfrac{L^2}{\gamma_1}Pu''_m & -(M_{xo} + Pv_m) - \dfrac{L^2}{\gamma_2}Pv''_m & 0 & GK_T + \bar{K} + \gamma_3\dfrac{\pi^2}{L^2}EI_\omega \end{bmatrix} \tag{7.17}$$

The matrix $[R]$ with a partially yielded cross section at mid-span is a function of the current state of forces and deformations as well as the properties of the material and the cross section. For an elastic-perfectly plastic section, the value of E is zero in the yield zone. Therefore, only the area of the elastic core will contribute to the integration in Eq. (7.17). This implies that further increment of external forces is resisted by the remaining elastic area of the section only. In the following numerical computations, the St. Venant torsional rigidity, GK_T, in (7.17) is assumed to be unaltered by the yielding of the mid-span cross section. The warping torsional rigidity, EI_ω, is assumed to be directly proportional to the ratio of the remaining elastic area to the total area of the section.

Equation (7.14) gives the relationship between the rate of change in the force vector and the rate of change in the deformation vector. Equation (7.14) is similar in form to that of Eq. (6.20), developed previously in Chap. 6 for a beam-column segment subjected to biaxial bending. The stiffness matrix $[R]$, accounting now for both the yielding and the instability effect, can be interpreted as the tangent of the load-deformation curve of the beam-column. The procedure of solving a long beam-column problem here is similar to that described in Chap. 6 for a short beam-column.

In general, the external loads acting on a beam-column can be categorized into two types: proportional and nonproportional. The formulation of the equations in this section is applicable to both cases.

7.3.2 Proportional Loading

In this case the applied moments are proportional to the axial force,

$$\dot{M}_{xo} = -\dot{P}e_y, \quad \dot{M}_{yo} = \dot{P}e_x \tag{7.18}$$

Thus the rate of change in the external force vector is from Eq. (7.15),

$$\{\dot{W}\} = \dot{P}\begin{Bmatrix} e_y - v_m - (e_x - u_m)\theta_m \\ e_x - u_m + (e_y - v_m)\theta_m \\ -1 \\ (-e_y + v_m)u''_m + (e_x - u_m)v''_m \end{Bmatrix} \tag{7.19}$$

The iteration procedure to compute the load-deformation relations of a proportionally loaded beam-column consists of the following steps (Santathadaporn and Chen, 1973), which are most suitable for a computer solution.

1. Input all necessary initial values for P, u_m, v_m, ε_m and θ_m. The corresponding curvatures u''_m, v''_m and θ''_m are computed by Eq. (7.11) in the similar form as that of the rate equations (7.12).

2. Assign the increment of curvature $\dot{v}''_m$. Since the rate Eq. (7.14) contains only four unknowns, one from Eq. (7.19), $\dot{P}$, and three from Eq. (7.16), $\dot{u}''_m$, $\dot{\varepsilon}_m$ and $\dot{\theta}''_m$, they can be solved. From the computed value of $\dot{P}$, revise the total axial force P.

3. Revise the deformation vector $\{\Delta\} = \{v''_m, u''_m, \varepsilon_m, \theta''_m\}$. Revise also the initial values for u_m, v_m and θ_m using Eq. (7.12). Compute the external force vector $\{F\}_{\text{ext}}$ from Eq. (7.8) and the internal force vector $\{F\}_{\text{int}}$ from Eq. (7.9). If the differences between the two force vectors are sufficiently small, transfer control to step 6, otherwise, proceed to step 4.

4. If the number of iteration cycles is too large, proceed to step 5. Otherwise, compute the unbalanced force $\{\dot{W}\} = \{F\}_{\text{ext}} - \{F\}_{\text{int}}$ and solve Eq. (7.14) for the corresponding correction deformation vector $\{\dot{\Delta}\} = [R]^{-1}\{\dot{W}\}$ in which $[R]^{-1}$ is the inverse of the new tangent stiffness matrix $[R]$ corresponding to the current state of axial load and deformations. This correction deformation vector will be added to the current deformations in order to eliminate the unbalanced force. The process is achieved by transferring back to step 3.

5. Make appropriate adjustment by either decreasing the increment of curvature $(\dot{v}''_m)$ or forcing the beam-column to undergo transformation from loading (stable equilibrium) to unloading (unstable equilibrium) portion of the load-displacement curve. Then, return to step 2.

6. Test the converged displacements. If they are smaller than the previous ones, return to step 5. Otherwise, print the solution. If the column has not weakened yet, return to step 2 for a new iteration. Otherwise terminate the program.

Step 5 is necessary when the load approaches its maximum value. At this point, the slope of the load-displacement curve is small and the determinant of the tangent stiffness matrix is nearly zero. The system of equations becomes ill-conditioned and requires a larger number of iteration cycles for a solution. The computational steps outlined above are summarized in the brief flow chart in Fig. 7.2.

7.3.3 Nonproportional Loading

This study will be restricted to one specific loading path only. The axial force, P, and the bending moment, M_{xo}, are applied first and maintained constant. The beam-column is then bent about the weak axis until it fails. The increment of external forces defined in Eq. (7.15) will differ for each step of the loading. For example, during the increment of axial force, the increments of the strong and weak axis bending moments are zero, i.e., $\dot{M}_{xo} = \dot{M}_{yo} = 0$. Thus,

$$P: \qquad \{\dot{W}\} = \dot{P}\begin{Bmatrix} -v_m + u_m\theta_m \\ -u_m - v_m\theta_m \\ -1 \\ v_m u''_m - u_m v''_m \end{Bmatrix} \tag{7.20}$$

Similar vectors due to the increments of moments $\dot{M}_{xo}$ and $\dot{M}_{yo}$ will be

$$M_{xo}: \qquad \{\dot{W}\} = \dot{M}_{xo}\begin{Bmatrix} -1 \\ -\theta_m \\ 0 \\ u''_m \end{Bmatrix} \tag{7.21}$$

and

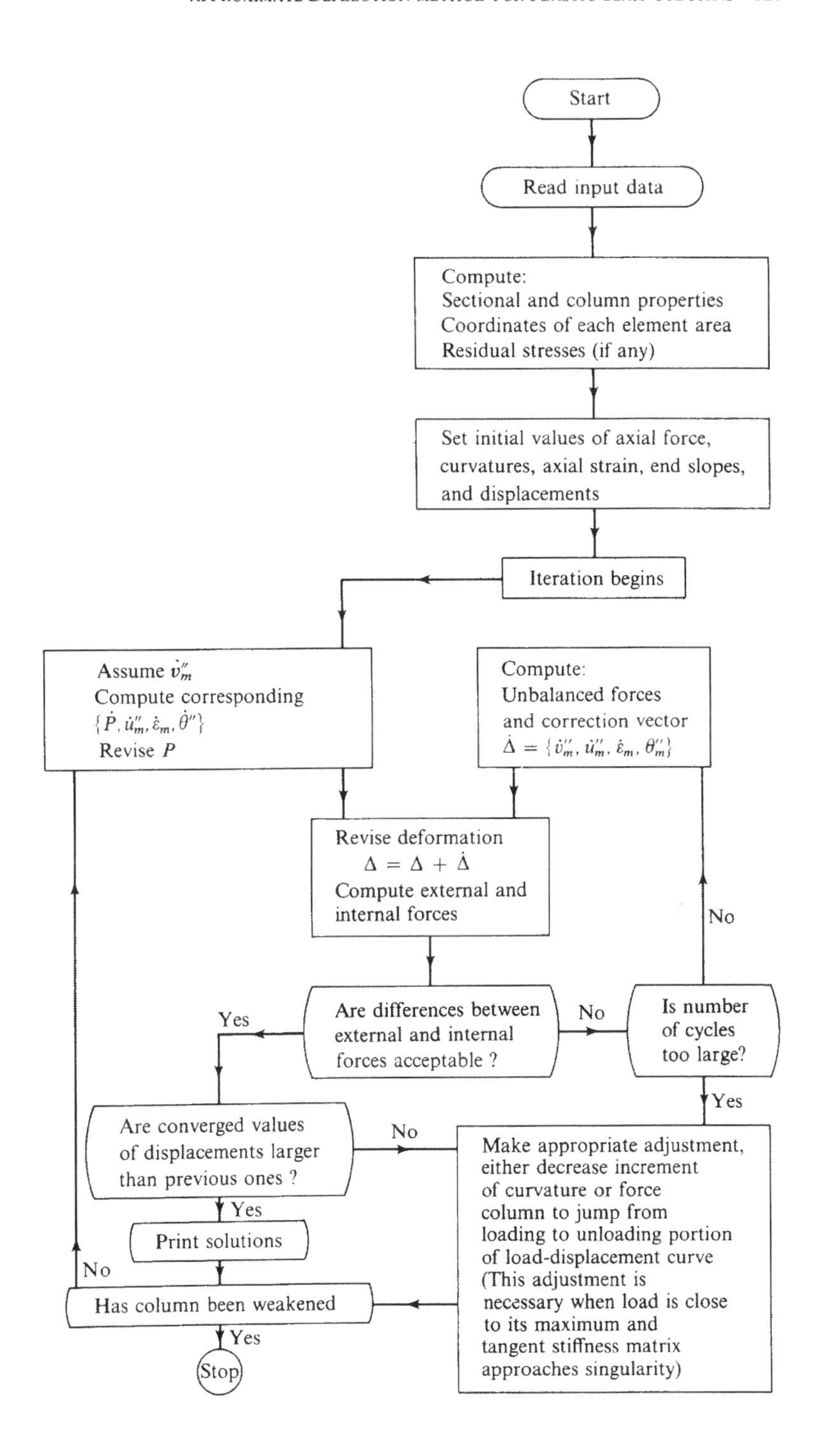

FIGURE 7.2
Brief flow chart for proportional loading case

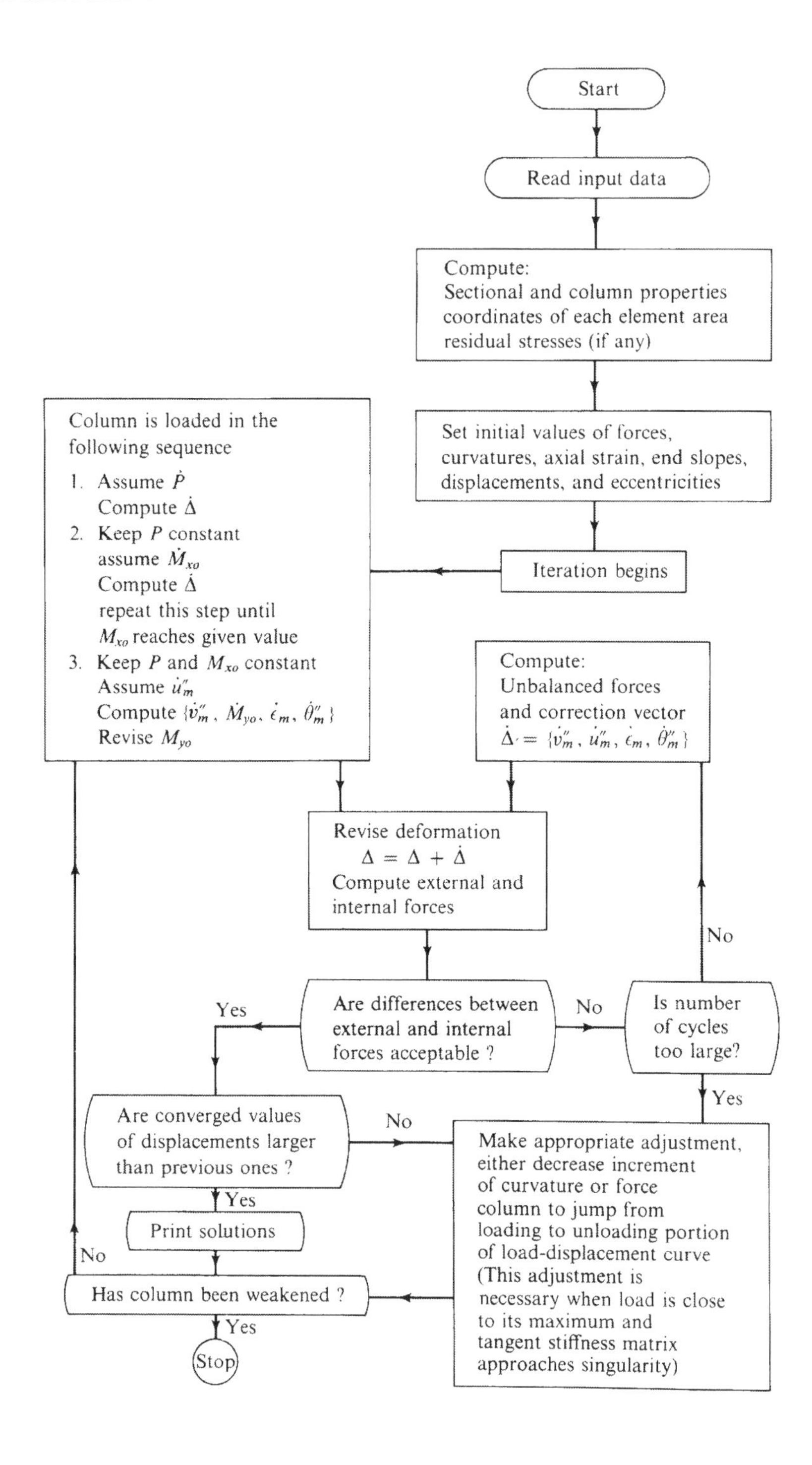

FIGURE 7.3
Brief flow chart for nonproportional loading case

$$M_{yo}: \qquad \{\dot{W}\} = \dot{M}_{yo} \begin{Bmatrix} -\theta_m \\ 1 \\ 0 \\ v''_m \end{Bmatrix} \tag{7.22}$$

The computational steps outlined previously are valid for the nonproportional loading case also. However, modification in step 2 is needed to account for the different loading sequence. All other steps remain the same. Figure 7.3 shows the brief flow chart for the nonproportional loading case.

7.3.4 Numerical Studies

An elastic beam-column of W14 × 43 section subjected to an equal biaxial eccentricity $e_x = e_y = 5.0$ in (127 mm) is solved twice here, using first the refined equations derived here and secondly the approximate Goodier's Eqs. (7.5), which neglect all the nonlinear terms. The solutions are compared in Fig. 7.4 for both warping-free (Fig. 7.4a) and warping-fully restrained (Fig. 7.4b) end conditions. Culver's exact solutions (1966) presented previously in Sec. 4.2, Chap. 4, using Goodier's approximate equations, are shown by the dots in the figure. Harstead, Birnstiel and Leu's (1968) solutions to be presented in Chap. 8, using a more refined numerical procedure which includes the nonlinear terms, are shown by the small open circles. As can be seen from Fig. 7.4, the present method of solutions using approximate equations agree well with those solved exactly by Culver. The difference in solutions between

FIGURE 7.4(a)
Elastic load-displacement curves [$L = 264.6$ in (6.7 m), $e_x = e_y = 5$ in (127 mm)] (1 in = 25.4 mm, 1 kip = 4.45 kN)

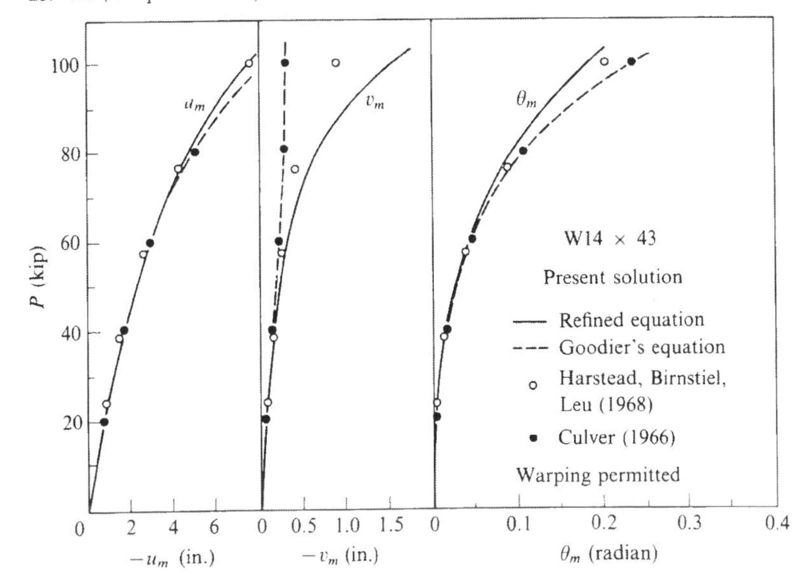

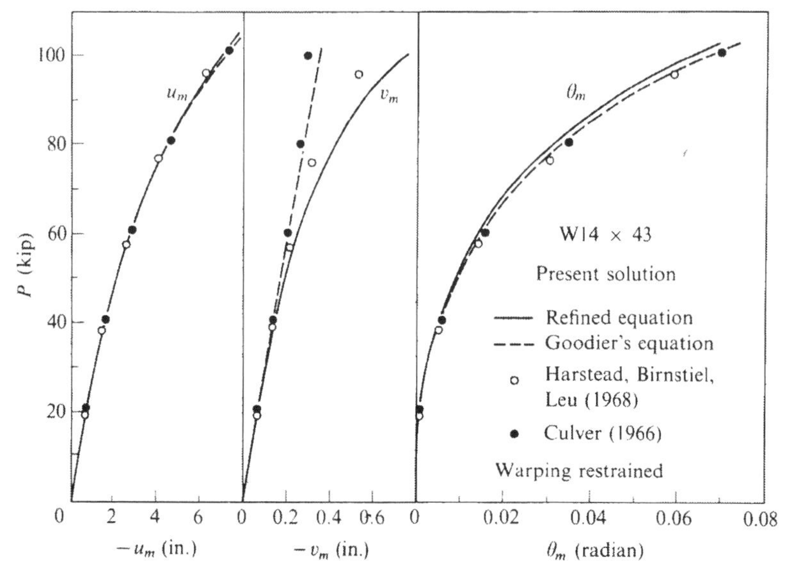

FIGURE 7.4(b)
Elastic load-displacement curves [$L = 264.6$ in (6.7 m), $e_x = e_y = 5$ in (127 mm)] (1 in = 25.4 mm, 1 kip = 4.45 kN)

the refined and Goodier's equations are clearly seen in the figures. This difference becomes large as the load goes higher. At this higher load range, Goodier's equations are not justified since the displacement v_m is considerably underestimated.

Figure 7.5 shows the effect of initial imperfections on the behavior of elastic beam-columns subjected to biaxially eccentric axial load. Both refined equations and

FIGURE 7.5(a)
Elastic load-displacement curves of an imperfect beam-column [$L = 280$ in (7 m), $E = 29{,}000$ ksi (200,000 MPa), $e_x = 0.75$ in (19 mm), $e_y = 10$ in (254 mm)] (1 in = 25.4 mm, 1 kip = 4.45 kN)

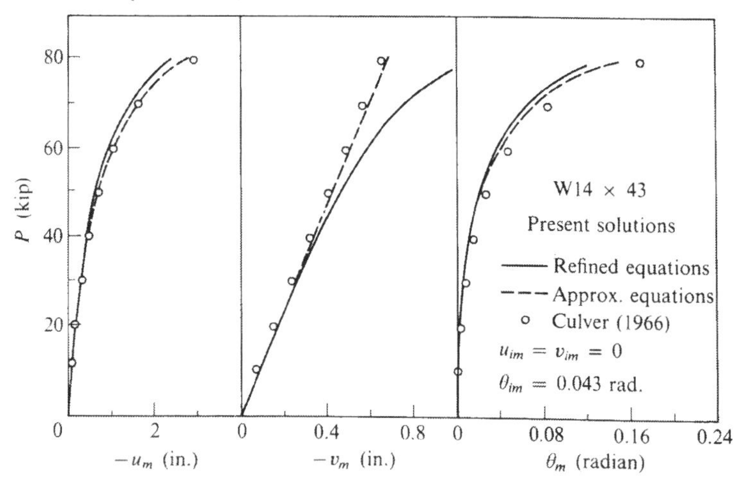

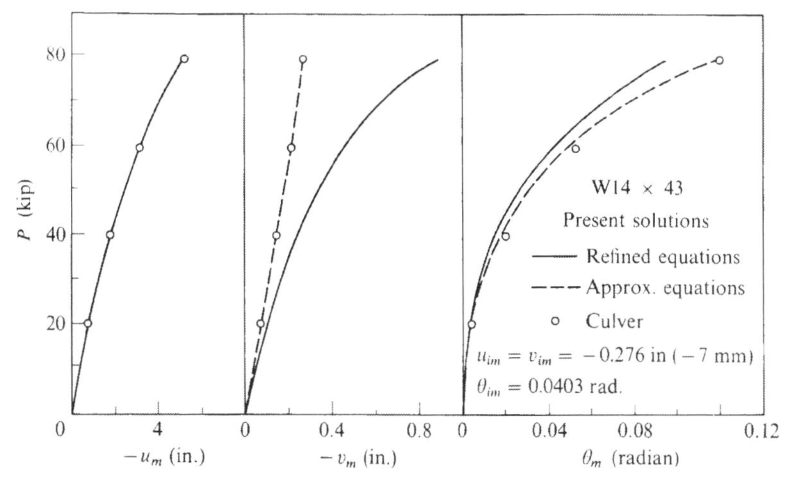

FIGURE 7.5(b)
Elastic load-displacement curves of an imperfect beam-column [$L = 264.6$ in (6.7 m), $E = 30{,}000$ ksi (207,000 MPa), $e_x = e_y = 5.0$ in (127 mm)] (1 in = 25.4 mm, 1 kip = 4.45 kN)

Goodier's approximate equations are solved. The initial imperfections are assumed to be sine functions for both lateral displacements and twisting angle and their maximum values at the mid-span are denoted by u_{im}, v_{im} and θ_{im}, respectively. In Fig. 7.5(a) the initial twisting angle θ_{im} is assumed to be 0.043 rad and the values of u_{im} and v_{im} are both zero. The beam-column is eccentrically loaded with $e_x = 0.75$ in (19 mm) and $e_y = 10$ in (254 mm). In Fig. 7.5(b), the initial imperfections are $u_{im} = v_{im} = -0.276$ in (−7 mm) and $\theta_{im} = 0.0403$ rad, and the beam-column is loaded by an equal biaxial eccentricity $e_x = e_y = 5.0$ in (127 mm). Again, note in these

FIGURE 7.6(a)
The effect of end warping on the behavior of a plastic beam-column [$L = 200$ in (5 m), $\sigma_y = 36$ ksi (248 MPa), $e_x = 0.5$ in (12.7 mm), $e_y = 5.0$ in (127 mm)] (1 in = 25.4 mm, 1 kip = 4.45 kN)

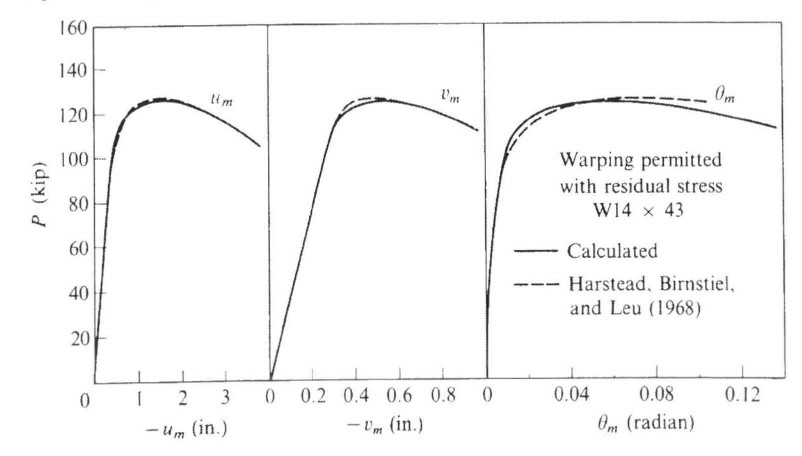

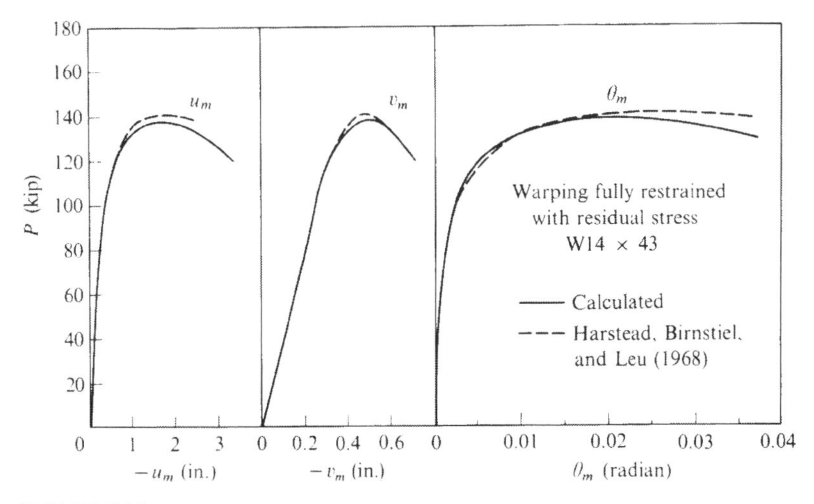

FIGURE 7.6(b)
The effect of end warping on the behavior of a plastic beam-column [$L = 220$ in (5 m), $\sigma_y = 36$ ksi (248 MPa), $e_x = 0.5$ in (12.7 mm), $e_y = 5.0$ in (127 mm)] (1 in = 25.4 mm, 1 kip = 4.45 kN)

figures that there is a large difference between the displacement v_m calculated from the approximate equations suggested by Goodier, and the refined results reported here. Solutions obtained by Culver are shown by the small circles in the figures.

Solutions to proportional loading case

Some numerical results are reported by Santathadaporn and Chen (1973) which are shown here in Fig. 7.6 to Fig. 7.14. First, as solutions of proportional loading case, Fig. 7.6 shows the effect of warping at the ends of a beam-column. Results indicate that the prevention of the beam-column ends from warping will strengthen the beam-column. The maximum loads computed from the present theory are slightly lower than those obtained more exactly by Harstead, Birnstiel and Leu (1968). This is probably due to the fact that the present formulation neglects elastic unloading of a plastic element, while the formulation in the reference has taken this effect into account. Details of the more refined numerical procedure will be presented in the following chapter.

Figures 7.7 compare the solutions of the present theory with the test data obtained by Birnstiel (1968). The theoretical curves in Figs. 7.7 are shown by solid lines and the small circles represent the results from experiments. Good agreement is generally observed from these figures. The twisting angle θ_m in Fig. 7.7(b) shows a large discrepancy at loads higher than 60 kip (267 kN). This disagreement probably resulted from the twist of the end fixture of the beam-column during the test (Birnstiel, 1968).

There have been few studies of the effect of initial imperfections on the space behavior of centrally loaded columns. The columns considered herein are made of

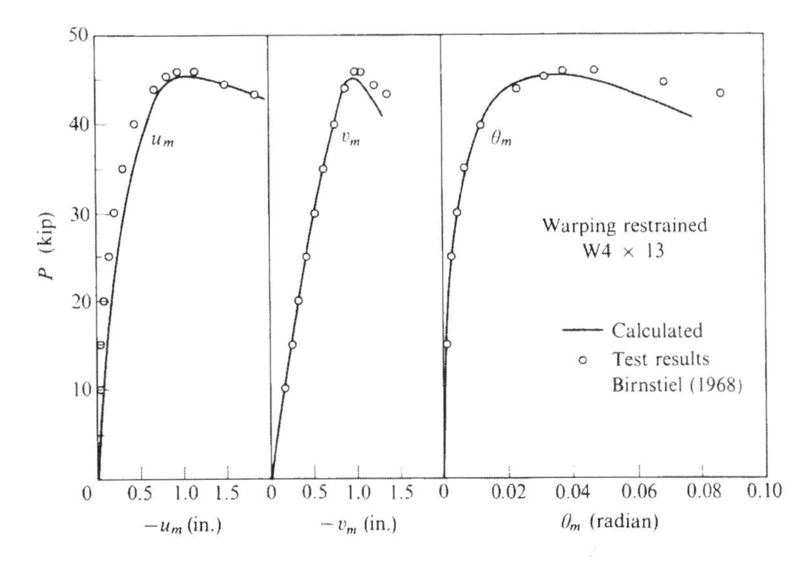

FIGURE 7.7(a)
Comparison of load-displacement curves with tests [$L = 120$ in (3 m), $\sigma_y = 65$ ksi (448 MPa), $e_x = 0.42$ in (10.7 mm), $e_y = 2.72$ in (69 mm)] (1 in = 25.4 mm, 1 kip = 4.45 kN)

W12 × 161 section and are initially twisted in a sine function shape with initial twisting angle $\theta_{im} = 0.05$ rad at mid-span. The initial displacements in the x and y directions are also assumed to be sine functions, and the maximum values at mid-

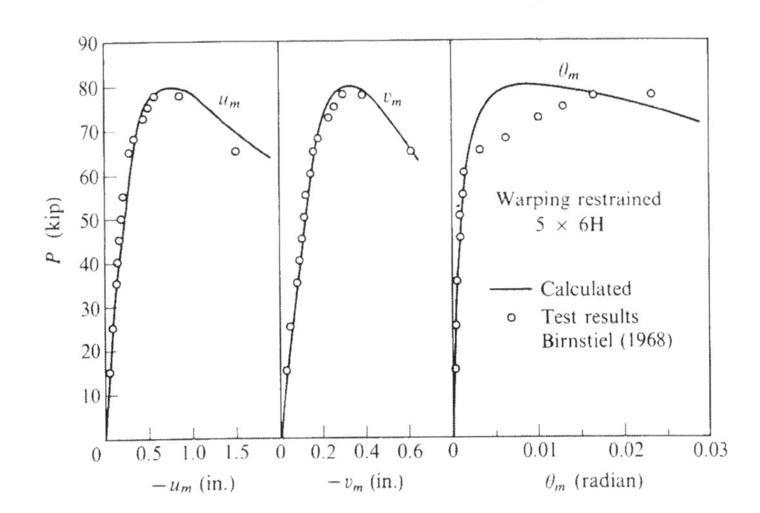

FIGURE 7.7(b)
Comparison of load-displacement curves with tests [$L = 96$ in (2.4 m), $\sigma_y = 36$ ksi (248 MPa), $e_x = 0.89$ in (22.6), $e_y = 2.82$ in (71.6 mm)] (1 in = 25.4 mm, 1 kip = 4.45 kN)

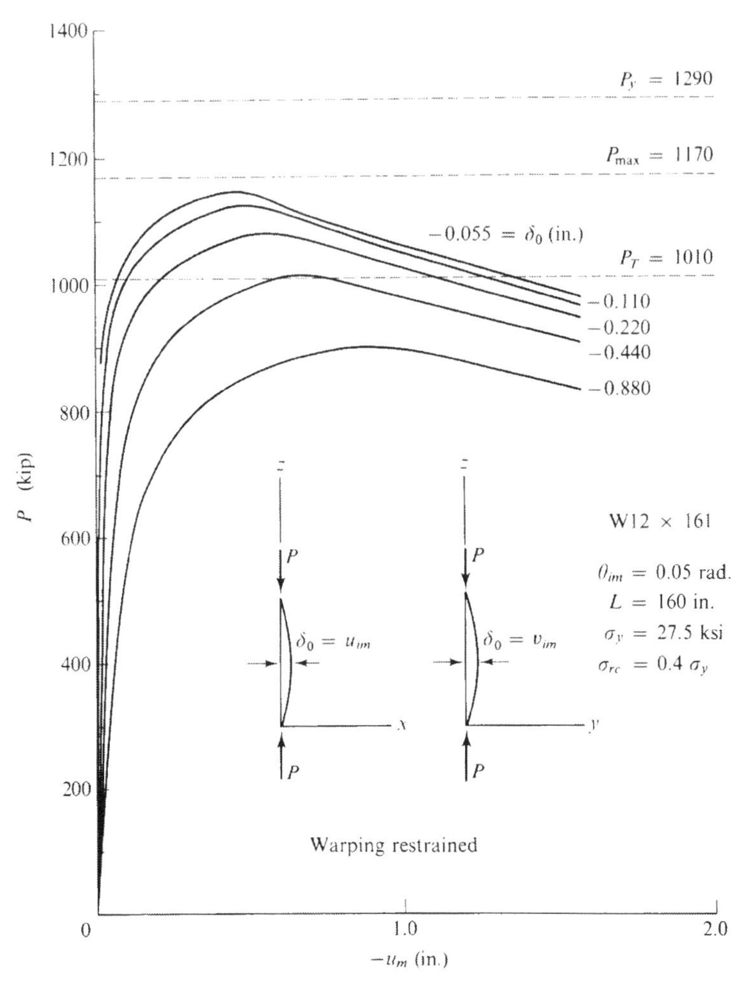

FIGURE 7.8
Load-displacement curves of a centrally loaded imperfect column (1 in = 25.4 mm, 1 kip = 4.45 kN, 1 ksi = 6.9 MPa)

span are equal and are arbitrarily assumed to vary from −0.055 in to −0.880 in (−1.4 mm to −22.4 mm). Results shown in Fig. 7.8 indicate a significant decrease of the maximum load of the column due to the initial imperfections. The yield load, P_y is 1290 kip (5740 kN) and the tangent modulus load, P_T, based on the measured residual stress is 1010 kip (4500 kN). The maximum loads, P_{max}, in the tests vary from 900 kip to 1170 kip (4000 to 5200 kN). Because of the lack of a measured value for the initial twist of the column, a comparison between the test data and the theoretical solution is not feasible. The technique of determining the true maximum strength of a centrally loaded column is shown in Fig. 7.9. The maximum strength values and the corresponding initial imperfections read from

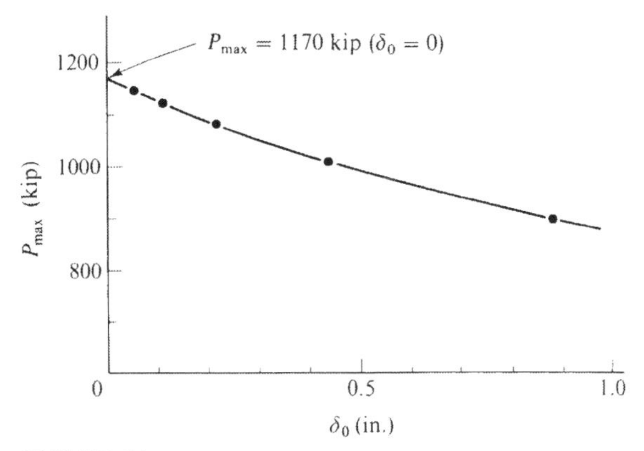

FIGURE 7.9
Relationship between the maximum strength and the initial imperfections of a centrally loaded column (1 in = 25.4 mm, 1 kip = 4.45 kN)

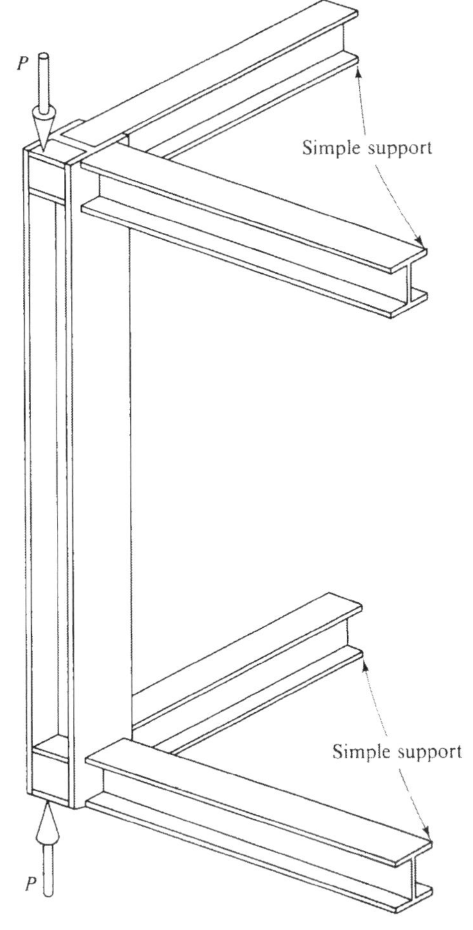

FIGURE 7.10
Space subassemblage under proportional loading

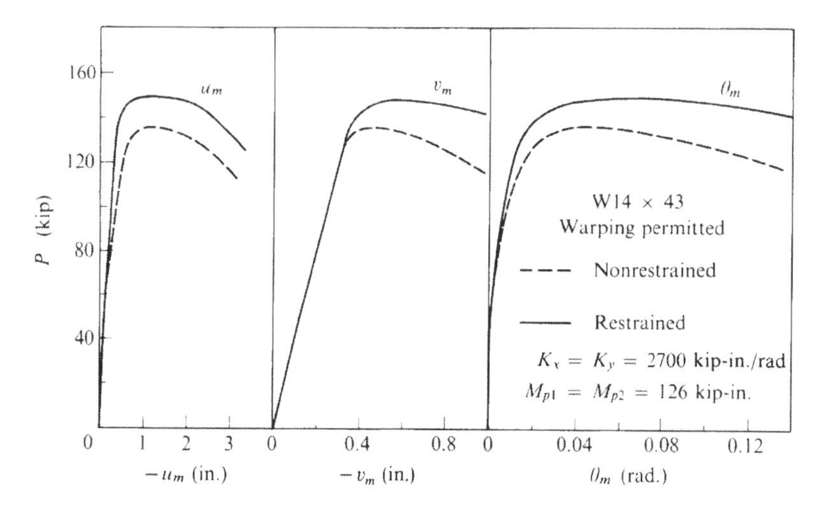

FIGURE 7.11
Load-displacement curves of the column in the space subassemblage shown in Fig. 7.10 [$L = 220$ in (5.6 m), $\sigma_y = 36$ ksi (248 MPa), $e_x = 0.5$ in (12.7 mm), $e_y = 5.0$ in (127 mm)] (1 in = 25.4 mm, 1 kip = 4.45 kN, 1 ksi = 6.9 MPa, 1 kip-in = 113 N-m)

Fig. 7.8 are plotted in this figure. The value of the true maximum strength is the load at which the initial displacement, δ_0, is zero.

The effect of beam bending restraint on the biaxially loaded column is studied for the space subassemblage sketched in Fig. 7.10. Assume that the beams connected to the column ends can resist only bending moment about their strong axis; bending moment about the weak axis and twisting moment in the beam are neglected. Furthermore, the effect of column shortening on the beam bending is also ignored. For the problem study, all the beams are made of S4 × 7.7 section and are 200 in (5.08 m) long. These beams which are simply supported at their other ends behave elastically until a plastic hinge forms at the joints. The rotational stiffness, K, and the plastic moment, M_p, of the two beams are calculated, respectively, from $3EI/l =$ 2700 kip/in/rad (3 MN-m/rad) and $\sigma_y Z = 126$ kip-in (14 N-m). In Fig. 7.11, the column solution, indicated by solid lines is compared with that of the nonrestrained column (dashed curves). Evidently, the maximum strength of the restrained column is higher than that of the nonrestrained column.

The problem of elastic-plastic lateral-torsional buckling of beam-column has been extensively investigated in the past and the present state-of-the-art is summarized in Chap. 3; however, the post-buckling behavior of this type of column has not been obtained. The purpose herein is to determine the failure load and the behavior of the column from the standpoint of deflection and biaxial bending by introducing the concept of initial imperfections. In Fig. 7.12 the initial displacement, u_{im}, is assumed to be -0.02 in (0.508 mm) and the values of v_{im} and θ_{im} are both zero. The column is eccentrically loaded about the strong axis ($e_y = 3.0$ in or 76.2 mm).

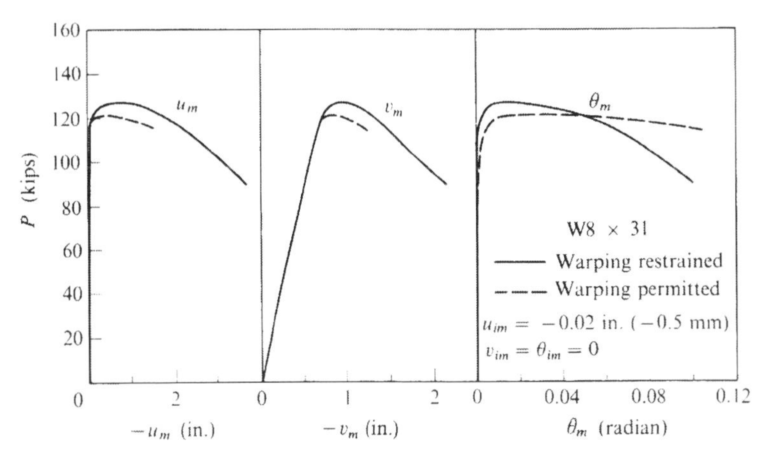

FIGURE 7.12
Lateral-torsional buckling behavior of an initially imperfect beam-column [L = 208 in (5.3 m), σ_y = 36 ksi (248 MPa), e_x = 0.01 in (0.25 mm), e_y = 3 in (76 mm)] (1 in = 25.4 mm, 1 kip = 4.45 kN)

FIGURE 7.13
Influence of the strong axis moment on the behavior of weak axis bending (1 in = 25.4 mm, 1 ksi = 6.9 MPa, 1 kip-in = 113 N-m)

θ_{im} = 0.001 rad.

$u_{im} = v_{im}$ = -0.001 in.

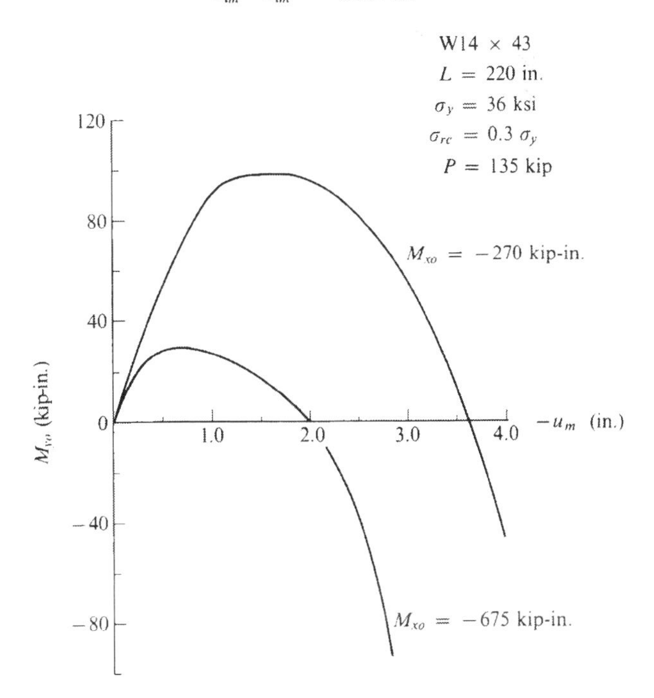

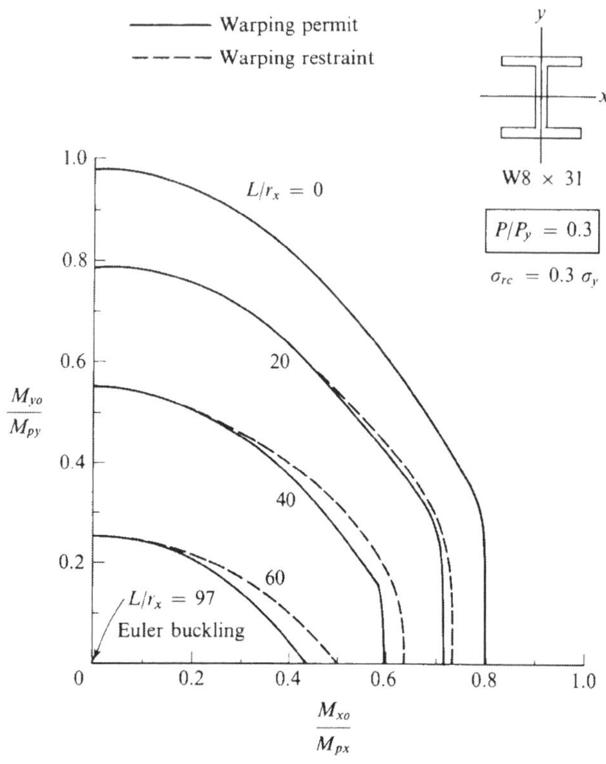

FIGURE 7.14
Maximum-strength interaction curves

The load acts slightly eccentrically about the weak axis ($e_x = 0.01$ in or 0.254 mm). The load-deflection curves are shown for two cases in Fig. 7.12. The solid curve indicates the case of warping fully restrained, and the dashed curve is for the case of warping free. It is observed that the lateral displacement, u_m, and the rotation, θ_m, are small until the axial force approaches a certain critical value, i.e., the value at the inception of lateral-torsional buckling. Again, the results indicate a higher strength of the column for the case with warping restrained.

Solutions of nonproportional loading case

Figure 7.13 investigates the influence of strong axis moment on weak axis bending. The axial force, P, remains constant at 135 kip (600 kN). Two different values of strong axis moments, −270 kip-in and −675 kip-in (−31 kN-m and −76 kN-m), are chosen for this study. The solutions differ significantly from each other.

Figure 7.14 shows the maximum strength interaction curves for biaxially loaded beam-column of W8 × 31 section. The ratios, for axial load, P/P_y and for residual stress, σ_{rc}/σ_y, are each equal to 0.3. The slenderness ratios, L/r_x, of the beam-column are 0, 20, 40 and 60. For $L/r_x = 97$, the column buckles elastically under axial load

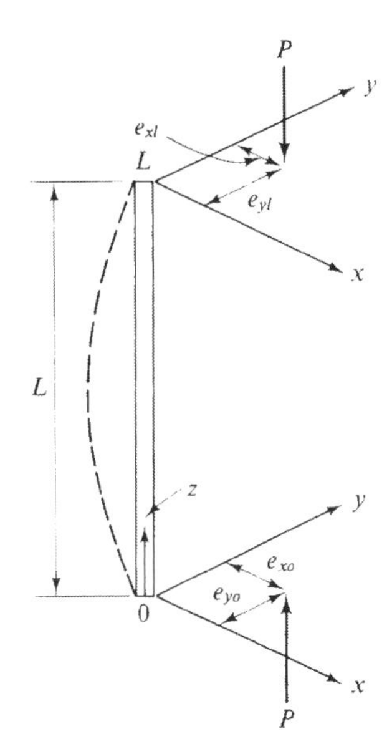

FIGURE 7.15
Beam-column under biaxially eccentric load with unequal end eccentricities

and, therefore, no bending moments can be applied to the column ends. In the case of $L/r_x = 0$, the solution is obtained from the limit analysis of column segment presented in Chap. 5. The solid curves represent the strength of the beam-column when its ends are free to warp, and the curves for the case of warping restrained are shown dashed in the figure. Note that the warping restraint does not affect the beam-column strength for small values of M_{xo}/M_{px}. However, its effect appears to be significant as the value of M_{xo}/M_{px} becomes higher. It is also observed from the figure that the effect of warping of the beam-column end becomes less significant as the slenderness ratio decreases. For the limiting case of $L/r_x = 0$, the warping restraint of the column ends has no influence at all on the results.

7.4 BEAM-COLUMN WITH UNEQUAL END ECCENTRICITIES

Consider an unsymmetrically loaded beam-column as shown in Fig. 7.15. Since this is a proportional loading case, the moments produced by the external forces at any section z of the beam-column are obtained from Eq. (2.162) by neglecting the effect of end twisting moment and by substituting

$$M_x = -P\left[e_{yo} + \frac{z}{L}(e_{yl} - e_{yo})\right]$$

$$M_y = P\left[e_{xo} + \frac{z}{L}(e_{xl} - e_{xo})\right]$$

into Eq. (2.162). Thus, we have

$$
\begin{aligned}
M_\eta &= P\left[e_{xo} + \frac{z}{L}(e_{xl} - e_{xo}) - u_0 - y_0\theta\right] \\
&\quad + P\theta\left[e_{yo} + \frac{z}{L}(e_{yl} - e_{yo}) - v_0 + x_0\theta\right] \\
M_\xi &= -P\left[e_{yo} + \frac{z}{L}(e_{yl} - e_{yo}) - v_0 + x_0\theta\right] \\
&\quad + P\theta\left[e_{xo} + \frac{z}{L}(e_{xl} - e_{xo}) - u_0 - y_0\theta\right] \\
M_\zeta &= P\left[e_{xo} + \frac{z}{L}(e_{xl} - e_{xo}) - u_0 - y_0\theta\right]v_0' \\
&\quad - P\left[e_{yo} + \frac{z}{L}(e_{yl} - e_{yo}) - v_0 + x_0\theta\right]u_0' \\
&\quad + P(y_0u_0' - x_0v_0') + \frac{P}{L}[(e_{yl} - e_{yo})u_0 - (e_{xl} - e_{xo})v_0]
\end{aligned}
\tag{7.23}
$$

while the internal forces are given by

$$
\begin{aligned}
P &= -\int_A \sigma\, dA \\
M_\eta &= -\int_A \sigma\xi\, dA \\
M_\xi &= \int_A \sigma\eta\, dA \\
M_\zeta &= -EI_\omega\theta''' + (GK_T + \bar{K})\theta'
\end{aligned}
\tag{7.24}
$$

in which

$$
\sigma = \begin{cases} \sigma_y & \text{for} \quad \varepsilon > \varepsilon_y \\ E\varepsilon & \text{for} \quad -\varepsilon_y \leq \varepsilon \leq \varepsilon_y \\ -\sigma_y & \text{for} \quad \varepsilon < -\varepsilon_y \end{cases}
\tag{7.25}
$$

and the strain at any point on the cross section is given by

$$
\varepsilon = \varepsilon_o - \xi u'' - \eta v'' + \omega\theta''
\tag{7.26}
$$

The equations of equilibrium are obtained by equating the external forces Eq. (7.23) to the corresponding internal forces Eq. (7.24). Syal and Sharma (1971) solved these equations in the elastic range by assuming deflection functions u, v, and θ as power series. The coefficients in the series are determined by satisfying the equilibrium equations at a sufficient number of points along the length of the beam-column including boundaries.

In the plastic range, in order to avoid difficulty involved due to nonlinearity,

Table 7.1 SPECIMENS AND END ECCENTRICITIES (SYAL AND SHARMA, 1971)

Specimen number (1)	Nominal size of column (2)	Length of column in. (3)	Yield stress of material, ksi (4)	Eccentricities of loading, in.			
				e_{xo} (5)	e_{xl} (6)	e_{yo} (7)	e_{yl} (8)
1	6 × 6H	96.0	33	1.61	1.61	2.78	2.78
2	5 × 5H	96.0	36	1.60	1.60	3.21	3.21
3	5 × 5H	120.0	36	0.80	0.80	2.63	2.63
4	6 × 6H	96.0	36	1.66	1.66	2.95	2.95
5	5 × 5H	96.0	36	2.36	2.36	3.17	3.17
6	5 × 5H	120.0	36	2.38	2.38	2.51	2.51
7	6 × 5W	96.0	36	−0.89	−0.89	2.82	2.82
8	6 × 5W	96.0	36	0.34	0.34	1.87	1.87
9	8 × 4W	96.0	36	0.38	0.39	1.72	−1.66
10	8 × 4W	96.0	36	0.19	0.19	2.60	2.60
11	5 × 5H	96.0	36	0.83	−0.78	1.73	−1.74
12	5 × 5H	120.0	36	−0.77	−0.77	2.78	2.78
13	4 × 4H	120.0	65	0.42	0.42	2.72	2.72
14	4 × 4H	120.0	65	0.83	0.83	2.35	2.35
15	5 × 5H	120.0	65	1.75	−1.47	2.96	−2.96
16	5 × 5H	120.0	65	0.14	0.35	−0.11	−2.80
B1	W12 × 79	180.0	33	6.00	6.00	12.0	12.0
H1	W14 × 43	264.6	33	0.50	0.00	5.00	0.00

1 in = 25.4 mm, 1 ksi = 6.9 MPa

Table 7.2 ULTIMATE LOAD IN SYMMETRIC LOADING (SYAL AND SHARMA, 1971)

Specimen number in Table 7.1 (1)	Ultimate loads, in kip						Residual strains considered (7)	Distance z/L to critical section (8)
			With nominal yield stress					
	Experiment (2)	Sharma and Gaylord's interaction curves* (3)	Linear interaction Eq. (13.17) (4)	Harstead *et al.* (1968) (5)	Syal and Sharma (1971) (6)			
1	92.80	93.4	80.6	92.8	93.1		Yes	0.5
2	54.1	49.9	42.8	52.6	50.75		Yes	0.5
3	62.7	58.3	52.3	62.8	60.84		Yes	0.5
4	86.3	83.6	67.4	83.3	84.35		Yes	0.5
5	49.6	51.4	42.5	52.7	50.20		Yes	0.5
6	47.9	49.2	42.0	49.4	47.76		Yes	0.5
7	76.6	70.4	65.9	79.3	80.16		Yes	0.5
8	109.4	98.0	92.8	110.3	110.1		No	0.5
10	85.0	75.7	71.6	80.5	78.7		No	0.5
12	51.0	51.5	50.3	55.7	56.23		Yes	0.5
13	46.1	42.7	31.4	45.0	47.24		No	0.5
14	38.7	37.2	29.2	41.2	40.12		No	0.5
		Sharma and Gaylord (1969)	Ringo (1964)	Birnstiel (1968)	Syal and Sharma (1971)			
B1		155.0	154.0	150.0	150.95			

* Fig. 7.17
1 kip = 4.45 kN

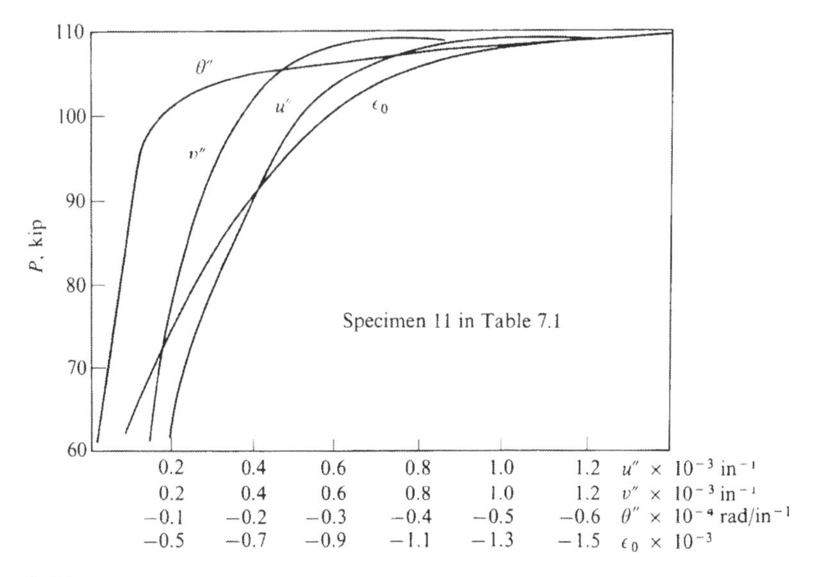

FIGURE 7.16
Load-deformation curves at critical section (Syal and Sharma, 1971) (1 kip = 4.45 kN)

it is assumed that the deflected shape of the beam-column in the elastic range remains qualitatively valid, i.e., the ratios among deflections u, v, and θ and the corresponding curvatures u'', v'' and θ'' remain constant with further application of load. Further the shifting of shear center with the penetration of yielding is neglected. This enables the four unknowns ε_o, u'', v'' and θ'' to be determined from the four equilibrium equations corresponding to each increment of load.

Numerical studies were made for eighteen beam-columns as shown in Table 7.1, five of them, namely, specimens 9, 11, 15, 16 and H1 were loaded unsymmetrically. In the numerical solutions, nonlinear terms $(u_0 + y_0\theta)\theta$, $(v_0 - x_0\theta)\theta$, $(v_0 - x_0\theta)u'$, and $(u_0 + y_0\theta)v'$ in Eq. (7.23) are neglected in the computation. As one of the numerical results, Fig. 7.16 shows relationship between load and deformations at the critical section of specimen 11 which is the case of biaxial bending into double curvatures. It can be observed that the twisting angle θ'' increases abruptly while the axial strain ε_o increases gradually. All the curves show their peak at the same P-value which is known to be the ultimate load of the beam-column. The ultimate loads computed for all the specimen are listed in Table 7.2 for the case of a symmetric loading and in Table 7.3 for the case of an unsymmetric loading. Comparison with results obtained by other investigators is also listed in the tables.

Numerical calculations were also made to investigate the effect of warping restraint at ends, the effect of cross-sectional shape, and the effect of residual stresses. It was concluded that restrained warping increases the ultimate loads by 7 percent for equal end eccentricity cases and by 4 percent for unequal eccentricity cases. As for the effect of cross-sectional shape, beam-columns of W12 × 65, W12 × 58,

Table 7.3 ULTIMATE LOAD IN UNSYMMETRIC LOADING (SYAL AND SHARMA, 1971)

	Ultimate load, in kip		
Specimen number in Table 7.1 (1)	Experiment (2)	Syal and Sharma (1971) (3)	Distance z/L to critical section (4)
9	83.00	80.16	0.33
11	108.00	109.50	0.02
15	103.50	101.75	0.02
16	93.00	92.10	0.69
	Harstead *et al.* (1968)	Syal and Sharma (1971)	
H1	152.00	144.19	0.34

1 kip = 4.45 kN

W12 × 50 and W12 × 36 were investigated by changing the length and eccentricities. It was found that the ratios of the ultimate load P_u obtained for the beam-column to P_{yu} the fully plastified section subjected to the same biaxial loading decreases with an increase in slenderness ratio L/r_y but it is only slightly affected with other parameters. It was also found that residual stress of magnitude $0.3\sigma_y$ can only decrease the ultimate load of beam-columns by a maximum of 7 percent.

7.5 ULTIMATE STRENGTH ANALYSES

Sharma and Gaylord (1969) have developed maximum strength interaction curves for symmetrically loaded beam-columns of W12 × 65 shape with no residual stress using the method described in the preceeding section. The material is elastic perfectly plastic with $\sigma_y = 36$ ksi (248 MPa). Deformations at midspan are solved by the stepwise procedure using equilibrium conditions at mid-span. The simplest deflected shapes are assumed as

$$\begin{aligned} u &= u_m \sin \frac{\pi z}{L} \\ v &= v_m \sin \frac{\pi z}{L} \\ \theta &= \theta_m \sin \frac{\pi z}{L} \end{aligned} \tag{7.27}$$

The results are summarized in design chart form in term of the slenderness ratio

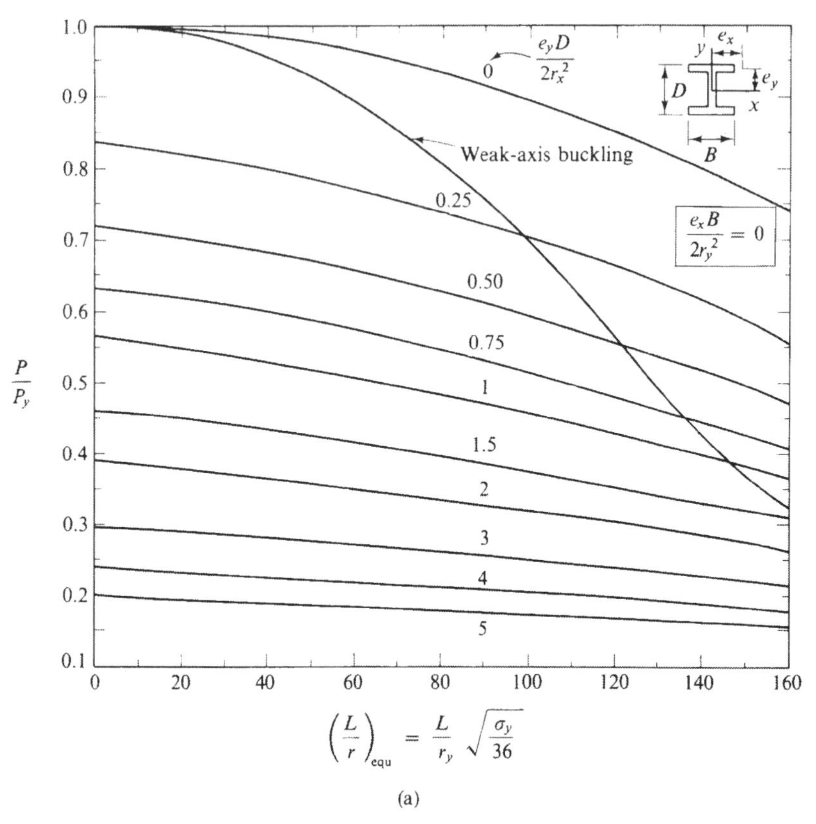

FIGURE 7.17 (a)
Interaction curves for biaxially loaded beam-column of I-sections (Sharma and Gaylord, 1969) $\sigma_y = 36$ ksi (248 MPa)

in weak axis bending L/r_y as shown in Fig. 7.17. The eccentricities e_x and e_y are nondimensionalized, using the overall dimensions of width B and depth D of the cross section and the radii of gyration r_x and r_y, into the two parameters,

$$\frac{e_x B}{2r_y^2} \quad \text{and} \quad \frac{e_y D}{2r_x^2} \tag{7.28}$$

Since it is known that the effect of cross-sectional shape on the nondimensionalized beam-column strength is small, these curves are considered to be applicable to most steel beam-columns of wide flange cross section. The curves can be used also for other yield stresses by using an *equivalent slenderness ratio* L/r_y according to the following formula:

$$\left(\frac{L}{r_y}\right)_{equ} = \frac{L}{r_y}\sqrt{\frac{\sigma_y}{36}} \tag{7.29}$$

The ultimate loads obtained by using the interaction curves in Fig. 7.17 are listed in

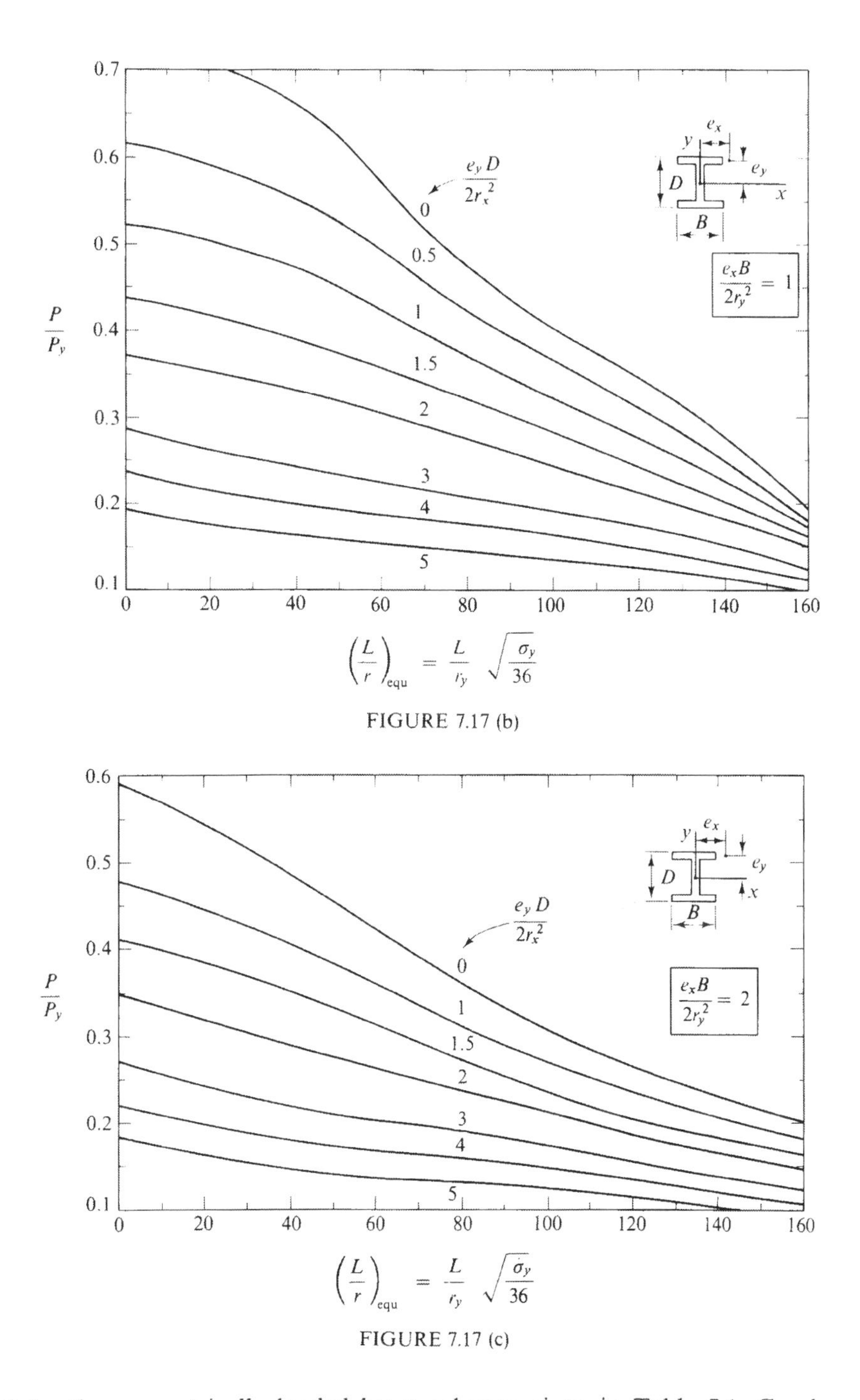

FIGURE 7.17 (b)

FIGURE 7.17 (c)

Table 7.2 for the symmetrically loaded beam-columns given in Table 7.1. Good agreement is generally observed when compared to the test results and other calculated values.

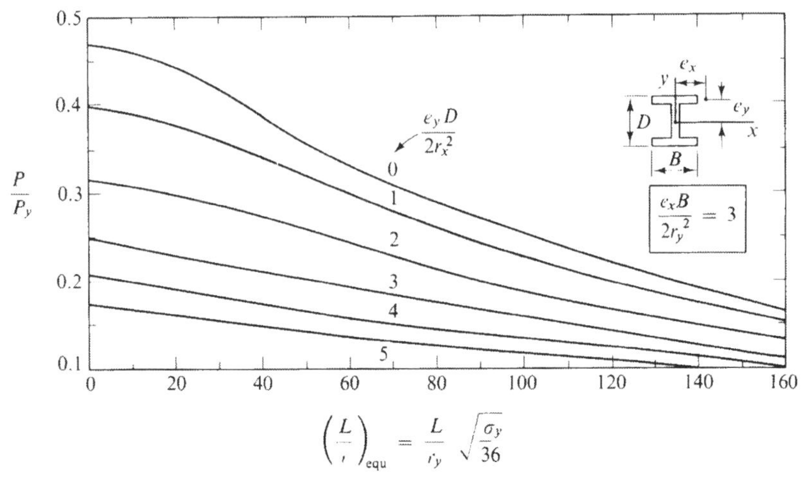

FIGURE 7.17 (d)

When the interaction curves in Fig. 7.17 are to be applied to beam-columns with unequal end eccentricities, e_{xo}, e_{xl} and e_{yo}, e_{yl}, the *equivalent moment parameters* C_{mx}, C_{my} can be used to modify them into the equivalent case of beam-columns with equal end eccentricities,

$$(e_x)_{eq} = C_{mx} e_{xo} \qquad (e_{xo} \geq e_{xl})$$
$$(e_y)_{eq} = C_{my} e_{yo} \qquad (e_{yo} \geq e_{yl}) \tag{7.30}$$

Table 7.4 COMPARISON OF ULTIMATE LOAD IN UNSYMMETRIC LOADING (SYAL AND SHARMA, 1971)

Specimen number in Table 7.1 (1)	Eccentricities of loading in inches e_{xo} (2)	e_{xl} (3)	e_{yo} (4)	e_{yl} (5)	P_u/P_{yu} Experiment Birnstiel (1968) (6)	Theory Syal and Sharma (1971) (7)	Interaction curves (8)
1	1.61	1.61	3.05	2.78	—	0.32	0.32
1[a]	1.61	1.61	3.33	2.78	—	0.316	0.304
9	0.38	0.39	1.72	−1.66	0.387	0.373	0.399
11	0.83	−0.78	1.73	−1.74	0.562	0.565	0.526
15	1.75	−1.47	2.96	−2.96	0.312	0.305	0.278
16	0.14	0.35	−0.11	−2.80	0.309	0.308	0.358
					Harstead *et al.* (1968)	Syal and Sharma (1971)	Interaction curves
H1	0.50	0.00	5.00	0.00	0.334	0.317	0.39

[a] Properties of this column cross section are the same as for test specimen 1
1 in = 25.4 mm

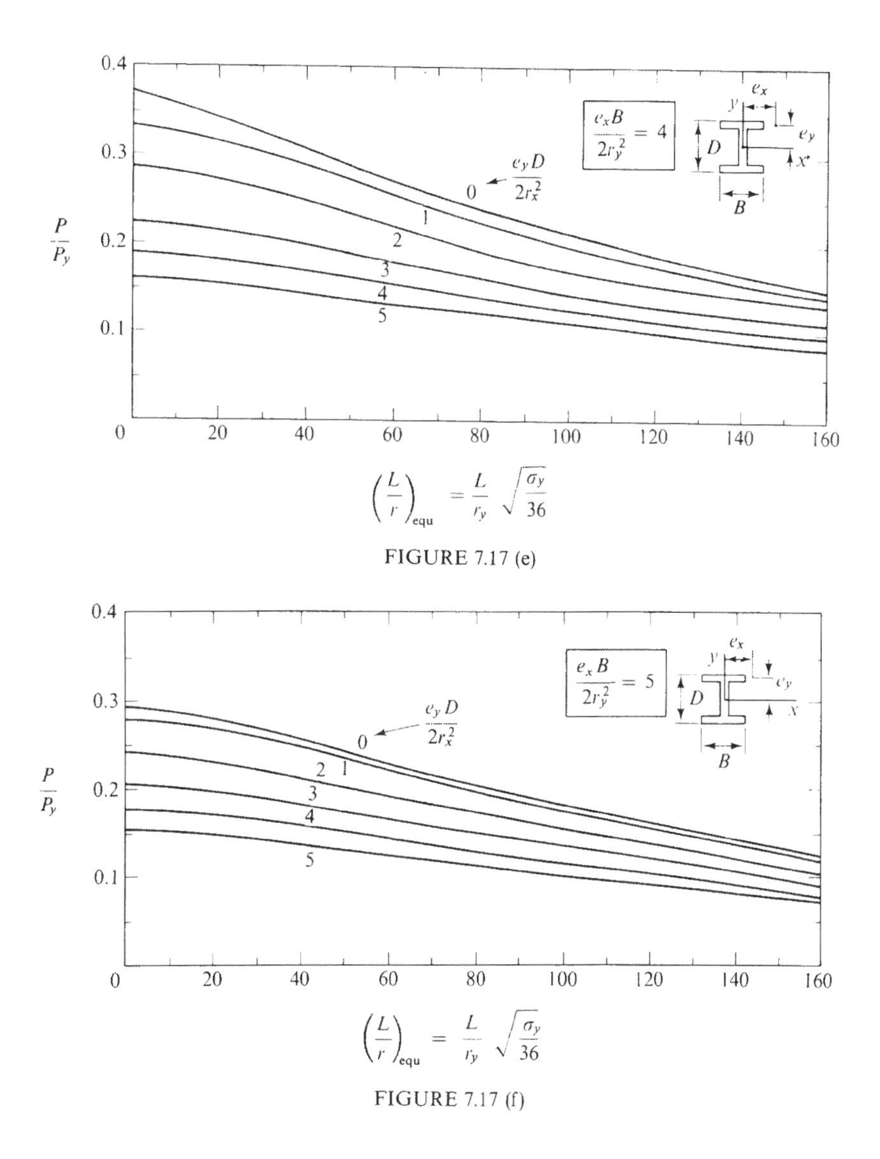

FIGURE 7.17 (e)

FIGURE 7.17 (f)

in which the simplest forms of C_{mx} and C_{my} as proposed by Austin (1961) are given by

$$C_{mx} = \left(0.6 + 0.4\frac{e_{xl}}{e_{xo}}\right) \geq 0.4$$
$$C_{my} = \left(0.6 + 0.4\frac{e_{yl}}{e_{yo}}\right) \geq 0.4 \tag{7.31}$$

Examples of the application of the above procedure is shown in Table 7.4 for the unsymmetrically loaded beam-columns given in Table 7.1. Although P_u/P_{yu} values obtained by the interaction curves are not very close to those obtained by tests and theory, it can be used in a rough estimation of beam-column strength at a preliminary design stage.

Lindner (1972) investigated biaxially loaded beam-columns with uniformly distributed lateral load, q_y, by using the Ritz method to the potential energy expression,

$$\prod = \frac{1}{2}\int_0^L \{EI_y(u'')^2 + EI_x(v'')^2 + EI_\omega(\theta'')^2 + (GK_T + \bar{K})(\theta')^2 + P[(v')^2 + (u')^2 + 2e_x v''\theta - 2e_y u''\theta] - 2q_y v\}\, dz \tag{7.32}$$

The deflected shapes are approximated by Hermite polynomials,

$$\begin{aligned} \theta &= \sum_i a_i \sum_j H_j(z) \\ v &= \sum_i b_i \sum_j H_j(z) \\ u &= \sum_i c_i \sum_j H_j(z) \end{aligned} \tag{7.33}$$

The bending moments and the twisting moment in the beam-column are found by solving Eqs. (7.32) and (7.33). From these moments, an improved curvatures u'', v'', θ''; and thus an improved displacements u, v, θ can be calculated by numerical integration along the member. This iteration will be repeated until no more modifications in the displacements are required. The ultimate

FIGURE 7.18(a)
Influence of residual stresses (Lindner, 1972)

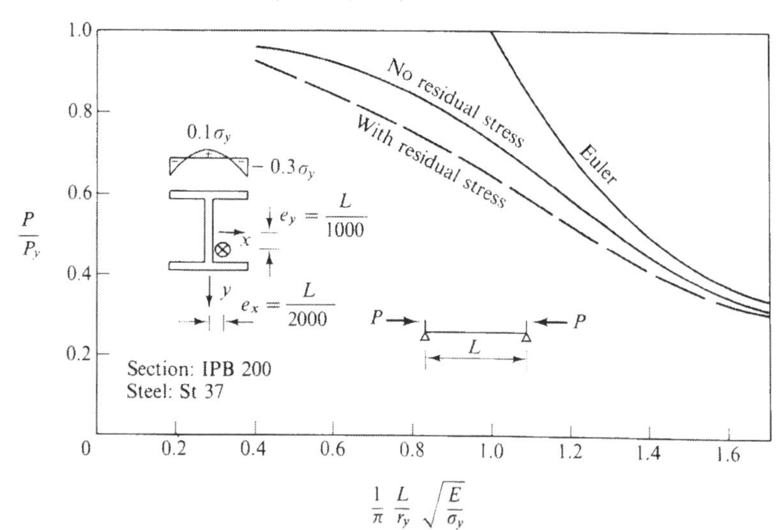

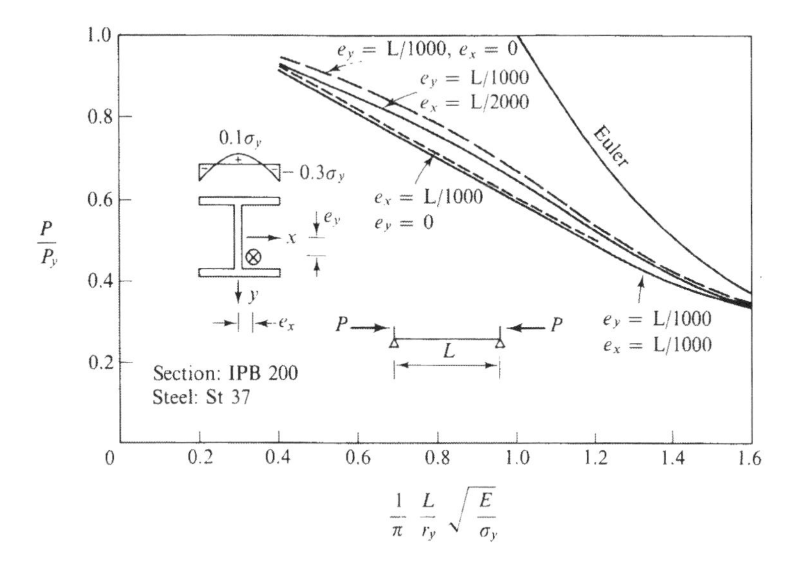

FIGURE 7.18(b)
Influence of end eccentricities (Lindner, 1972)

load is defined when the load-deflection curve cannot increase any further.

Numerical results are presented in the form of column strength curves in Fig. 7.18 comparing the influence of residual stresses [Fig. 7.18(a)], the influence of end eccentricities [Fig. 7.18(b)], the influence of the type of cross section [Fig. 7.18(c)],

FIGURE 7.18(c)
Influence of the type of cross-section (Lindner, 1972)

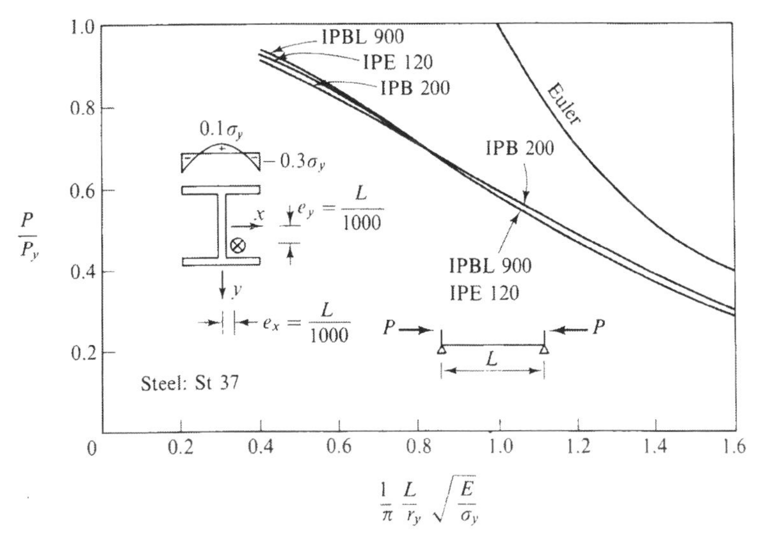

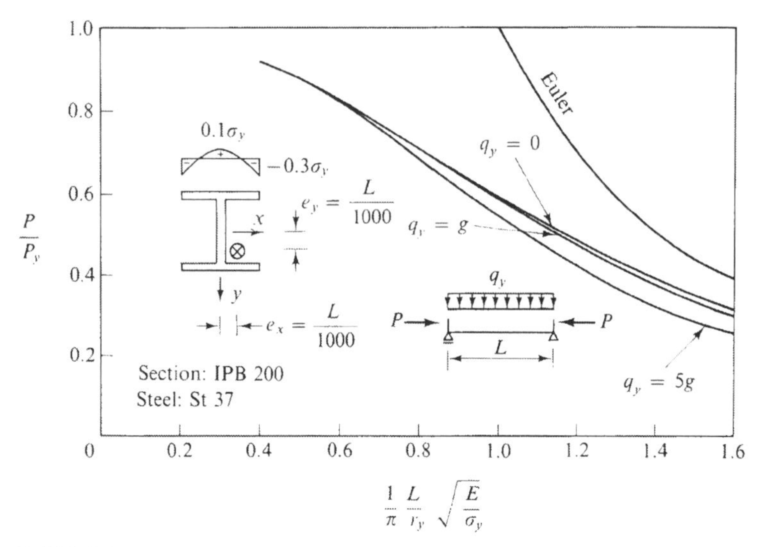

FIGURE 7.18(d)
Influence of lateral loads q_y (Lindner, 1972)

and the influence of lateral load q_y [Fig. 7.18(d)]. Similar conclusions made by Syal and Sharma (1971) are generally observed except that the existence of residual stresses can decrease the ultimate load of biaxially loaded beam-columns as much as 20 percent in the range of intermediate slenderness ratio [see Fig. 7.18(a)]. By comparing the curves in Fig. 7.18(b), we can notice a reduction of nearly 5 percent for the two upper curves with $e_y = L/1000$ and e_x varies from 0 to L/2000. If the eccentricity e_x increases to L/1000, an additional reduction is found of nearly 8 percent (lowest curve). By comparing the two lower curves with $e_x = L/1000$ and e_y varied from 0 to L/1000, we see that there is only a very small difference. The influence of three kinds of rolled shapes is shown in Fig. 7.18(c). The greatest differences among the three shapes are less than 5 percent. Figure 7.18(d) shows the results of a biaxially loaded beam-column with lateral load q_y acting in the plane of column web. If this lateral load is as great as the dead weight g of the beam-column, the influence is unimportant. If the load q_y reaches the value of $5g$, the maximum reduction of the beam-column strength can be as much as 20 percent.

7.6 AVERAGED FLOW MOMENTS FOR UNSYMMETRIC LOADING

In Chap. 13 of Vol. 1, a simple approximate analysis using the concept of averaged flow moment was developed, and simple interaction equations for the maximum strength of in-plane beam-columns were presented for various loading conditions. In that work, the moment-curvature-thrust relationship for a uniaxially loaded cross section was idealized to be elastic-perfectly plastic. The approximate average flow moment, M_{mc}, was assumed to lie between the initial yield moment, M_{yc}, and the plastic limit moment, M_{pc}.

Extension of this two-dimensional approximate approach for in-plane beam-columns to three-dimensional analysis for biaxially loaded beam-columns is attempted herein. The biaxial moment-curvature-thrust relationship is assumed to be elastic-perfectly plastic. The approximate averaged flow surface is assumed to lie between the initial yield surface and the fully plastic limit surface. Once the proper values for the averaged flow surface are selected, the maximum load carrying capacity of the biaxially loaded beam-column can be computed in a rather simple manner by an elastic analysis. This was done by Chen and Atsuta (1973) and their solutions are presented in what follows.

7.6.1 Elastic Moments at the Critical Section

Since the elastic solutions for the internal moments M_ξ, M_η and M_ζ in a biaxially loaded beam-column have already been obtained in Eq. (4.64) with the coefficients A_0, B_0 and C_0 defined in Eq. (4.63); C_1, C_2 in Eq. (4.40); and A_1, A_2, B_1, B_2 in Eq. (4.44), what must be done next is to search for the critical section. The critical section is the section where the initial yield condition

$$F_y(P, M_\xi, M_\eta, M_\zeta) = 0 \tag{7.34}$$

is reached first. Since in the elastic regime, $F_y < 0$, the location of the critical section ($z = z^*$) is determined by

$$\left.\frac{\partial F_y}{\partial z}\right|_{z = z^*} = 0 \tag{7.35}$$

As will be seen later, the expression for the initial yield function, F_y, is not simple enough such that Eq. (7.35) can be solved analytically. Herein, the simple linearized function nondimensionalized by the fully plastic states of the cross section

$$F_y = \left|\frac{P}{P_y}\right| + \left|\frac{M_\xi}{M_{p\xi}}\right| + \left|\frac{M_\eta}{M_{p\eta}}\right| - 1 = 0 \tag{7.36}$$

is used for the determination of the critical section. Substituting Eq. (4.64) into Eq. (7.36), then

$$\begin{aligned} F_y = &\left(\frac{P_{ex}}{M_{p\xi}} B_0 - \frac{P_{ey}}{M_{p\eta}} A_0\right) \sin\frac{\pi z}{L} + \frac{6}{\pi^2}\left(\frac{P_{ex}}{M_{p\xi}} B_2 - \frac{P_{ey}}{M_{p\eta}} A_2\right)\frac{z}{L} \\ &- \frac{2}{\pi^2}\left[\frac{P_{ey}}{M_{p\eta}}(A_1 - A_2) - \frac{P_{ex}}{M_{p\xi}}(B_1 - B_2)\right] + \frac{P}{P_y} - 1 \end{aligned} \tag{7.37}$$

in which P_{ex} and P_{ey} are Euler buckling loads defined as P_x and P_y in Eq. (4.54), respectively. From the condition Eq. (7.35), the location of the critical section is

$$z^* = \frac{L}{\pi}\cos^{-1}\left(-\frac{6}{\pi^3}\frac{\dfrac{P_{ex}}{M_{p\xi}}B_2 - \dfrac{P_{ey}}{M_{p\eta}}A_2}{\dfrac{P_{ex}}{M_{p\xi}}B_0 - \dfrac{P_{ey}}{M_{p\eta}}A_0}\right) \tag{7.38}$$

Thus the moments at the critical section are obtained from Eq. (4.64) as

$$\begin{aligned} M_\xi^* &= P_{ex}\left[B_0 \sin\frac{\pi z^*}{L} - \frac{2}{\pi^2}(B_2 - B_1) + \frac{6}{\pi^2}B_2\frac{z^*}{L}\right] \\ M_\eta^* &= -P_{ey}\left[A_0 \sin\frac{\pi z^*}{L} - \frac{2}{\pi^2}(A_2 - A_1) + \frac{6}{\pi^2}A_2\frac{z^*}{L}\right] \\ M_\zeta^* &= \frac{\pi}{L}\left(\frac{\pi^2 EI_\omega}{L^2} + GK_T + \bar{K}\right)C_0\cos\frac{\pi z^*}{L} \\ &\quad + \frac{1}{L}(GK_T + \bar{K})C_1\left(1 - 2\frac{z^*}{L}\right) \\ &\quad + \frac{1}{L}\left[\frac{6EI_\omega}{L^2} + (GK_T + \bar{K})\left(2\frac{z^*}{L} - 3\frac{z^{*2}}{L^2}\right)\right]C_2 \end{aligned} \tag{7.39}$$

The biaxially loaded beam-column reaches its ultimate strength when these moments on the critical section satisfy an averaged interaction relationship for the cross section ($F_A = 0$) which lies between the initial yield condition ($F_y = 0$) and the fully plastic limiting yield condition ($F_p = 0$). This averaged interaction relationship will be examined further later.

For the special case of a symmetric loading ($\gamma_{\omega o} = \gamma_{\omega l}$, $M_{xo} = M_{xl} = \bar{M}_x$, $M_{yo} = M_{yl} = \bar{M}_y$), the critical section is at the mid-span ($z^* = L/2$) and Eq. (7.39) reduces to

$$\begin{aligned} M_\xi^* &= P_{ex}B_0 + \bar{M}_x \\ M_\eta^* &= -P_{ey}A_0 + \bar{M}_y \\ M_\zeta^* &= 0 \end{aligned} \tag{7.40}$$

with

$$\begin{aligned} A_0 &= -\frac{\pi^2}{8}\frac{P}{P_{ey} - P}\frac{\bar{M}_y}{P_{ey}}\left[1 - \frac{\dfrac{4}{\pi}\dfrac{P_{ey} - P_{ex}}{P} - \left(\dfrac{P_{ex}}{P_{ey} - P} - \dfrac{P_{ey}}{P_{ex} - P}\right)\dfrac{\bar{M}_x^2}{P_{ex}}}{r_0^2\gamma(P_z - P) - \left(\dfrac{\bar{M}_y^2}{P_{ex} - P} + \dfrac{\bar{M}_x^2}{P_{ey} - P}\right)}\right] \\ B_0 &= \frac{\pi^2}{8}\frac{P}{P_{ex} - P}\frac{\bar{M}_x}{P_{ex}}\left[1 - \frac{\dfrac{4}{\pi}\dfrac{P_{ey} - P_{ex}}{P} - \left(\dfrac{P_{ex}}{P_{ey} - P} - \dfrac{P_{ey}}{P_{ex} - P}\right)\dfrac{\bar{M}_y^2}{P_{ey}}}{r_0^2\gamma(P_z - P) - \left(\dfrac{\bar{M}_y^2}{P_{ex} - P} + \dfrac{\bar{M}_x^2}{P_{ey} - P}\right)}\right] \end{aligned} \tag{7.41}$$

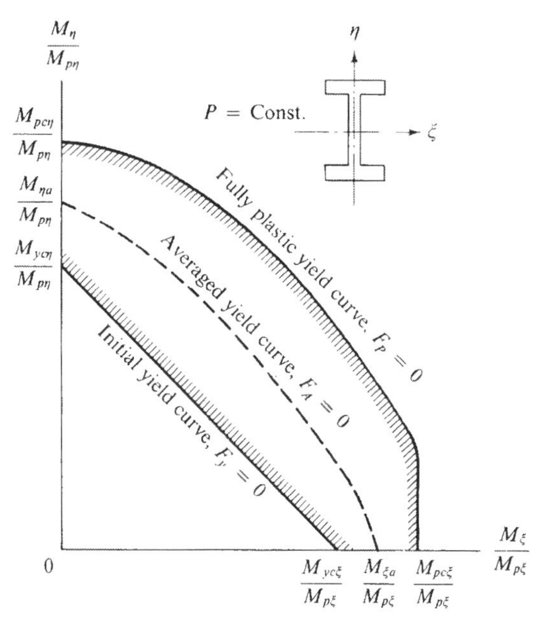

FIGURE 7.19
Averaged interaction curves of section

in which P_z is elastic torsional buckling load defined in Eq. (4.54). Further, for an uniaxial bending case ($\bar{M}_y = 0$), only M_ξ^* remains as

$$M_\xi^* = \left(\frac{\pi^2}{8}\frac{P}{P_{ex} - P} + 1\right)\bar{M}_x \tag{7.42}$$

7.6.2 Averaged Interaction Relationship for H-Section ($F_A = 0$)

Following the previous work on in-plane beam-columns (Chap. 13, Vol. 1), a certain averaged interaction relationship ($F_A = 0$) for a biaxially loaded cross section between initial yielding and the fully plastic state of the cross section can be found among the axial force (P/P_y) and the biaxial bending moments ($M_\xi/M_{p\xi}$, $M_\eta/M_{p\eta}$) at the critical section when the ultimate state of the beam-column is reached. The contribution of the twisting moment (M_ζ) to the yield condition is neglected herein.

This averaged interaction relationship ($F_A = 0$) is bounded by the initial yield condition ($F_y = 0$) and the fully plastic limiting yield condition ($F_p = 0$) as seen in Fig. 7.19.

$$F_y \leq F_A \leq F_p \tag{7.43}$$

In the following section, the wide flange shape of cross section is selected as an example for illustration. The initial yield function for a wide flange shape of cross section is

$$F_y = f_{s\xi}\frac{M_\xi}{M_{p\xi}} + f_{s\eta}\frac{M_\eta}{M_{p\eta}} + \frac{P}{P_y} - 1 = 0 \tag{7.44}$$

in which $f_{s\xi}, f_{s\eta}$ = shape factors of cross section.

The fully plastic yield function of a wide flange cross section has been derived approximately in Eq. (5.115), Chap. 5, as

$$F_p = \begin{cases} \dfrac{\left(\dfrac{M_\xi}{M_{p\xi}}\right)^a}{1-\left(\dfrac{P}{P_y}\right)^b} + \left(\dfrac{M_\eta}{M_{p\eta}}\right) + \left(\dfrac{P}{P_y}\right)^c - 1 & \left(\dfrac{M_\eta}{M_{p\eta}} \geq \dfrac{M_\eta^0}{M_{p\eta}}\right) \\ \dfrac{M_\xi}{M_{p\xi}} + \left(\dfrac{P}{P_y}\right)^d - 1 & \left(\dfrac{M_\eta}{M_{p\eta}} \leq \dfrac{M_\eta^0}{M_{p\eta}}\right) \end{cases} \tag{7.45}$$

in which

$$\frac{M_\eta^0}{M_{p\eta}} = 1 - \left(\frac{P}{P_y}\right)^c - \frac{[1-(P/P_y)^d]^a}{1-(P/P_y)^b} \tag{7.46}$$

FIGURE 7.20
Interaction curves of wide-flange section

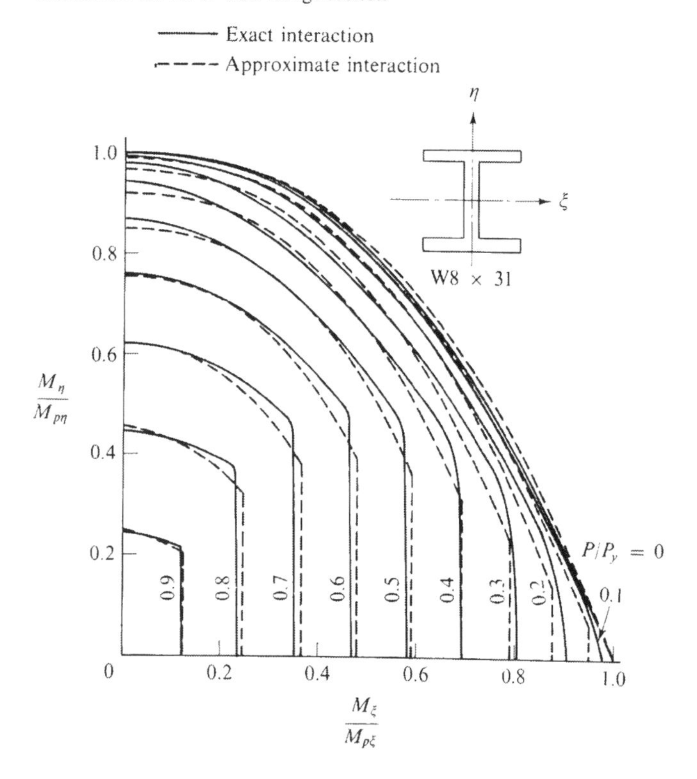

The four constants, a, b, c, and d, are determined according to the size of a particular wide flange cross section. For the cross section W8 × 31, e.g., these constants are found to be [see Eq. (5.118)]

$$a = 2.453, \quad b = 1.209, \quad c = 2.714, \quad d = 1.987 \tag{7.47}$$

The approximate fully plastic interaction curves ($F_p = 0$) calculated from Eq. (7.45) using the foregoing constants are compared in Fig. 7.20 with the exact curves reported in Sec. 5.3, Chap. 5. Good agreement is observed.

In order to obtain the averaged interaction relationship, the averaging factors, α and β, are introduced as shown in Fig. 7.19:

$$\begin{aligned} M_{\xi a} &= M_{pc\xi} - \alpha(M_{pc\xi} - M_{yc\xi}) \\ M_{\eta a} &= M_{pc\eta} - \beta(M_{pc\eta} - M_{yc\eta}) \end{aligned} \tag{7.48}$$

in which $M_{yc\xi}$, $M_{yc\eta}$ = initial yield moments in uniaxial bending including the effect of axial force P; and $M_{pc\xi}$, $M_{pc\eta}$ = fully plastic moments in uniaxial bending including the effect of axial force P. From Eqs. (7.44) and (7.45),

$$\begin{aligned} &\frac{M_{yc\xi}}{M_{p\xi}} = \frac{1}{f_{s\xi}}\left(1 - \frac{P}{P_y}\right), \quad \frac{M_{yc\eta}}{M_{p\eta}} = \frac{1}{f_{s\eta}}\left(1 - \frac{P}{P_y}\right) \\ &\frac{M_{pc\xi}}{M_{p\xi}} = 1 - \left(\frac{P}{P_y}\right)^d, \quad \frac{M_{pc\eta}}{M_{p\eta}} = 1 - \left(\frac{P}{P_y}\right)^c \end{aligned} \tag{7.49}$$

For simplicity, the averaged yield curve, F_A, is assumed to have the same form as that of the fully plastic yield curve, Eq. (7.45), but the biaxial bending moments, $M_\xi/M_{p\xi}$ and $M_\eta/M_{p\eta}$, are proportionally reduced by the factors $M_{\xi a}/M_{pc\xi}$ and $M_{\eta a}/M_{pc\eta}$, respectively. Using Eq. (7.48), then

$$F_A = \begin{cases} \dfrac{1}{1 - \left(\dfrac{P}{P_y}\right)^b}\left(\dfrac{\dfrac{M_\xi}{M_{p\xi}}}{1 - \alpha + \alpha\dfrac{M_{yc\xi}}{M_{pc\xi}}}\right)^a + \dfrac{\dfrac{M_\eta}{M_{p\eta}}}{1 - \beta + \beta\dfrac{M_{yc\eta}}{M_{pc\eta}}} + \left(\dfrac{P}{P_y}\right)^c - 1 & (M_\eta \geq M_\eta^{00}) \\ \dfrac{\dfrac{M_\xi}{M_{p\xi}}}{1 - \alpha + \alpha\dfrac{M_{yc\xi}}{M_{pc\xi}}} + \left(\dfrac{P}{P_y}\right)^d - 1 & (M_\eta \leq M_\eta^{00}) \end{cases} \tag{7.50}$$

in which

$$\frac{M_\eta^{00}}{M_{p\eta}} = \left(1 - \beta + \beta\frac{M_{yc\eta}}{M_{pc\eta}}\right)\left\{1 - \left(\frac{P}{P_y}\right)^c - \frac{\left[1 - \left(\dfrac{P}{P_y}\right)^d\right]^a}{1 - \left(\dfrac{P}{P_y}\right)^b}\right\} \tag{7.51}$$

When $\alpha = \beta = 0$, the averaged yield curve, $F_A = 0$, becomes identical to the fully plastic yield curve, $F_p = 0$ (upper bound), and when $\alpha = \beta = 1$, the curve passes through the two end points which represent the initial yield condition, $F_y = 0$, in the uniaxial bending cases ($M_\xi = 0$ or $M_\eta = 0$), but the two curves are not identical. For practical purposes, the values of α and β are assumed to be bounded by

$$0 \le \alpha \le 1, \quad 0 \le \beta \le 1 \tag{7.52}$$

The choice of a proper value for α and β for a biaxially loaded beam-column at its ultimate state will be considered in the following.

7.6.3 Interaction Equations for Long Beam-Columns

Using the averaged yield condition, Eq. (7.50), and the bending moments at the critical section of the beam-column, Eq. (7.39), the ultimate strength of the beam-column can be calculated as the combination of the applied end forces ($P, M_{xo}, M_{yo}, M_{xl}, M_{yl}$). This is the maximum strength of the biaxially loaded beam-column

$$F_c(P, M_{xo}, M_{yo}, M_{xl}, M_{yl}) = 0 \tag{7.53}$$

in which

$$F_c = \begin{cases} \dfrac{1}{1 - \left(\dfrac{P}{P_y}\right)^b} \left(\dfrac{\dfrac{M_\xi^*}{M_{p\xi}}}{1 - \alpha + \dfrac{\alpha}{f_{s\xi}}} \right)^a + \dfrac{\dfrac{M_\eta^*}{M_{p\eta}}}{1 - \beta + \dfrac{\beta}{f_{s\eta}}} + \left(\dfrac{P}{P_y}\right)^c - 1 & (M_\eta^* \ge M_\eta^{00}) \\ \dfrac{\dfrac{M_\xi^*}{M_{p\xi}}}{1 - \alpha + \dfrac{\alpha}{f_{s\xi}}} + \left(\dfrac{P}{P_y}\right)^d - 1 & (M_\eta^* \le M_\eta^{00}) \end{cases} \tag{7.54}$$

where $f_{s\xi} = M_{pc\xi}/M_{yc\xi}$ and $f_{s\eta} = M_{pc\eta}/M_{yc\eta}$.

Figure 7.21 shows the interaction curves for a beam-column (W8 × 31) under the symmetric loading condition. The open circles are the results reported by Santathadaporn and Chen (1973). These results are bounded by the two limiting curves (solid lines) and the averaging factors of $\alpha = \beta = 0.6$ are seen to give a good approximation to the more accurate results as shown by the dotted line.

After a number of trial calculations, it is found that the following formula gives a good estimation of the averaging flow factors

$$\begin{aligned} \alpha = \beta &= \frac{P_y}{P_{ey}} \quad &\text{if} \quad P_{ey} \ge P_y \\ \alpha = \beta &= 1.0 \quad &\text{if} \quad P_{ey} \le P_y \end{aligned} \tag{7.55}$$

in which P_{ey} = the Euler's buckling load given by

$$P_{ey} = \frac{\pi^2 E I_\eta}{L^2} \tag{7.56}$$

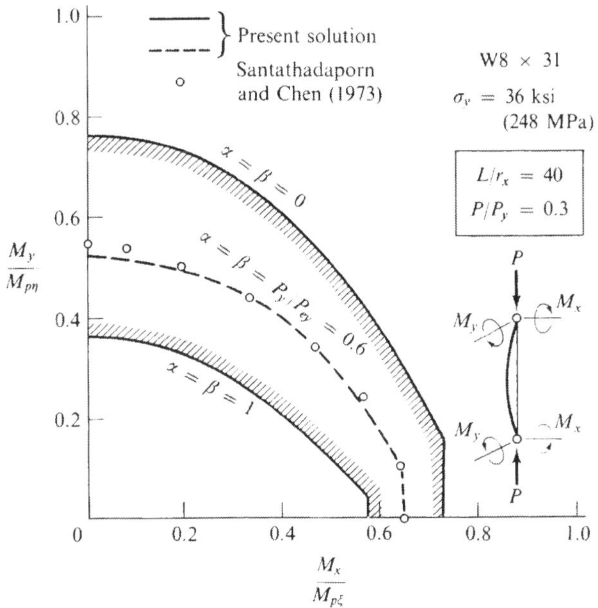

FIGURE 7.21
Bounded interaction curve of beam-column

FIGURE 7.22
Comparison in symmetric loading cases

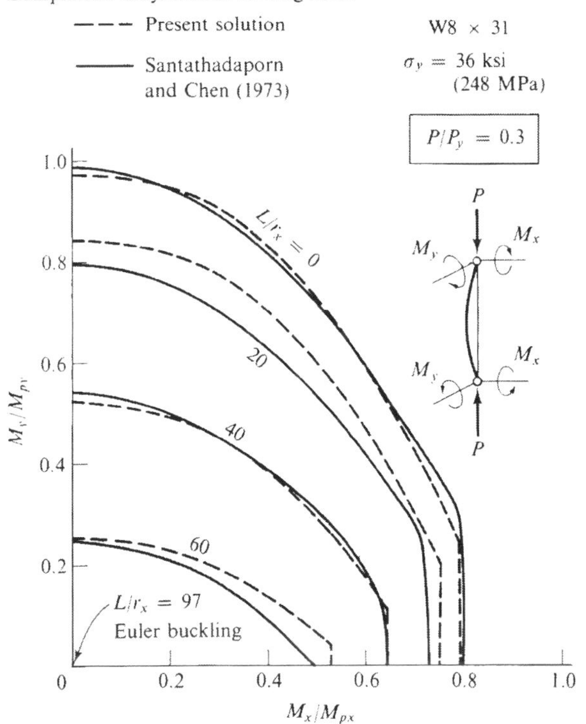

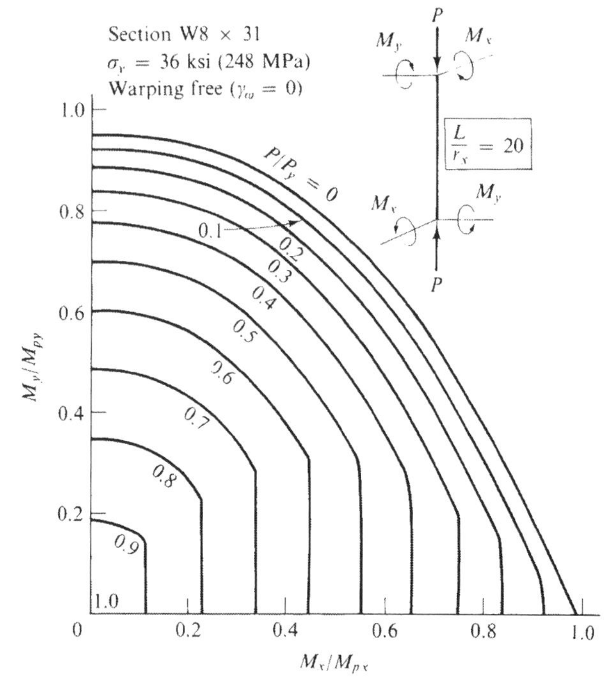

FIGURE 7.23(a)
Interaction curves of symmetrically loaded beam-column

FIGURE 7.23(b)
Interaction curves of symmetrically loaded beam-column

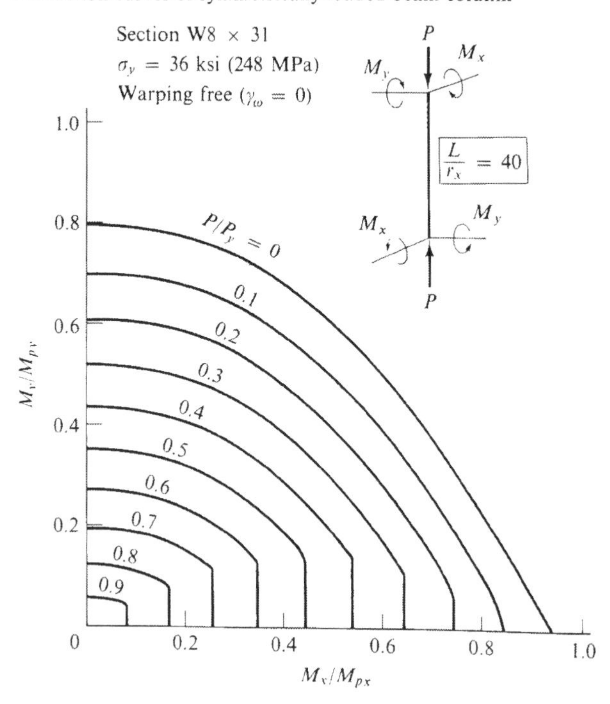

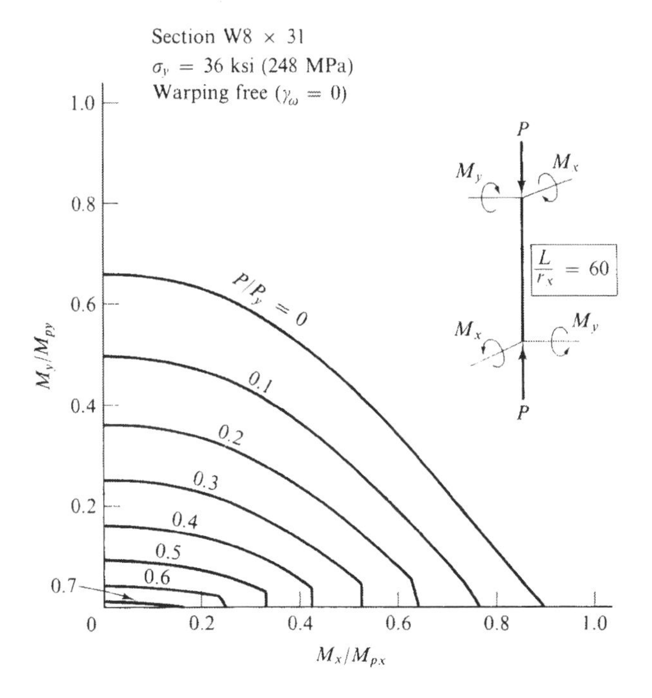

FIGURE 7.23(c)
Interaction curves of symmetrically loaded beam-column

FIGURE 7.23(d)
Interaction curves of symmetrically loaded beam-column

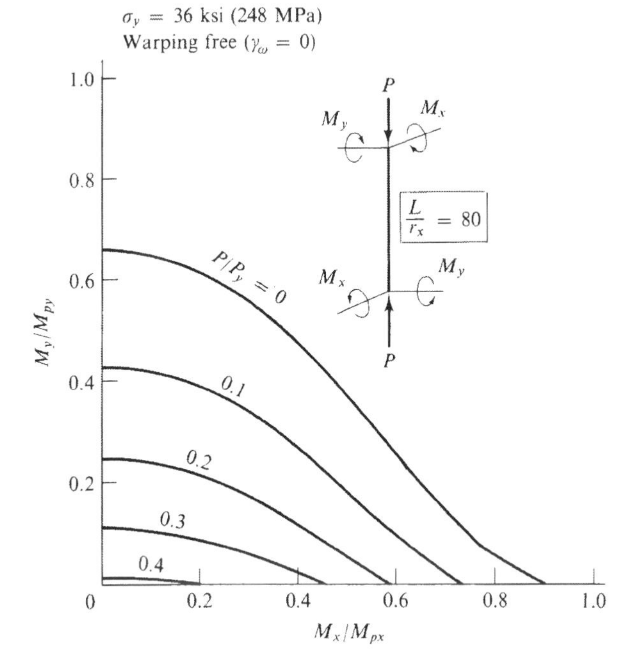

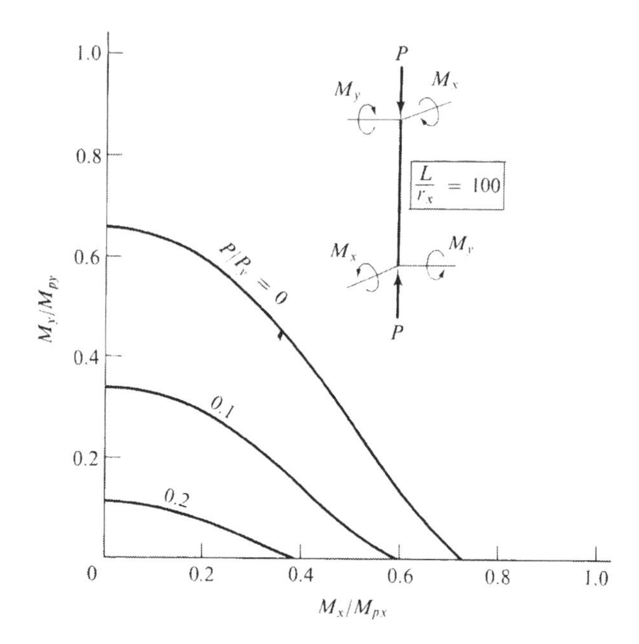

FIGURE 7.23(e)
Interaction curves of symmetrically loaded beam-column

This is a reasonable estimation because when $P_{ey} \ll P_y$, elastic buckling will occur first, thus the initial yield condition will govern ($\alpha = \beta = 1$). If $P_{ey} \gg P_y$, yielding of section will take place first and the ultimate state will be much closer to the fully plastic yield condition ($\alpha = \beta = 0$).

Using the formula [Eq. (7.55)], the interaction curves for the biaxially loaded column (W8 × 31) with different slenderness ratios were calculated where the axial force was kept constant ($P/P_y = 0.3$) as shown in Fig. 7.22. A good agreement is observed with the results reported by Santathadaporn and Chen (1973).

7.6.4 Numerical Examples

Figure 7.23 shows the numerical results for a symmetrically loaded beam-column of W8 × 31 cross section. The averaging factors, α and β are determined according to Eq. (7.55). Since the loading is symmetric, the critical section is at the mid-span ($z^* = L/2$). In such a case, a good agreement with more accurate results was usually observed (Fig. 7.22).

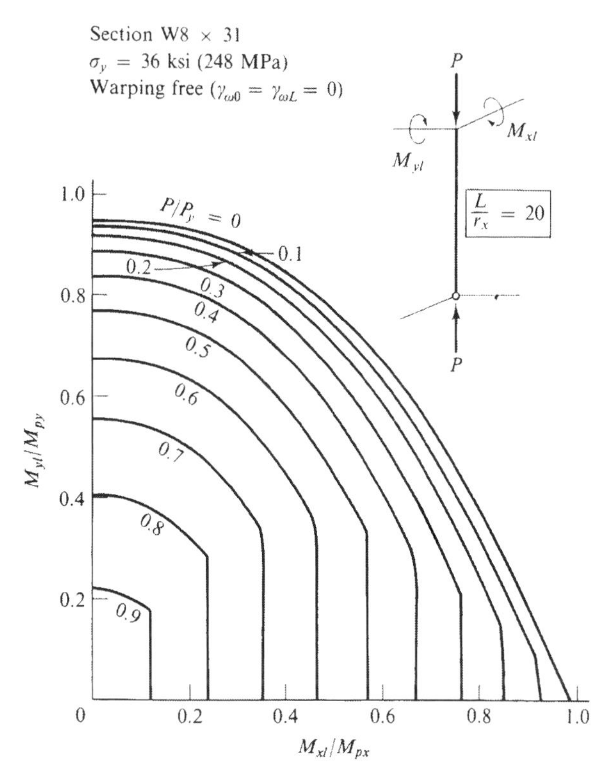

FIGURE 7.24(a)
Interaction curves of unsymmetrically loaded beam-column

Figure 7.24 shows results for an unsymmetrically loaded column with W8 × 31 section. The bending moments are applied only at the top end of the beam-column ($z = L$) and the critical section is located between the midspan and the top end ($L/2 < z^* < L$).

Harstead, Birnstiel and Leu (1968) reported experimental and theoretical analyses of biaxially loaded H-columns. Using the same specimens, Syal and Sharma (1971) calculated their ultimate strength. Details of the H-sections are listed in Table 7.1. The results of the ultimate strength are compared in Table 7.5 with the present approximate solutions. In Table 7.5, example number H1 is the only unsymmetric loading case. The present solution is seen to give good approximation comparing with the results reported by Syal and Sharma (1971).

7.7 SUMMARY

Direct solution of the governing differential equilibrium equations of a biaxially loaded beam-column in the plastic or nonlinear range is often intractable because

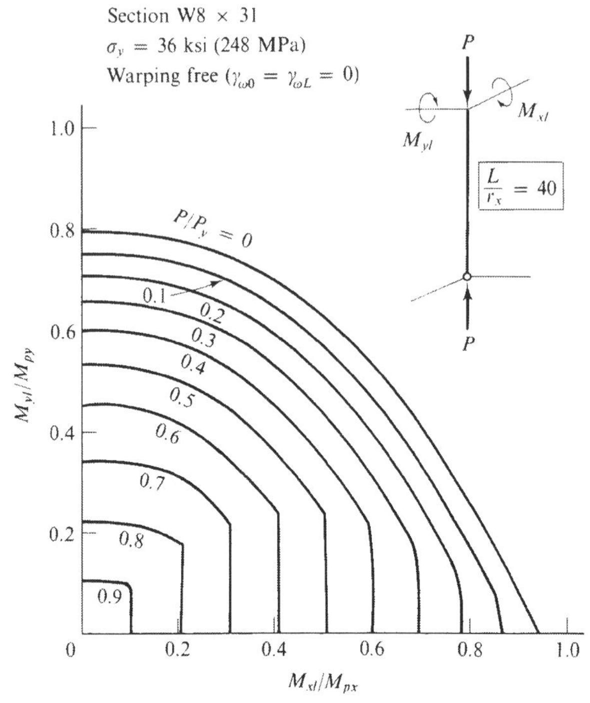

FIGURE 7.24(b)
Interaction curves of unsymmetrically loaded beam-column

Table 7.5 COMPARISON OF ULTIMATE LOAD (CHEN AND ATSUTA, 1973)

Specimen number in Table 7.1 (1)	Size (2)	Interaction constants of sections [Eq. (7.45)] a (3)	b (4)	c (5)	d (6)	Ultimate load: Experiment Birnstiel (1968) (7)	Theory Syal and Sharma (1971) (8)	Averaged interaction Eq. (7.54) (9)
1	6 × 6H	2.339	1.553	2.738	1.365	92.8	93.1	102.7
2	5 × 5H	2.372	1.539	2.660	1.308	54.1	50.75	49.4
3	5 × 5H	2.365	1.528	2.641	1.297	62.7	60.84	55.3
4	6 × 6H	2.432	1.331	2.767	1.329	86.3	84.35	84.9
5	5 × 5H	2.327	1.516	2.716	1.353	49.6	50.20	46.2
6	5 × 5H	2.327	1.516	2.716	1.353	47.9	47.76	39.6
7	6 × 5W	2.427	1.212	2.879	1.373	76.6	80.16	75.7
8	6 × 5W	2.427	1.212	2.879	1.373	109.4	110.10	110.6
10	8 × 4W	2.610	1.035	3.466	1.566	85.0	78.70	79.5
12	5 × 5H	2.365	1.491	2.667	1.314	51.0	56.23	53.3
13	4 × 4H	2.399	1.343	2.810	1.366	46.1	47.24	45.6
14	4 × 4H	2.399	1.343	2.810	1.366	38.7	40.12	44.1
B1	W12 × 79	2.427	1.313	2.689	1.282	—	150.95	155.9
H1	W14 × 43	2.566	1.070	3.128	1.392	152.0	144.19 (0.34)	143.8 (0.37)

Note: Material in parentheses indicates location of critical section from the bottom.

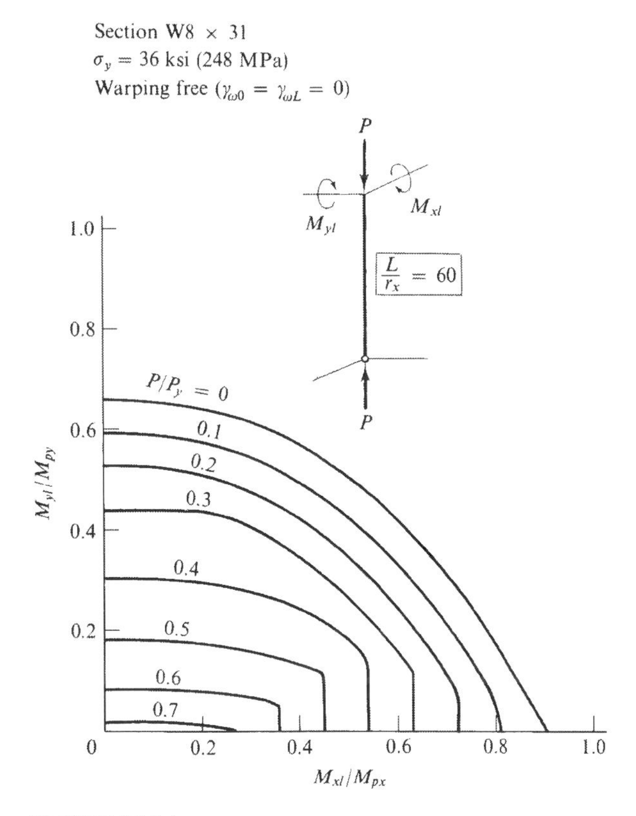

FIGURE 7.24(c)
Interaction curves of unsymmetrically loaded beam-column

of its nonlinearity and history dependency of cross section characteristics, and recourse must therefore be had to numerical methods or approximate methods to obtain solutions. All these methods are essentially based on numerical and iterative procedure, among which approximate deflection method is the most commonly used one.

Displacements in a biaxially loaded beam-column are the two perpendicular lateral deflection u, v and the twisting angle θ along the beam-column length (z-axis). In the approximate deflection method, these variables are assumed as fixed functions satisfying geometric boundary conditions and containing coefficients which represent the magnitude of deformations. The assumed deflection functions often used are

$$u = A_0 \sin \frac{\pi z}{L} \qquad \text{for simply supported beam-column}$$

$$u = A_0 \sin \frac{\pi z}{L} + A_1 z(L - z) \qquad \text{for symmetric loading}$$

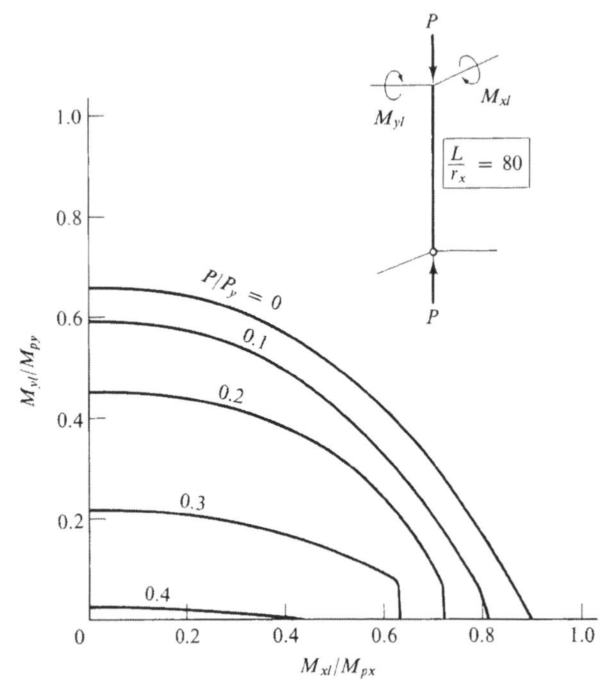

FIGURE 7.24(d)
Interaction curves of unsymmetrically loaded beam-column

$$u = A_0 \sin \frac{\pi z}{L} + z(L - z)(A_1 z - A_2) \qquad \text{for unsymmetric loading}$$

$$u = \sum A_i f_i(z) \qquad \text{for general loading case}$$

in the general loading case, some orthogonal functions are generally selected for the series function $f_i(z)$.

The coefficients of the deflection functions $\{A\}_i$ are determined from the governing equilibrium equations at the same number of stations selected along the beam-column as the number of unknown coefficients.

$$[K]_i\{A\}_i = \{F\}_i$$

Since the matrix $[K]_i$ is also a function of the deflected shape, namely, the deflection coefficient vector $\{A\}_i$, some iterative procedures must be applied to solve the deflection $\{A\}_i$. During the iteration cycles, how to adjust the current deflection vector $\{A\}_i$ for the subsequent iteration is a problem. An efficient procedure is the

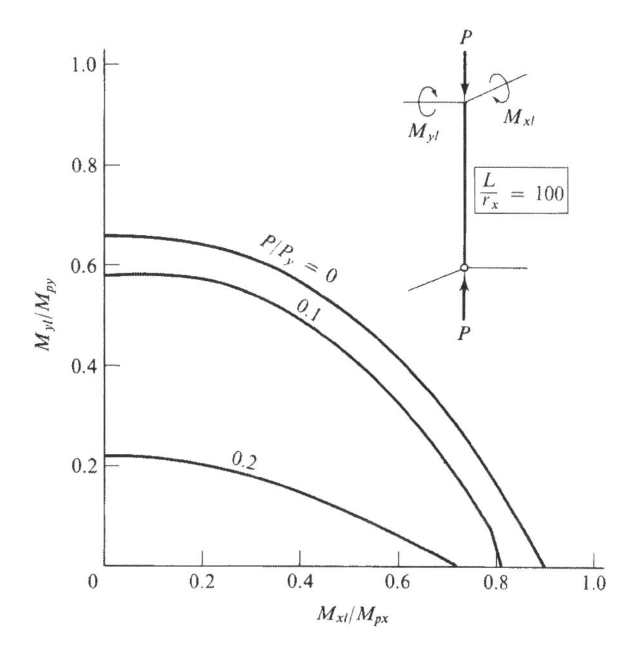

FIGURE 7.24(e)
Interaction curves of unsymmetrically loaded beam-column

use of the so-called tangent stiffness approach in which the deflection vector is adjusted so that the unbalanced force vector $\{\Delta F\}$, between the external force vector $\{F\}_{\text{ext}}$ and the internal force vector $\{F\}_{\text{int}}$, is diminished rapidly in the subsequent cycle and is reduced to within a tolerable limit in a number of cycles.

Generally, the ultimate strength of a beam-column is obtained as the peak point of a load-deflection curve and a great number of iterative cycles are required to reach this point. By the approximate deflection method, however, it is also possible to estimate directly the ultimate strength of a biaxially loaded beam-column from the viewpoint of elastic analysis and the concept of plastic hinge. In this approach, the internal forces in a beam-column calculated from an elastic analysis are assumed to satisfy a postulated yield condition $F_A = 0$ at the critical section at the ultimate state. This yield condition is bounded between the initial yield condition ($F_y = 0$) and the full plastic limiting condition ($F_p = 0$) of the cross section (Fig. 7.19).

From results of many numerical calculations, it has been found that the ultimate strength of beam-columns depends mainly on the slenderness ratio about the weak axis bending not on the cross sectional shape. Based on this information a set of maximum strength interaction curves using the slenderness ratio L/r_y have been plotted for a wide flange shape from which the ultimate strength of general H-section can be estimated (Fig. 7.17). These interaction curves are also applicable to unsymmetrically loaded steel beam-columns using the equivalent moment concept [Eq. (7.30)].

As for loading history dependency, the solutions available to date are only limited to proportional loading cases, such as an eccentrically loaded beam-column in which end moments are proportional to the axial force (Fig. 7.15), together with some very simple nonproportional loading cases.

7.8 PROBLEMS

7.1 From the equations of equilibrium for symmetric loading case, show that the biaxially loaded beam-column of circular cross section is reduced to uniaxial bending case or in-plane beam-column.

7.2 Comparing the maximum bending moment given in Eq. (7.42) to the exact expression for the in-plane beam-column, discuss the error of the approximate deflection method.

7.3 As was seen in this chapter, the assumed deflection function can be either an orthogonal function or not. Give the reason.

7.4 Determine the values of γ_1, γ_2 and γ_3 in Eq. (7.13).

7.9 REFERENCES

Atsuta, T., "Analyses of Inelastic Beam-Columns," Ph.D. Dissertation, Civil Engineering Department, Lehigh University, Bethlehem, Pennsylvania, (1972).

Austin, W. J., "Strength and Design of Metal Beam-Columns," *Journal of the Structural Division, ASCE*, Vol. 87, No. ST4, Proc. Paper 2802, April, pp. 1–29, (1961).

Birnstiel, C., "Experiments on H-Columns under Biaxial Bending," *Journal of the Structural Division, ASCE*, Vol. 94, No. ST10, Proc. Paper 6186, October, pp. 2429–2449, (1968).

Chen, W. F., "Approximate Solution of Beam-Columns," *Journal of the Structural Division, ASCE*, Vol. 97, No. ST2, Proc. Paper 7861, February, pp. 741–751, (1971).

Chen, W. F. and Atsuta, T., "Interaction Equations for Biaxially Loaded Sections," *Journal of the Structural Division, ASCE*, Vol. 98, No. ST5, Proc. Paper 8902, May, pp. 1035–1052, (1972).

Chen, W. F. and Atsuta, T., "Ultimate Strength of Biaxially Loaded Steel H-Columns," *Journal of the Structural Division, ASCE*, Vol. 99, No. ST3, Proc. Paper 9613, March, pp. 469–489, (1973).

Culver, C. G., "Exact Solution of the Biaxial Bending Equations," *Journal of the Structural Division, ASCE*, Vol. 92, No. ST2, Proc. Paper 4772, April, pp. 63–83, (1966).

Harstead, G. A., Birnstiel, C. and Leu, K. C., "Inelastic H-Columns Under Biaxial Bending," *Journal of the Structural Division, ASCE*, Vol. 94, No. ST10, Proc. Paper 6173, October, pp. 2371–2398, (1968).

Lindner, J., "Theoretical Investigations of Columns under Biaxial Loading," *Proceedings, The International Colloquium on Column Strength, IABSE-CRC-ECCS*, Paris, pp. 182–190, (1972).

Ringo, B. C., "Equilibrium Approach to the Ultimate Strength of a Biaxially Loaded Beam Column." Dissertation presented to the University of Michigan at Ann Arbor, Michigan, in partial fulfillment of the requirements for the degree of Doctor of Philosophy, (1964).

Santathadaporn, S. and Chen, W. F., "Analysis of Biaxially Loaded Steel H-Columns," *Journal of the Structural Division, ASCE*, Vol. 99, No. ST3, Proc. Paper 9621, March, pp. 491–509, (1973).

Sharma, S. S. and Gaylord, E. H., "Strength of Steel Columns with Biaxially Eccentric Load," *Journal of the Structural Division, ASCE*, Vol. 95, No. ST12, Proc. Paper 6960, December, pp. 2797–2812, (1969).

Syal, I. C. and Sharma, S. S., "Biaxially Loaded Beam-Column Analysis," *Journal of the Structural Division, ASCE*, Vol. 97, No. ST9, Proc. Paper 8384, September, pp. 2245–2259, (1971).

8

INFLUENCE COEFFICIENT METHODS FOR PLASTIC BEAM-COLUMNS

8.1 INTRODUCTION

The deflection methods described in the preceding chapter for biaxially loaded beam-columns are simple and easy to apply but the approximation is rather rough, since the deflected shapes of the member are assumed in simple functions and the equilibrium conditions are satisfied only at the mid-span for the case of symmetric loading, or at a critical point along the length of the member for the case of unsymmetrical loading. Here, an improved method called *influence coefficient method* is presented. In this method, a number of points are selected along the member and deflections are *solved* from the equilibrium conditions set up at these selected points.

The inter-relationships among the variables used in beam-column analysis are shown schematically in Fig. 8.1. The items enclosed by the circles, such as section shape, stress-strain relationship, boundary conditions and loads, are known. The variables in the rectangular boxes are unknowns but they are inter-related to each other between groups. In obtaining the solution indirectly, one of the groups of variables has to be assumed first and then the other groups of variables are determined numerically. In this process, new values for the initially assumed group of variables are re-calculated from which successively improved solutions can be obtained. The

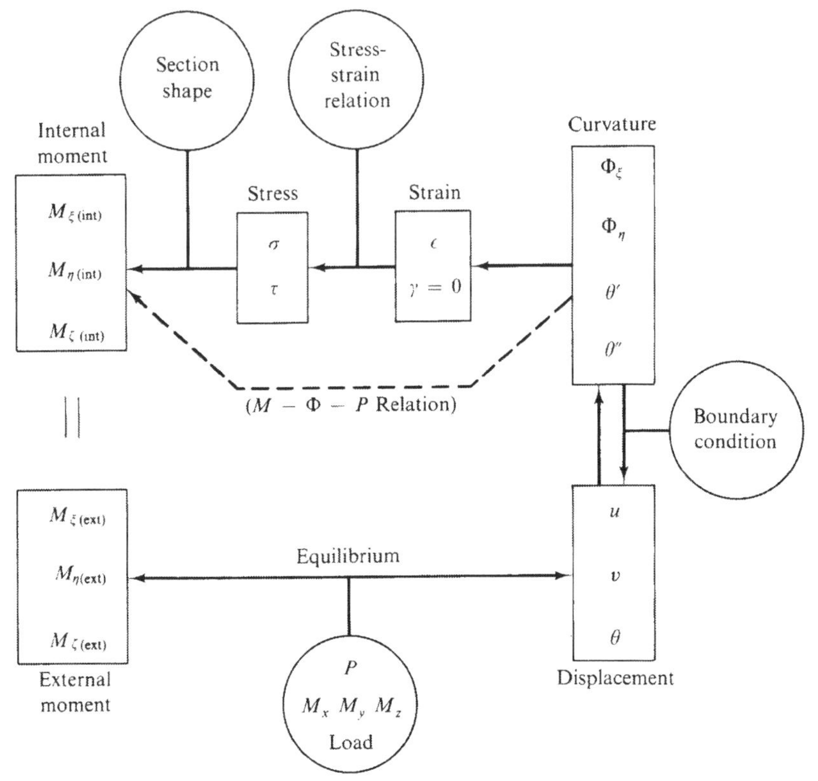

FIGURE 8.1
Cyclic relationship of variables

influence coefficient method is concerned with the development of efficient numerical procedures for this purpose.

There are two different approaches in the influence coefficient method (Fig. 8.1). One is to start from a group of variables, say $\{v_A\}$, and, after one cycle of calculation, it results in a new set of values for the same group of variables, $\{v'_A\}$. They are then compared and used to improve the initially assumed values. The other approach starts with a group of variables, $\{v_A\}$, and, after a half cycle of calculations, the values for the new variables, $\{v_B\}$, are obtained. Through the opposite route, a different set of values for the new variables, $\{v'_B\}$, is calculated. They are then compared and used to improve the previously assumed values.

In Sect. 8.2, the first approach is applied to solve a plastic steel-concrete composite beam-column. In the numerical procedure, the displacements (u_0, v_0 and θ_0) are first assumed at each point along the length of the member from which the corresponding curvatures ($\Phi_\xi, \Phi_\eta, \theta'_0$ and θ''_0), the strain (ε), the stresses (σ and τ) and the internal moments (M_ξ, M_η and M_ζ) are calculated following the upper route of Fig. 8.1. Using these internal moments, new values for the displacements ($\bar{u}, \bar{v}$ and $\bar{\theta}$) are calculated from the equilibrium conditions and the applied loads (P, M_x, M_y and M_z). However, the newly obtained displacements ($\bar{u}$, $\bar{v}$ and $\bar{\theta}$) are not used directly as the initial values for the second cycle of calculations. Rather, the modified values of displacements (u, v and θ) calculated through the *influence coefficient matrix* [ICM] equation as

$$\begin{Bmatrix} u - u_0 \\ v - v_0 \\ \theta - \theta_0 \end{Bmatrix} = \begin{bmatrix} \text{ICM} \end{bmatrix} \begin{Bmatrix} \bar{u} - u_0 \\ \bar{v} - v_0 \\ \bar{\theta} - \theta_0 \end{Bmatrix} \tag{8.1}$$

are used in order to ascertain the convergence of solution. The development of [ICM] is of major concern in Sec. 8.2.

In Sect. 8.3, the second approach is applied to solve plastic steel H-columns. In this case, values of curvatures (Φ_ξ, Φ_η, θ' and θ'') at each point along the length of the member are assumed first. The internal moments ($M_{\xi(\text{int})}$, $M_{\eta(\text{int})}$ and $M_{\zeta(\text{int})}$) are calculated through strain (ε) and stresses (σ and τ) as shown by the upper route of Fig. 8.1. Using proper boundary conditions, the assumed curvatures are also integrated numerically to obtain the displacements (u, v and θ) from which external moments ($M_{\xi(\text{ext})}$, $M_{\eta(\text{ext})}$ and $M_{\zeta(\text{ext})}$) are calculated from equilibrium conditions and the applied loads (P, M_x, M_y and M_z) as shown by the lower route of Fig. 8.1. The internal moments and the external moments at each station are compared and their differences $\{\Delta M\}$ are used to calculate the corrections $\{\Delta\Phi\}$ for the previously assumed curvatures through the influence coefficient matrix [ICM] equation as

$$\begin{Bmatrix} \Delta M_\xi \\ \Delta M_\eta \\ \Delta M_\zeta \end{Bmatrix} = \begin{bmatrix} \text{ICM} \end{bmatrix} \begin{Bmatrix} \Delta\Phi_\xi \\ \Delta\Phi_\eta \\ \Delta\theta' \\ \Delta\theta'' \end{Bmatrix} \tag{8.2}$$

Described above are general procedures for solving beam-column problems by the influence coefficient method. The actual combination of variables used in various example calculations may be different in each case for the sake of convenience in each numerical work.

The basic equations will be derived in what follows from the basic assumptions rather than taken directly from Chap. 2. This is because here we treat the beam-column not as a continuous member, but as a member represented by several discretized points (these are called stations hereafter). Thus, inter-stational relationships are needed rather than the differential equations of a continuous member. These inter-stational relationships, along with the equilibrium conditions of internal and external forces at each station, are used to develop the influence coefficient matrix of the system.

8.2 BEAM-COLUMN OF CLOSED OR SOLID CROSS SECTION

The influence coefficient method is utilized by Virdi and Dowling (1976) to obtain the ultimate strength of a pin-ended composite steel-concrete beam-column. In their analysis, all twisting effects are ignored because of the large twisting rigidity of the solid cross section. Thus, the method is quite simplified and applicable to many biaxially loaded beam-columns of solid cross sections such as reinforced concrete or steel-concrete composite sections as well as fabricated thin-walled steel tubular sections.

8.2.1 Moment-Curvature-Thrust Relationships

In order to calculate internal forces of a plastic section, the section is usually divided into finite elements and stresses and strains in each element are examined at every stage of deformation and loading as was discussed in detail in Chap. 6. In the case of steel beam-column, determination of the state of stress or strain is rather simple but in the case of general material, such as composite steel-concrete beam-column, the stress-strain characteristics of material must be considered for every stage of loading at every station along the length of beam-column for every element in the cross section. Thus, the number of cycles of repetitive calculation become very large. This is usually the most time consuming part of computer calculation. For these composite sections, the closed form moment-curvature-thrust expressions should be used to approximate the actual behavior of the cross section whenever possible. Results presented in Chap. 6 may be used for such purpose.

To obtain actual moment-curvature-thrust relationships for steel-concrete composite sections, numerical integration procedure must be applied. In order to reduce the computational time, these curvatures Φ_x and Φ_y may be combined to obtain the principal curvature $\Phi = \sqrt{\Phi_x^2 + \Phi_y^2}$.

With the neutral axis rotated by the angle

$$\theta = \tan^{-1}(\Phi_y/\Phi_x)$$

at the distance d_n from the origin O (Fig. 8.2), the strain distribution can be expressed in terms of the distance d_n as

$$\begin{aligned} \varepsilon &= \Phi(y \cos\theta - x \sin\theta - d_n) \\ &= \Phi\left(\frac{y - x(\Phi_y/\Phi_x)}{\sqrt{1 + (\Phi_y/\Phi_x)^2}} - d_n\right) \\ &= y\Phi_x - x\Phi_y - d_n\sqrt{\Phi_x^2 + \Phi_y^2} \end{aligned}$$

Using the stress-strain relations of material, the stress σ corresponding to the strain ε, thus, the resultant forces

$$\begin{aligned} N &= \int \sigma \, dA \\ M_x &= \int \sigma y \, dA \\ M_y &= -\int \sigma x \, dA \end{aligned} \tag{8.3}$$

are obtained.

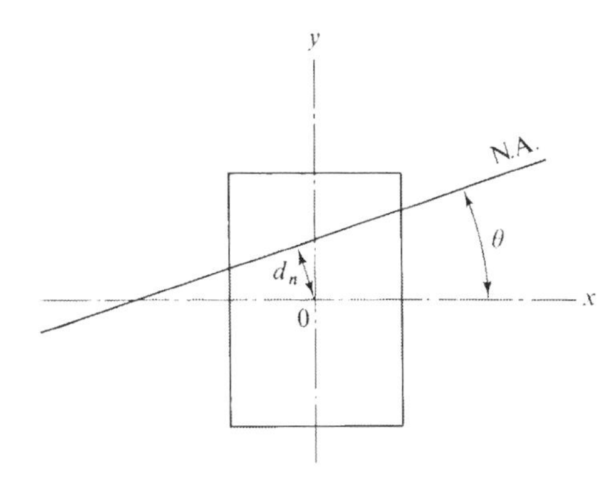

FIGURE 8.2
Neutral axis of principal curvature

Since the axial compression $P(= -N)$ is generally applied first and held constant, the first equation in (8.3) can be used to express the distance of the neutral axis location d_n in terms of P, Φ_x and Φ_y. The biaxial bending moment M_x and M_y are then obtained in terms of the biaxial curvatures Φ_x, Φ_y and the axial thrust, P

$$\begin{aligned} M_x &= M_x(\Phi_x, \Phi_y, P) \\ M_y &= M_y(\Phi_x, \Phi_y, P) \end{aligned} \tag{8.4}$$

These are the moment-curvature-thrust relationships which must be established numerically for every stage of computation.

8.2.2 Mathematical Formulation

Consider a beam-column of length L which is subjected to an axial compression P and biaxial end moments M_{xA}, M_{yA} at end A and M_{xB}, M_{yB} at end B (Fig. 8.3). Along the length of the beam-column we assign the stations 1 to $n+1$ from end A to end B with a constant spacing λ. We denote the lateral displacements of the centroid of the cross section at station i by u_i and v_i. The equilibrium condition at station i can be expressed by

$$\begin{aligned} M_{xi} &= Pv_i + \left(1 - \frac{i-1}{n}\right)M_{xA} + \frac{i-1}{n}M_{xB} \\ M_{yi} &= -Pu_i + \left(1 - \frac{i-1}{n}\right)M_{yA} + \frac{i-1}{n}M_{yB} \end{aligned} \tag{8.5}$$

from which the deflections u_i and v_i are solved

$$\begin{aligned} u_i &= -\frac{1}{P}\left[M_{yi} - \left(1 - \frac{i-1}{n}\right)M_{yA} - \frac{i-1}{n}M_{yB}\right] \\ v_i &= \frac{1}{P}\left[M_{xi} - \left(1 - \frac{i-1}{n}\right)M_{xA} - \frac{i-1}{n}M_{xB}\right] \end{aligned} \tag{8.6}$$

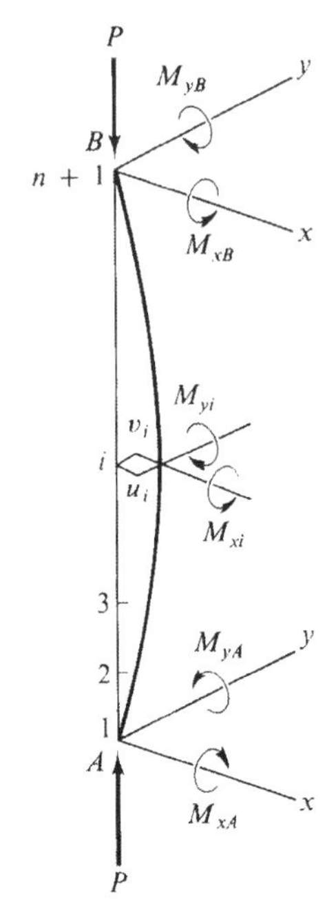

FIGURE 8.3
Biaxially loaded beam-column

The internal moments M_{xi} and M_{yi} are calculated next from the moment-curvature-thrust relations

$$
\begin{aligned}
M_{xi} &= M_x(\Phi_{xi}, \Phi_{yi}, P) \\
M_{yi} &= M_y(\Phi_{xi}, \Phi_{yi}, P)
\end{aligned}
\tag{8.7}
$$

In which the axial thrust P has a constant value and the curvatures are calculated from the assumed deflections by

$$
\begin{aligned}
\Phi_{xi} &= -\left.\frac{\partial^2 v}{\partial z^2}\right|_i = -\frac{1}{\lambda^2}(v_{i-1} - 2v_i + v_{i+1}) \\
\Phi_{yi} &= \left.\frac{\partial^2 u}{\partial z^2}\right|_i = \frac{1}{\lambda^2}(u_{i-1} - 2u_i + u_{i+1})
\end{aligned}
\tag{8.8}
$$

The end moments M_{xA}, M_{yA}, M_{xB} and M_{yB} are related to the end slopes θ_{xA}, θ_{yA}, θ_{xB} and θ_{yB} by the end constraint functions $\overline{M}_{xA}$, $\overline{M}_{yA}$, $\overline{M}_{xB}$ and $\overline{M}_{yB}$

$$
\begin{aligned}
M_{xA} &= \tilde{M}_{xA} - \overline{M}_{xA}(\theta_{xA}), \quad M_{xB} = \tilde{M}_{xB} - \overline{M}_{xB}(\theta_{xB}) \\
M_{yA} &= \tilde{M}_{yA} - \overline{M}_{yA}(\theta_{yA}), \quad M_{yB} = \tilde{M}_{yB} - \overline{M}_{yB}(\theta_{yB})
\end{aligned}
\tag{8.9}
$$

in which $\tilde{M}_{xA}$, $\tilde{M}_{yA}$, $\tilde{M}_{xB}$ and $\tilde{M}_{yB}$ are externally applied end moments. The applied moments may be produced by the end eccentricities of thrust P. In the case shown in Fig. 8.4, for example, we have

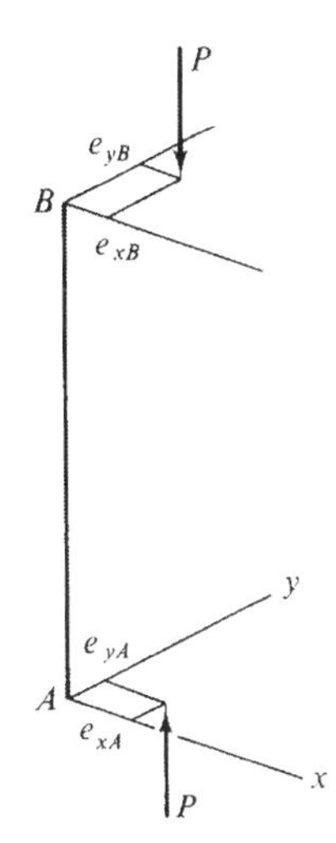

FIGURE 8.4
Eccentric loading on beam-column

$$\tilde{M}_{xA} = -Pe_{yA}, \quad \tilde{M}_{xB} = -Pe_{yB}$$
$$\tilde{M}_{yA} = Pe_{xA}, \quad \tilde{M}_{yB} = Pe_{xB} \tag{8.10}$$

The nonlinear moment-rotation relationships for end moments, as shown schematically in Fig. 8.5, are assumed known. For example, in the special case when the ends are supported by linear elastic springs with spring constants k's, we have the simple relations,

$$\bar{M}_{xA} = -k_{xA}\theta_{xA}, \quad \bar{M}_{xB} = k_{xB}\theta_{xB}$$
$$\bar{M}_{yA} = -k_{yA}\theta_{yA}, \quad \bar{M}_{yB} = k_{yB}\theta_{yB} \tag{8.11}$$

Further, if the ends are pin-connected, all these end restraint functions are zero. The end rotation angles are calculated by

FIGURE 8.5
Moment slope relation of end

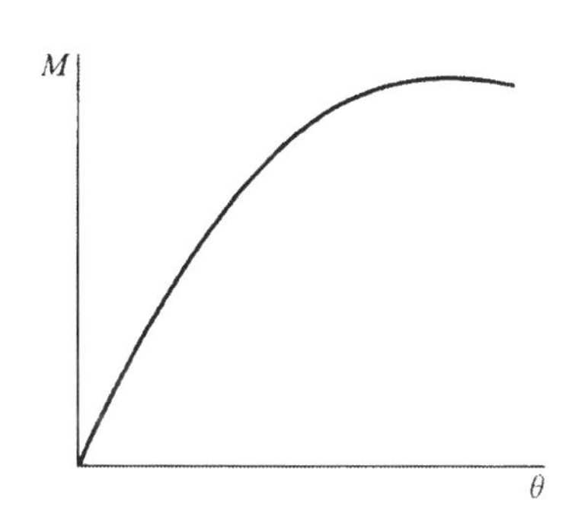

$$\theta_{xA} = -\left.\frac{\partial v}{\partial z}\right|_A = -\frac{1}{2\lambda}(v_3 - 4v_2), \quad \theta_{xB} = -\left.\frac{\partial v}{\partial z}\right|_B = \frac{1}{2\lambda}(v_{n-1} - 4v_n)$$

$$\theta_{yA} = \left.\frac{\partial u}{\partial z}\right|_A = \frac{1}{2\lambda}(u_3 - 4u_2), \quad \theta_{yB} = \left.\frac{\partial u}{\partial z}\right|_B = -\frac{1}{2\lambda}(u_{n-1} - 4u_n) \tag{8.12}$$

All the terms in the right hand side of Eq. (8.6) can now be determined using the assumed deflections. Note that for a beam-column with initial deflections, only the additional deflections produced by loads are used for the right hand side of Eq. (8.6) and the resulting new deflections in the left hand side are, however, the total deflections including the initial deflections. Equation (8.6) may now be written in the form

$$u_i = U_i(u_1 \ldots u_{n+1}; \quad v_1 \ldots v_{n+1})$$

$$v_i = V_i(u_1 \ldots u_{n+1}; \quad v_1 \ldots v_{n+1}) \tag{8.13}$$

Denoting the deflections at all stations by

$$\{w\} = \{u_1, u_2, \ldots u_{n+1}; \quad v_1, v_2, \ldots v_{n+1}\} \tag{8.14}$$

Equation (8.13) implies that the current deflection at station i is a function of previous deflections at all stations

$$w_i = W_i[\{w\}] \tag{8.15}$$

this relation corresponds to a cycle of calculation in Fig. 8.1 starting from an assumed set of displacements and ending up with a new set of displacement values.

This type of iteration problem as given in Eq. (8.15) may be readily solved by the application of Newton-Raphson iterative technique. This is described in the following. Supposing an approximate deflection $\{w^k\}$ has been obtained after the k-th cycle of iteration. By substituting this deflection in the right hand side of Eq. (8.15), we obtain a new value of deflection at station i

$$\bar{w}_i = W_i[\{w^k\}] \tag{8.16}$$

If the two deflections $\{w^k\}$ *and* $\{\bar{w}\}$ are very close, then the deflection $\{\bar{w}\}$ is considered to be the correct value. However, if significant differences exist between the two, $\{w^k - \bar{w}\}$, then the new value for the $(k+1)$-th iteration, $\{w^{k+1}\}$, is to be calculated by

$$\{w^{k+1}\} = \{w^k\} - ([I] - [J])^{-1}\{w^k - \bar{w}\} \tag{8.17}$$

in which $[I]$ is a unit matrix and $[J]$ is called influence coefficient matrix defined by

$$\{d\bar{w}\} = [J]\{dw\} \tag{8.18}$$

$[J]$ is also known as a Jacobian matrix whose element is given by

$$J_{ij} = \frac{\partial w_i}{\partial w_j} \tag{8.19}$$

Using the deflections given in Eq. (8.13), the influence coefficient matrix Eq. (8.18) may be written in the form as

$$\left\{\begin{array}{c} d\bar{u}_2 \\ d\bar{u}_i \\ d\bar{u}_n \\ \hline d\bar{v}_2 \\ d\bar{v}_i \\ d\bar{v}_n \end{array}\right\} = \left[\begin{array}{ccccccc|ccccccc} \frac{\partial U_2}{\partial u_2} & \frac{\partial U_2}{\partial u_3} & & & & & & \frac{\partial U_2}{\partial v_2} & \frac{\partial U_2}{\partial v_3} & & & & & \\ & & \frac{\partial U_i}{\partial u_{j-1}} & \frac{\partial U_i}{\partial u_j} & \frac{\partial U_i}{\partial u_{j+1}} & & & & & \frac{\partial U_i}{\partial v_{j-1}} & \frac{\partial U_i}{\partial v_j} & \frac{\partial U_i}{\partial v_{j+1}} & & \\ & & & & & \frac{\partial U_n}{\partial u_{n-1}} & \frac{\partial U_n}{\partial u_n} & & & & & & \frac{\partial U_n}{\partial v_{n-1}} & \frac{\partial U_n}{\partial v_n} \\ \hline \frac{\partial V_2}{\partial u_2} & \frac{\partial V_2}{\partial u_3} & & & & & & \frac{\partial V_2}{\partial v_2} & \frac{\partial V_2}{\partial v_3} & & & & & \\ & & \frac{\partial V_i}{\partial u_{j-1}} & \frac{\partial V_i}{\partial u_j} & \frac{\partial V_i}{\partial u_{j+1}} & & & & & \frac{\partial V_i}{\partial v_{j-1}} & \frac{\partial V_i}{\partial v_j} & \frac{\partial V_i}{\partial v_{j+1}} & & \\ & & & & & \frac{\partial V_n}{\partial u_{n-1}} & \frac{\partial V_n}{\partial u_n} & & & & & & \frac{\partial V_n}{\partial v_{n-1}} & \frac{\partial V_n}{\partial v_n} \end{array}\right] \left\{\begin{array}{c} du_2 \\ du_j \\ du_n \\ \hline dv_2 \\ dv_j \\ dv_n \end{array}\right\} \tag{8.20}$$

in which the deflections at both ends of the member are deleted because they are always zero

$$u_1 = v_1 = u_{n+1} = v_{n+1} = 0 \tag{8.21}$$

8.2.3 Elements of Influence Coefficient Matrix

The elements of the influence coefficient matrix can be obtained from Eq. (8.6) as

$$\begin{aligned}
\frac{\partial U_i}{\partial u_j} &= -\frac{1}{P}\left[\frac{\partial M_{yi}}{\partial u_j} - \left(1 - \frac{i-1}{n}\right)\frac{\partial M_{yA}}{\partial u_j} - \frac{i-1}{n}\frac{\partial M_{yB}}{\partial u_j}\right] \\
\frac{\partial U_i}{\partial v_j} &= -\frac{1}{P}\left[\frac{\partial M_{yi}}{\partial v_j} - \left(1 - \frac{i-1}{n}\right)\frac{\partial M_{yA}}{\partial v_j} - \frac{i-1}{n}\frac{\partial M_{yB}}{\partial v_j}\right] \\
\frac{\partial V_i}{\partial u_j} &= \frac{1}{P}\left[\frac{\partial M_{xi}}{\partial u_j} - \left(1 - \frac{i-1}{n}\right)\frac{\partial M_{xA}}{\partial u_j} - \frac{i-1}{n}\frac{\partial M_{xB}}{\partial u_j}\right] \\
\frac{\partial V_i}{\partial v_j} &= \frac{1}{P}\left[\frac{\partial M_{xi}}{\partial v_j} - \left(1 - \frac{i-1}{n}\right)\frac{\partial M_{xA}}{\partial v_j} - \frac{i-1}{n}\frac{\partial M_{xB}}{\partial v_j}\right]
\end{aligned} \tag{8.22}$$

The first term in the right hand side of Eq. (8.22) can be solved using the moment-curvature-thrust relationship in Eq. (8.7). Since M_{xi} is a function of the curvatures Φ_{xi} and Φ_{yi},

$$\frac{\partial M_{xi}}{\partial u_j} = \frac{\partial M_{xi}}{\partial \Phi_{xi}}\frac{\partial \Phi_{xi}}{\partial u_j} + \frac{\partial M_{xi}}{\partial \Phi_{yi}}\frac{\partial \Phi_{yi}}{\partial u_j} \tag{8.23}$$

Substituting the curvature expression of Eq. (8.8), we find $\partial \Phi_{xi}/\partial u_j = 0$ and hence Eq. (8.23) can be expressed in the form

$$\begin{aligned}
\frac{\partial M_{xi}}{\partial u_j} &= \frac{\partial M_{xi}}{\partial \Phi_{yi}}\left(\frac{1}{\lambda^2}\right)\frac{\partial}{\partial u_j}(u_{i-1} - 2u_i + u_{i+1}) \\
&= \frac{1}{\lambda^2}\frac{\partial M_{xi}}{\partial \Phi_{yi}}\begin{cases} 0 & (j < i-1) \\ 1 & (j = i-1) \\ -2 & (j = i) \\ 1 & (j = i+1) \\ 0 & (j > i+1) \end{cases}
\end{aligned} \tag{8.24}$$

which implies that there are only three non-zero terms, i.e.,

$$\frac{\partial M_{xi}}{\partial u_{i-1}} = -\frac{1}{2}\frac{\partial M_{xi}}{\partial u_i} = \frac{\partial M_{xi}}{\partial u_{i+1}} = \frac{1}{\lambda^2}\frac{\partial M_{xi}}{\partial \Phi_{yi}}$$

Similarly,

$$\begin{aligned}
\frac{\partial M_{xi}}{\partial v_{i-1}} &= -\frac{1}{2}\frac{\partial M_{xi}}{\partial v_i} = \frac{\partial M_{xi}}{\partial v_{i+1}} = -\frac{1}{\lambda^2}\frac{\partial M_{xi}}{\partial \Phi_{xi}} \\
\frac{\partial M_{yi}}{\partial u_{i-1}} &= -\frac{1}{2}\frac{\partial M_{yi}}{\partial u_i} = \frac{\partial M_{yi}}{\partial u_{i+1}} = \frac{1}{\lambda^2}\frac{\partial M_{yi}}{\partial \Phi_{yi}} \\
\frac{\partial M_{yi}}{\partial v_{i-1}} &= -\frac{1}{2}\frac{\partial M_{yi}}{\partial v_i} = \frac{\partial M_{yi}}{\partial v_{i+1}} = -\frac{1}{\lambda^2}\frac{\partial M_{yi}}{\partial \Phi_{xi}}
\end{aligned} \tag{8.25}$$

The last two terms in the right hand side of Eq. (8.22) can be solved using the end constraint function given in Eq. (8.9),

$$\begin{aligned}
\frac{\partial M_{xA}}{\partial u_j} &= -\frac{\partial \bar{M}_{xA}}{\partial \theta_{xA}}\frac{\partial \theta_{xA}}{\partial u_j}, \quad \frac{\partial M_{xB}}{\partial u_j} = -\frac{\partial \bar{M}_{xB}}{\partial \theta_{xB}}\frac{\partial \theta_{xB}}{\partial u_j} \\
\frac{\partial M_{xA}}{\partial v_j} &= -\frac{\partial \bar{M}_{xA}}{\partial \theta_{xA}}\frac{\partial \theta_{xA}}{\partial v_j}, \quad \frac{\partial M_{xB}}{\partial v_j} = -\frac{\partial \bar{M}_{xB}}{\partial \theta_{xB}}\frac{\partial \theta_{xB}}{\partial v_j} \\
\frac{\partial M_{yA}}{\partial u_j} &= -\frac{\partial \bar{M}_{yA}}{\partial \theta_{yA}}\frac{\partial \theta_{yA}}{\partial u_j}, \quad \frac{\partial M_{yB}}{\partial u_j} = -\frac{\partial \bar{M}_{yB}}{\partial \theta_{yB}}\frac{\partial \theta_{yB}}{\partial u_j} \\
\frac{\partial M_{yA}}{\partial v_j} &= -\frac{\partial \bar{M}_{yA}}{\partial \theta_{yA}}\frac{\partial \theta_{yA}}{\partial v_j}, \quad \frac{\partial M_{yB}}{\partial v_j} = -\frac{\partial \bar{M}_{yB}}{\partial \theta_{yB}}\frac{\partial \theta_{yB}}{\partial v_j}
\end{aligned} \tag{8.26}$$

Substituting the end slope expressions in Eq. (8.12) into Eq. (8.26), we have

$$\frac{\partial M_{xA}}{\partial u_j} = \frac{\partial M_{xB}}{\partial u_j} = \frac{\partial M_{yA}}{\partial v_j} = \frac{\partial M_{yB}}{\partial v_j} = 0 \tag{8.27}$$

and

$$\begin{aligned}
\frac{\partial M_{xA}}{\partial v_j} &= -\frac{\partial \bar{M}_{xA}}{\partial \theta_{xA}}\left(-\frac{1}{2\lambda}\right)\frac{\partial}{\partial v_j}(v_3 - 4v_2) \\
&= \frac{1}{2\lambda}\frac{\partial \bar{M}_{xA}}{\partial \theta_{xA}}\begin{cases} -4 & (j = 2) \\ 1 & (j = 3) \\ 0 & (j > 3) \end{cases}
\end{aligned} \tag{8.28}$$

$$\begin{aligned}
\frac{\partial M_{xB}}{\partial v_j} &= -\frac{\partial \bar{M}_{xB}}{\partial \theta_{xB}}\left(\frac{1}{2\lambda}\right)\frac{\partial}{\partial v_j}(v_{n-1} - 4v_n) \\
&= -\frac{1}{2\lambda}\frac{\partial \bar{M}_{xB}}{\partial \theta_{xB}}\begin{cases} 0 & (j < n-1) \\ 1 & (j = n-1) \\ -4 & (j = n) \end{cases}
\end{aligned} \tag{8.29}$$

which implies that only two terms are non-zero

$$\begin{aligned}
-\frac{1}{4}\frac{\partial M_{xA}}{\partial v_2} &= \frac{\partial M_{xA}}{\partial v_3} = \frac{1}{2\lambda}\frac{\partial \bar{M}_{xA}}{\partial \theta_{xA}} \\
\frac{1}{4}\frac{\partial M_{xB}}{\partial v_n} &= -\frac{\partial M_{xB}}{\partial v_{n-1}} = \frac{1}{2\lambda}\frac{\partial \bar{M}_{xB}}{\partial \theta_{xB}}
\end{aligned} \tag{8.30}$$

Similarly,

$$
\begin{aligned}
\frac{1}{4}\frac{\partial M_{yA}}{\partial u_2} &= -\frac{\partial M_{yA}}{\partial u_3} = \frac{1}{2\lambda}\frac{\partial \overline{M}_{yA}}{\partial \theta_{yA}} \\
-\frac{1}{4}\frac{\partial M_{yB}}{\partial u_n} &= \frac{\partial M_{yB}}{\partial u_{n-1}} = \frac{1}{2\lambda}\frac{\partial \overline{M}_{yB}}{\partial \theta_{yB}}
\end{aligned}
\tag{8.31}
$$

From the above derivation, it is known that the non-zero elements of the influence coefficient matrix are not many and they can be evaluated numerically from moment-curvature-thrust relations of cross section and moment-rotation relations at ends of the member. If we denote

$$
\begin{aligned}
\alpha_{xxi} &= \frac{\partial M_{xi}}{\partial \Phi_{xi}}, \qquad \alpha_{xyi} = \frac{\partial M_{xi}}{\partial \Phi_{yi}} \\
\alpha_{yxi} &= \frac{\partial M_{yi}}{\partial \Phi_{xi}}, \qquad \alpha_{yyi} = \frac{\partial M_{yi}}{\partial \Phi_{yi}}
\end{aligned}
\tag{8.32}
$$

and

$$
\begin{aligned}
\beta_{xA} &= \frac{\partial \overline{M}_{xA}}{\partial \theta_{xA}}, \qquad \beta_{xB} = \frac{\partial \overline{M}_{xB}}{\partial \theta_{xB}} \\
\beta_{yA} &= \frac{\partial \overline{M}_{yA}}{\partial \theta_{yA}}, \qquad \beta_{yB} = \frac{\partial \overline{M}_{yB}}{\partial \theta_{yB}}
\end{aligned}
\tag{8.33}
$$

which correspond to bending rigidities of the cross section and spring constants of the end constraint of the member, respectively, then the influence coefficient matrix

$$
\{d\overline{w}\} = [J]\{dw\} \qquad \text{or} \qquad \begin{Bmatrix} d\bar{u} \\ d\bar{v} \end{Bmatrix} = \begin{bmatrix} \dfrac{\partial U}{\partial u} & \dfrac{\partial U}{\partial v} \\ \dfrac{\partial V}{\partial u} & \dfrac{\partial V}{\partial v} \end{bmatrix} \begin{Bmatrix} du \\ dv \end{Bmatrix}
\tag{8.34}
$$

defined in Eq. (8.34) are determined from Eq. (8.20) by

$$
\left[\frac{\partial U}{\partial u}\right] = -\frac{1}{P\lambda^2}
$$

$$
\begin{bmatrix}
\alpha_{yy2} & & & & & (0) \\
& \alpha_{yy3} & & & & \\
& & \cdot\ \cdot & & & \\
& & & \alpha_{yyi} & & \\
& & & & \alpha_{yy(n-1)} & \\
(0) & & & & & \alpha_{yyn}
\end{bmatrix}
\begin{bmatrix}
-2 & 1 & & & & (0) \\
1 & -2 & 1 & & & \\
& \cdot & & & & \\
& & 1 & -2 & 1 & \\
& & & \cdot & \cdot & \\
& & & 1 & -2 & 1 \\
(0) & & & & 1 & -2
\end{bmatrix}
-\frac{1}{2P\lambda}
$$

$$\begin{bmatrix} 4\left(1-\frac{1}{n}\right) & -\left(1-\frac{1}{n}\right) & -\frac{1}{n} & 4\frac{1}{n} \\ 4\left(1-\frac{2}{n}\right) & -\left(1-\frac{2}{n}\right) & -\frac{2}{n} & 4\frac{2}{n} \\ \cdot & \cdot & \cdot & \cdot \\ 4\left(1-\frac{i-1}{n}\right) & -\left(1-\frac{i-1}{n}\right) \; (0) & -\frac{i-1}{n} & 4\frac{i-1}{n} \\ \cdot & & \cdot & \\ 4\left(1-\frac{n-2}{n}\right) & -\left(1-\frac{n-2}{n}\right) & -\frac{n-2}{n} & 4\frac{n-2}{n} \\ 4\left(1-\frac{n-1}{n}\right) & -\left(1-\frac{n-1}{n}\right) & -\frac{n-1}{n} & 4\frac{n-1}{n} \end{bmatrix} \begin{bmatrix} \beta_{yA} & & & & & & (0) \\ & \beta_{yA} & & & & & \\ & & (0) & & & & \\ & & & \cdot & & & \\ & & & & (0) & & \\ & & & & & \beta_{yB} & \\ (0) & & & & & & \beta_{yB} \end{bmatrix} \tag{8.35}$$

$$\left[\frac{\partial U}{\partial v}\right] = \frac{1}{P\lambda^2}$$

$$\begin{bmatrix} \alpha_{yx2} & & & & & (0) \\ & \alpha_{yx3} & & & & \\ & & \cdot & & & \\ & & & \alpha_{yxi} & & \\ & & & & \alpha_{yx(n-1)} & \\ (0) & & & & & \alpha_{yxn} \end{bmatrix} \begin{bmatrix} -2 & 1 & & & & & (0) \\ 1 & -2 & 1 & & & & \\ & & \cdot & & & & \\ & & 1 & -2 & 1 & & \\ & & & & \cdot & & \\ & & & & 1 & -2 & 1 \\ (0) & & & & & 1 & -2 \end{bmatrix} \tag{8.36}$$

$$\left[\frac{\partial V}{\partial u}\right] = \frac{1}{P\lambda^2}$$

$$\begin{bmatrix} \alpha_{xy2} & & & & & (0) \\ & \alpha_{xy3} & & & & \\ & & \cdot & & & \\ & & & \alpha_{xyi} & & \\ & & & & \cdot & \\ & & & & \alpha_{xy(n-1)} & \\ (0) & & & & & \alpha_{xyn} \end{bmatrix} \begin{bmatrix} -2 & 1 & & & & & (0) \\ 1 & -2 & 1 & & & & \\ & \cdot & \cdot & & & & \\ & & 1 & -2 & 1 & & \\ & & & \cdot & \cdot & & \\ & & & & 1 & -2 & 1 \\ (0) & & & & & 1 & -2 \end{bmatrix} \tag{8.37}$$

$$
\left[\frac{\partial V}{\partial v}\right] = -\frac{1}{P\lambda^2}
\begin{bmatrix}
\alpha_{xx2} & & & & & (0) \\
& \alpha_{xx3} & & & & \\
& & \cdot & & & \\
& & & \alpha_{xxi} & & \\
& & & & \cdot & \\
& & & & \alpha_{xx(n-1)} & \\
(0) & & & & & \alpha_{xxn}
\end{bmatrix}
\begin{bmatrix}
-2 & 1 & & & & & (0) \\
1 & -2 & 1 & & & & \\
& \cdot & \cdot & & & & \\
& & 1 & -2 & 1 & & \\
& & & \cdot & & & \\
& & & & 1 & -2 & 1 \\
(0) & & & & & 1 & -2
\end{bmatrix}
- \frac{1}{2P\lambda}
$$

$$
\begin{bmatrix}
4\left(1-\frac{1}{n}\right) & -\left(1-\frac{1}{n}\right) & & -\frac{1}{n} & 4\frac{1}{n} \\
4\left(1-\frac{2}{n}\right) & -\left(1-\frac{2}{n}\right) & & -\frac{2}{n} & 4\frac{2}{n} \\
\cdot & & & \cdot & \\
4\left(1-\frac{i-1}{n}\right) & -\left(1-\frac{i-1}{n}\right) & (0) & -\frac{i-1}{n} & 4\frac{i-1}{n} \\
\cdot & \cdot & & & \cdot \\
4\left(1-\frac{n-2}{n}\right) & -\left(1-\frac{n-2}{n}\right) & & -\frac{n-2}{n} & 4\frac{n-2}{n} \\
4\left(1-\frac{n-1}{n}\right) & -\left(1-\frac{n-1}{n}\right) & & -\frac{n-1}{n} & 4\frac{n-1}{n}
\end{bmatrix}
\begin{bmatrix}
\beta_{xA} & & & & & (0) \\
& \beta_{xA} & & & & \\
& & (0) & & & \\
& & & (0) & & \\
& & & & \beta_{xB} & \\
(0) & & & & & \beta_{xB}
\end{bmatrix}
\tag{8.38}
$$

8.2.4 Computation Procedure

The actual computations of load deflection behavior of biaxially loaded beam-column is to be carried out by the following steps.

Step 1 Read in initial deflections $\{u_0\}$ and $\{v_0\}$.

Step 2 Read in the initial value of applies loads $\{P, \tilde{M}_{xA}, \tilde{M}_{yA}, \tilde{M}_{xB}, \tilde{M}_{yB}\}$.

Step 3 Assume trial values of deflection $\{u\}$ and $\{v\}$.

Step 4 Compute end slopes, θ_A, θ_B [Eq. (8.12)] and effective end moments $M_{xA}, M_{xB}, M_{yA}, M_{yB}$ [Eq. (8.9)].

Step 5 Start with station $i = 2$.

Step 6 Compute net curvatures Φ_{xi} and Φ_{yi} [Eq. (8.8)].

Step 7 Compute internal bending moments M_{xi} and M_{yi} from the moment-curvature-thrust relations [Eq. (8.7)].

Step 8 Compute deflections $\bar{u}_i$ and $\bar{v}_i$ [Eq. (8.6)].

Step 9 Repeat steps 6 to 8 for every station $i = 2$ to n.

Step 10 If the solutions are close to the previously obtained or assumed values, that is,

$$\sum_{i=1}^{n} \left[(\bar{u}_i - u_i)^2 + (\bar{v}_i - v_i)^2\right] < \varepsilon,$$

a small quantity, the deflections have been found and proceed to step 15, otherwise,

Step 11 Change the value of curvatures Φ_{xi}, Φ_{yi} by a small amount and obtain the tangent rigidities α's from moment-curvature-thrust relations [Eq. (8.32)].

Step 12 Obtain the tangent spring constants β's using the current slope angles $\theta_{xA}, \theta_{yA}, \theta_{xB}, \theta_{yB}$ [Eq. (8.33)].

Step 13 Construct the influence coefficient matrix $[J]$ from Eqs. (8.35) to (8.38).

Step 14 Solve Eq. (8.17) for the new deflections $\{w\} = \{u, v\}$ and repeat steps 4 to 13 until the convergence criterion in step 10 is satisfied. If convergence is not obtained within a specified number of cycles, go to step 16.

Step 15 Increase the loads $\{P, \tilde{M}_{xA}, \tilde{M}_{yA}, \tilde{M}_{xB}, \tilde{M}_{yB}\}$ and select a trial value for deflection by an extrapolation of previously obtained results. Repeat steps 4 to 14.

Step 16 Decrease the loads $\{P, \tilde{M}_{xA}, \tilde{M}_{yA}, \tilde{M}_{xB}, \tilde{M}_{yB}\}$ and repeat steps 4 to 14 until the load increment becomes smaller than a specified small value. The last values of loads obtained in step 16 are the ultimate strength of the beam-column.

8.2.5 Numerical Results

Utilizing the method described here, Virdi and Dowling (1976) solved the composite steel-concrete beam-column problem shown in Fig. 8.6 in which a concrete encased steel I section in biaxial bending was considered.

They examined first the speed of convergence and the accuracy of solution for the composite steel-concrete problem; even though this is a second order nonlinear analysis problem, they found that the convergence of this problem is obtained extremely rapidly. This is because the twisting effect under biaxial bending for the composite section can be neglected. Further, the elimination of calculations for zero terms in the influence coefficient matrix also contributes to the speed of convergence. The most time consuming part of the computer solution is the generation of moment-curvature-thrust relationships for the composite section. If approximate

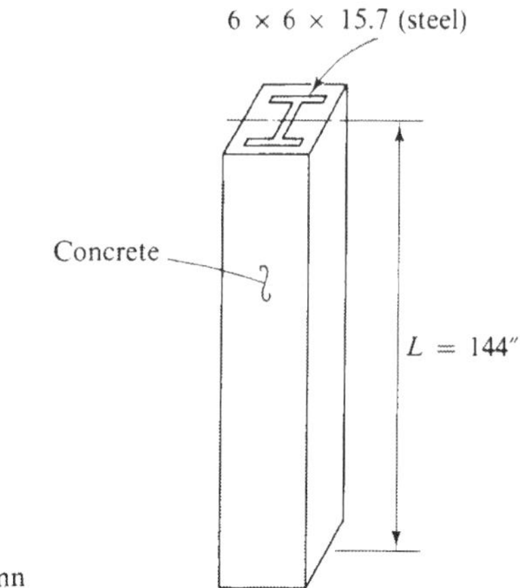

FIGURE 8.6
Composite steel-concrete beam-column

closed form expressions for moment-curvature-thrust relationships can be developed and used, computer time will be considerably reduced.

Accuracy of the solution depends, of course, on the fineness of element used for cross section, number of stations along the length of beam-column and incremental values of applied loads. The influence of the number of stations on the accuracy of the solution is examined for two extreme cases; a single curvature case and a double curvature case as shown in Fig. 8.7. The results are shown in Fig. 8.8. It can be seen that choosing eight segments in the single curvature case gives results within 0.1 percent of the exact results. The error is on the safe side. However, for the double curvature case, the optimum number is found to be about 20 and the errors are on the unsafe side. The error with 16 segments is 5.5 percent, which may be regarded

FIGURE 8.7
Single and double curvature bending

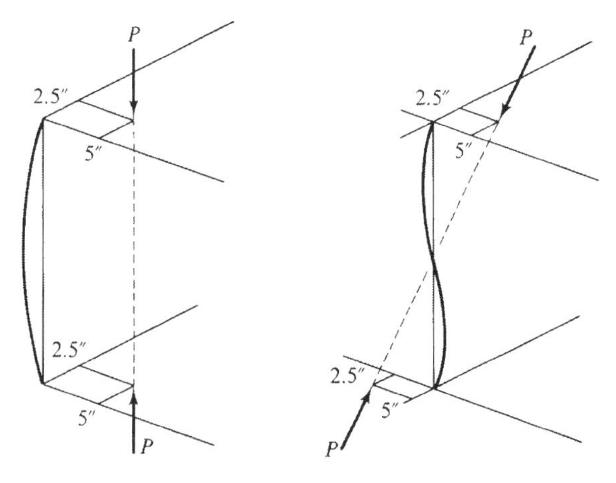

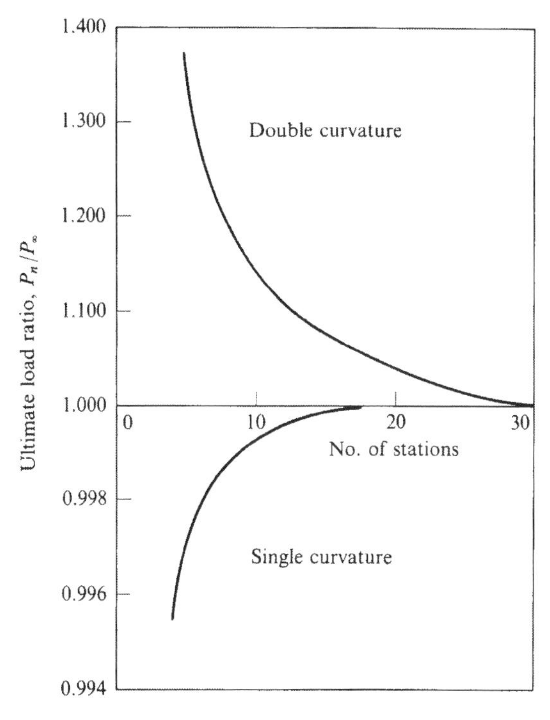

FIGURE 8.8
Convergence of solution

as acceptable, but errors obtained by having fewer than 16 segments may be considered unacceptable. The need for more stations for double curvature case is obviously caused by the sharp moment gradient near the ends of the double curvature bent beam-column.

Virdi and Dowling (1976) examined also the validity of the theory by comparing their theoretical solutions with available test results. Table 8.1 shows the comparison

Table 8.1 TESTS ON CONCRETE ENCASED STEEL SECTIONS (VIRDI AND DOWLING, 1976)

Column	L, in.	e_x, in.	e_y, in.	P_{test} tonf	P_{theory} tonf	P_{theory}/P_{test}
A	72	2.50	1.45	126	133.31	1.058
B	72	5.00	2.90	65	68.65	1.056
C	72	7.50	4.35	47.5	46.01	0.969
D	144	2.50	1.45	93	107.39	1.155
E	144	5.00	2.90	57.5	58.24	1.013
F	144	7.50	4.35	42	40.69	0.969
G	288	2.50	1.45	67	52.84	0.789
H	288	5.00	2.90	35.5	37.76	1.035
I	288	7.50	4.35	29.5	28.73	0.974

(1 in. = 25.4 mm, 1 tonf = 8.9 kN)

Mean 1.001907
Standard Deviation 9.95%

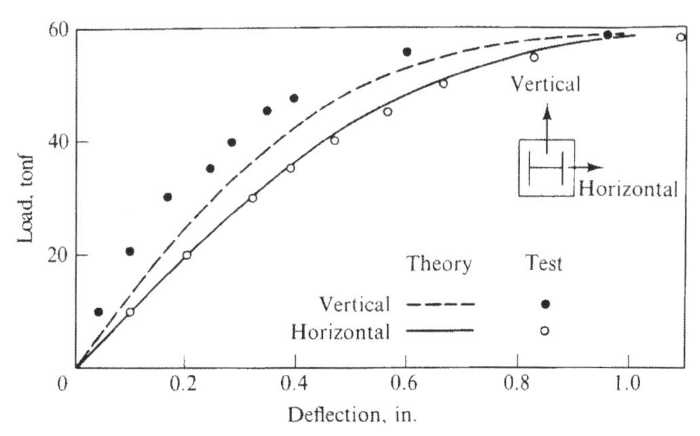

FIGURE 8.9
Comparison of deflection at mid-span (1 in = 25.4 mm, 1 tonf = 8.9 kN)

of ultimate load on nine pin-ended composite steel-concrete beam-columns. Good agreement is observed in all cases. (The average value of P_{theory}/P_{test} is 1.002 with a standard deviation of 9.95 percent.) Figure 8.9 shows the comparison in load-deflection behaviors in the direction of strong axis bending. The agreement is again very good. The difference in the minor axis bending may be caused by the actual initial deflection which may be less than the assumed value of L/1000 used in the theoretical calculation.

8.3 BEAM-COLUMN OF OPEN CROSS SECTION

In analysis of beam-columns of thin-walled open cross sections, the effect of column twisting upon the overall geometry must be considered. Harstead, Birnstiel and Leu (1968) solved biaxially loaded plastic beam-columns of wide flange sections making

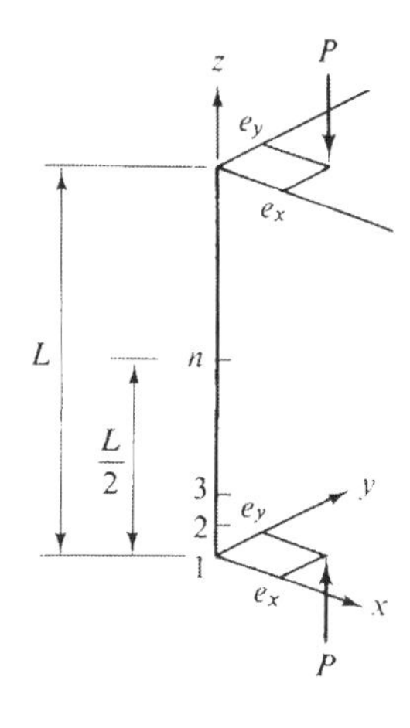

FIGURE 8.10
Eccentric symmetric loading

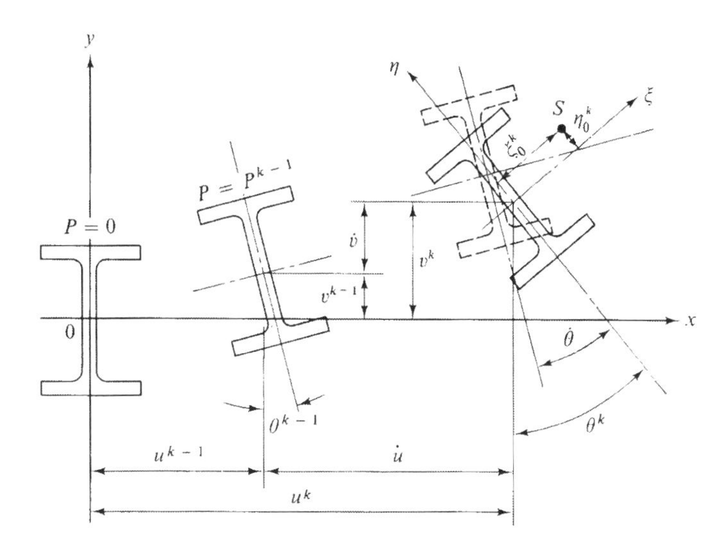

FIGURE 8.11
Displacement of cross section

use of the influence coefficient approach. Their method of solution is also applicable to beam-columns of general thin-walled open cross section. This is described here.

Herein, we consider a beam-column of wide-flange shape subjected to equal end moments due to biaxial eccentricities e_x and e_y of axial thrust P as shown in Fig. 8.10. Since the deformation is symmetric about the mid-span, the n-stations are set only along the half length of the beam-column such that station 1 and station n coincide with the end and the mid-span of the beam-column, respectively.

8.3.1 Coordinates and Displacements

The x, y and z axes are the fixed coordinate system in space where the x and y coincide with the principal axes ξ and η of the cross section at the end (station 1). The z is the axis passing through centroid of the cross section before deformation. After the application of axial force P with eccentricities e_x and e_y at both ends, the centroid of cross section at the station i translated laterally by the displacements u_i and v_i in x and y directions, respectively, and the principal axes ξ and η of the cross section rotate by an angle θ_i about ζ-axis which is tangent to the centroidal axis in the deflected state.

Figure 8.11 illustrates the displacements of a cross section at station i when the axial force P is increased by an amount $\dot{P}$ from P^{k-1} to P^k, i.e.,

$$P^k = P^{k-1} + \dot{P} \tag{8.39}$$

Accordingly, displacements increase and the current values are given by

$$\begin{Bmatrix} u^k \\ v^k \\ \theta^k \end{Bmatrix} = \begin{Bmatrix} u^{k-1} \\ v^{k-1} \\ \theta^{k-1} \end{Bmatrix} + \begin{Bmatrix} \dot{u} \\ \dot{v} \\ \dot{\theta} \end{Bmatrix} \tag{8.40}$$

As shown in Fig. 8.11, the increments in centroidal displacements $\dot{u}$ and $\dot{v}$ of the cross section have two components: one results from parallel translation $(\dot{\xi}, \dot{\eta})$ and the other from additional rotation $\dot{\theta}$ about the current torsion center $S(\xi_0^k, \eta_0^k)$.

$$\begin{aligned} \dot{u} &= \dot{\xi} + (\eta\dot{\theta} + \dot{\eta}\theta) = \dot{\xi} + 0.5\dot{\eta}_0(\theta^{k-1} + \theta^k) + \eta_0^k\dot{\theta} \\ \dot{v} &= \dot{\eta} - (\xi\dot{\theta} + \dot{\xi}\theta) = \dot{\eta} - 0.5\dot{\xi}_0(\theta^{k-1} + \theta^k) - \xi_0^k\dot{\theta} \end{aligned} \tag{8.41}$$

in which

$$\xi_0^k = \xi_0^{k-1} + \dot{\xi}_0 \qquad \text{and} \qquad \eta_0^k = \eta_0^{k-1} + \dot{\eta}_0$$

In the following discussion, the superscript k is dropped as it is understood that all the variables treated are in the current state.

8.3.2 External Moments

Using these current displacements, the external moments and torque can be expressed in x, y, z coordinate system as

$$\begin{Bmatrix} M_x \\ M_y \\ M_z \end{Bmatrix} = \begin{Bmatrix} -P(e_y - v) \\ P(e_x - u) \\ 0 \end{Bmatrix} \tag{8.42}$$

Using the directional cosine matrix between the two coordinate systems [see Eq. (2.154)]

$$\begin{Bmatrix} \xi \\ \eta \\ \zeta \end{Bmatrix} = \begin{bmatrix} 1 & \theta & -u' \\ -\theta & 1 & -v' \\ u' & v' & 1 \end{bmatrix} \begin{Bmatrix} x \\ y \\ z \end{Bmatrix} \tag{8.43}$$

The external moments and torque produced at any section z are given by

$$\begin{Bmatrix} M_{\xi\text{ext}} \\ M_{\eta\text{ext}} \\ M_{\zeta(1)} \end{Bmatrix} = P \begin{Bmatrix} -(e_y - v) + (e_x - u)\theta \\ (e_x - u) + (e_y - v)\theta \\ (e_x - u)v' - (e_y - v)u' \end{Bmatrix} \tag{8.44}$$

Although the twisting torque M_z about z-axis is zero, the twisting torque M_ζ about ζ-axis is not zero because the centroidal axis ζ which is fixed on the cross section is inclined by an angle v' and u' about x and y axes, respectively.

In addition to the torque $M_{\zeta(1)}$, the axial force P produces another torque $M_{\zeta(2)}$ which is caused by the shifting of torsion center $S(\xi_0, \eta_0)$ due to partial yielding of cross section

$$M_{\zeta(2)} = P(\eta_0 u' - \xi_0 v') \tag{8.45}$$

The total external torque produced at section after the application of the k-th increment of load is

$$M_{\zeta ext} = M_{\zeta(1)} + M_{\zeta(2)} \tag{8.46}$$

It is to be noted that this total torque is acting about the torsion center, and Eq. (8.45) is simply the correction of the pole from centroid to torsion center.

8.3.3 Internal Moments

The increment of the internal or resisting moments, axial force and torque corresponding to the changes of longitudinal strain distribution over the cross section at a station of the member can be conveniently determined by dividing the column cross section into many small elements. Details of this has been presented in Chap. 6.

In the following numerical procedure, account is kept of the longitudinal strain at the centroid of each of the elemental areas as the loading is incremented. The increment of longitudinal strain at an element due to axial deformation $\dot{\varepsilon}_o$ and bending curvatures $\dot{\xi}''$ and $\dot{\eta}''$ resulting from the k-th increment of load is

$$\dot{\varepsilon}_B = \dot{\varepsilon}_o - \eta\dot{\eta}'' - \xi\dot{\xi}'' \tag{8.47}$$

which is based on the assumption that plane before deformation remains plane after deformation. In addition to this axial and bending strain $\dot{\varepsilon}_B$, the longitudinal strain due to warping of the cross section $\dot{\varepsilon}_w$ exists which is caused by nonuniform torsion. The total longitudinal strain ε of the section is the sum of ε_B, ε_w and ε_r where ε_r denotes residual strain.

Since the warping stress has no contribution to axial force and bending moments, the increment of the internal forces are given by

$$\begin{aligned} \dot{N}_{int} &= \int_A E_t\dot{\varepsilon}_B \, dA \\ \dot{M}_{\xi int} &= \int_A E_t\dot{\varepsilon}_B\eta \, dA \\ \dot{M}_{\eta int} &= -\int_A E_t\dot{\varepsilon}_B\xi \, dA \end{aligned} \tag{8.48}$$

in which E_t is the tangent modulus of the material. In the present analysis, the material is assumed to be elastic-perfectly plastic, considering elastic unloading. The value of E_t is either E, the elastic modulus or zero, depending upon the state of stress and strain and strain rate of each element as was defined in Chap. 5, i.e.,

$$E_t = \begin{cases} E & |\varepsilon| < \varepsilon_y \quad \text{or} \quad \varepsilon\dot{\varepsilon} < 0 \\ 0 & |\varepsilon| \geq \varepsilon_y \quad \text{and} \quad \varepsilon\dot{\varepsilon} > 0 \end{cases} \tag{8.49}$$

in which ε_y is the initial yield strain of the material.

The internal resisting torque results from shearing stresses and from the longitudinal stresses caused by restraint of warping of the cross sections. The first part is the St. Venant torsion: this torque is directly proportional to the twist of the member, i.e., the first derivation of the rotation angle θ of the member,

$$M_{\zeta(1)} = GK_T\theta' \tag{8.50}$$

with

$$K_T = \frac{A_E}{A}\frac{1}{3}\sum_k b_k t_k^3 \tag{8.51}$$

in which b_k, t_k denotes the width and thickness of the k-th plate element of the H-column. In Eq. (8.51) we have assumed that the St. Venant torsional rigidity of an elastic-plastic section is directly proportional to the ratio of the elastically remained area A_E to the total cross sectional area A. This approximation was adopted by Harstead *et al.* (1968). It is to be noted that in their earlier work, Birnstiel and Michalos (1963), this torsional rigidity was calculated only for elastic elements as was done in Chap. 6 of this book.

$$GK_T = \frac{1}{3}\int G_t t^2\, dA \tag{8.52}$$

in which G_t is the shear modulus taking the value either G or zero similar to that of E_t. The actual value of G_t may lie somewhere between these two extremes.

The second part of the resisting torque results from restraint of warping of the cross section. The increment of warping strain $\dot{\varepsilon}_w$ at an element on the cross section is obtained by

$$\dot{\varepsilon}_w = \dot{\varepsilon}_{wo} - \dot{\theta}''\omega_0 = \dot{\varepsilon}_{wo} + \dot{\theta}''(\xi\eta - \xi\eta_0 + \xi_0\eta) \tag{8.53a}$$

because the unit warping for an I-shape is given from Eq. (2.94) as

$$\omega_0 = \int \rho_0\, ds = -\xi\eta + \xi\eta_0 - \xi_0\eta \tag{8.53b}$$

In Eq. (8.53a), $\dot{\varepsilon}_{wo}$ is the increment of warping strain at the centroid 0.
The corresponding increment of warping normal stress $\dot{\sigma}_w$ can be obtained by

$$\dot{\sigma}_w = E_t\dot{\varepsilon}_w \tag{8.54}$$

in which E_t is the tangent modulus discussed previously.

Now the location of the instantaneous torsion center during the application of load $\dot{P}$ can be determined from the condition that the warping normal stress σ_w has no contribution to the resultants of axial force in ζ direction and bending moments about ξ and η axes. Therefore

$$\begin{aligned} \dot{N}_w &= \int_A \dot{\sigma}_w\, dA = 0 \\ \dot{M}_{\xi w} &= \int_A \dot{\sigma}_w\eta\, dA = 0 \\ \dot{M}_{\eta w} &= -\int_A \dot{\sigma}_w\xi\, dA = 0 \end{aligned} \tag{8.55}$$

If these are not zero, the location of the instantaneous torsion center (ξ_0, η_0) and the warping strain at the centroid $(\dot{\varepsilon}_{wo})$ are to be adjusted by the following approximate equations (Prob. 8.1)

$$\Delta\xi_0 = \frac{\Delta\dot{M}_{\xi w}}{EI_x\dot{\theta}''}$$

$$\Delta\eta_0 = \frac{\Delta\dot{M}_{\eta w}}{EI_y\dot{\theta}''} \tag{8.56}$$

$$\Delta\dot{\varepsilon}_{wo} = \frac{\Delta\dot{N}_w}{EA}$$

Equations (8.56) do not directly result in the necessary change of the location of torsion center as they are a function of the yielded zones of the cross section which, in turn, depends on the coordinates of the torsion center. These equations must therefore be solved iteratively for patterns of warping strain based on successively better values of ξ_0, η_0 and $\dot{\varepsilon}_{wo}$ until Eqs. (8.56) are satisfied.

The warping strains given in Eq. (8.53) generate axial force, shear forces and bending moments in the flanges as well as in the web of the H-column. However, as far as twisting moment resistance by the warping stress is concerned, it is known that the major contributions for the twisting moment come from warping stresses in flanges not from the web, hence, we can neglect the warping stress contribution from the column web. The increment of bending moment about the η-axis in flanges resulting from the warping stresses are

$$\dot{M}_{w\eta \mathrm{UF}} = \int_{\mathrm{UF}} \xi\dot{\sigma}_w\, dA, \quad \dot{M}_{w\eta \mathrm{LF}} = \int_{\mathrm{LF}} \xi\dot{\sigma}_w\, dA \tag{8.57}$$

in which the integrations are made for the upper flange (UF) and for the lower flange (LF) separately.

The second part of the resisting torque at any station of the member results from shear forces in both flanges which is directly proportional to the moment change in flanges.

$$\dot{M}_{\zeta(2)} = \left(\frac{D-t}{2} - \eta_0\right)\dot{M}'_{w\eta \mathrm{UF}} + \left(\frac{D-t}{2} + \eta_0\right)\dot{M}'_{w\eta \mathrm{LF}} \tag{8.58}$$

In Eq. (8.58), D is the depth and t is the flange thickness of the cross section. The second part of the incremental resisting torque may be written as the first derivative of the bimoment M_ω similar to that of elastic case, Eq. (2.128b)

$$\dot{M}_{\zeta(2)} = \dot{M}'_\omega \tag{8.59}$$

because, from Eq. (2.118) and Eq. (8.53b), assuming $\omega_n = -\omega_0$ [Eq. (2.102)] and no warping on web-plate, we have

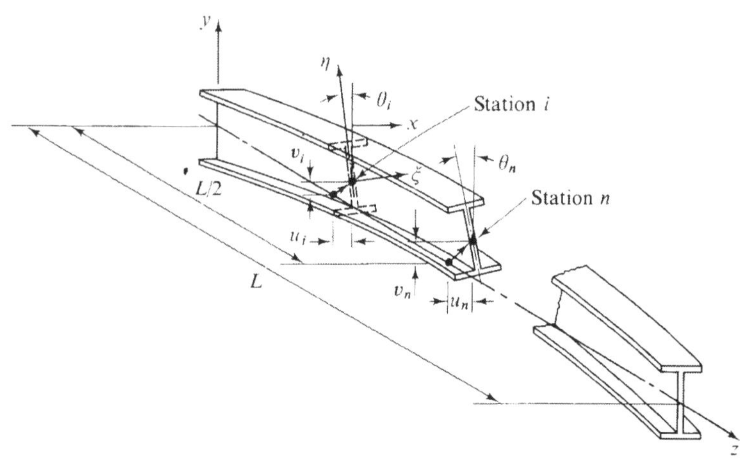

FIGURE 8.12
Beam-column in its deflected position

$$\dot{M}_\omega = \int \sigma_w \omega_n \, dA = -\int \sigma_w \omega_0 \, dA = \left(\frac{D-t}{2} - \eta_0\right)\dot{M}_{w\eta\mathrm{UF}} + \left(\frac{D-t}{2} + \eta_0\right)\dot{M}_{w\eta\mathrm{LF}} \tag{8.60}$$

For observation of this equation, see Prob. 8.2.

The third part of the internal torque is due to warping of cross section known as the Wagner's effect which was explained in Eq. (2.134), Chap. 2.

$$M_{\zeta(3)} = \theta'\bar{K} = \theta' \int_A E_t \varepsilon[(\eta - \eta_0)^2 + (\xi - \xi_0)^2] \, dA \tag{8.61}$$

Thus, the total internal torque is obtained by

$$M_{\zeta\,\mathrm{int}} = (GK_T + \bar{K})\theta' + M'_\omega \tag{8.62}$$

which is acting about torsion center $S(\xi_0, \eta_0)$.

Based on the equations derived above, Harstead *et al.* (1968) studied the behavior of eccentrically loaded beam-columns as shown in Fig. 8.12 for which the following boundary conditions were considered at $z = 0$ and L

$$u = u'' = 0 \qquad \text{and} \qquad v = v'' = 0$$

$$\theta = \theta' = 0 \qquad \text{or} \qquad \theta = \theta'' = 0$$

8.3.4 Method of Solution

Since section properties change depending upon loading history, the load must be incremented step by step. The computational procedure in k-th increment of loading is described in what follows along with the development of influence coefficient matrix.

It is assumed that the equilibrium position under the $(k-1)$-th increment of loading has been previously determined.

Step 1 Assume trial values of the deformations $\dot{\theta}''$, $\dot{\xi}''$, $\dot{\eta}''$ and $\dot{\varepsilon}_o$ at each station along the length of the beam-column. The first trial values for $k = 1$ are selected in the elastic range which satisfy specified boundary conditions to the problem. The trial values for $k = 2$ are proportional to the final values converged in $k = 1$ stage. The trial values for $k > 2$ are selected equal to $\dot{\theta}''^{(k-1)}$, $\dot{\xi}''^{(k-1)}$, $\dot{\eta}''^{(k-1)}$ and $\dot{\varepsilon}_o^{(k-1)}$ which are the converged values in the $(k-1)$-th increment of loading.

Step 2 Integrate numerically the curvatures $\dot{\theta}''$, $\dot{\xi}''$ and $\dot{\eta}''$ assumed in Step 1 for $\dot{\theta}'$, $\dot{\xi}'$ and $\dot{\eta}'$. Then, integrate once more for $\dot{\theta}$, $\dot{\xi}$ and $\dot{\eta}$. Next, compute u and v using Eq. (8.41) or

$$
\begin{aligned}
\dot{u} &= \dot{\xi} + \dot{\eta}_0\left(\theta + \frac{1}{2}\dot{\theta}\right) + \eta_0\dot{\theta}, \quad u = u^{k-1} + \dot{u} \\
\dot{v} &= \dot{\eta} - \dot{\xi}_0\left(\theta + \frac{1}{2}\dot{\theta}\right) - \xi_0\dot{\theta}, \quad v = v^{k-1} + \dot{v}
\end{aligned}
\tag{8.63}
$$

for which the most recently determined values of η_0 and ξ_0 are used.

Step 3 From the trial values in Step 1, compute increment of the internal forces $\dot{N}_{\text{int}}$, $\dot{M}_{\xi\text{int}}$ and $\dot{M}_{\eta\text{int}}$ from Eqs. (8.47) and (8.48)

$$
\dot{\varepsilon}_B = \dot{\varepsilon}_o - \eta\dot{\eta}'' - \xi\dot{\xi}'' \tag{8.64}
$$

$$
\begin{aligned}
\dot{N}_{\text{int}} &= \int_A E_t\dot{\varepsilon}_B\,dA \\
\dot{M}_{\xi\text{int}} &= \int_A E_t\dot{\varepsilon}_B\eta\,dA \\
\dot{M}_{\eta\text{int}} &= -\int_A E_t\dot{\varepsilon}_B\xi\,dA
\end{aligned}
\tag{8.65}
$$

where E_t is given by Eq. (8.49). Because the values of $\dot{N}_{\text{int}}$ will be different at each station, select $\dot{P} = -\dot{N}_{\text{int}}$ at mid-span as the control value. Compute the total force and moments by adding the increments to the previous total values

$$
\begin{aligned}
P &= P^{k-1} + \dot{P} \\
M_{\xi\text{int}} &= M_{\xi\text{int}}^{k-1} + \dot{M}_{\xi\text{int}} \\
M_{\eta\text{int}} &= M_{\eta\text{int}}^{k-1} + \dot{M}_{\eta\text{int}}
\end{aligned}
\tag{8.66}
$$

Step 4 Determine the values of the external bending moments from Eq. (8.44) or

$$
\begin{aligned}
M_{\xi\text{ext}} &= -P(e_y - v) + P(e_x - u)\theta \\
M_{\eta\text{ext}} &= P(e_x - u) + P(e_y - v)\theta
\end{aligned}
\tag{8.67}
$$

Step 5 At each station, compare the values of total internal forces computed in Step 3 with the corresponding external values. Also, compare the incremental value of axial force at all stations with the control value $\dot{P}$. If the difference exceeds the specified tolerable limit, then the trial values in Step 1 must be corrected by the amount $\Delta\dot{\xi}''$, $\Delta\dot{\eta}''$ and $\Delta\dot{\varepsilon}_o$. The corrections are computed through the influence coefficient matrix defined in the matrix equations

$$\begin{Bmatrix} \dot{M}_\eta(n) \\ \dot{M}_\xi(n) \\ \dot{P}(n) \end{Bmatrix}_{\text{ext}} - \begin{Bmatrix} \dot{M}_\eta(n) \\ \dot{M}_\xi(n) \\ N(n) \end{Bmatrix}_{\text{int}} = \begin{bmatrix} G(nxn) + H_{11}(nxn) & H_{12}(nxn) & H_{13}(nxn) \\ H_{21}(nxn) & G(nxn) + H_{22}(nxn) & H_{23}(nxn) \\ H_{31}(nxn) & H_{32}(nxn) & H_{33}(nxn) \end{bmatrix} \begin{Bmatrix} \Delta\dot{\xi}''(n) \\ -\Delta\dot{\eta}''(n) \\ \Delta\dot{\varepsilon}_o(n) \end{Bmatrix} \tag{8.68}$$

in which the symbols (n) and (nxn) indicate sizes of the submatrices involved. The submatrix $G(nxn)$ is the effect of unit changes in curvatures on the external moments and the submatrices $H(nxn)$ are the effect of unit changes in curvatures on the internal moments and axial force.

Elements of influence coefficient matrix

The submatrix $[G(nxn)]$ is determined from

$$\begin{aligned} \{\Delta\dot{M}_\eta(n)\}_{\text{ext}} &= -P\{\Delta\dot{\xi}(n)\} = [G(nxn)]\{\Delta\dot{\xi}''(n)\} \\ \{\Delta\dot{M}_\xi(n)\}_{\text{ext}} &= P\{\Delta\dot{\eta}(n)\} = [G(nxn)]\{-\Delta\dot{\eta}''(n)\} \end{aligned} \tag{8.69}$$

Since

$$\begin{aligned} \xi_i &= \xi_{i-1} + \frac{\lambda}{2}(\xi'_{i-1} + \xi'_i) \quad \text{and} \quad \xi'_i = \xi'_{i-1} + \frac{\lambda}{2}(\xi''_{i-1} + \xi''_i) \\ \eta_i &= \eta_{i-1} + \frac{\lambda}{2}(\eta'_{i-1} + \eta'_i) \quad \text{and} \quad \eta' = \eta'_{i-1} + \frac{\lambda}{2}(\eta''_{i-1} + \eta''_i) \end{aligned} \tag{8.70}$$

in which λ is the distance between stations. The matrix $[G(nxn)]$ is given by

$$[G(nxn)] = -\lambda^2 P \left[\begin{array}{cccccc|c} 0 & 0 & 0 & 0 & \cdots & 0 & 0 \\ 0 & 1 & 1 & 1 & \cdots & 1 & 1/2 \\ 0 & 1 & 2 & 2 & \cdots & 2 & 2/2 \\ 0 & 1 & 2 & 3 & \cdots & 3 & 3/2 \\ \vdots & \vdots & \vdots & \vdots & & \vdots & \vdots \\ 0 & 1 & 2 & 3 & \cdots & n-2 & (n-2)/2 \\ \hline 0 & 1 & 2 & 3 & \cdots & n-2 & (n-1)/2 \end{array}\right] \tag{8.71}$$

For the derivation of Eq. (8.71), see Prob. 8.3.

The submatrices $[H_{ij}(nxn)]$ are determined from Eqs. (8.47) and (8.48).

$$\begin{aligned}\{\dot{M}_\eta(n)\}_{\text{int}} &= \int E_t[\xi^2\dot{\xi}'' + \xi\eta\dot{\eta}'' - \xi\dot{\varepsilon}_o]_n\,dA \\ \{\dot{M}_\xi(n)\}_{\text{int}} &= -\int E_t[\xi\eta\dot{\xi}'' + \eta^2\dot{\eta}'' - \eta\dot{\varepsilon}_o]_n\,dA \\ \{\dot{N}(n)\}_{\text{int}} &= -\int E_t[\xi\dot{\xi}'' + \eta\dot{\eta}'' - \dot{\varepsilon}_o]_n\,dA\end{aligned} \tag{8.72}$$

The elements of $[H_{ij}]$ are given by

$$[H_{11}(nxn)] = \begin{bmatrix} (\int E_t\xi^2\,dA)_1 & & (0) \\ & (\int E_t\xi^2\,dA)_2 & \\ (0) & & (\int E_t\xi^2\,dA)_n \end{bmatrix}$$

$$[H_{12}(nxn)] = [H_{21}(nxn)] = -\begin{bmatrix} (\int E_t\xi\eta\,dA)_1 & & (0) \\ & (\int E_t\xi\eta\,dA)_2 & \\ (0) & & (\int E_t\xi\eta\,dA)_n \end{bmatrix}$$

$$[H_{13}(nxn)] = [H_{31}(nxn)] = -\begin{bmatrix} (\int E_t\xi\,dA)_1 & & (0) \\ & (\int E_t\xi\,dA)_2 & \\ (0) & & (\int E_t\xi\,dA)_n \end{bmatrix}$$

$$[H_{22}(nxn)] = \begin{bmatrix} (\int E_t\eta^2\,dA)_1 & & (0) \\ & (\int E_t\eta^2\,dA)_2 & \\ (0) & & (\int E_t\eta^2\,dA)_n \end{bmatrix} \tag{8.73}$$

$$[H_{23}(nxn)] = [H_{32}(nxn)] = \begin{bmatrix} (\int E_t\eta\,dA)_1 & & (0) \\ & (\int E_t\eta\,dA)_2 & \\ (0) & & (\int E_t\eta\,dA)_n \end{bmatrix}$$

$$[H_{33}(nxn)] = \begin{bmatrix} (\int E_t\,dA)_1 & & (0) \\ & (\int E_t\,dA)_2 & \\ (0) & & (\int E_t\,dA)_n \end{bmatrix}$$

Step 6 Repeat steps 2 through 5 until convergence is attained in step 5.

Step 7 Determine the warping strain distribution and the values of η_0 and ξ_0 from Eq. (8.56) by iteration

$$\xi_0 = \frac{\dot{M}_{\xi w}}{EI_x\dot{\theta}''} \qquad \eta_0 = \frac{\dot{M}_{\eta w}}{EI_y\dot{\theta}''} \tag{8.74}$$

Compute the incremental value of the external torque from Eqs. (8.44) and (8.45). Determine the value of the total external torque at each station as

$$M_{\zeta\text{ext}} = P[(e_x - u)v' - (e_y - v)u'] + \Sigma[\dot{P}(\eta_0 u' - \xi_0 v') + P^{k-1}(\eta_0\dot{u}' - \xi_0\dot{v}')] \tag{8.75}$$

Step 8 Compute the value of total internal torque from Eqs. (8.50), (8.60) and (8.61) or Eq. (8.62).

$$M_{\zeta\,\text{int}} = (GK_T + \bar{K})\theta' + M'_\omega \tag{8.76}$$

Step 9 Compare the values of total external torque $M_{\zeta\text{ext}}$ at each station in Step 7 with the value of internal torque $M_{\zeta\text{int}}$. If the difference $\{\Delta M_\zeta(n)\}$ is greater than the prescribed limit, then adjust the trial deformations by $\{\Delta\dot{\theta}''(n)\}$,

$$\{\Delta\dot{M}_\zeta(n)\} = ([R(nxn)] + [X(nxn)])\{\Delta\dot{\theta}''(n)\} \tag{8.77}$$

where $[R]$ are the effects on the first part and third part of the internal torque of unit changes of $\dot{\theta}''$. The matrix $[X]$ contains the effects of unit changes of $\dot{\theta}''$ on the second part of the internal torque. Changes on the external torque are not considered in the correction procedure since these quantities are affected only secondarily by variations in θ. The matrices $[R(nxn)]$ and $[X(nxn)]$ are derived separately for the two different warping restraint conditions at both ends.

(a) *Warping Restraint Case at Both Ends* ($\theta'_1 = 0$) The submatrix $[R(nxn)]$ is determined from Eqs. (8.50) and (8.61) as

$$\{\Delta\dot{M}_\zeta(n)\} = \{(GK_T + \bar{K})\Delta\dot{\theta}'(n)\} = [R(nxn)]\{\Delta\dot{\theta}''(n)\} \tag{8.78}$$

Since

$$\begin{aligned}\theta'_i &= \theta'_{i-1} + \frac{\lambda}{2}(\theta''_{i-1} + \theta''_i)\\ &= \theta'_1 + \frac{\lambda}{2}(\theta''_1 + 2\theta''_2 + 2\theta''_3 + \cdots + 2\theta''_{i-1} + \theta''_i)\end{aligned} \tag{8.79}$$

then we have

$$[R(nxn)] = \frac{\lambda}{2}\left[\begin{array}{cccccc|c} 0 & 0 & 0 & 0 & \cdots & 0 & 0\\ (K_T+\bar{K})_2 & (K_T+\bar{K})_2 & 0 & 0 & \cdots & 0 & 0\\ (K_T+\bar{K})_3 & 2(K_T+\bar{K})_3 & (K_T+\bar{K})_3 & 0 & \cdots & 0 & 0\\ & & & & & 0 & \\ (K_T+\bar{K})_{n-1} & 2(K_T+\bar{K})_{n-1} & 2(K_T+\bar{K})_{n-1} & \cdots & 2(K_T+\bar{K})_{n-1} & (K_T+\bar{K})_{n-1} & 0\\ \hline 10^{-4} & 2\times10^{-4} & 2\times10^{-4} & \cdots & 2\times10^{-4} & 2\times10^{-4} & 10^{-4}\end{array}\right] \tag{8.80}$$

The constant values 10^{-4} and 2×10^{-4} given in the last row of matrix $[R]$ force the corrections $\{\Delta\dot{\theta}''(n)\}$ to be such that $\theta' = 0$ at the column end (boundary condition) and at mid-span (because of symmetry).

The submatrix $[X(nxn)]$ is determined from Eq. (8.59)

$$\{\Delta\dot{M}_{\zeta(2)\,\text{int}}(n)\} = \{\Delta\dot{M}'_\omega(n)\} = [X(nxn)]\{\Delta\dot{\theta}''(n)\} \tag{8.81}$$

the bi-moment M_ω is approximately given by the in-plane bending moments in the upper flange (UF) and the lower flange (LF) due to the warping strain ε_w times their distance (Prob. 8.2),

$$\begin{aligned}\dot{M}_\omega &= \left(\frac{D-t}{2} - \eta_0\right)\int_{UF} E_t\dot{\varepsilon}_w\xi\, dA + \left(\frac{D-t}{2} + \eta_0\right)\int_{LF} E_t\dot{\varepsilon}_w\xi\, dA \\ &= \int E_t\dot{\varepsilon}_w\xi(\eta - \eta_0)\, dA\end{aligned} \tag{8.82}$$

The warping strain $\dot{\varepsilon}_w$ is from Eq. (8.53a) with $\dot{\varepsilon}_{wo} = \xi_o = 0$.

$$\dot{\varepsilon}_w = \dot{\theta}''(\xi\eta - \xi\eta_0) \tag{8.83}$$

From Eq. (8.59) and the result of Prob. 8.2, we have

$$\dot{M}_{\zeta(2)\text{int}} = \dot{M}'_\omega = \dot{\theta}''' \int E_t\xi^2(\eta - \eta_0)^2\, dA \tag{8.84}$$

Denoting this relationship by

$$\dot{M}_{(\zeta\text{int})i} = B_i\dot{\theta}'''_i \tag{8.85}$$

and using the numerical differentiation,

$$\theta'''_i = \frac{1}{2\lambda}(\theta''_{i+1} - \theta''_{i-1}) \tag{8.86}$$

the submatrix $[X(nxn)]$ is determined as

$$[X(nxn)] = \frac{1}{2\lambda}\begin{bmatrix} -3B_1 & 4B_1 & -B_1 & 0 & \dots & -0 & \dots & 0 \\ -B_2 & 0 & B_2 & 0 & \dots & 0 & 0 & 0 \\ 0 & -B_3 & 0 & B_3 & \dots & 0 & 0 & 0 \\ \dots & \dots & \dots & \dots & \dots & \dots & \dots & \dots \\ \dots & \dots & \dots & \dots & \dots & \dots & \dots & \dots \\ 0 & 0 & 0 & 0 & \dots & -B_{n-1} & 0 & B_{n-1} \\ 0 & 0 & 0 & 0 & \dots & 0 & 0 & 0 \end{bmatrix} \tag{8.87}$$

The elements in the first row are derived using the end conditions

$$\frac{1}{\lambda}(\theta''_2 - \theta''_1) = \frac{1}{2}(\theta'''_2 + \theta'''_1) \qquad \text{and} \qquad \theta'''_2 = \frac{1}{2\lambda}(\theta''_3 - \theta''_1) \tag{8.88}$$

Thus

$$\theta'''_1 = \frac{1}{2\lambda}(-3\theta''_1 + 4\theta''_2 - \theta''_3)$$

(b) *Warping Permitted Case at Both Ends* ($\theta''_1 = 0$) The first element in the correction vector $[\Delta\dot{\theta}''(n)]$ need not be corrected because it is desired to force the correction at station 1 to have a zero value and the size of the submatrices $[R]$ and $[X]$ is $(n-1) \times (n-1)$

Since

$$\begin{aligned}\theta'_i &= \theta'_{i+1} - \frac{\lambda}{2}(\theta''_i + \theta''_{i-1}) \\ &= \theta'_n - \frac{\lambda}{2}(\theta''_i + 2\theta''_{i+1} + \cdots + 2\theta''_{n-1} + \theta''_n)\end{aligned} \tag{8.89}$$

in which $\theta'_n = 0$.

Then the submatrix $[R(n-1, n-1)]$ is determined as

$$[R(n-1,n-1)] = \frac{\lambda}{2}\begin{bmatrix} 2(GK_T+\bar{K})_1 & 2(GK_T+\bar{K})_1 & 2(GK_T+\bar{K})_1 & \cdots & 2(GK_T+\bar{K})_1 & (GK_T+\bar{K})_1 \\ (GK_T+\bar{K})_2 & 2(GK_T+\bar{K})_2 & 2(GK_T+\bar{K})_2 & \cdots & 2(GK_T+\bar{K})_2 & (GK_T+\bar{K})_2 \\ & (GK_T+\bar{K})_3 & 2(GK_T+\bar{K})_3 & \cdots & 2(GK_T+\bar{K})_3 & (GK_T+\bar{K})_3 \\ & & (GK_T+\bar{K})_4 & \cdots & 2(GK_T+\bar{K})_4 & (GK_T+\bar{K})_4 \\ & (0) & & (GK_T+\bar{K})_{n-2} & 2(GK_T+\bar{K})_{n-2} & (GK_T+\bar{K})_{n-2} \\ & & & & (GK_T+\bar{K})_{n-1} & (GK_T+\bar{K})_{n-1} \end{bmatrix} \tag{8.90}$$

Similar to Eq. (8.87), the submatrix $[X(n-1, n-1)]$ has the following form

$$[X(n-1,n-1)] = \frac{1}{2\lambda}\begin{bmatrix} 4B_1 & -B_1 & 0 & 0 & \cdots & 0 & 0 & 0 \\ 0 & B_2 & 0 & 0 & \cdots & 0 & 0 & 0 \\ -B_3 & 0 & B_3 & 0 & \cdots & 0 & 0 & 0 \\ 0 & -B_4 & 0 & B_4 & \cdots & 0 & 0 & 0 \\ 0 & 0 & 0 & 0 & \cdots & 0 & B_{n-2} & 0 \\ 0 & 0 & 0 & 0 & \cdots & -B_{n-1} & 0 & B_{n-1} \end{bmatrix} \tag{8.91}$$

Step 10 Repeat steps 7 through 9 until adequate convergence is attained.

Step 11 Repeat step 5 to examine if the internal and external values of bending moments are still in adequate agreement.

Step 12 Repeat steps 6 through 10 until convergence is attained in steps 5 and 9.

Step 13 Repeat steps 1 through 12 until the ultimate load carrying capacity or the desired displacement has been attained.

8.3.5 Numerical Studies

Utilizing the numerical procedure described above Harstead *et al.* (1968) solved six biaxially loaded beam-column problems listed in Table 8.2. The problems are selected so as to make possible comparisons with solutions of the elastic case (Chap. 4) and to

Table 8.2 PROBLEMS FOR NUMERICAL STUDIES (HARSTEAD ET AL., 1968)

Problem Number	Length, in	Eccentricity		Warping of End Cross Sections	Residual Thermal Strains
		e_x, in	e_y, in		
1	264.6	5.0	5.0	Permitted	—
2	264.6	5.0	5.0	Restrained	—
3	220.0	0.5	5.0	Permitted	None
4	220.0	0.5	5.0	Restrained	None
5	220.0	0.5	5.0	Permitted	Yes
6	220.0	0.5	5.0	Restrained	Yes

(1 in. = 25.4 mm, 1 ksi = 6.9 MPa)

Section Properties (*W14* × *43*)
$A = 12.65$ in^2
$I_\xi = 429.0$ in^4
$I_\eta = 45.1$ in^4
$D = 13.68$ in
$t = 0.528$ in
$w = 0.308$
$E = 30\,000$ ksi
$\varepsilon_y = 0.0012$ in/in
$B = 8.00$ in
$G = 11\,500$ ksi

Other Data
Number of columns in flange grid = 24
Number of rows in flange grid = 8
Number of columns in web grid = 4
Number of rows in web grid = 20

examine the effects of warping restraint and residual strains on plastic behavior of beam-columns under biaxial bending. The wide-flange shape W14 × 43 was chosen for all problems. Problems 1 and 2 in Table 8.2 were solved on the basis of completely elastic behavior. For Prob. 3 to 6, the material of the beam-columns is assumed to be elastic-perfectly plastic with the yield stress level of 36 ksi (248 MPa). The partitioning

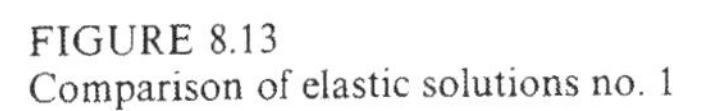
FIGURE 8.13
Comparison of elastic solutions no. 1

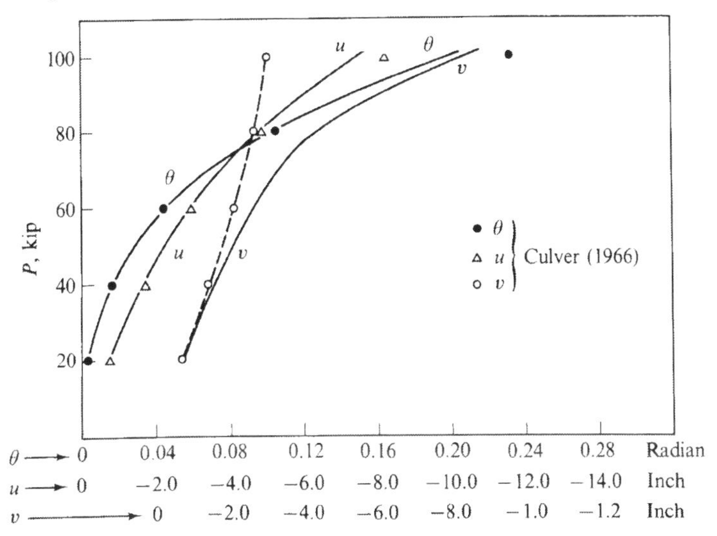

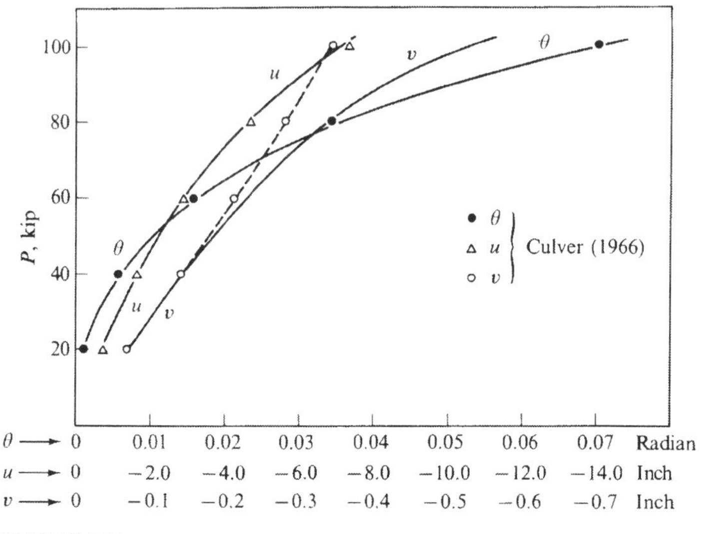

FIGURE 8.14
Comparison of elastic solutions no. 2

of the cross section at each station along the beam-column is 8×24 for each flange and 4×20 for the web plate.

(a) *Elastic solutions*

Problems 1 and 2 were solved to compare the numerical results by the present computational procedure with reported elastic solutions by Culver (1966) (see Chap. 4). Plots of load P versus the midheight displacements u, v and θ of Probs. 1 and 2 are shown in Fig. 8.13 and Fig. 8.14, respectively. It is seen that the twisting angle θ for the warping restrained beam-column (Fig. 8.14) is much less than that of warping permitted beam-column (Fig. 8.13).

In both Figs. 8.13 and 8.14 solutions by Culver (1966) are also plotted which are almost identical to the present solution except for the deflection v. Harstead *et al.*, (1968) pointed out that the discrepancy in v was due mainly to the fact that the differential equations which Culver solved exactly were derived by Goodier (1941) under the assumptions of

$$\begin{aligned} u'' = \xi'' \quad &\text{and} \quad v'' = \eta'' \\ \theta u \ll v \quad &\text{and} \quad \theta v \ll u \end{aligned} \tag{8.92}$$

These assumptions are acceptable when the twisting angle θ is small. As can be seen in Figs. 8.13 and 8.14, when the twisting angle gets large, the assumptions of Eq. (8.92) result in an unrealistic solution for the deflection v as indicated by the dotted lines.

(b) *Effect of warping restraint*

Figure 8.15 illustrates the effect of warping restraint at the end of the beam-column on the plastic deformation of Probs. 3 and 4 in Table 8.2. It is again observed that the

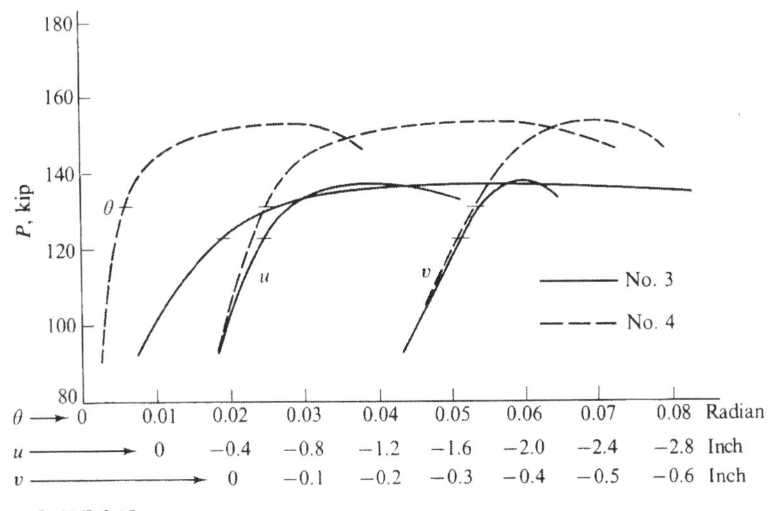

FIGURE 8.15
Effect of warping restraint no. 3 and no. 4

twisting angle θ differs largely. The ultimate strength of the beam-column is raised by 12 percent due to the warping restraint.

(c) *Effect of residual strains*

Effect of residual strains are illustrated in Fig. 8.16 for warping permitted beam-columns (Probs. 3 and 5) and in Fig. 8.17 for warping restrained beam-columns (Probs. 4 and 6). The residual stress pattern is assumed to have a uniform tension of 6.19 ksi (42.7 MPa) in the web and a linear distribution of − 11.88 ksi (− 82 MPa) at the flange

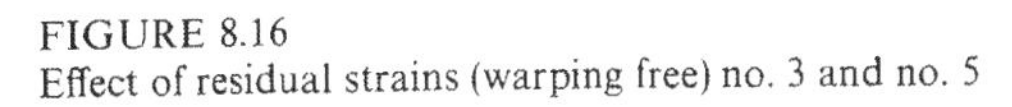
FIGURE 8.16
Effect of residual strains (warping free) no. 3 and no. 5

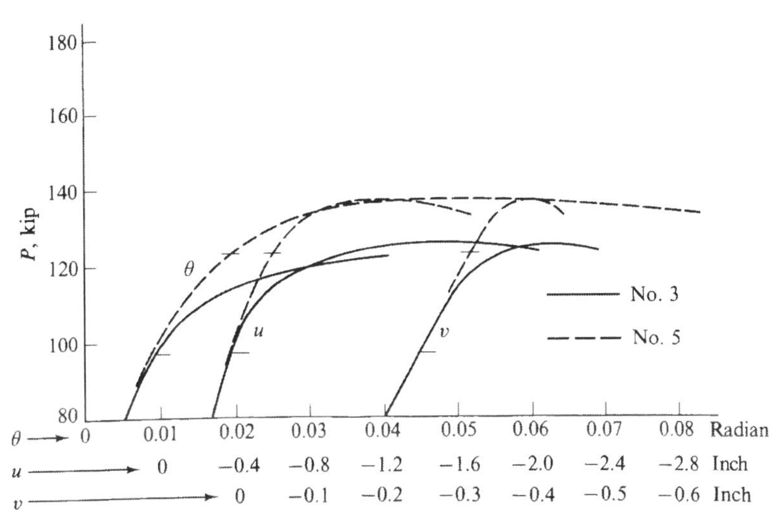

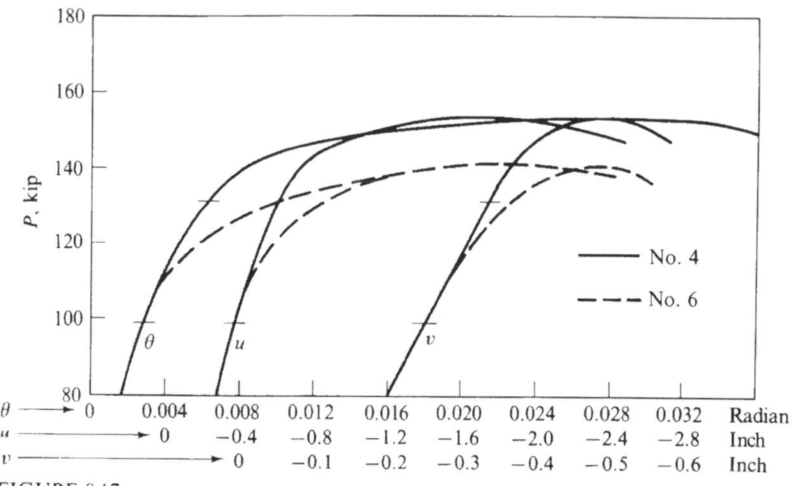

FIGURE 8.17
Effect of residual strains (warping restrained) no. 4 and no. 6

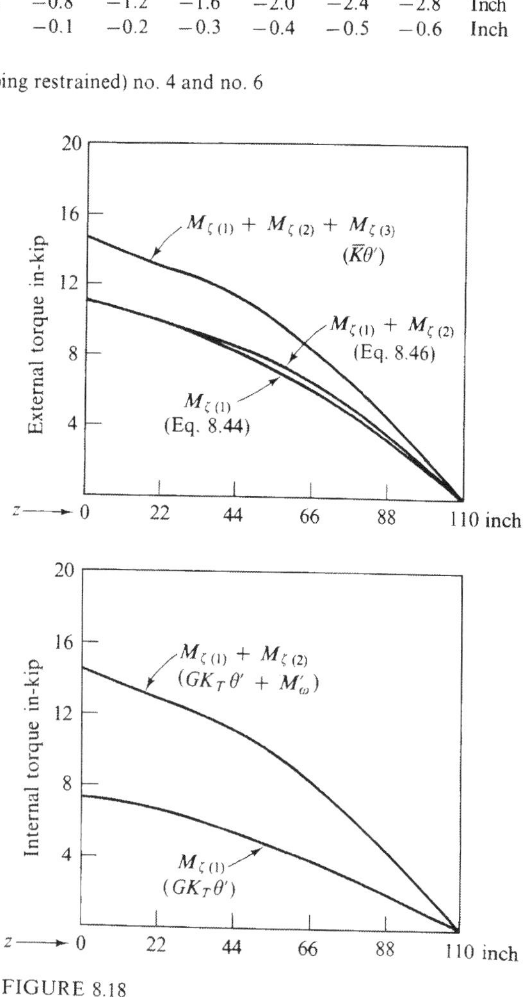

FIGURE 8.18
Twisting moment distribution at $P = 136.9$ kip, no. 3

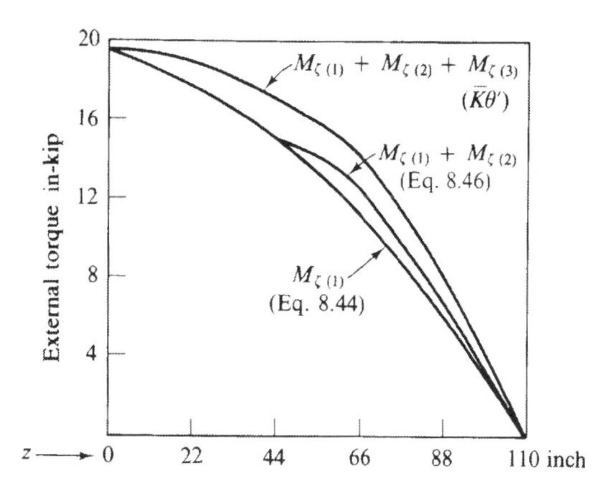

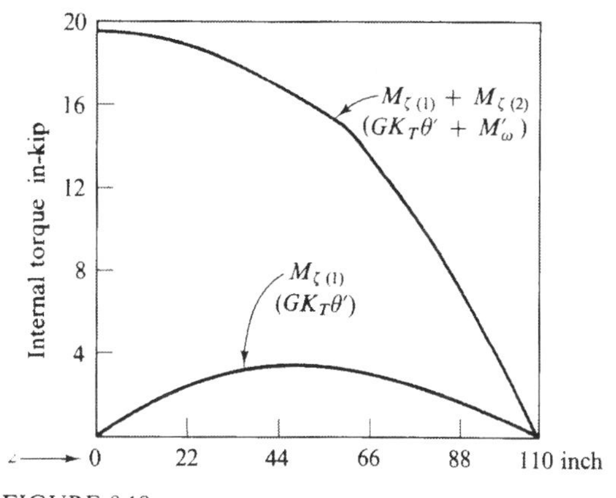

FIGURE 8.19
Distribution of twisting moment at $P = 153$ kip, no. 4

tips to 6.19 ksi (42.7 MPa) at the middle of the flange. From these two figures, it is known that the existence of residual strain reduces the computed value of the ultimate load by about 8.5 percent for both warping permitted and warping restrained conditions.

(d) *Twisting moment distribution*

The effect of warping restraint on the variation of external and internal torque along the column length at loads close to the ultimate is compared in Fig. 8.18 for Prob. 3 at $P = 136.9$ kip (609 kN) and in Fig. 8.19 for Prob. 4 at $P = 153$ kip (681 kN). In these illustrations, $M_{\zeta(3)} = \bar{K}\theta'$ is treated as a part of external moment. In the case of a beam-column with warping of the end cross sections permitted (Fig. 8.18), major part of the external twisting moment is due to $M_{\zeta(1)}$ [Eq. (8.44)] while the internal twisting moment is composed of almost equal magnitudes of

$$M_{\zeta(1)} = GK_T\theta' \qquad \text{and} \qquad M_{\zeta(2)} = \dot{M}'_\omega$$

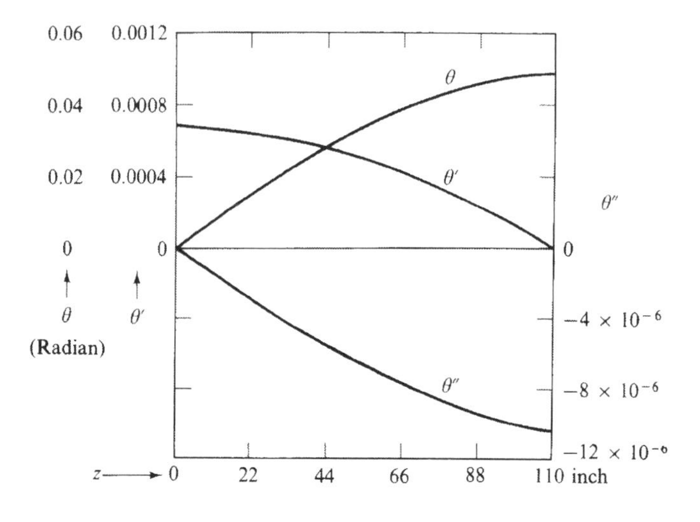

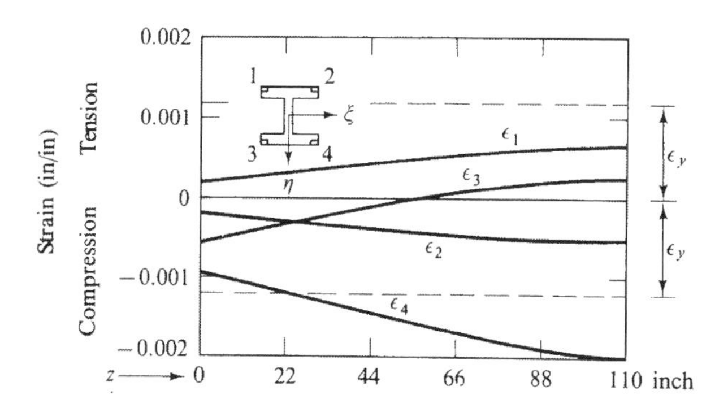

FIGURE 8.20
Distribution of twist and strains at $P = 136.9$ kip, no. 3

In the case of the warping restrained beam-column (Fig. 8.19), however, $M_{\zeta(1)}$ [Eq. (8.44)] has the most part of the contribution to the external moment while the internal twisting moment results mainly from $M_{\zeta(2)} = M'_{\omega}$.

(e) *Distribution of twist and strains*

The variation of twist and longitudinal strain at selected points on the cross section along the length are shown in Figs. 8.20 and 8.21 for Probs. 3 and 4, respectively. The amount of twist is considerably larger for Prob. 3 than for Prob. 4. It is known from the stress distribution that the yielded zone extends for nearly the entire length of the beam-columns.

(f) *Comparison with test results*

The deformational response predicted for the 12 beam-columns listed in Table 8.3 was compared with their experimental results due to Birnstiel (1968). As a

Table 8.3 COLUMN TEST SPECIMEN DATA (BIRNSTIEL, 1968)

Column No.	Nominal Size	Type[1]	Alloy	Length (in.)	Average Cross Sectional Dimensions: Flange Width (in.)	Flange Thickness (in.)	Depth (in.)	Web Thickness (in.)	Eccentricity of Loading (in.): e_x	e_y	Residual Strains Considered in Computations
1	6 × 6H	A	A7	96.0	6.01	0.60	6.01	0.51	1.61	2.78	Yes
2	5 × 5H	C	A36	96.0	5.10	0.42	5.13	0.29	1.60	3.21	Yes
3	5 × 5H	C	A36	120.0	5.11	0.42	5.13	0.28	0.80	2.63	Yes
4	6 × 6H	C	A36	96.0	6.11	0.45	6.45	0.33	1.66	2.95	Yes
5	5 × 5H	A	A36	96.0	5.00	0.51	5.01	0.41	2.36	3.17	Yes
6	5 × 5H	A	A36	120.0	5.02	0.51	5.02	0.41	2.38	2.51	Yes
7	6 × 5W	B	A36	96.0	5.01	0.48	6.29	0.33	−0.89	2.82	No
8	6 × 5W	B	A36	96.0	5.01	0.47	6.28	0.35	0.34	1.87	No
10	8 × 4W	B	A36	96.0	4.00	0.45	8.00	0.34	0.19	2.60	No
12	5 × 5H	C	A36	120.0	5.04	0.42	5.02	0.29	−0.77	2.78	Yes
13	4 × 4H	D	V65	120.0	4.01	0.35	4.12	0.30	0.42	2.72	No
14	4 × 4H	D	V65	120.0	4.01	0.35	4.12	0.31	0.83	2.35	No

1 in. = 25.4 mm

[1] Type A — Weldment, stress relieved and machined
Type B — Rolled shape, stress relieved and machined
Type C — As-rolled shape, not straightened
Type D — As-delivered shape, rotarized

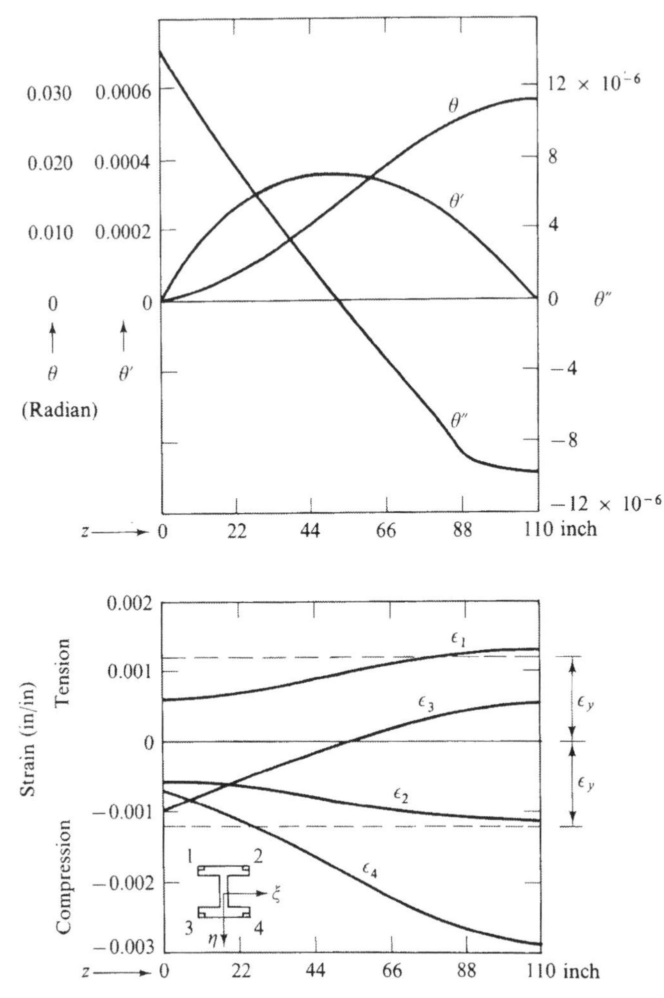

FIGURE 8.21
Distribution of twist and strains at $P = 153$ kip, no. 4

representative one, Fig. 8.22 shows the comparison for the specimen number 5 in Table 8.3. In most cases, adequate agreements were observed between computed curve and experimental results at load less than 90 percent of the measured ultimate loads. The difference in the ultimate load are less than 9 percent. The comparison of the ultimate loads are listed in Table 8.4. The largest differences are observed in the value of twisting angle which is considered due to initial twist of the specimens or, more likely, to twist of the end loading fixtures.

In order to further verify the computational procedure in the elastic-plastic range, the longitudinal strains were compared at four points of the midheight cross section for each beam-column listed in Table 8.3. As a representative one, the comparison is made in Fig. 8.23 for specimen number 5. The correspondence between

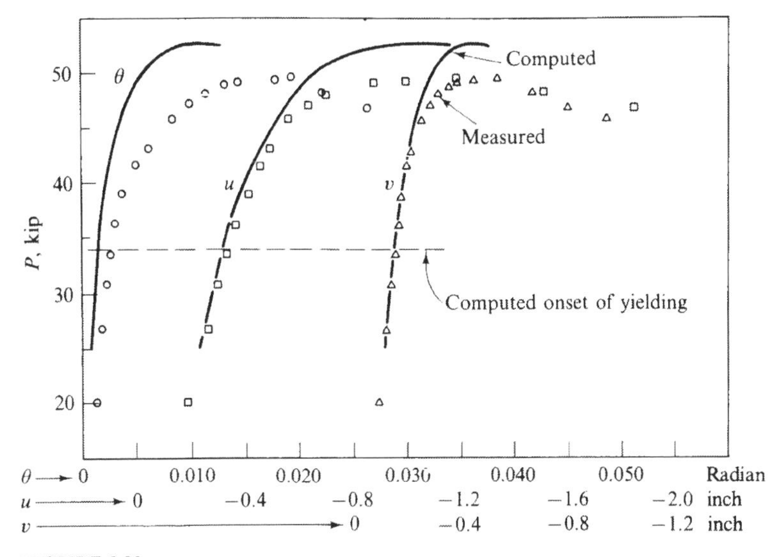

FIGURE 8.22
Comparison of deflections with test results, no. 5

measured and computed values of longitudinal strain is quite good for most beam-columns at loads less than 95 percent of the measured ultimate load except for some columns.

8.4 SUMMARY

Two numerical methods to solve space behavior of plastic beam-columns were explained following the works by Virdi and Dowling (1976) and Birnstiel, Harstead

Table 8.4 COMPARISON OF ULTIMATE LOAD—PREDICTED AND OBSERVED (BIRNSTIEL, 1968)

Specimen No.	Ultimate Load (kip)		Ratio: $\frac{P_{computed}}{P_{test}}$
	Computed	Test	
1	92.8	92.8	1.00
2	52.6	54.1	0.97
3	62.8	62.7	1.00
4	83.3	86.3	0.97
5	52.7	49.6	1.06
6	49.4	47.9	1.03
7	79.3	76.6	1.04
8	110.3	109.4	1.01
10	80.5	85.0	0.95
12	55.7	51.0	1.09
13	45.0	46.1	0.98
14	41.2	38.7	1.06

1 kip = 4.45 kN

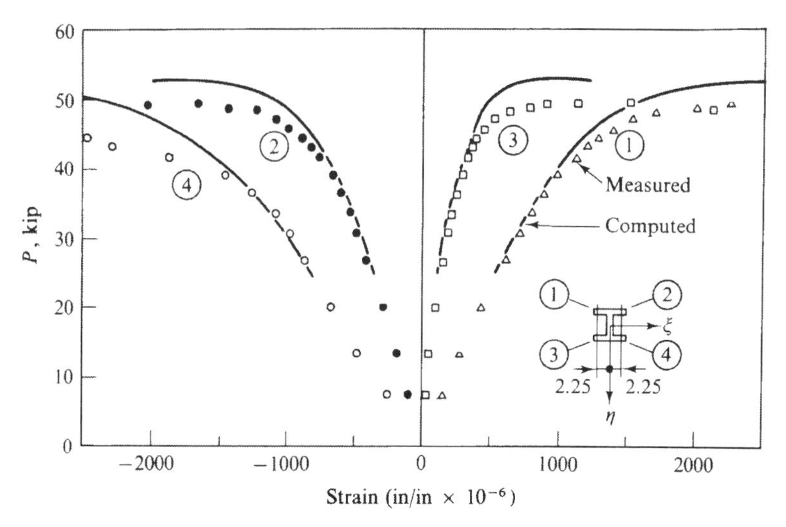

FIGURE 8.23
Comparison of strains with test results, no. 5

and Leu (1968). The approaches are categorized as the influence coefficient method because variables at stations which are set up along the length of the beam-column are iteratively solved starting from initially assumed values by making use of influence coefficient matrix.

When the effects of twisting on the overall geometry of the beam-column is small such as the cases for beam-columns of solid cross sections or thin-walled closed cross sections, the first method can be applied. The numerical procedures are summarized as follows,

Given: external loads $P, \tilde{M}_{xA}, \tilde{M}_{yA}, \tilde{M}_{xB}, \tilde{M}_{yB}$

end restraining function $\bar{M}_{xA}(\theta), \bar{M}_{yA}(\theta), \bar{M}_{xB}(\theta), \bar{M}_{yB}(\theta)$

moment-curvature-thrust relations $M_x = M_x(\Phi_x, \Phi_y, P)$

$M_y = M_y(\Phi_x, \Phi_y, P)$

1. Assume trial values of deflection

$$
\begin{aligned}
\{u_o\} &= \{u_2, u_3 \ldots u_n\} \\
\{v_o\} &= \{v_2, v_3 \ldots v_n\}
\end{aligned}
\tag{8.93}
$$

2. Compute end slopes

$$
\begin{aligned}
\theta_{xA} &= -\frac{1}{2\lambda}(v_3 - 4v_2), \quad \theta_{xB} = \frac{1}{2\lambda}(v_{n-1} - 4v_n) \\
\theta_{yA} &= \frac{1}{2\lambda}(u_3 - 4u_2), \qquad \theta_{yB} = -\frac{1}{2\lambda}(u_{n-1} - 4u_n)
\end{aligned}
\tag{8.94}
$$

3. Compute end moments

$$M_{xA} = \tilde{M}_{xA} - \bar{M}_{xA}(\theta_{xA}), \quad M_{xB} = \tilde{M}_{xB} - \bar{M}_{xB}(\theta_{xB})$$
$$M_{yA} = \tilde{M}_{yA} - \bar{M}_{yA}(\theta_{yA}), \quad M_{yB} = \tilde{M}_{yB} - \bar{M}_{yB}(\theta_{yB}) \tag{8.95}$$

4. Compute curvatures

$$\Phi_{xi} = -\frac{1}{\lambda^2}(v_{i-1} - 2v_i + v_{i+1})$$
$$\Phi_{yi} = \frac{1}{\lambda^2}(u_{i-1} - 2u_i + u_{i+1}) \tag{8.96}$$

5. Compute internal bending moments from moment-curvature-thrust relations

$$M_{xi} = M_x(\Phi_{xi}, \Phi_{yi}, P)$$
$$M_{yi} = M_y(\Phi_{xi}, \Phi_{yi}, P) \tag{8.97}$$

6. Compute new values for deflections from equilibrium condition

$$\bar{u}_i = -\frac{1}{P}\left[M_{yi} - \left(1 - \frac{i-1}{n}\right)M_{yA} - \frac{i-1}{n}M_{yB}\right]$$
$$\bar{v}_i = \frac{1}{P}\left[M_{xi} - \left(1 - \frac{i-1}{n}\right)M_{xA} - \frac{i-1}{n}M_{xB}\right] \tag{8.98}$$

7. Compare

$$\Sigma[(\bar{u}_i - u_i)^2 + (\bar{v}_i - v_i)^2] < \varepsilon$$

if yes, $\bar{u}_i$ and $\bar{v}_i$ are the solutions, otherwise,

8. Compute tangent rigidities

$$\alpha_{xxi} = \frac{\partial M_{xi}}{\partial \Phi_{xi}}, \quad \alpha_{xyi} = \frac{\partial M_{xi}}{\partial \Phi_{yi}}$$
$$\alpha_{yxi} = \frac{\partial M_{yi}}{\partial \Phi_{xi}}, \quad \alpha_{yyi} = \frac{\partial M_{yi}}{\partial \Phi_{yi}} \tag{8.99}$$

9. Compute tangent spring constants

$$\beta_{xA} = \frac{\partial \bar{M}_{xA}}{\partial \theta_{xA}}, \quad \beta_{xB} = \frac{\partial \bar{M}_{xB}}{\partial \theta_{xB}}$$
$$\beta_{yA} = \frac{\partial \bar{M}_{yA}}{\partial \theta_{yA}}, \quad \beta_{yB} = \frac{\partial \bar{M}_{yB}}{\partial \theta_{yB}} \tag{8.100}$$

10. Construct influence coefficient matrix

$$\begin{Bmatrix} d\bar{u} \\ \\ d\bar{v} \end{Bmatrix} = \begin{bmatrix} \dfrac{\partial U}{\partial u} & \dfrac{\partial U}{\partial v} \\ \\ \dfrac{\partial V}{\partial u} & \dfrac{\partial V}{\partial v} \end{bmatrix} \begin{Bmatrix} du \\ \\ dv \end{Bmatrix} \tag{8.101}$$

11. Compute new values for deflections on Newton-Raphson equations

$$\begin{Bmatrix} u^{k+1} \\ \\ v^{k+1} \end{Bmatrix} = \begin{Bmatrix} u^{k} \\ \\ v^{k} \end{Bmatrix} - \left([\mathrm{I}] - \begin{bmatrix} \dfrac{\partial U}{\partial u} & \dfrac{\partial U}{\partial v} \\ \\ \dfrac{\partial V}{\partial u} & \dfrac{\partial V}{\partial v} \end{bmatrix} \right)^{-1} \begin{Bmatrix} u^{k} - \bar{u} \\ \\ v^{k} - \bar{v} \end{Bmatrix} \tag{8.102}$$

12. Using the newly obtained values for deflections, repeat above procedures until convergence is reached.

Applying this method, composite steel-concrete beam-columns were solved and good agreement was reported by Virdi and Dowling (1976).

In order to solve beam-columns of thin-walled open cross section, the effects of twisting on the overall geometry of the beam-column must be taken into consideration for which the second method is proposed. The procedures are summarized as follows:

Problem

Known: curvatures in previous state (ξ'', η'', ε_o, θ'', P)
Given: load increment ($\dot{P}$)
Solve: curvature increments ($\dot{\xi}''$, $\dot{\eta}''$, $\dot{\varepsilon}_o$, $\dot{\theta}''$)

Procedure

Step 1 Assume curvature increments ($\dot{\xi}''$, $\dot{\eta}''$, $\dot{\varepsilon}_o$, $\dot{\theta}''$)

Step 2 Integrate for $\dot{\xi} = \int\int \dot{\xi}''(dz)^2$, $\dot{\eta} = \int\int \dot{\eta}''(dz)^2$, $\dot{\theta} = \int\int \dot{\theta}''(dz)^2$ (8.103)

Compute

$$\begin{aligned} u &= u^{k-1} + \dot{\xi} + \dot{\eta}_0(\theta + 0.5\dot{\theta}) + \eta_0\dot{\theta} \\ v &= v^{k-1} + \dot{\eta} - \dot{\xi}_0(\theta + 0.5\dot{\theta}) - \xi_0\dot{\theta} \end{aligned} \tag{8.104}$$

Step 3 Compute $\dot{\varepsilon} = \dot{\varepsilon}_o - \xi\dot{\xi}'' - \eta\dot{\eta}''$, $\quad \varepsilon = \varepsilon^{k-1} + \dot{\varepsilon}$ (8.105)

Determine $E_t = E$ or 0 for each element (Eq. 8.49)

Compute

$$\dot{N}_{\mathrm{int}} = \int E_t\dot{\varepsilon}\,dA, \quad \dot{M}_{\xi\mathrm{int}} = \int E_t\dot{\varepsilon}\eta\,dA, \quad \dot{M}_{\eta\mathrm{int}} = -\int E_t\dot{\varepsilon}\xi\,dA \tag{8.106}$$

$$P = P^{k-1} + \dot{P}, \quad M_{\xi\mathrm{int}} = M^{k-1}_{\xi\mathrm{int}} + \dot{M}_{\xi\mathrm{int}}, \quad M_{\eta\mathrm{int}} = M^{k-1}_{\eta\mathrm{int}} + \dot{M}_{\eta\mathrm{int}} \tag{8.107}$$

Step 4 Compute

$$M_{\xi ext} = -P[e_y - v - (e_x - u)\theta]$$
$$M_{\eta ext} = P[e_x - u + (e_y - v)\theta] \qquad (8.108)$$
$$M_{\zeta ext} = P[(e_x - u)v' - (e_y - v)u']$$

Step 5 Compare

$$\Delta\dot{P} = \dot{N}_{int} - \dot{P}, \ \Delta\dot{M}_{\xi} = \dot{M}_{\xi ext} - \dot{M}_{\xi int}, \ \Delta\dot{M}_{\eta} = \dot{M}_{\eta ext} - \dot{M}_{\eta int} \qquad (8.109)$$

If $(\Delta\dot{P}, \Delta\dot{M}_{\xi}, \Delta\dot{M}_{\eta})$ are small, go to Step 7.

Step 6 Compute corrections

$$\begin{Bmatrix} \Delta\dot{\xi}'' \\ -\Delta\dot{\eta}'' \\ \Delta\dot{\varepsilon}_o \end{Bmatrix} = \begin{bmatrix} G + H_{11} & H_{12} & H_{13} \\ H_{21} & G + H_{22} & H_{23} \\ H_{31} & H_{32} & G + H_{33} \end{bmatrix}^{-1} \begin{Bmatrix} \Delta\dot{M}_{\eta} \\ \Delta\dot{M}_{\xi} \\ \Delta\dot{P} \end{Bmatrix} \qquad (8.110)$$

Go to Step 1.

Step 7 Iterate for

$$\dot{\varepsilon}_w = \dot{\varepsilon}_{wo} + \dot{\theta}''(\xi\eta - \xi\eta_0 + \xi_0\eta)$$
$$\dot{N}_w = \int E_t\dot{\varepsilon}_w dA, \quad \dot{M}_{\xi w} = \int E_t\dot{\varepsilon}_w\eta\, dA, \quad \dot{M}_{\eta w} = -\int E_t\dot{\varepsilon}_w\xi\, dA$$
$$\eta_0 = \frac{\dot{M}_{\eta w}}{EI_y\dot{\theta}''}, \quad \xi_0 = \frac{\dot{M}_{\xi w}}{EI_x\dot{\theta}''} \qquad (8.111)$$

Compute

$$M_{\zeta(1)ext} = P[(e_x - u)v' - (e_y - v)u']$$
$$\dot{M}_{\zeta(2)ext} = \dot{P}(\eta_0 u' - \xi_0 v') + P(\eta_0\dot{u}' - \xi_0\dot{v}'), M_{\zeta(2)ext} = M^{k-1}_{\zeta(2)ext} + \dot{M}_{\zeta(2)ext}$$
$$M_{\zeta(3)ext} = -\theta'\overline{K} \quad \text{where} \quad \overline{K} = \int E_t\varepsilon[(\xi - \xi_0)^2 + (\eta - \eta_0)^2]\, dA \qquad (8.112)$$
$$M_{\zeta ext} = M_{\zeta(1)ext} + M_{\zeta(2)ext} + M_{\zeta(3)ext}$$

Step 8 Compute

$$M_{\zeta(1)int} = GK_T\theta' \quad \text{where} \quad K_T = \frac{A_E}{3A}\sum_i b_i t_i^3$$
$$\dot{M}_{\zeta(2)int} = \dot{\theta}''' \int E_t\xi^2(\eta - \eta_0)^2\, dA, M_{\zeta(2)int} = M^{k-1}_{\zeta(2)int} + \dot{M}_{\zeta(2)int} \qquad (8.113)$$
$$M_{\zeta int} = M_{\zeta(1)int} + M_{\zeta(2)int}$$

Step 9 Compare $\Delta M_{\zeta} = M_{\zeta ext} - M_{\zeta int}$ (8.114)

If ΔM_{ζ} is small, go to step 5.

Step 10 Correct $\dot{\theta}'' = (\dot{\theta}'')^{k-1} + \Delta\dot{\theta}''$ where $\Delta\dot{\theta}'' = [R + X]^{-1}\Delta\dot{M}_{\zeta}$ (8.115)

Go to step 7.

In the procedure summarized above, there are two iterative loops. The first nesting loop (Steps 6 to 1) are for corrections of the curvatures (ξ'', η'', ε_o) with respect to the corresponding bending moments and axial force (M_η, M_ξ, P). The second nesting loop (Steps 10 to 7) are for corrections of the twisting curvature (θ'') with respect to the twisting moment (M_ζ). For the calculation in the second loop, the location of center of twist is calculated at every stage of iteration.

The member treated is an H-shaped straight steel beam-column which is subjected to a proportional loading by axial force with biaxial eccentricity. The boundary conditions are the same at both ends including warping restraint condition. A half part of the beam-column is partitioned at several stations. At each station, the cross section is divided into small elements. On each element, stress and strain are calculated using the prescribed material characteristics of elastic-perfectly plastic. Residual strain can be read into each element.

Some numerical results were presented to show effects of warping restraint and existence of residual strain and it was seen that these have significant effect on the ultimate strength of the beam-column. The solutions were also compared with experimental results and good agreement was observed, except for the smaller computed twisting angle near the ultimate state which may be due to the initial twist or rotation of the end loading fixture.

The presented numerical procedure is quite general and can be applicable to other cross sections, other loading cases, other material properties and other beam-columns with initial imperfections.

8.5 PROBLEMS

8.1 Derive the approximate equations for the center of twist and the centroidal warping strain during the increment of load $\dot{P}$, Eq. (8.56), or

$$\Delta\xi_0 = \frac{\Delta \dot{M}_{\xi w}}{EI_x\theta''}, \quad \Delta\eta_0 = \frac{\Delta \dot{M}_{\eta w}}{EI_y\theta''}, \quad \Delta\dot{\varepsilon}_{wo} = \frac{\Delta\dot{N}_w}{EA}$$

Discuss the assumptions made.

8.2 Derive the following approximate equations for H-shapes

$$K_T = \frac{1}{3}\int G_t t^2 \, dA$$

$$M_\omega = \theta'' \int E_t \xi^2 (\eta - \eta_0)^2 \, dA$$

in which E_t and G_t are tangent moduli of the material.

8.3 Derive the submatrix $[G(nxn)]$ in Eq. (8.71).

8.6 REFERENCES

Birnstiel, C., "Experiments on H-Columns Under Biaxial Bending," *Journal of the Structural Division, ASCE*, Vol. 94, No. ST10, Proc. paper 6186, October, pp. 2429–2449, (1968).

Birnstiel, C. and Michalos, J., "Ultimate Load of H-Columns Under Biaxial Bending," *Journal of the Structural Division, ASCE*, Vol. 89, No. ST2, Proc. Paper 3503, April, pp. 161–197, (1963).

Culver, C. G., "Exact Solution of the Biaxial Bending Equations," *Journal of the Structural Division, ASCE*, Vol. 92, No. ST2, Proc. Paper 4772, April, pp. 63–83, (1966).
Goodier, J. N., "Buckling of Compressed Bars by Torsion and Flexure," *Cornell University Engineering Experiment Station Bulletin*, No. 27, December, (1941).
Goodier, J. N., "Flexural-Torsional Buckling of Bars of Open Section Under Bending, Eccentric Thrust or Torsional Loads," *Cornell University Engineering Experiment Station Bulletin* No. 28, January, (1942).
Harstead, G. A., Birnstiel, C. and Leu, K.-C., "Inelastic H-Columns Under Biaxial Bending," *Journal of the Structural Division, ASCE*, Vol. 94, No. ST10, Proc. Paper 6173, October, pp. 2371–2398, (1968).
Virdi, K. S. and Dowling, P. J., "The Ultimate Strength of Composite Columns in Biaxial Bending," *Proceedings, Institute of Civil Engineers*, No. 55, March, pp. 251–272, (1973).
Virdi, K. S. and Dowling, P. J., "The Ultimate Strength of Biaxially Restrained Columns," *Proceedings, Institute of Civil Engineers*, Part 2, No. 61, March, pp. 41–58, (1976).

9

NUMERICAL INTEGRATION METHOD FOR PLASTIC BEAM-COLUMNS

9.1 INTRODUCTION

The biaxially loaded beam-column problem can be approached from the standpoint of numerical integration for the Column Deflection Curves (CDCs). The concept of CDCs in two dimensions has been discussed thoroughly in Chap. 8 of Vol. 1 for the case of in-plane analysis of beam-columns in which the deflected axis of any beam-column problem can be represented by a portion of the CDCs. The CDC approach was employed years ago at Lehigh University to give needed solutions to various in-plane beam-column problems of interest in plastic design of multi-story steel frames (*ASCE Manual*, **41**, 1971). The application of the concept to three-dimensional beam-column problems is presented here.

The first part of the chapter discusses the particular application of the CDC concept for a biaxially loaded, thin-walled beam-column of *closed* cross section or solid cross section for which the three-dimensional column-deflection-curves can be constructed in space by numerical integration technique neglecting the effect of torsional stress and twisting deformation. A brief description of the numerical procedure follows a rigorous development of the relationship between moments and curvatures for a square and thin-walled box. Studies made of the ultimate capacity of hollow box and circular tube loaded biaxially using this analysis are presented.

The numerical integration method is then extended to solve biaxially loaded beam-columns of thin-walled *open* cross section for which the effect of torsional stress and twisting deformation upon the overall geometry of the beam-column must be considered. The method is used to predict the behavior and strength of beam-columns of wide flange sections.

The CDC method is developed further and applied to solve lateral torsional buckling of beam-columns. In this further development, strain hardening of material is considered in the analysis. Comparisons are then made with other theoretical solutions.

Although the technique of applying the CDC method to obtain solutions for in-plane beam-columns has been highly developed for many years, its extension to three-dimensional space situations is just beginning. Perhaps the delay of this development is a result of not being able to obtain the fundamental analytical relationship between moments and curvatures for a commonly used structural section under combined axial force, biaxial bending and twisting moment. This development requires a complete history for the distribution pattern of yielded zone in the cross section during the entire history of loading, which is very complicated when the residual stresses are included in the formulation.

In the CDC approach, a series of three-dimensional CDCs are generally constructed first. Solutions corresponding to a beam-column with any set of end moments and end restraints can then be obtained. The CDC method is most effective in developing design charts such as the ultimate strength interaction diagrams. It is generally not suitable for solving a specific beam-column problem. This is because of the nature of the column-deflection-curves which contain information for many beam-column problems. This was demonstrated clearly in the case of in-plane beam-column analysis (Chap. 8, Vol. 1) and will be demonstrated again in the case of biaxially loaded beam-columns of *closed* cross section. However, for thin-walled *open* cross section for which the effect of twist of the beam-column must be taken into consideration, the present method of CDC analysis can only deal with a specific beam-column problem. In other words, an ultimate strength interaction curve for a given beam-column can be constructed only by solving many individual problems with different loading combinations for the same beam-column. The volume of data obtained in one loading combination can not be used directly in generating solutions for other loading combinations. Each loading combination must be treated as a new problem.

9.2 ULTIMATE STRENGTH OF BEAM-COLUMN OF CLOSED SECTION

Since the effect of twisting deformation upon the overall geometry of a beam-column of closed section can be ignored, the basic approach here is essentially the same as that of the CDC method described in Chap. 8, Vol. 1 for in-plane beam-columns. A direct extension of the in-plane CDC method to the three-dimensional biaxially loaded beam-column is described here.

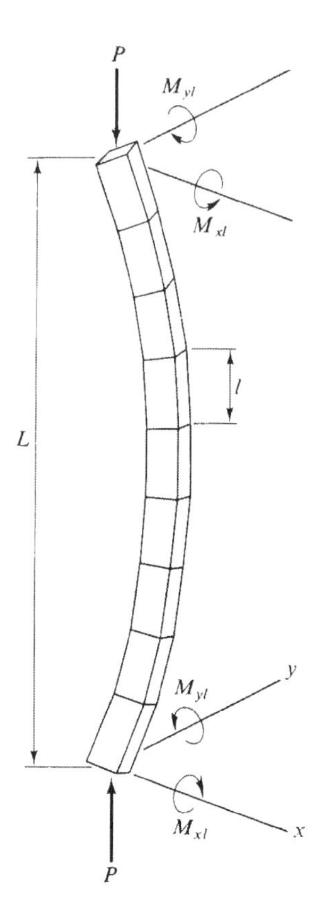

FIGURE 9.1
Beam-column consisting of segments

Consider a beam-column of arbitrary closed or solid section which is subjected to a set of symmetric end loads, i.e., the axial compression P and end moments M_{xl} and M_{yl} as shown in Fig. 9.1. In order to integrate the deflections along the length of the member, the beam-column is divided into a number of segments. The length of each segment (l) can, of course, be arbitrary but it is reported (Ellis, Jury and Kirk, 1964) that four times the radius of gyration of its cross section gives sufficiently accurate results for practical applications.

9.2.1 Moment-Curvature-Thrust Relationships

In the following analysis, we further assume that only the upper face of a segment moves downward by a displacement Δ in z-direction and rotates by angles θ_x and θ_y about the axes in x- and y-directions, respectively, using right-hand screw rule on front face as shown in Fig. 9.2. Thus, the average normal strain ε_o and the average curvatures Φ_x and Φ_y of the segment are obtained by

$$\varepsilon_o = -\Delta/l, \quad \Phi_x = \theta_x/l, \quad \Phi_y = \theta_y/l \tag{9.1}$$

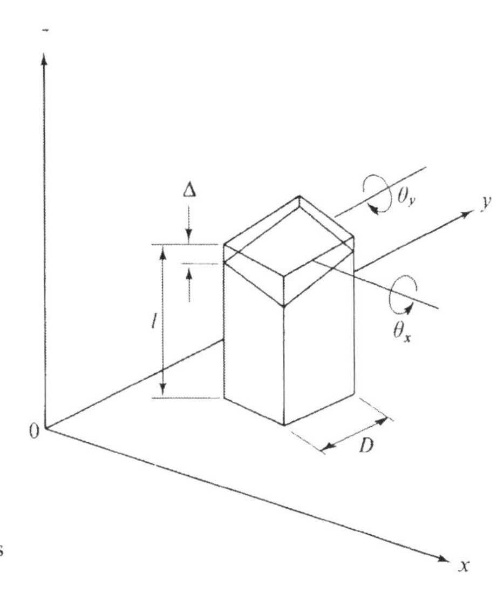

FIGURE 9.2
Segment and its deformations

From these generalized strains of a segment, the normal strain ε at any point (ξ, η) on the cross section can be calculated by

$$\varepsilon = \varepsilon_o + \eta\Phi_x - \xi\Phi_y \tag{9.2}$$

From this strain distribution, together with a stress strain relation of material, the stress distribution in the segment is known. For simplicity, we assume here an elastic-perfectly plastic material,

$$\sigma = \begin{cases} -\sigma_y & (\varepsilon < -\varepsilon_y) \\ E\varepsilon & (-\varepsilon_y \leq \varepsilon \leq \varepsilon_y) \\ \sigma_y & (\varepsilon_y < \varepsilon) \end{cases} \tag{9.3}$$

Integrating this normal stress over the entire cross section, the generalized stresses of a section, i.e., normal force N and bending moments M_x and M_y, are calculated numerically by

$$N = \int \sigma \, dA, \quad M_x = \int \sigma\eta \, dA, \quad M_y = -\int \sigma\xi \, dA \tag{9.4}$$

Summing up, when the deformations of a segment Δ, θ_x and θ_y are known, forces acting on the upper face of a segment N, M_x and M_y can be calculated. These are considered to be the moment-curvature-thrust relationships of the cross section in biaxial loading. An example of this relationship is shown in Fig. 9.3 for a hollow square cross section. The curves are expressed in terms of the nondimensional parameters

$$\frac{P}{P_y}, \quad \frac{M_x}{M_{px}}, \quad \frac{M_y}{M_{py}} \tag{9.5}$$

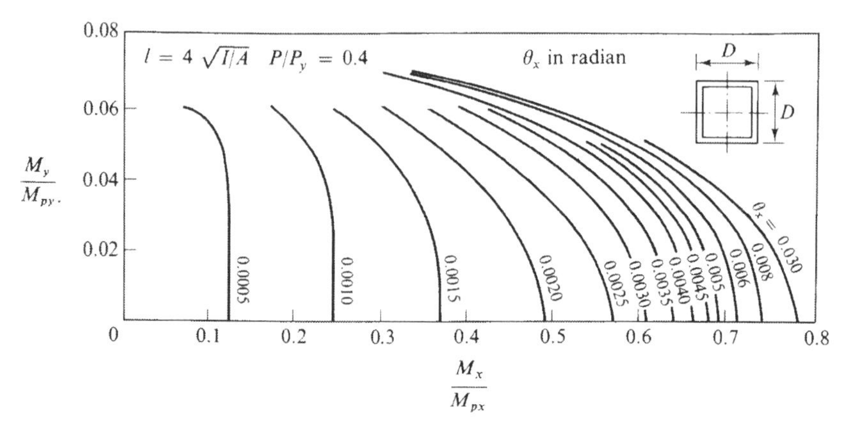

FIGURE 9.3
Moment curvature thrust relations

in which M_{px} and M_{py} are fully plastic moment of the cross section under zero axial thrust. In this example, the length l of a segment is selected to be four times the radius of gyration of the cross section r,

$$l = 4r = 4\sqrt{\frac{I}{A}}\left(= \frac{4}{\sqrt{6}}D\right) \tag{9.6}$$

in which D is the width of the square section. The biaxial moment curves shown in Fig. 9.3 are plotted for the constant axial force $P/P_y = 0.4$ with various constant values of θ_x. Since the section is doubly symmetric, this chart can also be used for the constant value of θ_y by simply exchanging x- and y-axes.

9.2.2 Column-Deflection-Curves in Space

The column-deflection-curve is a deflected shape of an axially loaded column after buckling due to a constant axial compression P Thus, the length of the column alters depending on the magnitude of the compressive force P Deflected configuration of any in-plane beam-column with end moments is known to be a part of the column-deflection-curve (Chap. 8, Vol. 1). In other words, one column-deflection-curve contains information for many beam-columns. The same concept is applied here to solve biaxially loaded beam-columns.

In order to construct the column-deflection-curves for a constant thrust P, set a global coordinate system x, y, z at the mid-span of the beam-column of length L as shown in Fig. 9.4. The number of segments is n for the half portion of the beam-column, $L/2 = nl$. The end moment M_{xl} and M_{yl} are represented by end eccentricities of the thrust P

$$M_{xl} = e_y P, \quad M_{yl} = -e_x P \tag{9.7}$$

The procedures to construct a column-deflection-curve are as follows:

Step 1 Place the first segment on the x-y plane at an arbitrarily selected lateral deflection point (u_m, v_m) as shown in Fig. 9.4.

Step 2 Since the bending moments in the segment are known by

$$M_{xm} = v_m P, \quad M_{ym} = -u_m P \tag{9.8}$$

The displacements of the upper face of the segment, $(\Delta_1, \theta_{x1}, \theta_{y1})$ are found from the moment-curvature-thrust relations of the segment prepared in Fig. 9.3. Now the location of the centroid of the upper face of the first segment is obtained

$$x_1 = u_m, \quad y_1 = v_m, \quad z_1 = l - \Delta_1 \tag{9.9}$$

Step 3 Place the second segment on the first one. Now the location of the centroid of the upper face of the second segment is found

$$\begin{aligned} x_2 &= x_1 + l\theta_{y1} \\ y_2 &= y_1 - l\theta_{x1} \\ z_2 &= z_1 + l - \Delta_2 \end{aligned} \tag{9.10}$$

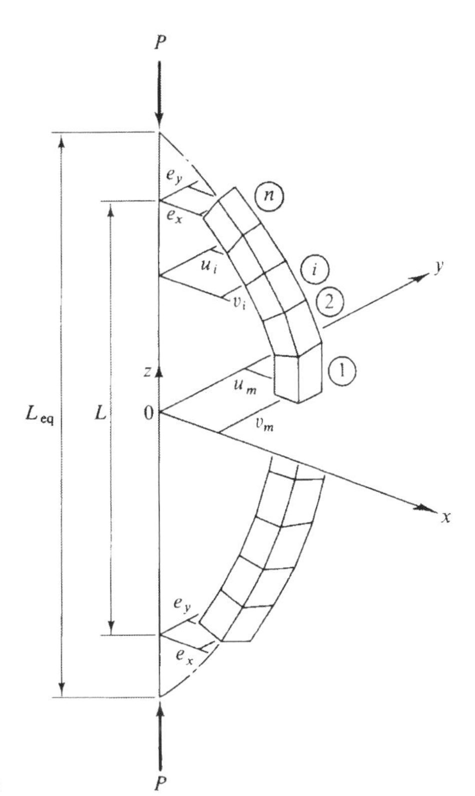

FIGURE 9.4
Construction of column deflection curve

Step 4 In general, for the ith-segment, we have

$$x_i = u_m + l \sum_{k=1}^{i-1} \theta_{yk}$$

$$y_i = v_m - l \sum_{k=1}^{i-1} \theta_{xk} \tag{9.11}$$

$$z_i = il - \sum_{k=1}^{i} \Delta_k$$

Repeat this calculation for segments $i = 1$ to n.

Step 5 At the end of the beam-column, we have

$$x_n = u_m + l \sum_{k=1}^{n-1} \theta_{yk}$$

$$y_n = v_m - l \sum_{k=1}^{n-1} \theta_{xk} \tag{9.12}$$

$$z_n = nl - \sum_{k=1}^{n} \Delta_k$$

Comparing these displacements with the end eccentricities e_x and e_y, if

$$x_n = e_x, \quad y_n = e_y \tag{9.13}$$

then, we know that the obtained configuration is the correct deflection of the beam-column. If Eqs. (9.13) are not satisfied, change the initially assumed mid-span deflection u_m and v_m and repeat above steps until Eqs. (9.13) are satisfied within a tolerable error. By the procedure described above, a half portion of the deflected shape of the beam-column can be solved. The other is simply a reflection of the first half with respect to the x-y plane for the symmetrically loading case.

9.2.3 Ultimate Strength of Beam-Columns

As previously mentioned, the column-deflection-curve method is very efficient in generating ultimate strength design chart for beam-column, rather than to solve a specific beam-column problem. A brief description of this is given below.

Consider a group of column deflection curves as shown in Fig. 9.5, where various mid-span deflections (u_m, v_m) of the axially loaded column under a constant thrust P are plotted. Noting that in constructing these column curves, no specific beam-column problems need be considered. When the column is entirely in the elastic range, many column-curves can pass through the same end points as shown by the dotted lines in Fig. 9.5. These are the elastic buckling curves for which the axial load P is the Euler's buckling load. The distance between the two fixed end points is Euler's buckling length. The existence of many such curves indicates the indefinite characteristics of deflection after buckling.

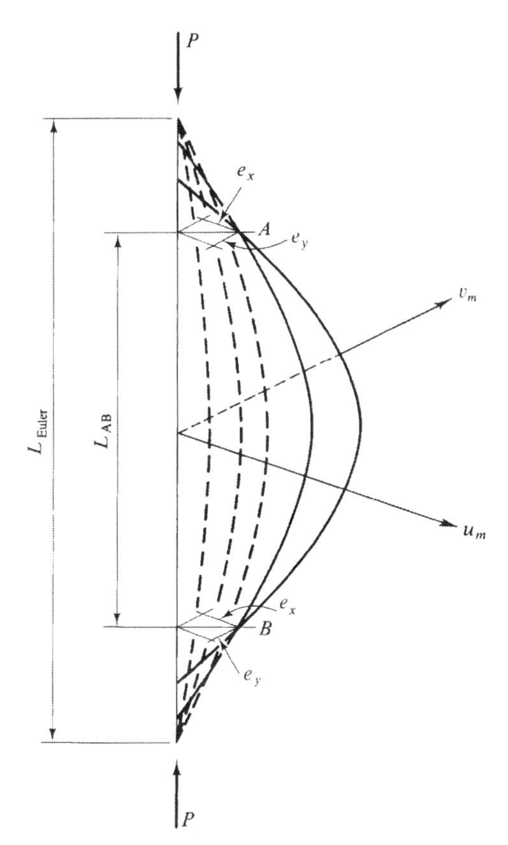

FIGURE 9.5
Ultimate strength criterion

This characteristic deflection can be used as the criterion for the ultimate state of plastic beam-columns. In general, the column-deflection-curves in the plastic range intersect one another such as the two solid curves shown in Fig. 9.5 which pass through the points A and B. This indicates that two beam-columns of the same length L_{AB} which are subjected to the same end loading of axial thrust P with end eccentricities e_x and e_y can have two different equilibrium configurations. If the difference between the two equilibrium configurations is infinitesimally small and yet it still has the same loading condition, we know that the system is unstable and the corresponding loading condition gives the ultimate state for the plastic beam-column (Prob. 9.1).

9.2.4 Numerical Examples

Using this approach, Ellis, Jury and Kirk (1964) computed ultimate strength of beam-column of a square hollow section (Fig. 9.6). In the first example, the biaxial end moments are the same for both axes, $M_{xl} = M_{yl}$. Thus, the plane of deflection lies in the plane of equal angle from the x and y axes as shown in Fig. 9.6(a). Utilizing the ultimate criterion described above, the relationship between the length

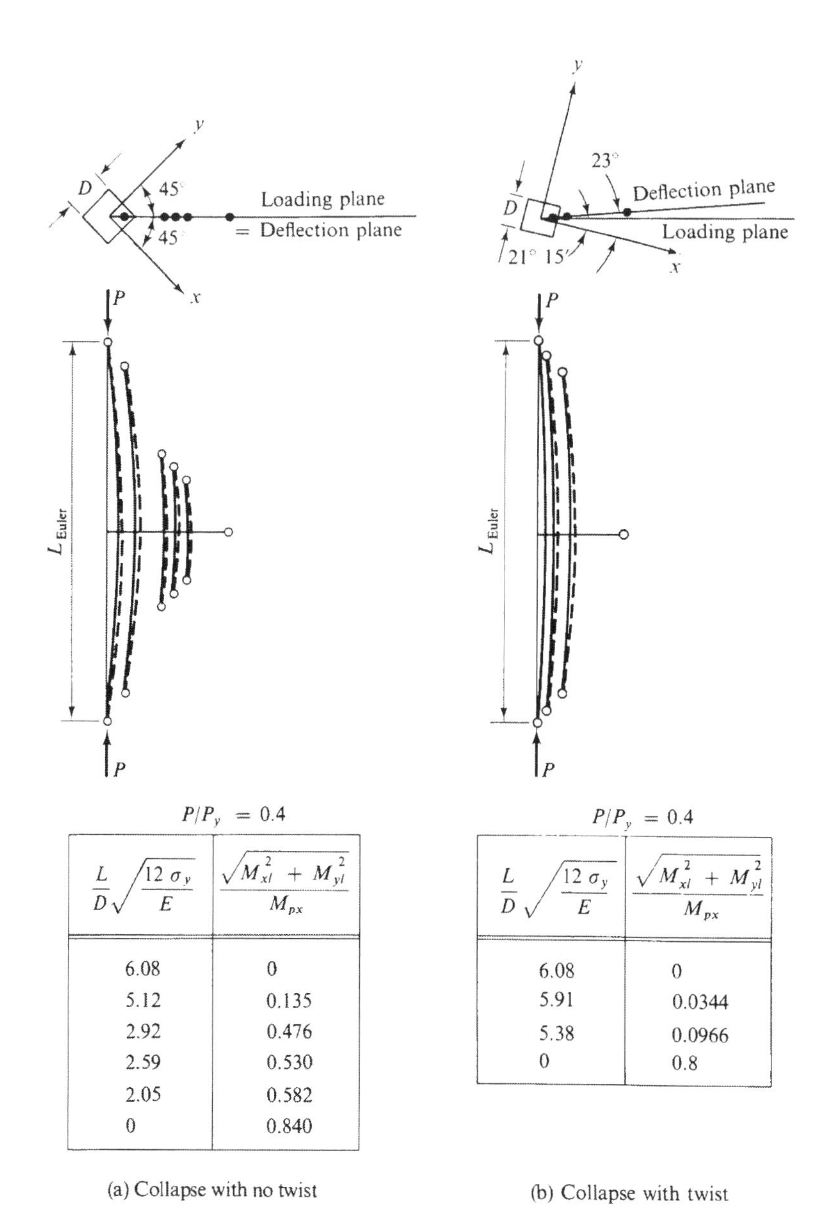

$P/P_y = 0.4$

$\frac{L}{D}\sqrt{\frac{12\,\sigma_y}{E}}$	$\frac{\sqrt{M_{xl}^2 + M_{yl}^2}}{M_{px}}$
6.08	0
5.12	0.135
2.92	0.476
2.59	0.530
2.05	0.582
0	0.840

(a) Collapse with no twist

$P/P_y = 0.4$

$\frac{L}{D}\sqrt{\frac{12\,\sigma_y}{E}}$	$\frac{\sqrt{M_{xl}^2 + M_{yl}^2}}{M_{px}}$
6.08	0
5.91	0.0344
5.38	0.0966
0	0.8

(b) Collapse with twist

FIGURE 9.6
Column deflection curves at ultimate state

L and the resultant moment $\sqrt{M_{xl}^2 + M_{yl}^2}$ at the ultimate state under a constant thrust $P = 0.4P_y$ is obtained as shown in the table of Fig. 9.6(a) in which the length and the resultant moment are nondimensionalized by

$$\frac{L}{D}\sqrt{\frac{12\sigma_y}{E}} \quad \text{and} \quad \frac{\sqrt{M_{xl}^2 + M_{yl}^2}}{M_{px}} \tag{9.14}$$

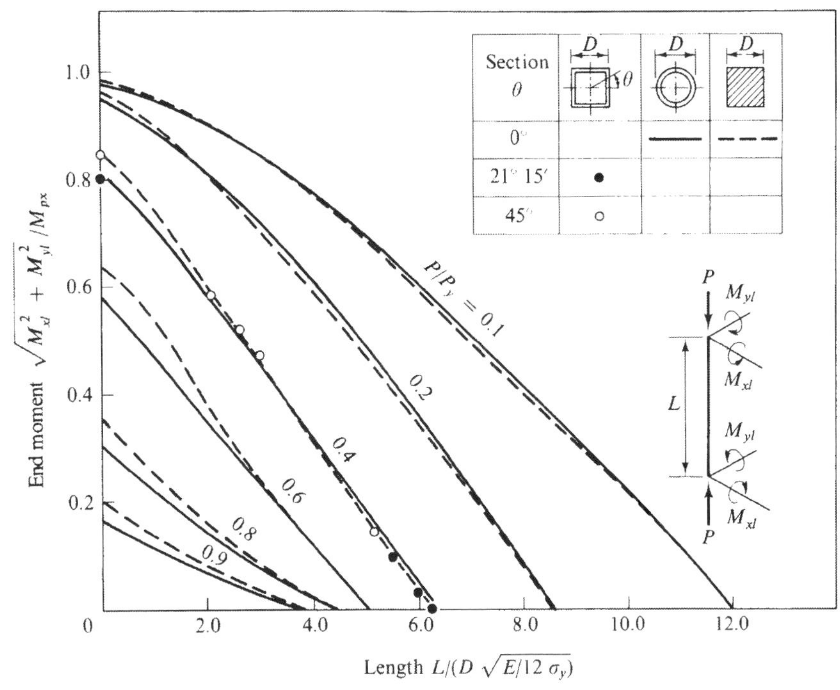

FIGURE 9.7
Numerical results of ultimate strength

When biaxial bending moments are applied differently about x- and y-axes so that the resultant moments are not in the plane of the symmetry of the cross section, the deflected configuration will not remain in the plane of the resultant moment (the loading plane). This was examined in Fig. 9.6(b) in which the loading plane is at 21° 15′ from the x-axis, but the maximum deflection solved by the numerical integration method occurs in the direction of 23° from the x-axis. It is known, in general, that the deviation of deflection plane from the loading plane is not great for beam-column of closed section.

The ultimate strength data obtained in the two examples of Figs. 9.6(a) and (b) are plotted in Fig. 9.7 by open circles and solid dots, respectively, together with numerical results of ultimate strength of beam-column of the rectangular cross section and circular tube cross section based on in-plane bending analysis. From Fig. 9.7, it is known that, if we plot the ultimate strength of beam-column of closed section in nondimensional parameters, large differences are not observed due to the difference in shape of the cross section and direction of loading plane.

9.2.5 Unsymmetric Loading Cases

The numerical integration technique described above is also applicable to unsymmetric loading cases, although the number of column-deflection-curves required

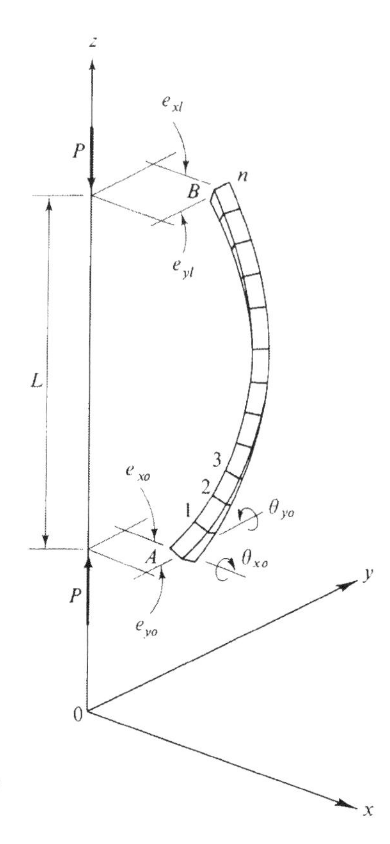

FIGURE 9.8
Column deflection curve for unsymmetric loading

to get an ultimate strength interaction curve becomes large. Since there exists no symmetry, the numerical integration must be done along the entire length of the beam-column starting from the segment at one end with all deflections and slopes specified.

The procedures to obtain the column-deflection-curves for a specified thrust P are as follows. First, the entire beam-column AB of length L is divided into n-segments of length l as shown in Fig. 9.8. At the starting point of integration, the bottom end A is selected at which eccentricities of thrust $e_{xo}(=u_0)$, $e_{yo}(=v_0)$ and slopes θ_{xo}, θ_{yo} are specified. Thus, the bending moments at this end are

$$M_{xo} = Pe_{yo} \qquad \text{and} \qquad M_{yo} = -Pe_{xo}$$

From these end loads on the first segment, the displacements at the upper face of this segment can be calculated using the moment-curvature-thrust relationships prepared in Fig. 9.3. The second segment is placed on the first one and deflections of its upper face are calculated in a similar manner as that of the symmetric loading cases. The procedures are repeated for all n-segments along the entire length L of the beam-column. Thus, the deflections at the top end B, $u_n(=e_{xl})$ and $v_n(=e_{yl})$ are obtained from which the end moments are

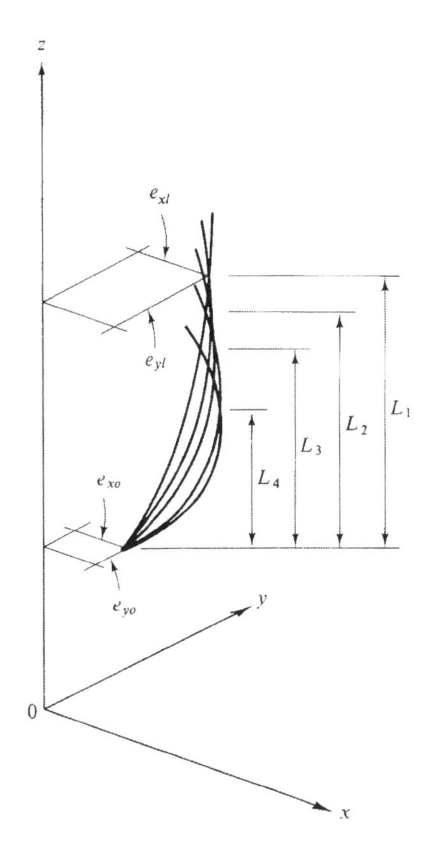

FIGURE 9.9
A set of column-deflection-curves in space

$$M_{xl} = Pe_{yl} \qquad \text{and} \qquad M_{yl} = -Pe_{xl}$$

The column-deflection-curve thus obtained is the solution of a beam-column. Changing the initially specified end slopes θ_{xo} and/or θ_{yo}, a set of column-deflection-curves is obtained as shown in Fig. 9.9. These curves do not give any specific information yet for a specific beam-column problem as they are constructed. As was done in symmetric loaaing cases, however, a pair of intersecting and adjacent column-deflection-curves which represents the ultimate condition of the member can now be found from this set of curves. The collapse information is then extracted from the intersection of the column-deflection-curves which are adjacent. This pair of ultimate column-deflection-curve is shown schematically in Fig. 9.10 for the cases of symmetric loading, unsymmetric single curvature loading and unsymmetric double curvature loading.

As can be seen in Fig. 9.9, these ultimate conditions are based on the intersection of adjacent column shapes which are functions of end moments, M_{xo}, M_{yo}, M_{xl}, M_{yl} and the length L. Plotting the results for collapse columns with lengths such as $L_1, L_2 \ldots$ as shown in Fig. 9.9 provides a set of ultimate strength interaction curves. Numerical calculations were made by Marshall and Ellis (1970) for unsymmetrically loaded beam-columns of a circular section and a box section which are shown in

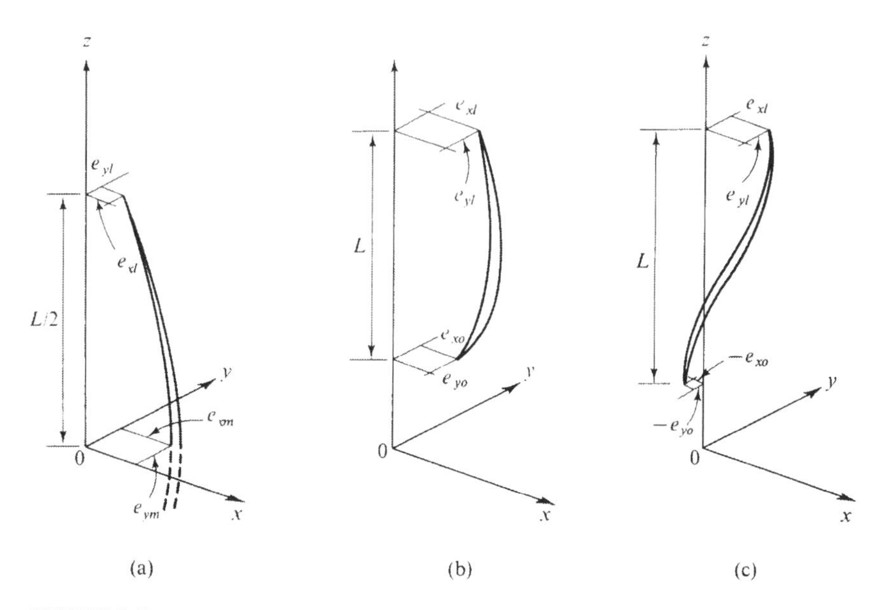

FIGURE 9.10
A pair of column-deflection curves representing ultimate state. (a) Symmetric loading. (b) Single curvature. (c) Double curvature

Fig. 9.11 where M_{px} denotes fully plastic moment, assuming no axial thrust, with respect to the x-axis of bending. For the interaction curves of circular cross section (solid lines in Fig. 9.11), in-plane analysis is available. The interaction curves of box section (indicated by circles in Fig. 9.11) were obtained by applying the bending moments in the plane of 26° 34′ from one of the principal axes under the constant axial thrust $P/P_y = 0.4$. In these results, significant differences between the two sections are not observed. This indicates that the ultimate strength of a box section is similar to that of a circular section. Hence, the complicated biaxial bending analysis for a box section may be substituted by an in-plane bending analysis.

9.2.6 Comparison with Experimental Results

Marshall and Ellis (1970) made a series of tests to examine the validity of the theory. Twenty-nine small scale steel beam-columns of box shape cross section were tested after the removal of residual stresses. Test results are listed in Table 9.1 and Table 9.2 in terms of ratios to the theoretical predictions. Table 9.1 shows all the symmetric loading cases in which specimens in series A, B and C were loaded in the plane of 26° 34′ and series G were in the plane of 45°. Table 9.2 shows the unsymmetric loading cases for both single and double curvature bendings. In all cases, good agreements were observed and the ratios of experimental results to theoretically predicted loads varied from 0.94 to 1.11 and had an average value of 1.03.

Table 9.1 EXPERIMENTAL RESULTS OF SYMMETRICAL SINGLE CURVATURE COLUMNS, NOMINAL SIZE 1 × 1 × 0.1 IN (MARSHALL AND ELLIS, 1970)

Column number	Lower yield stress σ_y psi	Slenderness ratio L/r	Nondimensional end moments		$\frac{P_{experimental}}{P_{predicted}}$	$\frac{P_{experimental}}{P_{yield}}$	$\frac{P_{experimental}}{P_{ultimate\,CRC^*}}$
			M_{xo}/M_{px}	M_{yo}/M_{py}			
S1	36 070	27	0.625	0.313	0.98	1.52	1.31
S2	34 600	27	0.621	0.310	1.05	1.61	1.40
S3	35 300	27	0.621	0.310	1.10	1.69	1.47
A1	34 300	50	0.508	0.254	1.09	1.53	1.36
A2	33 900	51	0.508	0.254	1.06	1.50	1.32
A3	35 190	50	0.508	0.254	1.05	1.48	1.32
B1	34 400	67	0.408	0.204	1.11	1.45	1.32
B2	35 260	67	0.408	0.204	1.09	1.43	1.31
B3	35 230	68	0.408	0.204	1.02	1.33	1.22
C1	37 700	92	0.244	0.122	1.09	1.28	1.28
C2	36 500	94	0.244	0.122	1.10	1.29	1.20
C3	36 300	94	0.244	0.122	1.01	1.28	1.15
G1	33 600	50	0.411	0.411	0.97	1.22	1.31
G2	33 900	50	0.411	0.411	1.00	1.24	1.33
G3	34 000	50	0.411	0.411	0.98	1.26	1.31
G4	33 000	64	0.375	0.375	0.98	1.20	1.30
G5	33 600	64	0.375	0.375	0.99	1.20	1.30
G6	32 900	64	0.375	0.375	0.96	1.17	1.26
G7	34 300	72	0.336	0.336	0.94	1.12	1.22
G8	34 000	72	0.336	0.336	0.96	1.14	1.24
G9	34 100	72	0.336	0.336	0.96	1.18	1.23

1 in = 25.4 mm,* Eq. (9.40), 1 psi = 6.9 kN/m²

Table 9.2 EXPERIMENTAL RESULTS OF NONSYMMETRICALLY LOADED COLUMNS, NOMINAL SIZE 1.5 × 1.5 × 0.125 IN (MARSHALL AND ELLIS, 1970)

Column number	Lower yield stress σ_y psi	Slenderness ratio L/r	Nondimensional end moments				$\frac{P_{experimental}}{P_{predicted}}$
			End A		End B		
			M_{xo}/M_{px}	M_{yo}/M_{py}	M_{xl}/M_{px}	M_{yl}/M_{py}	
D1	35 000	135	0.095	0.047	0	0	1.06
D2	36 200	135	0.070	0.035	0	0	1.11
D3	36 400	134	0.075	0.038	0	0	1.02
E1	33 100	80	0.582	0.497	−0.350	−0.175	1.03
E2	38 100	82	0.600	0.300	−0.149	−0.049	0.98
E3	36 400	84	0.300	0.150	0.105	−0.043	1.06
E4	33 100	41	0.625	0.395	0.400	0.100	1.04
E5	36 500	61	0.442	0.314	0.400	0.100	1.03

1 in = 25.4 mm, 1 psi = 6.9 kN/m²

For 21 of the 29 tests as listed in Table 9.1 the loads at which yielding first occurred were calculated and it was found that the ratios of experimental results to these computed yield loads varied from 1.12 to 1.69 and had an average value of 1.34. For these same 21 tests, the ratios of their experimental results to the values predicted by the Structural Stability Research Council (formerly Column Research

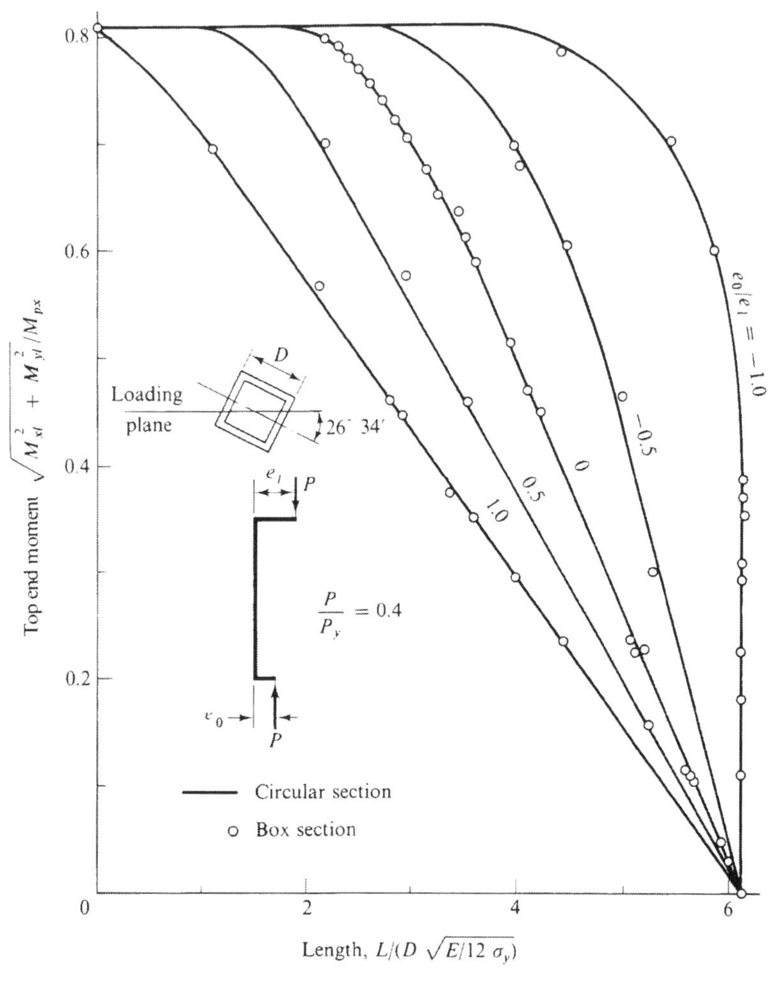

FIGURE 9.11
Interaction of unsymmetrically loaded beam-column

Council) formula [see Eq. (9.40)] for biaxial loading are also shown. Their ratios varied from 1.15 to 1.47 and had an average value of 1.29. Thus, it is known that the CRC equation underestimates by about 30 percent the ultimate capacities of biaxially loaded beam-columns of box cross section which are free from residual stresses. More accurate formulae for estimating the ultimate load carrying capacities of biaxially loaded beam-columns will be presented in Chap. 13.

9.3 ULTIMATE STRENGTH OF BEAM-COLUMN OF OPEN SECTION

Numerical integration method was applied by Aglan (1972) to obtain column-deflection-curves of a specific biaxially loaded beam-column of wide-flange shape of

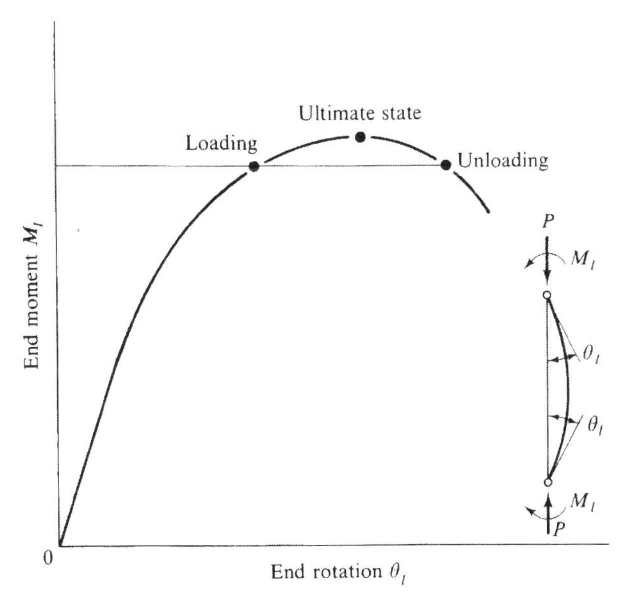

FIGURE 9.12
End moment vs. end rotation curve

cross section in which the effects of twist are included. This particular application of CDCs is described here. In the analysis, the effect of residual stresses is not taken into consideration; however, it can be handled using the same procedure with some modification. The procedure can also be modified to handle beam-columns of general open cross section.

In order to examine the ultimate strength of biaxially loaded beam-column, the end moment vs. end rotation ($M_l - \theta_l$) relations are searched for such as the curve shown schematically in Fig. 9.12. From the peak point of the load-rotation curve, conditions of the ultimate strength are known. The $M_l - \theta_l$ curves are readily obtained once the column-deflection-curves (CDCs) have been traced. Here, as in Chap. 8, Vol. 1, the CDCs in space are constructed by numerical integration of curvatures at each station along the length of beam-column. To solve the curvatures in terms of moments, the moment-curvature-thrust-twist relationships of the cross section must be utilized.

Since an open cross section such as the wide-flange shape is weak in resisting twisting moment, the effect of this torsion on the overall geometry of the beam-column must be taken into consideration. Torsional characteristics of an open cross section depend highly on the pattern of distribution of yielded zones of an elastic-plastic cross section. For simplicity, however, we neglect the shifting of shear center due to partial yielding of a cross section. Further, the axial shortening due to compression force P is also neglected. The boundary conditions considered in the present case are that both ends are simply supported with warping completely free:

$$u = v = \theta = o, \quad \theta'' = 0 \tag{9.15}$$

9.3.1 Basic Equations

From the assumptions of small displacements and symmetric loading denoted by the end moments M_{xl} and M_{yl}, the differential equations, Eq. (2.173), are applicable which is now written in terms of the shear center displacements u, v and θ as

$$\begin{aligned} &EI_x v'' + (v - \theta x_0 - \theta u)P + M_{xl} + \theta M_{yl} = 0 \\ &EI_y u'' + (u + \theta y_0 + \theta v)P - M_{yl} + \theta M_{xl} = 0 \\ &EI_\omega \theta''' - (GK_T + \bar{K})\theta' + u'(M_{xl} + Py_0 + Pv) + v'(M_{yl} - Px_0 - Pu) = 0 \end{aligned} \tag{9.16}$$

in which the sectional parameters, $EI_x, EI_y, EI_\omega, GK_T, x_0, y_0$ and $\bar{K}$ change according to the distribution of yielded portion of cross section during each stage of loading. The stress-strain relationship of material is assumed as:

$$\sigma = \begin{cases} E\varepsilon & (|\varepsilon| \le \varepsilon_y) \\ \pm\sigma_y & (|\varepsilon| \ge \varepsilon_y) \end{cases} \tag{9.17}$$

Since the location of shear center is assumed unaltered after partial yielding of the cross section, i.e.,

$$x_0 = 0, \quad y_0 = 0 \tag{9.18}$$

Further, neglecting all nonlinear terms in Eq. (9.16) and using the assumption $x_0 = y_0 = 0$ in the plastic range, one obtains

$$\begin{aligned} &EI_x v'' + Pv + M_{xl} + \theta M_{yl} = 0 \\ &EI_y u'' + Pu - M_{yl} + \theta M_{xl} = 0 \\ &EI_\omega \theta''' - (GK_T + \bar{K})\theta' + u'M_{xl} + v'M_{yl} = 0 \end{aligned} \tag{9.19}$$

Differentiate the last equation once and we have

$$EI_\omega \theta^{\mathrm{IV}} - (GK_T + \bar{K})\theta'' = \Phi_x M_{yl} - \Phi_y M_{xl} \tag{9.20}$$

in which Φ_x and Φ_y are bending curvatures,

$$\Phi_x = -v'' \qquad \text{and} \qquad \Phi_y = u'' \tag{9.21}$$

Next, assume the angle of rotation along the length of the member having the form

$$\theta = \theta_m \sin\frac{\pi z}{L} \tag{9.22}$$

which satisfies the end boundary conditions, Eq. (9.15). Considering the equilibrium at the mid-span, the maximum rotation angle θ_m is obtained from Eq. (9.20), using Eq. (9.22)

$$\theta_m = \frac{M_{yl}\Phi_{xm} - M_{xl}\Phi_{ym}}{EI_\omega\left(\dfrac{\pi}{L}\right)^4 + (GK_T + \bar{K})\left(\dfrac{\pi}{L}\right)^2} \tag{9.23}$$

in which Φ_{xm} and Φ_{ym} are curvatures at the mid-span of the beam-column.

The torsional coefficients of a plastic section EI_ω, GK_T and $\bar{K}$ are not constant but depend on the current stress distribution over the section, thus they must be calculated numerically after the strain distribution over the section has been determined from bending curvatures, axial strain and warping curvature. This will be described in the following section.

The St. Venant torsional rigidity GK_T is assumed to be equal to the elastic torsional rigidity of the section. This is because the contribution resulted from St. Venant torsional rigidity GK_T is not large compared with that of warping torsional rigidity EI_ω for an open cross section such as the wide-flange shape considered here. The warping rigidity, EI_ω, is calculated by considering only the elastic portion of the elastic-plastic cross section. Wagner's coefficient, $\bar{K} = \int \sigma a^2 \, dA$, is calculated numerically by dividing each flange and web area into finite element areas.

9.3.2 Moment-Curvature-Thrust-Twist Relationship

The normal strain of any point (ξ, η) on the cross section may be written as

$$\varepsilon = \varepsilon_o + \Phi_\xi \eta - \Phi_\eta \xi + \varepsilon_w \tag{9.24}$$

in which ε_o is the average normal strain and ε_w is the warping strain caused by warping restraint. From Eq. (2.90) we have

$$\varepsilon_w = \frac{\partial w}{\partial z} = w'_A - \theta''\omega_0 = -\theta''\omega_0 \tag{9.25}$$

in which the reference point A is selected on the web plate where no warping occurs. Since the shifting of shear center (x_0, y_0) is neglected, the double sectorial area for the wide-flange section (Fig. 9.13) is

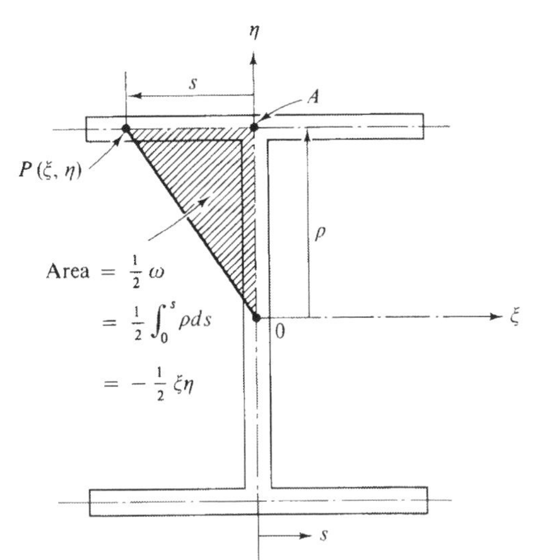

FIGURE 9.13
Coordinates on wide flange section

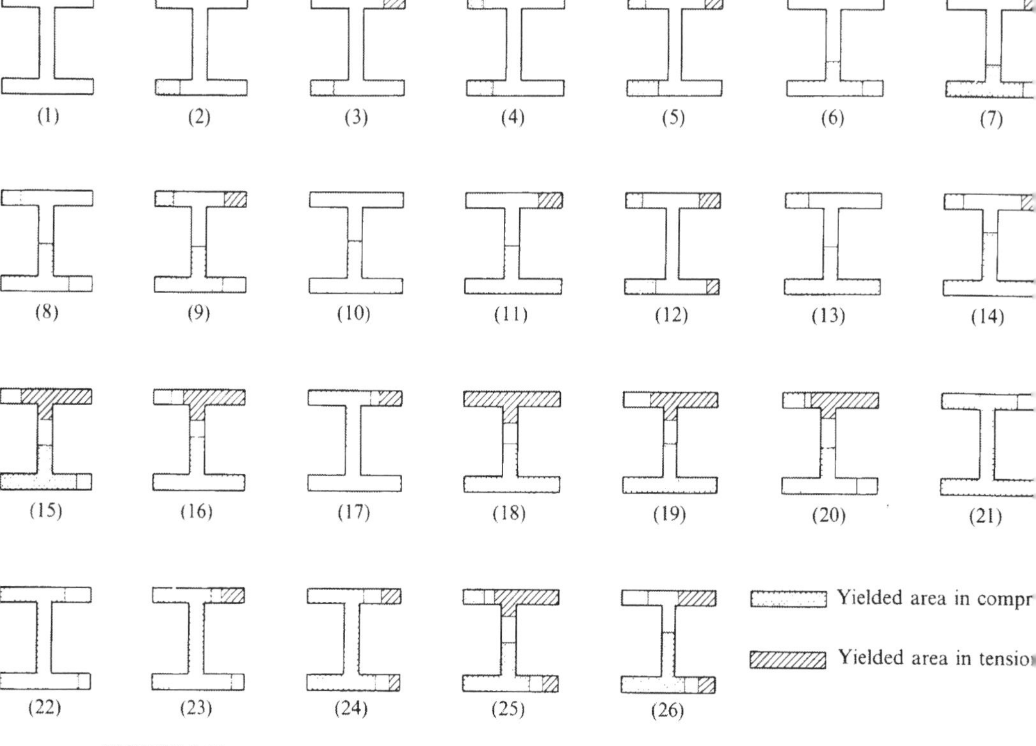

FIGURE 9.14
Possible strain-configurations for a wide flange section

$$\omega_0 = \omega = \int_0^s \rho \, ds = -\xi\eta \tag{9.26}$$

Thus, Eq. (9.24) is reduced to

$$\varepsilon = \varepsilon_o + \eta\Phi_\xi - \xi\Phi_\eta + \xi\eta\theta'' \tag{9.27}$$

From this, the yield condition, Eq. (9.17), is checked and the yield pattern of cross section is determined. For a biaxially loaded wide-flange shape, there are 26 strain configurations possible as shown in Fig. 9.14. For each strain configuration, stress distribution is determined from Eq. (9.17) and the internal resultant forces are obtained as

$$\begin{aligned} P &= \int_A \sigma \, dA \\ M_\xi &= \int_A \sigma\eta \, dA \\ M_\eta &= -\int_A \sigma\xi \, dA \end{aligned} \tag{9.28}$$

If all the generalized strains, i.e., ε_o, Φ_ξ, Φ_η and θ'' are given, the strain distribution $\varepsilon(\xi, \eta)$ is known from Eq. (9.27), thus the corresponding strain configurations can be selected from Fig. 9.14. From the strain distribution $\varepsilon(\xi, \eta)$, the corresponding stress distribution $\sigma(\xi, \eta)$ is determined using Eq. (9.17). Integrating the stress over the entire cross section, the internal forces are obtained by Eq. (9.28). Therefore, they may be written in general functional forms as

$$
\begin{aligned}
P &= f_o(\varepsilon_o, \Phi_\xi, \Phi_\eta, \theta'') \\
M_\xi &= f_\xi(\varepsilon_o, \Phi_\xi, \Phi_\eta, \theta'') \\
M_\eta &= f_\eta(\varepsilon_o, \Phi_\xi, \Phi_\eta, \theta'')
\end{aligned}
\tag{9.29}
$$

These functions, f_o, f_ξ and f_η can be determined for each strain configuration of Fig. 9.14. Since it is not practical in an actual iterative calculation to improve all the elements of the generalized strain $(\varepsilon_o, \Phi_\xi, \Phi_\eta, \theta'')$ at the same time, the average normal strain ε_o is usually used as the control of the iterative calculation. For this purpose, the first equation of (9.29) is solved for ε_o as

$$
\varepsilon_o = g_0(P, \Phi_\xi, \Phi_\eta, \theta'') \tag{9.30}
$$

The function g_0 can also be determined for each strain configuration.

Since the strain distribution is linear on each of the cross section plate elements (flanges and webs), the average strain function ε_o of Eq. (9.30) is a solution of the quadratic equation

$$
Q\varepsilon_o^2 + R\varepsilon_o + S = 0 \tag{9.31}
$$

in which Q, R and S are functions of P, Φ_ξ, Φ_η and θ''.

This relationship may be understood by considering a typical strain distribution in a plate element Bt as shown in Fig. 9.15. The strain at the center of the cross section can be written in the quadratic form as

$$
3\varepsilon_o^2 + 2\left(\varepsilon_y + \frac{B}{2}\Phi_\xi\right)\varepsilon_o - \left(\varepsilon_y - \frac{B}{2}\Phi_\xi\right)^2 - 2\frac{P}{Et}\Phi_\xi = 0 \tag{9.32}
$$

which is a particular case of Eq. (9.31) [for derivation of Eq. (3.32), see Prob. 9.2]. Aglan (1972) has derived expressions for Q, R and S for each strain configuration shown in Fig. 9.14.

The procedures in obtaining the moment-curvature-thrust-twist relationships for a cross section are now briefly described in what follows. Such typical curves are shown in Fig. 9.16.

Step 1.1 Specify values of P and θ''.

Step 1.2 Assign values of Φ_ξ and Φ_η and select a strain configuration for which the coefficient functions Q, R and S are calculated.

Step 1.3 Solve Eq. (9.31) for ε_o.

Step 1.4 Calculate strain distribution from Eq. (9.27).

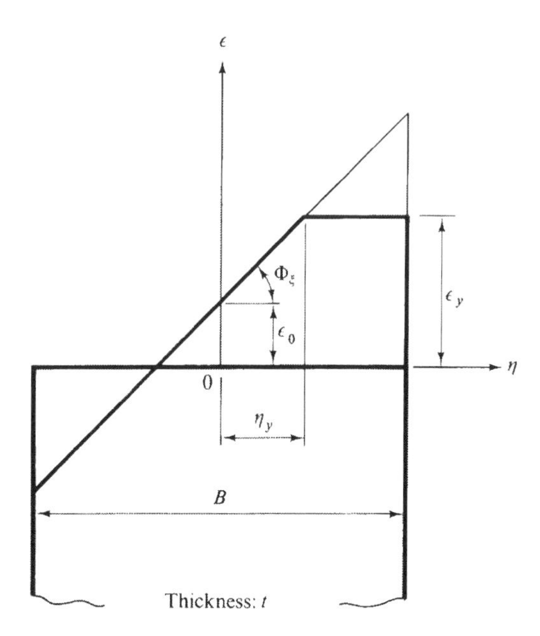

FIGURE 9.15
Strain distribution of a plate element

Step 1.5 Repeat Steps 1.2 to 1.4 until the calculated strain configuration becomes the same as that of the assumed one. The moments M_ξ and M_η are calculated using the last two equations of (9.28).

The procedures described above provide the moment-curvature-thrust-twist relationship in the form given in the last two equations of Eq. (9.29).

In the numerical integration procedure for determining the column-deflection-curves, different forms of moment-curvature-thrust-twist relationship are needed, i.e., the curvatures (Φ_ξ, Φ_η) are also required to be expressed explicitly in terms of the moments (M_ξ, M_η), thrust P and warping curvature θ'' as

$$\begin{aligned} \Phi_\xi &= g_\xi(M_\xi, M_\eta, \theta'', P) \\ \Phi_\eta &= g_\eta(M_\xi, M_\eta, \theta'', P) \end{aligned} \tag{9.33}$$

These relationships can be obtained from the relationships of Eq. (9.29) by the following procedures:

Step 2.1 By changing each value of Φ_ξ and Φ_η over the desired range, compute M_ξ and M_η and plot the curves shown in Figs. 9.16(a) and (b).

Step 2.2 Similar charts are obtained by varying the warping curvature θ''.

Step 2.3 For a specified value of M_ξ (for example, $M_{\xi 0}$), the curvatures Φ_ξ are found from the intersections of $M_{\xi 0}$ with the curves in Fig. 9.16(a). A plot of these intersections makes the curve A in Fig. 9.17.

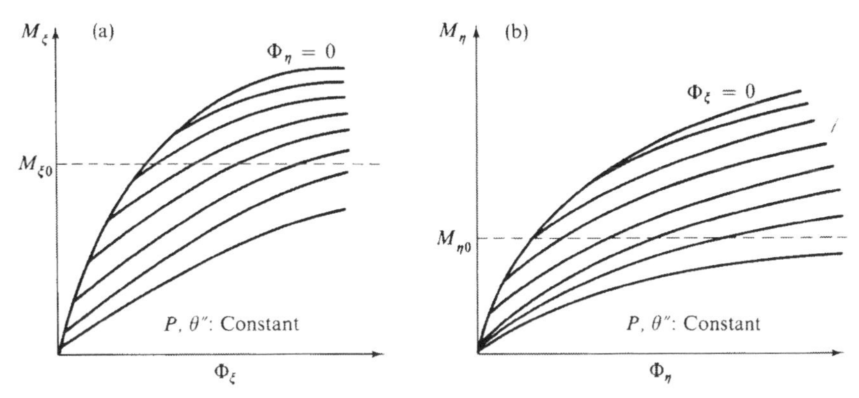

FIGURE 9.16
Typical moment-curvature curves

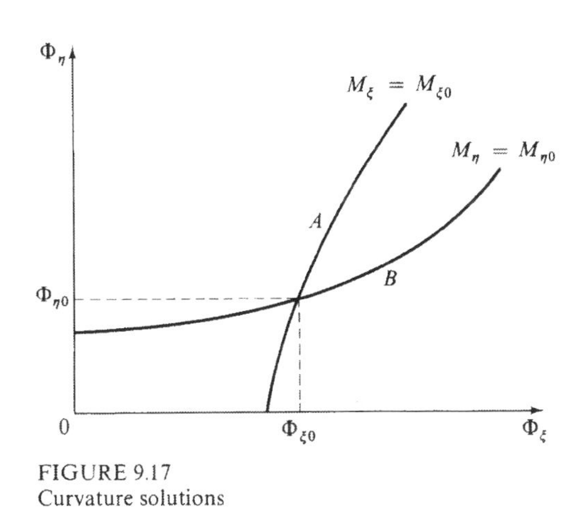

FIGURE 9.17
Curvature solutions

Step 2.4 In a similar manner, obtain curve B for $M_\eta = M_{\eta 0}$ in Fig. 9.17.

Step 2.5 The intersection of curve A with B determines the curvatures $\Phi_{\xi 0}$ and $\Phi_{\eta 0}$.

Thus, the relationships of Eq. (9.33) are developed.

9.3.3 Column-Deflection-Curves in Space

Figure 9.18 shows a typical column-deflection-curve for a biaxially loaded beam-column with equal eccentricities, e_x and e_y, at both ends in each of the principal directions. The ratio of the two end moments M_{yl} and M_{xl} is designated as γ

$$\gamma = \frac{-M_{yl}}{M_{xl}} = \frac{Pe_x}{Pe_y} = \frac{e_x}{e_y} \tag{9.34}$$

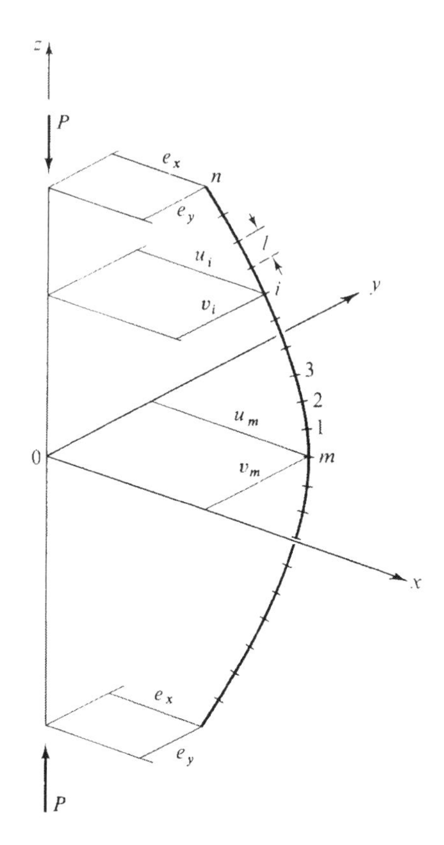

FIGURE 9.18
Column deflection curve in space

Since the beam-column is loaded symmetrically, only the upper half portion of the beam-column is considered and divided into n-small segments of equal length l as shown in Fig. 9.18. The displacements and curvatures at station i are denoted by (u_i, v_i) and (Φ_{xi}, Φ_{yi}), respectively. Assuming that the projection of the segment deflection onto each of the x-z and y-z planes is a flat circular arc, the following relations are obtained (see Chap. 8, Vol. 1).

$$
\begin{aligned}
v_{i+1} &= v_i + l v_i' - \frac{1}{2} l^2 \Phi_{xi} \\
u_{i+1} &= u_i + l u_i' + \frac{1}{2} l^2 \Phi_{yi} \\
\theta_{x(i+1)} &= v_{i+1}' = v_i' - l\Phi_{xi} \\
\theta_{y(i+1)} &= u_{i+1}' = u_i' + l\Phi_{yi}
\end{aligned}
\tag{9.35}
$$

in which the prime (′) indicates derivatives with respect to z.

At station i, bending moments about the axes in x and y directions are calculated by

$$\begin{Bmatrix} M_{xi} \\ M_{yi} \end{Bmatrix} = P \begin{bmatrix} 0 & 1 \\ -1 & 0 \end{bmatrix} \begin{Bmatrix} u_i \\ v_i \end{Bmatrix} \tag{9.36}$$

Denoting the rotation angle of the section at station i by θ_i, the bending moments about the principal axes ξ and η of the section are transformed from M_{xi} and M_{yi} using the simplified rotation matrix,

$$\begin{Bmatrix} M_{\xi i} \\ M_{\eta i} \end{Bmatrix} = \begin{bmatrix} 1 & \theta_i \\ -\theta_i & 1 \end{bmatrix} \begin{Bmatrix} M_{xi} \\ M_{yi} \end{Bmatrix} = -P \begin{bmatrix} \theta_i & -1 \\ 1 & \theta_i \end{bmatrix} \begin{Bmatrix} u_i \\ v_i \end{Bmatrix} \tag{9.37}$$

The curvatures in the two coordinate systems are also related by the rotation matrix,

$$\begin{Bmatrix} \Phi_{xi} \\ \Phi_{yi} \end{Bmatrix} = \begin{bmatrix} 1 & \theta_i \\ -\theta_i & 1 \end{bmatrix}^{-1} \begin{Bmatrix} \Phi_{\xi i} \\ \Phi_{\eta i} \end{Bmatrix} = \begin{bmatrix} 1 & -\theta_i \\ \theta_i & 1 \end{bmatrix} \begin{Bmatrix} \Phi_{\xi i} \\ \Phi_{\eta i} \end{Bmatrix} \tag{9.38}$$

in which inverse of the rotation matrix is equated to the transverse of it.

To construct the CDCs in space for a specified value of γ, the mid-span of the beam-column is selected as the starting point for the numerical integration process, since the slopes u'_m and v'_m at this point are both equal to zero (Fig. 9.18). The procedures are now outlined in the following steps:

Step 3.1 The displacements u_m and v_m are specified and the angle of twist, θ_m, is assumed. At the beginning of the calculations, it is convenient to assume θ_m to be equal to zero. Later, this value will be corrected by trial and error.

Step 3.2 The moments (M_{xm}, M_{ym}) and curvatures (Φ_{xm}, Φ_{ym}) are then calculated using Eqs. (9.36) to (9.38), together with moment-curvature-thrust-twist relationships.

Step 3.3 Deflections (u_1, v_1) and rotations $(\theta_{x1}, \theta_{y1})$ at the next station are found using Eqs. (9.35) where the subscript i refers to the mid-span. Then the process is repeated to cover the full length of the column ($i = 1$ to n).

Step 3.4 At the column end, the ratio $-M_{yl}/M_{xl}$ is calculated and compared to the specified value γ.

Step 3.5 If the calculated ratio is equal to γ plus or minus, a permissible error, the procedure is followed immediately by Step 3.6. If not, u_m is corrected.

Step 3.6 The angle of twist θ_m is then calculated using Eq. (9.23).

Step 3.7 The calculated angle of twist, θ_m, is used as the assumed one and the procedure is repeated from Step 3.3 until the difference between the calculated angle of twist and the assumed one comes within a tolerable error. This would give a column-curvature-curve of the beam-column.

Step 3.8 The value of u_m, v_m and θ_m are incremented by trial and error until sufficient number of column-curvature-curves are obtained.

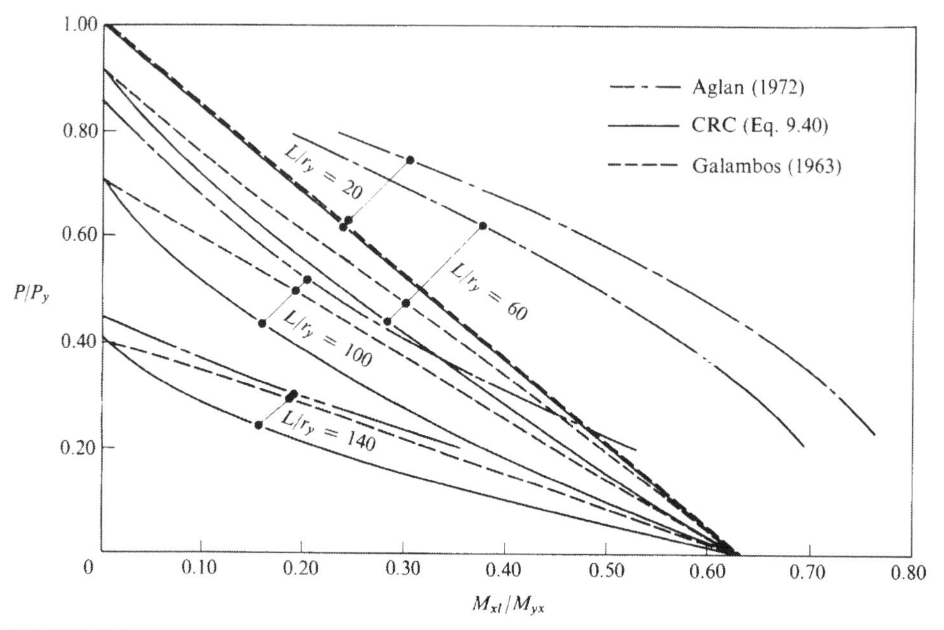

FIGURE 9.19
Interaction curves for W8 × 31 ($\gamma = e_x/e_y = 0.2$)

The end moment vs. end rotation curves can then be plotted from a set of the column-deflection-curves obtained by the procedures explained above.

9.3.4 Numerical Results

An example of a beam-column with W8 × 31 shape cross section is investigated by Aglan (1972) in which the slenderness ratio about the weak axis, L/r_y, was considered in the range from 20 to 140 and P/P_y from 0.2 to 0.8 with the values of $\gamma = e_x/e_y = 0.2$, 0.4 and 0.6.

The interaction curves, relating the axial force and the ultimate end moment, were presented in nondimensionalized form as shown in Figs. 9.19 to 9.21. The nondimensionalizations were made by using the initial yield quantities,

$$P_y = \sigma_y A$$
$$M_{yx} = 2\frac{EI_x}{D}\varepsilon_y \tag{9.39}$$

in which D is the depth of the wide flange shape.

The Second and Third Editions of the Column Research Council's *Guide to Design Criteria for Metal Compression Members* has the interaction equation (Johnston, 1966, 1976) [see Eq. (13.17) page 630]

$$\frac{P}{P_u} + \frac{M_x}{M_{ux}\left(1 - \frac{P}{P_{ex}}\right)} + \frac{M_y}{M_{uy}\left(1 - \frac{P}{P_{ey}}\right)} = 1.0 \tag{9.40}$$

where M_{ux} and M_{uy} are the ultimate bending moments of a *beam* about the x- and y-axis, respectively, reduced for the possible presence of lateral torsional buckling, if necessary, and P_{ex} and P_{ey} are Euler buckling loads with respect to the x- and y-axis, and P_u is the ultimate load of an axially loaded column. The CRC axially loaded column strength equation has the form [Eq. (14.15), Chap. 14, Vol. 1].

$$\frac{P_u}{P_y} = 1.0 - \frac{\sigma_y}{4\pi^2 E}\left(\frac{L}{r_y}\right)^2 \tag{9.41}$$

If the initial yield moments M_{yx} and M_{yy} are used for M_{ux} and M_{uy}, Eq. (9.40) can be rewritten in the form

$$\frac{P}{P_y}\left[\frac{1}{(P_u/P_y)} + \frac{\bar{e}_y}{(1 - P/P_{ex})} + \frac{\bar{e}_x}{(1 - P/P_{ey})}\right] = 1.0 \tag{9.42}$$

where the eccentricity ratios $\bar{e}_x$ and $\bar{e}_y$ are defined as

$$\bar{e}_y = \frac{e_y D}{2r_x^2}, \quad \bar{e}_x = \frac{e_x B}{2r_y^2} \tag{9.43}$$

in which r_x and r_y are the radius of gyration of the cross section about the x- and y-axis, respectively.

FIGURE 9.20
Interaction curves for W8 × 31 ($\gamma = e_x/e_y = 0.4$)

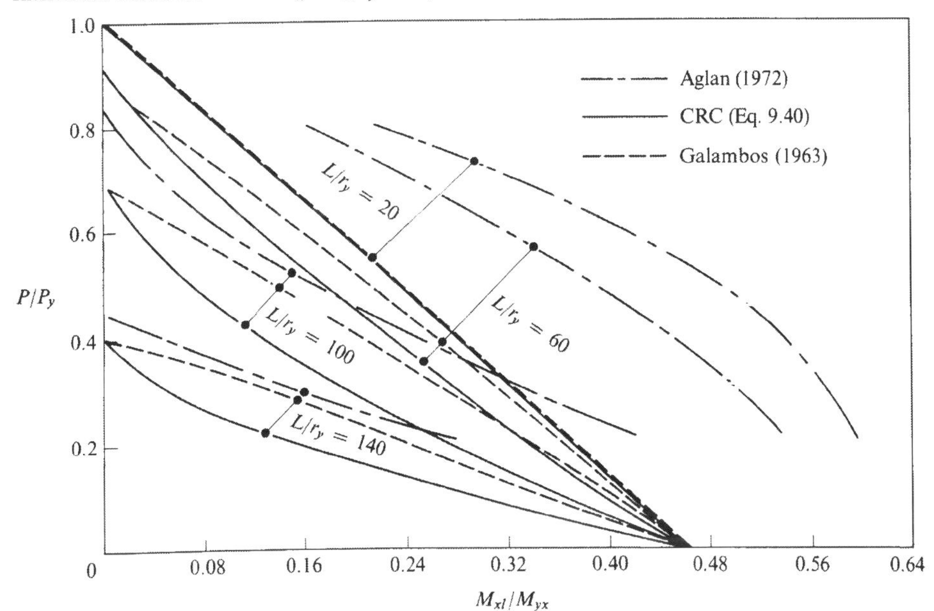

The results of Eq. (9.42) or Eq. (9.40) are also plotted on the same curves of Figs. 9.19 to 9.21. Galambos (1963) proposed a simplification of Eq. (9.42) as

$$\frac{P}{P_y} = \frac{1.0}{\dfrac{1}{(P_u/P_y)} + \bar{e}_y + \bar{e}_x} \tag{9.44}$$

This equation neglects the amplification factors $(1 - P/P_{ex})$ and $(1 - P/P_{ey})$ in Eq. (9.42). The interaction curves calculated by Eq. (9.44) are also shown in Figs. 9.19 to 9.21. From these figures, the following results are observed:

1. For a constant value of the axial load P and the slenderness ratio L/r_y, the ultimate end moment about the x-axis decreases appreciably with the increase in the ratio $\gamma = e_x/e_y$.
2. For a constant value of γ and L/r_y, the ultimate end moments decrease considerably with the increase in the value of P.
3. For a constant value of γ and P, there is a *significant* decrease in the values of the ultimate end moments as the slenderness ratio L/r_y increases.
4. The comparison between the interaction curves developed in this analysis with the interaction curves of the modified CRC Equation (9.42), show that the latter equation is too conservative, especially for short columns. This deviation is attributed to the fact that here the CRC equation is modified and based only on

FIGURE 9.21
Interaction curves for W8 × 31 ($\gamma = e_x/e_y = 0.6$)

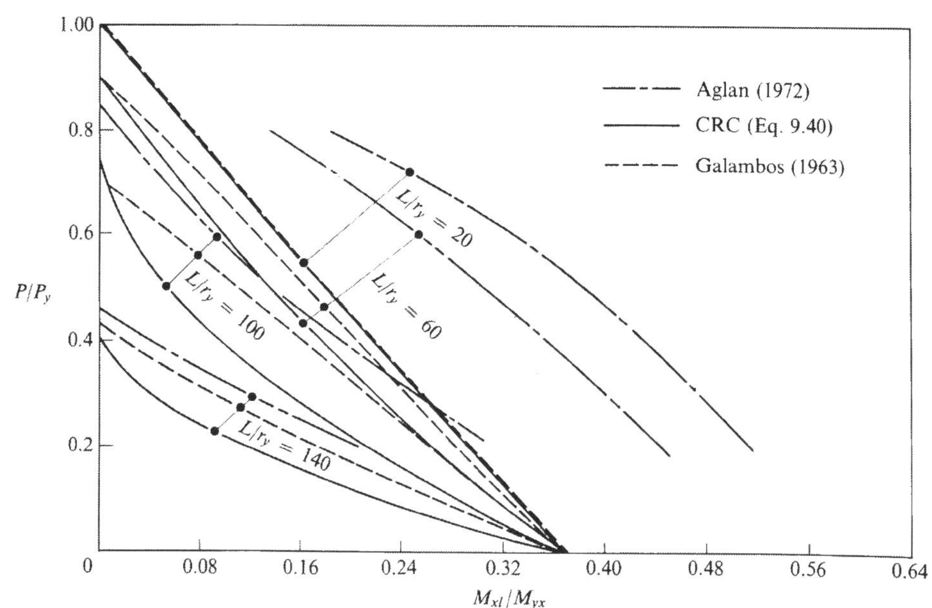

elastic analysis for moments M_{ux} and M_{uy}; whereas, for short beam-columns, a large part of the section yields before the ultimate carrying capacity of the beam-column is reached. Further discussions on various proposed interaction equations are given in Chap. 13.

5. The comparison between the interaction equation (9.44) with the present analysis shows a better agreement. Despite the unconservative assumptions, this equation gives conservative values for the ultimate load.

Table 9.3 shows the comparison between the ultimate strength predicted by the present analysis with those of the experimental works conducted by Chubkin (1959) and Klöppel and Winkelmann (1962), which are summarized by Galambos (1963) (see Sect. 13.7, Chap. 13, p. 658). Also, it shows a comparison between the present solution with the theoretical results by Harstead, Birnstiel and Leu (1968). The comparison shows good agreement between these results. A better agreement is observed when the slenderness ratio L/r_y increases since the effect of the residual stresses, which is neglected in this analysis, is predominant for beam-columns with smaller L/r_y ratios.

9.4 LATERAL TORSIONAL BUCKLING OF BEAM-COLUMNS

A wide flange beam-column subjected to an axial force and end moments only about its major axis may fail in lateral torsional buckling before it collapses by excessive bending as shown in Fig. 9.22. A detailed discussion on lateral torsional buckling of beam-columns has been presented in Chap. 3. Extending the column-deflection-curve method developed previously in Sect. 9.3 for analysis of biaxially loaded beam-column, Abdel-Sayed and Aglan (1973) obtain solutions for lateral torsional buckling of plastic beam-columns taking into consideration the strain

FIGURE 9.22
Lateral torsional buckling

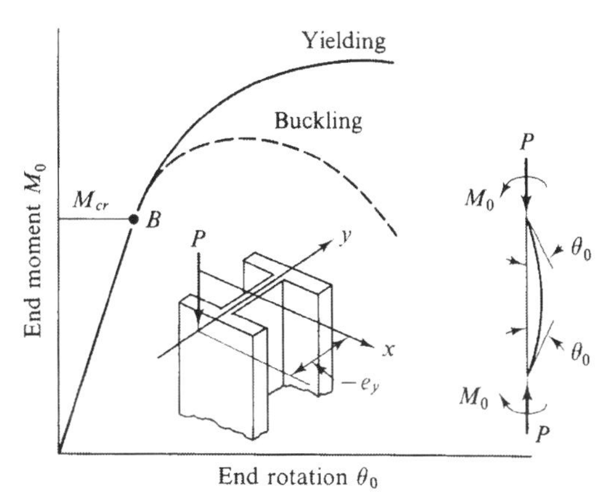

Table 9.3 COMPARISON OF ULTIMATE STRENGTH

Specimen number	Section size	$\frac{P}{P_y}$	Yield stress σ_y ksi	$\frac{L}{r_y}$	$\gamma = \frac{e_x}{e_y}$	Ultimate strength M_{xt}/M_{yx} Previous work	Present work	% Difference	Type of previous work
1	$B = 16$ cm	0.549	38.2	57	0.365	0.34	0.366	7.7	Experimental
	$D = 16$ cm								Chubkin (1959)
2	$t = 1.5$ cm	0.390	37.2	57	0.365	0.48	0.474	1.33	
	$w = 0.8$ cm								
3		0.214	37.4	114	0.365	0.338	0.350	3.5	
4	$B = 9.4$ cm	0.521	35.6	50	0.285	0.260	0.280	7.8	Experimental
	$D = 18$ cm								Klöppel and Winkelmann
	$t = 0.6$ cm								(1962)
5	$A = 30.6\ \text{cm}^2$	0.343	35.4	100	0.285	0.170	0.180	5.9	
6	W5 × 18.5	0.280	35.7	75	0.500	0.495	0.560	13.2	Theoretical
7	W5 × 18.5	0.280	35.7	94	0.305	0.525	0.525	9.0	Harstead *et al.* (1968)

1 ksi = 6.89 MPa

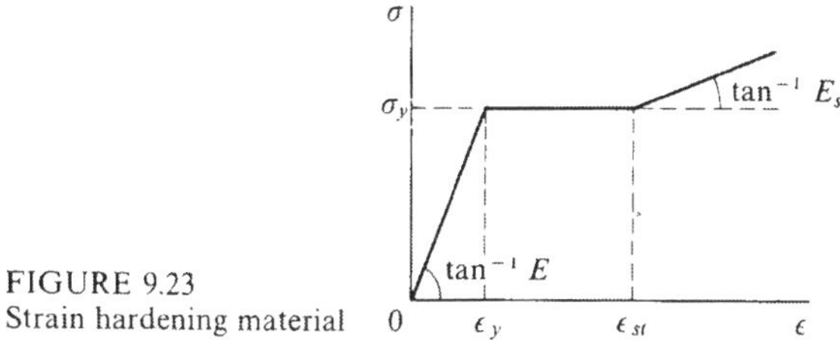

FIGURE 9.23
Strain hardening material

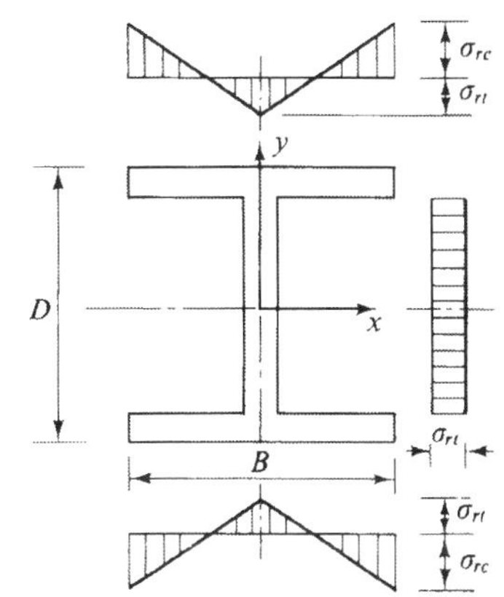

FIGURE 9.24
Residual stress

hardening in material (Fig. 9.23), residual stress in cross section (Fig. 9.24) and prebuckling deformation in beam-column.

9.4.1 Differential Equations

The differential equations governing the lateral torsional buckling of a beam-column which is subjected to axial compressive force P and end moment M_0 about the major axis at both end are obtained from Eq. (9.16) by substituting

$$\begin{aligned} M_{xl} &= M_0, \quad M_{yl} = 0 \\ x_0 &= 0, \qquad v \text{ By } v + v_i \end{aligned} \tag{9.45}$$

Thus,

$$\begin{aligned} &EI_x(v'' + v_i'') + (v + v_i)P = -M_0 \\ &EI_y u'' + Pu + [M_0 + P(v_i + y_0)]\theta = 0 \\ &EI_\omega \theta''' - (GK_T + \bar{K})\theta' + [M_0 + P(v_i + y_0)]u' = 0 \end{aligned} \tag{9.46}$$

in which v_i is the prebuckling deflection in the plane of loading. In the derivation of the first equation in Eq. (9.46), higher order terms related with θ and u are neglected so that this equation can solve the in-plane deflection $v + v_i$ independently to the other two equations. Lateral torsional buckling strength can be solved from the last two equations of Eq. (9.46).

9.4.2 Method of Solution

The method of solution is similar to those for biaxial loading analysis given previously in Sec. 9.3. In the present problem, the end moment vs. end rotation curve in Fig. 9.22 represents a stable uniaxially loaded beam-column deflected in the plane of applied end moments. The deflection curve and bending moment distribution are obtained from the first equation in Eq. (9.46) by the numerical integration of column deflection curve using the moment-curvature-thrust relations. With the specified thrust and moment, one can select one of the strain configurations in Fig. 9.25 which are similar to symmetric ones in Fig. 9.14 but adding the strain hardening and residual stress effects. For each of the strain configurations, the plastic section properties are calculated.

FIGURE 9.25
Sequence for checking the strain configurations

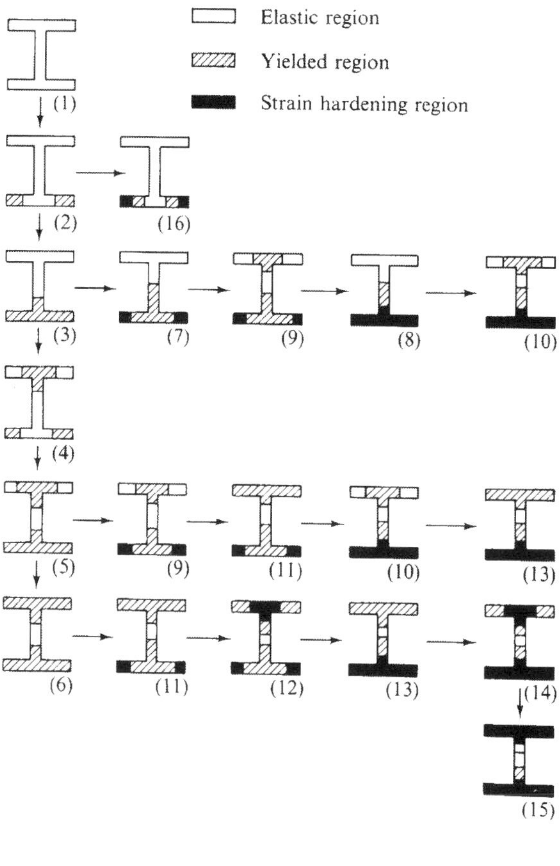

The lateral torsional buckling (bifurcation point in Fig. 9.22) is examined next using the last two equations of Eqs. (9.46). The coefficients in Eqs. (9.46) are variable with respect to z. Therefore, a direct solution is rather difficult and a finite difference approximation is applied. This leads to a set of simultaneous algebraic equations in terms of lateral displacement u_i and rotation angle θ_i at each station along the length of the beam-column. Further applications of finite difference method to plastic beam-columns will be presented in Chap. 10.

This system of equations can be written in matrix form as follows:

$$([C] - \lambda[I])\{X\} = 0 \tag{9.47}$$

in which

$$[C] = \text{real nonsymmetric coefficient matrix}$$

$$\lambda = \text{eigenvalues}$$

$$[I] = \text{unit matrix}$$

$$\{X\} = \text{eigenvectors}$$

The point of onset of lateral torsional buckling and the plastic section properties are interdependent. Therefore, a trial and error approach is applied to solve the problem.

A critical end moment, M^*_{cr}, is first assumed using an approximate method for a beam-column with specified length and axial load P The corresponding deflection and the moment at evenly spaced stations along the length of the beam-column can be calculated using the column-deflection-curves. The plastic section properties are calculated for the section at each station. The coefficient matrix $[C]$ can then be calculated and also the corresponding minimum critical end moment M_{cr} is obtained from the eigenvalues of matrix $[C]$. The calculated critical end moment M_{cr} is then compared with the assumed value M^*_{cr}. If the differences exceed the tolerable limit, then the procedure is repeated using the newly obtained value for M_{cr} until the difference becomes small enough to determine the critical end moment of the beam-column.

9.4.3 Numerical Results

A numerical example calculation is made for beam-column of W8 × 31 shape cross section. Stress-strain model of the material is (Fig. 9.23)

$$\sigma_y = 36 \text{ ksi (248 MPa)}, \quad \varepsilon_{st}/\varepsilon_y = 12, \quad E_{st}/E = 0.022$$

The moment-curvature-thrust relationship for this section is shown in Fig. 9.26 in which the domains of different strain configuration specified in Fig. 9.25 are also shown. The lateral torsional buckling strength curves are presented in Fig. 9.27 in which effects of residual stresses and prebuckling deflections are compared by four different curves denoted by numbers 1 to 4:

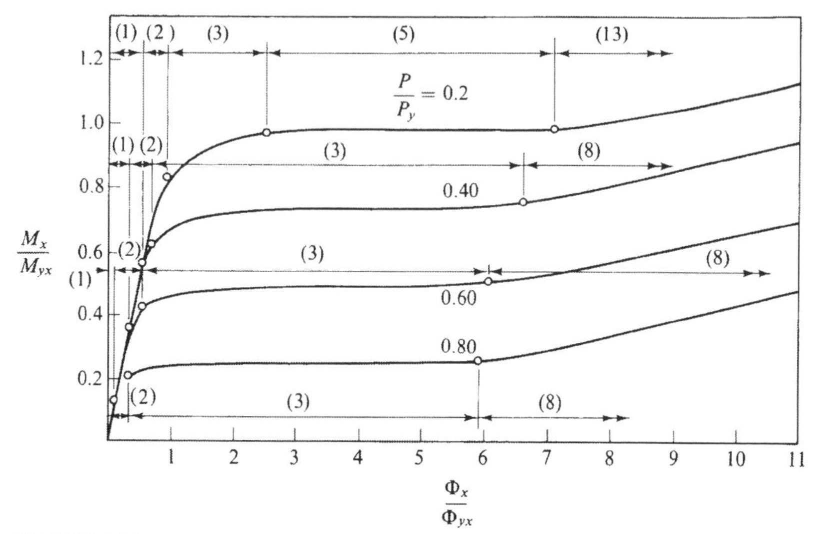

FIGURE 9.26
Moment-curvature-thrust relationships for W8 × 31 section

FIGURE 9.27
Numerical results for lateral torsional buckling

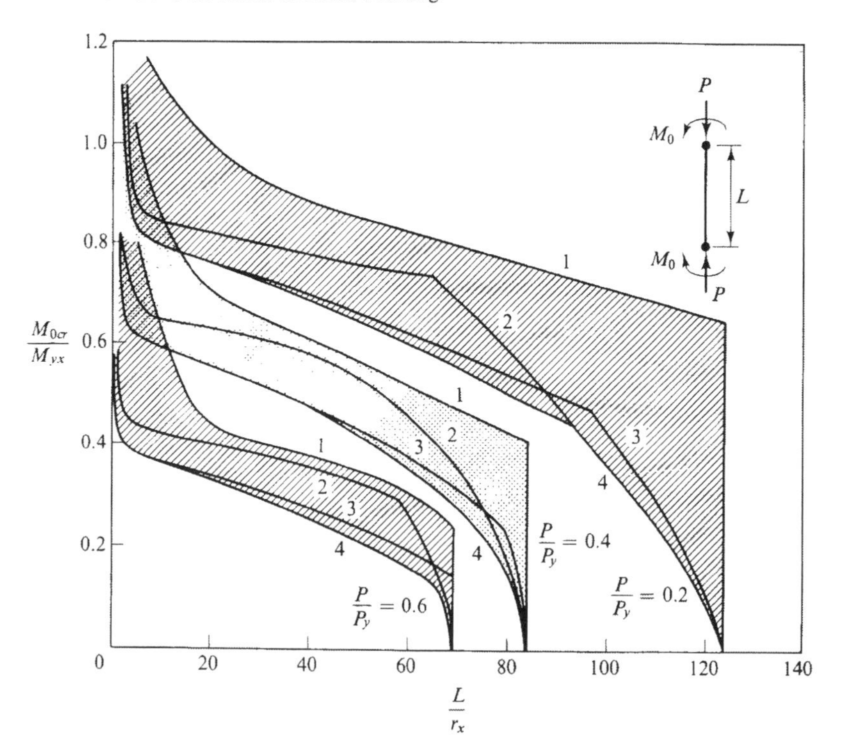

1. maximum moment by excessive yielding considering residual stresses;
2. lateral torsional buckling with no residual stresses but considering prebuckling deflections;
3. lateral torsional buckling with no prebuckling deflections but considering residual stresses;
4. lateral torsional buckling with residual stresses and prebuckling deflections.

It is seen that the residual stress has a negligible effect on the lateral torsional buckling strength in the elastic range but it has considerable effect within the plastic range (curves 2 and 4) because it extends the yielding zone in the cross section leading to a reduction in the bending and warping rigidities.

The effect of prebuckling deflections on the lateral torsional buckling strength of beam-columns is relatively small compared with that of residual stresses (curves 3 and 4). Especially for a shorter beam-column, the effect is negligible. This is

FIGURE 9.28
Effect of strain hardening excluding residual stresses

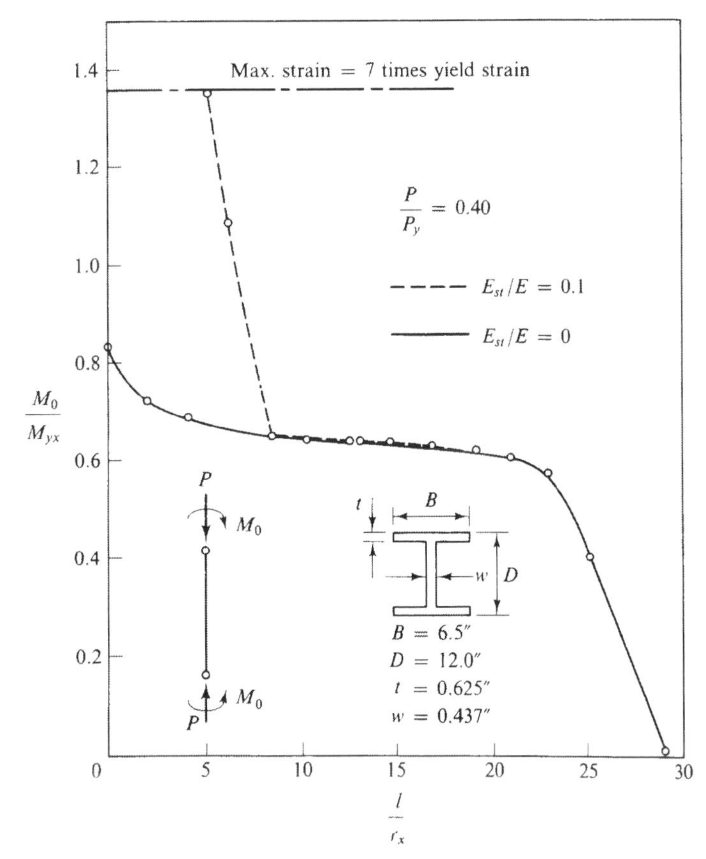

because the prebuckling displacements are much smaller than the shifting of shear center y_0 due to partial yielding of cross section for a short beam-column.

In order to examine the effect of strain hardening of material on the lateral torsional buckling strength of a beam-column, an aluminum alloy section (Alcan No. 28021) is examined considering $E_{st}/E = 0.1$ and $\varepsilon_{st}/\varepsilon_y = 1.0$. A comparison of two curves in Fig. 9.28 shows that the strain hardening has appreciable effect on the lateral torsional buckling strength of a relatively short beam-column (where $L/r_x < 8$).

9.5 SUMMARY

Two numerical integration methods to solve biaxially loaded beam-columns were studied. One for closed cross section and the other for open cross section.

The former is simple because, with closed cross section, the twisting deformation is small and the effect of this upon the overall geometry can be ignored. The construction of column-deflection-curves follows, essentially, the same procedure as that for in-plane beam-columns (Chap. 8, Vol. 1). Using biaxial moment-curvature-thrust relationships of segments, the column-deflection-curves are constructed in space by numerical integration of curvatures starting from one end and proceeding to the other end. When a symmetrically loaded beam-column is to be solved, the numerical integration can be started from the mid-span section where the slopes in both x and y axes are zero. Thus, considerable computer time can be saved.

From the set of column-deflection-curves thus obtained, a pair of intersecting and adjacent curves which represent the ultimate state are selected numerically from which the collapse information can be extracted. Numerical calculations were shown both for symmetric loading and for unsymmetric loading.

In open cross section, the twisting of a beam-column must be taken into consideration. In order to avoid the difficulty in determining twisting deformation, the rotation angle is assumed in a simple function,

$$\theta = \theta_m \sin \frac{\pi z}{L} \tag{9.48}$$

The rotation angle at the mid-span θ_m is related to end moments and mid-span curvatures by

$$\theta_m = \frac{M_{yl}\Phi_{xm} - M_{xl}\Phi_{ym}}{EI_\omega\left(\frac{\pi}{L}\right)^4 + (GK_T + \bar{K})\left(\frac{\pi}{L}\right)^2} \tag{9.49}$$

First, the moment-curvature-thrust-twist relations are prepared in the form

$$\begin{aligned} M_\xi &= f_\xi(\Phi_\xi, \Phi_\eta, \theta'', P) \\ M_\eta &= f_\eta(\Phi_\xi, \Phi_\eta, \theta'', P) \end{aligned} \tag{9.50}$$

These relationships are obtained by the following procedures under a given thrust P.

1. Specify curvatures Φ_ξ, Φ_η and θ''.
2. Select a strain configuration in Fig. 9.14.
3. Compute coefficients, Q, R and S.
4. Solve ε_o from the quadratic equation

$$Q\varepsilon_0^2 + R\varepsilon_0 + S = 0 \tag{9.51}$$

5. Determine strain distribution by

$$\varepsilon = \varepsilon_o + \eta\Phi_\xi - \xi\Phi_\eta + \xi\eta\theta'' \tag{9.52}$$

6. Compare the strain configuration with the selected one in Step 2. Repeat Steps 3, 4 and 5 until it falls in the same strain configuration.
7. Compute stress distribution by

$$\sigma = \begin{cases} -\sigma_y & (\varepsilon < -\varepsilon_y) \\ E\varepsilon & (-\varepsilon_y \le \varepsilon \le \varepsilon_y) \\ \sigma_y & (\varepsilon_y < \varepsilon) \end{cases} \tag{9.53}$$

8. Integrate the stress over the section and obtain internal moments

$$M_\xi = \int_A \sigma\eta \, dA, \quad M_\eta = -\int_A \sigma\xi \, dA \tag{9.54}$$

Thus, the moment-curvature-thrust-twist relationships in Eq. (9.50) are obtained. In an actual application, these relationships are prepared in charts and can be used to obtain curvatures explicitly in the form

$$\begin{aligned} \Phi_\xi &= g_\xi(M_\xi, M_\eta, \theta'', P) \\ \Phi_\eta &= g_\eta(M_\xi, M_\eta, \theta'', P) \end{aligned} \tag{9.55}$$

In order to obtain the ultimate strength of beam-column which is subjected to end moments due to thrust eccentricities,

$$M_{xl} = Pe_y, \quad M_{yl} = -Pe_x \qquad \text{with} \qquad \gamma = \frac{e_x}{e_y} \tag{9.56}$$

the column deflection curves are constructed in space by the following numerical integration procedures:

1. Specify the mid-span displacements, u_m and v_m where

$$u'_m = v'_m = 0$$

2. Assume the mid-span twisting angle θ_m.

3. Compute bending moments

$$\begin{Bmatrix} M_{\xi m} \\ M_{\eta m} \end{Bmatrix} = -P \begin{bmatrix} \theta_m & -1 \\ 1 & \theta_m \end{bmatrix} \begin{Bmatrix} u_m \\ v_m \end{Bmatrix} \tag{9.57}$$

4. Compute the twisting curvature θ''_m at mid-span

$$\theta''_m = -\left(\frac{\pi}{L}\right)^2 \theta_m \tag{9.58}$$

5. From the moment-curvature-thrust-twist relationships, Eq. (9.55), we obtain curvatures

$$\begin{aligned} \Phi_{\xi m} &= g_\xi(M_{\xi m}, M_{\eta m}, \theta''_m, P) \\ \Phi_{\eta m} &= g_\eta(M_{\xi m}, M_{\eta m}, \theta''_m, P) \end{aligned} \tag{9.59}$$

6. Numerically integrate the curvatures for deflection at the next point

$$\begin{Bmatrix} v_1 \\ u_1 \end{Bmatrix} = \begin{Bmatrix} v_m \\ u_m \end{Bmatrix} + \frac{l^2}{2} \begin{bmatrix} -1 & \theta_m \\ \theta_m & 1 \end{bmatrix} \begin{Bmatrix} \Phi_{\xi m} \\ \Phi_{\eta m} \end{Bmatrix} \tag{9.60}$$

7. Repeat the same procedures at every station. When deflection u_i and v_i at station i are known, deflections at the next station $i + 1$ are obtained by

$$\begin{Bmatrix} v_{i+1} \\ u_{i+1} \end{Bmatrix} = \begin{Bmatrix} v_i \\ u_i \end{Bmatrix} + \frac{l^2}{2} \begin{bmatrix} -1 & \theta_i \\ \theta_i & 1 \end{bmatrix} \begin{Bmatrix} \Phi_{\xi i} \\ \Phi_{\eta i} \end{Bmatrix} + l \begin{Bmatrix} v'_{i-1} \\ u'_{i-1} \end{Bmatrix} + l^2 \begin{bmatrix} -1 & \theta_{i-1} \\ \theta_{i-1} & 1 \end{bmatrix} \begin{Bmatrix} \Phi_{\xi(i-1)} \\ \Phi_{\eta(i-1)} \end{Bmatrix} \tag{9.61}$$

in which

$$\begin{aligned} \Phi_{\xi i} &= g_\xi(M_{\xi i}, M_{\eta i}, \theta''_i, P) \\ \Phi_{\eta i} &= g_\eta(M_{\xi i}, M_{\eta i}, \theta''_i, P) \end{aligned} \tag{9.62}$$

$$\begin{Bmatrix} M_{\xi i} \\ M_{\eta i} \end{Bmatrix} = -P \begin{bmatrix} \theta_i & -1 \\ 1 & \theta_i \end{bmatrix} \begin{Bmatrix} u_i \\ v_i \end{Bmatrix} \tag{9.63}$$

$$\theta_i = \theta_m \sin \frac{\pi z_i}{L}, \quad \theta''_i = -\left(\frac{\pi}{L}\right)^2 \theta_i \tag{9.64}$$

8. At the top of beam-column, $z = L/2$, calculate the ratio

$$-\frac{M_{yn}}{M_{xn}} = \frac{P u_n}{P v_n} = \frac{u_n}{v_n} \tag{9.65}$$

and compare with the specified value $\gamma = e_x/e_y$. If these two values are satisfactorily close, go to Step 9, otherwise, correct the assumed mid-span displacement u_m and repeat Steps 3 to 8.

9. Calculate curvatures at mid-span

$$\begin{Bmatrix} -\Phi_{xm} \\ \Phi_{ym} \end{Bmatrix} = \frac{2}{l^2} \begin{Bmatrix} v_1 - v_m \\ u_1 - u_m \end{Bmatrix} \tag{9.66}$$

10. Calculate angle of twist at mid-span

$$\theta_m = \frac{M_{yl}\Phi_{xm} - M_{xl}\Phi_{ym}}{EI_\omega\left(\frac{\pi}{L}\right)^4 + (GK_T + \bar{K})\left(\frac{\pi}{L}\right)^2} \tag{9.67}$$

Compare this with the assumed value in Step 2. If the difference exceeds the tolerable limit, repeat Steps 2 to 10 until a satisfactory agreement is attained.

From the procedures described above, a column-deflection curve is obtained. In order to search the ultimate strength, the end moment vs. end rotation curve is plotted from the column-deflection curves, the peak of which gives the information of the ultimate state of the beam-column.

There are three iterative calculation loops. One is for determining the correct strain configuration in the moment-curvature-thrust-twist relationships. The second loop is to adjust the loading plane by changing the mid-span displacements. The third loop is to obtain the correct twisting angle at mid-span. In the case of a closed section, the third loop is not needed and computational time is much reduced.

Ultimate strength of biaxially loaded beam-columns for both closed and open cross sections was calculated and compared with the experimental results. In both cases, good agreement was observed although the theoretical results tend to overestimate the ultimate loads which are considered, due to the neglect or approximation of twisting angle function, and neglect of residual stresses and initial deflections.

Finally, the method of solving biaxially loaded beam-columns of open cross sections is extended to analyze the problem of lateral torsional buckling of beam-columns and the effects of residual stresses, prebuckling deformations and strain hardening of materials are also studied.

9.6 PROBLEMS

9.1 In the construction of column-deflection-curves for a beam-column, if the same loading condition is obtained for two infinitesimally adjacent deflection configurations, it is known that this loading condition is the ultimate or collapse state of the beam-column. Give the reason why the difference must be infinitesimally small.

9.2 Derive Eq. (9.32) which is the quadratic equation for solving the normal strain ε_o at the center of a plate element shown in Fig. 9.15.

9.7 REFERENCES

Abdel-Sayed, G. and Aglan, A. A., "Inelastic Lateral Torsional Buckling of Beam-Columns," *IABSE Publications* 33-II, pp. 1–16, (1973).

Aglan, A. A., "The Ultimate Carrying Capacity of Beam-Columns." Thesis presented to the University of Windsor in partial fulfillment of the requirements for the degree of Doctor of Philosophy, Windsor, Ontario, Canada, (1972).

ASCE Manual No. 41, "Commentary on Plastic Design in Steel," 2nd edn, Welding Research Council and the American Society of Civil Engineers, New York, (1971).

Chubkin, G. M., "Experimental Research on the Stability of Thin Plate Steel Members with Biaxial Eccentricity," Paper No. 6 in *Analysis of Spatial Structures*, Vol. 5, Moscow, GILS, (1959).

Ellis, J. S., Jury, E. J. and Kirk, D. W., "Ultimate Capacity of Steel Columns Loaded Biaxially," EIC-64-BR and STR2, *Transactions, Engineering Institute of Canada*, Vol. 7, No. A-2, February, pp. 3–11, (1964).

Galambos, T. V., "Review of Tests on Biaxially Loaded Steel Wide-Flange Beam-Columns," *Report to CRC Task Group No. 3, Fritz Engineering Laboratory Report* No. 287.4, Lehigh University, Bethlehem, Pa., April, (1963).

Harstead, G. A., Birnstiel, C. and Leu, K. C., "Inelastic H-Columns Under Biaxial Bending," *Journal of the Structural Division, ASCE*, Vol. 94, No. ST10, Proc. Paper 6173, October, pp. 2371–2398, (1968).

Johnston, B. G., Editor, *Guide to Design Criteria for Metal Compression Members*, John Wiley & Sons, New York, 2nd edn, (1966), 3rd edn, (1976).

Klöppel, K. and Winkelmann, E., "Experimentalle und Theoretisch Untersuchungen über die Traglast von Zweiachsig Aussermittig Gedrückten Stahlstäben," *Der Stahlbau*, Vol. 31, No. 2, February and March, p. 33, (1962).

Marshall, P. J. and Ellis, J. S., "Ultimate Biaxial Capacity of Box Steel Columns," *Journal of the Structural Division, ASCE*, Vol. 96, No. ST9, Proc. Paper 7515, September, pp. 1873–1887, (1970).

Pillai, S. U. and Ellis, J. S., "Hollow Tubular Beam-Columns in Biaxial Bending," *Journal of the Structural Division, ASCE*, Vol. 97, No. ST5, Proc. Paper 8094, pp. 1399–1406, May, (1971).

10

FINITE DIFFERENCE METHOD FOR PLASTIC BEAM-COLUMNS

S. Vinnakota

10.1 INTRODUCTION

Among the most severe limitations of theoretical *closed form* solutions, presented in Chap. 4, are the idealized conditions, such as ideal elasticity, homogeneity and simple end-conditions, that are often assumed to facilitate the mathematical solution. These idealized conditions are rarely realized in natural phenomena, and their departure from real conditions is often considerable. The great advantage of numerical methods, such as the finite difference method to be described in the present chapter, or the numerical integration method, finite segment method and finite element method presented in Chapters 9, 11 and 12 lies in their ability to solve problems with conditions that are far too complicated for analytical methods. This enables us to introduce into the problem of flexural torsional stability some of the complexities like plasticity, residual stresses, initial deformations and general end-conditions and evaluate their effects on the result. Numerical methods have the additional advantage that they can be handled by electronic computers. The practical significance is that this makes it feasible to investigate a large number of variables such as loading conditions, end conditions, etc., and make parametric studies. The engineer can thus systematically analyze the effect of these possible conditions on the final result—

flexural-torsional stability of columns or lateral buckling of beams as the case may be.

For uniform members, generally encountered in civil engineering practice, the finite difference formulation to be described in this chapter is very powerful and economical in computer time. For more complicated structures and where local buckling preceeds overall stability, the method of finite elements to be described in Chap. 12 should be considered.

10.2 FINITE DIFFERENCES

The finite difference method for solution of differential equations is a technique for the reduction of a continuum to a system with finite number of degrees of freedom. The basis of the method is that the derivatives of functions at a point can be approximated by an algebraic expression consisting of the value of the function at that point and at several nearby points.

Finite difference methods have been discussed extensively in textbooks on numerical analysis [see, for example, Collatz (1966) and Forsythe and Wassov (1960)]. In this section several finite difference expressions are derived, and their application to boundary value problems is described with the help of some very simple examples. Additional examples can be found in Sect. 12.4, Chap. 12, Vol. 1 where the method has been illustrated for the problem of in-plane beam-columns.

Let us consider a function $u = f(z)$, plotted in Fig. 10.1, whose value is known at $z = z_i$ and at several evenly spaced points to the right and to the left of z_i. We designate points to the right of z_i as z_{i+1}, z_{i+2}, and so on, and those to the left of z_i as z_{i-1}, z_{i-2}, and identify the corresponding ordinates as $u_{i+1}, u_{i+2}, u_{i-1}, u_{i-2}$, respectively, u_i being the ordinate at z_i.

FIGURE 10.1 Pivotal and grid points

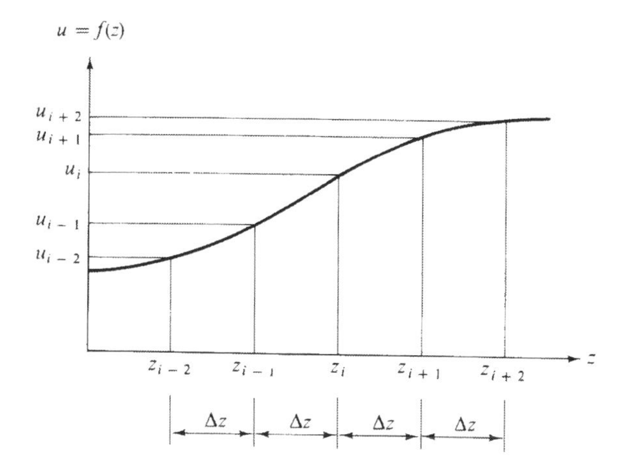

The Taylor series for the function $u = f(z)$ at $z_{i+1} = (z_i + \Delta z)$ expanded about z_i is

$$u_{i+1} = u_i + \Delta z u_i' + \frac{(\Delta z)^2}{2!} u_i'' + \frac{(\Delta z)^3}{3!} u_i''' + \frac{(\Delta z)^4}{4!} u_i'''' + \cdots \tag{10.1}$$

where a prime denotes differentiation with respect to z. The function at $z_{i-1} = (z_i - \Delta z)$ is similarly given by

$$u_{i-1} = u_i - \Delta z u_i' + \frac{(\Delta z)^2}{2!} u_i'' - \frac{(\Delta z)^3}{3!} u_i''' + \frac{(\Delta z)^4}{4!} u_i'''' - \cdots \tag{10.2}$$

By subtracting Eq. (10.2) from Eq. (10.1), we get

$$u_i' = \frac{1}{2\Delta z}(u_{i+1} - u_{i-1}) - \frac{(\Delta z)^2}{3!} u_i''' - \cdots \tag{10.3}$$

If the interval Δz is made sufficiently small, the value of

$$\frac{(\Delta z)^2}{3!} u_i'''$$

and of all succeeding terms are very small. If we neglect these terms, we obtain

$$u_i' = \frac{1}{2\Delta z}(u_{i+1} - u_{i-1}) \tag{10.4}$$

Similarly, on adding Eq. (10.1) and Eq. (10.2) and neglecting the higher order terms, we have

$$u_i'' = \frac{1}{(\Delta z)^2}(u_{i+1} - 2u_i + u_{i-1}) \tag{10.5}$$

Point i, at which finite difference approximations such as Eqs. (10.4) and (10.5) are expressed is called *pivotal* point. For distinction, points $i + 1$, $i - 1$, etc., at which the function is defined, are called *grid* points.

Equations (10.4) and (10.5) may also be obtained if we consider the points $i - 1$, $i, i + 1$ to be close enough that u may be taken as linear between these points. Then we have

$$u_i' = \frac{u_{i+1} - u_i}{\Delta z} \qquad \text{from } i \text{ to } i + 1 \tag{10.6}$$

$$u_i' = \frac{u_i - u_{i-1}}{\Delta z} \qquad \text{from } i - 1 \text{ to } i \tag{10.7}$$

or

$$u_i' = \frac{1}{2}\left(\frac{u_{i+1} - u_i}{\Delta z} + \frac{u_i - u_{i-1}}{\Delta z}\right) = \frac{1}{2}\frac{u_{i+1} - u_{i-1}}{\Delta z} \qquad \text{from } i - 1 \text{ to } i + 1 \tag{10.8}$$

They are called, respectively, the forward, backward, and central differences. Forward-difference expressions for the various derivatives are entirely in terms of values of the function at the pivotal point z_i and points to the right of z_i. In a similar manner, derivative expressions which are entirely in terms of values of the function at z_i and points to the left of z_i are known as backward-finite-differences. Expressions for the derivatives that involve values of the function on both sides of the z value at which the derivatives of the function is desired are called central-differences. In numerical differentiation, forward-difference-expressions are used when data to the left of a point at which a derivative is desired are not available; such as left-end of a beam-column. Backward-difference-expressions are used when data to the right of the desired point are not available. The right-end of a beam-column is an example. Central-difference-expressions, however, are more accurate than either forward- or backward-difference-expressions for a given spacing Δz. With a view to using central-differences throughout, fictitious grid points are generally added beyond the support points.

Error bounds for the finite difference approximations are determined from the lowest order term (in Δz), which is not included in the expression. This is easily obtained if a Taylor series representation of the function values is substituted back into the finite difference expression. For instance, if the expressions (10.1) and (10.2) are substituted into the expressions for the first-order derivatives (10.6) to (10.8) we find

$$u_i' = \frac{u_{i+1} - u_i}{\Delta z} - \frac{\Delta z}{2} u_i'' \tag{10.6a}$$

$$u_i' = \frac{u_i - u_{i-1}}{\Delta z} + \frac{\Delta z}{2} u_i'' \tag{10.7a}$$

$$u_i' = \frac{u_{i+1} - u_{i-1}}{2\Delta z} - \frac{\Delta z^2}{6} u_i''' \tag{10.8a}$$

It can be shown that if the second-order derivative in Eqs. (10.6a) and (10.7a) is represented by its maximum value within the interval, the term $\frac{1}{2}\Delta z u''$ represents a bound on the error. We say, therefore, that the first-order derivative in Eqs. (10.6) and (10.7) is of first-order accuracy; the error term E is $0(\Delta z)$. In the same way, in the finite difference expression given by Eq. (10.8a), $E = 0(\Delta z^2)$. That is, the first-order derivative in Eq. (10.8) is of second-order accuracy. Thus, the central differences lead to more accurate formulation than the forward or backward differences. The accuracy of these could however be increased if more terms are used in the Taylor series and more function values at grid points are included as shown below. We have

$$\begin{aligned} u_{i+1} &= u_i + \Delta z u_i' + \frac{\Delta z^2}{2!} u_i'' + \frac{\Delta z^3}{3!} u_i''' + \cdots \\ u_{i+2} &= u_i + 2\Delta z u_i' + \frac{(2\Delta z)^2}{2!} u_i'' + \frac{(2\Delta z)^3}{3!} u_i''' + \cdots \end{aligned} \tag{10.9}$$

From these relations, we obtain

$$u_i' = \frac{1}{2\Delta z}(-3u_i + 4u_{i+1} - u_{i+2}) + \frac{1}{3}\Delta z^2 u_i''' \tag{10.10}$$

That is, the first-order forward-difference-expression in Eq. (10.10), using three grid points, has the same order of accuracy as the first-order central-difference-expression in Eq. (10.8).

In stability problems, the equilibrium equations that define the deformed shape of the bar are, in general, fourth-order differential equations. First-, second-, third- and fourth-order central derivatives given below result in a set of expressions that is homogeneous in accuracy.

$$\begin{aligned}
u_i' &= \frac{1}{2\Delta z}(u_{i+1} - u_{i-1}) \\
u_i'' &= \frac{1}{\Delta z^2}(u_{i+1} - 2u_i + u_{i-1}) \\
u_i''' &= \frac{1}{2\Delta z^3}(u_{i+2} - 2u_{i+1} + 2u_{i-1} - u_{i-2}) \\
u_i'''' &= \frac{1}{\Delta z^4}(u_{i+2} - 4u_{i+1} + 6u_i - 4u_{i-1} + u_{i-2})
\end{aligned} \tag{10.11}$$

From these relations we see that the approximate finite difference equation corresponding to a fourth order differential equation will contain five adjacent grid points. In particular the application of this equation at the boundary point O will involve the two "unwanted" values of u_{-1} and u_{-2}. We therefore need two extra equations at this point to obtain a unique solution, and these are provided by the two boundary conditions corresponding to this end, which when applied at O will each involve both u_{-1} and u_{-2}, but no more external values. We see that if there are m segments in the range, then the application of the difference equations at all these points introduces these values at four extra points, two outside each end of the range. There are therefore $m + 1$ equations connecting $m + 5$ unknowns, and we need four extra conditions to give a unique solution. These four conditions are supplied in the form of two boundary conditions at each end.

EXAMPLE 10-1 *Application to Elastic In-Plane Buckling of a Column (Fourth-Order Differential Equations)* The finite difference method will be used to determine the elastic critical load of the hinged-hinged column shown in Fig. 10.2. The differential equation for a hinged-hinged column is

$$u'''' + \frac{P}{EI}u'' = 0 \tag{10.12}$$

After the finite difference expressions taken from Eq. (10.11) are substituted for the derivatives, the equation corresponding to pivotal point i is

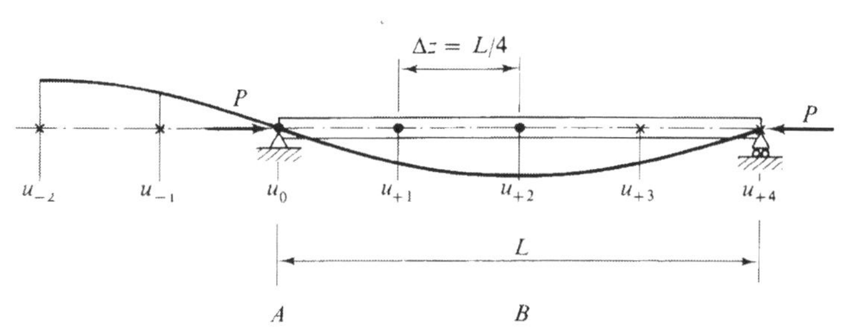

FIGURE 10.2
Finite difference grid for a hinged-hinged column (Fourth-order differential equation).

$$u_{i+2} - 4u_{i+1} + 6u_i - 4u_{i-1} + u_{i-2} + \frac{P\Delta z^2}{EI}(u_{i+1} - 2u_i + u_{i-1}) = 0 \qquad (10.13)$$

Due to the fact that the buckling mode of a hinged-hinged column is symmetrical, only half of the column length need be considered. If the member is divided into four equal parts of length $\Delta z = L/4$, the grid points to be considered are as shown in Fig. 10.2. The boundary conditions at A are

$$u = u'' = 0 \qquad (10.14)$$

and at B,

$$u' = u''' = 0 \qquad (10.15)$$

Applying Eq. (10.13) at pivotal points 0, +1 and +2, and substituting the finite difference expressions taken from Eq. (10.11) in the boundary conditions (10.14) and (10.15), we obtain the following equation system

$$\begin{aligned} u_0 &= 0 \\ u_{-1} - 2u_0 + u_1 &= 0 \\ (u_{-2} - 4u_{-1} + 6u_0 - 4u_1 + u_2) + \lambda(u_{-1} - 2u_0 + u_1) &= 0 \\ (u_{-1} - 4u_0 + 6u_1 - 4u_2 + u_3) + \lambda(u_0 - 2u_1 + u_2) &= 0 \\ (u_0 - 4u_1 + 6u_2 - 4u_3 + u_4) + \lambda(u_1 - 2u_2 + u_3) &= 0 \\ u_3 - u_1 &= 0 \\ -u_0 + 2u_1 - 2u_3 + u_4 &= 0 \end{aligned} \qquad (10.16)$$

where $\lambda = P\Delta z^2/EI$. Equations (10.16) are homogeneous, and the buckling load is given by the value of P for which the coefficient determinant equals zero. Thus

$$P_{cr} = 9.376\frac{EI}{L^2} \qquad (10.17)$$

Comparison of this result with the exact solution, 9.87 EI/L^2 shows the finite difference approximation to be in error by about 5 percent. More accurate values are obtained for the buckling load if a closer grid spacing is used. However, as the number of grid points increases so does the number of simultaneous equations that must be solved. The amount of numerical work could be reduced, however, if the finite difference method is applied to the corresponding second-order differential equation of the problem. This is illustrated with the help of the following example.

EXAMPLE 10-2 *Elastic In-Plane Buckling of a Column (Second-Order Differential Equation)* Let us again consider the buckling of the hinged-hinged column which was analyzed in Example 10-1. The second order differential equation and boundary conditions for this column are

$$u'' + \frac{P}{EI}u = 0 \tag{10.18}$$

$$u(0) = u(L) = 0 \tag{10.19}$$

If the finite difference expression given in Eq. (10.11) is substituted for the second derivative in Eq. (10.18), one obtains

$$u_{i+1} - 2u_i + u_{i-1} + \frac{P\Delta z^2}{EI}u_i = 0 \tag{10.20}$$

As shown in Fig. 10.3 the member is again divided into four equal parts of length $\Delta z = L/4$. There are three interior points at which the difference equation can be written. Due to symmetry of the buckling mode $u_1 = u_3$ and the number of equations that must be written is reduced to two. At $i = 1$, Eq. (10.20) leads to

$$u_2 - 2u_1 + u_0 + \lambda u_1 = 0 \tag{10.21}$$

and at $i = 2$ we obtain

$$u_3 - 2u_2 + u_1 + \lambda u_2 = 0 \tag{10.22}$$

where $\lambda = PL^2/16EI$. Making use of the boundary conditions and symmetry, these equations can be rewritten as

$$\begin{aligned} u_1(\lambda - 2) + u_2 &= 0 \\ u_1(2) + u_2(\lambda - 2) &= 0 \end{aligned} \tag{10.23}$$

FIGURE 10.3
Finite difference grid for a hinged-hinged column (Second-order differential equation).

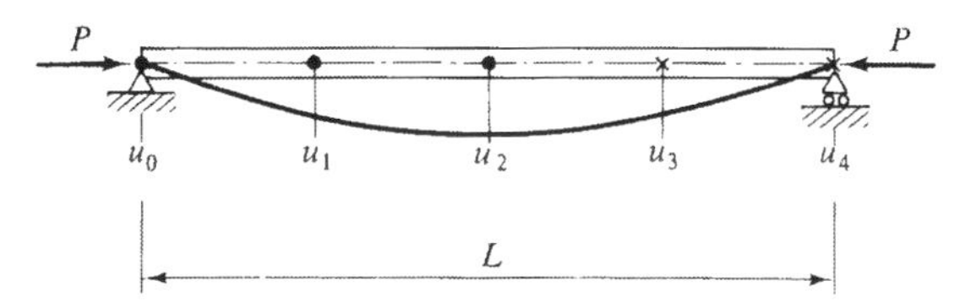

Setting the determinant of Eqs. (10.23) equal to zero gives the quadratic equation

$$\lambda^2 - 4\lambda + 2 = 0 \tag{10.24}$$

whose smallest root is $\lambda = 0.59$. Hence

$$P_{cr} = 9.376 \frac{EI}{L^2} \tag{10.25}$$

The same result obtained in Example 10-1.

In this section we developed finite difference expressions for the special case in which the grid points are equally spaced at distances Δz. Finite difference expressions with variable grid spacing could be obtaned by an expansion of the function $f(z)$ in a Taylor series about the pivotal point. Experience shows that variable-spacing grids can often be used to expedite the analysis. Such is the case with the plastic beam-columns studied in this chapter (for details see Sect. 10.5.1).

10.3 SCOPE

The basic equations governing the behavior of thin-walled open-section members were formulated in Chap. 2, using the coordinate system x-y-z fixed in space. Lateral buckling of beams and beam-columns is treated in Chap. 3, elastic solutions of eccentrically loaded beam-columns are discussed in Chap. 4, while plastic analysis of beam-columns is considered in Chaps. 7 to 9. The coordinate system (x-y-z) is usually defined in accordance with the principal axes of the cross section having its origin at the centroid CG; and the displacements (u, v, θ) of a cross section are taken with respect to its shear center, S. Here, as in Chap. 2, the governing differential equations are derived for an arbitrary open section member, without making use of the notions of center of gravity, principal axes or shear center. However, for the convenience of the present method, the equilibrium equations are written for a rectangular coordinate system X-Y-Z which displaces itself parallel to the arbitrary

FIGURE 10.4
Types of buckling phenomenon. (a) Bifurcation type. (b) Stability type.

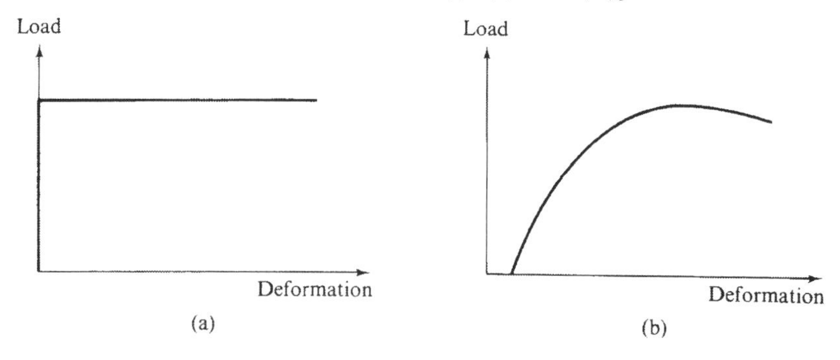

fixed coordinate system x-y-z. In addition, the displacements (u, v, θ) of a cross section are given for an arbitrary point 0 taken as pole of the sectorial areas. In the formulation, the deformation field is considered nonlinear (the displacements u and v are still assumed to be small, but θ is considered to be sufficiently large that the simplifications $\cos\theta = 1$ and $\sin\theta = \theta$ are no longer valid). The equilibrium equations derived for an initially straight beam-column are later corrected for the presence of initial geometrical imperfections. Inclusion of the geometrical imperfections will allow us to study the bifurcation phenomenon (lateral buckling of beam-columns, for instance) as a limiting case of the stability phenomenon (load-deflection of beam-columns) as the geometrical imperfections are made small (Fig. 10.4).

10.4 THEORY

10.4.1 General

We consider a beam-column of an arbitrary open cross section of length L, shown in Figs. 10.5 and 10.6. The member is provided with two rotational restraints at each end. End displacements and end twist are prevented. Warping at the ends could be either free or completely restrained. The external loads are shown in Fig. 10.5, while the deformed configuration is indicated in Fig. 10.7. In the development of the theory, it is assumed that the cross section retains its original shape during deformation. The stress-strain diagram of the material is considered ideally elastic-perfectly plastic. Strain hardening and strain reversal are neglected. The deformations are considered small and the shearing strain is neglected. It is also assumed that yielding is governed by normal stress only; and loads are conservative.

The reference axes x-y-z are right-handed rectangular coordinate system, stationary in space. The z axis is directed along the beam length and passes through an arbitrary point C of the cross section. At each section, a local coordinate

FIGURE 10.5
Beam-column under consideration and external loads

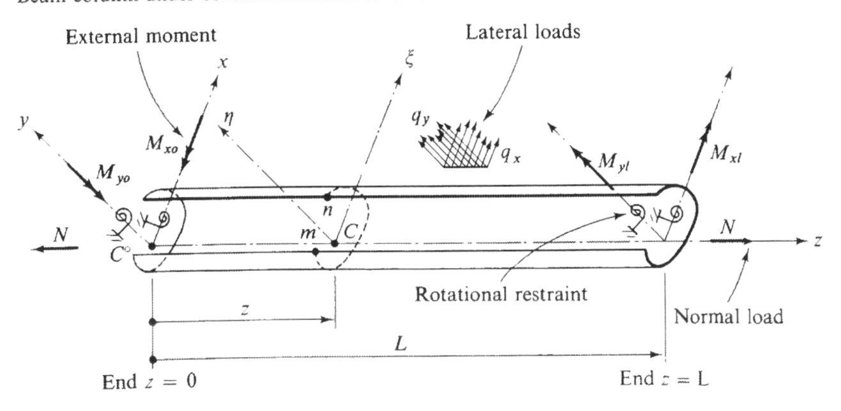

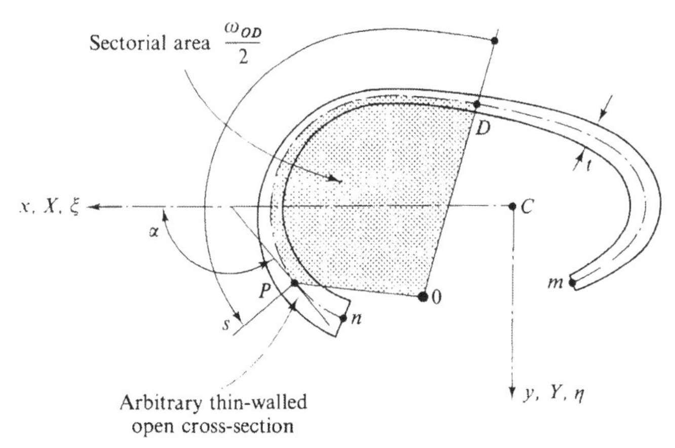

FIGURE 10.6
Cross-section under consideration and definitions

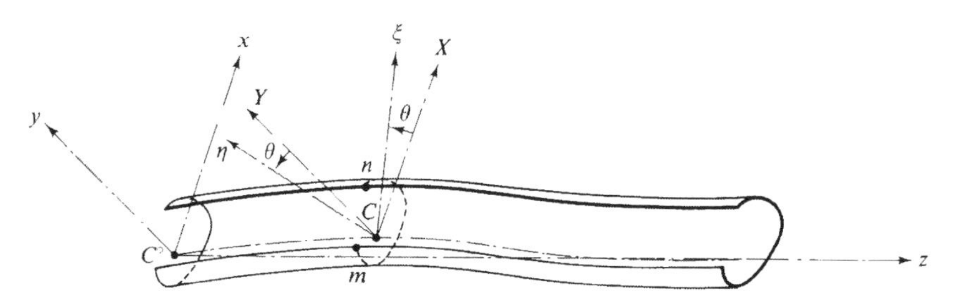

FIGURE 10.7
Deformed configuration of the bar and definitions of the ξ-η-ζ and X-Y-Z coordinate systems

system ξ-η-ζ fixed to the cross section and deforming with it is considered. Its origin coincides with the reference point C of the cross section and, in the unloaded position, the $\xi\zeta$ and $\eta\zeta$ planes coincide with the xz and yz planes respectively. A second local system of axes X-Y-Z whose origin coincides with the reference point C and whose axes are always parallel to the x-y-z axes is also considered in the analysis (Fig. 10.7). Finally, the position of any point on the middle line of the cross section can also be defined by its distance s measured from an arbitrary point D of the cross-section (Fig. 10.6).

10.4.2 Kinematics of Deformation

Let $P(\xi, \eta)$ be a general point on the middle line of the cross section (Fig. 10.6). α is the angle between the ξ axis and the tangent to the middle line of the cross section at P. Let $0(\xi_0, \eta_0)$ denote an arbitrary reference point of the cross section and let ω_{0D} be the warping function or double sectorial area of the point P with 0 as the pole and $0D$ as the initial vector.

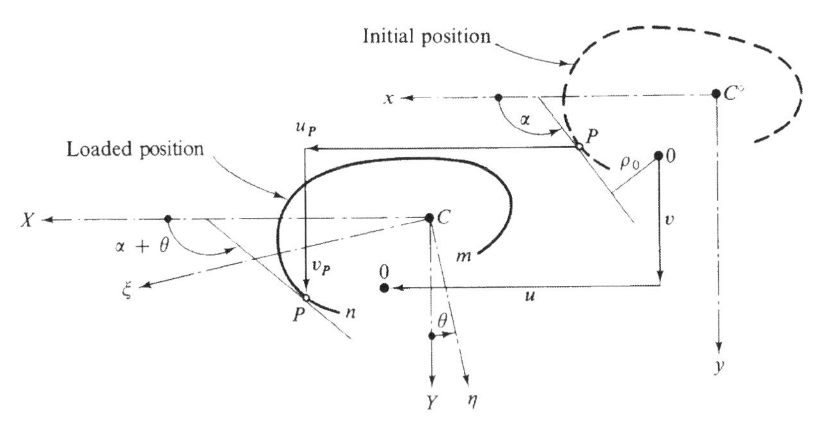

FIGURE 10.8
Displacements of the pole 0 and a general point P of a cross-section

We now consider a section mn at a distance z from the origin (Fig. 10.5). The deformation of this section is defined by the displacement components u, v of the pole 0, and by the angle of twist θ of the cross section (Fig. 10.8). The angle of twist θ is taken positive about the z axis according to the right hand rule and the displacements are positive in the positive directions of the corresponding axes. In the deformed position of the cross section, let (X_0, Y_0) and (X, Y) represent the coordinates of the points 0 and P of the cross section mn respectively. Also, let u_P, v_P be the displacements of the point P.

The following relations could be written with the help of Fig. 10.8.

$$X = \xi \cos\theta - \eta \sin\theta \qquad Y = \xi \sin\theta + \eta \cos\theta$$

$$X_0 = \xi_0 \cos\theta - \eta_0 \sin\theta \quad Y_0 = \xi_0 \sin\theta + \eta_0 \cos\theta$$

$$\frac{d\omega_{0D}}{ds} = \rho_0 = (\xi - \xi_0)\sin\alpha - (\eta - \eta_0)\cos\alpha$$

$$= (X - X_0)\sin(\alpha + \theta) - (Y - Y_0)\cos(\alpha + \theta) \tag{10.26}$$

in which ρ_0 is the distance from point 0 to the tangent line shown in Fig. 10.8. As the section is considered indeformable, the displacement u_P, v_P of the point P could be expressed as a function of the displacement u, v of the pole 0, by expressing the fact that the vector $\overline{OP}$ does not vary in length (see Prob. 10.4). Thus we obtain

$$u_P = u - (\eta - \eta_0)\sin\theta - (\xi - \xi_0)(1 - \cos\theta)$$

$$v_P = v + (\xi - \xi_0)\sin\theta - (\eta - \eta_0)(1 - \cos\theta) \tag{10.27}$$

By replacing the cosines of small angles by unity and the sines of small angles by the angles themselves, these relations reduce to

$$u_P = u - (\eta - \eta_0)\theta \quad v_P = v + (\xi - \xi_0)\theta \tag{10.28}$$

In a majority of the earlier studies (Bleich, 1952; Vlasov, 1961; Timoshenko and Gere, 1961; Galambos, 1968) the deformation field is used in this linearized form. In the present chapter these simplifying assumptions will not be made and the finite deformation field defined by Eq. (10.27) will be retained.

With the help of Eqs. (10.26) and (10.27), the first derivatives of u_P and v_P can be written as (Prob. 10.4)

$$u'_P = u' - (Y - Y_0)\theta' \quad v'_P = v' + (X - X_0)\theta' \tag{10.29}$$

where the primes indicate differentiation with respect to z. Noting that X' is equal to $-Y\theta'$ and Y' is equal to $X\theta'$ from Eq. (10.26), the above relations can also be written in the form

$$u'_P = (u + X - X_0)' \quad v'_P = (v + Y - Y_0)' \tag{10.30}$$

10.4.3 Equilibrium Equations

Three equilibrium equations for the member in its deformed position may be written in terms of the three displacement variables u, v, and θ as shown in the following paragraphs.

As the section is considered thin-walled, only normal stresses σ and shearing stresses need be considered in the analysis. As mentioned in Chap. 2, these shearing stresses develop due to two distinct modes of deformation: a pure twisting of the member during which all sections are free to warp and a transverse bending of the bar coupled with nonuniform axial deformation. The first mode of deformation, known as St. Venant's torsion, leads to shearing stresses which vary linearly over the thickness of the wall and are indicated by τ_{sv}. For open sections under consideration, the additional shearing stresses developed during the second mode of deformation are essentially uniform over the wall thickness, and are indicated by the

FIGURE 10.9
Forces acting on the portion of the beam-column to the left of mn.

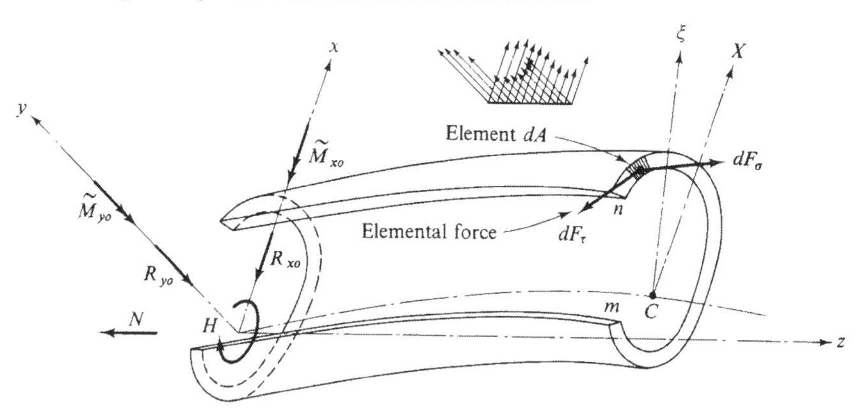

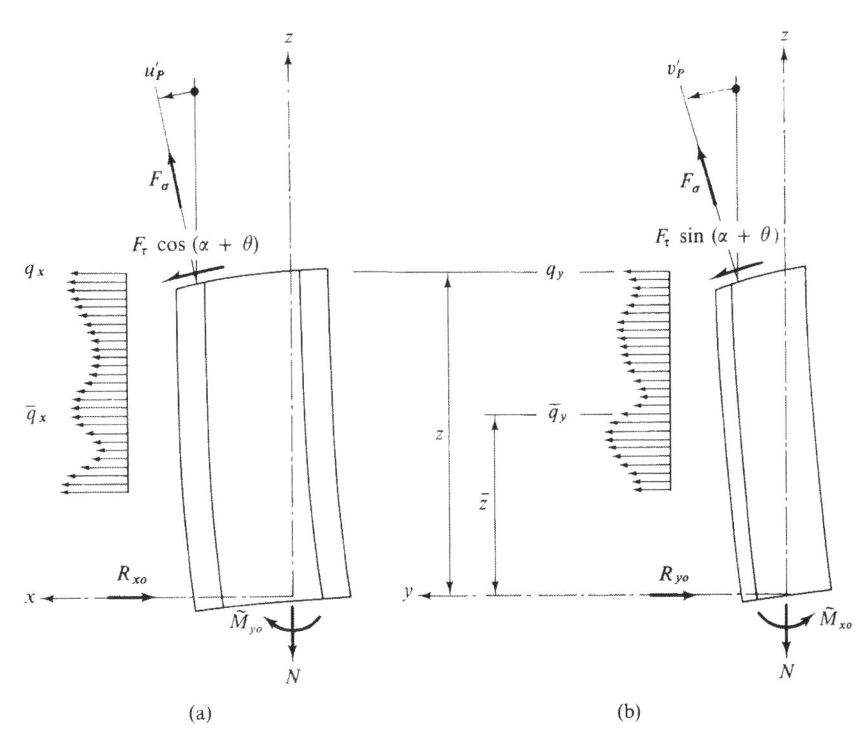

FIGURE 10.10
Projection of the forces acting on the portion of the beam-column to the left of the section *mn*. (a) Projection on the *xz* plane. (b) Projection on the *yz* plane.

symbol τ. The linear shearing stresses τ_{sv} do not influence the equilibrium conditions in the longitudinal direction.

The portion of the member to the left of section *mn* situated at a distance z is shown in a deformed configuration in Fig. 10.9 while Fig. 10.10 gives the projection, on the xz and yz planes, of the forces acting on this portion. Let us consider an elemental area $dA(= t\,ds)$ situated at the point P. In the deformed state the stress element $dF_\sigma = \sigma\,dA$ is inclined to the z axis by the angle du_P/dz (Fig. 10.10a) and by the angle dv_P/dz in the yz plane (Fig. 10.10b). The stress element $dF_\tau = \tau\,dA$ is inclined by the angle $\alpha + \theta$ to the x axis (Fig. 10.11). The angle of twist θ and the angles of rotation u' and v' are assumed to be small as before. The components of the stress elements along the x and y axes are therefore

$$\begin{aligned} dF_{\sigma X} &= \sigma\,dA\,u'_P & dF_{\sigma Y} &= \sigma\,dA\,v'_P \\ dF_{\tau X} &= \tau\,dA\cos(\alpha + \theta) & dF_{\tau Y} &= \tau\,dA\sin(\alpha + \theta) \end{aligned} \tag{10.31}$$

The resultants of these stress elements integrated over the cross section *mn*, the distributed lateral loads q_x, q_y and the reactions at the end $z = 0$, keep the beam-column to the left of the section *mn* in equilibrium.

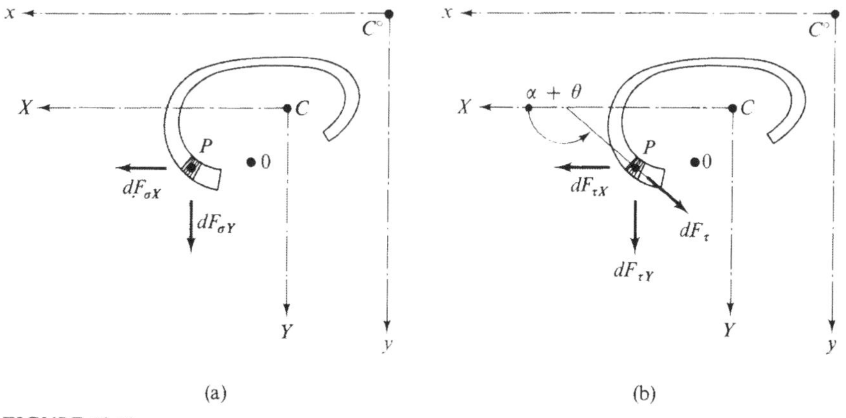

FIGURE 10.11
Components of elemental forces along X and Y axes, acting at section mn. (a) Forces resulting from normal stress. (b) Forces resulting from tangential stress.

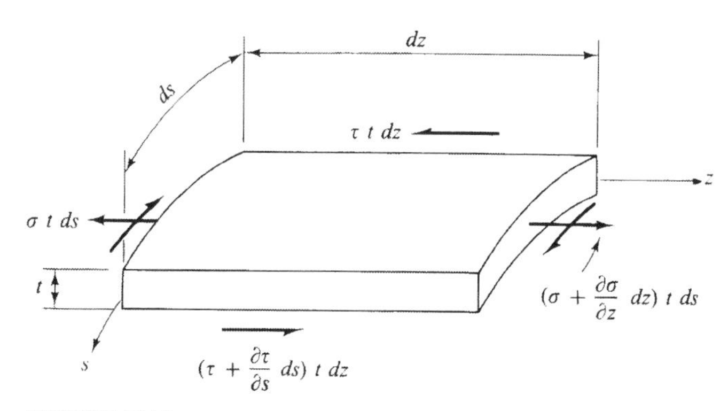

FIGURE 10.12
Element of beam-column illustrating equilibrium of forces in the longitudinal direction

The equilibrium condition of a typical wall element $ds\,dz$, such as that shown in Fig. 10.12, by summing forces in the z direction and dividing by $ds\,dz$, gives

$$\frac{\partial(\tau t)}{\partial s} + t\frac{\partial \sigma}{\partial z} = 0 \tag{10.32}$$

Let us now consider the equilibrium of the member in the xz plane. By taking the components $dF_{\sigma X}, dF_{\tau X}$ of the stress elements (Fig. 10.11) and integrating over the whole cross section, we have

$$F_x = \int_A (dF_{\tau X} + dF_{\sigma X}) = \int_A [\tau \cos(\alpha + \theta) + \sigma u'_P]\, dA \tag{10.33a}$$

Noting that there are no applied shear stresses along the lateral edges m and n of the open cross section, we obtain with the help of Eqs. (10.26), (10.30) and (10.32)

$$F_x = -M'_Y + (u - X_0)'N \tag{10.33b}$$

where N and M_Y are the stress resultants defined by

$$N = \int_A \sigma\, dA, \quad M_Y = -\int_A \sigma X\, dA \tag{10.34}$$

We denote by $\bar{z}$ a section situated to the left of mn; so that $0 \leq \bar{z} \leq z$. Let $\bar{q}_x$ and $\bar{q}_y$ be the intensities of the transverse load at this point. Also, let R_{xo} be the reaction at end $z = 0$ in the x-direction. The equilibrium equation of the portion of the member to the left of mn is (Fig. 10.10a):

$$-M'_Y + (u - X_0)'N + \int_0^z \bar{q}_x\, d\bar{z} = R_{xo} \tag{10.35}$$

Similarly, the equilibrium of the member in the yz plane results in the second equilibrium condition

$$M'_X + (v - Y_0)'N + \int_0^z \bar{q}_y\, d\bar{z} = R_{yo} \tag{10.36}$$

where

$$M_X = \int_A \sigma Y\, dA \tag{10.37}$$

Remarks

From the relations (10.26) it could be seen that the coordinates X_0, Y_0 of the pole 0 appearing in the equilibrium equations (10.35) and (10.36) are not constants, but functions of the twist θ, of the section z. Also, from the definitions for X, Y given in Eq. (10.26), the following relations could be written with the help of Eqs. (10.34) and (10.37):

$$\begin{aligned} M_Y &= -\int_A \sigma(\xi \cos\theta - \eta \sin\theta)\, dA = +M_\eta \cos\theta + M_\xi \sin\theta \\ M_X &= \int_A \sigma(\xi \sin\theta + \eta \cos\theta)\, dA = -M_\eta \sin\theta + M_\xi \cos\theta \end{aligned} \tag{10.38}$$

where M_ξ, M_η are the moments with respect to the axes ξ and η respectively. In the derivation of the equilibrium equations (10.35) and (10.36), the longitudinal force N is considered tensile. If the beam-column is subjected to a compressive longitudinal force P, we replace N in the equilibrium equations by $-P$.

The third equilibrium equation is obtained by considering the twisting of the beam. For this, moments are taken about the pole 0 of the section mn in its deformed position.

The twisting moment $M_{T(z)}$ acting at the section mn has three components: the first one is the St. Venant torsion moment T_{sv} which is the resultant of the shear stresses τ_{sv}; the second one is the warping torsion moment T_w which is the resultant of shear stresses τ. The third component T_σ is the additional moment due to the Wagner effect. Thus

$$M_{T(z)} = T_{sv} + T_w + T_\sigma \tag{10.39}$$

From Eq. (2.50), the relation between T_{sv} and the angle of twist θ is simply
simply

$$T_{sv} = GK_T\theta' \tag{10.40}$$

where G is the shear modulus and K_T is the torsion constant of the section.

The force components $dF_{\tau X}$ and $dF_{\tau Y}$ acting on the elemental area dA result in a twisting moment which is equal to

$$dF_{\tau Y}(X - X_0) - dF_{\tau X}(Y - Y_0) \tag{10.41}$$

Integration over the cross section mn and simplification leads to the relation

$$T_w = \frac{d}{dz}\int_A \sigma\omega \, dA = M'_\omega \tag{10.42}$$

where M_ω is the bimoment

$$M_\omega = \int_A \sigma\omega \, dA \tag{10.43}$$

The force components $dF_{\sigma X}$ and $dF_{\sigma Y}$ acting on the elemental area dA, also result in a twisting moment about the point 0 which is equal to

$$dF_{\sigma Y}(X - X_0) - dF_{\sigma X}(Y - Y_0)$$

Introduction of the expressions for $dF_{\sigma X}$ and $dF_{\sigma Y}$ and integration over the cross sectional area gives after some simplification

$$T_\sigma = v'(-M_Y - X_0N) - u'(M_X - Y_0N) + \bar{K}\theta' \tag{10.44}$$

where

$$\bar{K} = \int_A \sigma[(X - X_0)^2 + (Y - Y_0)^2] \, dA = \int_A \sigma[(\xi - \xi_0)^2 + (\eta - \eta_0)^2] \, dA \tag{10.45}$$

It is to be noted that in Chap. 2, $T_\sigma = \bar{K}\theta'$ because it was the twisting moment about ζ axis. Now it is about an axis parallel to z axis, passing through the pole.

Replacing T_{sv}, T_w and T_σ in Eq. (10.39) by their values given by Eqs. (10.40), (10.42) and (10.44), we obtain

$$M_{T(z)} = M'_\omega + (GK_T + \bar{K})\theta' - v'(M_Y + X_0N) - u'(M_X - Y_0N) \tag{10.46}$$

Next we consider the torsion moments resulting from the forces acting at the end $z = 0$ of the member. These moments are again calculated with respect to an axis parallel to the axis z, passing through the pole 0 of the section mn in its deformed

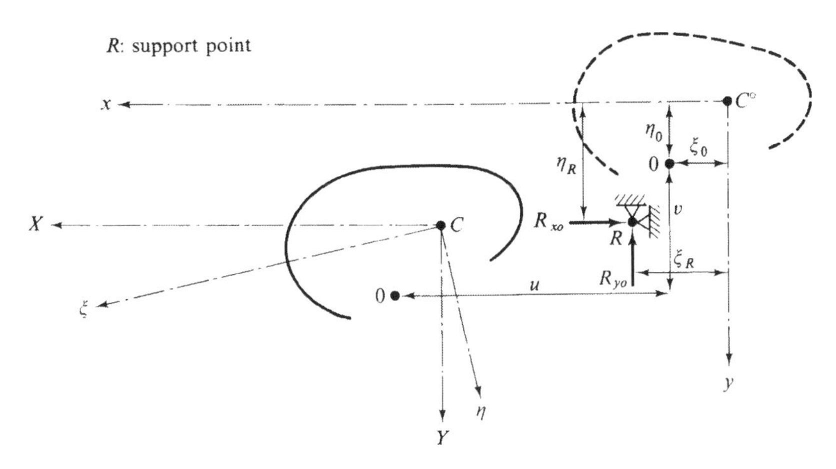

FIGURE 10.13
Torsion moment resulting from end-reactions

position. First, the end reactions R_{x0} and R_{y0}, acting at the end $z = 0$, create a torsion moment which has a value (Fig. 10.13):

$$R_{x0}[v - (\eta_R - \eta_0)] - R_{y0}[u - (\xi_R - \xi_0)]$$

where ξ_R, η_R are the coordinates of the point of application of the reactions ($\xi_R = \eta_R = 0$ in Fig. 10.9). The end reactions R_{x0}, R_{y0} and the end shears F_{x0}, F_{y0} are always in equilibrium, but do not necessarily have the same line of action (this will be especially true when the end section becomes progressively plastic). So a torsional moment is introduced at the end, designated as M_R. Let M_{z0} be the external torsional moment, if any, applied at the end $z = 0$. The net torsional moment resulting from the forces at this end is therefore

$$M_{T(0)} = R_{x0}[v - (\eta_R - \eta_0)] - R_{y0}[u - (\xi_R - \xi_0)] + M_R + M_{z0} = R_{x0}v - R_{y0}u + H \tag{10.47}$$

where the quantity H is, for a given load level, a constant independent of z.

$$H = - R_{x0}(\eta_R - \eta_0) + R_{y0}(\xi_R - \xi_0) + M_R + M_{z0} \tag{10.48}$$

Finally, we consider the torsional moment caused by the distributed lateral forces q_x, q_y. We denote again by $\bar{z}$ a section situated to the left of mn; so that $0 \leq \bar{z} \leq z$. Let $\bar{q}_x$ and $\bar{q}_y$ be the intensities of the transverse load at this point and $\bar{X}_q, \bar{Y}_q$ the point of application of the load. Also let $\bar{X}_0, \bar{Y}_0$, the coordinates, $\bar{u}, \bar{v}$ the displacements of the pole 0 of the section $\bar{z}$, in the loaded position. The torsional moment, $M_{T(q)}$ acting on the portion of the beam-column to the left of mn is therefore

$$M_{T(q)} = \int_0^z [\bar{q}_y(\bar{X}_q - \bar{X}_0 - u + \bar{u}) - \bar{q}_x(\bar{Y}_q - \bar{Y}_0 - v + \bar{v})]\, d\bar{z} \tag{10.49}$$

The third equilibrium equation is now obtained by writing $M_{T(z)} + M_{T(q)} = M_{T(0)}$. Thus we have, from Eqs. (10.46), (10.47) and (10.49)

$$M'_\omega + (GK_T + \bar{K})\theta' - v'(M_Y + X_0 N) - u'(M_X - Y_0 N)$$
$$= R_{x0}v - R_{y0}u + H + \int_0^z [\bar{q}_x(\bar{Y}_q - \bar{Y}_0 - v + \bar{v}) - \bar{q}_y(\bar{X}_q - \bar{X}_0 - u + \bar{u})]\, d\bar{z} \tag{10.50}$$

Equations (10.35), (10.36) and (10.50) are the three equilibrium equations for the transversely loaded, rectilinear thin-walled, open section beam-column shown in Fig. 10.5. Taking derivatives once, we obtain an alternative form of these equilibrium equations

$$[-M_Y + (u - X_0)N]'' + q_x = 0 \tag{10.51a}$$

$$[M_X + (v - Y_0)N]'' + q_y = 0 \tag{10.51b}$$

$$M''_\omega + [(GK_T + \bar{K})\theta']' - v''(M_Y + X_0 N) - u''(M_X - Y_0 N)$$
$$+ [q_y(X_q - X_0) - q_x(Y_q - Y_0)] = 0 \tag{10.51c}$$

Note that these equations are valid for a small element dz of the beam-column at section mn situated at a distance z from the end $z = 0$. In deriving Eq. (10.51c) from Eq. (10.50), we made use of the relations (10.35) and (10.36). q_x and q_y in Eqs. (10.51) are the intensities of the transverse load at section z. We also note that these equations are written with reference to the coordinate system X, Y which translates (but does not rotate) in connection with the fixed coordinate system x-y-z as the beam-column deforms. Equations (2.173) of Chap. 2 are, however, written with regard to the coordinate system ξ-η-ζ.

10.4.4 Initial Crooked Beam-Column

In the previous section the three equilibrium equations are derived for a bar, initially straight and untwisted. We now consider a bar with initial geometrical imperfections. Let u_i, v_i, θ_i represent the initial deformations of the pole 0 of the cross section mn (Fig. 10.14 shows different shapes of geometrical imperfections utilized in the computer program to study the behavior of I-beams). In the present chapter we assume that these initial deformations are nil at both ends. For the bar with initial imperfections, the three equilibrium equations (10.51a), (10.51b) and (10.51c) become

$$[-M_Y + (u + u_i - X_0)N]'' + q_x = 0 \tag{10.52a}$$

$$[M_X + (v + v_i - Y_0)N]'' + q_y = 0 \tag{10.52b}$$

$$M''_\omega + (GK_T\theta')' + [\bar{K}(\theta + \theta_i)']' - (v + v_i)''(M_Y + X_0 N)$$
$$- (u + u_i)''(M_X - Y_0 N) + [q_y(X_q - X_0) - q_x(Y_q - Y_0)] = 0 \tag{10.52c}$$

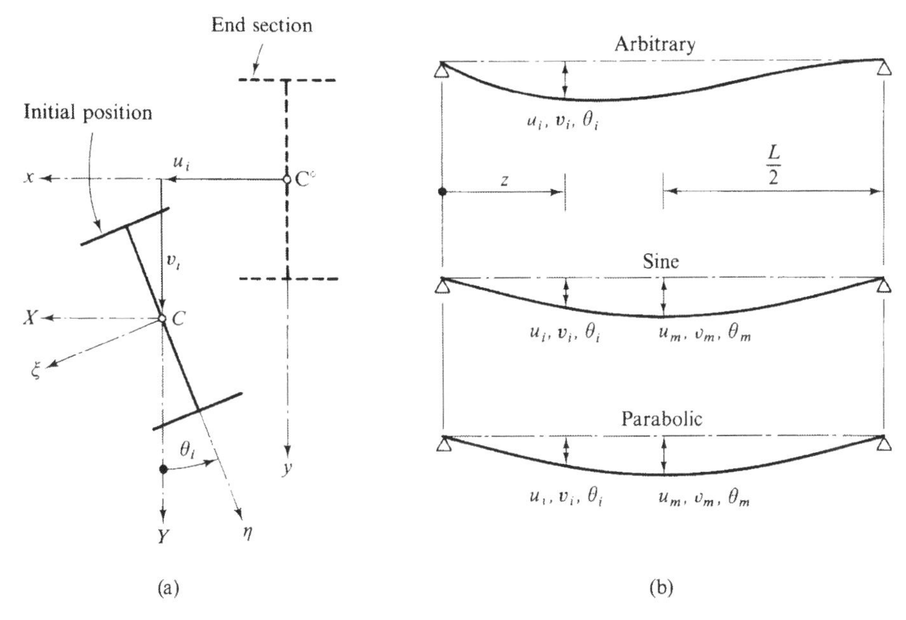

FIGURE 10.14
Geometrical imperfections. (a) Intial position of the section at distance z. (b) Longitudinal variation.

where

$$X = \xi \cos(\theta + \theta_i) - \eta \sin(\theta + \theta_i)$$
$$Y = \xi \sin(\theta + \theta_i) + \eta \cos(\theta + \theta_i) \tag{10.53}$$

10.4.5 Second Order Equilibrium Equations

If the equilibrium equations (10.52a), (10.52b) and (10.52c) are developed as a function of deformations, fourth-order differential equations will be obtained. In order to increase the accuracy of the finite difference method of solution used in this chapter, we integrate these equilibrium equations twice and reduce the order of differential equations by two.

Integration of Eq. (10.52a) twice, results in

$$-M_Y + (u + u_i - X_0)N + M_{qX} = C_1 z + C_2 \tag{10.54}$$

where C_1 and C_2 are the integration constants and M_{qX} is a function of z that satisfies the following two conditions (see Sect. 10.5 for further details):

$$M''_{qX} = q_x \quad M_{qX(0)} = M_{qX(L)} = 0 \tag{10.55}$$

The constants C_1 and C_2 are determined from the end conditions (Fig. 10.9)

$$u = 0 \quad \text{and} \quad M_Y = \tilde{M}_{y0} \quad \text{at} \quad z = 0$$
$$u = 0 \quad \text{and} \quad M_Y = \tilde{M}_{yl} \quad \text{at} \quad z = L \tag{10.56}$$

which result in

$$-M_Y + uN = -(\tilde{M}_{yl} - \tilde{M}_{y0})\frac{z}{L} - \tilde{M}_{y0} + (X_0 - X_{0(0)} - u_i)N - M_{qX} \tag{10.57}$$

where $X_{0(0)}$ is the X coordinate of the pole 0 at the section $z = 0$. Similarly, Eq. (10.52b) becomes

$$M_X + vN = (\tilde{M}_{xl} - \tilde{M}_{x0})\frac{z}{L} + \tilde{M}_{x0} + (Y_0 - Y_{0(0)} - v_i)N - M_{qY} \tag{10.58}$$

where:

$$M''_{qY} = q_y \quad M_{qY(0)} = M_{qY(L)} = 0 \tag{10.59}$$

and $Y_{0(0)}$ is the Y coordinate of the pole at the section $z = 0$. Finally, integrating Eq. (10.52c) twice results in

$$M_\omega + GK_T\theta + \bar{K}(\theta + \theta_i) - M_m = C_1 z + C_2 \tag{10.60}$$

where

$$M''_m = (v + v_i)''(M_Y + X_0N) + (u + u_i)''(M_X - Y_0N)$$

and

$$-[q_y(X_q - X_0) - q_x(Y_q - Y_0)] + [\bar{K}'(\theta + \theta_i)]' \tag{10.61a}$$

$$M_{m(0)} = M_{m(L)} = 0 \tag{10.61b}$$

Note that in the development of the above equation the term $(GK_T)'$ is neglected. In the elastic range, and for bars of constant cross section $(GK_T)'$ is equal to zero. In the plastic range—that is for the majority of (short) beams encountered in civil engineering practice—warping torsion predominates St. Venant's torsion and the influence of $(GK_T)'$ could be neglected without error.

The constants of integration C_1 and C_2 [in Eq. (10.60)] are again determined from the end conditions

$$\theta = 0 \quad \text{and} \quad M_\omega = M_{\omega 0} \quad \text{at} \quad z = 0$$
$$\theta = 0 \quad \text{and} \quad M_\omega = M_{\omega l} \quad \text{at} \quad z = L$$

which results in the relation

$$M_\omega + (GK_T + \bar{K})\theta = (M_{\omega l} - M_{\omega 0})\frac{z}{L} + M_{\omega 0} - \bar{K}\theta_i + M_m \tag{10.62}$$

10.4.6 Generalized Stress-Strain Relations

The normal strain ε at a point $P(X, Y, \omega)$ of the cross section mn situated at a distance z from the origin, and the deformations u, v, θ of the pole of this cross section are related by the following formula

$$\varepsilon = \bar{\varepsilon} - u''X - v''Y - \theta''\omega + \varepsilon_r \tag{10.63}$$

where $\bar{\varepsilon}$ is a constant, for a given z. ε_r is the residual strain, if any, existing at the point P. Note that for an initially deformed bar, X and Y are defined by Eq. (10.53). If the strain ε is inferior to the yield strain of the material, the fiber is still elastic and

$$\sigma = E[\bar{\varepsilon} - u''X - v''Y - \theta''\omega] + \sigma_r \tag{10.64}$$

and if it is equal to or superior to the yield strain

$$\sigma = \pm\sigma_y \tag{10.65}$$

where E is the modulus of elasticity σ_y, the yield stress of the material, and σ_r, the residual stress at point P. In this chapter, we assume that the residual stress distribution satisfies the following conditions (Vinnakota and Aysto, 1976):

$$\int_A \sigma_r\,dA = \int_A \sigma_r X\,dA = \int_A \sigma_r Y\,dA = \int_A \sigma_r \omega\,dA = 0 \tag{10.66}$$

For a self-equilibrating system of residual stresses the first three conditions are necessary, while the fourth condition is added to simplify the final equations.

In what follows, we use the subscripts e and p to define the elastic and plastic parts of a partially plastified section. We define several geometrical characteristics based on the elastic core. These are A_e, the area of the elastic core; S_{Xe}, S_{Ye}, static moments of area about X and Y axes; I_{Xe}, I_{Ye}, moments of inertia about X and Y axes; I_{XYe}, product of inertia; $S_{\omega e}$, sectorial static moment; $I_{\omega e}$, sectorial moment of inertia; $I_{\omega Xe}, I_{\omega Ye}$, sectorial products of inertia with respect to the Y and X axes. These quantities are defined as follows:

$$A_e = \int_{A_e} dA$$

$$\begin{aligned}
S_{Xe} &= \int_{A_e} Y\,dA & S_{Ye} &= \int_{A_e} X\,dA & S_{\omega e} &= \int_{A_e} \omega\,dA \\
I_{Xe} &= \int_{A_e} Y^2\,dA & I_{Ye} &= \int_{A_e} X^2\,dA & I_{\omega e} &= \int_{A_e} \omega^2\,dA \\
I_{\omega Ye} &= \int_{A_e} \omega Y\,dA & I_{\omega Xe} &= \int_{A_e} \omega X\,dA & I_{XYe} &= \int_{A_e} XY\,dA
\end{aligned} \tag{10.67}$$

In addition we introduce the following notations:

$$\begin{aligned}
\tilde{I}_{Xe} &= I_{Xe} - \frac{S_{Xe}^2}{A_e} & \tilde{I}_{\omega Xe} &= I_{\omega Xe} - \frac{S_{Ye}S_{\omega e}}{A_e} \\
\tilde{I}_{Ye} &= I_{Ye} - \frac{S_{Ye}^2}{A_e} & \tilde{I}_{\omega Ye} &= I_{\omega Ye} - \frac{S_{Xe}S_{\omega e}}{A_e} \\
\tilde{I}_{\omega e} &= I_{\omega e} - \frac{S_{\omega e}^2}{A_e} & \tilde{I}_{XYe} &= I_{XYe} - \frac{S_{Xe}S_{Ye}}{A_e}
\end{aligned} \tag{10.68}$$

From their definitions given in Eq. (10.53), we note that the coordinates X, Y are not constants but functions of the twist of the cross section, $\theta + \theta_i$. The quantities $S_{Xe}, S_{Ye}, I_{Xe}\ldots$ defined in Eqs. (10.67) and (10.68) are therefore not constants, even in the elastic range.

With the help of the relation $N = \int_A \sigma \, dA$, the term $\bar{\varepsilon}$ could be eliminated from Eq. (10.64) giving

$$\sigma = E\left[\left(\frac{S_{Ye}}{A_e} - X\right)u'' + \left(\frac{S_{Xe}}{A_e} - Y\right)v'' + \left(\frac{S_{\omega e}}{A_e} - \omega\right)\theta''\right] + \frac{N_e}{A_e} + \sigma_r \tag{10.69}$$

where N_e is the axial load resisted by the elastic part of the cross section $(= \int_{A_e}(\sigma - \sigma_r)\, dA)$. Introducing this value of σ in Eqs. (10.34), (10.37) and (10.43), and making use of the relations (10.67) and (10.68), the generalized stresses M_Y, M_X and M_ω can be expressed as a function of generalized strains u'', v'' and θ''. Thus, we have

$$\begin{aligned} M_Y &= \quad E(\tilde{I}_{Ye}u'' + \tilde{I}_{XYe}v'' + \tilde{I}_{\omega Xe}\theta'') - f_Y \\ M_X &= -E(\tilde{I}_{XYe}u'' + \tilde{I}_{Xe}v'' + \tilde{I}_{\omega Ye}\theta'') + f_X \\ M_\omega &= -E(\tilde{I}_{\omega Xe}u'' + \tilde{I}_{\omega Ye}v'' + \tilde{I}_{\omega e}\theta'') + f_\theta \end{aligned} \tag{10.70}$$

where

$$\begin{aligned} f_Y &= \frac{N_e S_{Ye}}{A_e} + \int_{A_p} (\sigma - \sigma_r) X \, dA \\ f_X &= \frac{N_e S_{Xe}}{A_e} + \int_{A_p} (\sigma - \sigma_r) Y \, dA \\ f_\theta &= \frac{N_e S_{\omega e}}{A_e} + \int_{A_p} (\sigma - \sigma_r) \omega \, dA \end{aligned} \tag{10.71}$$

The integrals in Eq. (10.71) represent the part of the moments M_Y, M_X and M_ω resisted by the plasticized part A_p of the cross section.

10.4.7 Differential Equations

For the biaxially restrained member shown in Fig. 10.5 compatibility requires that the member ends and the attached rotational springs (which furnish only bending restraint) rotate through the same angle. In an actual structure the restraints offered by the springs (adjacent members) are complex nonlinear functions of deformations. In the present chapter, we consider that these springs are linearly elastic, with the spring constants given as $\alpha_{x0}, \alpha_{xl}, \alpha_{y0}$ and α_{yl}. Equilibrium at each end requires that:

$$\begin{aligned} M_{y0} &= \tilde{M}_{y0} - \alpha_{y0}u_0', \quad M_{yl} = \tilde{M}_{yl} + \alpha_{yl}u_l' \\ M_{x0} &= \tilde{M}_{x0} + \alpha_{x0}v_0' \quad M_{xl} = \tilde{M}_{xl} - \alpha_{xl}v_l' \end{aligned} \tag{10.72}$$

Regarding the end conditions for warping, only the following two extreme cases are considered: either the member end is prevented from warping and

$$\theta' = 0 \tag{10.73}$$

or it is free to warp, and

$$M_\omega = 0 \tag{10.74}$$

By replacing the bending moments M_Y, M_X and the bimoment M_ω in Eqs. (10.72) and (10.74) by their values given in Eq. (10.70), the end conditions become

$$\begin{aligned}
&E\tilde{I}_{Ye}u_0'' + E\tilde{I}_{XYe}v_0'' + E\tilde{I}_{\omega Xe}\theta_0'' - \alpha_{y0}u_0' = M_{y0} + f_{Y0}\\
&E\tilde{I}_{Ye}u_l'' + E\tilde{I}_{XYe}v_l'' + E\tilde{I}_{\omega Xe}\theta_l'' + \alpha_{yl}u_l' = M_{yl} + f_{Yl}\\
&E\tilde{I}_{XYe}u_0'' + E\tilde{I}_{Xe}v_0'' + E\tilde{I}_{\omega Ye}\theta_0'' - \alpha_{x0}v_0' = -M_{x0} + f_{X0}\\
&E\tilde{I}_{XYe}u_l'' + E\tilde{I}_{Xe}v_l'' + E\tilde{I}_{\omega Ye}\theta_l'' + \alpha_{xl}v_l' = -M_{xl} + f_{Xl}\\
&\theta_0' = 0 \quad \text{or} \quad E\tilde{I}_{\omega Xe}u_0'' + E\tilde{I}_{\omega Ye}v_0'' + E\tilde{I}_{\omega e}\theta_0'' = f_{\theta 0}\\
&\theta_l' = 0 \quad \text{or} \quad E\tilde{I}_{\omega Xe}u_l'' + E\tilde{I}_{\omega Ye}v_l'' + E\tilde{I}_{\omega e}\theta_l'' = f_{\theta l}\\
&u_0 = v_0 = \theta_0 = u_l = v_l = \theta_l = 0
\end{aligned} \tag{10.75}$$

The section properties $\tilde{I}$ in any of these equations are the values corresponding to that end. If we substitute for the generalized stresses M_Y, M_X, M_ω their values given by Eqs. (10.70) and replace the internal end-moments by their values obtained from the Eq. (10.72), the equilibrium equations (10.57), (10.58) and (10.62) become

$$\begin{aligned}
&E\tilde{I}_{Ye}u'' + E\tilde{I}_{XYe}v'' + E\tilde{I}_{\omega Xe}\theta'' - Nu - \left(1 - \frac{z}{L}\right)\alpha_{y0}u_0' + \frac{z}{L}\alpha_{yl}u_l'\\
&\quad = +\left(1 - \frac{z}{L}\right)M_{y0} + \frac{z}{L}M_{yl} - (X_0 - X_{0(0)} - u_i)N + M_{qX} + f_Y\\
&E\tilde{I}_{XYe}u'' + E\tilde{I}_{Xe}v'' + E\tilde{I}_{\omega Ye}\theta'' - Nv - \left(1 - \frac{z}{L}\right)\alpha_{x0}v_0' + \frac{z}{L}\alpha_{xl}v_l'\\
&\quad = -\left(1 - \frac{z}{L}\right)M_{x0} - \frac{z}{L}M_{xl} - (Y_0 - Y_{0(0)} - v_i)N + M_{qY} + f_X\\
&E\tilde{I}_{\omega Xe}u'' + E\tilde{I}_{\omega Ye}v'' + E\tilde{I}_{\omega e}\theta'' - (GK_T + \bar{K})\theta\\
&\quad = -\left(1 - \frac{z}{L}\right)M_{\omega 0} - \frac{z}{L}M_{\omega l} + \bar{K}\theta_i - M_m + f_\theta
\end{aligned} \tag{10.76}$$

These are the three nonlinear, nonhomogeneous, coupled differential equations of equilibrium for a beam-column subjected to end moments and transverse loads, and loaded into the plastic range. The equilibrium is written for the portion to the left of section mn situated at a distance z from the origin, in the deformed position of the bar. Note that the origin C is not the center of gravity, but an arbitrary point of the cross section. The pole 0, whose displacements u and v appear in the above equations is again an arbitrary point of the cross section, not the shear center. The ordinates X, Y are measured in accord with a system of axes X-Y moving parallel to the arbitrary coordinate system x-y-z, fixed in space (Fig. 10.7). The deformation field considered is not linearized as is the case with the majority of the studies [see Eqs. (10.27) and (10.28)]. u_i, v_i and θ_i define the initial geometrical imperfections of the bar at section z. When the section enters the plastic range, the rigidities are calculated for the remaining elastic core. The quantities f_X, f_Y and f_θ are functions of the part of the bending moments and bi-moment resisted by the plasticized part of the cross section. M_{qX}, M_{qY} and M_m represent the influence of the external transverse loads. The influence of residual stresses appears in the functions f_X, f_Y and f_θ; indirectly in the calculation of the elastic core; and in the calculation of the term $\bar{K}$ [Eq. (10.45)].

As we considered an arbitrary open cross section, the equilibrium equations are valid, in particular, to the I, Z, T, U, double-angle and hat-shaped sections generally encountered in civil engineering application of steel structures.

The exact solution of these three coupled differential equations is not feasible. To obtain a numerical solution, a finite difference technique is used, which will form the subject of the next section.

10.5 NUMERICAL TECHNIQUE

10.5.1 Finite Differences and Finite Integrals

The finite difference grid used in the calculations is shown in Fig. 10.15. The member is first divided into m equal parts of $\Delta z = L/m$, defining $m + 1$ grid points; in addition we consider two half segments at each end, thus defining two additional grid points. The deformed shape of the member is completely defined by the three unknown components u, v and θ at these $(m + 3)$ grid points, making a total of 3 $(m + 3)$ unknowns. The three equilibrium equations at each of the pivotal points 2 to m, and the six boundary conditions at each end furnish a total of 3 $(m + 3)$ equations containing the 3 $(m + 3)$ unknowns.

For the pivotal point i, with $3 \leq i \leq m - 1$, we use central-differences with equal grid spacing. From Eq. (10.11), we have

FIGURE 10.15
Finite difference grid used in the program

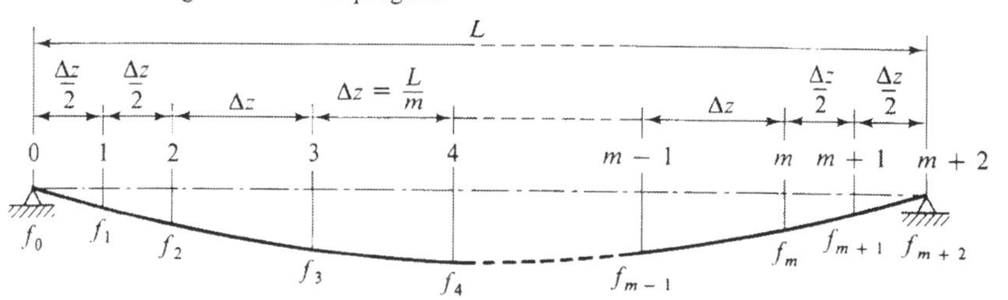

$$f_i'' = \frac{f_{i-1} - 2f_i + f_{i+1}}{(\Delta z)^2} \tag{10.77}$$

where f stands for one of the three deformations u, v or θ.

For the pivotal points 2 and m, we use central-differences with variable grid spacing ($\Delta z/2$ and Δz). The corresponding expressions are

$$f_2'' = \frac{2f_1 - 3f_2 + f_3}{\frac{3}{4}(\Delta z)^2} \qquad f_m'' = \frac{f_{m-1} - 3f_m + 2f_{m+1}}{\frac{3}{4}(\Delta z)^2} \tag{10.78}$$

For the pivotal point O we use forward-differences, while for the pivotal point $m + 2$ we use backward-differences. Using four grid points, we have the following difference expressions:

$$f_O'' = \frac{7f_O - 16f_1 + 10f_2 - f_3}{(\Delta z)^2}$$

$$f_{m+2}'' = \frac{-f_{m-1} + 10f_m - 16f_{m+1} + 7f_{m+2}}{(\Delta z)^2} \tag{10.79}$$

Finally, for the first derivatives at the pivotal points O and $m + 2$, we obtain the following relations with the help of Eq. (10.10)

$$f_O' = \frac{-3f_O + 4f_1 - f_2}{(\Delta z)} \qquad f_{m+2}' = \frac{f_m - 4f_{m+1} + 3f_{m+2}}{(\Delta z)} \tag{10.80}$$

Note that the usual procedure of considering fictitious grid points beyond the supports—which will permit the use of central differences throughout—is not utilized here. For members whose support sections are prone to large plastification, such a procedure is likely to lead to inaccuracies and convergence difficulties. Note also that only two adjacent grid points are used for central differences, while four points are used for forward and backward differences, to increase the accuracy.

In Sec. 10.4.5 we introduced the quantities M_{qX}, M_{qY} and M_m which are functions of the transverse loads q_x, q_y acting on the member. These quantities satisfy the following homogeneous conditions:

$$M_{qX}'' = q_x \qquad M_{qX(0)} = M_{qX(L)} = 0 \tag{10.81}$$

$$M_{qY}'' = q_y \qquad M_{qY(0)} = M_{qY(L)} = 0 \tag{10.82}$$

$$\begin{aligned} M_m'' &= (v + v_i)''(M_Y + X_0 N) + (u + u_i)''(M_X - Y_0 N) \\ &\quad - [q_y(X_q - X_0) - q_x(Y_q - Y_0)] + [\bar{K}'(\theta + \theta_i)]' = q_t \\ &\qquad M_{m(0)} = M_{m(L)} = 0 \end{aligned} \tag{10.83}$$

From these relations it could be seen that M_{qX}, M_{qY} and M_m could be assimilated to the bending moments of a simply supported beam of span L with distributed loads q_x, q_y and q_t respectively. For the integration of q_x, q_y, q_t to obtain M_{qX}, M_{qY}, M_m, these distributed loads are first replaced by equivalent concentrated loads acting

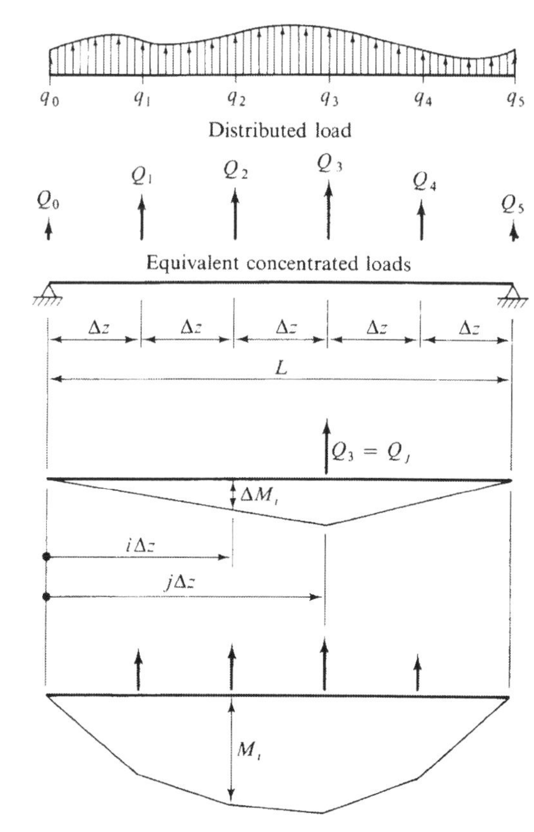

FIGURE 10.16
Calculation of the terms M_{qX}, M_{qY} and M_{qt}

at the grid points. A continuous variable loading of some higher order or perhaps unknown order (this is generally so with q_t) is best treated by considering it as parabolic over each three adjacent grid points in turn. The procedure is shown in Fig. 10.16 for the case of equal grid spacing. Thus

$$Q_0 = \frac{\Delta z}{24}(7q_0 + 6q_1 - q_2) \qquad Q_1 = \frac{\Delta z}{12}(q_0 + 10q_1 + q_2) \qquad \text{etc.} \tag{10.84}$$

The bending moment ΔM_i at a pivotal point i due to a concentrated load Q_j acting at a grid point j is (Fig. 10.16):

$$\begin{aligned} \Delta M_i &= Q_j\left(1 - \frac{i\Delta z}{L}\right)j\Delta z \qquad \text{for} \qquad 0 \le j \le i \\ \Delta M_i &= Q_j\left(1 - \frac{j\Delta z}{L}\right)i\Delta z \qquad \text{for} \qquad j > i \end{aligned} \tag{10.85}$$

The bending moment at the pivotal point i due to the continuously variable transverse load q is therefore

$$M_i = \Sigma_j \Delta M_i \tag{10.86}$$

Calculation of M_{qX}, M_{qY} and M_m could be simplified if the transverse load has a simpler form. Thus, in the case of a single concentrated load Q acting at a distance a from the origin

$$\begin{aligned} M &= QL\left(1 - \frac{a}{L}\right)\frac{z}{L} \quad \text{for} \quad 0 \le z \le a \\ M &= QL\left(1 - \frac{z}{L}\right)\frac{a}{L} \quad \text{for} \quad a \le z \le L \end{aligned} \tag{10.87}$$

In the case of a trapezoidal loading with intensities q_1 at the end $z = 0$ and q_2 at the end $z = L$ of the member, we have

$$M = \frac{q_1 L^2}{6}\left[2\left(\frac{z}{L}\right) - 3\left(\frac{z}{L}\right)^2 + \left(\frac{z}{L}\right)^3\right] + \frac{q_2 L^2}{6}\left[\left(\frac{z}{L}\right) - \left(\frac{z}{L}\right)^3\right] \tag{10.88}$$

10.5.2 Method of Solution

Replacing the derivatives in the three equilibrium equations (10.76) at each of the pivotal points 2 to m and in the twelve boundary conditions (10.75), by the appropriate central, forward and backward differences mentioned in the previous section, results in a system of simultaneous algebraic equations. This system can be expressed in the form

$$[A]\{U\} = \{W\} \tag{10.89}$$

where

$\{U\}$ is a vector of 3 $(m + 3)$ pivotal deformations.

$$\{U\}^T = \{u_O, v_O, \theta_O, u_1, v_1, \theta_1, \ldots, u_{m+2}, v_{m+2}, \theta_{m+2}\}$$

$[A]$ is a square matrix of size 3 $(m + 3) \times 3\,(m + 3)$ whose elements are geometrical characteristics $E\tilde{I}_{Ye}, E\tilde{I}_{XYe}$, etc., of the elastic core at these pivotal points; the axial tension N (or the axial thrust $P = -N$) and the spring stiffnesses α.

$\{W\}$ is a vector of size $3\,(m + 3)$ which consists of the right hand sides of the equilibrium equations and of the appropriate boundary conditions. In particular they contain the external end-moments; the terms resulting from the transverse loads q_x, q_y; the influence of initial deformations; and also the nonlinear terms resulting from plastification and twisting deformation.

The set of simultaneous equations (10.89) cannot be solved directly because the vector $\{W\}$ is an unknown function of $\{U\}$. Furthermore, the geometric characteristics vary due to the twisting deformation and due to the yielding. So the matrix $[A]$ is also an unknown function of $\{U\}$. Therefore, to solve the problem an iteration procedure, described below, is used.

1. The input data are read and all the initial values that are necessary for computation are generated.

   The beam-column is divided into m intervals of equal length resulting in $m+3$ pivotal points as shown in Fig. 10.15. In the numerical examples presented later in the Sec. 10.6, m varied from 8 to 12.

2. A convenient load level is chosen to start the analysis. The value adopted is such that the bar is in the elastic state and elastic geometrical characteristics can be utilized for first trial calculations.

3. Trial values for deformations $\bar{\varepsilon}, u, v$ and θ at each pivotal point are considered as follows:

   (a) For the first cycle of the first load level the deformations are assumed to be the initial deformations.

   (b) For the first cycle at any other load level, the deformations obtained from a previously converged calculation are used.

   (c) For subsequent cycles at any load level, the deformations obtained from the previous cycle are used.

4. From these deformations the normal strain distribution across the cross section at each pivotal point is computed using Eq. (10.63). For details see Sect. 10.5.3.)

5a. From this strain distribution the geometric characteristics of the unyielded core ($A_e, S_{Xe}, I_{Xe}, \ldots, \tilde{I}_{Xe}, \tilde{I}_{Ye}$, etc.) defined by Eqs. (10.67) and (10.68) are computed. The normal stress, σ, is then computed using Eqs. (10.69) and (10.65).

5b. The generalized stresses N, M_X, M_Y, M_ω; the coefficient $\bar{K}$ defined by Eq. (10.45) and the quantities f_X, f_Y, f_θ defined by Eq. (10.71) are then evaluated with the normal stress distribution computed above.

5c. The quantities M_{qX}, M_{qY} and M_m at each pivotal point are also calculated using the procedure described in Sec. 10.5.1.

5d. The right hand sides of the equilibrium equations (10.76), and of the appropriate boundary conditions are then evaluated.

5e. The quantities determined in 5a, 5b, 5c, 5d, above, permit to establish the elements of the matrix $[A]$ and the vector $\{W\}$.

6. A new set of values for deformations $\{U\}$ is then obtained from the solution of the system (10.89).

7. The deformations calculated in two successive cycles are compared for convergence. The following rule of convergence is used:

$$\frac{\delta_i - \delta_{i-1}}{\delta_{i-1}} \leq \mu \qquad (10.90)$$

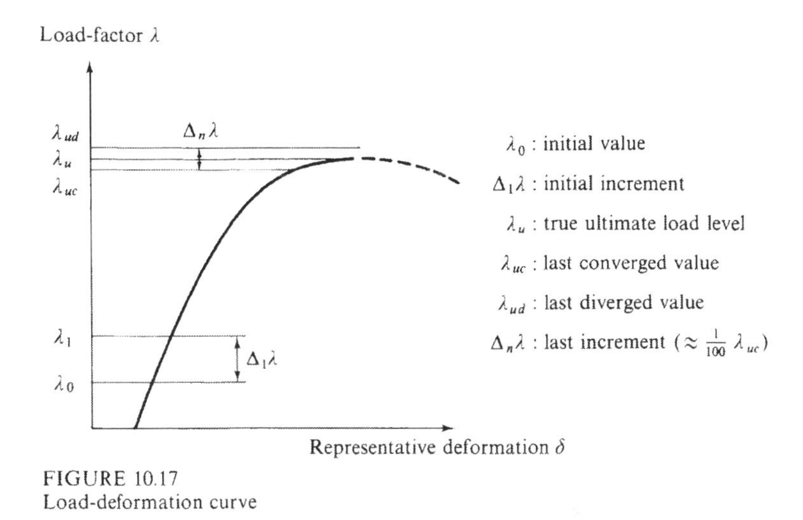

FIGURE 10.17
Load-deformation curve

where δ_{i-1} and δ_i are the deformations at a chosen point in the $i-1$ and i-th cycle, and μ represents the tolerance ratio of convergence. The value of μ in this analysis is generally taken as 1 percent.

(a) If the test is satisfied, the member is still stable. So the load is increased and the calculations start from Step 3.

(b) If the test is not satisfied and if the number of iteration cycles is less than a predefined number (NCYMAX), a new cycle of calculations with the same load starts from Step 3.

(c) If the test is not satisfied within the predefined number of iteration cycles NCYMAX, the calculations restart from the previously converged load level with reduced load increment at Step 3.

The calculations are repeated until the ratio between this reduced load increment and the last converged load becomes less than 1 percent. The ultimate load is considered as the last converged load. The procedure is schematically represented in Fig. 10.17.

10.5.3 Computer Program for I-Section Members

A computer program based on the theory presented in this chapter is developed in order to study the behavior of I-section members. The aim of the program is to study the influence of various parameters on the ultimate strength of structural steel I-section members (in-plane or space behavior of beams, columns or beam-columns).

For carrying out the numerical calculations the I-section is divided into small elements as shown in Fig. 10.18(a). Three patterns of residual stress distributions, generally found in rolled sections, are included in the program (Fig. 10.19). As shown in Fig. 10.18(b), the actual surface of strain distribution across an element, under

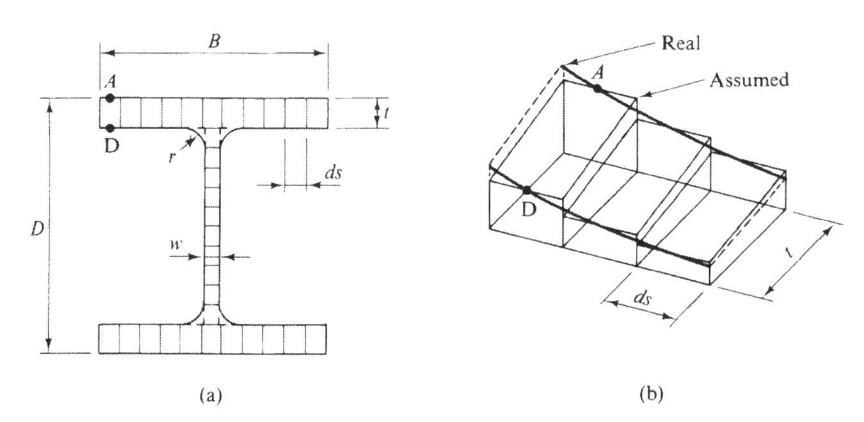

FIGURE 10.18
Partitioning of cross-section for numerical calculations. (a) Elemental areas. (b) Real and assumed strain distribution

the combined influence of bending, warping and residual strains (parabolic type, for example), may not be a plane. Also shown in this figure is the trapezoidal type of strain distribution across each element, assumed in the calculation of the zones of plasticity, geometrical properties of the elastic core, etc. Note that the influence of the flange web fillet is included in the calculations.

Three types of geometrical imperfections included in the program are already shown in Fig. 10.14. The external forces considered in the program are non-proportional. In effect, each of them can be composed of a constant part and a variable part. Thus

$$\begin{aligned} N &= N_c + \lambda N_v \qquad q_y = q_{cy} + \lambda q_{vy} \qquad q_x = q_{cx} + \lambda q_{vx} \\ M_{xo} &= M_{cxo} + \lambda M_{vxo} \qquad M_{xl} = M_{cxl} + \lambda M_{vxl} \\ M_{yo} &= M_{cyo} + \lambda M_{vyo} \qquad M_{yl} = M_{cyl} + \lambda M_{vyl} \end{aligned} \tag{10.91}$$

where λ represents the load factor or load level; subscript c stands for a constant part of a force; and subscript v for the variable part of a force corresponding to

FIGURE 10.19
Patterns of residual stresses included in the program

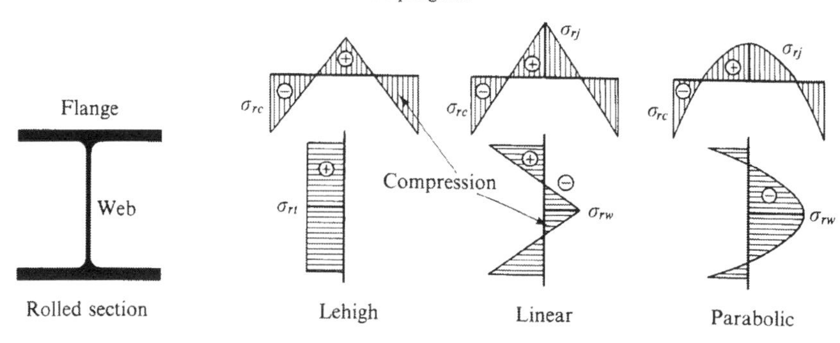

the values of $\lambda = 1$. In the program, λ is increased monotonically and the corresponding deformations are calculated.

10.6 NUMERICAL EXAMPLES

In this section several numerical examples are given to illustrate the validity of the theory developed to I-section members. For clarity, these examples are divided into isolated beam-columns under biaxial loading (Sect. 10.6.1), buckling of isolated beams and beam-columns (Sect. 10.6.2) and restrained beam-columns under biaxial loading (Sect. 10.6.3). In each section, the examples selected include those for which experimental or analytical results are available in the literature.

10.6.1 Isolated Beam-Columns under Biaxial Loading

The isolated beam-columns studied include two biaxially loaded beam-columns tested by Birnstiel (1968); one biaxially loaded beam-column analyzed by Soltis and Christiano (1972) and several parametric studies made by Vinnakota (1976a).

FIGURE 10.20
Test specimen of isolated beam-column used by Birnstiel, 1968

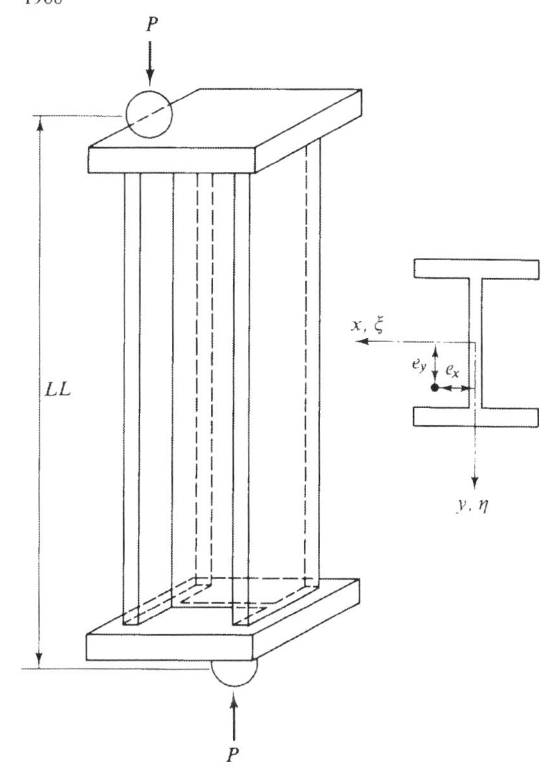

Table 10.1 DATA FOR EXAMPLE 10.3

	Specimen Number 4	Specimen Number 15
Profile	6 × 6H	5 × 5H
Length, L	96 in	120 in
Yield-stress, σ_y	36 ksi	65 ksi
e_{xo}	1.67 in	1.75 in
e_{xl}	1.66 in	−1.47 in
e_{yo}	2.95 in	2.96 in
e_{yl}	2.95 in	−2.96 in

1 in = 25.4 mm; 1 ksi = 6.9 MN/m².

EXAMPLE 10-3 *Biaxially Loaded Beam-Columns: Comparison with Test Results of Birnstiel (1968)* The two columns selected are the specimen numbers 4 and 15—isolated steel H columns subjected to biaxially eccentric load—tested at the New York University (see Sect. 13.7.3, Chap. 13, p. 663). They were held in position against horizontal displacement at the ends, which were also prevented from twisting about the longitudinal axis. The end sections were permitted to rotate about any axis lying in the x-y plane (Fig. 10.20). Warping at the ends was prevented by welding very thick plates as shown in this figure. For specimen number 4 the eccentricities of loading were substantially alike at both ends resulting in simple curvature with respect to both principal axes. On the other hand, for specimen number 15 the

FIGURE 10.21
Example of a symmetrically bent isolated column (specimen no. 4 of Birnstiel, 1968)

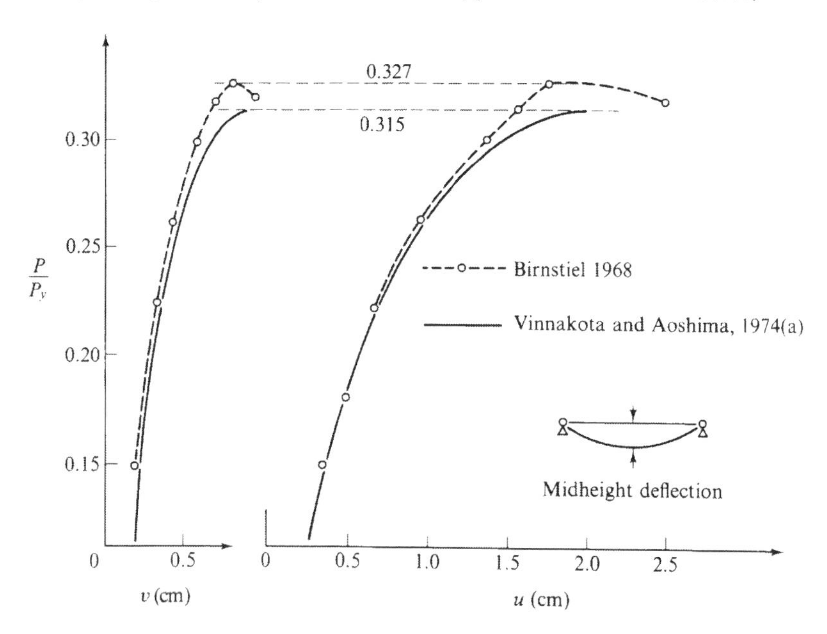

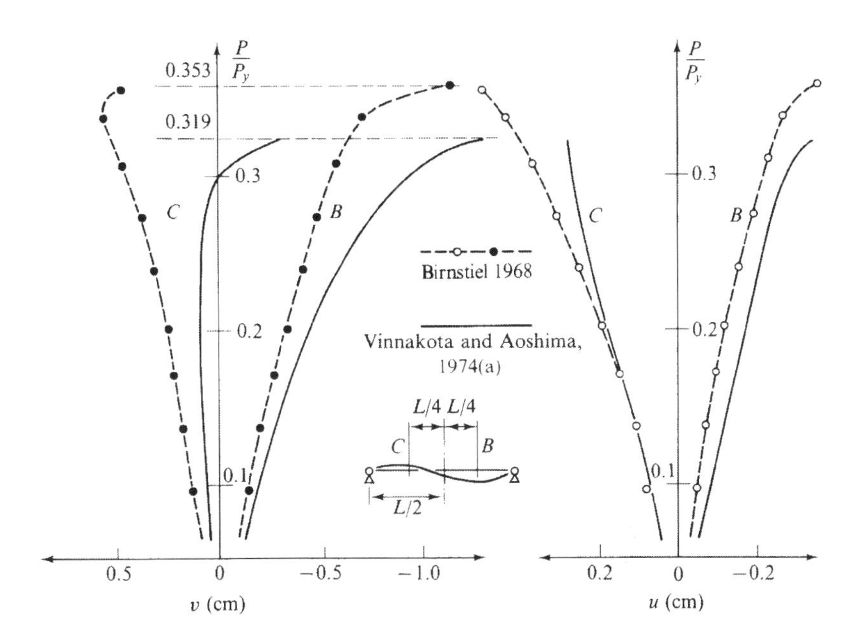

FIGURE 10.22
Example of an antimetrically bent isolated column (specimen no. 15 of Birnstiel, 1968)

eccentricities of loading were unlike at both ends. The relevant data is given in Table 10.1.

The solutions of the present theory are compared, in Fig. 10.21 for specimen number 4 and in Fig. 10.22 for specimen number 15, with the data obtained by Birnstiel. The theoretical curves are shown by thick solid lines and the small circles connected by thin lines represent the results from experiments. The axial load, P, divided by the yield load $P_y(=\sigma_y A)$ is used in these curves. A reasonable agreement can be observed for specimen number 4 but the correspondence between the computed and the experimentally determined values of displacements for specimen number 15 is not good. The discrepancy probably results from the fact that the column length L is taken in the analysis as the length LL between the bearings, see Fig. 10.20. In reality, however, very thick plates (rigid in their plane) exist at the ends. It is observed in the calculation that, near the critical load, about 20 percent of the area of the cross section at the ends has yielded for the specimen number 4 and 45 percent for the specimen number 15. The numerical results are thus based on a flexible model compared to the test, more so for specimen number 15 than for specimen number 4.

EXAMPLE 10-4 *Biaxially Loaded Beam-Column: Comparison with the Results of Third-Order Analysis* The example selected—a W14 × 43 beam-column having slenderness ratio $L/r_y = 140$—is the one studied by Soltis and Christiano (1972) using a third-order analysis, described already in Sect. 4.6. The beam-column is loaded by a

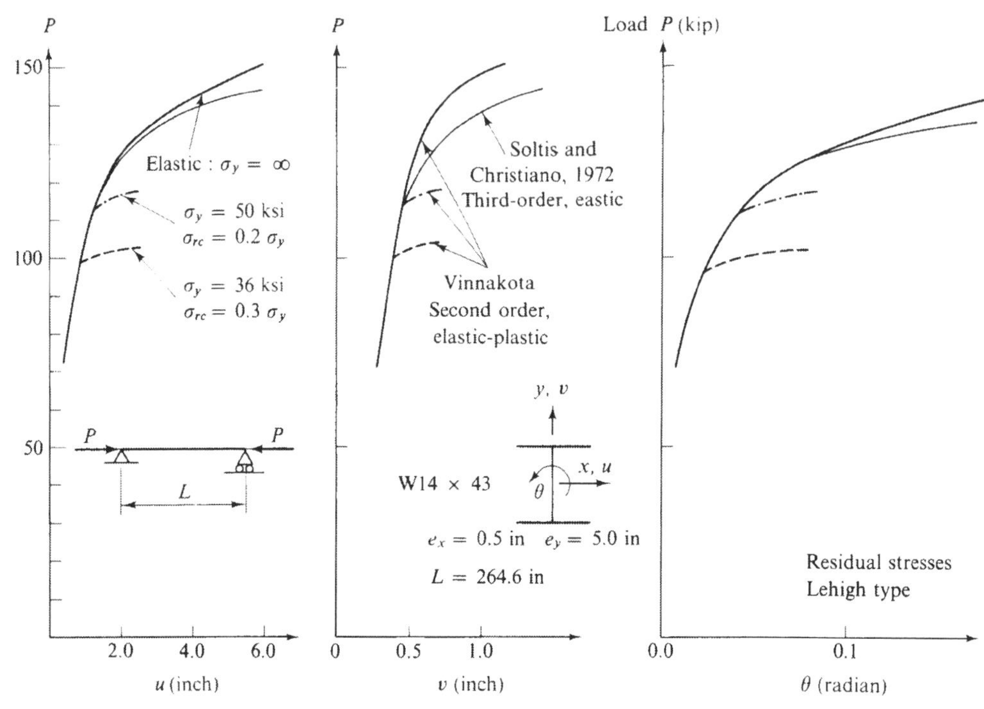

FIGURE 10.23
Midspan deformations of beam-column (Soltis and Christiano, 1972) (1 kip = 4.45 kN, 1 in = 2.54 cm)

longitudinal thrust P with biaxial eccentricity of $e_x = 0.5$ in (1.27 cm) and $e_y = 5.0$ in (12.7 cm). Simply supported, free to warp conditions are assumed in the analysis. Mid-height deformations u, v and θ computed through third-order, elastic theory obtained by Soltis and Christiano (1972) are shown in Fig. 10.23. Also given in this figure are the values obtained by the present analysis, for three different values of yield stress σ_y, namely, $\sigma_y = 36$ ksi; 50 ksi and infinity. For the first two calculations, Lehigh type residual stresses having $\sigma_{rc} = 0.3\sigma_y$ and $0.2\sigma_y$ respectively, are included in the analysis.

Note that the present analysis neglects terms such as $[1 + (u')^2]^{1/2}$ and $[1 + (v')^2]^{1/2}$ which indicate the degree of inclination of the ξ-η plane to the X-Y plane [see Fig. 10.7 and Eq. (4.89)] and certain products of derivatives of u and v which are considered small. Both theories are based on nonlinear deformation field, as far as the angle of twist θ is concerned. (Thus the usual simplications such as $\sin\theta = \theta$ and $\cos\theta = 1$ are not made in both the analysis.) The present theory, however, includes the influence of plasticity and residual stresses.

The difference between the two elastic solutions ($\sigma_y = \infty$) indicates the difference between the second- and third-order theories. From the results given for $\sigma_y = 36$ ksi and 50 ksi, it could be concluded that, for practical columns, the influence of plasticity is much more important than the influence of third-order terms.

EXAMPLE 10-5 *Influence of Residual Stresses on the Strength of Biaxially Loaded Beam-Columns* The numerical examples worked out to find the influence of the variations in the level and pattern of residual stresses on the maximum strength of biaxially loaded beam-columns are all based on columns made of W8 × 31 section. Three doubly symmetric patterns of residual stress distributions are considered. They are the Lehigh type, Linear type and Parabolic type shown in Fig. 10.19. Four levels of maximum compressive residual stresses, that occur at flange tips for rolled shapes, are considered in the present study. They are $\sigma_{rc} = 0.0, 0.1, 0.3$ and 0.5 of σ_y. For the linear and parabolic types the residual stress intensity at the center of the web (σ_{rw}, in Fig. 10.19) is taken as 2/3 of σ_{rc}.

The slenderness ratio, L/r_x, of all the columns considered is 30. They are initially twisted, with a twist angle θ_m at mid-height equal to 0.001 radian. In addition, they have small initial crookedness in the x and y directions with lateral deflections at mid-span u_m and v_m equal to $L/10\,000$. The ends are simply supported, with warping permitted at both ends. The axial force, P, remains constant at 0.3 of the yield load of the section, P_y. In addition, the columns are subjected to symmetric, single curvature moments in the two principal planes. The maximum value of the minor axis bending moments M_y, corresponding to various chosen values of the major axis bending moments, M_x, is calculated with the help of the computer program and given in the Table 10.2. Also given there is the maximum value of the major axis bending moment when the minor axis bending moment is kept equal to zero (lateral buckling load of the beam-column, M_{xcr}). For the Lehigh pattern with $\sigma_{rc} = 0.3\sigma_y$ the maximum strengths obtained by Tebedge and Chen (1974) are also indicated for comparative purposes (values given in brackets). Note, however, that their calculations are based on a lower level of initial deformations ($L/10^6$).

Figure 10.24 contains interaction curves for the two intermediate values of

Table 10.2 INFLUENCE OF RESIDUAL STRESSES ON THE ULTIMATE STRENGTH OF BIAXIALLY LOADED BEAM-COLUMNS

Pattern of Residual Stress Distribution	σ_{rc}/σ_y	Maximum Value of $m_y = M_y/M_{py}$, ($m_x = M_x/M_{px}$)							$m_{x\,max} = m_{xcr}$
		$m_x = 0.0$	$m_x = 0.1$	$m_x = 0.2$	$m_x = 0.3$	$m_x = 0.4$	$m_x = 0.5$	$m_x = 0.6$	$m_y = 0.0$
Annealed	0.0	0.725	0.712	0.672	0.591	0.456	0.278	0.096	0.637
Lehigh	0.1	0.719	0.703	0.666	0.584	0.453	0.272	0.079	0.664
	0.3	0.697	0.684	0.647	0.575	0.441	0.253	0.002	0.616
	0.3*	(0.693)	(0.681)	(0.649)	(0.597)	(0.514)	(0.390)	(0.292)	—
	0.5	0.675	0.662	0.625	0.550	0.425	0.228	—	0.591
Linear	0.1	0.722	0.709	0.666	0.584	0.447	0.259	0.074	0.634
	0.3	0.709	0.697	0.647	0.559	0.409	0.203	0.001	0.609
	0.5	0.691	0.669	0.616	0.516	0.350	0.138	—	0.575
Parabolic	0.1	0.728	0.716	0.672	0.591	0.453	0.269	0.085	0.641
	0.3	0.728	0.716	0.666	0.581	0.434	0.241	0.060	0.628
	0.5	0.722	0.703	0.647	0.553	0.403	0.197	—	0.584

Initial deformations: $u_m = -v_m = L/10\,000$; $\theta_m = 0.001$ radian.
* Values in the brackets are obtained by Tebedge and Chen, (1974), corresponding to $u_m = v_m = L/10^6$, $\theta_m = 1/10^6$ rad.

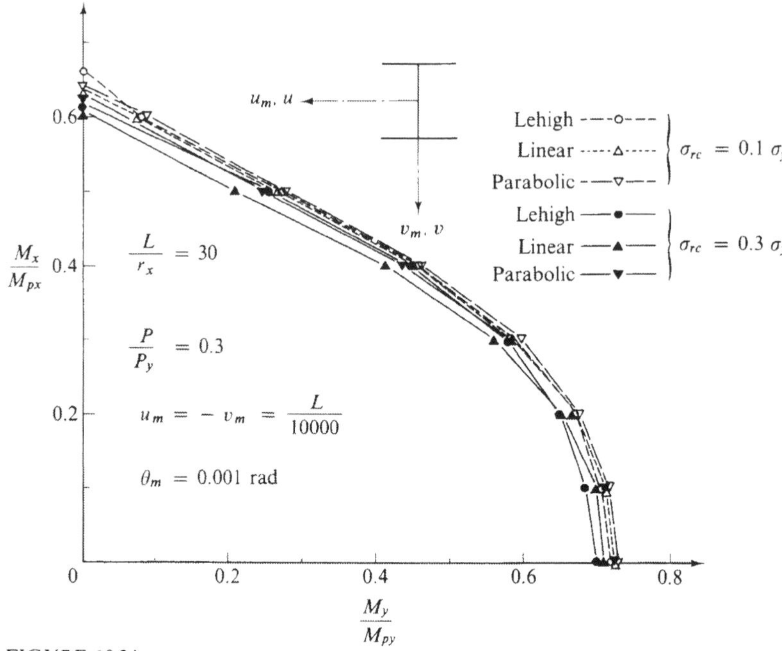

FIGURE 10.24
Influence of residual stresses on the strength of biaxially bent columns ($\sigma_{rc}/\sigma_y = 0.1$ and 0.3)

FIGURE 10.25
Influence of residual stresses on the strength of biaxially bent columns ($\sigma_{rc}/\sigma_y = 0.0$ and 0.5)

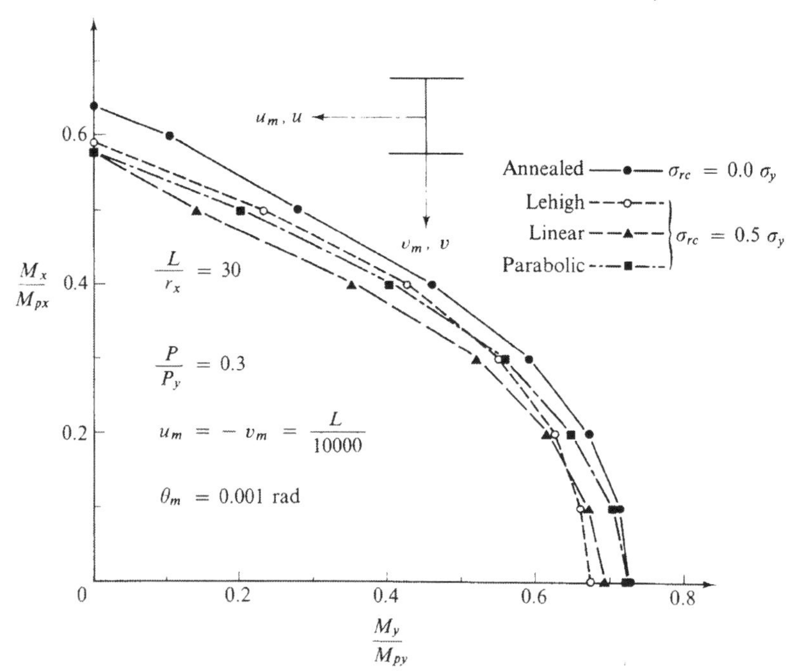

Table 10.3 INFLUENCE OF THE SIGN OF INITIAL DEFLECTIONS ON THE STRENGTH OF BIAXIALLY LOADED BEAM-COLUMNS ($|u_m| = |v_m| = L/500$)

Column Designation	$m_x = \frac{M_x}{M_{px}}$	Maximum Value of $m_y = M_y/M_{py}$ — Coordinates at Midheight Defining Initial Deformation: $(+u_m, +v_m)$	$(-u_m, +v_m)$	$(+u_m, -v_m)$	$(-u_m, -v_m)$
DS 330-0	0.0	0.728	0.656	0.728	0.656
DS 330-1	0.1	0.719	0.647	0.716	0.641
DS 330-2	0.2	0.687	0.609	0.678	0.600
DS 330-3	0.3	0.619	0.531	0.603	0.512
DS 330-4	0.4	0.506	0.397	0.478	0.366
DS 330-5	0.5	0.341	0.200	0.303	—
DS 330-6	0.6	0.146	—	—	—

	m_y	Maximum Value of m_x			
DS 330-10	0.0	0.584	0.587	0.556	0.559

residual stress intensities, namely, $\sigma_{rc}/\sigma_y = 0.1$ and 0.3: while Fig. 10.25 contains results for two extreme values of 0.0 and 0.5. The width of the band into which the results shown in Fig. 10.24 fall, will be typical of columns of slenderness ratio and axial load intensity considered in these examples and suggest that variations in the pattern and intensity of residual stresses do not have a pronounced effect on the maximum strength of biaxially loaded beam-columns.

EXAMPLE 10-6 *Influence of the Sign of Initial Deflections on the Strength of Biaxially Loaded Beam-Columns* In this example the influence of the sign of initial deflections on the maximum strength of biaxially loaded, pin-ended columns is studied. θ_m is taken as 0.001 radian for all columns. The shape of the initial crookedness is assumed to be an half wave of a sine function. All the columns considered are made of W8 × 31 section and are of length $L = 30r_x$. They are subjected to a constant axial force of $P = 0.3P_y$ and to symmetric, single curvature moments in the two principal planes. Residual stresses are considered to be of Lehigh type with $\sigma_{rc} = 0.3\sigma_y$.

For the purpose of comparison, columns with the four possible combinations of signs of initial deflections in the x and y plane are considered. These are: $(+u_m, +v_m)$; $(+u_m, -v_m)$; $(-u_m, +v_m)$ and $(-u_m, -v_m)$. The calculated results are given in Table 10.3 for an intensity of $L/500$ and in Table 10.4 for an intensity of $L/1000$. In addition, these results are shown graphically in Figs. 10.26 and 10.27. It could be seen that for a given value of the minor axis bending moment, the maximum value of M_x is obtained with $(+u_m, +v_m)$ and the minimum value of M_x with $(-u_m, -v_m)$. We note that for the combinations $(+u_m, +v_m)$ and $(+u_m, -v_m)$ the absolute maximum value of M_x (lateral buckling load) does not correspond to

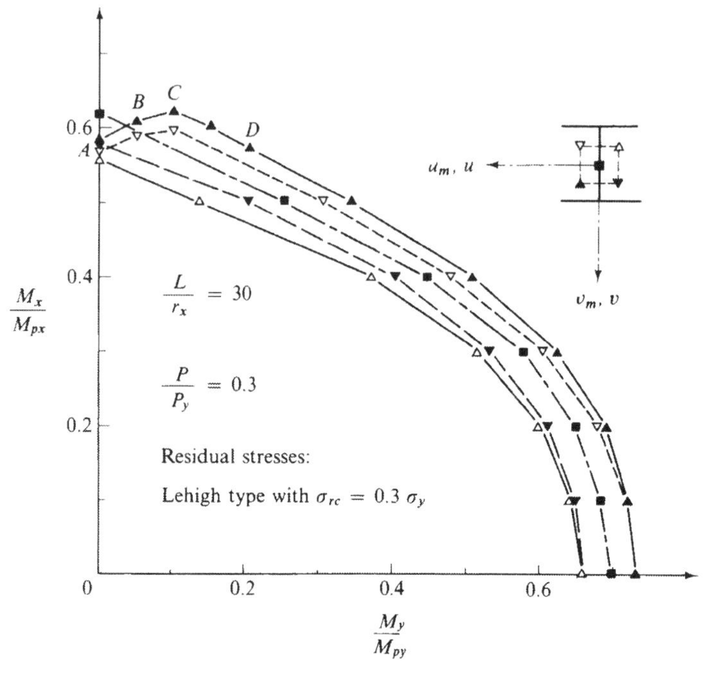

FIGURE 10.26
Influence of the sign of initial deflections on the strength of biaxially bent columns ($|u_m| = |v_m| = L/500$)

$M_y = 0$, but with a definite value of M_y. The answer could be obtained by observing the deformations in the minor axis plane, shown in Fig. 10.28, for the columns A, B, C and D marked on Fig. 10.26. For the columns A and B the deflection

Table 10.4 INFLUENCE OF THE SIGN OF INITIAL DEFLECTIONS ON THE STRENGTH OF BIAXIALLY LOADED BEAM-COLUMNS ($|u_m| = |v_m| = L/1000$)

Column Designation	$m_x = \frac{M_x}{M_{px}}$	Maximum Value of $m_y = M_y/M_{py}$			
		Coordinates at Midheight Defining Initial Deformation			
		$(+u_m, +v_m)$	$(-u_m, +v_m)$	$(+u_m, -v_m)$	$(-u_m, -v_m)$
DS 330-0	0.0	0.712	0.675	0.712	0.675
DS 330-1	0.1	0.700	0.666	0.700	0.662
DS 330-2	0.2	0.666	0.628	0.662	0.622
DS 330-3	0.3	0.594	0.550	0.584	0.541
DS 330-4	0.4	0.475	0.419	0.459	0.403
DS 330-5	0.5	0.297	0.222	0.272	0.197
DS 330-6	0.6	0.000	—	—	—
	m_y	Maximum Value of m_x			
DS 330-10	0.0	0.600	0.603	0.584	0.587

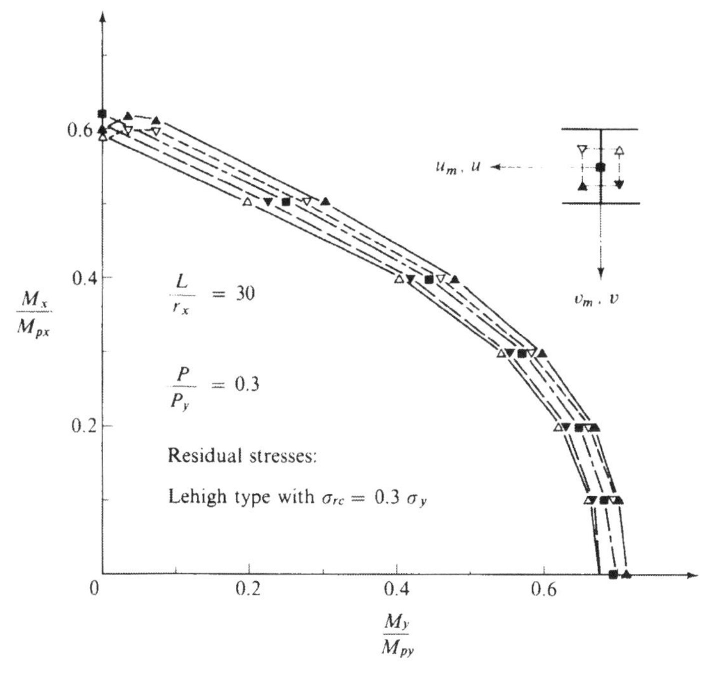

FIGURE 10.27
Influence of the sign of initial deflections on the strength of biaxially bent columns ($|u_m| = |v_m| = L/1000$)

at the ultimate load is of the same shape as the initial crookedness. For the column D, the deflection changes sign before the ultimate strength is attained. In the case of column C, the deformation in the y plane is virtually nil at the ultimate load. Though the phenomenon is more complex, we could consider that the moment $M_y = 0.1M_{py}$, initially applied, makes the initially crooked column straight. The results indicate that the strength of a biaxially loaded beam-column could be increased by imposing a suitable initial camber.

EXAMPLE 10-7 *Influence of the Magnitude of Initial Deflections on the Strength of Biaxially Loaded Beam-Columns* The influence of the magnitude of initial deformations on the maximum strength is studied in this example. For these calculations the sign of the initial deflections is taken to be the most severe one ($-u_m$, $-v_m$). The intensities considered are $L/500$, $L/750$, $L/1000$, $L/5000$ and $L/10\,000$. All other data for the columns is that considered in Example 10-6. The results are given in Table 10.5 and the interaction curves are shown in Fig. 10.29. These results indicate that, for the slenderness and axial load considered, a column with an initial deformation of $L/5000$ or less could be considered as straight. The decrease in maximum strength for the usually stipulated initial deformation of $L/1000$ is not much.

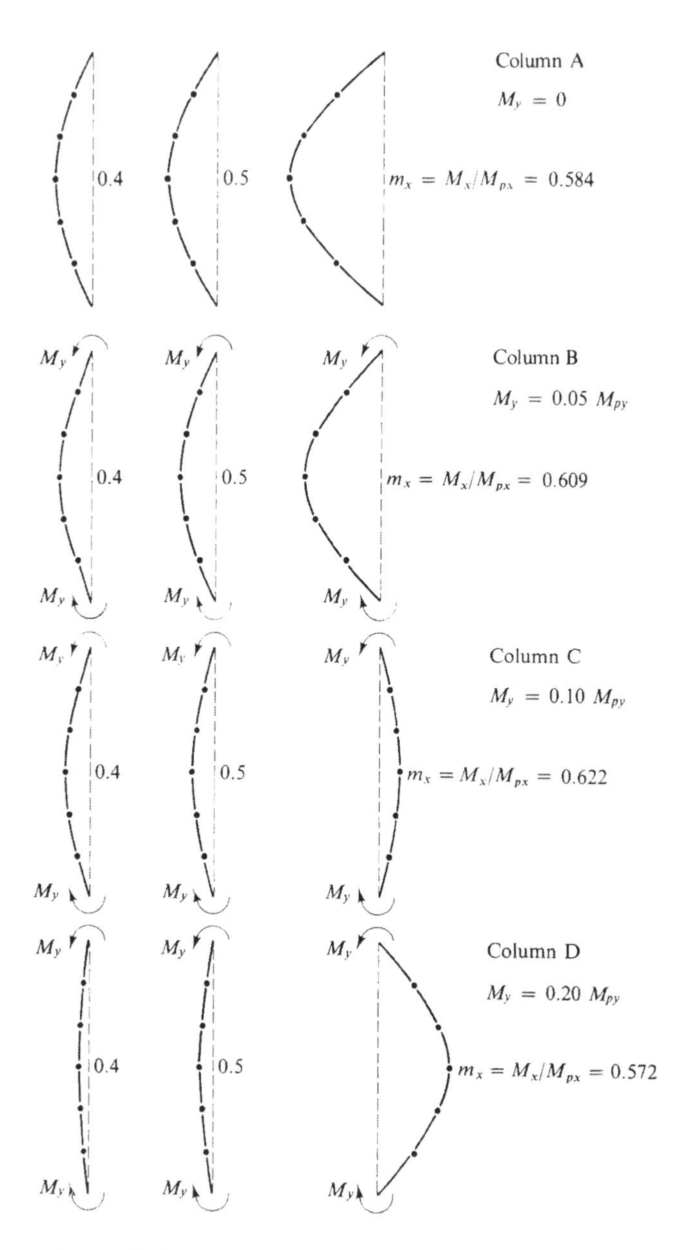

FIGURE 10.28
Deformations u of columns A, B, C and D with increasing M_x.

EXAMPLE 10-8 *Variation of the Influence of Imperfections, with Slenderness, on the Strength of Biaxially Loaded Beam-Columns* The variation of the influence of initial deformations and residual stresses as a function of slenderness is studied now, by making a series of calculations with columns of slenderness ratio, L/r_x equal to 30

Table 10.5 INFLUENCE OF THE MAGNITUDE OF INITIAL DEFLECTIONS ON THE STRENGTH OF BIAXIALLY LOADED BEAM-COLUMNS ($L/r_x = 30$, $P/P_y = 0.3$, RESIDUAL STRESSES: LEHIGH TYPE, $\sigma_{rc} = 0.3\sigma_y$)

Column Designation	$m_x = \frac{M_x}{M_{px}}$	Maximum Value of $m_y = M_y/M_{py}$				
		Initial Deformation at Midheight $-u_m = -v_m$				
		$L/10\,000$	$L/5000$	$L/1000$	$L/750$	$L/500$
DM 330-0	0.0	0.691	0.691	0.675	0.669	0.656
DM 330-1	0.1	0.681	0.678	0.662	0.656	0.641
DM 330-2	0.2	0.644	0.641	0.622	0.616	0.600
DM 330-3	0.3	0.566	0.562	0.541	0.531	0.512
DM 330-4	0.4	0.438	0.434	0.403	0.394	0.366
DM 330-5	0.5	0.247	0.241	0.197	0.181	—
DM 330-51	0.51	0.222	0.216	0.175	0.159	—
DM 330-53	0.53	0.178	0.172	—	—	—

	m_y	Maximum Value of $m_x = m_{x\,cr}$				
DM 330-10	0.0	0.615	0.612	0.587	0.578	0.559

and 50. These columns, of W8 × 31 section, are subjected to an axial force $P = 0.3P_y$, that is kept constant. In addition, they are subjected to symmetric, single curvature moments in the two principal planes.

FIGURE 10.29
Influence of the magnitude of initial deflections on the strength of biaxially bent columns

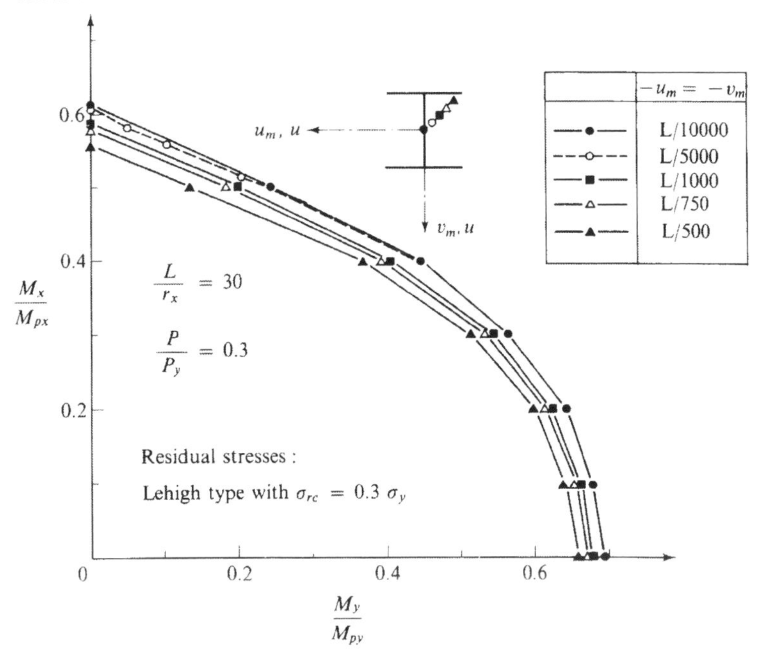

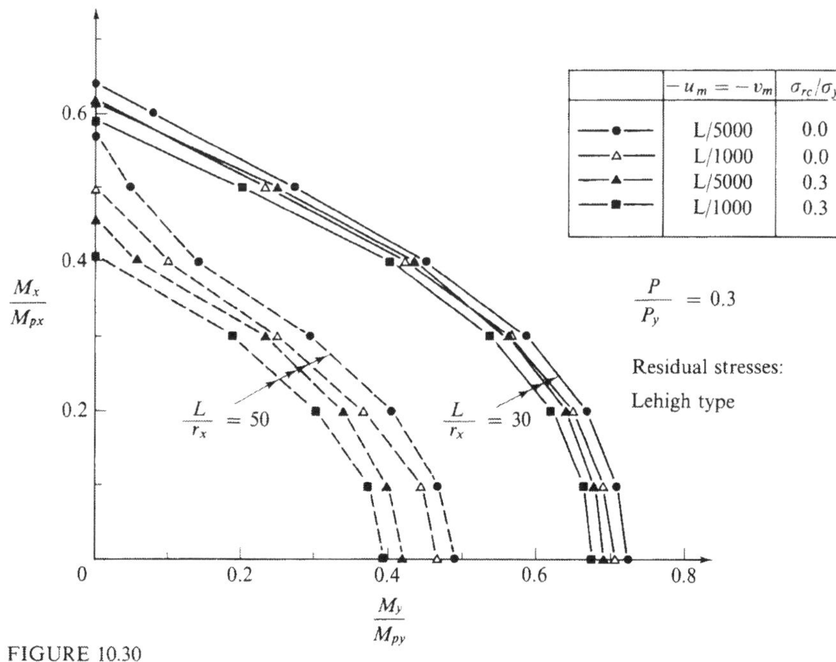

FIGURE 10.30
Variation of the influence of imperfections with slenderness on the strength of biaxially bent columns

A slenderness ratio, L/r_x, equal to 50 (or $L/r_y = 86$) corresponds to the slenderness function

$$\lambda = \frac{L}{r_y}\frac{1}{\lambda_p} = \frac{L}{r_y}\frac{1}{\pi}\sqrt{\frac{\sigma_y}{E}} = 0.95$$

This value is in a region where the planar strength of isolated, pin-ended columns is influenced to the maximum extent by the imperfections (Beer and Schulz, 1975).

Table 10.6 VARIATION OF THE INFLUENCE OF IMPERFECTIONS, WITH SLENDERNESS, ON THE STRENGTH OF BIAXIALLY LOADED BEAM-COLUMNS (W8 × 31 SHAPE, RESIDUAL STRESSES: LEHIGH TYPE, $P/P_y = 0.3$)

$\frac{L}{r_x}$	$-u_m = -v_m$	$\frac{\sigma_{rc}}{\sigma_y}$	Maximum Value of $m_y = M_y/M_{py}$ ($m_x = M_x/M_{px}$)							$m_{x\,max} = m_{x\,cr}$
			$m_x = 0.0$	$m_x = 0.1$	$m_x = 0.2$	$m_x = 0.3$	$m_x = 0.4$	$m_x = 0.5$	$m_x = 0.6$	$m_y = 0.0$
	$L/1000$	0.3	0.675	0.662	0.622	0.541	0.403	0.197	—	0.587
	$L/1000$	0.0	0.706	0.691	0.650	0.562	0.419	0.228	—	0.616
30	$L/5000$	0.3	0.691	0.678	0.641	0.562	0.434	0.241	0.020	0.612
	$L/5000$	0.0	0.722	0.706	0.666	0.584	0.447	0.269	0.082	0.634
	$L/10^6$*	0.3	(0.693)	(0.681)	(0.649)	(0.597)	(0.514)	(0.390)	(0.292)	—
	$L/1000$	0.3	0.394	0.372	0.303	0.188	—	—	—	0.409
	$L/1000$	0.0	0.463	0.441	0.372	0.250	0.097	—	—	0.497
50	$L/5000$	0.3	0.419	0.397	0.338	0.231	0.059	—	—	0.459
	$L/5000$	0.0	0.488	0.466	0.403	0.294	0.141	0.047	—	0.575
	$L/10^6$*	0.3	(0.409)	(0.402)	(0.364)	(0.306)	(0.217)	(0.076)	—	—

* Values in the brackets are obtained by Tebedge and Chen, 1974.

For each slenderness, four different combinations of imperfections are considered. They are: (a) columns with initial imperfections and residual stresses; (b) columns with initial imperfections, but no residual stresses; (c) straight columns with residual stresses and (d) straight, annealed columns. As mentioned earlier, for numerical purposes, columns with an initial deflection of $L/5000$ could be considered as straight.

The results for $L/r_x = 30$ and $L/r_x = 50$ are summarized in Table 10.6 and in Fig. 10.30. From these results the effects of initial crookedness and/or residual stresses may be compared with the idealized strength where both crookedness and residual stresses were absent. It is seen that the influence of initial deformations and residual stresses is relatively more important for $L/r_x = 50$ than for $L/r_x = 30$, as is to be expected.

10.6.2 Buckling of Isolated Beams and Beam-Columns

The application of the theory and the program developed in this chapter for predicting the plastic lateral buckling strength of laterally unsupported I beams will now be considered with the help of three examples. In all these examples, the eigenvalue problem considered is treated as a load-deflection problem of a beam with slight initial out-of-plane imperfections. The first two examples selected are for transversely load beams, for which earlier theoretical or experimental results are available in the literature. The lateral buckling strength of beam-columns subjected to symmetric, single curvature major axis end-moments is treated in the last example.

EXAMPLE 10-9 *Buckling of Plastic I Beams* Kitipornchai and Trahair (1975b) presented results of six carefully conducted lateral buckling tests on full scale simply supported 10 UB 29 beams with central concentrated loads. The point of application of the load was 3.5 in (89 mm) above the top flange. Four beams were tested in the as-rolled condition with the residual stresses resulting from the cooling process. After stress relieving two of these four beams, two more tests were carried out. The maximum loads observed from the tests are compared in Table 10.7 with those calculated by the computer program described in Sect. 10.5. For these calculations,

Table 10.7 LATERAL BUCKLING LOAD OF BEAMS—COMPARISON WITH TEST RESULTS OF KITIPORNCHAI AND TRAHAIR (1975b)

Beam Condition	Beam	Length (in)	Maximum Beam Load (kip)	
			Experimental	Present Study
As-rolled	S1-20	240	12.8	12.02
As-rolled	S2-10	120	41.6	42.50
As-rolled	S3-12	144	32.6	31.87
As-rolled	S4-8	96	52.8	58.28
Stress-relieved	S2-10	120	43.6	45.70
Stress-relieved	S2-12	144	31.5	34.23

1 in = 25.4 mm; 1 kip = 4.45 kN

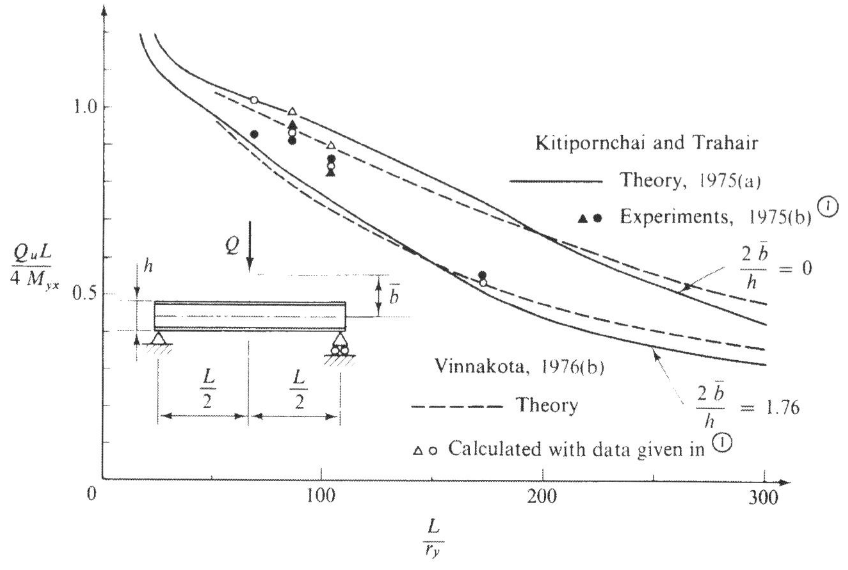

FIGURE 10.31
Comparison with experimental and theoretical results (Kitipornchai and Trahair)

the yield stress is taken as 43.8 ksi (302 MN/m^2). The measured residual stress distribution is replaced by a cosine pattern with compressive stresses of $0.3\sigma_y$ and $0.5\sigma_y$ at the flange edges and web center, respectively. The beams are considered to have an initial crookedness in the x-x plane of sine shape with a maximum ordinate of $u_m = L/1000$ at midspan.

For the longer beams S1-20 and S3-12, the initial deformations considered in the calculations are likely to be more severe than the combined effect of initial deformations and load eccentricities of test pieces, while the reverse is likely to be true for the short beam S2-10 and more so for S4-8. This explains partly the difference between the calculated and experimental results. For the shortest beam, S4-8, the large difference between the calculated and experimental result is considered to result from the assumption in the present analysis that contribution of shear stress on yielding was neglected. As is to be expected, the annealed beams have a higher calculated strength compared to the corresponding rolled beams. In general, the agreement between the theoretical and experimental values is quite good.

Numerical results obtained by Kitipornchai and Trahair, 1975a, for simply supported 10UB29 beams with central concentrated loads Q acting at the geometrical axis ($2\bar{b}/h = 0$) and above the top flanges ($2\bar{b}/h = 1.76$, which corresponds to the value used in the tests) are shown in Fig. 10.31. Also shown in this figure are the values obtained by the present ana'ysis. M_{yx} in Fig. 10.31 is the initial yield moment of the section about x-axis.

EXAMPLE 10-10 *Buckling of Plastic I Beams—Influence of Load Position* The influence of load position on the lateral buckling strength of simply supported I

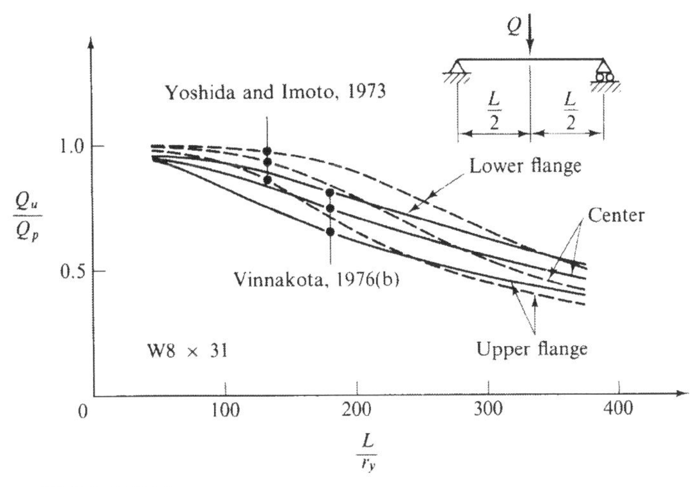

FIGURE 10.32
Lateral buckling curves for simply supported beam with concentrated load

beams is considered in this example. Figure 10.32 shows a comparison of the results for W8 × 31 beams under a concentrated load at the center of the span using the present analysis, with those given by Yoshida and Imoto (1973). The load positions considered are: load at lower flange, load at the upper flange and load at the center of web. Linear pattern of residual stress distribution [Fig. 10.19] with $\sigma_{rj} = -\sigma_{rc} = -\sigma_{rw} = 0.3\sigma_y$ is assumed in both the calculations. Initial deformations, in the x-x plane, of sine shape and of intensity $u_m = L/1000$ are included in the present calculations. Q_p is the load Q corresponding to the formation of a plastic hinge at the midspan. Yoshida and Imoto neglected the effect of the shift in the shear center due to plasticity and the influence of prebuckling deformations on plasticity. This, in addition to the effect of initial deformations considered in the present calculations, explains the difference between the two results in the plastic range. In the elastic range, the values given by the present theory are higher than those of Yoshida and Imoto due to the influence of nonlinear deformation field included in the present theory.

EXAMPLE 10-11 *Buckling of Plastic I-Section Beam-Columns* In this example, we study the lateral buckling strength of laterally unsupported I-section beam-columns, subjected to symmetric, single curvature major axis end-moments M_x and axial compressive loads, P. For the numerical calculations W8 × 31 section having Lehigh type residual stress pattern (with $\sigma_{rc} = 0.3\sigma_y$) is considered. The ends are simply supported for the lateral displacements and rotations, and warping is permitted. The beams are assumed to have small initial geometrical imperfections of sine shape, defined by $u_m = v_m = -L/1000$ and $\theta_m = -0.001$ radian. Note that the signs of u_m and v_m are chosen to obtain the worst effect of the initial imperfections. In the calculations, GK_T is assumed to remain constant during plastification of

Table 10.8 LATERAL BUCKLING STRENGTH m_{xcr} OF BEAM-COLUMNS

L/r_x	20	30	40	50	60	70	80	90	100
p \ L/r_y	34.5	51.8	69.0	86.3	103.5	120.8	138.0	155.3	172.5
0.0	0.937	0.916	0.872	0.822	0.759	0.716	0.678	0.644	0.609
0.1	0.844	0.828	0.759	0.678	0.609	0.553	0.503	0.447	0.394
0.2	0.781	0.706	0.628	0.541	0.466	0.397	0.325	0.250	0.159
0.3	0.653	0.587	0.503	0.409	0.323	0.242	0.136	(0.284)	(0.238)
0.4	0.553	0.472	0.384	0.288	0.192	0.067	(0.344)		
0.5	0.441	0.363	0.272	0.166	0.041	(0.422)			
0.6	0.334	0.256	0.159	0.034	(0.513)				
0.7	0.228	0.153	0.034	(0.614)					
0.8	0.125	0.045	(0.716)						
0.9	0.023	(0.828)							
1.0	(0.919)								

NOTE: $p = P/P_y$, $m_x = M_x/M_{px}$, $m_y = M_y/M_{py}$.
The values in the brackets are the values of p corresponding to $m_x = m_y = 0$.
DATA: Profil W8 × 31. Residual stresses, Lehigh type with $\sigma_{rc} = 0.3\sigma_y$.
Warping permitted. GK_T constant.
Initial deformations $u_m = v_m = -L/1000$, $\theta_m = -0.001$ radian.

sections. The results obtained with the help of computer program are given in Table 10.8 in nondimensional form. Also given in this table, in brackets, are the plastic buckling strengths of laterally unsupported columns with geometrical imperfections.

10.6.3 Restrained Beam-Columns under Biaxial Loading

The three examples of restrained columns presented are selected from the experimental program of Gent and Milner (1968), from the analytical results presented by Santathadaporn and Chen (1973) and by Vinnakota (1974).

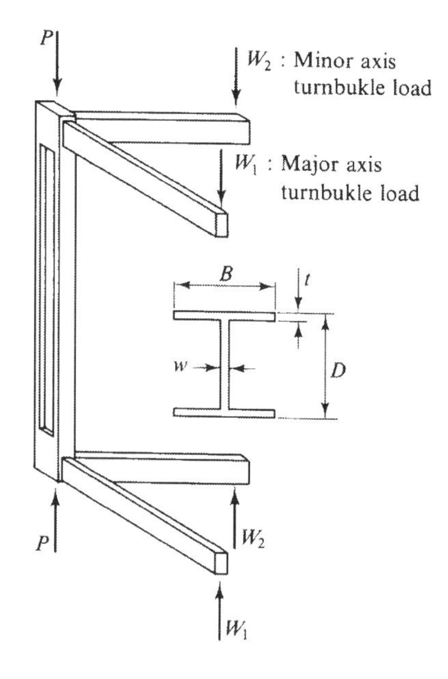

FIGURE 10.33
Test set-up of Gent and Milner (1968)

Table 10.9 DATA FOR EXAMPLE 10-12 [GENT AND MILNER (1968)]

	Specimen Number A4	Specimen Number B3
Column data (H section)		
Length	16.5 in	18.0 in
B	1.000 in	0.750 in
D	1.000 in	0.750 in
t	0.101 in	0.055 in
w	0.060 in	0.035 in
Yield stress, σ_y	35 ksi	35 ksi
Beam data (rectangular section)		
Major axis: length	11.7 in	12.1 in
B	0.375 in	0.375 in
D	0.500 in	0.850 in
Minor axis: length	11.7 in	12.1 in
B	0.375 in	0.375 in
D	0.850 in	0.850 in
Initially applied beam moments		
Major axis	0.500 in-kip	0.210 in-kip
Minor axis	0.510 in-kip	0.170 in-kip

1 in = 25.4 mm, 1 ksi = 6.9 MN/m^2, 1 in-kip = 113 N-m.

EXAMPLE 10-12 *Rotationally Restrained Beam-Columns* The validity of the theory described in this chapter to study the behavior of restrained columns is verified by considering the specimen numbers A4 and B3 tested by Gent and Milner (1968). The test set up used is shown in Fig. 10.33. The column was first bent by turnbuckle loads W_1 and W_2. Then, clamping these turnbuckles rigidly, the axial load P was increased to failure. With the increase of the axial load, the column deformed increasing its joint rotations and relaxing its end moments which were controlled by the beam stiffness. Table 10.9 gives a summary of the relevant data for the H-section columns and the major and minor axes rectangular beams tested.

Milner and Gent (1971) also developed a computer program to study the behavior of these restrained columns. In their analysis, based on finite difference method, the calculations were first carried out ignoring strain hardening, strain reversal and torsion, but the effect of including these on the collapse load was subsequently explored independently for cases of interest.

Variations of the column end moments and of the mid-height displacements, as a function of the axial loads, measured in the experiments and those calculated using the numerical analysis presented in this chapter are shown in Figs. 10.34 and 10.35. Residual stresses and torsion are not included in the analysis. Owing to the symmetry of the problem, the numerical calculations are carried out on one half of the column with the following end conditions: $u = v = 0$, moments given at the end $z = 0$, and $u' = v' = 0$, $u''' = v''' = 0$ at the end $z = L/2$.

The correlation between computed and experimental results is good. It was also found that if torsion was included in the analysis the collapse load was reduced by less than 2 percent.

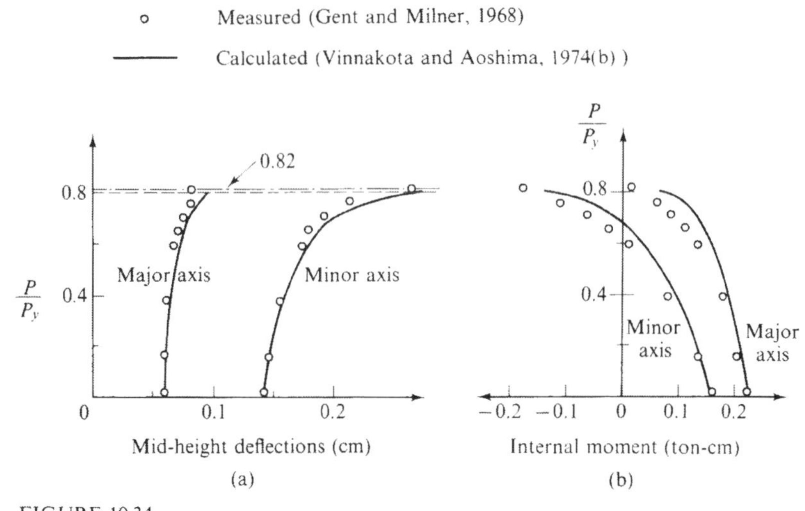

FIGURE 10.34
Rotationally restrained beam-column. Specimen B3 of Gent and Milner (1968). (a) Load deflection curve. (b) Variation of internal moment with load

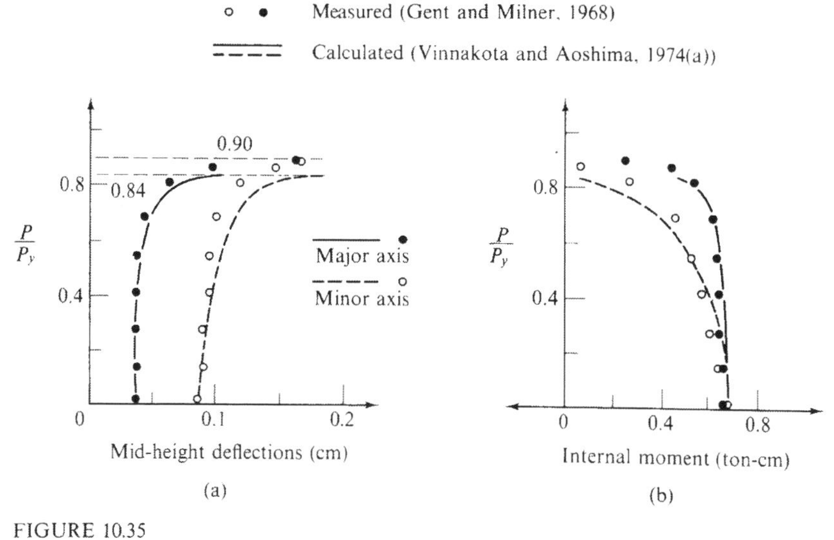

FIGURE 10.35
Rotationally restrained beam-column. Specimen A4 of Gent and Milner (1968). (a) Load deflection curve. (b) Variation of internal moment with load

EXAMPLE 10-13 *Rotationally Restrained Column* The second example considered in this series is the symmetric and symmetrically loaded restrained beam-column (Fig. 10.36) studied by Santathadaporn and Chen (1973). In the analysis a W14 × 43 ($\approx$ IPE 380) column of length 220 in (559 cm) was subjected to a double eccentric longitudinal load, P ($e_y = 5.0$ in or 12.7 cm, $e_x = 0.5$ in or

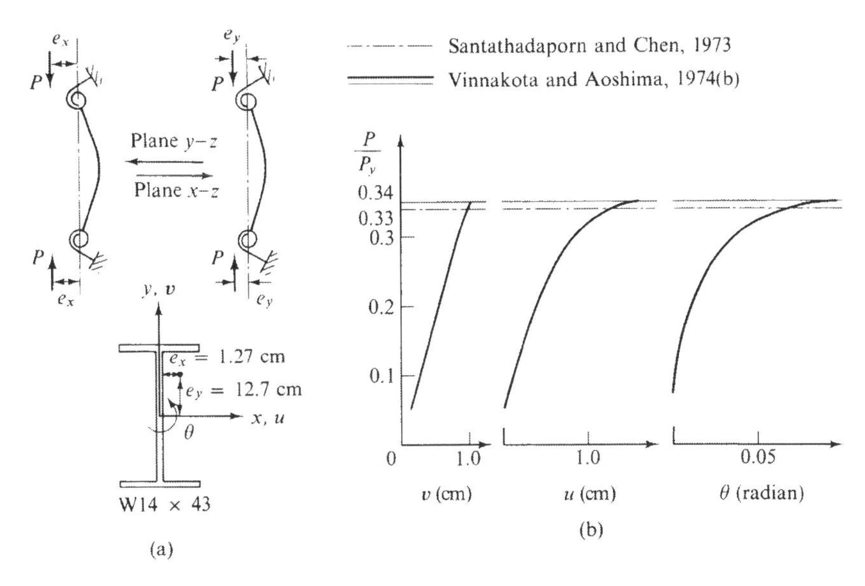

FIGURE 10.36
Rotationally restrained beam-column of Santathadaporn and Chen (1973). (a) Problem definition. (b) Load versus mid-height deformations

1.3 cm). The stiffness α of the rotational restraints about x and y axes was 2700 kip in/radian (305 MN-m/radian). The yield stress of the material was 36 ksi (248.4 MN/m^2). The sway and twist at the ends as well as warping of column ends were prevented. Residual stresses and geometrical imperfections were not considered.

The mid-height deformations calculated by the present analysis are given in Fig. 10.36. The ultimate load calculated (= 0.34 of the yield load) compares favorably with the value (0.33) given by Santathadaporn and Chen (1973).

EXAMPLE 10-14 *Unsymmetrically Loaded, Rotationally Restrained Beam-Column* The last example selected is the one designed in the *Handbook of Steel Construction* of the Canadian Institute of Steel Construction (1970), by the allowable stress interaction formulas given by the CSA Standard S16 (1969). The column, of length $L = 144$ in (3.66 m), is subjected to an axial load $P = 350$ kip (1.56 MN). In addition, under working loads, rigid beam columns about both axes induce the following moments: $\tilde{M}_{xo} = 100$ ft-kip (136 kN-m), $\tilde{M}_{xl} = 135$ ft-kip (183 kN-m) and $\tilde{M}_{yo} = \tilde{M}_{yl} = 50$ ft-kip (68 kN-m). The direction of the moments is such that double curvature is induced in the column. The effective lengths K_x and K_y were taken as 0.9. A section W12 × 133 was selected and found to be adequate ($L/r_x = 25.8$, $L/r_y = 45.5$).

This column is analyzed with the help of the program developed for two cases as shown in Fig. 10.37(a). In case 1, $K_x = K_y = 1.0$ (no end-restraints). In case 2, $K_x = K_y = 0.85$. [This value, instead of the values 0.9 used in design, is chosen so as to make use of the curves given by Kavanagh (1962), to calculate

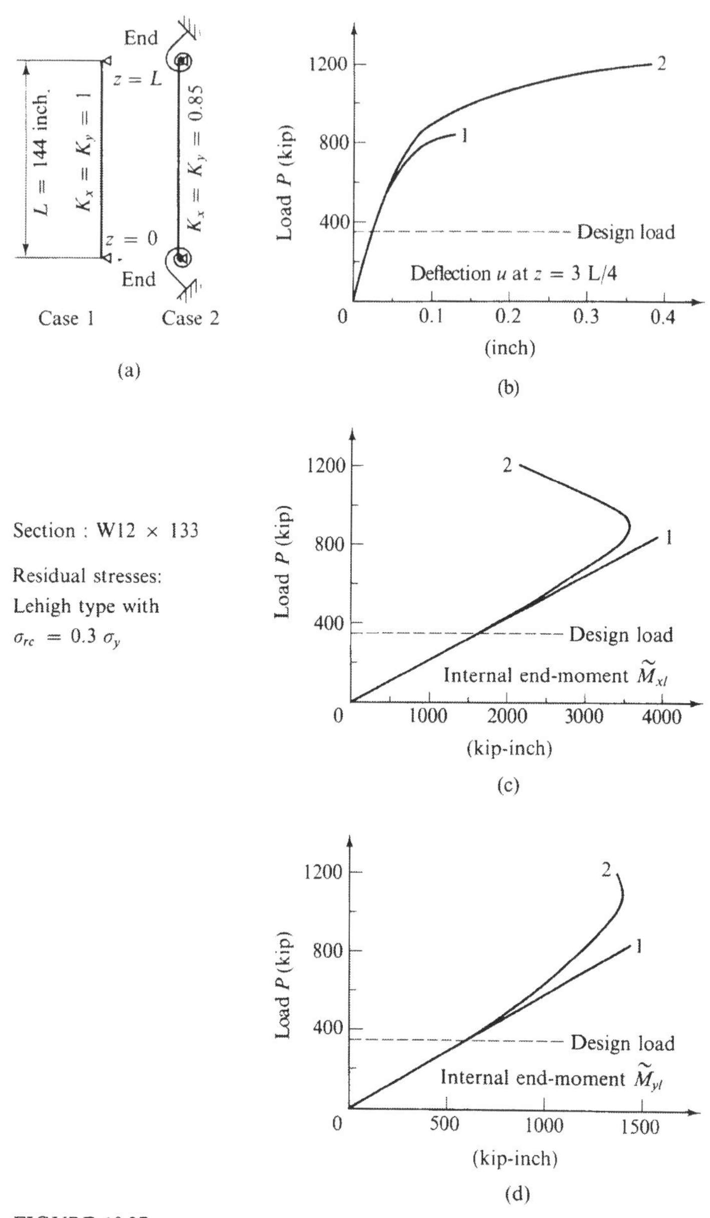

FIGURE 10.37
Unsymmetrically loaded, rotationally restrained beam-column (Vinnakota, 1974). (a) Effective length factors. (b) Load-deflection curve. (c) Variation of $\tilde{M}_{xl}$ with load. (d) Variation of $\tilde{M}_{yl}$ with load

the spring constants.] The external moments of the restrained column are chosen such that, at working load, the internal moments acting on the column ends are same as those in case 1. The resulting eccentricities of the load are given in

Table 10.10 DATA FOR EXAMPLE 10.14

	Case 1	Case 2
e_{yo}	3.43 in	3.85 in
e_{yl}	−4.63 in	−5.72 in
e_{xo}	1.71 in	2.04 in
e_{xl}	−1.71 in	−2.04 in
$K_y = K_x$	1.00	0.85
α_y	0	91 143 kip-in/radian
α_x	0	283 070 kip-in/radian

1 in = 25.4 mm, 1 in-kip = 113 N-m.

Table 10.10. In the analysis the biaxially eccentric load is increased proportionally. Warping at ends is completely restrained. The calculations are based on a residual stress pattern of Lehigh type with $\sigma_{rc} = 0.3\sigma_y$, and a yield stress of 40 ksi (276 MN/m^2).

The load displacement curves for the two cases are shown in Fig. 10.37(b) where u is the displacement at the quarter height point from end $z = L$. The relation between the axial load and the internal moments at end $z = L$ are shown in Figs. 10.37(c) and (d). In case 1 (column with no restraints), the relation is a straight line, as the loading is proportional. In the case of the restrained column, while the axial load and the external moments increase continuously the internal moments increase at first and after reaching a maximum value diminish (phenomena of "Shedding down of end-moments"). The calculated factor of safety is ≈ 2.4 in case 1 and ≈ 3.4 in case 2. The calculated factor of safety with $K_x = K_y = 0.9$ will be between these two values and probably near towards the higher value. (Note: the factor of safety for axially compressed members, assumed in the standards, is between 1.67 and 1.92.)

The influence of initial deflections, which are usually present, are not taken into account. But this should be more than compensated by the strain hardening and strain reversal which are also not included in the analysis.

10.7 SUMMARY

The use of finite difference method to solve biaxially loaded beam-columns is presented. For this, a general theory that describes the space behavior of arbitrary open section members subjected to end moments, axial load and transverse loads is developed. The equilibrium equations (10.76) are written, for the deformed position of the bar, with respect to an arbitrary system of coordinate axes, and no reference is made to the shear center, center of gravity or principal axes. The deformation field considered is nonlinear [see Eq. (10.27)]. Nonlinearity due to material plasticity, and the influence of residual stresses and initial geometrical imperfections are included in the analysis. Strain hardening and strain reversal are neglected in the theory. Based on this theory, a computer program is developed to study the behavior of I-section members. The validity of the theory and the computer program to calculate the strength of biaxially loaded beam-columns and the lateral buckling load of I beams and beam-columns is controlled by comparison with several experimental and theoretical results, available in the existing literature. In most cases, good agreement was observed. The problem of lateral buckling of beams and beam-columns is treated as a load-deflection problem with slight out-of-plane imperfections. From the variety of numerical examples treated it could be concluded that, for the types of members generally encountered in civil engineering structures, the application of finite difference method leads to efficient and economic procedures.

10.8 PROBLEMS

10.1 Solve Example 10-1 without making use of fictitious grid points beyond the supports, but using forward and backward differences. Comment on the result obtained.

10.2 Develop an expression for the elastic lateral buckling load of an I beam, subjected to symmetric single curvature major axis end moments, using Eqs. (10.51). Compare the result with that given in Chap. 3, using linearized deformation field and discuss the solutions.

10.3 Prepare a computer program to determine the elastic deformations of a simply supported, I-section beam-column. Assume that the beam-column is initially straight and untwisted. Take the principal axes as the coordinate system and the center of gravity as the pole.

10.4 Derive the expressions given in Eqs. (10.27) to (10.30) using the fact that the shape of section is indeformable.

10.9 REFERENCES

Beer, H. and Schulz, G., "The European Column Curves," *Proceedings, International Colloquium on Column Strength*, Paris, November 23–24, 1972, published by the International Association for Bridge and Structural Engineering, Zurich, pp. 385–398, (1975).

Birnstiel, C., "Experiments on H-Columns Under Biaxial Bending," *Journal of the Structural Division, ASCE*, Vol. 94, No. ST10, Proc. paper 6186, pp. 2429–2449, October, (1968).

Bleich, F., *Buckling Strength of Metal Structures*, McGraw-Hill Book Co., Inc., New York, NY, (1952).

Canadian Institute of Steel Construction, *Hand Book of Steel Construction*, (1970).

Collatz, L., *Functional Analysis and Numerical Mathematics*, Academic Press, New York, (1966).

Forsythe, G. E. and Wassov, W. R., *Finite Difference Methods for Partial Differential Equations*, Wiley, New York, (1960).

Galambos, T. V., *Structural Members and Frames*, Prentice Hall, Inc., Englewood Cliffs, NY, (1968).

Gent, A. R. and Milner, H. R., "The Ultimate Load Capacity of Elastically Restrained H-Columns Under Biaxial Bending," *Proceedings of the Institution of Civil Engineers*, Vol. 41, pp. 685–704, London, England, December, (1968).

Kavanagh, T. C., "Effective Length of Framed Columns," *Transactions, ASCE*, Vol. 127, pp. 88–101, (1962).

Kitipornchai, S. and Trahair, N. S., "Buckling of Inelastic I-Beams Under Moment Gradient," *Journal of the Structural Division, ASCE*, Vol. 101, No. ST5, Proc. paper 11295, pp. 991–1004, May, (1975a).

Kitipornchai, S. and Trahair, N. S., "Inelastic Buckling of Simply Supported Steel I-Beams," *Journal of the Structural Division, ASCE*, Vol. 101, No. ST7, Proc. paper 11419, pp. 1333–1347, July, (1975b).

Milner, H. R. and Gent, A. R., "Ultimate Load Calculations for Restrained H-Columns Under Biaxial Bending," *Civil Engineering Transactions, Institution of Engineers*, Australia, pp. 35–44, (1971).

Santathadaporn, S. and Chen, W. F., "Analysis of Biaxially Loaded Steel H-Columns," *Journal of the Structural Division, ASCE*, Vol. 99, No. ST3, Proc. paper 9621, pp. 491–509, March, (1973).

Soltis, L. A. and Christiano, P., "Finite Deformation of Biaxially Loaded Columns," *Journal of the Structural Division, ASCE*, Vol. 98, No. ST12, Proc. paper 9407, pp. 2647–2662, December, (1972).

Tebedge, N. and Chen, W. F., "Design Criteria for H-Columns Under Biaxial Bending," *Journal of the Structural Division, ASCE,* Vol. 100, No. ST3, Proc. paper 10400, pp. 579–598, March, (1974).

Timoshenko, S. P. and Gere, J. M., *Theory of Elastic Stability,* 2nd edn, McGraw Hill Book Co., Inc., New York, NY, (1961).

Trahair, N. S. and Kitipornchai, S., "Buckling of Inelastic I-Beams Under Uniform Moment," *Journal of the Structural Division, ASCE,* Vol. 98, No. ST11, Proc. paper 9339, pp. 2551–2566, November, (1972).

Vinnakota, S., *Design and Analysis of Restrained Columns Under Biaxial Bending,* Stavebnicky Casopis, SAV, XXII, 4, Bratislava, Czechoslovakia, pp. 182–206, (1974).

Vinnakota, S., "The Influence of Imperfections on the Maximum Strength of Biaxially Bent Columns," *Canadian Journal of Civil Engineering,* Vol. 3, No. 2, pp. 186–197, June, (1976a).

Vinnakota, S., "Inelastic Stability of Laterally Unsupported I-Beams," paper presented at the Second National Symposium on Computerized Structural Analysis and Design at the School of Engineering and Applied Science, George Washington University, Washington DC, March, (1976b).

Vinnakota, S. and Aoshima, Y., "Inelastic Behavior of Rotationally Restrained Columns Under Biaxial Bending," *The Structural Engineer,* London, England, Vol. 52, No. 7, pp. 245–255, July, (1974a).

Vinnakota, S. and Aoshima, Y., "Spatial Behavior of Rotationally and Directionally Restrained Beam-Columns," *Publications, International Association of Bridge and Structural Engineering,* Zurich, Vol. 34, II, pp. 169–194, (1974b).

Vinnakota, S. and Aysto, P., "Inelastic Spatial Stability of Restrained Beam-Columns," *Journal of the Structural Division, ASCE,* Vol. 100, ST11, Proc. paper 10919, pp. 2235–2254, November, (1974).

Vinnakota, S. and Aysto, P., Closure to "Inelastic Spatial Stability of Restrained Beam-Columns," *Journal of the Structural Division, ASCE,* Vol. 102, No. ST6, pp. 1257–1260, June (1976).

Vinnakota, S., Badoux, J. C., and Aoshima, Y., "Equations fondamentales du comportement des poutres-colonnes à section ouverte et parois minces," *Bulletin Technique de la Suisse Romande,* Lausanne, Vol. 101, No. 26, pp. 437–445, Décembre, (1975).

Vlasov, V. Z., "Thin Walled Elastic Beams," 2nd edn, National Science Foundation, Washington, DC, (1961).

Yoshida, H. and Imoto, Y., "Inelastic Lateral Buckling of Restrained Beams," *Journal of the Engineering Mechanics Division, ASCE,* Vol. 99, No. EM 2, Proc. paper 9666, pp. 343–366, April, (1973).

11

FINITE SEGMENT METHOD FOR PLASTIC BEAM-COLUMNS

11.1 INTRODUCTION

In the preceding chapter, the governing differential equations of a biaxially loaded beam-column have been solved approximately by the well-known numerical technique of *finite difference method*. Here, the actual beam-column is physically replaced by an assembly of finite segments. The elastic-plastic behavior of these segments throughout the entire range of loading up to ultimate load has been presented in Chap. 6. As a result, the beam-column problem can now be formulated and solved approximately in terms of the behavior of these segments without recourse to complex differential equations. In a sense, the *finite segment approach* may be considered as a physical interpretation of the finite difference method as applied numerically to solve differential equations. The systematic development and utilization of a more refined, discretized approximation of an actual beam-column known as the *finite element method* will be presented in the following chapter. The finite element method may be considered as a generalization of the finite segment method described herein.

The general analysis of a biaxially loaded beam-column by finite segment method is essentially the same as that of a space structure. In this method, the beam-column

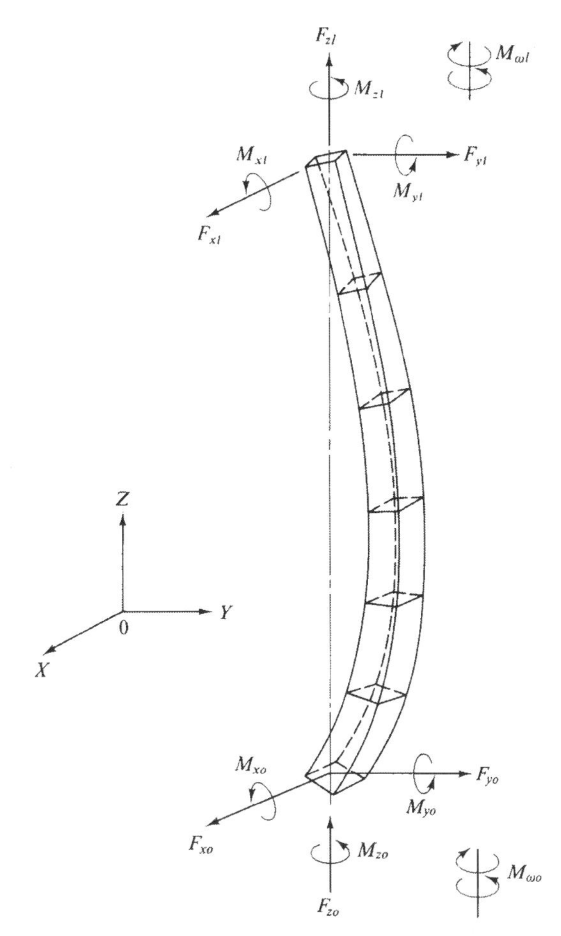

FIGURE 11.1
Segmentation of biaxially loaded beam-column

is assumed to consist of a number of segments and is treated as a space structure as shown in Fig. 11.1. Three dimensional displacements (u_x, u_y, u_z), rotations ($\theta_x, \theta_y, \theta_z$) and a warping ($\bar{w}$) are taken into account at each node cross section. In computation of the segment stiffness matrices, the *direct stiffness method* is utilized. The elastic-plastic section properties are calculated making use of the method discussed in Chap. 6. Geometric and material nonlinearities are handled by application of a modified tangent stiffness approach. The following assumptions are made in the analysis:

1. Cross-sectional shapes considered herein are wide-flange sections.
2. Material properties governing stress-strain behavior are that of an elastic-perfectly plastic material.
3. Yielding is considered to be a function of normal stress only.

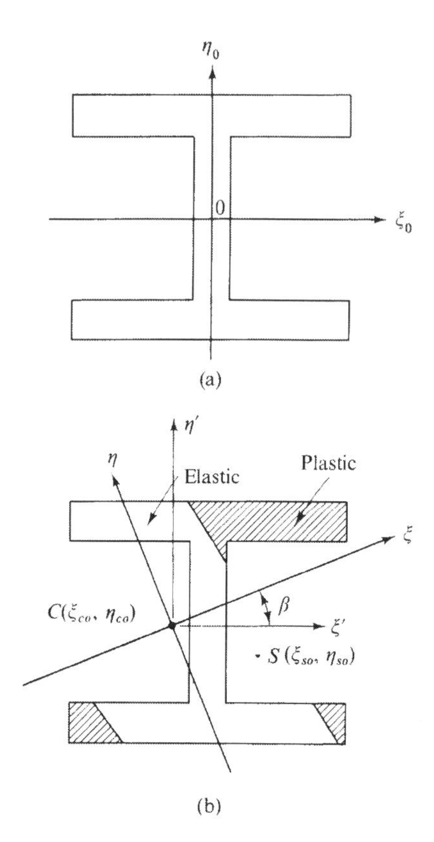

FIGURE 11.2
Wide flange and its principle axes. (a) Initial elastic state. (b) Elastic plastic state

4. Segmental stiffnesses are based on undeformed straight segment behavior.
5. Section properties for the tangent stiffness of a segment are invariants within the segment and are calculated using the elastic portion of the cross section.
6. In computation of the warping moment of inertia, yielding is considered to occur only in the flange plates.
7. Warping $\bar{w} = \theta'_\zeta$ and bimoment m_ω are considered as one of the generalized strains and stresses of a cross-section, respectively. Note that $\bar{w}$ denoting warping in this chapter is the average warping on the cross section (cf. $w = \omega_n \theta'_\zeta$).
8. The directional change of warpings and bimoments between two adjacent segments are neglected because of the small angle changes between segments. Therefore, these are treated like scalar quantities.

11.2 ELASTIC-PLASTIC SECTION PROPERTIES

Consider a wide-flange section which has its centroid and shear center initially at the origin O. The initial principal axes are represented by ξ_o and η_o as shown in Fig. 11.2(a).

At a moment after application of some loading, the section may be partially plastified as shown by the shaded portions in Fig. 11.2(b). The instantaneous centroid and shear center move to point C and S, respectively. The instantaneous principal axes ξ and η are also determined for the remaining elastic portion of the section.

As generalized strain rates on the cross section, the following four deformation rates are considered:

$$\dot{\varepsilon}_c = \text{normal strain rate at the centroid } C$$

$$\dot{\Phi}_\xi = \text{bending curvature rate about the } \xi\text{-axis } (= \dot{\theta}'_\xi)$$

$$\dot{\Phi}_\eta = \text{bending curvature rate about the } \eta\text{-axis } (= \dot{\theta}'_\eta)$$

$$\dot{\Phi}_\omega = \text{warping curvature rate about the shear center } S\ (= \dot{w}')$$

from which the normal strain rate distribution on the entire section is given by Eq. (2.116), on

$$\dot{\varepsilon} = \dot{\varepsilon}_c + \eta\dot{\Phi}_\xi - \xi\dot{\Phi}_\eta + \omega_n\dot{\Phi}_\omega \tag{11.1}$$

where ω_n is normalized warping on the section which will be calculated by Eq. (11.15) as explained below. The current total strain distribution is then obtained by addition of the strain rate $\dot{\varepsilon}$ to the strain ε_o in the previous state,

$$\varepsilon = \varepsilon_o + \dot{\varepsilon} \tag{11.2}$$

Since the material is assumed to be elastic-perfectly plastic and only normal stress contributes to yielding, the tangent modulus E_t at a point on the section is uniquely determined from the total strain ε and the strain rate $\dot{\varepsilon}$ using the rule discussed in Table 6.1, in Chap. 6 (page 284).

$$E_t = \begin{cases} E & \text{for elastic or unloading portion} \\ O & \text{for plastic loading portion} \end{cases} \tag{11.3}$$

The shear modulus G is used only in elastic regime,

$$G = \frac{E}{2(1 + \nu)} \tag{11.4}$$

where ν is Poisson's ratio.

Sectional properties of the elastic portion are calculated first about the initial principal axes ξ_0 and η_0,

$$\begin{aligned} EA_0 &= \int E_t\,dA & EI_{\xi_0} &= \int E_t\eta_0^2\,dA \\ ES_{\eta 0} &= \int E_t\eta_0\,dA & EI_{\eta_0} &= \int E_t\xi_0^2\,dA \\ ES_{\xi 0} &= \int E_t\xi_0\,dA & EI_{\xi\eta_0} &= \int E_t\xi_0\eta_0\,dA \end{aligned} \tag{11.5}$$

The above integrations are carried out using finite elements cut from the section as was done in Chap. 6.

The location of the instantaneous centroid $C(\xi_{co}, \eta_{co})$ and rotation angle of the instantaneous principal axes β are calculated by

$$\xi_{co} = \frac{ES_{\xi 0}}{EA_0}, \quad \eta_{co} = \frac{ES_{\eta 0}}{EA_0} \tag{11.6}$$

and

$$\beta = \tfrac{1}{2}\tan^{-1}\left[\frac{2I'_{\xi\eta}}{I'_{\xi} - I'_{\eta}}\right] \tag{11.7}$$

in which

$$\begin{aligned} EI'_{\xi} &= EI_{\xi_0} - \eta_{co}^2 EA_0 \\ EI'_{\eta} &= EI_{\eta_0} - \xi_{co}^2 EA_0 \\ EI'_{\xi\eta} &= EI_{\xi\eta_0} - \xi_{co}\eta_{co} EA_0 \end{aligned} \tag{11.8}$$

are sectional properties of the elastic portion about the ξ' and η' axes as shown in Fig. 11.2.

Rigidities of the section against force rates with respect to the instantaneous principal axes ξ and η are now obtained as follows:

(a) *Tensile Rigidity* (EA)

$$EA = EA_0 \tag{11.9}$$

(b) *Bending Rigidities* (EI_{ξ}, EI_{η})

$$\left.\begin{matrix} EI_{\xi} \\ EI_{\eta} \end{matrix}\right\} = \frac{EI'_{\xi} + EI'_{\eta}}{2} \pm \frac{EI'_{\xi\eta}}{\sin 2\beta} \tag{11.10}$$

(c) *St. Venant Torsion Rigidities* (GK_T)

$$GK_T = \frac{1}{3} G\Sigma b_i t_i^3 \tag{11.11}$$

where b_i and t_i are the breadth and thickness of the elastic portion of the i-th plate element of the thin-wall section.

(d) *Warping Torsion Rigidity* (EI_{ω})

Location of the instantaneous shear center $S(\xi_{so}, \eta_{so})$ is determined so that the warping moment of inertia I_{ω} of the elastic portion of the section about the point may be minimum. For this purpose, it is assumed herein that the web plate remains elastic and the elastic portion of the section is idealized as shown in Fig. 11.3. The elastic length b_i of the i-th flange plate in each quadrant may be calculated approximately by

$$b_i = \frac{1}{Et_i}\int_{A_i} E_t\, dA \quad (i = 1, \ldots 4) \tag{11.12}$$

First, the location of the shear center $S(\xi_{so}, \eta_{so})$ is assumed using the distances ρ_1 and ρ_2 from the upper flange and the web plate as shown in Fig. 11.3(a).

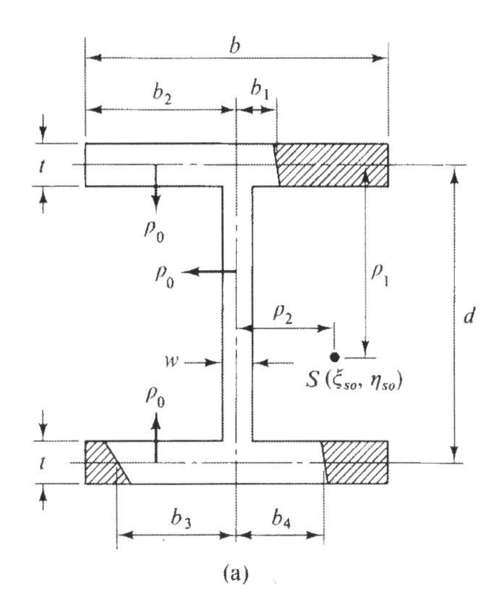

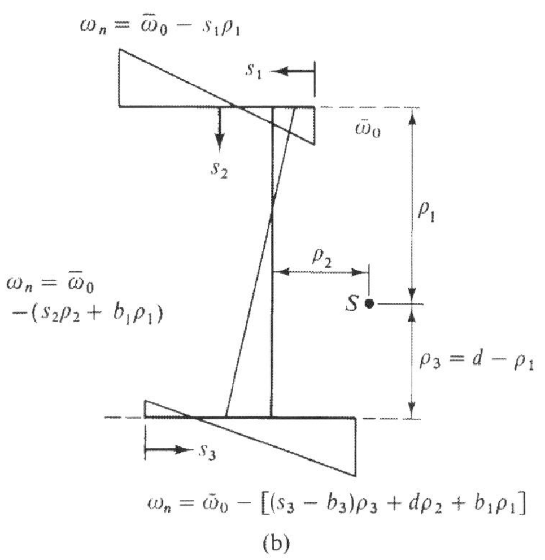

FIGURE 11.3
Warping distribution on elastic portion of section. (a) Idealization of elastic portion. (b) Normalized warping about point S

$$\xi_{s0} = \rho_2, \quad \eta_{s0} = \frac{d}{2} - \rho_1 \tag{11.13}$$

Warping about the assumed shear center S is expressed by

$$\omega_0 = \begin{cases} s_1\rho_1 & \text{for upper flange} \\ s_2\rho_2 + b_1\rho_1 & \text{for web plate} \\ (s_3 - b_3)(d - \rho_1) + b_1\rho_1 + d\rho_2 & \text{for lower flange} \end{cases} \tag{11.14}$$

in which s_1, s_2 and s_3 are coordinates on each plate element along its length from an edge as shown in Fig. 11.3(b). The warping is normalized as [see Eq. (2.102)],

$$\omega_n = \bar{\omega}_0 - \omega_0 \tag{11.15}$$

where

$$\bar{\omega}_0 = \frac{1}{A}\int \omega_0 \, dA = e_1\rho_1 + e_2\rho_2 + e_3 \tag{11.16}$$

and

$$e_1 = \frac{1}{2A}\left[(b_1 + b_2)^2 t + 2b_1(b_3 + b_4)t - (b_4^2 - b_3^2)t + 2b_1\, dw\right]$$

$$e_2 = \frac{1}{2A} d\left[dw + 2(b_3 + b_4)t\right] \tag{11.17}$$

$$e_3 = \frac{1}{2A} dt(b_4^2 - b_3^2)$$

in which A is the elastic area of the cross section, t thickness of the flange and w thickness of the web. The distribution of ω_n is shown in Fig. 11.3(b). This normalized warping is used for computation of normal strain rate $\dot{\varepsilon}$ in Eq. (11.1).

The warping moment of inertia about the assumed shear center is now calculated by [see Eq. (2.109)]

$$I_\omega = \int \omega_n^2 \, dA = C_1\rho_1^2 + C_2\rho_1\rho_2 + C_3\rho_2^2 + C_4\rho_1 + C_5\rho_2 + C_6 \tag{11.18}$$

in which

$$C_1 = e_1^2(b_1 + b_2)t - e_1(b_1 + b_2)^2 t + (e_1 - b_1)^2\, dw + \frac{1}{3}(b_1 + b_2)^3 t$$
$$+ (e_1 - b_1 - b_3)^2(b_3 + b_4)t + (e_1 - b_1 - b_3)(b_3 + b_4)^2 t + \frac{1}{3}(b_3 + b_4)^3 t$$

$$C_2 = 2e_1e_2(b_1 + b_2)t - e_2(b_1 + b_2)^2 t + 2(e_1 - b_1)bdw - (e_1 - b_1)d^2w$$
$$+ 2(e_1 - b_1 - b_3)(e_2 - d)(b_3 + b_4)t + (e_2 - d)(b_3 + b_4)^2 t$$

$$C_3 = e_2^2(b_1 + b_2)t + e_2^2\, dw - bd^2w + (b - d)^2(b_3 + b_4)t + \frac{1}{3}d^3w$$

$$C_4 = 2e_1e_3(b_1 + b_2)t - e_3(b_1 + b_2)^2 t + 2(e_1 - b_1)e_3\, dw \tag{11.19}$$
$$+ 2(e_1 - b_1 - b_3)(e_3 + b_3 d)(b_3 + b_4)t + (e_3 + b_3 d)(b_3 + b_4)^2 t$$
$$- (e_1 - b_1 - b_3)(b_3 + b_4)^2\, dt - \frac{2}{3}(b_3 + b_4)^3\, dt$$

$$C_5 = 2e_2e_3(b_1 + b_2)t + 2e_2e_3\, dw - e_3\, d^2w$$
$$+ 2(e_2 - d)(e_3 + b_3 d)(b_3 + b_4)t - (e_2 - d)(b_3 + b_4)^2\, dt$$

$$C_6 = e_3^2(b_1 + b_2)t + e_3^2\, dw + (e_3 + b_3 d)^2(b_3 + b_4)t$$
$$- (e_3 + b_3 d)(b_3 + b_4)^2\, dt + \frac{1}{3}d^2(b_3 + b_4)^3 t$$

Minimize I_ω using the conditions,

$$\frac{\partial I_\omega}{\partial \rho_1} = 0 \qquad \text{and} \qquad \frac{\partial I_\omega}{\partial \rho_2} = 0 \tag{11.20}$$

and obtain the location of shear center

$$\begin{aligned} \rho_1 &= (C_2C_5 - 2C_3C_4)/(4C_1C_3 - C_2^2) \\ \rho_2 &= (C_2C_4 - 2C_1C_5)/(4C_1C_3 - C_2^2) \end{aligned} \tag{11.21}$$

Substitution of Eq. (11.21) into Eq. (11.18) gives the warping moment of inertia I_ω of the elastic portion of the section.

Contribution of axial force to twisting moment ($\bar{K}$)

When warping $\bar{w} = \theta_\zeta'$ exists on a section, axial stress σ has some contribution to twisting moment m_ζ. As explained previously in Eq. (2.133), this is the Wagner effect expressed by

$$m_\zeta = \bar{K}\bar{w} \tag{11.22}$$

and

$$\bar{K} = \int \sigma a^2 \, dA \tag{11.23}$$

where a is the distance from the point on the middle line of the cross section to the current *torsion center* which does not necessarily coincide with the instantaneous shear center S. In the present analysis, the torsion center is assumed to be at the origin O for simplicity and the average value of $\sigma = -P/A$ for the entire cross section is used, thus

$$\bar{K} = -\frac{P}{A}(I_{\xi_0} + I_{\eta_0}) \tag{11.24}$$

in which P is axial compressive force acting on the section.

11.3 SEGMENT STIFFNESS MATRIX

Consider an undeformed segment AB of length L as shown in Fig. 11.4. The local coordinates (ξ, η, ζ) are defined as principal axes of the elastic portion of the section according to the right-hand screw rule,

ξ = strong axis of the elastic section

η = weak axis of the elastic section

ζ = longitudinal axis from A to B

Forces and displacements at ends A and B are related by the stiffness relationship. Since warping is considered only in the longitudinal direction of the segment, there are seven forces and seven displacements at each end, namely,

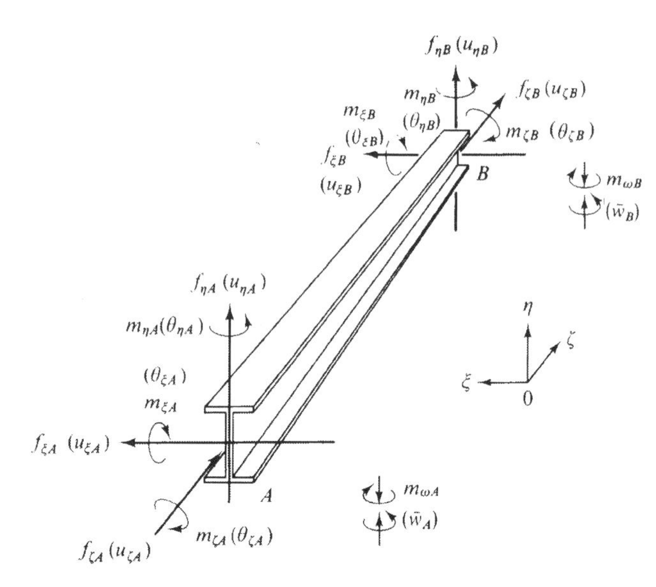

FIGURE 11.4
Forces and displacements of a segment

three forces and three displacements

$$\{f(3)\} = \begin{Bmatrix} f_\xi \\ f_\eta \\ f_\zeta \end{Bmatrix} \qquad \{u(3)\} = \begin{Bmatrix} u_\xi \\ u_\eta \\ u_\zeta \end{Bmatrix} \tag{11.25}$$

three moments and three rotations

$$\{m(3)\} = \begin{Bmatrix} m_\xi \\ m_\eta \\ m_\zeta \end{Bmatrix} \qquad \{\theta(3)\} = \begin{Bmatrix} \theta_\xi \\ \theta_\eta \\ \theta_\zeta \end{Bmatrix} \tag{11.26}$$

and a bi-moment and a warping

$$m_\omega(1) = m_\omega \qquad \bar{w}(1) = \theta'_\zeta \tag{11.27}$$

where the superscript prime indicates derivative with respect to ζ.

In this chapter, the parentheses { } and [ ] indicate an array of numbers as vector and matrix respectively and the integer in ( ) shows the size of the array. The lower case of f, m and $\theta, u, \bar{w}$ denote the internal forces and corresponding displacements at the ends of a segment with respect to the local coordinates (ξ, η, ζ), while the upper case of their counterparts refers to the global coordinates (X, Y, Z) (Fig. 11.1).

The bimoment or warping moment m_ω (moment distance product due to warping having the dimension of force multiplied by area) is defined in Eq. (2.118) and it is related to warping ω by Eq. (2.123) or

$$m_\omega = -EI_\omega\theta''_\zeta = -EI_\omega\bar{w}' \tag{11.28}$$

The segment stiffness matrix $[k(14 \times 14)]$ relates the fourteen forces $\{f(14)\}$ to the fourteen displacements $\{u(14)\}$.

$$\begin{Bmatrix} f_A(3) \\ m_A(3) \\ m_{\omega A}(1) \\ f_B(3) \\ m_B(3) \\ m_{\omega B}(1) \end{Bmatrix} = \begin{bmatrix} \\ \\ k(14 \times 14) \\ \\ \\ \end{bmatrix} \begin{Bmatrix} u_A(3) \\ \theta_A(3) \\ \bar{w}_A(1) \\ u_B(3) \\ \theta_B(3) \\ \bar{w}_B(1) \end{Bmatrix}. \tag{11.29}$$

The subscripts A and B indicate the quantities at the end points A and B, respectively.

The segment stiffness matrix $[k(14 \times 14)]$ is computed considering axial deformation, bending deformation, and torsion-warping deformation, separately. Further, the section properties are assumed to be constant within the segment. Thus the elastic analysis is available. In the following, the segment stiffness matrix $[k(14 \times 14)]$ is derived by solving separately the axial, flexural, and torsional problems, using the following force-displacement relations

$$\begin{aligned} &P = -EAu'_\zeta \\ &m_\xi = -EI_\xi u''_\eta, \qquad && m_\eta = EI_\eta u''_\xi \\ &m_\zeta = -EI_\omega \theta'''_\zeta + (GK_T + \bar{K})\theta'_\zeta, \qquad && m_\omega = -EI_\omega \theta''_\zeta \end{aligned} \tag{11.30}$$

Let us note that the relations (11.30) imply the orthogonality of coordinates ξ, η and $\bar{w}$, or that in other words the axial force P and bending moments m_ξ and m_η are reduced to the center of gravity while the twist moment m_ζ and transverse shear forces f_ξ and f_η are reduced on the shear center. The equations of equilibrium of the segment in terms of componental deformations for the linear static problem are due to the orthogonality of ξ, η and $\bar{w}$ uncoupled.

$$\begin{aligned} EAu''_\zeta &= 0 \\ EI_\xi u''''_\eta + Pu''_\eta &= 0 \\ EI_\eta u''''_\xi + Pu''_\xi &= 0 \\ EI_\omega \theta''''_\zeta - (GK_T + \bar{K})\theta''_\zeta &= 0 \end{aligned} \tag{11.31}$$

where primes denote differentiation with respect to ζ. The first equation in (11.31) governs the problem of an axially stressed segment while the next two govern the beam-column segment of flexure in two principal planes. The last equation governs the problem of warping torsion of a thin-walled segment. A set of fourteen boundary conditions (seven for each terminal cross section) should be attached to system (11.31) in order to define the boundary value problem.

(a) *Axial Stiffness*

The segment AB is subjected to the axial end forces $f_{\zeta A}$ and $f_{\zeta B}$ and the end displacements are $u_{\zeta A}$ and $u_{\zeta B}$ as shown in Fig. 11.5(a). The governing differential equation is

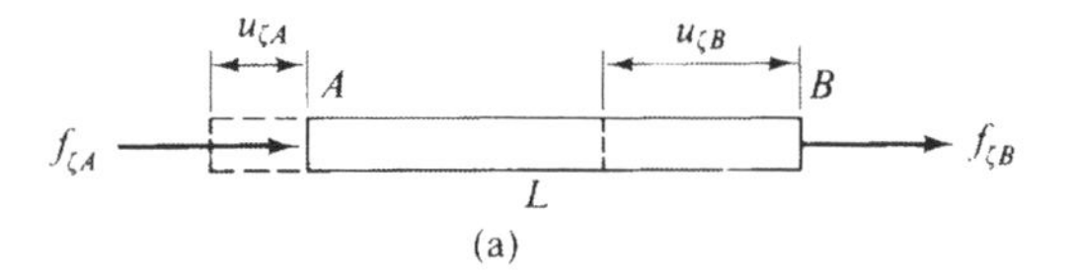

(a)

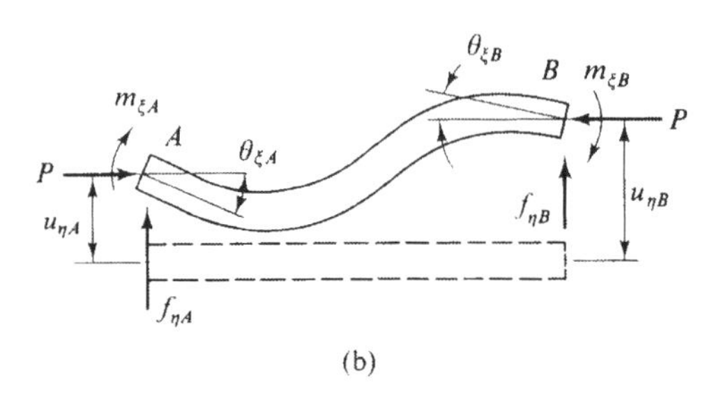

(b)

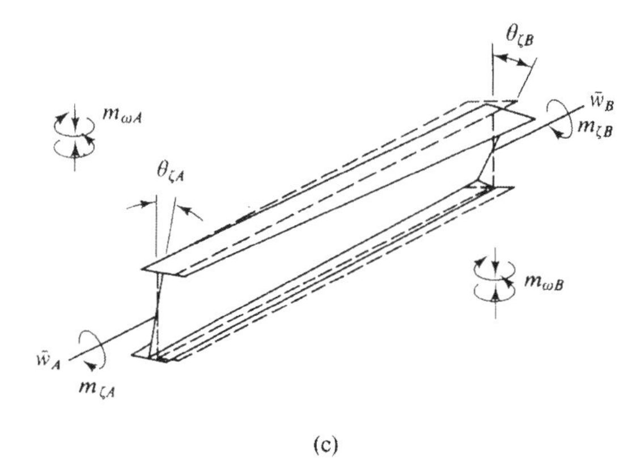

(c)

FIGURE 11.5
Independent deformation of a segment. (a) Axial deformation. (b) Bending and shear deformation. (c) Torsion and warping deformation

$$EAu_\zeta'' = 0 \tag{11.32}$$

The general solution of Eq. (11.32) is

$$u_\zeta(\zeta) = D_1 + D_2\zeta \tag{11.33}$$

where D_1 and D_2 are integration constants. At the boundaries A and B,

$$\begin{Bmatrix} u_{\zeta A} \\ u_{\zeta B} \end{Bmatrix} = \begin{Bmatrix} u_\zeta(0) \\ u_\zeta(L) \end{Bmatrix} = \begin{bmatrix} 1 & 0 \\ 1 & L \end{bmatrix} \begin{Bmatrix} D_1 \\ D_2 \end{Bmatrix} \tag{11.34}$$

and

$$\begin{Bmatrix} f_{\zeta A} \\ f_{\zeta B} \end{Bmatrix} = EA \begin{Bmatrix} -u_\zeta'(0) \\ u_\zeta'(L) \end{Bmatrix} = EA \begin{bmatrix} 0 & -1 \\ 0 & 1 \end{bmatrix} \begin{Bmatrix} D_1 \\ D_2 \end{Bmatrix} \tag{11.35}$$

Eliminating D_1 and D_2 between Eqs. (11.34) and (11.35), we obtain the axial stiffness relationship,

$$\begin{Bmatrix} f_{\zeta A} \\ f_{\zeta B} \end{Bmatrix} = \frac{EA}{L} \begin{bmatrix} 1 & -1 \\ -1 & 1 \end{bmatrix} \begin{Bmatrix} u_{\zeta A} \\ u_{\zeta B} \end{Bmatrix} \tag{11.36}$$

(b) *Bending and Shear Stiffness*

The segment AB is subjected to end forces $f_{\eta A}, f_{\eta B}, m_{\xi A}$ and $m_{\xi B}$. The end displacements are $u_{\eta A}$, $u_{\eta B}$, $\theta_{\xi A}$ and $\theta_{\xi B}$ as shown in Fig. 11.5(b).

The governing differential equation is

$$EI_{\xi} u_{\eta}'''' + P u_{\eta}'' = 0 \tag{11.37a}$$

or

$$u_{\eta}'''' + k_{\xi}^2 u_{\eta}'' = 0 \tag{11.37b}$$

where P is the axial compressive force,

$$P = f_{\zeta A} = -f_{\zeta B} \tag{11.38}$$

and

$$k_{\xi} = \sqrt{\frac{P}{EI_{\xi}}} \tag{11.39}$$

The general solution of Eq. (11.37) is

$$u_{\eta}(\zeta) = D_1 \cos k_{\xi}\zeta + D_2 \sin k_{\xi}\zeta + D_3\zeta + D_4 \tag{11.40}$$

where $D_1 \ldots D_4$ are the integration constants.

At the boundaries A and B,

$$\begin{Bmatrix} u_{\eta A} \\ \theta_{\xi A} \\ u_{\eta B} \\ \theta_{\xi B} \end{Bmatrix} = \begin{Bmatrix} u_{\eta}(0) \\ -u_{\eta}'(0) \\ u_{\eta}(L) \\ -u_{\eta}'(L) \end{Bmatrix} = \begin{bmatrix} 1 & 0 & 0 & 1 \\ 0 & -k_{\xi} & -1 & 0 \\ C_{\xi} & S_{\xi} & L & 1 \\ k_{\xi}S_{\xi} & -k_{\xi}C_{\xi} & -1 & 0 \end{bmatrix} \begin{Bmatrix} D_1 \\ D_2 \\ D_3 \\ D_4 \end{Bmatrix} \tag{11.41}$$

and

$$\begin{Bmatrix} f_{\eta A} \\ m_{\xi A} \\ f_{\eta B} \\ m_{\xi B} \end{Bmatrix} = EI_{\xi} \begin{Bmatrix} u_{\eta}'''(0) + k_{\xi}^2 u_{\eta}'(0) \\ u_{\eta}''(0) \\ -u_{\eta}'''(L) - k_{\xi}^2 u_{\eta}'(L) \\ -u_{\eta}''(L) \end{Bmatrix}$$
$$= EI_{\xi}k_{\xi}^2 \begin{bmatrix} 0 & 0 & 1 & 0 \\ -1 & 0 & 0 & 0 \\ 0 & 0 & -1 & 0 \\ C_{\xi} & S_{\xi} & 0 & 0 \end{bmatrix} \begin{Bmatrix} D_1 \\ D_2 \\ D_3 \\ D_4 \end{Bmatrix} \tag{11.42}$$

where

$$C_\xi = \cos k_\xi L \qquad \text{and} \qquad S_\xi = \sin k_\xi L \tag{11.43}$$

Eliminating $D_1 \ldots D_4$ between Eqs. (11.41) and (11.42) we obtain the bending and shear stiffness relation about the ξ-axis,

$$\begin{Bmatrix} f_{\eta A} \\ m_{\xi A}k_\xi \\ f_{\eta B} \\ m_{\xi B}k_\xi \end{Bmatrix} = \frac{EI_\xi k_\xi^2}{2 - 2C_\xi - k_\xi LS_\xi} \begin{bmatrix} S_\xi & C_\xi - 1 & -S_\xi & C_\xi - 1 \\ & S_\xi - k_\xi LC_\xi & 1 - C_\xi & k_\xi L - S_\xi \\ & & S_\xi & 1 - C_\xi \\ \text{SYMM.} & & & S_\xi - k_\xi LC_\xi \end{bmatrix} \begin{Bmatrix} u_{\eta A}k_\xi \\ \theta_{\xi A} \\ u_{\eta B}k_\xi \\ \theta_{\xi B} \end{Bmatrix} \tag{11.44}$$

In a similar way, the bending and shear stiffness relationship about the η-axis is given by,

$$\begin{Bmatrix} f_{\xi A} \\ m_{\eta A}k_\eta \\ f_{\xi B} \\ m_{\eta B}k_\eta \end{Bmatrix} = \frac{EI_\eta k_\eta^2}{2 - 2C_\eta - k_\eta LS_\eta} \begin{bmatrix} S_\eta & 1 - C_\eta & -S_\eta & 1 - C_\eta \\ & S_\eta - k_\eta LC_\eta & C_\eta - 1 & k_\eta L - S_\eta \\ & & S_\eta & C_\eta - 1 \\ \text{SYMM.} & & & S_\eta - k_\eta LC_\eta \end{bmatrix} \begin{Bmatrix} u_{\xi A}k_\eta \\ \theta_{\eta A} \\ u_{\xi B}k_\eta \\ \theta_{\eta B} \end{Bmatrix} \tag{11.45}$$

where

$$k_\eta = \sqrt{\frac{P}{EI_\eta}} \tag{11.46}$$

and

$$C_\eta = \cos k_\eta L \qquad \text{and} \qquad S_\eta = \sin k_\eta L \tag{11.47}$$

(c) *Torsion and Warping Stiffness*

The segment AB is subjected to the end forces $m_{\zeta A}$, $m_{\zeta B}$, $m_{\omega A}$ and $m_{\omega B}$. The corresponding end displacements are $\theta_{\zeta A}$, $\theta_{\zeta B}$, $\bar{w}_A$ and $\bar{w}_B$ as shown in Fig. 11.5(c). The governing differential equation is

$$EI_\omega \theta_\zeta'''' - (GK_T + \bar{K})\theta_\zeta'' = 0 \tag{11.48a}$$

or

$$\theta_\zeta'''' - k_\zeta^2 \theta_\zeta'' = 0 \tag{11.48b}$$

where

$$k_\zeta = \sqrt{\frac{GK_T + \bar{K}}{EI_\omega}} \tag{11.49}$$

The general solution of Eq. (11.48) is

$$\theta_\zeta(\zeta) = D_1 \cosh k_\zeta \zeta + D_2 \sinh k_\zeta \zeta + D_3 \zeta + D_4 \tag{11.50}$$

where $D_1 \ldots D_4$ are integration constants. At the boundaries A and B,

$$\begin{Bmatrix} \theta_{\zeta A} \\ \bar{w}_A \\ \theta_{\zeta B} \\ \bar{w}_B \end{Bmatrix} = \begin{Bmatrix} \theta_\zeta(0) \\ \theta'_\zeta(0) \\ \theta_\zeta(L) \\ \theta'_\zeta(L) \end{Bmatrix} = \begin{bmatrix} 1 & 0 & 0 & 1 \\ 0 & k_\zeta & 1 & 0 \\ C_\zeta & S_\zeta & L & 1 \\ k_\zeta S_\zeta & k_\zeta C_\zeta & 1 & 0 \end{bmatrix} \begin{Bmatrix} D_1 \\ D_2 \\ D_3 \\ D_4 \end{Bmatrix} \tag{11.51}$$

and

$$\begin{Bmatrix} m_{\zeta A} \\ m_{\omega A} \\ m_{\zeta B} \\ m_{\omega B} \end{Bmatrix} = EI_\omega \begin{Bmatrix} \theta'''_\zeta(0) - k_\zeta^2 \theta'_\zeta(0) \\ -\theta''_\zeta(0) \\ -\theta'''_\zeta(L) + k_\zeta^2 \theta'_\zeta(L) \\ \theta''_\zeta(L) \end{Bmatrix}$$

$$= k_\zeta^2 EI_\omega \begin{bmatrix} 0 & 0 & -1 & 0 \\ -1 & 0 & 0 & 0 \\ 0 & 0 & 1 & 0 \\ C_\zeta & S_\zeta & 0 & 0 \end{bmatrix} \begin{Bmatrix} D_1 \\ D_2 \\ D_3 \\ D_4 \end{Bmatrix} \tag{11.52}$$

where $$S_\zeta = \sinh k_\zeta L \text{ and } C_\zeta = \cosh k_\zeta L. \tag{11.53}$$

Eliminating $D_1 \ldots D_4$ between Eqs. (11.51) and (11.52), we obtain the torsion and warping stiffness relationship,

$$\begin{Bmatrix} m_{\zeta A} \\ m_{\omega A} k_\zeta \\ m_{\zeta B} \\ m_{\omega B} k_\zeta \end{Bmatrix} = \frac{k_\zeta^2 EI_\omega}{2 - 2C_\zeta + k_\zeta L S_\zeta} \begin{bmatrix} S_\zeta & C_\zeta - 1 & -S_\zeta & C_\zeta - 1 \\ & k_\zeta L C_\zeta - S_\zeta & 1 - C_\zeta & S_\zeta - k_\zeta L \\ & & S_\zeta & 1 - C_\zeta \\ \text{SYMM.} & & & k_\zeta L C_\zeta - S_\zeta \end{bmatrix} \begin{Bmatrix} \theta_{\zeta A} k_\zeta \\ \bar{w}_A \\ \theta_{\zeta B} k_\zeta \\ \bar{w}_B \end{Bmatrix} \tag{11.54}$$

The stiffness matrix for a thin-walled member was first derived by Krahula (1967) and the present form of Eq. (11.54) was obtained by Krajcinovic (1969). Similar developments for solid members are given elsewhere (see for instance, Archer, 1965, and Hartz, 1965). Further discussions on this subject will be taken up again in Chap. 12 when the detail formulation on the finite element method is presented.

Summing up the above stiffness relationships together, the segment stiffness matrix $[k]$ is obtained (Atsuta, 1972),

$$[k(14 \times 14)] = \begin{bmatrix} k_{AA}(7 \times 7) & k_{AB}(7 \times 7) \\ k_{BA}(7 \times 7) & k_{BB}(7 \times 7) \end{bmatrix} \tag{11.55}$$

which is a matrix of 14-th order. While the 7-th order submatrices $[k_{AA}]$, $[k_{BB}]$ on the main diagonal are evidently symmetric, $[k_{BA}]$ is the transpose of $[k_{AB}]$ which is an anti-symmetric submatrix of 7-th order. These submatrices have the forms as given in Eqs. (11.56) to (11.59). One notes that the elements in the submatrices $[k_{AA}]$ and $[k_{BB}]$ are identical except the sign of off-diagonal elements. Note also that the three non-zero off-diagonal elements of $[k_{AB}]$ as given by Eq. (11.58) are the same as the corresponding elements of $[k_{BB}]$ in Eq. (11.57), and the other three anti-symmetric off-diagonal elements of $[k_{AB}]$ are apparently the same as those of $[k_{AA}]$.

$$
[k_{AA}] = \begin{bmatrix}
\dfrac{EI_\eta k_\eta^3 S_\eta}{2 - 2C_\eta - k_\eta L S_\eta} & & & & & & \\
0 & \dfrac{EI_\xi k_\xi^3 S_\xi}{2 - 2C_\xi - k_\xi L S_\xi} & & & & \text{SYMM.} & \\
0 & 0 & \dfrac{EA}{L} & & & & \\
0 & -\dfrac{EI_\xi k_\xi^2 (1 - C_\xi)}{2 - 2C_\xi - k_\xi L S_\xi} & 0 & \dfrac{EI_\xi k_\xi (S_\xi - k_\xi L C_\xi)}{2 - 2C_\xi - k_\xi L S_\xi} & & & \\
\dfrac{EI_\eta k_\eta^2 (1 - C_\eta)}{2 - 2C_\eta - k_\eta L S_\eta} & 0 & 0 & 0 & \dfrac{EI_\eta k_\eta (S_\eta - k_\eta L C_\eta)}{2 - 2C_\eta - k_\eta L S_\eta} & & \\
0 & 0 & 0 & 0 & 0 & \dfrac{EI_\omega k_\zeta^3 S_\zeta}{2 - 2C_\zeta + k_\zeta L S_\zeta} & \\
0 & 0 & 0 & 0 & 0 & \dfrac{EI_\omega k_\zeta^2 (C_\zeta - 1)}{2 - 2C_\zeta + k_\zeta L S_\zeta} & \dfrac{EI_\omega k_\zeta (k_\zeta L C_\zeta - S_\zeta)}{2 - 2C_\zeta + k_\zeta L S_\zeta}
\end{bmatrix} \tag{11.56}
$$

$$
[k_{BB}] = \begin{bmatrix}
\dfrac{EI_\eta k_\eta^3 S_\eta}{2 - 2C_\eta - k_\eta L S_\eta} & & & & & & \\
0 & \dfrac{EI_\xi k_\xi^3 S_\xi}{2 - 2C_\xi - k_\xi L S_\xi} & & & & \text{SYMM.} & \\
0 & 0 & \dfrac{EA}{L} & & & & \\
0 & \dfrac{EI_\xi k_\xi^2 (1 - C_\xi)}{2 - 2C_\xi - k_\xi L S_\xi} & 0 & \dfrac{EI_\xi k_\xi (S_\xi - k_\xi L C_\xi)}{2 - 2C_\xi - k_\xi L S_\xi} & & & \\
\dfrac{-EI_\eta k_\eta^2 (1 - C_\eta)}{2 - 2C_\eta - k_\eta L S_\eta} & 0 & 0 & 0 & \dfrac{EI_\eta k_\eta (S_\eta - k_\eta L C_\eta)}{2 - 2C_\eta - k_\eta L S_\eta} & & \\
0 & 0 & 0 & 0 & 0 & \dfrac{EI_\omega k_\zeta^3 S_\zeta}{2 - 2C_\zeta + k_\zeta L S_\zeta} & \\
0 & 0 & 0 & 0 & 0 & \dfrac{-EI_\omega k_\zeta^2 (C_\zeta - 1)}{2 - 2C_\zeta + k_\zeta L S_\zeta} & \dfrac{EI_\omega k_\zeta (k_\zeta L C_\zeta - S_\zeta)}{2 - 2C_\zeta + k_\zeta L S_\zeta}
\end{bmatrix}
\tag{11.57}
$$

$$[k_{AB}] = \begin{bmatrix} \dfrac{-EI_\eta k_\eta^3 S_\eta}{2-2C_\eta - k_\eta L S_\eta} & & & & & & \\ 0 & \dfrac{-EI_\xi k_\xi^3 S_\xi}{2-2C_\xi - k_\xi L S_\xi} & & & & \text{ANTI-SYMM.} & \\ 0 & 0 & \dfrac{-EA}{L} & & & & \\ 0 & \dfrac{EI_\xi k_\xi^2(1-C_\xi)}{2-2C_\xi - k_\xi L S_\xi} & 0 & \dfrac{EI_\xi k_\xi(k_\xi L - S_\xi)}{2-2C_\xi - k_\xi L S_\xi} & & & \\ \dfrac{-EI_\eta k_\eta^2(1-C_\eta)}{2-2C_\eta - k_\eta L S_\eta} & 0 & 0 & 0 & \dfrac{EI_\eta k_\eta(k_\eta L - S_\eta)}{2-2C_\eta - k_\eta L S_\eta} & & \\ 0 & 0 & 0 & 0 & 0 & \dfrac{-EI_\omega k_\zeta^3 S_\zeta}{2-2C_\zeta + k_\zeta L S_\zeta} & \\ 0 & 0 & 0 & 0 & 0 & \dfrac{-EI_\omega k_\zeta^2(C_\zeta - 1)}{2-2C_\zeta + k_\zeta L S_\zeta} & \dfrac{-EI_\omega k_\zeta(k_\zeta L - S_\zeta)}{2-2C_\zeta + k_\zeta L S_\zeta} \end{bmatrix} \tag{11.58}$$

and

$$[k_{BA}] = [k_{AB}]^T \tag{11.59}$$

In computation of the segment stiffness, plastic section properties are constant throughout the segment as given by Eqs. (11.8) to (11.24). The sectional deformation rates which are necessary in computation of strain rate distribution in Eq. (11.1) are calculated as averages of the end displacement rates of the segment.

$$\dot{\varepsilon}_c = \dot{u}'_\zeta = \frac{\dot{u}_{\zeta B} - \dot{u}_{\zeta A}}{L}, \quad \dot{\Phi}_\eta = \dot{\theta}'_\eta = \frac{\dot{\theta}_{\eta B} - \dot{\theta}_{\eta A}}{L}$$
$$\dot{\Phi}_\xi = \dot{\theta}'_\xi = \frac{\dot{\theta}_{\xi B} - \dot{\theta}_{\xi A}}{L}, \quad \dot{\Phi}_\omega = \dot{\bar{w}}' = \frac{\dot{\bar{w}}_B - \dot{\bar{w}}_A}{L} \tag{11.60}$$

The warping curvature Φ_ω, the warping $\bar{w}$, and the twisting angle θ_ζ are related by

$$\Phi_\omega = \bar{w}' = \theta''_\zeta \tag{11.61}$$

11.4 ROTATION MATRIX AND MASTER STIFFNESS MATRIX

In order to make the master stiffness matrix for the whole structure by stacking up the segment stiffness matrices, a rotation matrix $[R(3 \times 3)]$ is required which transforms a vector from the global system (X, Y, Z) to the local system (ξ, η, ζ)

$$\underset{\text{local}}{\{V(3)\}} = [R(3 \times 3)]\underset{\text{global}}{\{V(3)\}} \tag{11.62}$$

or

$$\begin{Bmatrix} V_\xi \\ V_\eta \\ V_\zeta \end{Bmatrix} = \begin{bmatrix} (\xi X) & (\xi Y) & (\xi Z) \\ (\eta X) & (\eta Y) & (\eta Z) \\ (\zeta X) & (\zeta Y) & (\zeta Z) \end{bmatrix} \begin{Bmatrix} V_X \\ V_Y \\ V_Z \end{Bmatrix} \tag{11.63}$$

where (ξX) is the direction cosine between the local ξ-axis and the global X-axis, etc.

In computation of the rotation matrix, $[R(3 \times 3)]$, total displacements of segments rather than their rates must be taken into account. Let us consider a segment AB deformed from the original points $A_O B_O$ in the global system as shown in Fig. 11.6 The location of the original points A_O and B_O and the unit vectors in the local axes $\mathbf{v}_{\xi O}$, $\mathbf{v}_{\eta O}$ and $\mathbf{v}_{\zeta O}$ are known,

$$\mathbf{v}_{\xi O}: \begin{Bmatrix} (\xi_0 X) \\ (\xi_0 Y) \\ (\xi_0 Z) \end{Bmatrix}, \quad \mathbf{v}_{\eta O}: \begin{Bmatrix} (\eta_0 X) \\ (\eta_0 Y) \\ (\eta_0 Z) \end{Bmatrix}, \quad \mathbf{v}_{\zeta O}: \begin{Bmatrix} (\zeta_0 X) \\ (\zeta_0 Y) \\ (\zeta_0 Z) \end{Bmatrix} \tag{11.64}$$

After the deformation, the end points move by $\{U_A\}$ and $\{U_B\}$ thus

$$\text{A}: \begin{Bmatrix} X_A \\ Y_A \\ Z_A \end{Bmatrix} = \begin{Bmatrix} X_{AO} \\ Y_{AO} \\ Z_{AO} \end{Bmatrix} + \begin{Bmatrix} U_{XA} \\ U_{YA} \\ U_{ZA} \end{Bmatrix} \tag{11.65}$$

and

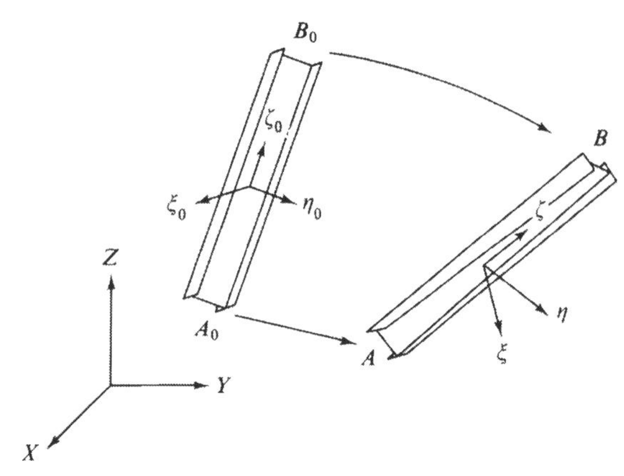

FIGURE 11.6
Displacements of a segment

$$\text{B}: \begin{Bmatrix} X_B \\ Y_B \\ Z_B \end{Bmatrix} = \begin{Bmatrix} X_{BO} \\ Y_{BO} \\ Z_{BO} \end{Bmatrix} + \begin{Bmatrix} U_{XB} \\ U_{YB} \\ U_{ZB} \end{Bmatrix} \tag{11.66}$$

The length AB is now

$$L = \sqrt{(X_B - X_A)^2 + (Y_B - Y_A)^2 + (Z_B - Z_A)^2} \tag{11.67}$$

and the new ζ-axis is determined by its unit vector

$$\mathbf{v}_\zeta: \begin{Bmatrix} (\zeta X) \\ (\zeta Y) \\ (\zeta Z) \end{Bmatrix} = \begin{Bmatrix} (X_B - X_A)/L \\ (Y_B - Y_A)/L \\ (Z_B - Z_A)/L \end{Bmatrix} \tag{11.68}$$

Determination of the new η-axis is not direct. First, the η_O-axis rotates about the ζ_O-axis by the average angle of rotation at ends A_O and B_O plus the rotation of principal axes due to yielding of the section given in Eq. (11.7)

$$\bar{\theta}_{\zeta O} = \frac{1}{2}(\theta_{\zeta OA} + \theta_{\zeta OB}) + \beta \tag{11.69}$$

Thus the $\bar{\eta}$-axes is obtained as shown in Fig. 11.7.

$$\mathbf{v}_{\bar{\eta}}: \begin{Bmatrix} (\bar{\eta}X) \\ (\bar{\eta}Y) \\ (\bar{\eta}Z) \end{Bmatrix} = \begin{Bmatrix} (\eta_0 X)\cos\bar{\theta}_{\zeta 0} - [(\eta_0 Y)(\zeta_0 Z) - (\eta_0 Z)(\zeta_0 Y)]\sin\bar{\theta}_{\zeta 0} \\ (\eta_0 Y)\cos\bar{\theta}_{\zeta 0} - [(\eta_0 Z)(\zeta_0 X) - (\eta_0 X)(\zeta_0 Z)]\sin\bar{\theta}_{\zeta 0} \\ (\eta_0 Z)\cos\bar{\theta}_{\zeta 0} - [(\eta_0 X)(\zeta_0 Y) - (\eta_0 Y)(\zeta_0 X)]\sin\bar{\theta}_{\zeta 0} \end{Bmatrix} \tag{11.70}$$

Since the direction of the η-axis coincides with the projection of $\mathbf{v}_{\bar{\eta}}$ on the plane perpendicular to the ζ-axis,

$$\mathbf{v}_\eta: \begin{Bmatrix} (\eta X) \\ (\eta Y) \\ (\eta Z) \end{Bmatrix} = \frac{1}{L_\eta} \begin{Bmatrix} (\bar{\eta}X) - M(\zeta X) \\ (\bar{\eta}Y) - M(\zeta Y) \\ (\bar{\eta}Z) - M(\zeta Z) \end{Bmatrix} \tag{11.71}$$

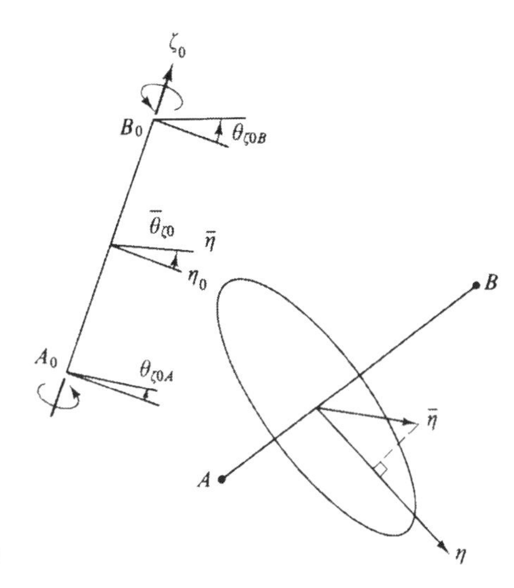

FIGURE 11.7
Transformation of axis

where M is the dot-product of $\mathbf{v}_{\bar{\eta}}$ and $\mathbf{v}_\zeta$,

$$M = \mathbf{v}_{\bar{\eta}} \cdot \mathbf{v}_\zeta = (\bar{\eta}X)(\zeta X) + (\bar{\eta}Y)(\zeta Y) + (\bar{\eta}Z)(\zeta Z) \tag{11.72}$$

and the length of the projected $\mathbf{v}_{\bar{\eta}}$ is

$$L_\eta = \sqrt{[(\bar{\eta}X) - M(\zeta X)]^2 + [(\bar{\eta}Y) - M(\zeta Y)]^2 + [(\bar{\eta}Z) - M(\zeta Z)]^2} \tag{11.73}$$

The unit vector in the new ξ-axis, $\mathbf{v}_\xi$, is obtained as the cross-product of $\mathbf{v}_\eta$ and $\mathbf{v}_\zeta$,

$$\mathbf{v}_\xi : \begin{Bmatrix} (\xi X) \\ (\xi Y) \\ (\xi Z) \end{Bmatrix} = \begin{Bmatrix} (\eta Y)(\zeta Z) - (\eta Z)(\zeta Y) \\ (\eta Z)(\zeta X) - (\eta X)(\zeta Z) \\ (\eta X)(\zeta Y) - (\eta Y)(\zeta X) \end{Bmatrix} \tag{11.74}$$

Now, the rotation matrix of the segment AB after the deformation is determined from Eqs. (11.68), (11.71) and (11.74).

$$[R(3 \times 3)] = \begin{bmatrix} (\xi X) & (\xi Y) & (\xi Z) \\ (\eta X) & (\eta Y) & (\eta Z) \\ (\zeta X) & (\zeta Y) & (\zeta Z) \end{bmatrix} \tag{11.75}$$

It is to be emphasized here that this rotation matrix has been obtained taking both the geometric change and material yielding into consideration.

Since warping is considered only in the longitudinal direction of the beam-column and treated like a scalar quantity (assumption 8), the force vector $\{f(14)\}$ and the displacement vector $\{u(14)\}$ in the local system are transformed into the global system $\{F(14)\}$ and $\{U(14)\}$ by using the rotation matrix $[R(14 \times 14)]$.

$$\begin{aligned} \{F(14)\} &= [R^T(14 \times 14)]\{f(14)\} \\ \{U(14)\} &= [R^T(14 \times 14)]\{u(14)\} \end{aligned} \tag{11.76}$$

$$\begin{Bmatrix} F_A(3) \\ M_A(3) \\ M_{\omega A}(1) \\ \cdots \\ F_B(3) \\ M_B(3) \\ M_{\omega B}(1) \end{Bmatrix} = \left[\begin{array}{ccc|ccc} R^T(3\times 3) & & & & & \\ & R^T(3\times 3) & & & 0 & \\ & & 1 & & & \\ \hline & & & R^T(3\times 3) & & \\ & 0 & & & R^T(3\times 3) & \\ & & & & & 1 \end{array}\right] \begin{Bmatrix} f_A(3) \\ m_A(3) \\ m_{\omega A}(1) \\ \cdots \\ f_B(3) \\ m_B(3) \\ m_{\omega B}(1) \end{Bmatrix} \tag{11.77}$$

and

$$\begin{Bmatrix} U_A(3) \\ \Theta_A(3) \\ W_A(1) \\ \cdots \\ U_B(3) \\ \Theta_B(3) \\ W_B(1) \end{Bmatrix} = \left[\begin{array}{ccc|ccc} R^T(3\times 3) & & & & & \\ & R^T(3\times 3) & & & 0 & \\ & & 1 & & & \\ \hline & & & R^T(3\times 3) & & \\ & 0 & & & R^T(3\times 3) & \\ & & & & & 1 \end{array}\right] \begin{Bmatrix} u_A(3) \\ \theta_A(3) \\ \bar{w}_A(1) \\ \cdots \\ u_B(3) \\ \theta_B(3) \\ \bar{w}_B(1) \end{Bmatrix} \tag{11.78}$$

where $[R^T(3 \times 3)]$ is the transpose of matrix $[R(3 \times 3)]$ which is given by Eq. (11.75).

Now the segment stiffness relationship Eq. (11.29) is also transformed into the global system by use of Eq. (11.76).

$$\{F(14)\} = [R^T(14 \times 14)][k(14 \times 14)][R(14 \times 14)]\{U(14)\} \tag{11.79}$$

Stacking up this relationship for all segments, the stiffness relationship of the whole structure is obtained

$$\{F(N)\} = [K(N \times N)]\{U(N)\} \tag{11.80}$$

where

$$\begin{aligned} [K(N \times N)] &= \text{master stiffness matrix} = \Sigma[R^T][k][R] \\ N &= 7 \times (\text{number of nodal points}) \end{aligned} \tag{11.81}$$

Since the master stiffness matrix $[K(N \times N)]$ is symmetric and diagonally populated, the banded stiffness matrix $[K(N \times NB)]$ is used to save computer storage. The band width NB is given by

$$NB = 7 \times (\text{the maximum number of connected nodes})$$

In the case of beam-column analyses, only two nodes are connected by a segment, thus $NB = 14$ is sufficient.

11.5 TANGENT STIFFNESS APPROACH FOR NUMERICAL SOLUTIONS

The master stiffness matrix $[K]$ obtained in Eq. (11.80) is a tangent stiffness of the system, because the current geometry $[R]$ and the instantaneous section properties were used in its computation.

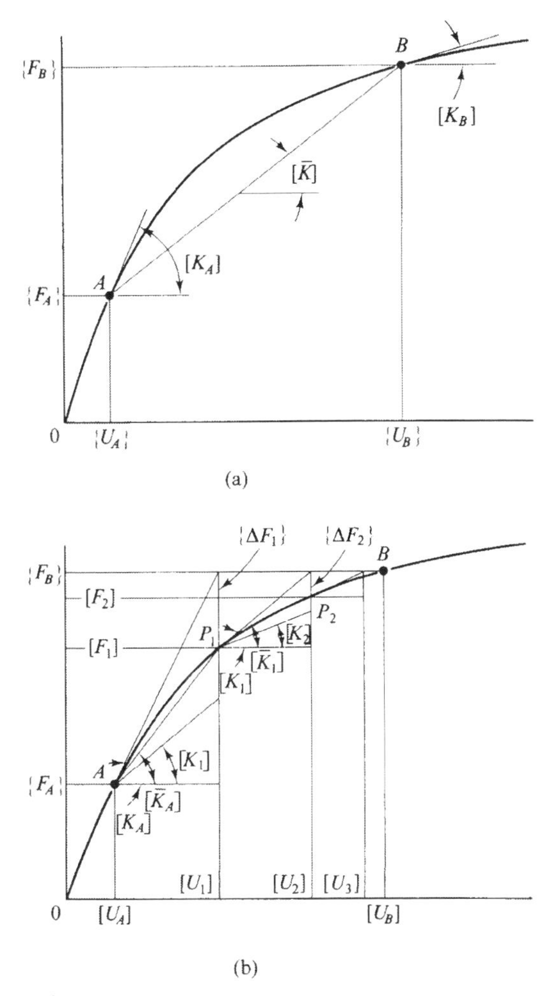

FIGURE 11.8
Modified tangent stiffness approach. (a) Stiffness change and second stiffness. (b) Iterative procedure for displacements

Consider a structure which is subjected to forces $\{F_A\}$ and has displacements $\{U_A\}$. This state is shown by point A in Fig. 11.8(a). When the forces are increased from $\{F_A\}$ to $\{F_B\}$, the displacements become $\{U_B\}$. The correct values of $\{U_B\}$ cannot be obtained directly because of non-linearity of the system. Thus some iterative procedures must be applied. The tangent stiffness approach described previously in Chap. 6 is one of the techniques which is used herein being somewhat modified as presented in what follows.

Since stiffness of a system is a function of the current deformation, the tangent stiffness matrix $[K]$ changes from $[K_A]$ to $[K_B]$ during the deformation change from $\{U_A\}$ to $\{U_B\}$. In Fig. 11.8(a), the stiffnesses are represented by the slopes of the load-deflection curve. In a loading state of a system, the stiffness may be assumed to decrease monotonically. Thus the secant stiffness $[\bar{K}]$ between the two states A and B is bounded by

$$|K_B| \leq |\bar{K}| \leq |K_A| \tag{11.82}$$

or using an averaging factor, α, it is expressed as

$$[\bar{K}] = \alpha[K_A] + (1 - \alpha)[K_B] \tag{11.83}$$

The averaging factor α is bounded by

$$0 \leq \alpha \leq 1 \tag{11.84}$$

In the extreme cases,

$$\begin{aligned} \alpha = 1 \quad [K] = [K_A] \quad \text{upper limit} \\ \alpha = 0 \quad [K] = [K_B] \quad \text{lower limit} \end{aligned} \tag{11.85}$$

If the factor α could be correctly estimated, the deformation $\{U_B\}$ would be solved directly without using laborious iterative techniques. Since it is not possible, a successive approximation approach is applied using a fixed value for α. Referring to Fig. 11.8(b), the initial force difference is

$$\{\Delta F_0\} = \{F_B\} - \{F_A\} \tag{11.86}$$

Using the tangent stiffness at point A, $[K_A]$, the new deformations $\{U_1\}$ are calculated by

$$\{U_1\} = [K_A]^{-1}\{\Delta F_0\} + \{U_A\} \tag{11.87}$$

From the deformations, the tangent stiffness $[K_1]$ is obtained. The corresponding forces $\{F_1\}$ are calculated using Eq. (11.83) as

$$\{F_1\} = [\alpha K_A + (1 - \alpha)K_1]\{U_1 - U_A\} + \{F_A\} \tag{11.88}$$

This state is shown by point P_1 in Fig. 11.8(b). There is an unbalanced force of

$$\{\Delta F_1\} = \{F_B\} - \{F_1\} \tag{11.89}$$

Then the same procedure is repeated to obtain the new state P_2 by

$$\begin{aligned} \{U_2\} &= [K_1]^{-1}\{\Delta F_1\} + \{U_1\} \\ \{F_2\} &= [\alpha K_1 + (1 - \alpha)K_2]\{U_2 - U_1\} + \{F_1\} \\ \{\Delta F_2\} &= \{F_B\} - \{F_2\} \end{aligned} \tag{11.90}$$

After n-cycles of iteration the above relations yield

$$\begin{aligned} \{U_n\} &= [K_{n-1}]^{-1}\{\Delta F_{n-1}\} + \{U_{n-1}\} \\ \{F_n\} &= [\alpha K_{n-1} + (1 - \alpha)K_n]\{U_n - U_{n-1}\} + \{F_{n-1}\} \\ \{\Delta F_n\} &= \{F_B\} - \{F_n\} \end{aligned} \tag{11.91}$$

or

$$\{U_n\} = \sum_{i=1}^{n} [K_{i-1}]^{-1}\{\Delta F_{i-1}\} + \{U_A\}$$
$$\{\Delta F_n\} = (1-\alpha)^n \prod_{i=1}^{n} ([I] - [K_i][K_{i-1}]^{-1})\{\Delta F_o\} \tag{11.92}$$

where $[I]$ = unit matrix, Σ = summation symbol, Π = multiplication symbol. This procedure is repeated until the unbalanced forces $\{\Delta F_n\}$ become less than the prescribed tolerable limit.

The convergence and accuracy of this method are examined using Eq. (11.92). Since in the loading state, the stiffness decreases monotonically,

$$0 < |[I] - [K_i][K_{i-1}]^{-1}| < 1 \tag{11.93}$$

thus the unbalanced force $\{\Delta F_n\}$ converges to zero in the order of at least $(1-\alpha)^n$. Accuracy of the solution can be improved as much as needed by decreasing the loading steps. In an extreme case as

$$[K_i][K_{i-1}]^{-1} = [I] \tag{11.94}$$

the upper solution and the lower solution coincide with each other.

11.6 NUMERICAL EXAMPLES

11.6.1 Large Deflection of an Elastic Column

In order to examine the effect of segment size and load steps on convergence and accuracy of the solution in a non-linear problem, the large deflection of an elastic column with an initial imperfection is discussed below.

As shown in Fig. 11.9, the column is simply supported at its ends and has an initial deflection of sinusoidal form,

$$U = U_0 \sin \frac{\pi Z}{L} \tag{11.95}$$

The exact solution of this problem with a parabolic imperfection case is obtained by Atsuta (1972).

(a) *Effect of segment size*

Using a column with initial deflection of $U_0/L = 0.02$, the effect of segment size on the load deflection relation is examined in three examples with different numbers of segments NS.

Example A-1: two segments ($NS = 2$)

Example A-2: four segments ($NS = 4$)

Example A-3: eight segments ($NS = 8$)

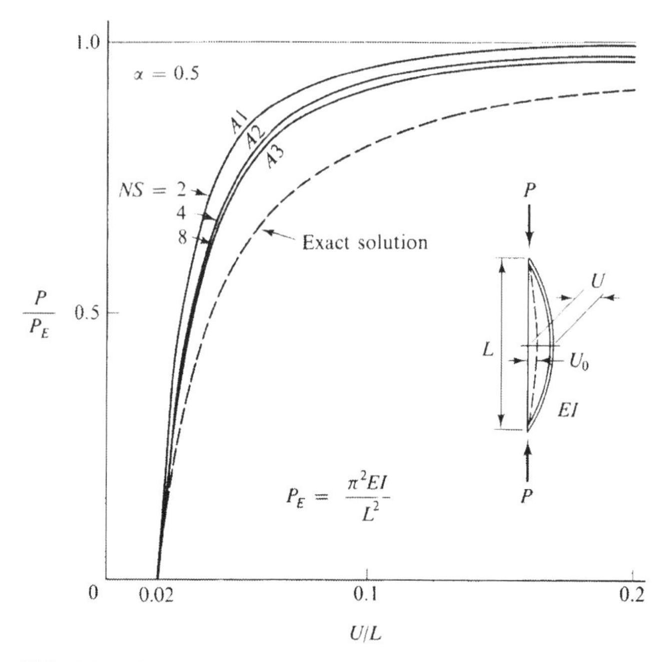

FIGURE 11.9
Averaged solution and number of segments

The solutions obtained by using the averaged stiffness matrices ($\alpha = 0.5$) are plotted in Fig. 11.9. The broken line is the exact solution. The columns with more than four segments give almost the same results, thus the use of four segments is considered to be small enough to analyze elastic columns.

(b) *Effect of averaging factor*

Since all above solutions are not close to the exact one as shown in Fig. 11.9, the averaging factor α of the stiffness matrix must be examined. The upper and lower solutions for the same column were examined in six different examples,

Example B-1: $NS = 2$
Example B-2: $NS = 4$ $\Big\}$ $\alpha = 0$ (lower solution)
Example B-3: $NS = 8$
Example C-1: $NS = 2$
Example C-2: $NS = 4$ $\Big\}$ $\alpha = 1$ (upper solution)
Example C-3: $NS = 8$

The results are plotted in Fig. 11.10. Refinement of segment size has little effect on the upper bound solution but it improves the lower bound solutions very much, and the lower bound solution approaches the exact solution.

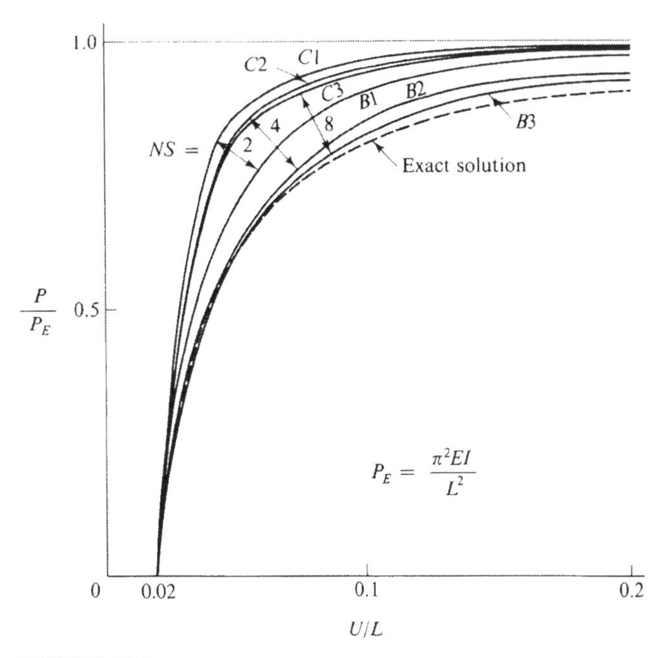

FIGURE 11.10
Number of segments and boundaries of solutions

(c) *Effect of load steps*

Using the same column with an initial deflection of $U_0/L = 0.02$, six examples were calculated with different load steps (ΔP),

Example D: $P/P_E = 0, 0.5, 0.8, 0.9, 0.95, 0.98, 0.99$

$$\left.\begin{array}{ll} \text{D-1} & \alpha = 0.0 \\ \text{D-2} & \alpha = 0.5 \\ \text{D-3} & \alpha = 1.0 \end{array}\right\} NS = 8$$

Example E: $P/P_E = 0, 0.25, 0.5, 0.65, 0.8, 0.85, 0.9, 0.925, 0.95, 0.965, 0.98, 0.985, 0.99, 0.995$

$$\left.\begin{array}{ll} \text{E-1} & \alpha = 0.0 \\ \text{E-2} & \alpha = 0.5 \\ \text{E-3} & \alpha = 1.0 \end{array}\right\} NS = 8$$

The results are plotted in Fig. 11.11 which shows that load steps affect only the upper bound solution. The lower bound solutions are almost identical in both Examples D-1 and E-1.

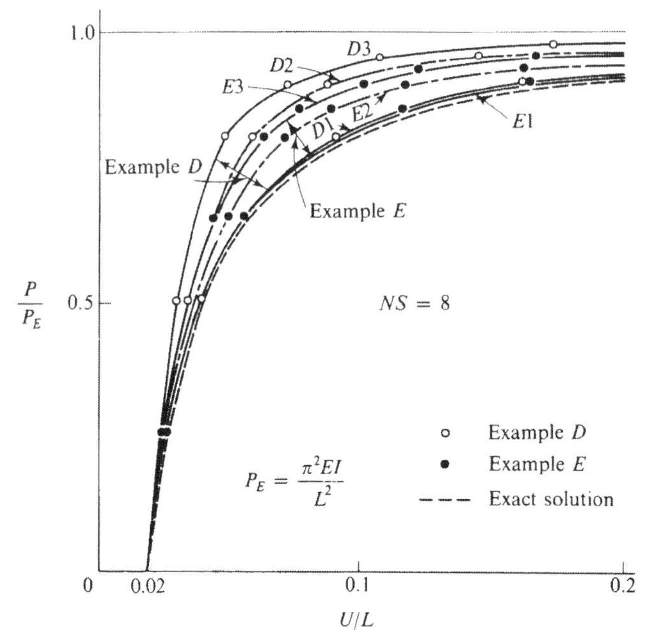

FIGURE 11.11
Effect of load steps

FIGURE 11.12
Lower bound solution with initial deflections

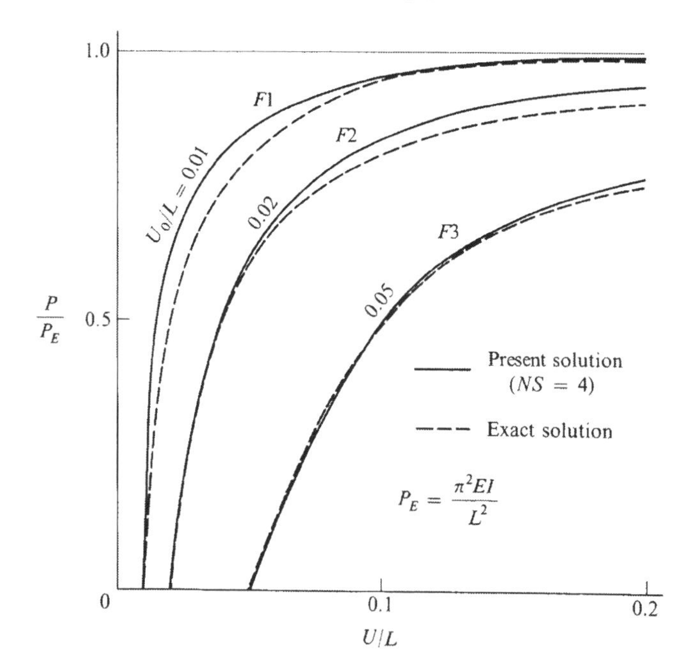

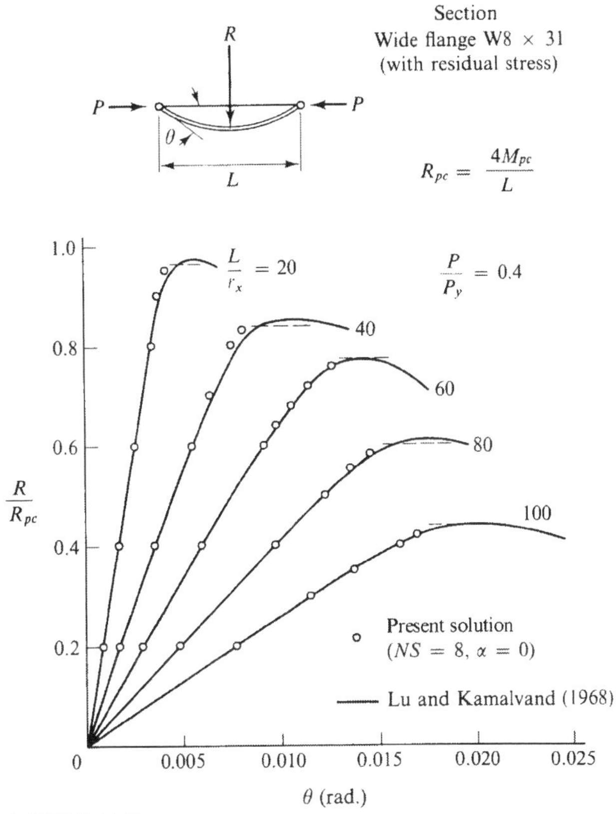

FIGURE 11.13
Example of an in-plane beam-column

(d) *Effect of initial imperfection*

From the above discussion it may be concluded that the lower bound solution ($\alpha = 0$) with four segments ($NS = 4$) gives good results in the analysis of elastic columns. In order to check this conclusion, three other examples were calculated for a column with different initial imperfections.

Example F-1: $U_0/L = 0.01$

Example F-2: $U_0/L = 0.02 \quad NS = 4, \quad \alpha = 0$

Example F-3: $U_0/L = 0.05$

As shown in Fig. 11.12, good agreement was observed with the exact solutions in all three cases.

11.6.2 In-Plane Beam-Column with a Lateral Load

As a plastic beam-column example, an in-plane beam-column with a concentrated lateral load at its mid-span was studied as shown in Fig. 11.13.

The wide-flange section W8 × 31 with residual stress was used. Five different slenderness ratios were selected:

$$\left.\begin{array}{ll} \text{Example E-1:} & L/r_x = 20 \\ \text{Example E-2:} & L/r_x = 40 \\ \text{Example E-3:} & L/r_x = 60 \\ \text{Example E-4:} & L/r_x = 80 \\ \text{Example E-5:} & L/r_x = 100 \end{array}\right\} \quad NS = 8, \quad \alpha = 0$$

The beam-column was cut into eight segments two of which were short segments ($l = 1$ inch) located under the lateral load. The axial force P was kept constant ($P/P_y = 0.4$) while the lateral load R was increased, which is nondimensionalized by the fraction to

$$R_{pc} = \frac{4M_{pc}}{L} \tag{11.96}$$

Figure 11.13 shows numerical results of an end rotation angle θ in radians. Open circles represent the present solutions and the solid lines are the results reported by Lu and Kamalvand (1968), using Column Deflection Curve method described in Chap. 8 of Vol. 1. Although only loading states were calculated, good agreement was observed in the two analyses.

FIGURE 11.14
Example of a biaxially loaded beam-column (1 in = 25.4 mm, 1 kip = 4.45 kN, 1 ksi = 6.9 MPa)

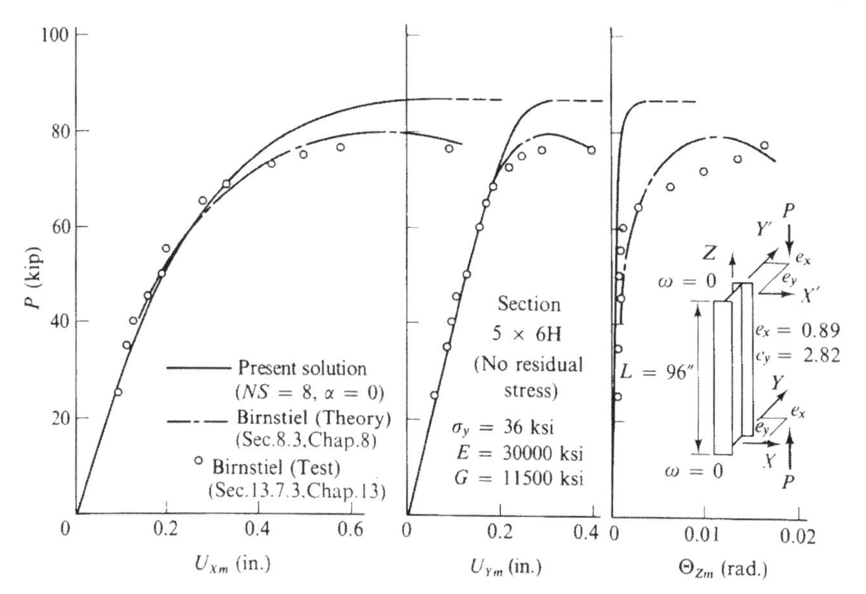

11.6.3 Biaxially Loaded Beam-Column

Figure 11.14 shows numerical results of a beam-column which is subjected to axial thrust P with symmetric eccentricities at both ends.

Example F

Section:	$5 \times 6H$ (without residual stress)
Length:	$L = 96$ in (2.44 m)
Material:	$\sigma_y = 36$ ksi (248 MN/m^2)
	$E = 30\,000$ ksi (207 000 MN/m^2)
	$G = 11\,500$ ksi (79 350 MN/m^2)
Eccentricities:	$e_x = 0.89$ in (22.6 mm)
	$e_y = 2.82$ in (71.6 mm)
Warping:	restrained at both ends

The mid-height displacement U_{Xm}, U_{Ym} and the rotation angle Θ_{Zm} were computed using eight segments ($NS = 8$) and the lowest stiffness factor ($\alpha = 0$). Since the loading was symmetric, two small segments ($l = 1$ in) were installed at the mid-height. In Fig. 11.14, the solutions by present analysis are plotted by solid lines. Open circles and broken lines show experimental results (see Chap. 13) and theoretical results (see Chap. 8) respectively, reported by Birnstiel (1968).

The experiment shows a sudden increase in rotation angle Θ_{Zm} after onset of yielding. The present solution does not follow this large rotation and thus gives a higher value for the ultimate strength. This is considered to result from the assumption in the present analysis that contribution of shear stress on yielding was neglected.

11.6.4 Space Frame

The present method is directly applicable to the analysis of space structures if warping deformation is perfectly free or perfectly restrained at all junctions of members. This condition is satisfied at joints of a space frame at which a maximum of three members meet at right angles, because no bimoments exist at those joints.

As a numerical example, the space frame which had been studied by Morino (1970) was selected here. The frame is composed of four columns and four beams as shown in Fig. 11.15(a).

Example G

Columns:	W10 $\times$ 60, $L = 144$ in (3.7 m)
Beams:	W18 $\times$ 60, $L = 144$ in (3.7 m)

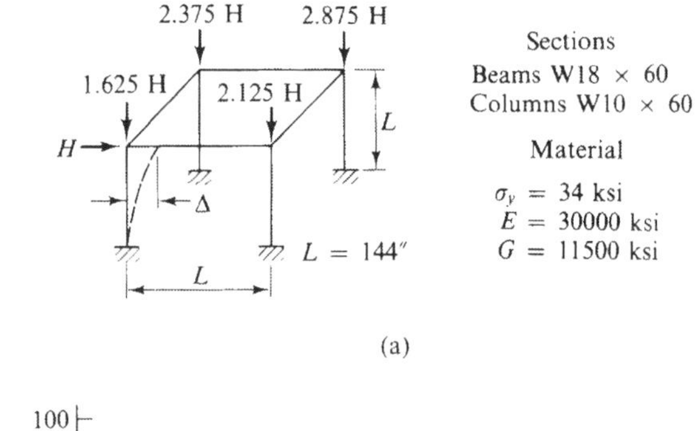

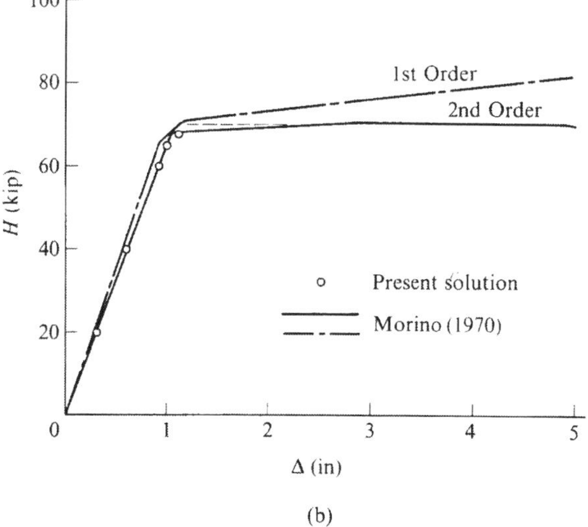

FIGURE 11.15
Example of a space frame (1 in = 25.4 mm, 1 kip = 4.45 kN, 1 ksi = 6.9 MPa). (a) Space frame and load. (b) Numerical results for horizontal deflection

Material: $\sigma_y = 34$ ksi (235 MN/m^2)

$E = 30\,000$ ksi (207 000 MN/m^2)

$G = 11\,500$ ksi (79,400 MN/m^2)

Eighteen segments and eighteen nodal points were used. Nodal points were numbered so that band width NB of the master stiffness matrix $[K]$ might be the smallest (in this example $NB = 7 \times 4 = 28$). Eight short segments ($l = 1$ inch) were selected where plastic hinges were expected to be formed.

Four vertical loads and a horizontal load were applied proportionally as shown in Fig. 11.15(a). Figure 11.15(b) shows horizontal displacement Δ at point where the horizontal force H was applied. Open circles are the solutions obtained by the present analysis. The broken horizontal line represents the lowest value of H for which the system was found to be unstable. The solid lines are results reported by Morino

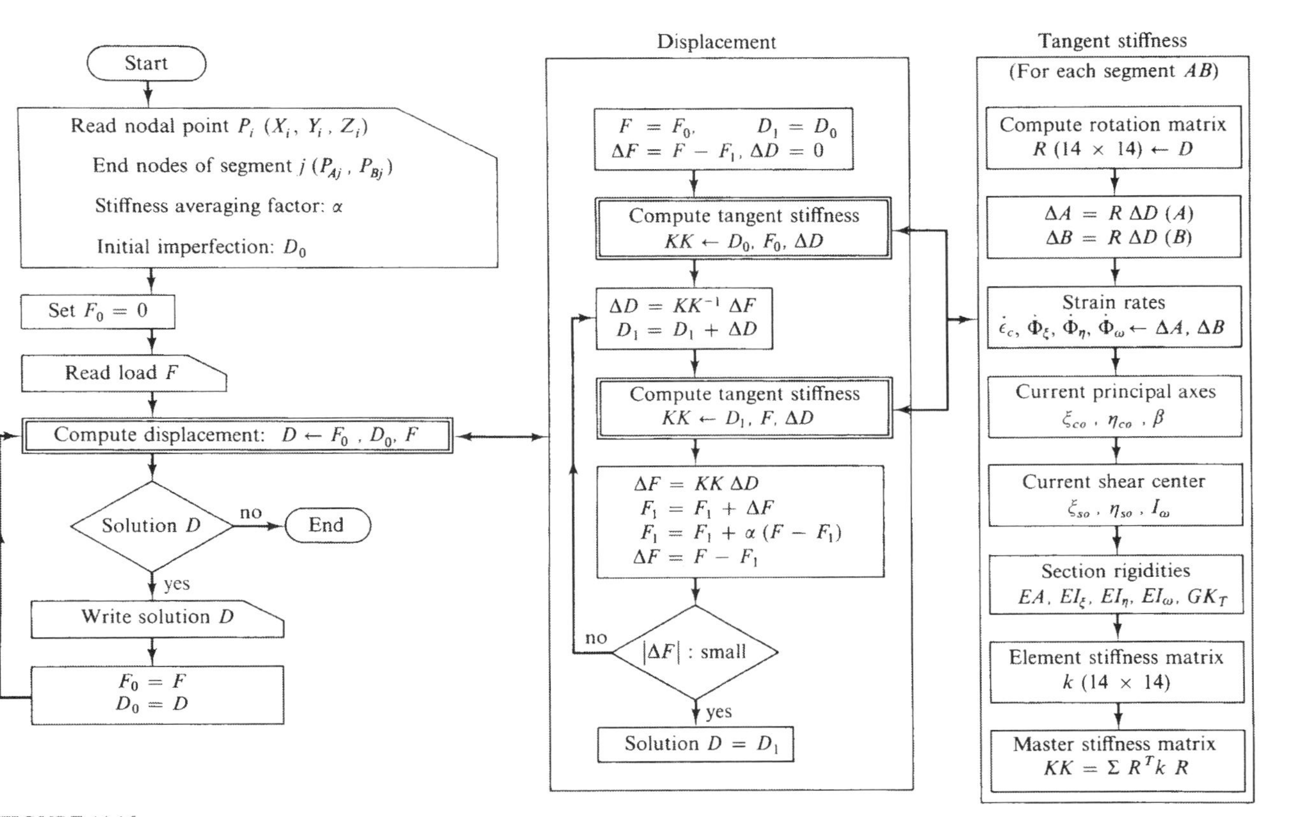

FIGURE 11.16
Flow diagram of finite segment method

(1970). Although a solution was not obtained in the large deflection portion, good agreement was observed in elastic portion and for the ultimate strength.

11.7 SUMMARY

The finite segment method to solve biaxially loaded beam-columns is presented. The beam-column was treated as a space structure after segmentation. The sections were cut into finite element areas only in the segments where yielding was expected. The segment stiffness relationships were computed using the direct stiffness method. Nonlinearity due to material plasticity and geometrical change was overcome by a modified tangent stiffness approach. Numerical examples were presented on large deflection of an elastic column, an in-plane beam-column with a lateral load, a biaxially loaded beam-column and a space frame.

All numerical results were compared with other reported values. In most cases, good agreement was observed. For a predominant torsional deformation, this analysis procedure indicated a higher value for the ultimate strength. This indication was considered to be related to the simpler yield criterion assumed. This method would be improved by accounting for shear stress in the yield condition. Further, space structures could be solved more generally by treating warping and bimoment as three dimensional vectors at intersections of members.

As a summary of the computational procedure, a flow diagram of the finite segment method is shown in Fig. 11.16 which is the one that the authors used for their computer programing. A generalization of the finite segment method, known as the finite element method, will be presented in the following chapter.

11.8 PROBLEMS

11.1 Show that when the axial thrust and St. Venant torsion rigidity are small, the segment stiffness matrices as given by Eq. (11.56) through (11.59) are reduced to the usual stiffness matrices for a beam, i.e.,

$$[k_{AA}] = \begin{bmatrix} \frac{12EI_\eta}{L^3} & & & & & & \\ 0 & \frac{12EI_\xi}{L^3} & & & & \text{SYMM.} & \\ 0 & 0 & \frac{EA}{L} & & & & \\ 0 & -\frac{6EI_\xi}{L^2} & 0 & \frac{4EI_\xi}{L} & & & \\ \frac{6EI_\eta}{L^2} & 0 & 0 & 0 & \frac{4EI_\eta}{L} & & \\ 0 & 0 & 0 & 0 & 0 & \frac{12EI_\omega}{L^3} & \\ 0 & 0 & 0 & 0 & 0 & \frac{6EI_\omega}{L^2} & \frac{4EI_\omega}{L} \end{bmatrix} \tag{11.97}$$

$$[k_{AB}] = \begin{bmatrix} -\frac{12EI_\eta}{L^3} & & & & & & \\ 0 & -\frac{12EI_\xi}{L^3} & & & \text{ANTI-SYMM.} & & \\ 0 & 0 & -\frac{EA}{L} & & & & \\ 0 & \frac{6EI_\xi}{L^2} & 0 & \frac{2EI_\xi}{L} & & & \\ -\frac{6EI_\eta}{L^2} & 0 & 0 & 0 & \frac{2EI_\eta}{L} & & \\ 0 & 0 & 0 & 0 & 0 & -\frac{12EI_\omega}{L^3} & \\ 0 & 0 & 0 & 0 & 0 & -\frac{6EI_\omega}{L^2} & \frac{2EI_\omega}{L} \end{bmatrix} \quad (11.98)$$

$$[k_{BB}] = \begin{bmatrix} \frac{12EI_\eta}{L^3} & & & & & & \\ 0 & \frac{12EI_\xi}{L^3} & & & & \text{SYMM.} & \\ 0 & 0 & \frac{EA}{L} & & & & \\ 0 & \frac{6EI_\xi}{L^2} & 0 & \frac{4EI_\xi}{L} & & & \\ -\frac{6EI_\eta}{L^2} & 0 & 0 & 0 & \frac{4EI_\eta}{L} & & \\ 0 & 0 & 0 & 0 & 0 & \frac{12EI_\omega}{L^3} & \\ 0 & 0 & 0 & 0 & 0 & -\frac{6EI_\omega}{L^2} & \frac{4EI_\omega}{L} \end{bmatrix} \quad (11.99)$$

and

$$[k_{BA}] = [k_{AB}]^T \quad (11.100)$$

11.2 Show that the location of shear center (x_s, y_s) for the elastic portion of a partially yielded cross section can be found from the condition, Eq. (11.20), or

$$\frac{\partial I_\omega}{\partial x_s} = 0 \quad \text{and} \quad \frac{\partial I_\omega}{\partial y_s} = 0 \quad (11.101)$$

Using this condition, calculate the shear center and the warping moment of inertia of a channel section.

11.3 Show that rotation angle of the instantaneous principal axes β is calculated by Eq. (11.7).

11.4 In this chapter, the warping deformation and the bimoment are treated as a scalar value. Discuss the physical meaning of this assumption for a space frame joint where members meet each other with right angles.

11.9 REFERENCES

Archer, J. S., "Consistent Matrix Formulations for Structural Analysis Using Finite-Elements Techniques," *AIAA Journal*, Vol. 3, No. 10, October, pp. 1910–1918, (1965).

Atsuta, T., "Analyses of Inelastic Beam-Columns," thesis presented to Department of Civil Engineering, Lehigh University, Bethlehem, Pa., in partial fulfilment of the requirements for the degree of Doctor of Philosophy, (1972).

Birnstiel, C., "Experiments on H-columns under biaxial bending," *Journal of the Structural Division, ASCE*, Vol. 94, No. ST10, Proc. Paper 6186, October, pp. 2429–2449, (1968).

Hartz, B. J., "Matrix Formulation of Structural Stability Problems," *Journal of the Structural Division, ASCE*, Vol. 91, No. ST6, Proc. Paper 4572, December, pp. 141–157, (1965).

Krahula, J. L., "Analysis of Bent and Twisted Bars Using Finite Elements Method," *AIAA Journal*, Vol. 5, No. 6, June, pp. 1194–1197, (1967).

Krajcinovic, D., "A Consistent Discrete Elements Technique for Thin-Walled Assemblages," *International Journal of Solids and Structures*, Vol. 5, No. 7, Pergamon Press, Oxford, England, pp. 639–662, (1969).

Lu, L. W. and Kamalvand, H., "Ultimate Strength of Laterally Loaded Columns," *Journal of the Structural Division, ASCE*, Vol. 94, No. ST6, Proc. Paper 6009, June, pp. 1505–1523, (1968).

Morino, S., "Analysis of Space Frames," thesis presented to Department of Civil Engineering, Lehigh University, Bethlehem, Pa., in partial fulfilment of the requirements for the degree of Doctor of Philosophy, (1970).

12

FINITE ELEMENT METHOD FOR PLASTIC BEAM-COLUMNS

S. Rajasekaran

12.1 INTRODUCTION

Numerical solutions for the governing differential equilibrium equations for elastic and plastic beam-columns have been achieved in Chap. 10 by the finite difference technique in which the differential equation is approximated by discrete values of the variables at selected points. On the other hand, finite-segment method is applied in Chap. 11 for the elastic-plastic beam-column by replacing it with an assembly of finite segments. Segment stiffness matrices have been evaluated using the exact displacement field of elastic beam-column along the length of the segment. In this chapter, the beam-column is physically replaced by an assembly of discrete elements and the element stiffness will be evaluated using the approximate displacement field along the element length. Hence one can consider the finite difference scheme described in Chap. 10 as point-wise approximation, and the finite segment method described in the preceding chapter and the finite element method described here as element-wise approximations.

Basic concept

The basic concept underlying the finite element method is that a real structure can be modelled analytically by its subdivisions into finite number of elements. In each of the finite element regions, the behavior of the element can be studied independently of the behavior of other elements in the ensemble by a set of assumed functions approximating the stresses or displacements in that region. The process of connecting elements together to form a complete model is purely a topological one and is independent of the physical nature of the problem. The set of functions for displacements (or stresses) are so chosen that they ensure continuity (or equilibrium) throughout the entire system. For the numerical solutions for practical problems using finite element method, it is required to formulate and solve a large system of linear algebraic equations.

Although finite element formulations can be based on either stress fields or displacement fields, most often displacement based finite element formulation is applied in practice since it can easily be programed for digital computer. Hence in this chapter, displacement model is used to arrive at the force displacement relationship for a beam-column element by considering the principle of virtual work. Correspondingly, the treatment for a complete structure is then accomplished with the use of stiffness equations of the element. The steps involved in the finite element method are

1. Discretization of the body into finite elements.
2. Evaluation of element stiffness by deriving nodal force-displacement relationship.
3. Assemblage of the stiffness and force matrices for the system of elements and nodes.
4. Introduction of boundary conditions.
5. Solution of the resulting equations for nodal displacements.
6. Calculation of strains and stresses based on nodal displacements.

The advantages of the finite element method are

1. The method is able to cope with irregularities in loading, material, geometry and complex boundary conditions. Sizes and shapes of the elements may be selected in such a way that highly irregular geometric forms may be approximated to an almost arbitrary degree of accuracy.
2. The method is suitable for automation of the equation formation process.

The limitations of the finite element method are

1. The method requires relatively large amounts of computer memory and time even for most efficient computer codes. Hence, it requires expensive computation time.
2. The engineer's judgement is of primary importance in interpreting the large volume of output generated by the finite element method.

12.2 FORMULATION OF BASIC EQUILIBRIUM EQUATIONS

In this section, total and incremental equilibrium equations are developed using the principle of virtual work. The principle of virtual work is less restrictive than a potential energy formulation as the equations developed by the principle of virtual work are valid for both elastic and plastic responses. For a beam-column element a relationship between the generalized stress and stress using equilibrium condition and a relationship between the generalized strain and strain using geometric condition along with kinematic assumption are established. The stresses and strains are substituted in terms of generalized stress and generalized strain in the virtual work equation to arrive at the total equilibrium equation for the beam-column element. Incremental equilibrium equation is then developed as explained in Sect. 12.2.2.

12.2.1 Total Equilibrium Equation

A thin-walled beam of open section is shown in Fig. 12.1(a). Points "C" and "O" represent an origin of coordinates and an arbitrary point, respectively. ξ, η and ζ, represent the coordinate axes in the right hand coordinate system having the origin at "C." The equation of virtual work for the thin-walled member neglecting body forces can be written as

$$\int_v (\sigma\,\delta\varepsilon + \tau_{\xi\zeta}\,\delta\gamma_{\xi\zeta} + \tau_{\eta\zeta}\,\delta\gamma_{\eta\zeta})\,dv = \int_s \sum_i t_i\,\delta u_i\,ds \tag{12.1}$$

since $\sigma_\xi = \sigma_\eta = \tau_{\xi\eta} = 0$ for a thin-walled beam. In Eq. (12.1), σ is the normal stress and $\tau_{\xi\zeta}$ and $\tau_{\eta\zeta}$ are components of shearing stress, $\tau_{t\zeta}$ as shown in Fig. 12.1(b) and $\delta\varepsilon$, $\delta\gamma_{\xi\zeta}$ and $\delta\gamma_{\eta\zeta}$ are corresponding virtual strains at an arbitrary point "N" on the

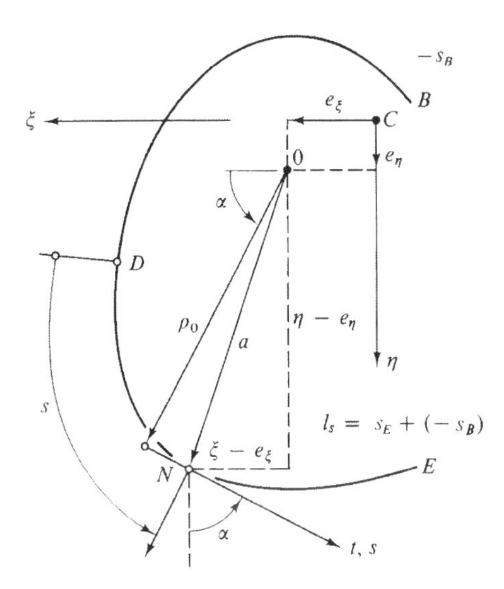

FIGURE 12.1(a)
Section geometry and coordinates

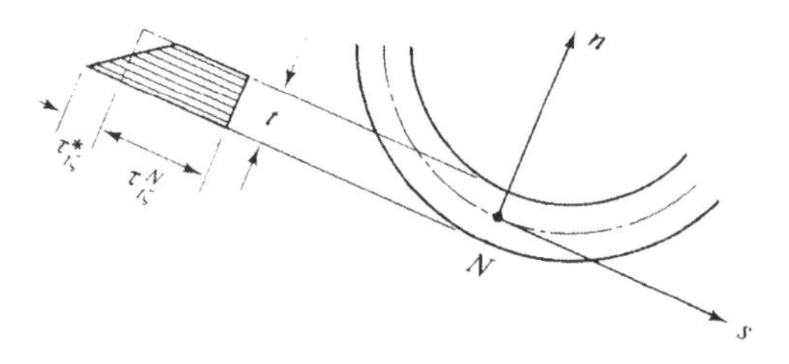

$\tau^N_{i\zeta}$ – Uniform warping shear

$\tau^*_{i\zeta}$ – St. Venant shear (max)

$$\tau_{i\zeta} = \tau^N_{i\zeta} - \tau^*_{i\zeta}\left(\frac{2n}{t}\right)$$

FIGURE 12.1(b)
Shear stresses in a thin-walled beam

contour of the thin-walled beam shown in Fig. 12.1(a). In Eq. (12.1), t_i and δu_i denote surface forces and corresponding virtual displacements respectively.

The resultant stress quantities $m_\xi, m_\eta, m_\omega, m_\zeta, f_\xi, f_\eta$ and f_ζ are considered positive in the directions indicated in Fig. 12.2. Here, for the convenience of present formulation, the positive bending moment m_η does not follow the usual right hand screw rule for moment. It is to be noted that the normal force f_ζ and bending moments m_ξ and m_η are reduced to "C" axis while twisting moments m_ζ and transverse shear

FIGURE 12.1(c)
Displacement of arbitrary point N

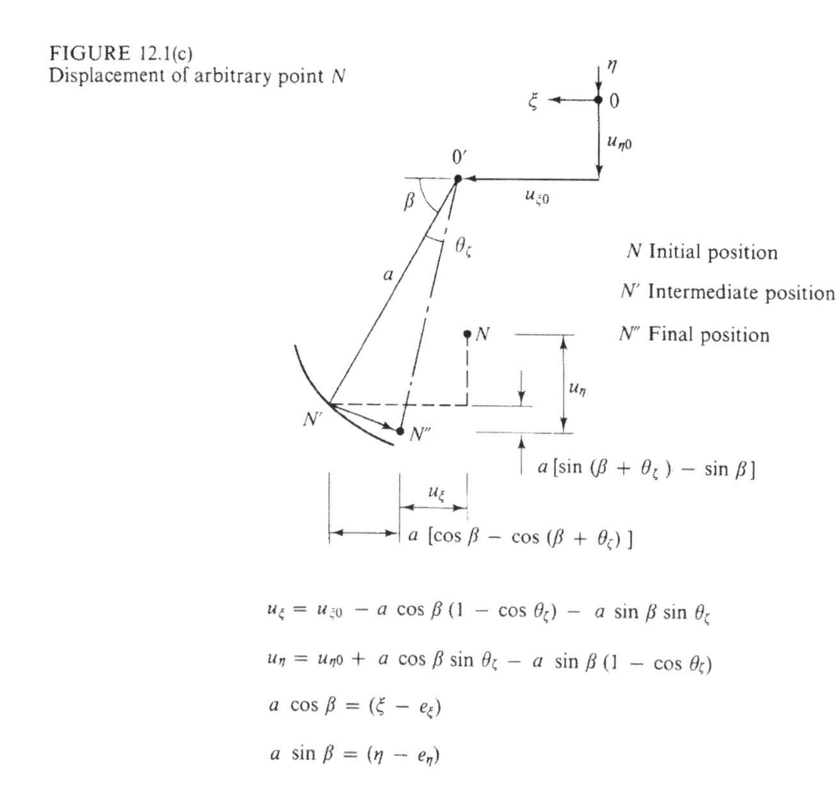

$$u_\xi = u_{\xi 0} - a\cos\beta\,(1 - \cos\theta_\zeta) - a\sin\beta\sin\theta_\zeta$$

$$u_\eta = u_{\eta 0} + a\cos\beta\sin\theta_\zeta - a\sin\beta\,(1 - \cos\theta_\zeta)$$

$$a\cos\beta = (\xi - e_\xi)$$

$$a\sin\beta = (\eta - e_\eta)$$

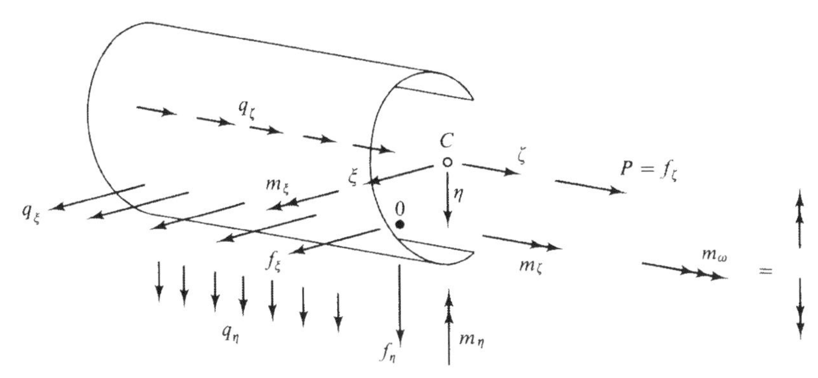

FIGURE 12.2
Distributed loading and conventional stress resultants

forces f_ξ and f_η are reduced to the "O" axis. One also notes that the transverse shear forces f_ξ and f_η, and twisting moment m_ζ, cannot be defined in terms of deformations as a result of imposed thin-walled beam assumptions about the deformation of the cross section to be described in what follows. They are not considered as *generalized stresses*. They are called *reactions* because they are only needed for equilibrium (see Chap. 1, Vol. 1). The transverse distributed loads q_ξ, q_η and q_ζ, and support reactions pass through "O" axis as shown in Fig. 12.2. The relationship between stress and stress resultant is established using the equations of equilibrium as

$$
\begin{aligned}
f_\xi &= \int_A \tau_{\xi\zeta}\, dA = \frac{dm_\eta}{d\zeta} = m'_\eta \\
f_\eta &= \int_A \tau_{\eta\zeta}\, dA = \frac{dm_\xi}{d\zeta} = m'_\xi \\
f_\zeta &= \int_A \sigma\, dA = P \\
m_\xi &= \int_A \sigma\eta\, dA \\
m_\eta &= \int_A \sigma\xi\, dA \\
m_\zeta &= \int_A \left[\tau^N_{\eta\zeta}(\xi - e_\xi) - \tau^N_{\xi\zeta}(\eta - e_\eta)\right] dA + T_{sv} = T_w + T_{sv} \\
m_\omega &= \int_A \sigma\omega^0_{Ds}\, dA
\end{aligned}
\tag{12.2}
$$

where m_ω is the bimoment having the dimension of force multiplied by area and $\omega^0_{Ds} = \omega_{nD}$ is known as the sectorial coordinate of the point "N" with respect to origin "C," pole "O," and sectorial origin "D" [see Fig. 12.1(a)] and is given by

$$\omega_{Ds}^{0} = \omega_{nD} = -\int_{D}^{s} \rho_0 \, ds \tag{12.3a}$$

where ρ_0 is the perpendicular distance from point O to the tangent at N, and s is positive in the direction indicated in Fig. 12.1(a). m_ζ is the total twisting moment being a vector sum of the St. Venant torsion and warping torsion $T_{sv} + T_w$ [see Eq. (2.131), Chap. 2]. The warping moment T_w is defined as the torque due to the uniform warping shear stress $\tau_{t\zeta}^{N}$ [Fig. 12.1(b)] along the midsurface contour and can be evaluated by Eq. (2.128b) or

$$T_w = -\frac{d}{d\zeta}\left(\int_A \sigma \omega_{Ds}^{0} \, dA\right) = -m'_\omega \tag{12.3b}$$

In the case of problems involving stability, changes in geometry due to deformations must be considered in the equilibrium equations. Hence, the nonlinear form of the strain-displacement relationship is adopted as

$$\begin{aligned} \varepsilon &= u_{\zeta,\zeta} + \frac{1}{2}(u_{\xi,\zeta}^2 + u_{\eta,\zeta}^2 + u_{\zeta,\zeta}^2) \\ \gamma_{\xi\zeta} &= u_{\zeta,\xi} + u_{\xi,\zeta} + (u_{\xi,\xi}u_{\xi,\zeta} + u_{\eta,\xi}u_{\eta,\zeta} + u_{\zeta,\xi}u_{\zeta,\zeta}) \\ \gamma_{\eta\zeta} &= u_{\zeta,\eta} + u_{\eta,\zeta} + (u_{\xi,\eta}u_{\xi,\zeta} + u_{\eta,\eta}u_{\eta,\zeta} + u_{\zeta,\eta}u_{\zeta,\zeta}) \end{aligned} \tag{12.4}$$

where u_ξ, u_η and u_ζ are the displacements of the arbitrary point "N" in the ξ, η and ζ directions, respectively, and $u_{\zeta,\zeta} = \partial u_\zeta/\partial_\zeta$, etc.

In order to establish a relationship between the displacement of a point on the middle surface of cross section to the displacement and rotation of the cross section, it is now necessary to make the assumptions (Vlasov, 1961) normally associated with the behavior of thin-walled beams which is a more general hypothesis governing the kinematics of the cross section than that of Bernoulli's hypothesis of plane sections for solid beams. The assumptions are:

(a) the shearing deformation in the mid-surface of the thin-walled plate is extremely small and can be neglected;

(b) the shape of the projected cross section is unaltered (rigid) during deformation.

We shall show in the following that the displacement of a point on the middle surface of the cross section can now be described by three displacement components $u_{\xi 0}$, $u_{\eta 0}$ and $u_{\zeta c}$, at the points "O" and "C" of the cross section, respectively, in the direction of ξ, η and ζ, and the angle of rotation θ_ζ about a longitudinal axis. The displacements u_ξ, u_η of an arbitrary point "N" may be written as [see Fig. 12.1(c)]

$$\begin{aligned} u_\xi &= u_{\xi 0} - (\eta - e_\eta)\sin\theta_\zeta - (\xi - e_\xi)(1 - \cos\theta_\zeta) \\ u_\eta &= u_{\eta 0} + (\xi - e_\xi)\sin\theta_\zeta - (\eta - e_\eta)(1 - \cos\theta_\zeta) \end{aligned} \tag{12.5a}$$

for analyzing thin-walled beams in finite deformation range. However, for "reasonably small" twisting rotations, i.e., ($\sin\theta_\zeta \approx \theta_\zeta$; $\cos\theta_\zeta \approx 1$), the displacements u_ξ and u_η of Eq. (12.5a) can be reduced to

$$u_\xi = u_{\xi 0} - (\eta - e_\eta)\theta_\zeta$$
$$u_\eta = u_{\eta 0} + (\xi - e_\xi)\theta_\zeta \tag{12.5b}$$

where $u_{\xi 0} = u_{\xi 0}(\zeta)$ and $u_{\eta 0} = u_{\eta 0}(\zeta)$ denote the displacements of the axis passing through "O" in ξ and η directions, respectively, and e_ξ, e_η locate "O" with respect to the origin "C" and $\theta_\zeta = \theta_\zeta(\zeta)$ is the rotation of the cross section about any longitudinal axis.

The displacement u_ζ at an arbitrary point "N" can be derived by noting that the tangential displacement of N in the direction of the mid-surface contour [coordinate t of Fig. 12.1(a)] may be written as

$$u_t = u_{\eta 0}\cos\alpha - u_{\xi 0}\sin\alpha + \rho_0\theta_\zeta \tag{12.6}$$

Since $u_\zeta(\xi, \eta, \zeta)$ represents the longitudinal displacement of point N, assumption (a) requires that

$$\frac{\partial u_t}{\partial \zeta} + \frac{\partial u_\zeta}{\partial s} = 0 \tag{12.7}$$

Substituting Eq. (12.6) into Eq. (12.7) yields

$$\frac{\partial u_\zeta}{\partial s} = -u'_{\eta 0}\cos\alpha + u'_{\xi 0}\sin\alpha - \rho_0\theta'_\zeta \tag{12.8}$$

in which primes denote differentiation with respect to ζ or $(\ )' = d(\ )/d\zeta$. Integrating Eq. (12.8) yields

$$u_\zeta\Big|_D^s = -u'_{\eta 0}\int_D^s \cos\alpha\, ds + u'_{\xi 0}\int_D^s \sin\alpha\, ds - \theta'_\zeta\int_D^s \rho_0\, ds \tag{12.9}$$

From Fig. 12.1(a), we have

$$\frac{d\xi}{ds} = -\sin\alpha$$
$$\frac{d\eta}{ds} = \cos\alpha \tag{12.10}$$

so that Eq. (12.9) becomes

$$u_\zeta(\xi, \eta, \zeta) = u_{\zeta D} - (\eta - \eta_D)u'_{\eta 0} - (\xi - \xi_D)u'_{\xi 0} + \omega^0_{Ds}\theta'_\zeta \tag{12.11}$$

where the term ω^0_{Ds} defined in Eq. (12.3a) gives rise to a longitudinal "warping" of the midsection of the contour. Although "C" is not on the contour, the longitudinal displacement at "C" may be obtained by visualizing a connection to it from any point on the contour and by substituting the coordinates in Eq. (12.11) to those of the origin. This results

$$u_{\zeta c} = u_{\zeta D} + \eta_D u'_{\eta 0} + \xi_D u'_{\xi 0} + \omega^0_{Dc}\theta'_\zeta \tag{12.12a}$$

using Eq. (12.12a), Eq. (12.11) can now be written in the form

$$u_\zeta = u_{\zeta c} - \eta u'_{\eta 0} - \xi u'_{\xi 0} + (\omega^0_{Ds} - \omega^0_{Dc})\theta'_\zeta \tag{12.12b}$$

The imaginary path connecting D to C may always be chosen such that ω^0_{Dc} vanishes and hence

$$u_\zeta = u_{\zeta c} - \eta u'_{\eta 0} - \xi u'_{\xi 0} + \omega^0_{Ds}\theta'_\zeta \tag{12.13}$$

Equations (12.5) and (12.13) give the displacements of an arbitrary point N on the section contour in terms of $u_{\xi 0}, u_{\eta 0}, u_{\zeta c}$ and θ_ζ or their derivatives, for any location of O, C, and D consistent with the assumptions of thin-walled beam theory.

Substituting nonlinear strain-displacement relationship, i.e., Eq. (12.4) in Eq. (12.1), the virtual work equation reduces to

$$\begin{aligned}\int_v \{&[\sigma\delta(u_{\zeta,\zeta}) + \tau_{\xi\zeta}\delta(u_{\zeta,\xi} + u_{\xi,\zeta}) + \tau_{\eta\zeta}\delta(u_{\zeta,\eta} + u_{\eta,\zeta})] \\ &+ [\sigma(u_{\xi,\zeta}\delta u_{\xi,\zeta} + u_{\eta,\zeta}\delta u_{\eta,\zeta}) + \tau_{\xi\zeta}\delta(u_{\xi,\xi}u_{\xi,\zeta} + u_{\eta,\xi}u_{\eta,\zeta}) \\ &+ \tau_{\eta\zeta}\delta(u_{\xi,\eta}u_{\xi,\zeta} + u_{\eta,\eta}u_{\eta,\zeta})]\}\, dV = \int_s \sum_i t_i \delta u_i \, ds\end{aligned} \tag{12.14}$$

In arriving at Eq. (12.14), the terms containing the products of derivatives of u_ζ are neglected.

Now, apply the virtual work equation, Eq. (12.14), to the beam-column segment, a variational form of the total equilibrium equation is obtained by substituting the displacements u_ξ, u_η and u_ζ in Eqs. (12.5) and (12.13) into Eq. (12.14) using Eqs. (12.2) for stress resultants as

$$\begin{aligned}&\int_0^l [P\,\delta u'_{\zeta c} - m_\xi\,\delta u''_{\eta 0} - m_\eta\,\delta u''_{\xi 0} + m_\omega\,\delta\theta''_\zeta + T_{sv}\,\delta\theta'_\zeta]\, d\zeta \\ &+ \int_0^l \Big[f_\xi\,\delta(\theta_\zeta u'_{\eta 0}) - f_\eta\,\delta(\theta_\zeta u'_{\xi 0}) + \frac{P}{2}\delta(u'^2_{\xi 0} + u'^2_{\eta 0} + 2e_\eta u'_{\xi 0}\theta'_\zeta - 2e_\xi u'_{\eta 0}\theta'_\zeta) \\ &+ m_\eta\,\delta(u'_{\eta 0}\theta'_\zeta) - m_\xi\,\delta(u'_{\xi 0}\theta'_\zeta) + \frac{\bar{K}}{2}\delta(\theta'_\zeta)^2 \Big] d\zeta \\ &= \int_0^l (q_\xi\,\delta u_{\xi 0} + q_\eta\,\delta u_{\eta 0} + q_\zeta\,\delta u_{\zeta c})\, d\zeta \\ &+ \Big[P^*\,\delta u_{\zeta c} + f^*_\xi\,\delta u_{\xi 0} + f^*_\eta\,\delta u_{\eta 0} - m^*_\xi\,\delta u'_{\eta 0} - m^*_\eta\,\delta u'_{\xi 0} + m^*_\zeta\,\delta\theta_\zeta + m^*_\omega\,\delta\theta'_\zeta \Big]_0^l\end{aligned} \tag{12.15a}$$

in which

$$\bar{K} = \int_A \sigma a^2\, dA \tag{12.15b}$$

Table 12.1 NATURAL BOUNDARY CONDITIONS

Displacement Variable	Associated Generalized Force (Natural Boundary Condition)
$\delta u_{\zeta c}$	P
$\delta u_{\xi 0}$	$f_\xi + \{-f_\eta \theta_\zeta + Pe_\eta \theta'_\zeta - m_\xi \theta'_\zeta + Pu'_{\xi 0}\}$
$\delta u_{\eta 0}$	$f_\eta + \{f_\xi \theta_\zeta - Pe_\xi \theta'_\zeta + m_\eta \theta'_\zeta + Pu'_{\eta 0}\}$
$\delta \theta_\eta = \delta u'_{\xi 0}$	$-m_\eta$
$\delta \theta_\xi = \delta u'_{\eta 0}$	$-m_\xi$
$\delta \theta_\zeta$	$T_{sv} - m'_\omega + \{Pe_\eta u'_{\xi 0} - Pe_\xi u'_{\eta 0} + m_\eta u'_{\eta 0} - m_\xi u'_{\xi 0} + \bar{K}\theta'_\zeta\}$
$\delta \theta'_\zeta$	m_ω

In Eq. (12.15a), $u''_{\xi 0}$ and $u''_{\eta 0}$ are the bending curvatures of the cross section about η and ξ axes passing through point "O." These curvatures are denoted in the previous chapters by Φ_y and Φ_x.

The term a is the distance of point "N" from "O." Equation (12.15a) is a variational form of virtual work total equilibrium equation which is valid for arbitrary reference axes "O" and "C" and arbitrary sectorial origin D. It is approximate only to the extent imposed by the assumptions of thin-walled beam theory and the higher order terms neglected. Since no assumptions have been made with respect to the origin of stress, it is equally valid for both elastic and plastic analyses. The stress resultants contained in the last brackets of Eq. (12.15a) are the specified values of resultant stress at the ends of the members.

Integrating Eq. (12.15a) by parts yields the differential equilibrium equations as the Euler-Lagrange equations of the functional, namely

$$
\begin{aligned}
P' &= -q_\zeta \\
m''_\eta + [-(m_\xi \theta_\zeta)'' + (Pu'_{\xi 0})' + (Pe_\eta \theta'_\zeta)'] &= -q_\xi \\
m''_\xi + [(m_\eta \theta_\zeta)'' + (Pu'_{\eta 0})' - (Pe_\xi \theta'_\zeta)'] &= -q_\eta \qquad (12.16) \\
m''_\omega - T'_{sv} + [m_\xi u''_{\xi 0} - m_\eta u''_{\eta 0} - (\bar{K}\theta'_\zeta)' - (Pe_\eta u'_{\xi 0})' + (Pe_\xi u'_{\eta 0})'] &= 0
\end{aligned}
$$

and the associated boundary conditions are given in Table 12.1. In arriving at Eq. (12.16) it is necessary to relate the transverse shearing force f_ξ and f_η to the derivative of moments.

12.2.2 Incremental Equilibrium Equation

Referring to Fig. 12.3, let A represent the deformed equilibrium configuration (the initial configuration) of the beam-column under surface loads t, and B represent the deformed configuration under the loads t. Let σ, ε and u denote stress, strain and displacements, respectively, in configuration A and $\bar{\sigma}$, $\bar{\varepsilon}$ and $\bar{u}$ the corresponding quantities in configuration B. $\dot{\sigma}$, $\dot{u}$ and $\dot{t}$ denote the increments in stress, displacements and surface loads, respectively. Let

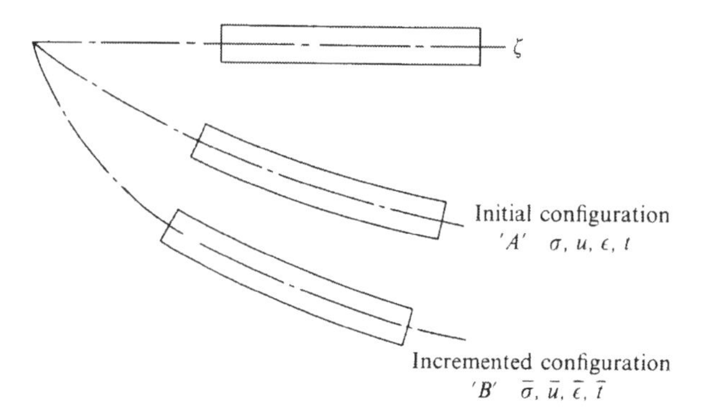

FIGURE 12.3
Stress strain and displacement nomenclature in initial and incremented configuration

$$
\begin{aligned}
\bar{\sigma} &= \sigma + \dot{\sigma} \\
\bar{\tau}_{\eta\zeta} &= \tau_{\eta\zeta} + \dot{\tau}_{\eta\zeta} \\
\bar{\tau}_{\xi\zeta} &= \tau_{\xi\zeta} + \dot{\tau}_{\xi\zeta} \\
\bar{u}_{\xi} &= u_{\xi} + \dot{u}_{\xi} \\
\bar{u}_{\eta} &= u_{\eta} + \dot{u}_{\eta} \\
\bar{u}_{\zeta} &= u_{\zeta} + \dot{u}_{\zeta} \\
\bar{t} &= t + \dot{t}
\end{aligned}
\tag{12.17}
$$

The principle of virtual work applied as before to the B configuration, regarding strain and displacement increments as variables yield

$$
\int_v [(\sigma + \dot{\sigma})\,\delta\bar{\varepsilon} + (\tau_{\xi\zeta} + \dot{\tau}_{\xi\zeta})\,\delta\bar{\gamma}_{\xi\zeta} + (\tau_{\eta\zeta} + \dot{\tau}_{\eta\zeta})\,\delta\bar{\gamma}_{\eta\zeta}]\,dv = \int_s \sum_i (t_i + \dot{t}_i)\,\delta\bar{u}_i\,ds \tag{12.18}
$$

In the above equation, virtual strains $\delta\bar{\varepsilon}$, $\delta\bar{\gamma}_{\xi\zeta}$ and $\delta\bar{\gamma}_{\eta\zeta}$ may be substituted in terms of the virtual displacements using the nonlinear strain-displacement relationship of Eq. (12.4). The equilibrium equation in configuration B is

$$
\begin{aligned}
\int_v \Big\{ &(\sigma + \dot{\sigma})\left[\delta\bar{u}_{\zeta,\zeta} + \frac{1}{2}\delta(\bar{u}^2_{\xi,\zeta} + \bar{u}^2_{\eta,\zeta} + \bar{u}^2_{\zeta,\zeta})\right] \\
&+ (\tau_{\xi\zeta} + \dot{\tau}_{\xi\zeta})[\delta\bar{u}_{\zeta,\xi} + \delta\bar{u}_{\xi,\zeta} + \delta(\bar{u}_{\xi,\xi}\bar{u}_{\xi,\zeta} + \bar{u}_{\eta,\xi}\bar{u}_{\eta,\zeta} + \bar{u}_{\zeta,\xi}\bar{u}_{\zeta,\zeta})] \\
&+ (\tau_{\eta\zeta} + \dot{\tau}_{\eta\zeta})[\delta\bar{u}_{\zeta,\eta} + \delta\bar{u}_{\eta,\zeta} + \delta(\bar{u}_{\xi,\eta}\bar{u}_{\xi,\zeta} + \bar{u}_{\eta,\eta}\bar{u}_{\eta,\zeta} + \bar{u}_{\zeta,\eta}\bar{u}_{\zeta,\zeta})]\Big\}\,dv \\
= \int_s &\sum_i (t_i + \dot{t}_i)\,\delta u_i\,ds
\end{aligned}
\tag{12.19}
$$

It is to be noted that when taking variation in displacements in configuration B, u_i are treated as constants. Thus $\delta \bar{u}_i = \delta \dot{u}_i$ or $\delta \bar{u}_{\zeta,\xi} = \delta \dot{u}_{\zeta,\xi}$.

The equilibrium equation in configuration A is obtained by letting $\dot{\sigma} = \dot{\tau}_{\xi\zeta} = \dot{\tau}_{\eta\zeta} = \dot{t} = \dot{u}_\xi = \dot{u}_\eta = \dot{u}_\zeta = 0$ in Eq. (12.19) which results in

$$\int_v [\sigma(\delta\dot{u}_{\zeta,\zeta} + u_{\xi,\zeta}\,\delta\dot{u}_{\xi,\zeta} + u_{\eta,\zeta}\,\delta\dot{u}_{\eta,\zeta}) + \tau_{\xi\zeta}(\delta\dot{u}_{\xi,\zeta} + \delta\dot{u}_{\zeta,\xi} + u_{\xi,\xi}\,\delta\dot{u}_{\xi,\zeta} + u_{\xi,\zeta}\,\delta\dot{u}_{\xi,\xi}$$
$$+ u_{\eta,\xi}\,\delta\dot{u}_{\eta,\zeta} + u_{\eta,\zeta}\,\delta\dot{u}_{\eta,\xi}) + \tau_{\eta\zeta}(\delta\dot{u}_{\zeta,\eta} + \delta\dot{u}_{\eta,\zeta} + u_{\xi,\eta}\,\delta\dot{u}_{\xi,\zeta} + u_{\xi,\zeta}\,\delta\dot{u}_{\xi,\eta} + u_{\eta,\eta}\,\delta\dot{u}_{\eta,\zeta}$$
$$+ u_{\eta,\zeta}\,\delta\dot{u}_{\eta,\eta})]\,dv = \int_s \sum_i t_i\,\delta u_i\,ds \tag{12.20}$$

The incremental equilibrium equation can be obtained using Biot's (1965) procedure by finding the difference between the equilibrium equations when the configuration changes from A to B. Neglecting higher order terms, approximate incremental equilibrium equation is obtained by subtracting Eq. (12.20) from Eq. (12.19) and written as

$$\int_v \{[\dot{\sigma}\,\delta(\dot{u}_{\zeta,\zeta}) + \dot{\tau}_{\xi\zeta}\,\delta(\dot{u}_{\zeta,\xi} + \dot{u}_{\xi,\zeta}) + \dot{\tau}_{\eta\zeta}\,\delta(\dot{u}_{\zeta,\eta} + \dot{u}_{\eta,\zeta})]$$
$$+ [\sigma(\dot{u}_{\xi,\zeta}\,\delta\dot{u}_{\xi,\zeta} + \dot{u}_{\eta,\zeta}\,\delta\dot{u}_{\eta,\zeta}) + \tau_{\xi\zeta}\,\delta(\dot{u}_{\xi,\xi}\dot{u}_{\xi,\zeta} + \dot{u}_{\eta,\xi}\dot{u}_{\eta,\zeta}) \tag{12.21}$$
$$+ \tau_{\eta\zeta}\,\delta(\dot{u}_{\xi,\eta}\dot{u}_{\xi,\zeta} + \dot{u}_{\eta,\eta}\dot{u}_{\eta,\zeta})]\}\,dv = \int_s \sum_i \dot{t}_i\,\delta u_i\,ds$$

The approximate incremental equilibrium equation in terms of stress resultants and displacements may be obtained in a similar manner as that of the total equilibrium Eq. (12.15a) by substituting the displacements u_ξ, u_η and u_ζ in Eqs. (12.5) and (12.13) into Eq. (12.21) using Eqs. (12.2) for stress resultants. The incremental equilibrium equation has the form

$$\int_0^l [\dot{P}\,\delta\dot{u}'_{\zeta c} - \dot{m}_\xi\,\delta\dot{u}''_{\eta 0} - \dot{m}_\eta\,\delta\dot{u}''_{\xi 0} + \dot{m}_\omega\,\delta\dot{\theta}''_\zeta + \dot{T}_{sv}\,\delta\dot{\theta}'_\zeta]\,d\zeta$$
$$+ \int_0^l \Bigg[f_\xi\,\delta(\dot{\theta}_\zeta\dot{u}'_{\eta 0}) - f_\eta\,\delta(\dot{\theta}_\zeta\dot{u}'_{\xi 0}) + \frac{P}{2}\,\delta(\dot{u}'^2_{\xi 0} + \dot{u}'^2_{\eta 0} + 2e_\eta\dot{u}'_{\xi 0}\dot{\theta}'_\zeta - 2e_\xi\dot{u}'_{\eta 0}\dot{\theta}'_\zeta)$$
$$+ m_\eta\,\delta(\dot{u}'_{\eta 0}\dot{\theta}'_\zeta) - m_\xi\,\delta(\dot{u}'_{\xi 0}\dot{\theta}'_\zeta) + \frac{\bar{K}}{2}\,\delta(\dot{\theta}'_\zeta)^2\Bigg]\,d\zeta \tag{12.22}$$
$$= \int_0^l (\dot{q}_\xi\,\delta\dot{u}_{\xi 0} + \dot{q}_\eta\,\delta\dot{u}_{\eta 0} + \dot{q}_\zeta\,\delta\dot{u}_{\zeta c})\,d\zeta$$
$$+ \Big[\dot{P}^*\,\delta\dot{u}_{\zeta c} + \dot{f}^*_\xi\delta\dot{u}_{\xi 0} + \dot{f}^*_\eta\,\delta\dot{u}_{\eta 0} - \dot{m}^*_\xi\,\delta\dot{u}'_{\eta 0} - \dot{m}^*_\eta\,\delta\dot{u}'_{\xi 0} + \dot{m}^*_\zeta\,\delta\dot{\theta}_\zeta + \dot{m}^*_\omega\,\delta\dot{\theta}'_\zeta\Big]_0^l$$

where $\dot{P}$, $\dot{m}_\xi$, $\dot{m}_\eta$ and $\dot{m}_\omega$ are increments in axial force, bending moments and bimoment which are considered here as the generalized stresses of the cross section. $\dot{T}_{sv}$ increment in St. Venant torsional moment, and $\dot{u}_{\xi 0}$, $\dot{u}_{\eta 0}$, $\dot{u}_{\zeta c}$ and $\dot{\theta}_\zeta$ are increments in displacements and rotation of the cross section at the reference point or axis. Equation (12.22) is also applicable for both elastic and plastic analysis for a thin-walled beam with arbitrary reference axis O and C and arbitrary sectorial origin D.

12.3 GENERALIZED STRESS-STRAIN RELATIONSHIPS

For the case of thin-walled beams as it has been discussed in Chap. 11, only normal stress contributes to the yielding of the material and the tangent modulus E_t at a point on the section is uniquely determined by the total strain ε and the strain rate $\dot{\varepsilon}$. Since plastic behavior is load path dependent, it usually requires step-by-step calculations that follow the history of loading. The relationship between the total strain ε and the generalized strains $u'_{\zeta c}$, $u''_{\eta 0}$, $u''_{\xi 0}$ and θ''_ζ is from Eq. (12.13) (neglecting higher order terms), or,

$$\varepsilon = u'_{\zeta c} - \eta u''_{\eta 0} - \xi u''_{\xi 0} + \omega^0_{Ds}\theta''_\zeta \tag{12.23}$$

The stress increment $\dot{\sigma}$ in terms of strain increment for any thin-walled beam of elastic-plastic material is (see Sec. 12.7.1)

$$\dot{\sigma} = E_t\dot{\varepsilon} \tag{12.24a}$$

$$\dot{\varepsilon} = \dot{u}'_{\zeta c} - \eta\dot{u}''_{\eta 0} - \xi\dot{u}''_{\xi 0} + \omega^0_{Ds}\dot{\theta}''_\zeta \tag{12.24b}$$

Substituting Eq. (12.24) in Eq. (12.2), the generalized stress increments $\dot{P}$, $\dot{m}_\xi$, $\dot{m}_\eta$ and $\dot{m}_\omega$ for the cross section of a thin-walled beam can be expressed in terms of generalized strain increments $\dot{u}'_{\zeta c}$, $\dot{u}''_{\eta 0}$, $\dot{u}''_{\xi 0}$ and $\dot{\theta}''_\zeta$ about the reference axes as

$$\begin{Bmatrix} \dot{P} \\ \dot{m}_\eta \\ \dot{m}_\xi \\ \dot{m}_\omega \end{Bmatrix} = \begin{bmatrix} \int_A E_t\,dA & \int_A E_t\xi\,dA & \int_A E_t\eta\,dA & \int_A E_t\omega^0_{Ds}\,dA \\ & \int_A E_t\xi^2\,dA & \int_A E_t\xi\eta\,dA & \int_A E_t\omega^0_{Ds}\xi\,dA \\ & & \int_A E_t\eta^2\,dA & \int_A E_t\omega^0_{Ds}\eta\,dA \\ \text{SYMMETRIC} & & & \int_A E_t\omega^{02}_{Ds}\,dA \end{bmatrix} \begin{Bmatrix} \dot{u}'_{\zeta c} \\ -\dot{u}''_{\xi 0} \\ -\dot{u}''_{\eta 0} \\ \dot{\theta}''_\zeta \end{Bmatrix} \tag{12.25}$$

and St. Venant's torsional moment has the value

$$\dot{T}_{sv} = GK_T\dot{\theta}'_\zeta \tag{12.26}$$

where G is taken as the shear modulus in the elastic regime and K_T is the St. Venant torsional constant defined by

$$K_T \approx \frac{1}{3}\sum_i \frac{E_t}{E} b_i t_i^3 \tag{12.27}$$

where b_i and t_i are the breadth and thickness of the i-th plate.

It is to be noted that the generalized stress-strain relationship presented in Chap. 6 for the plastic response of metal column segments subjected to varying loading histories by considering the elastic unloading of yielded fibers, is of the same form as Eq. (12.25). However, in the present chapter, the elastic unloading of yielded fibers will be ignored and the error from this assumption is minimized if the loads are increased monotonically.

12.4 FINITE ELEMENT MODEL FOR ELASTIC RESPONSE

The analysis of thin-walled structural system was first approached using the method of transfer matrices by Vlasov (1961). Force method was applied for the analysis of thin-walled beams by Bazant (1965) and Przemieniecki (1968). These methods are, however, found to be inferior to the displacement method. The first published work of Gallagher and Padlog (1963) for the beam-column element is significant. This has enabled subsequent application of the finite element method to the thin-walled beams by Arghyris (1964), Krahula (1967) and Renton (1967).

Consider the prismatic element shown in Fig. 12.4. The element undergoes

FIGURE 12.4
Finite element nodal displacements and stress resultants

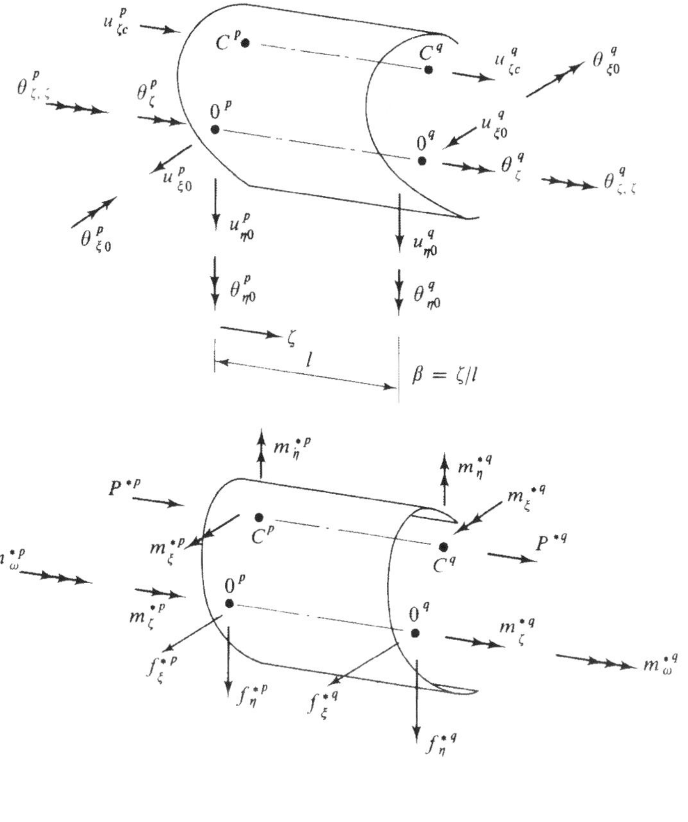

axial, flexural and torsional displacements under the action of joint forces P^*, f_ξ^*, f_η^*, m_ξ^*, m_η^*, m_ζ^* and m_ω^*. For the elastic case the tangent modulus E_t of the material is equal to E, i.e.,

$$E_t = E \tag{12.28}$$

Choosing the coordinates ξ and η as the principal coordinates of the cross section and selecting C and O as centroid and shear center and D as sectorial pole, the off-diagonal terms in the coefficient matrix of the generalized stress-strain relationship (12.25) obviously vanish. Hence the generalized stresses P, m_ξ, m_η and m_ω can be written as

$$\begin{aligned}
P &= E\left(\int_A dA\right)u'_{\zeta c} = EAu'_{\zeta c} \\
m_\xi &= -E\left(\int_A \eta^2\, dA\right)u''_{\eta 0} = -EI_\xi u''_{\eta 0} \\
m_\eta &= -E\left(\int_A \xi^2\, dA\right)u''_{\xi 0} = -EI_\eta u''_{\xi 0} \\
m_\omega &= E\left(\int_A \omega_{Ds}^{0^2}\, dA\right)\theta''_\zeta = EI_\omega \theta''_\zeta
\end{aligned} \tag{12.29}$$

and
$$T_{sv} = G\left(\frac{1}{3}\sum_i b_i t_i^3\right)\theta'_\zeta = GK_T\theta'_\zeta$$

where A is the area of the cross section and I_ξ and I_η are principal moments of inertia and I_ω is the sectorial or warping moment of inertia. They are defined in Eq. (12.29). Substituting the generalized stress-strain relations of Eq. (12.29) in Eq. (12.15a), the virtual work total equilibrium equation for elastic case is obtained as

$$\begin{aligned}
&\int_0^l [EAu'_{\zeta c}\,\delta u'_{\zeta c} + EI_\xi u''_{\eta 0}\,\delta u''_{\eta 0} + EI_\eta u''_{\xi 0}\,\delta u''_{\xi 0} + EI_\omega \theta''_\zeta\,\delta\theta''_\zeta + GK_T\theta'_\zeta\,\delta\theta'_\zeta]\,d\zeta \\
&\quad + \int_0^l \Bigg[f_\xi\,\delta(\theta_\zeta u'_{\eta 0}) - f_\eta\,\delta(\theta_\zeta u'_{\xi 0}) + \frac{P}{2}\,\delta(u'^2_{\xi 0} + u'^2_{\eta 0} + 2e_\eta u'_{\xi 0}\theta'_\zeta - 2e_\xi u'_{\eta 0}\theta'_\zeta) \\
&\quad + m_\eta\,\delta(u'_{\eta 0}\theta'_\zeta) - m_\xi\,\delta(u'_{\xi 0}\theta'_\zeta) + \frac{\bar{K}}{2}\,\delta(\theta'_\zeta)^2 \Bigg] d\zeta \\
&= \int_0^l (q_\xi\,\delta u_{\xi 0} + q_\eta\,\delta u_{\eta 0} + q_\zeta\,\delta u_{\zeta c})\,d\zeta + \Big[P^*\,\delta u_{\zeta c} + f_\xi^*\,\delta u_{\xi 0} + f_\eta^*\,\delta u_{\eta 0} - m_\xi^*\,\delta u'_{\eta 0} \\
&\quad - m_\eta^*\,\delta u'_{\xi 0} + m_\zeta^*\,\delta\theta_\zeta + m_\omega^*\,\delta\theta'_\zeta \Big]_0^l
\end{aligned} \tag{12.30}$$

The above equation states that δU (internal virtual work) $= \delta W_e$ (external virtual work) or

$$\delta\Pi = 0 \tag{12.31a}$$

where Π is known as total potential given by

$$\Pi = U - W_e \tag{12.31b}$$

In order to establish the force displacement relationship for the beam-column element, the continuous displacement in Eq. (12.30) such as $u_{\zeta c}$, $u_{\xi 0}$, $u_{\eta 0}$ and θ_ζ are to be written in terms of nodal displacements at the ends and the integration is carried out throughout the length of the element. This will be described later.

An alternative way of representing Eq. (12.31a) is

$$\delta\Pi = \frac{\partial \Pi}{\partial\{\underline{u}\}}\,\delta\{\underline{u}\} = 0 \tag{12.32a}$$

Since $\delta\{\underline{u}\}$ is arbitrary,

$$\frac{\partial \Pi}{\partial\{\underline{u}\}} = 0 \tag{12.32b}$$

where u is the nodal displacement. Hence the principle of virtual displacement is the same as the statement that the first variation of total potential energy of the system is zero for any virtual displacement consistent with the constraints.

12.4.1 Choice of the Displacement Function

For any acceptable numerical formulation, the numerical solution must converge or tend to the exact solution of the problem, if the subdivision of the beam-column is made finer. It has been shown for the finite element formulation that under certain circumstances the displacement formulation provides an upper bound to the true stiffness of the beam-column and hence such a finite element formulation will converge to the exact displacement solution from below. In order for this convergence to be rigorously assured, three conditions must be met for the choice of the displacement function [Zienkiewicz (1971)].

(a) The displacement function must be continuous within the element and the displacement must be compatible between adjacent elements. This is ensured if the displacement field be continuous on the inter-element boundary up to the derivative of one order less than the highest derivative appearing in strain-displacement relation. This condition in the finite element literature is known as *compatible* or *conforming condition.*

(b) The displacement function must include rigid body displacements of the element.

(c) The displacement function must include constant strain state of the element.

The last two conditions are known as *completeness criteria.* To satisfy the completeness criteria, the displacement expansion must be at least a complete

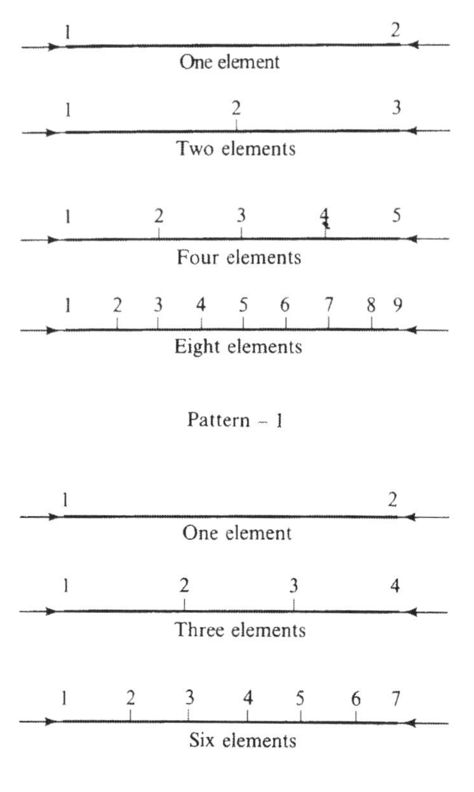

FIGURE 12.5
Discretization for a minimizing sequence

polynomial of order equal to the highest derivative occurring in the strain-displacement relationship. Although all three conditions must be met to ensure convergence, sometimes nonconformable elements (which relax the conforming condition) converge more quickly only if they pass Iron's patch test. For more details the reader may refer to Strang and Fix (1973).

It has been proved that the finite element solution gives an upper bound for the total potential energy when the inter-element compatibility is satisfied and the solution will converge to the true solution, provided all uniform strain states can be represented by the expansion. The solution will converge monotonically only if the subdivisions comprise a minimizing sequence, i.e., nth discretization must contain all the previous nodes and the displacement expansions (Brebbia and Connor, 1973). As an example for a beam-column element, discretization shown in Fig. 12.5 is a minimizing sequence.

For the beam-column element the strain displacement relationship contains second derivatives in lateral displacements and twist and the first derivative in the axial displacement. Hence it is necessary to choose the displacement function such that $u_{\zeta c}$, $u_{\xi 0}$, $\theta_{\eta 0} = u'_{\xi 0}$, $u_{\eta 0}$, $\theta_{\xi 0} = u'_{\eta 0}$ and θ_{ζ}, θ'_{ζ} must be continuous at the nodes. This can be achieved by adopting a linear displacement field for $u_{\zeta c}$ and cubic displacement field for other degrees of freedom. Using the notation $\langle\ \rangle$ for a row vector and $\{\ \}$ for a column vector, $u_{\xi 0}$ is written as

$$u_{\xi 0} = \langle 1 \ \ \zeta \ \ \zeta^2 \ \ \zeta^3 \rangle \begin{Bmatrix} \alpha_1 \\ \alpha_2 \\ \alpha_3 \\ \alpha_4 \end{Bmatrix} \tag{12.33}$$

$$\theta_{\eta 0} = \frac{du_{\xi 0}}{d\zeta} = \langle 0 \ \ 1 \ \ 2\zeta \ \ 3\zeta^2 \rangle \begin{Bmatrix} \alpha_1 \\ \alpha_2 \\ \alpha_3 \\ \alpha_4 \end{Bmatrix} \tag{12.34}$$

Substituting the corresponding values of ζ at the ends p and q of the beam element shown in Fig. 12.4, we get

$$\begin{Bmatrix} u^p_{\xi 0} \\ \theta^p_{\eta 0} \\ u^q_{\xi 0} \\ \theta^q_{\eta 0} \end{Bmatrix} = \begin{bmatrix} 1 & 0 & 0 & 0 \\ 0 & 1 & 0 & 0 \\ 1 & l & l^2 & l^3 \\ 0 & 1 & 2l & 3l^2 \end{bmatrix} \begin{Bmatrix} \alpha_1 \\ \alpha_2 \\ \alpha_3 \\ \alpha_4 \end{Bmatrix} = [A_\alpha]\{\alpha\} \tag{12.35}$$

Hence

$$u_{\xi 0} = \langle 1 \ \ \zeta \ \ \zeta^2 \ \ \zeta^3 \rangle [A_\alpha]^{-1} \begin{Bmatrix} u^p_{\xi 0} \\ \theta^p_{\eta 0} \\ u^q_{\xi 0} \\ \theta^q_{\eta 0} \end{Bmatrix} \tag{12.36}$$

or

$$u_{\xi 0} = \langle n_3 \rangle \{\underline{u}_{\xi 0}\} \tag{12.37}$$

where n_3 is the cubic *interpolation function* given by

$$\langle n_3 \rangle = \langle (1 - 3\beta^2 + 2\beta^3), (\beta - 2\beta^2 + \beta^3), (3\beta^2 - 2\beta^3), (\beta^3 - \beta^2) \rangle \tag{12.38}$$

and $\{\underline{u}_{\xi 0}\}$ the nodal displacement as

$$\langle \underline{u}_{\xi 0} \rangle = \langle u^p_{\xi 0}, l\theta^p_{\eta 0}, u^q_{\xi 0}, l\theta^q_{\eta 0} \rangle \tag{12.39}$$

where

$$\beta = \zeta / l \tag{12.40}$$

Derivations for other quantities $u_{\eta 0}$, $u_{\zeta c}$ and θ_ζ can be made in a similar manner.

The displacement function expressed in Eq. (12.38) is the exact displacement field for a flexural element without axial load. It is of interest to note that Krajcinovic (1969) developed a finite element formulation for thin-walled members based on the use of hyperbolic functions to represent twist θ_ζ. Barsoum and Gallagher (1970), using the principle of minimum potential energy developed a finite element formulation for the linear stability analysis of structural systems subjected to conservative and nonconservative forces using both cubic and fifth degree polynomial. Similar treatment to the linear stability of thin-walled beams by finite element analysis using different formulations was carried out by Powell and Klinger (1970), Mei (1970), Rajasekaran and Murray (1971) and Tebedge and Tall (1973). However, Rajasekaran and Murray (1973) have extended finite element method to plastic member buckling and coupled local and member buckling problems.

In summary, the displacement in an element is represented by the nodal displacements at the ends as

$$\begin{aligned} u_{\xi 0} &= \langle n_3 \rangle \{\underline{u}_{\xi 0}\} \\ u_{\eta 0} &= \langle n_3 \rangle \{\underline{u}_{\eta 0}\} \\ u_{\zeta c} &= \langle n_1 \rangle \{\underline{u}_{\zeta c}\} \\ \theta_\zeta &= \langle n_3 \rangle \{\underline{\theta}_\zeta\} \end{aligned} \tag{12.41}$$

where

$$\langle n_1 \rangle = \langle (1-\beta), \beta \rangle$$

and the nodal displacements of the element at the ends p and q are given by

$$\begin{aligned} \langle \underline{u}_{\xi 0} \rangle &= \langle u^p_{\xi 0}, l\theta^p_{\eta 0}, u^q_{\xi 0}, l\theta^q_{\eta 0} \rangle \\ \langle \underline{u}_{\eta 0} \rangle &= \langle u^p_{\eta 0}, l\theta^p_{\xi 0}, u^q_{\eta 0}, l\theta^q_{\xi 0} \rangle \\ \langle \underline{\theta}_\zeta \rangle &= \langle \theta^p_\zeta, l\theta^p_{\zeta,\zeta}, \theta^q_\zeta, l\theta^q_{\zeta,\zeta} \rangle \\ \langle \underline{u}_{\zeta c} \rangle &= \langle u^p_{\zeta c}, u^q_{\zeta c} \rangle \end{aligned} \tag{12.42}$$

Hence at each node of a beam-column element there are seven degrees of freedom $(u_{\xi 0}, \theta_{\xi 0}, u_{\eta 0}, \theta_{\eta 0}, u_{\zeta c}, \theta_\zeta, \theta_{\zeta,\zeta})$ as shown in Fig. 12.4.

12.4.2 Stiffness Matrices for Elastic Beam-Column

Total equilibrium equation for an elastic thin-walled beam-column element can now be obtained in terms of nodal displacements at the ends by substituting Eq. (12.41) in Eq. (12.30). For example, the first term of Eq. (12.30) $(EA\,\delta u'_{\zeta c} u'_{\zeta c})$ can be written in matrix form as

$$EA(\delta u'_{\zeta c})^T(u'_{\zeta c}) = \text{Scalar}$$

where

$$\begin{aligned} (u'_{\zeta c}) &= \frac{d}{d\zeta}(u_{\zeta c}) = \frac{d}{d\zeta}\langle n_1 \rangle \{\underline{u}_{\zeta c}\} \\ &= \frac{1}{l}\frac{d}{d\beta}\langle n_1 \rangle \{\underline{u}_{\zeta c}\} = \frac{1}{l}\langle n'_1 \rangle \{\underline{u}_{\zeta c}\} \end{aligned} \tag{12.43}$$

and the superscript T denotes the transpose of a vector. Hereafter the prime associated with *shape functions* denotes the differential with respect to nondimensional parameter $\beta = \zeta/l$

$$(\delta u'_{\zeta c})^T = \frac{1}{l}\langle \delta \underline{u}_{\zeta c}\rangle\{n'_1\}$$

Hence, (12.44)

$$\int_0^l EAu'_{\zeta c}\,\delta u'_{\zeta c}\,d\zeta = \langle \delta \underline{u}_{\zeta c}\rangle \frac{EA}{l}\left(\int_0^1 \{n'_1\}\langle n'_1\rangle\,d\beta\right)\{\underline{u}_{\zeta c}\}$$

Similarly all the other terms of Eq. (12.30) are written in terms of nodal displacements and grouping terms together we get the virtual work total equilibrium equation as

$$\begin{aligned}
&\langle \delta \underline{u}_{\xi 0}\rangle\left[\left(\int_0^1 \frac{EI_\eta}{l^3}\{n''_3\}\langle n''_3\rangle\,d\beta + \int_0^1 \frac{P}{l}\{n'_3\}\langle n'_3\rangle\,d\beta\right)\{\underline{u}_{\xi 0}\}\right.\\
&\left.+\left(\int_0^1 \frac{(Pe_\eta - m_\xi)}{l}\{n'_3\}\langle n'_3\rangle\,d\beta - \int_0^1 f_\eta\{n'_3\}\langle n_3\rangle\,d\beta\right)\{\underline{\theta}_\zeta\}\right]\\
&+\langle \delta \underline{u}_{\eta 0}\rangle\left[\left(\int_0^1 \frac{EI_\xi}{l^3}\{n''_3\}\langle n''_3\rangle\,d\beta + \int_0^1 \frac{P}{l}\{n'_3\}\langle n'_3\rangle\,d\beta\right)\{\underline{u}_{\eta 0}\}\right.\\
&\left.+\left(\int_0^1 \frac{(-Pe_\xi + m_\eta)}{l}\{n'_3\}\langle n'_3\rangle\,d\beta + \int_0^1 f_\xi\{n'_3\}\langle n_3\rangle\,d\beta\right)\{\underline{\theta}_\zeta\}\right]\\
&+\langle \delta \underline{\theta}_\zeta\rangle\left[\left(\int_0^1 \frac{EI_\omega}{l^3}\{n''_3\}\langle n''_3\rangle\,d\beta + \int_0^1 \frac{\bar{K}}{l}\{n'_3\}\langle n'_3\rangle\,d\beta + \int_0^1 \frac{GK_T}{l}\{n'_3\}\langle n'_3\rangle\,d\beta\right)\right.\\
&\times\{\underline{\theta}_\zeta\} + \left(\int_0^1 \frac{(Pe_\eta - m_\xi)}{l}\{n'_3\}\langle n'_3\rangle\,d\beta - \int_0^1 f_\eta\{n_3\}\langle n'_3\rangle\,d\beta\right)\{\underline{u}_{\xi 0}\}\\
&\left.+\left(\int_0^1 \frac{(-Pe_\xi + m_\eta)}{l}\{n'_3\}\langle n'_3\rangle\,d\beta + \int_0^1 f_\xi\{n_3\}\langle n'_3\rangle\,d\beta\right)\{\underline{u}_{\eta 0}\}\right]\\
&+\langle \delta \underline{u}_{\zeta c}\rangle\left(\int_0^1 \frac{EA}{l}\{n'_1\}\langle n'_1\rangle\,d\beta\right)\{\underline{u}_{\zeta c}\}\\
&= l\int_0^1 [\langle \delta \underline{u}_{\xi 0}\rangle\{n_3\}q_\xi + \langle \delta \underline{u}_{\eta 0}\rangle\{n_3\}q_\eta + \langle \delta \underline{u}_{\zeta c}\rangle\{n_1\}q_\zeta]\,d\beta\\
&+ [P^*\,\delta u^*_{\zeta c} + f^*_\xi\,\delta u^*_{\xi 0} + f^*_\eta\,\delta u^*_{\eta 0} - m^*_\xi\,\delta\theta^*_\xi - m^*_\eta\,\delta\theta^*_\eta + m^*_\zeta\,\delta\theta^*_\zeta + m^*_\omega\,\delta\theta'^*_\zeta]_0^l
\end{aligned}$$

(12.45)

Here we have used the notations $\theta_{\xi 0} = \theta_\xi$ and $\theta_{\eta 0} = \theta_\eta$ in consistent with the notation used for rotation θ_ζ. The terms contained inside the first bracket of Eq. (12.30) [which can be identified in Eq. (12.45)] constitute the usual flexural stiffness matrix $[k_S]$ and the remaining terms in the integral on the left hand side of Eq. (12.30) represent the "initial stress" or "geometric stiffness" matrix $[k_G]$. It is of interest to note that Turner *et al.*, (1960) first showed that a "new class (geometric)" stiffness matrices should be introduced into the equilibrium equations while investigating large deflection and stability problems.

Assuming constant axial load and a linear variation of moments along the length of the prismatic element, and integrating with respect to β, and recognizing the arbitrary nature of the virtual displacement, the total equilibrium equation for the elastic beam-column element can be written as the following four equations

$$\frac{EI_\eta}{l^3}[K_{33}^{220}]\{\underline{u}_{\xi 0}\} + \frac{P}{l}[K_{33}^{110}]\{\underline{u}_{\xi 0}\} + \left\{\frac{(Pe_\eta - m_\xi^p)}{l}[K_{33}^{110}] - f_\eta([K_{33}^{111}] + [K_{33}^{100}])\right\}\{\underline{\theta}_\zeta\}$$

$$= \begin{Bmatrix} f_\xi^p \\ -m_\eta^p/l \\ f_\xi^q \\ -m_\eta^q/l \end{Bmatrix} \tag{12.46a}$$

$$\frac{EI_\xi}{l^3}[K_{33}^{220}]\{\underline{u}_{\eta 0}\} + \frac{P}{l}[K_{33}^{110}]\{\underline{u}_{\eta 0}\}$$

$$+ \left\{\frac{(-Pe_\xi + m_\eta^p)}{l}[K_{33}^{110}] + f_\xi([K_{33}^{100}] + [K_{33}^{111}])\right\}\{\underline{\theta}_\zeta\}$$

$$= \begin{Bmatrix} f_\eta^p \\ -m_\xi^p/l \\ f_\eta^q \\ -m_\xi^q/l \end{Bmatrix} \tag{12.46b}$$

$$\left(\frac{EI_\omega}{l^3}[K_{33}^{220}] + \frac{GK_T}{l}[K_{33}^{110}]\right)\{\underline{\theta}_\zeta\}$$

$$+ \left\{\frac{(Pe_\eta - m_\xi^p)}{l}[K_{33}^{110}] - f_\eta([K_{33}^{010}] + [K_{33}^{111}])\right\}\{\underline{u}_{\xi 0}\}$$

$$+ \left\{\frac{(-Pe_\xi + m_\eta^p)}{l}[K_{33}^{110}] + f_\xi([K_{33}^{010}] + [K_{33}^{111}])\right\}\{\underline{u}_{\eta 0}\}$$

$$+ \left(\frac{\bar{K}^p}{l}[K_{33}^{110}] + \frac{(\bar{K}^q - \bar{K}^p)}{l}[K_{33}^{111}]\right)\{\underline{\theta}_{\zeta 0}\}$$

$$= \begin{Bmatrix} m_\zeta^p \\ m_\omega^p/l \\ m_\zeta^q \\ m_\omega^q/l \end{Bmatrix} \tag{12.46c}$$

$$\frac{EA}{l}[K_{11}^{110}]\{\underline{u}_{\zeta c}\} = \begin{Bmatrix} P^p \\ P^q \end{Bmatrix} \tag{12.46d}$$

where $$[K_{gh}^{stj}] = \int_0^1 \beta^j \{n_g^s\} \langle n_h^t \rangle \, d\beta \tag{12.47}$$

The subscripts g and h are degree of interpolation vectors and s and t are the order of differentiation and j is the exponent of the multiplying factor β. Various matrices occurring in Eq. (12.46) are given in Table 12.2.

Total equilibrium equation (12.46) in local coordinate system ξ, η and ζ for the elastic beam-column element can be written in a compact form as

$$[k_S]\{u_E\}_L + [k_G]\{u_E\}_L = \{F_E\}_L \tag{12.48}$$

$$\{u_E\}_L^T = \langle \{\underline{u}_{\xi 0}\}^T, \{\underline{u}_{\eta 0}\}^T, \{\underline{\theta}_\zeta\}^T, \{\underline{u}_{\zeta c}\}^T \rangle \tag{12.49}$$

and $\{F_E\}_L$ denote the nodal forces given in the right hand side of Eq. (12.46). Equation (12.46) or (12.48) express the total equilibrium requirement implied by Eq. (12.1). $[k_S]$ is the usual elastic element flexural stiffness matrix, and $[k_G]$ is the geometric stiffness matrix. The geometric stiffness matrix derives the name from the fact that it depends on the geometry of the displaced element. It is interesting to note that the element stiffness matrix contemplates all the possible fourteen degrees of freedom for each element. The matrix $[k_G]$ must be evaluated using the stress resultants in the final equilibrium position which are directly dependent on the displacements $\{u_E\}$.

Table 12.2 FINITE ELEMENT MATRICES

$$30[K_{33}^{110}] = \begin{bmatrix} 36 & 3 & -36 & 3 \\ 3 & 4 & -3 & -1 \\ -36 & -3 & 36 & -3 \\ 3 & -1 & -3 & 4 \end{bmatrix} \qquad 30[K_{33}^{111}] = \begin{bmatrix} 18 & 3 & -18 & 0 \\ 3 & 1 & -3 & -1/2 \\ -18 & -3 & 18 & 0 \\ 0 & -1/2 & 0 & 3 \end{bmatrix}$$

$$30[K_{33}^{100}] = \begin{bmatrix} -15 & -3 & -15 & 3 \\ 3 & 0 & -3 & 1/2 \\ 15 & 3 & 15 & -3 \\ -3 & -1/2 & 3 & 0 \end{bmatrix} \qquad [K_{23}^{120}] = \begin{bmatrix} 4 & 3 & -4 & 1 \\ -8 & -4 & 8 & -4 \\ 4 & 1 & -4 & 3 \end{bmatrix}$$

$$3[K_{22}^{110}] = \begin{bmatrix} 7 & -8 & 1 \\ -8 & 16 & -8 \\ 1 & -8 & 7 \end{bmatrix} \qquad [K_{33}^{220}] = \begin{bmatrix} 12 & 6 & -12 & 6 \\ 6 & 4 & -6 & 2 \\ -12 & -6 & 12 & -6 \\ 6 & 2 & -6 & 4 \end{bmatrix}$$

$$[K_{33}^{221}] = \begin{bmatrix} 6 & 2 & -6 & 4 \\ 2 & 1 & -2 & 1 \\ -6 & -2 & 6 & -4 \\ 4 & 1 & -4 & 3 \end{bmatrix} \qquad 210[K_{33}^{112}] = \begin{bmatrix} 72 & 15 & -72 & -6 \\ 15 & 4 & -15 & -3 \\ -72 & -15 & 72 & 6 \\ -6 & -3 & 6 & 18 \end{bmatrix}$$

$$[K_{23}^{121}] = \begin{bmatrix} 1 & 2/3 & -1 & 1/3 \\ -4 & -4/3 & 4 & -8/3 \\ 3 & 2/3 & -3 & 7/3 \end{bmatrix} \qquad 6[K_{22}^{111}] = \begin{bmatrix} 3 & -4 & 1 \\ -4 & 16 & -12 \\ 1 & -12 & 11 \end{bmatrix}$$

$$[K_{11}^{110}] = \begin{bmatrix} 1 & -1 \\ -1 & 1 \end{bmatrix} \qquad [K_{33}^{010}] = [K_{33}^{100}]^T$$

$$[K_{32}^{210}] = [K_{23}^{120}]^T \qquad [K_{32}^{211}] = [K_{23}^{121}]^T$$

In order to establish the incremental virtual work equation for the elastic case, incremental displacements $\dot{u}_{\xi 0}, \dot{u}_{\eta 0}, \dot{u}_{\zeta c}$ and $\dot{\theta}_{\zeta}$ must be substituted instead of $u_{\xi 0}, u_{\eta 0}$, $u_{\zeta c}$ and θ_{ζ}. The incremental equilibrium equation Eq. (12.22) can be written in the finite element form as

$$[k_S]\{\dot{u}_E\}_L + [k_G]\{\dot{u}_E\}_L = \{\dot{F}_E\}_L \tag{12.50}$$

where $\{\dot{u}_E\}_L$ and $\{\dot{F}_E\}_L$ are the increments in nodal displacements and forces of an element. $[k_S]$ and $[k_G]$ in Eq. (12.50) have exactly the same form as Eq. (12.48). However, a distinction occurs in that for Eq. (12.50) the geometric stiffness is evaluated from the stress resultants at the beginning of the increment.

The flexural stiffness matrix $[k_S]$ and the geometric stiffness $[k_G]$ are given in Table 12.3 for an elastic beam-column element.

12.4.3 Assembly of Finite Element Equations

The element stiffness matrices of Eq. (12.48) and Eq. (12.50) have been evaluated with respect to nodal displacements referred to local coordinate systems of ξ, η and ζ axes. In many cases it is convenient to select a different set of reference axes x, y and z for the global system of nodal displacements.

Let the lateral loads be applied through an axis passing through arbitrary point $\bar{O}$ of the cross section, and the axial load be applied through $\bar{C}$ and the

Table 12.3(a) FLEXURAL STIFFNESS MATRIX $[k_S]$

$[k_S]_{14 \times 14} =$

u_ξ^p	u_η^p	u_ζ^p	θ_ζ^p	$l\theta_\eta^p$	$l\theta_\xi^p$	$l\theta_\zeta'^p$	u_ξ^q	u_η^q	u_ζ^q	θ_ζ^q	$l\theta_\eta^q$	$l\theta_\xi^q$	$l\theta_\zeta'^q$
a				b			$-a$				b		
	e				f			$-e$				f	
		m							$-m$				
			i			j				$-i$			j
				c			$-b$				d		
					g			$-f$				h	
						k				$-j$			o
							a				$-b$		
								e				$-f$	
									m				
										i			$-j$
											c		
												g	
SYMMETRIC													k

$a = 12EI_\eta/l^3$; $\quad e = 12EI_\xi/l^3$; $\quad i = \dfrac{12EI_\omega}{l^3} + \dfrac{36}{30}\dfrac{GK_T}{l}$; $\quad m = EA/l$.

$b = 6EI_\eta/l^3$; $\quad f = 6EI_\xi/l^3$; $\quad j = \dfrac{6EI_\omega}{l^3} + \dfrac{3}{30}\dfrac{GK_T}{l}$;

$c = 4EI_\eta/l^3$; $\quad g = 4EI_\xi/l^3$; $\quad k = \dfrac{4EI_\omega}{l^3} + \dfrac{4}{30}\dfrac{GK_T}{l}$;

$d = 2EI_\eta/l^3$; $\quad h = 2EI_\xi/l^3$; $\quad o = \dfrac{2EI_\omega}{l^3} - \dfrac{GK_T}{30l}$;

Table 12.3(b) GEOMETRIC STIFFNESS MATRIX $[k_G]$

$$
\underset{14\times 14}{[k_G]} =
\begin{array}{cccccccccccccc}
u_\xi^p & u_\eta^p & u_\zeta^p & \theta_\zeta^p & l\theta_\eta^p & l\theta_\xi^p & l\theta_\zeta'^p & u_\xi^q & u_\eta^q & u_\zeta^q & \theta_\zeta^q & l\theta_\eta^q & l\theta_\xi^q & l\theta_\zeta'^q \\
\hline
a & & & e & b & & f & -a & & & -k & b & & g \\
 & a & & -e' & & b & -f' & & -a & & k' & & b & -g' \\
 & & & & & & & & & & & & & \\
 & & & m & -i & i' & n & -e & e' & & -m & s & -s' & q \\
 & & & & c & & h & -b & & & i & d & & j \\
 & & & & & c & -h' & & -b & & -i' & & d & -j' \\
 & & & & & & o & -f & f' & & -n & t & -t' & p \\
 & & & & & & & a & & & k & -b & & -g \\
 & & & & & & & & a & & -k' & & -b & g' \\
 & & & & & & & & & & & & & \\
 & & & & & & & & & & m & -s & s' & -q \\
 & & & & & & & & & & & c & & w \\
 & & & & & & & & & & & & c & -w' \\
\text{SYMMETRIC} & & & & & & & & & & & & & r
\end{array}
$$

$$a = \frac{36}{30}\frac{P}{l} \qquad b = \frac{3}{30}\frac{P}{l} \qquad c = \frac{4}{30}\frac{P}{l} \qquad d = -\frac{1}{30}\frac{P}{l}$$

$$e = \frac{36}{30l}(Pe_\eta - m_\xi^p) - \frac{3}{30}f_\eta$$

$$f = \frac{3}{30l}(Pe_\eta - m_\xi^p)$$

$$g = \frac{3}{30l}(Pe_\eta - m_\xi^p) - \frac{3}{30}f_\eta$$

$$h = \frac{4}{30l}(Pe_\eta - m_\xi^p) - \frac{f_\eta}{30}$$

$$i = -\frac{3}{30l}(Pe_\eta - m_\xi^p) + \frac{6f_\eta}{30}$$

$$j = -\frac{(Pe_\eta - m_\xi^p)}{30l}$$

$$k = \frac{36}{30l}(Pe_\eta - m_\xi^p) - \frac{33}{30}f_\eta$$

$$w = \frac{4}{30l}(Pe_\eta - m_\xi^p) - \frac{3}{30}f_\eta$$

$$m = \frac{36}{30}\frac{\bar{K}^p}{l} + \frac{18}{30l}(\bar{K}^q - \bar{K}^p)$$

$$n = \frac{3}{30}\frac{\bar{K}^p}{l} + \frac{3}{30l}(\bar{K}^q - \bar{K}^p)$$

$$o = \frac{4}{30}\frac{\bar{K}^p}{l} + \frac{1}{30l}(\bar{K}^q - \bar{K}^p)$$

$$p = -\frac{1}{30}\frac{\bar{K}^p}{l} - \frac{1}{60l}(\bar{K}^q - \bar{K}^p)$$

$$q = \frac{3}{30}\frac{\bar{K}^p}{l}$$

$$r = \frac{4}{30}\frac{\bar{K}^p}{l} + \frac{3}{30l}(\bar{K}^q - \bar{K}^p)$$

$$s = \frac{3}{30l}(Pe_\eta - m_\xi^p) + \frac{3}{30}f_\eta$$

$$t = -\frac{1}{30l}(Pe_\eta - m_\xi^p) + \frac{f_\eta}{30}$$

$e', f', g', h', i', j', k', w'$ are obtained by replacing e_η by e_ξ and m_ξ by m_η and f_η by f_ξ in the expressions $e, f, g, h, i, j, k,$ and w.

principal axis of the cross section be at an angle of α to the global axes as shown in Fig. 12.6. The element displacements with respect to the local coordinate system can be related to those in the global system by the transformation

$$\{u_E\}_L = [T]\{u_E\}_G \tag{12.51}$$

where $[T]$ is given by

$$[T] = \begin{bmatrix} T_{11} & 0 \\ 0 & T_{11} \end{bmatrix} \tag{12.52}$$

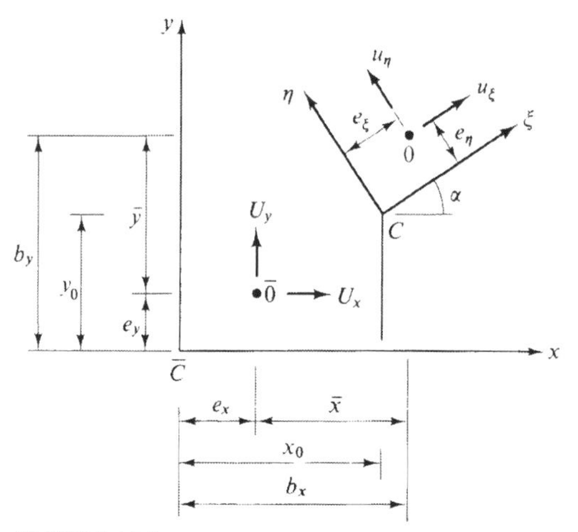

FIGURE 12.6
Thin-walled beam cross section in local and global orientation

$$[T_{11}] =$$

$$\begin{bmatrix} \cos\alpha & \sin\alpha & 0 & [(b_x - e_x)\sin\alpha - (b_y - e_y)\cos\alpha] & 0 & 0 & 0 \\ -\sin\alpha & \cos\alpha & 0 & [(b_y - e_y)\sin\alpha + (b_x - e_x)\cos\alpha] & 0 & 0 & 0 \\ 0 & 0 & 1 & 0 & -x_0 & -y_0 & 0 \\ 0 & 0 & 0 & 1 & 0 & 0 & 0 \\ 0 & 0 & 0 & 0 & l\cos\alpha & l\sin\alpha & l[(b_x - e_x)\sin\alpha - (b_y - e_y)\cos\alpha] \\ 0 & 0 & 0 & 0 & -l\sin\alpha & l\cos\alpha & l[(b_y - e_y)\sin\alpha + (b_x - e_x)\cos\alpha] \\ 0 & 0 & 0 & 0 & 0 & 0 & l \end{bmatrix} \tag{12.53}$$

and

$$\{u_E\}_G^T = \langle U_x^p, U_y^p, U_z^p, \theta_z^p, \theta_y^p, \theta_x^p, \theta_z'^p, U_x^q, U_y^q, U_z^q, \theta_z^q, \theta_y^q, \theta_x^q, \theta_z'^q \rangle \tag{12.54}$$

Substituting for $\{u_E\}_L$ in Eq. (12.48) in terms of $\{u_E\}_G$ and premultiplying by $[T]^T$ yields

$$[T]^T[k_S][T]\{u_E\}_G + [T]^T[k_G][T]\{u_E\}_G = \{F_E\}_G \tag{12.55}$$

or

$$([k_S]_G + [k_G]_G)\{u_E\}_G = \{F_E\}_G \tag{12.56}$$

The overall unconstrained stiffness matrix is generated following the rules that govern the assembly process used in the matrix analysis of framed structure. The matrix is generated by a simple summation of the individual stiffness and loads at the nodes using the compatibility of displacements and slopes. The flexural and geometric stiffness matrices for the total structure are written as

$$[K_S] = \sum_{i=1}^{N} [k_S]_{Gi} \tag{12.57a}$$

$$[K_G] = \sum_{i=1}^{N} [k_G]_{Gi} \tag{12.57b}$$

$$\{F\} = \sum_{i=1}^{N} \{F_E\}_{Gi} \tag{12.57c}$$

and total equilibrium equation for the structure

$$[K_S]\{u\} + [K_G]\{u\} = \{F\} \tag{12.57d}$$

N denotes total number of the elements.

Structural members are subjected to force and displacement boundary conditions. The force boundary conditions are incorporated automatically into the load vector. When imposing the displacement boundary condition the following procedure may be used. For example,

$$\begin{bmatrix} K_{11} & K_{12} & K_{13} \\ K_{21} & K_{22} & K_{23} \\ K_{31} & K_{32} & K_{33} \end{bmatrix} \begin{Bmatrix} U_1 \\ U_2 \\ U_3 \end{Bmatrix} = \begin{Bmatrix} F_1 \\ F_2 \\ F_3 \end{Bmatrix} \tag{12.58}$$

If U_2 is specified as δ make

$$F_i = F_i - K_{i2}\delta \qquad i \neq 2$$
$$F_i = U_i \qquad i = 2$$

and

$$K_{2j} = K_{j2} = 0 \qquad j \neq 2$$
$$K_{22} = 1$$

After this modification, Eq. (12.58) is now written as

$$\begin{bmatrix} K_{11} & 0 & K_{13} \\ 0 & 1 & 0 \\ K_{31} & 0 & K_{33} \end{bmatrix} \begin{Bmatrix} U_1 \\ U_2 \\ U_3 \end{Bmatrix} = \begin{Bmatrix} F_1 - K_{12}\delta \\ \delta \\ F_3 - K_{32}\delta \end{Bmatrix} \tag{12.59}$$

The above procedure keeps the size of the master stiffness matrix same and thus avoiding reorganization of the computer storage. Since the stiffness matrices are symmetric and diagonally populated, the banded stiffness matrix is used to save computer storage. The semi-band width B is given by

$$B = (D + 1)d \tag{12.60}$$

where D is the maximum largest difference occurring for all elements between two external nodes of the members of a single element and d denotes the degrees of freedom at each node.

In the case of beam-column analyses, only two nodes are connected by a segment and $d = 7$, thus the semi-band width $B = 14$ is sufficient.

The finite element form of the incremental equilibrium equation may be written as

$$[K_S]\{\dot{u}\} + [K_G]\{\dot{u}\} = \{\dot{F}\} \tag{12.61}$$

where $[K_S]$ and $[K_G]$ are identical to those of Eq. 12.57(d) but $[K_G]$ is determined from the initial stress prior to the load increment.

12.5 ELASTIC BUCKLING OF COLUMNS

With regard to stability problems, two different approaches are commonly used in analysis; both of them furnish two different kinds of valid and important information. One way of approaching these problems is by load-deflection approach which attempts to solve stability problems by predicting the load deflection behavior throughout the entire range of loading (including loading and unloading). This approach is difficult and time-consuming. On the other hand, eigenvalue approach attempts to determine the stability load in a direct manner without calculating the deflection. In this approach, an ideal or perfect member and loading condition are considered and deformations prior to the attainment of buckling load are ignored. As discussed in Chap. 1, Vol. 1, the buckling load determined by a load-deflection analysis for any real member is always less than bifurcation load (which always forms an upper bound for a real system) determined by an eigenvalue approach.

12.5.1 Buckling Formulation

A buckling problem is formulated by imposing a variation of displacement in the matrix equation (12.57d) under constant load. The result is identical to the homogeneous form of Eq. (12.61) is

$$[K_S]\{\dot{u}\} + [K_G]\{\dot{u}\} = \{0\} \tag{12.62}$$

The above equation is satisfied, for a non-zero displacement vector, only if the stiffness matrix is singular, and the *buckling criterion* can be represented by the vanishing of the determinant of the stiffness matrix, i.e.,

$$\det([K_S] + [K_G]) = 0 \tag{12.63}$$

Assuming that throughout the loading history the stress state can be represented in terms of some function whose intensity is governed by a single parameter λ, the initial state then takes the form

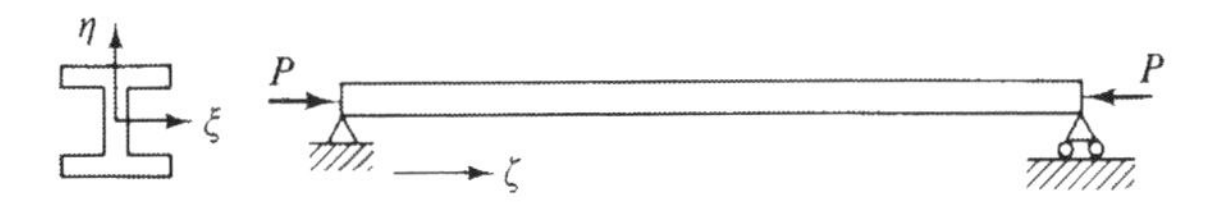

FIGURE 12.7
A thin-walled member with two axes of symmetry

$$[K_G] = \lambda[\bar{K}_G] \tag{12.64}$$

where $[\bar{K}_G]$ represents the initial stress stiffness matrix computed for some arbitrary stress distribution. Introducing Eq. (12.64) into Eq. (12.62) yields

$$[K_S]\{\dot{u}\} = -\lambda[\bar{K}_G]\{\dot{u}\} \tag{12.65a}$$

or

$$\frac{1}{\lambda}\{\dot{u}\} = -[K_S]^{-1}[\bar{K}_G]\{\dot{u}\} \tag{12.65b}$$

The critical buckling load is now obtained by solving Eq. (12.65b) for the unknown eigenvalue λ. The eigenvalue solution can be obtained using any one of the standard library programs. The buckling mode can be identified as the eigenvector corresponding to the lowest eigenvalue.

Before any finite element solution can be used with confidence, some ideas of its accuracy and convergence characteristics are required. To illustrate the validity and application of the method, finite element method is applied to the following example whose analytical solution is straight forward.

12.5.2 Torsional Buckling of a Thin-Walled Member with Two Axes of Symmetry

The buckling phenomenon in question involves torsional deformation of a thin-walled member shown in Fig. 12.7 about ζ axis. The derivative of the angle of twist θ'_ζ is included in order to account for possible warping effects. Thus, at each end of the member the nodal displacements are θ_ζ and θ'_ζ.

Under the conditions stated above, with the inclusion of torsional instability effects, the incremental virtual work equilibrium equation defined in Eq. (12.22) together with Eq. (12.29) for an element reduces to

$$\int_{0\,(e_\eta = e_\xi = 0)}^{l} (EI_\omega \dot{\theta}''_\zeta\, \delta\dot{\theta}''_\zeta + GK_T \dot{\theta}'_\zeta\, \delta\dot{\theta}'_\zeta + \bar{K}\dot{\theta}'_\zeta\, \delta\dot{\theta}'_\zeta)\, d\zeta = 0 \tag{12.66}$$

$$\bar{K} = \int_A \sigma a^2\, dA = -PI_p/A \tag{12.67}$$

where I_p is the polar moment of inertia. The basic flexural stiffness matrix is given by

$$[k_S] = \frac{EI_\omega}{l^3} \begin{bmatrix} 12 + \frac{36}{30}\kappa^2 l^2 & 6 + \frac{3}{30}\kappa^2 l^2 & -12 - \frac{36}{30}\kappa^2 l^2 & 6 + \frac{3}{30}\kappa^2 l^2 \\ & 4 + \frac{4}{30}\kappa^2 l^2 & -6 - \frac{3}{30}\kappa^2 l^2 & 2 - \frac{\kappa^2 l^2}{30} \\ & & 12 + \frac{36}{30}\kappa^2 l^2 & -6 - \frac{3}{30}\kappa^2 l^2 \\ \text{SYMMETRIC} & & & 4 + \frac{4}{30}\kappa^2 l^2 \end{bmatrix} \quad (12.68)$$

(columns: $\dot{\theta}_\zeta^p$, $l(\dot{\theta}'_\zeta)^p$, $\dot{\theta}_\zeta^q$, $l(\dot{\theta}'_\zeta)^q$)

where κ^2 is the ratio of St. Venant rigidity to warping rigidity, $\kappa^2 = GK_T/EI_\omega$. The explicit form for the matrix $[k_G]$ is

$$[k_G] = \frac{-PI_p}{30lA} \begin{bmatrix} 36 & 3 & -36 & 3 \\ & 4 & -3 & -1 \\ & & 36 & -3 \\ \text{SYMMETRIC} & & & 4 \end{bmatrix} \quad (12.69)$$

(columns: $\dot{\theta}_\zeta^p$, $l(\dot{\theta}'_\zeta)^p$, $\dot{\theta}_\zeta^q$, $l(\dot{\theta}'_\zeta)^q$)

As an example, the problem of torsional buckling of the thin-walled member for one element representation is solved

$$\dot{\theta}_\zeta^p = \dot{\theta}_\zeta^q = 0 \quad (12.70)$$

and $(\dot{\theta}'_\zeta)^p = (\dot{\theta}'_\zeta)^q \neq 0$ since warping is permitted. For $\kappa = 0$

$$\begin{bmatrix} 4 & 2 \\ 2 & 4 \end{bmatrix} \begin{Bmatrix} (\dot{\theta}'_\zeta)^p \\ (\dot{\theta}'_\zeta)^q \end{Bmatrix} = \lambda \begin{bmatrix} 4 & -1 \\ -1 & 4 \end{bmatrix} \begin{Bmatrix} (\dot{\theta}'_\zeta)^p \\ (\dot{\theta}'_\zeta)^q \end{Bmatrix} \quad (12.71)$$

where

$$\lambda = PI_p l^2/30EI_\omega A \quad (12.72)$$

Solving, one obtains

$$P_{cr} = 12EI_\omega A/I_p l^2 \quad (12.73)$$

where the exact solution is

$$\frac{9.87EI_\omega A}{I_p l^2} \quad (12.74)$$

Table 12.4 shows a comparison of the results of finite element formulation with that of the closed form solution for various values of κ. It can be seen that the four element approximation gives results closer to the exact solution. Convergence characteristics of the torsional buckling problem obtained with the use of derived formulation is shown in Fig. 12.8. The corresponding results for a finite difference

Table 12.4 TORSIONAL BUCKLING OF THIN-WALLED BEAM

Multiplicator Factor $\frac{EI_{\omega}A}{I_p l^2}$

κL	Exact Solution	Twist Represented by Hyperbolic Function Krajcinovic	Cubic Polynomial One Element	Cubic Polynomial Two Elements	Cubic Polynomial Four Elements	Cubic Polynomial Eight Elements
0	9.87	9.92	12.0	9.94385	9.87466	9.86993
2	13.87	14.01	16.0	13.9438	13.8747	13.8699
4	25.87	26.30	28.0	25.9438	25.8747	25.8699
6	45.87	46.99	48.0	45.9438	45.8747	45.8699
8	73.87	76.13	76.0	73.9438	73.8747	73.8699
10	109.87	113.65	112.0	109.944	109.875	109.87
20	409.87	423.20	412.0	409.944	409.875	409.87

solution are also shown for the purpose of comparison. The lessened accuracy of the finite difference model is due to the assumption of a linear variation of θ_{ζ} between joints. It is also to be noted that finite difference employs only a single parameter at a joint.

FIGURE 12.8
Convergence between finite difference and finite element solution

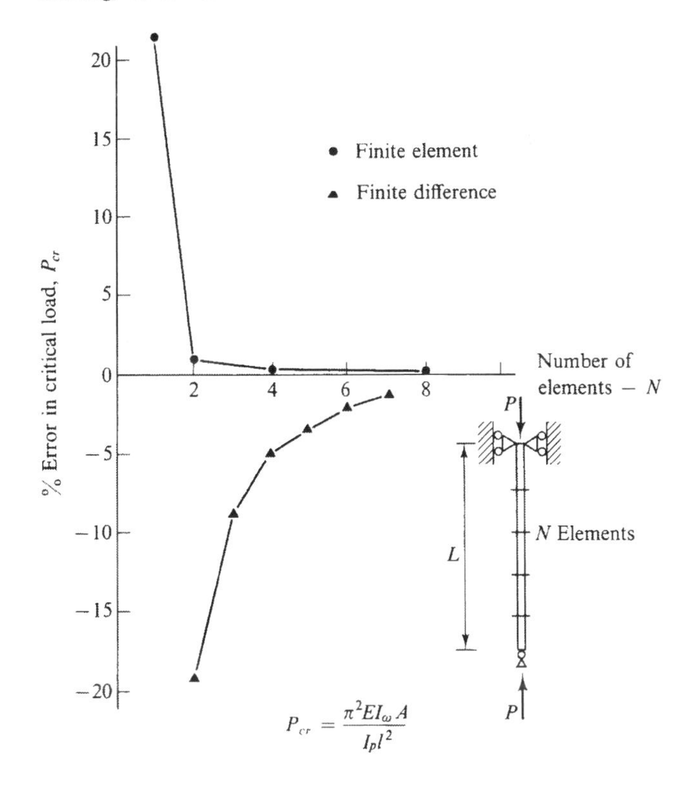

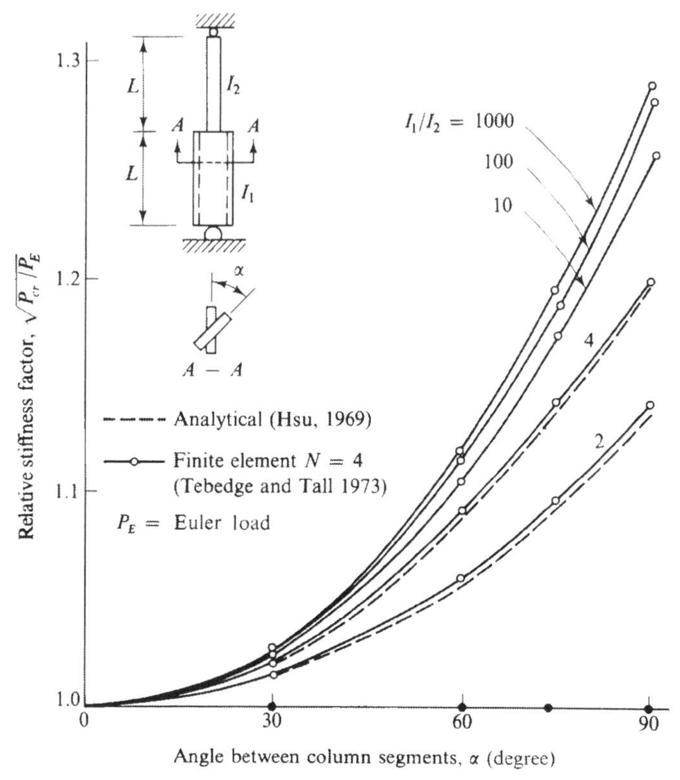

FIGURE 12.9
Buckling loads of two-component columns with pin ends

Of course, the finite element method is not intended as a procedure for calculation of the buckling load for problems given above which can be treated rather efficiently by classical and other methods such as finite segment, etc. However, there are numerous problems, whose solutions are not straight forward when conventional approaches are used, especially when there are irregularities in geometry and loading. The following examples are given to show the efficacy of the finite element method in the linear stability problem.

12.5.3 Piecewise Prismatic Columns

A piecewise prismatic column is composed of prismatic column segments joined together with an offset of the principal axes directions. The deformed configuration of such columns does not necessarily become perpendicular to the axis of the least moment of inertia. The solution of these problems by classical approach is difficult, since the governing differential equations are nonlinear. Element stiffnesses are formulated with respect to local coordinate system and must be transformed to a global system using the transformation matrix given in Eq. (12.52). The equation for a complete system may be formulated as

$$[K_S]\{\dot{u}\} + [K_G]\{\dot{u}\} = \{0\} \tag{12.75}$$

where $[K_S]$ and $[K_G]$ are given by Eq. (12.57). Figure 12.9 compares the critical loads of two component columns with arbitrary offset angle α supported on spherical pins for various values of the stiffness factor I_1/I_2. The results were obtained by Tebedge and Tall (1973) using the finite element method.

12.5.4 Frame Instability

In formulating the finite element relationship for the indeterminate frame, the internal forces in each member prior to buckling are first evaluated. Flexural and geometric stiffnesses are determined in terms of the local coordinate system, a displacement and force transformation is performed relating local and global systems and the eigenvalue solution is carried out. The application of the method to more complex problem is illustrated by solving the space frame shown in Fig. 12.10. The critical loads and the associated buckling modes are obtained in a single operation by solving Eq. (12.75) as an eigenvalue problem.

There are some practical engineering problems such as a pile embedded in a frictional medium, guyed towers, industrial racks whose stability analysis by the analytical solutions involve integrals that are difficult and essentially impossible to

FIGURE 12.10
Space frame buckling by finite element method (1 kip = 4.448 kN)

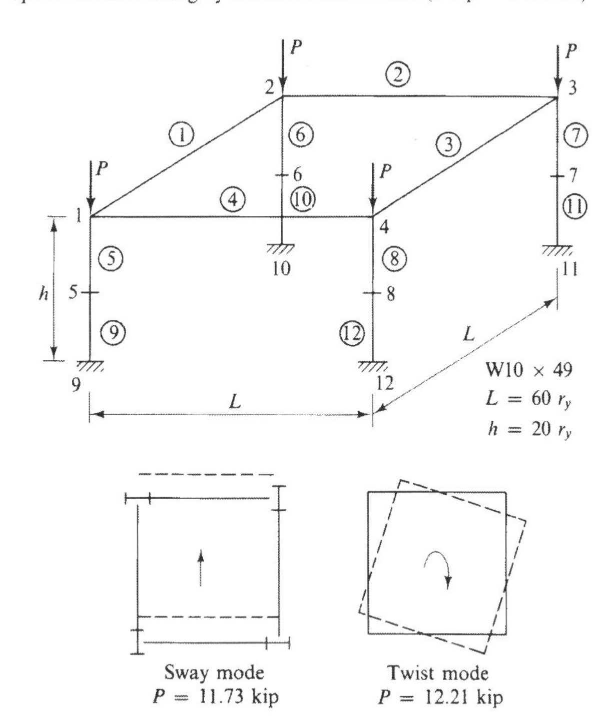

evaluate. For such problems, the formulation given in this chapter gives results with satisfactory accuracy by using few elements and with minimal computational effort. Examples of the finite element elastic instability problems are to be found in the work by Barsoum (1970), Powell and Klinger (1970), Rajasekaran (1971) and Tebedge (1972).

12.6 ELASTIC STABILITY OF BEAM-COLUMNS

In the case of beam-columns, the problem must be solved by "load-deflection approach." This involves the solution of Eq. (12.57d), which, because of the dependence of $[K_G]$ on $\{u\}$, is nonlinear. There are a number of methods of solving this set of nonlinear equations, and the survey of solution techniques utilizing the finite element concept for nonlinear problems is given elsewhere (Martin, 1969; Rajasekaran and Murray, 1975). Basically, most of the numerical procedures fall into two broad categories, namely, incremental and iterative methods. The incremental methods do not necessarily satisfy equilibrium conditions whereas the iterative methods tend to stay on the equilibrium path at all stages of loading.

12.6.1 Solution Techniques

(a) *Incremental Method*

The nonlinear behavior is determined by solving a sequence of linear problems. The load is applied as a sequence of sufficiently small increments so that during loading the structure is assumed to respond linearly. For each load increment the displacement increment is computed by solving Eq. (12.61). A subsequent increment in load is applied and the process is repeated until the desired number of load increments has been completed. The basic disadvantage of the method is that real

FIGURE 12.11(a)
Simple incremental technique

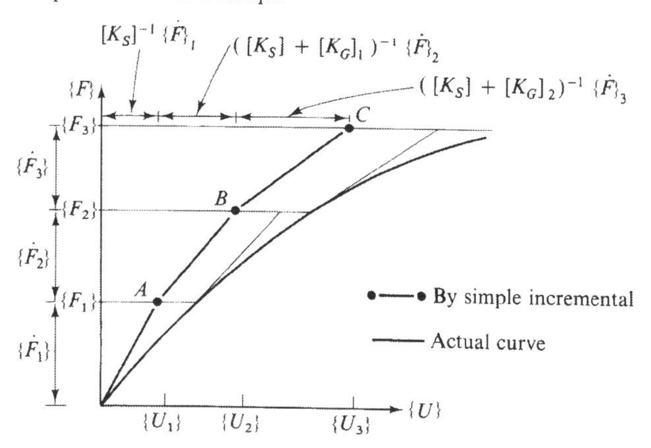

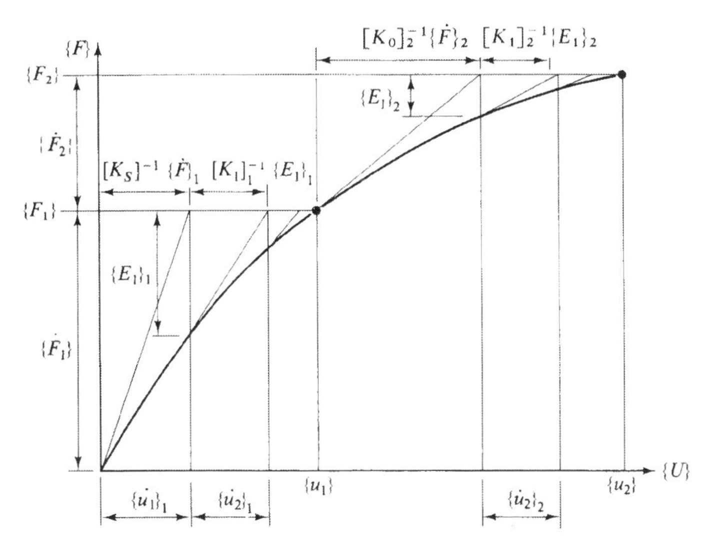

FIGURE 12.11(b)
Incremental and iterative technique (Newton-Raphson procedure)

estimate of solution accuracy is not possible since equilibrium is not satisfied at any given load level. The procedure is illustrated in Fig. 12.11(a). The algorithm is

$$\{u\}_r = [K]_{r-1}^{-1}\{\dot{F}\}_r + \{u\}_{r-1} \tag{12.76}$$

where

$$[K]_{r-1} = ([K_S] + [K_G]_{r-1}) \tag{12.77a}$$

and

$$[K_G]_0 = [0] \tag{12.77b}$$

and $\{\dot{F}\}_r$ is the increment in load vector at the rth load step and $\{u\}_r$ represents the total displacement vector.

(b) *Incremental and Iterative* (*Using Newton-Raphson's Approach*)

Initially, a small displacement solution of Eq. (12.57d) is obtained for the first load increment, with $[K_G]$ set to zero. The element stress resultants are computed from the known displacements and geometric stiffness for the structure is evaluated. The total resisting force at any node is then the sum of the resisting forces obtained from flexural stiffness and geometric stiffness. The difference between the resisting forces and applied forces represents a set of unbalanced forces for the given configuration of the model. Knowing the set of unbalanced forces on the model in the configuration, one can solve for the increment in nodal displacements. The process is repeated until the configuration of the model, maintains equilibrium with the applied loads. The procedure is similar to that of tangent stiffness method described in Chaps. 6

and 7. The algorithm of Newton-Raphson method of solving beam-column problems is given below and illustrated in Fig. 12.11(b)

For any rth load step

1.
$$\{E_{n-1}\}_r = \{\dot{F}\}_r \qquad \text{for} \qquad n = 1$$
$$\{E_{n-1}\}_r = \{F\}_r - [K_{n-1}]_r\{u_{n-1}\}_r \qquad \text{for} \qquad n > 1 \tag{12.78a}$$

where

$$\{F\}_r = \sum_{i=1}^{r} \{\dot{F}\}_i \tag{12.78b}$$
$$[K_{(n-1)}]_r = [K_S] + [K_{G(n-1)}]_r \tag{12.78c}$$

and $\{E_{n-1}\}_r$ is known as the error vector after $(n-1)$ iterations in the rth load step

2.
$$\{\dot{u}_n\}_r = [K_{(n-1)}]_r^{-1}\{E_{(n-1)}\}_r \tag{12.79}$$

3.
$$\{u_n\}_r = \{u_{n-1}\}_r + \{\dot{u}_n\}_r \tag{12.80}$$

4. The steps 1 to 3 are repeated until the unbalanced force vector is less than the allowable tolerance value (i.e., until the ratio of norm of unbalanced forces to that of applied forces is less than the prescribed tolerance), i.e.,

$$\frac{\{E_n\}_r^T\{E_n\}_r}{\{F\}_r^T\{F\}_r} < \varepsilon \text{ (tolerance)} \tag{12.81}$$

FIGURE 12.11(c)
Incremental and iterative technique modified Newton-Raphson procedure

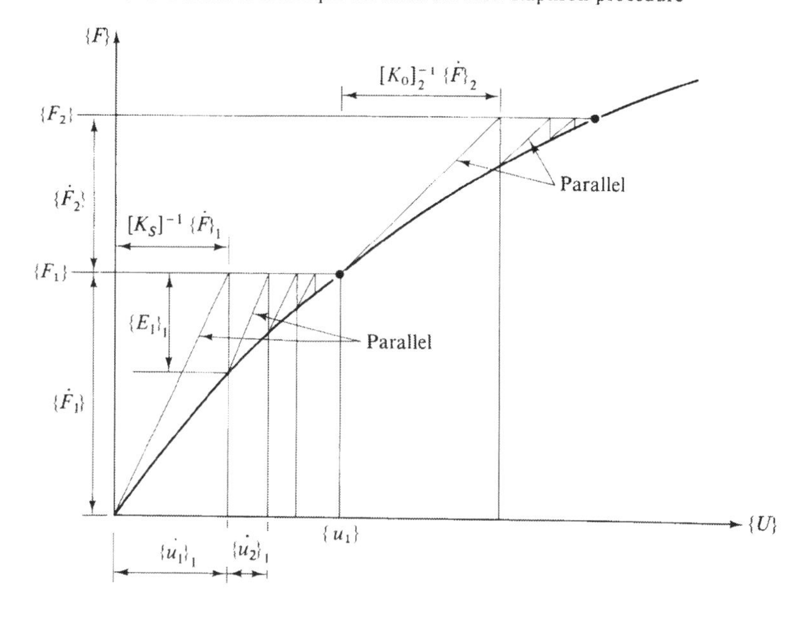

5. A new load increment is applied and the steps 1 to 4 are repeated. The principal disadvantage of this technique is that it is necessary to triangularize a new stiffness matrix for every iteration. This method is illustrated in Fig. 12.11(b).

(c) *Incremental and Iterative (Using Modified Newton-Raphson)*

1. For any rth load step

$$\{E_{n-1}\}_r = \{\dot{F}\}_r \qquad \text{for} \qquad n = 1 \tag{12.82a}$$

FIGURE 12.12(a)
Flow chart for simple incremental technique

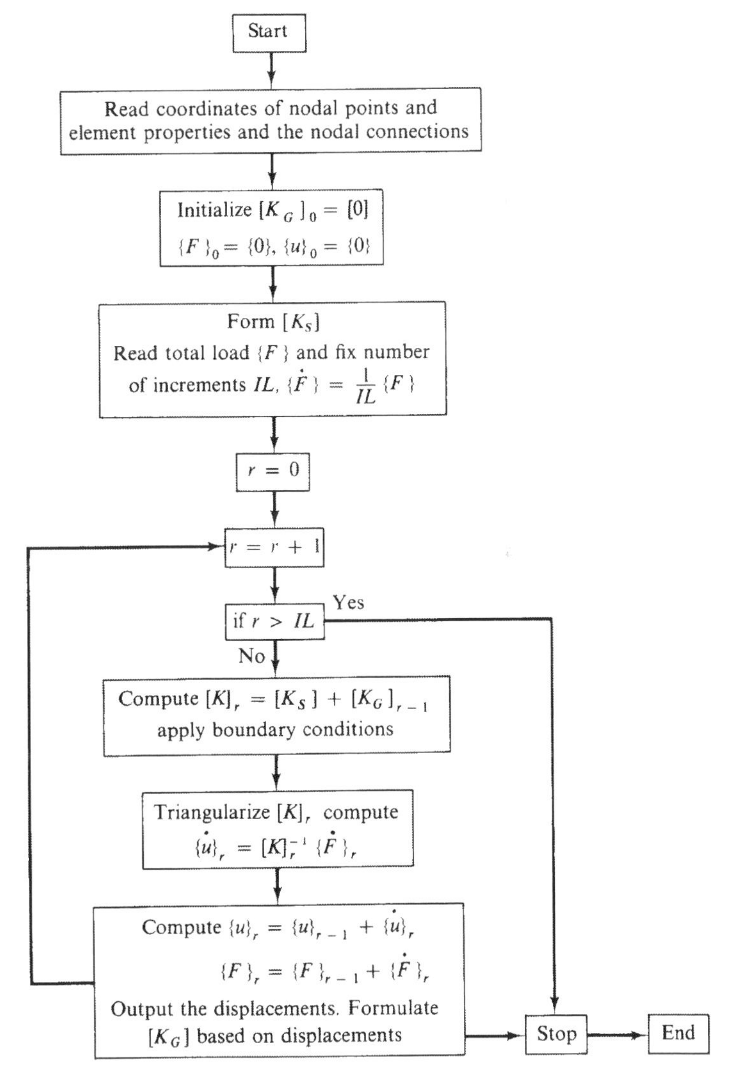

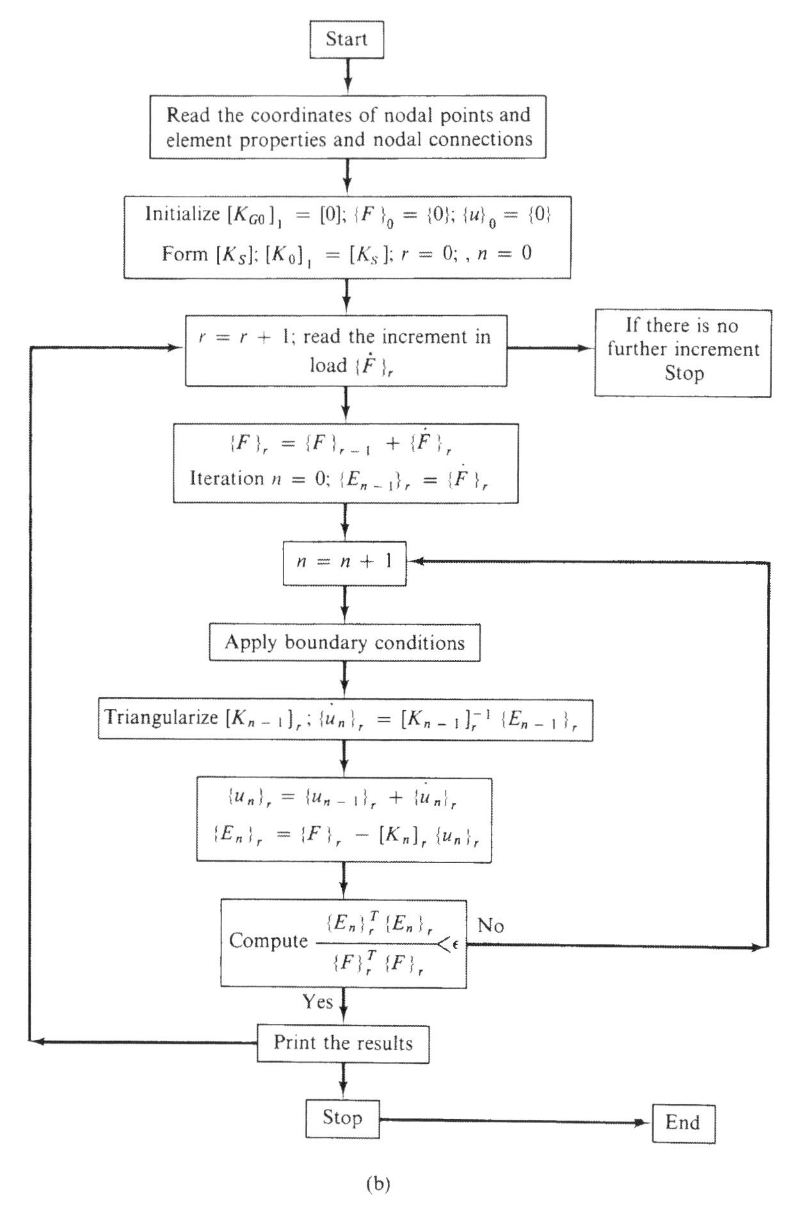

FIGURE 12.12(b)
Flow chart for incremental iterative technique using Newton-Raphson method

and

$$\{E_{n-1}\}_r = \{F\}_r - [K_{n-1}]_r\{u_{n-1}\}_r \qquad \text{for} \qquad n > 1 \tag{12.82b}$$

2.

$$\{\dot{u}_n\}_r = [K]_{r-1}^{-1}\{E_{n-1}\}_r = [K_0]_r^{-1}\{E_{n-1}\}_r \tag{12.83a}$$

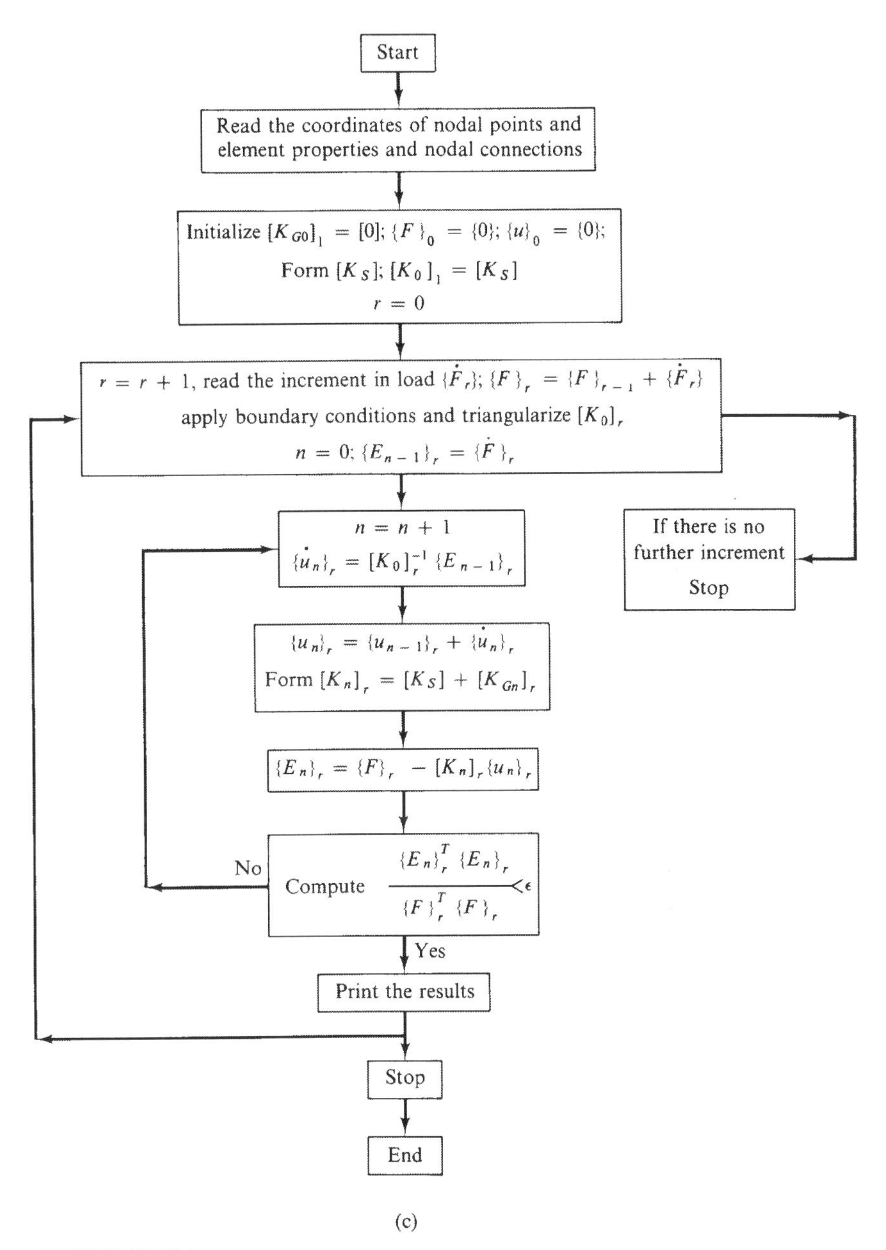

(c)

FIGURE 12.12(c)
Flow chart for incremental iterative technique using modified Newton-Raphson method

where

$$[K]_{r-1} = [K_S] + [K_G]_{r-1} \tag{12.83b}$$

3.

$$\{u_n\}_r = \{u_{n-1}\}_r + \{\dot{u}_n\}_r \tag{12.84}$$

The iteration is continued until the displacement for the nth iteration is the same as that for the $(n-1)$th iteration at which the unbalanced force is arbitrarily small. The advantage of this method is that the matrix $[K_0]_r$ is inverted only once for any rth load step by updating $[K_G]$ at the beginning of each load step. This method is illustrated in Fig. 12.11(c).

The flow charts for the solution techniques described above are given in Fig. 12.12. Efficient solution techniques such as Alpha-constant technique of Nayak and Zienkiewicz (1972) may also be applied. This technique essentially converts the displacement increment obtained in modified Newton-Raphson to that of Newton-Raphson.

12.6.2 Geometric Imperfection

When the beam has an initial imperfection of $\{u\}^*$, Eq. (12.57d) can be written as

$$[K_S]\{u\} + [K_G]\{u\} = \{F\} - [K_G]\{u\}^* \tag{12.85a}$$

and the unbalanced force vector after $(n-1)$ iterations is

$$\{E_n\} = \{F_n\} - [K_S]\{u_{n-1}\} - [K_{G(n-1)}]\{u_{n-1}\} - [K_{G(n-1)}]\{u\}^* \tag{12.85b}$$

The correction $\{\dot{u}_n\}$ is given by the solution to Eq. (12.85) either by Newton-Raphson or by modified Newton-Raphson method.

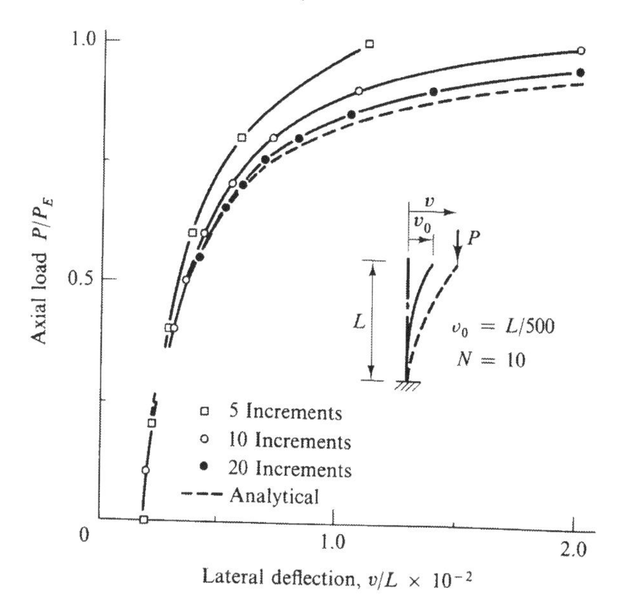

FIGURE 12.13
Effect of number of load increments (simple incremental technique) (Tebedge and Tall, 1972)

12.6.3 Numerical Examples

EXAMPLE 12-1 The first numerical example to illustrate the validity of the procedure described is that of a cantilever column with an initial out-of-straightness $U_{y0}^* = v_0$ at the free end. The out-of-straightness along the length of the column is assumed to be expressed by $U_y^* = v_0(1 - \cos \pi z/L)$. In this study by Tebedge (1972), direct incremental procedure without iteration is used in evaluating the load displacement relationship. The effect of the number of load increments in the final solution is shown in Fig. 12.13. When using direct incremental procedure it is observed that the results agree closely to the analytical solution as the load increments become small.

EXAMPLE 12-2 Figure 12.14 shows the moment and out-of-plane response of a W8 × 31 beam when there is an initial lateral deflection of 0.01 sin $\pi z/L$. U_x, U_y

FIGURE 12.14
Beam with initial imperfection (1 kip-in = 112 N-m; 1 in = 25.4 mm)

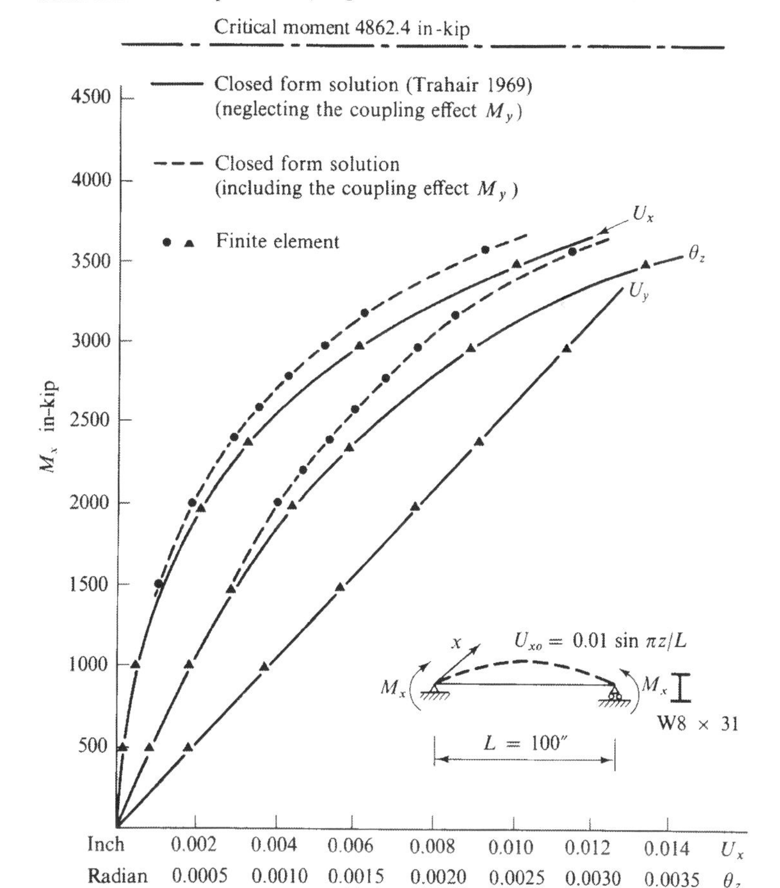

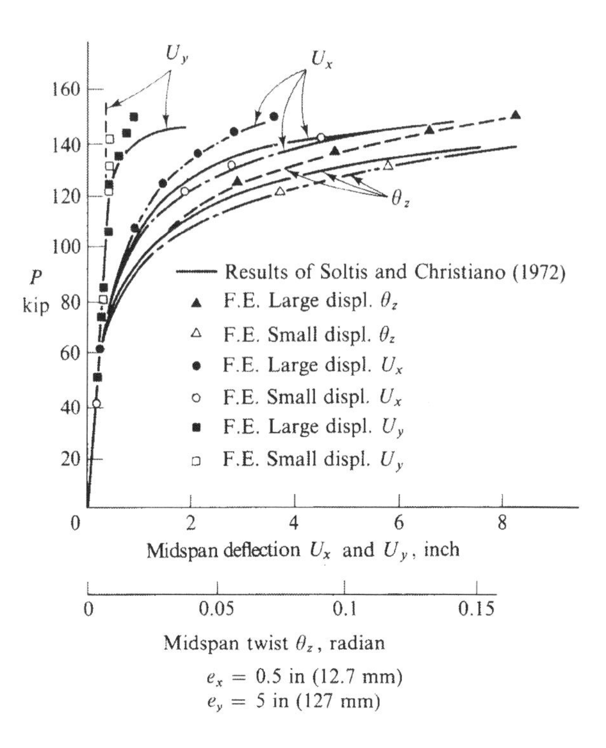

FIGURE 12.15
Beam-column under biaxial loading (1 kip = 4.45 kN; 1 in = 25.4 mm)

and θ_z in Fig. 12.14 correspond with displacements and rotations at mid-span. Moment corresponding to the eigenvalue solution neglecting the prebuckling displacements is also shown in the figure by a dotted line. Classical eigenvalue solution forms an upper bound since prebuckling deformations are neglected. The explicit form of the stiffness matrices are given in Table 12.3 (for wide flange shape $e_\xi = e_\eta = 0$).

EXAMPLE 12-3 A finite element small displacement as well as large displacement analyses have been carried out on a W14 × 43 beam-column having $L/r_y = 140$ (Rajasekaran and Murray, 1974), ($L = 264.6$ in $= 672$ cm) and subjected to biaxial bending ($e_x = 0.5$ in $= 12.7$ mm); ($e_y = 5$ in $= 127$ mm). Figure 12.15 shows the comparison between the results of the finite element analysis with that of Soltis and Christiano (1972). The small displacement finite element solution has yielded displacements in excess of those predicted by Soltis and Christiano in the range in which they detect no divergence between large and small displacement theory (see Sect. 4.6, Chap. 4).

12.7 FINITE ELEMENT MODEL FOR PLASTIC RESPONSE

12.7.1 Plastic Stress-Strain Relations

In the mathematical theory of plasticity two types of theories have been advanced. The one which is called *deformation (total) theory* is based on the premise that the final state of stress is uniquely determined by the final state of strain. In the other, which is referred to as *flow (incremental) theory*, it is stated that the plastic strain increments are related to the final state of stress, plastic strain, and the stress increment. In general, these relations are not integrable and the integral depends on the loading path. For this reason the term *non-holonomic* theory also has been suggested to designate the theory. It is now a fairly well established fact that the deformation theory is incapable of tracing of the totality of the loading-deformation history, especially during the reversal of loading. Using finite-element method and incremental techniques, the flow theory can easily be applied to solve practical problems. A brief review of the flow theory is given in Chap. 2, Vol. 1.

Flow theory

Yielding of material can occur only if the stress $\{\sigma\}$ satisfies the yield criteria

$$f(\{\sigma\}, k) = 0 \tag{12.86}$$

where k is a "hardening parameter." Using the flow rule (*normality principle*) of von Mises, plastic strain increment $d\{\varepsilon^p\}$ can be written as

$$d\{\varepsilon^p\} = \lambda \frac{\partial f}{\partial\{\sigma\}} \tag{12.87}$$

which is the same as Eq. (2.58), Vol. 1. λ is known as a proportionality constant which is a positive scalar function. During an infinitesimal increment of stress, changes of strain are assumed to be divisible into elastic and plastic parts as

$$d\{\varepsilon\} = d\{\varepsilon^e\} + d\{\varepsilon^p\} \tag{12.88}$$

Elastic strain increments are related to stress increments by the *elastic constitutive matrix* $[D]$ and plastic strain increment by normality principle, i.e.,

$$d\{\varepsilon\} = [D]^{-1} d\{\sigma\} + \lambda \frac{\partial f}{\partial\{\sigma\}} \tag{12.89}$$

when plastic yielding occurs, the stresses are on the yield surface given by Eq. (12.86). Differentiating

$$\frac{\partial f}{\partial\{\sigma\}} d\{\sigma\} + \frac{\partial f}{\partial k} dk = 0 \tag{12.90}$$

Combining Eqs. (12.89) and (12.90) in matrix form as

$$\begin{Bmatrix} d\varepsilon_1 \\ d\varepsilon_2 \\ \vdots \\ \hline 0 \end{Bmatrix} = \left[\begin{array}{cc|c} & & \dfrac{df}{\partial\sigma_1} \\ & [D]^{-1} & \dfrac{\partial f}{\partial\sigma_2} \\ & & \vdots \\ \hline \dfrac{\partial f}{\partial\sigma_1} & \dfrac{\partial f}{\partial\sigma_2} \quad - & \dfrac{1}{\lambda}\dfrac{\partial f}{\partial k}dk \end{array}\right] \begin{Bmatrix} d\sigma_1 \\ d\sigma_2 \\ \vdots \\ \hline \lambda \end{Bmatrix} \tag{12.91}$$

or

$$d\{\sigma\} = [D]_{ep}\, d\{\varepsilon\} \tag{12.92a}$$

i.e.,

$$\{\dot{\sigma}\} = [D]_{ep}\{\dot{\varepsilon}\} \tag{12.92b}$$

$[D]_{ep}$ is known as *elastoplastic matrix* (Zienkiewicz, 1971) which is symmetric, *positive definite* and is given by

$$[D]_{ep} = [D] - [D]\left\{\frac{\partial f}{\partial\{\sigma\}}\right\}\left\{\frac{\partial f}{\partial\{\sigma\}}\right\}^T [D]\left(\frac{1}{\lambda}\frac{\partial f}{\partial k}dk + \left\{\frac{\partial f}{\partial\{\sigma\}}\right\}^T [D]\left\{\frac{\partial f}{\partial\{\sigma\}}\right\}\right)^{-1} \tag{12.93}$$

The reader may refer to Chap. 2, Vol. 1, for more details of the flow theory.

Deformation theory

The basic assumption of this theory is that no strain reversal occurs (see Fig. 12.16) and therefore the relationship between increment in stress and increment in strain may be determined from the tangent modulus. Moreover, since the stress in the member is essentially uniaxial, the *effective strain* for the determination of tangent modulus is considered to be the longitudinal strain. Hence,

$$\dot{\sigma} = E_t\dot{\varepsilon} \tag{12.94}$$

By using the above assumptions, the deformation theory is essentially a *nonlinear elastic theory* and this is used in further development of the model for the plastic response of beam-columns.

Assumptions The following assumptions are made in addition to the usual assumptions of thin-walled beams.

1. Members are initially straight and prismatic.
2. The load is static and applied proportionately.
3. There will be no local buckling failure.

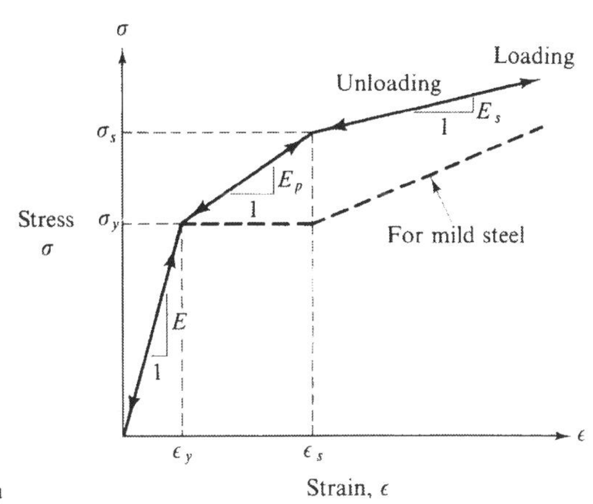

FIGURE 12.16
Trilinear stress-strain diagram

4. Residual stresses are present and they must satisfy the following equations of equilibrium

$$\int_A \sigma_r \, dA = \int_A \sigma_r \eta \, dA = \int_A \sigma_r \xi \, dA = 0 \tag{12.95a}$$

$$-m''_{\omega r} + T'_{svr} + (\bar{K}_r \theta'_\zeta)' = 0 \tag{12.95b}$$

5. Strain reversal does not occur.

12.7.2 Incremental Equilibrium Equation

Equation (12.15) and Eq. (12.22) represent the total and an approximate incremental equilibrium equation even for the plastic response except that the generalized stresses P, m_ξ, m_η and m_ω cannot explicitly be written in terms of displacements, instead they can be determined from Eq. (12.2). Since the usual elastic centroidal and shear center axis lose their significance after the initiation of plastic response and the instantaneous incremental centroidal and shear center axes are constantly shifting in both location and orientation, the incremental equilibrium equations are derived with respect to any two arbitrary axes which are fixed in relation to the cross section (preferably centroidal principal axes, shear center and sectorial centroid of the elastic section).

The assumption for incremental displacements for the plastic case differ from the elastic case only in the axial displacement term, where a quadratic rather a linear function is used. For a constant load increment of $\dot{P}, \dot{u}_{\zeta,\zeta}$, is not constant. Since

$$\left(\int_A E_t \, dA\right) \dot{u}_{\zeta,\zeta} = \dot{P} \tag{12.96a}$$

and

$$\left(\int_A E_t \, dA\right) \tag{12.96b}$$

varies along the length of the element. Hence in order to ensure convergence in the axial load, quadratic function is used as a shape function for the increment in axial displacement as

$$\dot{u}_{\zeta c} = \langle n_2 \rangle \begin{Bmatrix} \dot{u}^p_{\zeta c} \\ \dot{u}^{(p+q)/2}_{\zeta c} \\ \dot{u}^q_{\zeta c} \end{Bmatrix} \tag{12.97a}$$

$$\langle n_2 \rangle = \langle (2\beta - 1)(\beta - 1), -4\beta(\beta - 1), \beta(2\beta - 1) \rangle \tag{12.97b}$$

where $\dot{u}^{(p+q)/2}_{\zeta c}$ is the increment in axial displacement at the midsection of the element length. The incremental stiffness at any section may be determined from the instantaneous tangent modulus at all parts of the cross section. The incremental section properties may be determined by the transformed area concept and are given later by Eq. (12.103) where the $A^T \ldots I^T_\omega$ represent the properties of the transformed section with respect to O and C and ξ and η reference axes.

The incremental virtual work equation for the prismatic element for plastic case can be established by substituting for the stress resultants $\dot{P}$, $\dot{m}_\xi$, $\dot{m}_\eta$, $\dot{m}_\omega$ and $\dot{T}_{sv}$ in Eq. (12.22) in terms of displacement increments of Eqs. (12.25) and (12.26) as

$$\begin{aligned}
&\int_0^l \Bigg\{ \delta \dot{u}'_{\zeta c} \Bigg[\left(\int_A E_t \, dA\right) \dot{u}'_{\zeta c} - \left(\int_A E_t \xi \, dA\right) \dot{u}''_{\xi 0} - \left(\int_A E_t \eta \, dA\right) \dot{u}''_{\eta 0} \\
&\quad + \left(\int_A E_t \omega^0_{Ds} \, dA\right) \dot{\theta}''_\zeta \Bigg] + \delta \dot{u}''_{\xi 0} \Bigg[- \left(\int_A E_t \xi \, dA\right) \dot{u}'_{\zeta c} + \left(\int_A E_t \xi^2 \, dA\right) \dot{u}''_{\xi 0} \\
&\quad + \left(\int_A E_t \xi \eta \, dA\right) \dot{u}''_{\eta 0} - \left(\int_A E_t \omega^0_{Ds} \xi \, dA\right) \dot{\theta}''_\zeta \Bigg] + \delta \dot{u}''_{\eta 0} \Bigg[- \left(\int_A E_t \eta \, dA\right) \dot{u}'_{\zeta c} \\
&\quad + \left(\int_A E_t \xi \eta \, dA\right) \dot{u}''_{\xi 0} + \left(\int_A E_t \eta^2 \, dA\right) \dot{u}''_{\eta 0} - \left(\int_A E_t \omega^0_{Ds} \eta \, dA\right) \dot{\theta}''_\zeta \Bigg] \\
&\quad + \delta \dot{\theta}''_\zeta \Bigg[\left(\int_A E_t \omega^0_{Ds} \, dA\right) \dot{u}'_{\zeta c} - \left(\int_A E_t \omega^0_{Ds} \xi \, dA\right) \dot{u}''_{\xi 0} - \left(\int_A E_t \omega^0_{Ds} \eta \, dA\right) \dot{u}''_{\eta 0} \\
&\quad + \left(\int_A E_t \omega^{02}_{Ds} \, dA\right) \dot{\theta}''_\zeta \Bigg] \Bigg\} d\zeta + G \int_0 K_T \dot{\theta}'_\zeta \, \delta \dot{\theta}'_\zeta \, d\zeta \\
&\quad + \int_0^l \Bigg[f_\xi \, \delta(\dot{\theta}_\zeta \dot{u}'_{\eta 0}) - f_\eta \, \delta(\dot{\theta}_\zeta \dot{u}'_{\xi 0}) + \frac{P}{2} \delta(\dot{u}'^2_{\xi 0} + \dot{u}'^2_{\eta 0} + 2e_\eta \dot{u}'_{\xi 0} \dot{\theta}'_\zeta - 2e_\xi \dot{u}'_{\eta 0} \dot{\theta}'_\zeta) \\
&\quad + m_\eta \, \delta(\dot{u}'_{\eta 0} \dot{\theta}'_\zeta) - m_\xi \, \delta(\dot{u}'_{\xi 0} \dot{\theta}'_\zeta) + \frac{\overline{K}}{2} \delta(\dot{\theta}'_\zeta)^2 \Bigg] d\zeta \\
&= \int_0^l [\dot{q}_\xi \, \delta \dot{u}_{\xi 0} + \dot{q}_\eta \, \delta \dot{u}_{\eta 0} + \dot{q}_\zeta \, \delta \dot{u}_{\zeta c}] \, d\zeta + [\dot{P}^* \, \delta \dot{u}_{\zeta c} + \dot{f}^*_\xi \, \delta \dot{u}_{\xi 0} + \dot{f}^*_\eta \, \delta \dot{u}_{\eta 0} - \dot{m}^*_\xi \, \delta \dot{\theta}_\xi \\
&\quad - \dot{m}^*_\eta \, \delta \dot{\theta}_\eta + \dot{m}^*_\zeta \, \delta \dot{\theta}_\zeta + \dot{m}^*_\omega \, \delta \dot{\theta}'_\zeta]^l_0
\end{aligned} \tag{12.98}$$

where K_T is defined in Eq. (12.27). Substituting Eq. (12.41) and (12.97) for $\dot{u}_{\xi 0}$, $\dot{u}_{\eta 0}$, $\dot{u}_{\zeta c}$ and $\dot{\theta}_\zeta$ into the variational form of displacement equation of equilibrium (12.98) and carrying out the integration as explained in Sect. 12.4.2, yields the standard form of finite element incremental equilibrium equation for the plastic case of an arbitrary element in local coordinate system as

$$[k_S^T]\{\dot{u}_E\}_L + [k_G]\{\dot{u}_E\}_L = \{\dot{F}_E\}_L \tag{12.99}$$

$[k_S^T]$ is now a full matrix with submatrices denoted as

$$[k_S^T] = \begin{bmatrix} k_{u_\xi u_\xi} & \text{SYMMETRIC} & & \\ k_{u_\eta u_\xi} & k_{u_\eta u_\eta} & & \\ k_{\theta_\zeta u_\xi} & k_{\theta_\zeta u_\eta} & k_{\theta_\zeta \theta_\zeta} & \\ k_{u_\zeta u_\xi} & k_{u_\zeta u_\eta} & k_{u_\zeta \theta_\zeta} & k_{u_\zeta u_\zeta} \end{bmatrix} \tag{12.100}$$

and $[k_G]$ is of the form given by

$$[k_G] = \begin{bmatrix} g_{u_\xi u_\xi} & 0 & g_{u_\xi \theta_\zeta} & 0 \\ & g_{u_\eta u_\eta} & g_{u_\eta \theta_\zeta} & 0 \\ & & g_{\theta_\zeta \theta_\zeta} & 0 \\ \text{SYMMETRIC} & & 0 & 0 \end{bmatrix} \tag{12.101}$$

The element matrices are given in Table 12.5 where the numerical evaluation of plastic sectional properties $A^T \ldots I_\omega^T$ is given in Eq. (12.103).

Table 12.5(a) SUBMATRICES IN THE FLEXURAL STIFFNESS MATRIX $[k_S^T]$ OF EQ.(12.100)

$$\frac{l^3}{E}[k_{u_\xi u_\xi}] = I_{\eta p}[K_{33}^{220}] + (I_{\eta q} - I_{\eta p})[K_{33}^{221}]$$

$$\frac{l^3}{E}[k_{u_\xi u_\eta}] = I_{\xi\eta p}[K_{33}^{220}] + (I_{\xi\eta q} - I_{\xi\eta p})[K_{33}^{221}]$$

$$\frac{l^3}{E}[k_{u_\xi \theta_\zeta}] = -\{S_{\omega\eta p}[K_{33}^{220}] + (S_{\omega\eta q} - S_{\omega\eta p})[K_{33}^{221}]\}$$

$$\frac{l^2}{E}[k_{u_\xi u_\zeta}] = -\{Q_{\eta p}[K_{32}^{210}] + (Q_{\eta q} - Q_{\eta p})[K_{32}^{211}]\}$$

$$\frac{l^3}{E}[k_{u_\eta u_\eta}] = I_{\xi p}[K_{33}^{220}] + (I_{\xi q} - I_{\xi p})[K_{33}^{221}]$$

$$\frac{l^3}{E}[k_{u_\eta \theta_\zeta}] = -\{S_{\omega\xi p}[K_{33}^{220}] + (S_{\omega\xi q} - S_{\omega\xi p})[K_{33}^{221}]$$

$$\frac{l^2}{E}[k_{u_\eta u_\zeta}] = -\{Q_{\xi p}[K_{32}^{210}] + (Q_{\xi q} - Q_{\xi p})[K_{32}^{211}]$$

$$\frac{l^3}{E}[k_{\theta_\zeta \theta_\zeta}] = I_{\omega p}[K_{33}^{220}] + (I_{\omega q} - I_{\omega p})[K_{33}^{221}]$$

$$\frac{l^2}{E}[k_{\theta_\zeta u_\zeta}] = S_{\omega p}[K_{32}^{210}] + (S_{\omega q} - S_{\omega p})[K_{32}^{211}]$$

$$\frac{l}{E}[k_{u_\zeta u_\zeta}] = A_p[K_{22}^{110}] + (A_q - A_p)[K_{22}^{111}]$$

Table 12.5(b) SUBMATRICES IN THE GEOMETRIC STIFFNESS MATRIX $[k_G]$ OF EQ.(12.101)

$$l[g_{u_\xi u_\xi}] = l[g_{u_\eta u_\eta}] = P_p[K_{33}^{110}] + (P_q - P_p)[K_{33}^{111}]$$

$$l[g_{u_\xi \theta_\zeta}] = l[g_{\theta_\zeta u_\xi}]^T = -f_\eta l([K_{33}^{100}] + [K_{33}^{111}]) - m_{\xi p}[K_{33}^{110}] + P_p e_{\eta p}[K_{33}^{110}] + (P_q - P_p)e_{\eta p}[K_{33}^{111}]$$

$$l[g_{u_\eta \theta_\zeta}] = l[g_{\theta_\zeta u_\eta}]^T = f_\xi l([K_{33}^{100}] + [K_{33}^{111}]) + m_{\eta p}[K_{33}^{110}] - P_p e_{\xi p}[K_{33}^{110}] - (P_q - P_p)e_{\xi p}[K_{33}^{111}]$$

$$\frac{l}{G}[g_{\theta_\zeta \theta_\zeta}] = (K_T + \bar{K}/G)_p[K_{33}^{110}] + [(K_T + \bar{K}/G)_q - (K_T + \bar{K}/G)_p][K_{33}^{111}]$$

Linear variation can be assumed for all the section properties in Eq. (12.98) as well as for the stress resultants m_ξ, m_η and $\bar{K}$. Assembling the element equations by the direct stiffness procedure (using compatibility and equilibrium relationship as explained in Sect. 12.4.2) and applying boundary conditions yield the total set of incremental equilibrium equations which may be written as

$$[K_S^T]\{\dot{u}\} + [K_G]\{\dot{u}\} = \{\dot{F}\} \tag{12.102}$$

where $[K_S^T]$ and $[K_G]$ are tangent and geometric stiffness matrices, respectively.

12.7.3 Numerical Evaluation of Section Properties and Stress Resultants

Expressions for the evaluation of the section properties are given in terms of the transformed section for the tangent modulus approach for the trilinear stress-strain curve of Fig. 12.16, since this is the most complex case which has been considered.

FIGURE 12.17
Transformed section of a plate segment

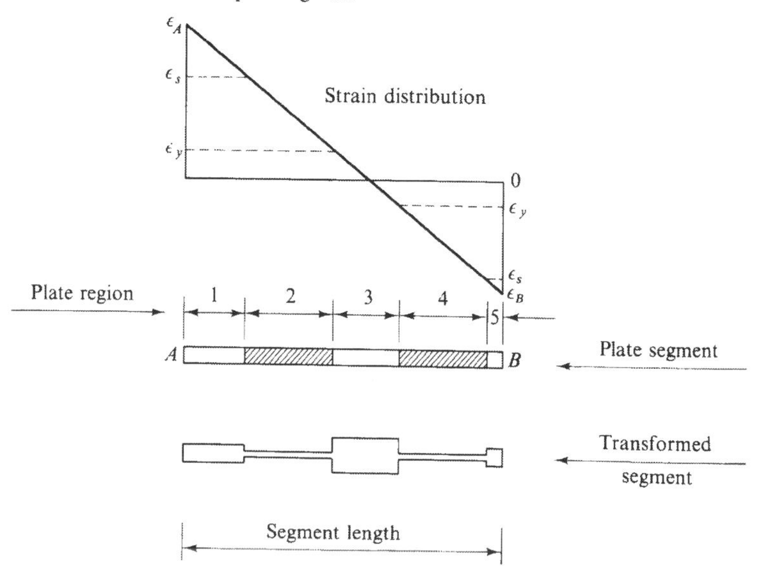

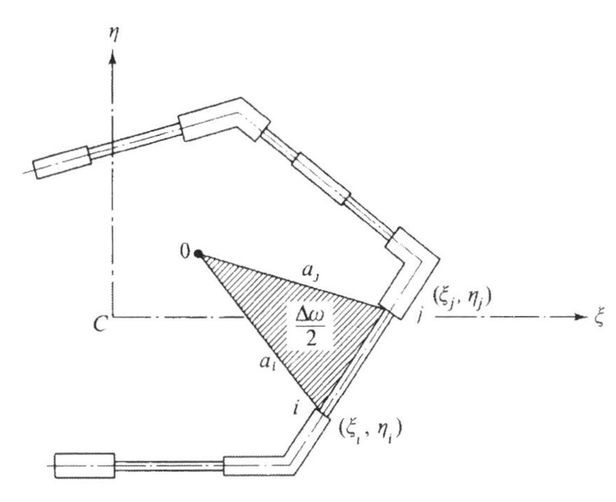

FIGURE 12.18
Arbitrary transformed section

Consider any straight segment of plate for which residual strain is assumed to vary linearly. By superposition, and the beam assumptions, the total strain will also vary linearly. For the stress-strain curve in Fig. 12.16, a linear variation of strain divides the plate segment into at most five regions, as shown in Fig. 12.17.

To evaluate the tangent modulus section properties, the plate thickness in each region can be modified by the modular ratio and the transformed section can be treated as though it possesses a constant modulus. Let Fig. 12.18 represent an arbitrary transformed section with reference axes C and O. The tangent section properties may be evaluated as

$$A^T = \int_A \frac{E_t}{E}\, dA = \sum_{k=1}^{N} A_k \tag{12.103a}$$

$$Q_\xi^T = \int_A \frac{E_t}{E} \eta\, dA = \sum_{k=1}^{N} A_k \eta_k \tag{12.103b}$$

$$Q_\eta^T = \int_A \frac{E_t}{E} \xi\, dA = \sum_{k=1}^{N} A_k \xi_k \tag{12.103c}$$

$$S_\omega^T = \int_A \frac{E_t}{E} \omega_{Ds}^0\, dA = \sum_{k=1}^{N} \sum_{r=1}^{5} \frac{1}{2} b_r t_r (\omega_{ir} + \omega_{jr}) \tag{12.103d}$$

$$I_\xi^T = \int_T \frac{E_t}{E} \eta^2\, dA = \sum_{k=1}^{N} I_{\xi\xi k} + \sum_{k=1}^{N} A_k \eta_k^2 \tag{12.103e}$$

$$I_\eta^T = \int_A \frac{E_t}{E} \xi^2\, dA = \sum_{k=1}^{N} I_{\eta\eta k} + \sum_{k=1}^{N} A_k \xi_k^2 \tag{12.103f}$$

$$I_{\xi\eta}^T = \int_A \frac{E_t}{E} \xi\eta\, dA = \sum_{k=1}^{N} I_{\xi\eta k} + \sum_{k=1}^{N} A_k \xi_k \eta_k \tag{12.103g}$$

$$S_{\omega\eta}^{T} = \int_{A} \frac{E_t}{E} \omega_{Ds}^{0} \xi \, dA = \sum_{k=1}^{N} \sum_{r=1}^{5} \frac{b_r t_r}{6} [\omega_{ir}(\xi_{jr} + 2\xi_{ir}) + \omega_{jr}(\xi_{ir} + 2\xi_{jr})] \tag{12.103h}$$

$$S_{\omega\xi}^{T} = \int_{A} \frac{E_t}{E} \omega_{Ds}^{0} \eta \, dA = \sum_{k=1}^{N} \sum_{r=1}^{5} \frac{b_r t_r}{6} [\omega_{ir}(\eta_{jr} + 2\eta_{ir}) + \omega_{jr}(\eta_{ir} + 2\eta_{jr})] \tag{12.103i}$$

$$I_{\omega}^{T} = \int_{A} \frac{E_t}{E} \omega_{Ds}^{02} \, dA = \sum_{k=1}^{N} \sum_{r=1}^{5} \frac{b_r t_r}{3} (\omega_{ir}^2 + \omega_{ir}\omega_{jr} + \omega_{jr}^2) \tag{12.103j}$$

In which

r = plate region
N = number of plate segments
k = plate segment index
t_r = plate region thickness
b_r = plate region length
ω_{ir}, ω_{jr} = warping coordinates at i and j.

The stress resultants required in Eq. (12.2) and (12.16), and for the evaluation of $[k_G]$, may be computed numerically as

$$P = \sum_{k=1}^{N} \sum_{r=1}^{5} \frac{b_r t_k}{2} (\sigma_{ir} + \sigma_{jr}) \tag{12.104a}$$

$$m_{\xi} = \sum_{k=1}^{N} \sum_{r=1}^{5} \frac{b_r t_k}{6} [\sigma_{ir}(\eta_{jr} + 2\eta_{ir}) + \sigma_{jr}(\eta_{ir} + 2\eta_{jr})] \tag{12.104b}$$

$$m_{\eta} = \sum_{k=1}^{N} \sum_{r=1}^{5} \frac{b_r t_k}{6} [\sigma_{ir}(\xi_{jr} + 2\xi_{ir}) + \sigma_{jr}(\xi_{ir} + 2\xi_{jr})] \tag{12.104c}$$

$$m_{\omega} = \sum_{k=1}^{N} \sum_{r=1}^{5} \frac{b_r t_k}{6} [\sigma_{ir}(\omega_{jr} + 2\omega_{ir}) + \sigma_{jr}(\omega_{ir} + 2\omega_{jr})] \tag{12.104d}$$

$$\bar{K} = \sum_{k=1}^{N} \sum_{r=1}^{5} \frac{b_r t_k}{12} [\sigma_{ir}(4a_{ir}^2 + 2a_{jr}^2 - b_r^2) + \sigma_{jr}(4a_{jr}^2 + 2a_{ir}^2 - b_r^2)] \tag{12.104e}$$

$$f_{\xi} = \frac{dm_{\eta}}{d\zeta} \tag{12.104f}$$

$$f_{\eta} = \frac{dm_{\xi}}{d\zeta} \tag{12.104g}$$

In which

a_{ir}, a_{jr} = distance of the i and j section points of the region with respect to 0.

The stress resultant T_{sv} must be evaluated from displacements as

$$T_{sv} = GK_T^T \theta'_\zeta \tag{12.105}$$

in which K_T^T may be computed from the transformed section as

$$K_T^T = \frac{1}{3} \sum_{k=1}^{N} \sum_{r=1}^{5} b_r^* t_k^3 \tag{12.106a}$$

in which

$$b_r^* = \frac{E_t b_r}{E} \tag{12.106b}$$

12.8 PLASTIC BUCKLING OF COLUMNS

In addition to the assumptions stated in Sec. 12.7.1, the following assumptions are made for the plastic member buckling problem.

1. Boundary conditions are such that the member in the plastic range is statically determinate.
2. The effect of prebuckling displacements on the equilibrium equation may be neglected.

12.8.1 Solution Technique

The condition required to satisfy the requirement of bifurcation loading is that the determinant of the coefficient matrix of Eq. (12.102) be equal to zero

$$\det([K_S^T] + [K_G]) = 0 \tag{12.107a}$$

or

$$[K_S^T]\{\dot{u}\} + [K_G]\{\dot{u}\} = \{0\} \tag{12.107b}$$

If $[K_S^T]$ were independent of load and $[K_G]$ were linearly dependent on loading, the buckled shape could be obtained from a standard eigenvalue analysis. However, for an arbitrary cross section subjected to arbitrary loads, the matrix is highly sensitive to the loading condition after plastic response has been initiated.

For any arbitrary statically determinate loading, equilibrium is established including the effect of plastic material response but neglecting the effect of displacements on equilibrium equations, by an iterative approach. Knowing the strains after augmenting the residual strains throughout the member, the tangent stiffness $[K_S^T]$ can be established. The prebuckling stress resultants such as P, m_ξ, m_η, m_ω and $\bar{K}$ can also be established by simple numerical integration using Eqs. (12.104).

An examination of stiffness matrix of Eq. (12.107b) shows that the equation may be written as

$$([K_1] + [K_2] + [K_3])\{\dot{u}\} + [K_G]\{\dot{u}\} = \{0\} \tag{12.108}$$

It is to be noted that the St. Venant torsional rigidity terms are included in $[K_G]$ instead of $[K_S^T]$ matrix. The matrices $[K_1]$, $[K_2]$ and $[K_3]$ contain terms proportional to $1/L$, $1/L^2$ and $1/L^3$, respectively, and $[K_G]$ is proportional to $1/L$. Defining the critical length $L_c = \lambda L$ and factoring λ; the condition for the existence of non-trivial solution of Eq. (12.108) is written as

$$\frac{1}{\lambda^3}\left([K_3] + \lambda[K_2] + \lambda^2[K_1]\right)\{\dot{u}\} = -\frac{1}{\lambda}[K_G]\{\dot{u}\} \tag{12.109a}$$

or

$$\frac{1}{\lambda_{i+1}^2}[K(\lambda_i)]\{\dot{u}\} = -[K_G]\{\dot{u}\} \tag{12.109b}$$

Assuming a value of λ_i, an eigenvalue solution will yield λ_{i+1}. Equation (12.109b) may be iterated until $\lambda_{i+1} = \lambda_i$, at which stage the critical length is determined. For all practical purposes two iterations may suffice. It is to be noted that the above procedure is only valid if the stress resultants remain constant at each node as the member length is scaled. A simultaneous scaling of the transverse loading is required to satisfy this condition and hence this is applicable to statically determinate loading. An algorithm for determining the plastic critical length is given below.

1. Determine static stress resultants:
   (a) Subdivide the member into elements of equal length.
   (b) Compute elastic section properties such as A, I_ξ, I_η, I_ω, K_T, with C and O located at shear center and centroid of the section.
   (c) Solve for displacements with $[K_G] = [0]$ and $[K_S]$ determined from elastic properties.
   (d) Compute stress resultants P, m_ξ, m_η and m_ω at each node from this small deflection solution.
2. Determine strains (iteratively) which produce static stress resultants:
   (a) Compute strains from initial approximate displacement as

$$\varepsilon_i = u'_{\zeta c} - \eta u''_{\eta 0} - \xi u''_{\xi 0} + \omega^0_{Ds}\theta''_\zeta + \varepsilon_r$$

   (b) Calculate stress resultants $\hat{P}$, $\hat{m}_\xi$, $\hat{m}_\eta$ and $\hat{m}_\omega$ (using stress-strain curve) for the given strain distribution.
   (c) Compute the unbalance in stress resultants as

$$\bar{P} = P - \hat{P}$$

$$\bar{m}_\xi = m_\xi - \hat{m}_\xi$$

$$\bar{m}_\eta = m_\eta - \hat{m}_\eta$$

$$\bar{m}_\omega = m_\omega - \hat{m}_\omega$$

(d) Estimate strain increment from the equation

$$\Delta\varepsilon_i = \frac{\bar{P}}{A} + \frac{\bar{m}_\xi}{I_\xi}\eta + \frac{\bar{m}_\eta}{I_\eta}\xi + \frac{\bar{m}_\omega}{I_\omega}\omega_{Ds}^0$$

and a new approximate strain is

$$\varepsilon_{i+1} = \varepsilon_i + \Delta\varepsilon_i$$

(e) Iterate on steps (2b) to (2d) until the unbalanced stress resultants are negligible. (A set of strains have now been determined which will provide stress resultants to equilibrate the external forces, assuming the prebuckling displacements have negligible effect on the equilibrium equations).

3. Computation of tangent and geometric stiffness matrices:

The tangent stiffness matrix $[K_S^T]$ and $[K_G]$ of Eq. (12.107b) may now be evaluated as explained in Sect. 12.7.3 and given in Table 12.5.

4. Solution for critical length

Equation (12.109b) is solved for λ to determine the critical length $L_c = \lambda L$.

It is seen that when the tangent properties are the same all along the length, or if the axial and flexural behavior representations are uncoupled, the first iteration will give exact solution for the critical length. To illustrate the accuracy of finite element formulation for plastic buckling, the following examples are solved and compared with the available results.

12.8.2 Numerical Examples

EXAMPLE 12-4 *Lateral Torsional Buckling of Double Angle Strut* Figure 12.19 shows a double angle strut subjected to axial load. It is seen that the coefficient matrix $[K_S^T]$ will be uncoupled by choosing the reference points at instantaneous centroid and shear center. Hence, the transformed section will be subjected to both axial load, P, and bending moment $m_\xi = Pe_\eta$. $[K_S^T]$ and $[K_G]$ matrices are given in Table 12.6. A plot of σ/σ_y against L/r_x for lateral torsional buckling and Euler buckling about x-axis is shown in Fig. 12.19. The influence of residual stress and the significance of the plastic response is at once apparent. Eight elements are used for the idealization.

EXAMPLE 12-5 *Plastic Lateral Buckling of a Wide Flange Beam Subjected to a Central Load* In Fig. 12.20, the critical moment vs. slenderness ratio curve is shown for W18 × 50 beam subjected to a central load. A large reduction in the buckling strength results when the presence of residual stress initiates plastic behavior. The curves are compared with Massey and Pitman (1966) who recommended the curve based on a theoretical analysis. In their analysis, residual stress distribution is not

Table 12.6 STIFFNESS MATRICES FOR PLASTIC TORSIONAL-FLEXURAL BUCKLING OF A PRISMATIC MEMBER

$$
\begin{array}{cccccccc}
u^p_{\xi 0} & \theta^p_\zeta & l\theta^p_\eta & l\theta^p_{\zeta,\zeta} & u^q_{\xi 0} & \theta^q_\zeta & l\theta^q_\eta & l\theta^q_{\zeta,\zeta}
\end{array}
$$

$$
[k^T_S] = \frac{E}{l^3}\begin{bmatrix}
12I^T_\eta & & 6I^T_\eta & & -12I^T_\eta & & 6I^T_\eta & \\
 & 12I^T_\omega & & 6I^T_\omega & & -12I^T_\omega & & 6I^T_\omega \\
 & & 4I^T_\eta & & -6I^T_\eta & & 2I^T_\eta & \\
 & & & 4I^T_\omega & & -6I^T_\omega & & 2I^T_\omega \\
 & & & & 12I^T_\eta & & -6I^T_\eta & \\
 & \text{SYMMETRIC} & & & & 12I^T_\omega & & -6I^T_\omega \\
 & & & & & & 4I^T_\eta & \\
 & & & & & & & 4I^T_\omega
\end{bmatrix}
$$

$$
[k_G] = \frac{1}{60l}\begin{bmatrix}
72P & -72m_\xi & 6P & -6m_\xi & -72P & 72m_\xi & 6P & -6m_\xi \\
 & 72GJ & -6m_\xi & 6GJ & 72m_\xi & -72GJ & -6m_\xi & 6GJ \\
 & & 8P & -8m_\xi & -6P & 6m_\xi & -2P & 2m_\xi \\
 & & & 8GJ & 6m_\xi & -6GJ & 2m_\xi & -2GJ \\
 & & & & 72P & -72m_\xi & -6P & 6m_\xi \\
 & \text{SYMMETRIC} & & & & 72GJ & 6m_\xi & -6GJ \\
 & & & & & & 8P & -8m_\xi \\
 & & & & & & & 8GJ
\end{bmatrix}
$$

$J = (K_T + \bar{K}/G)$

FIGURE 12.19
Plastic buckling of double angle strut

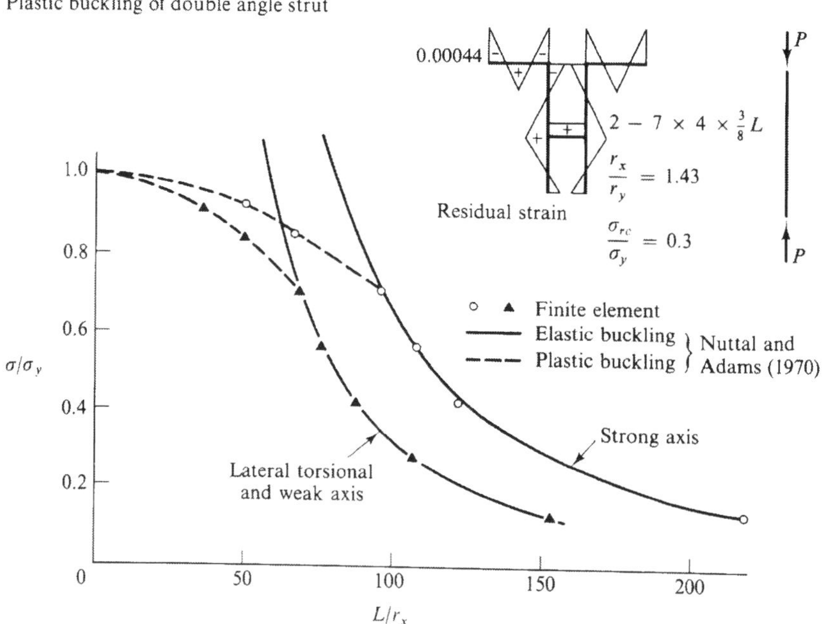

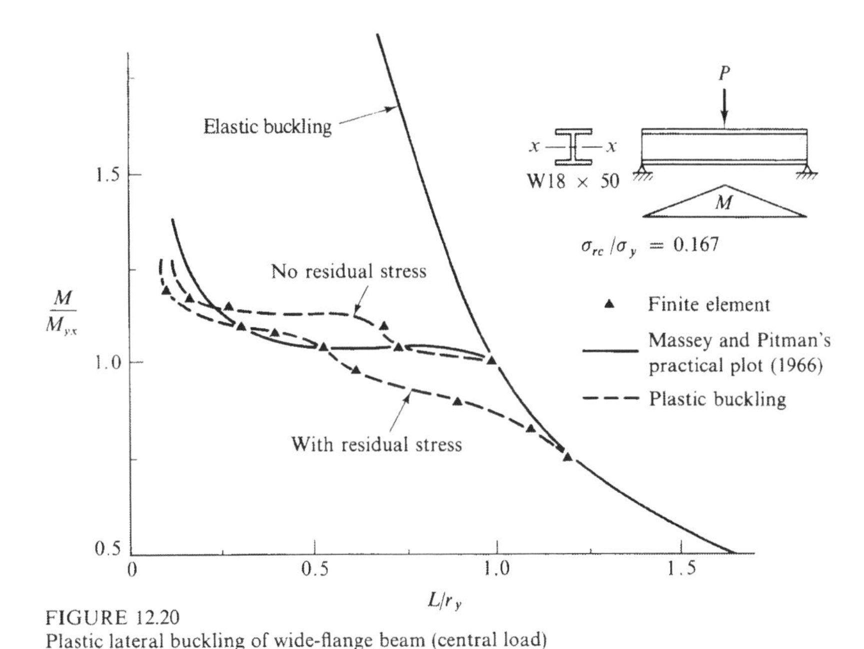

FIGURE 12.20
Plastic lateral buckling of wide-flange beam (central load)

taken into account. In the present analysis, eight elements are used and computed points assume the loads act through shear center.

EXAMPLE 12-6 *Plastic Lateral Torsional Buckling with Biaxial Bending* A plastic buckling analysis of W14 × 43 section is carried out for the beam-column shown

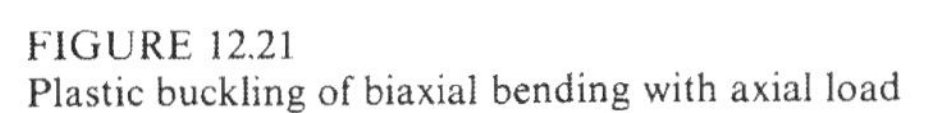

FIGURE 12.21
Plastic buckling of biaxial bending with axial load

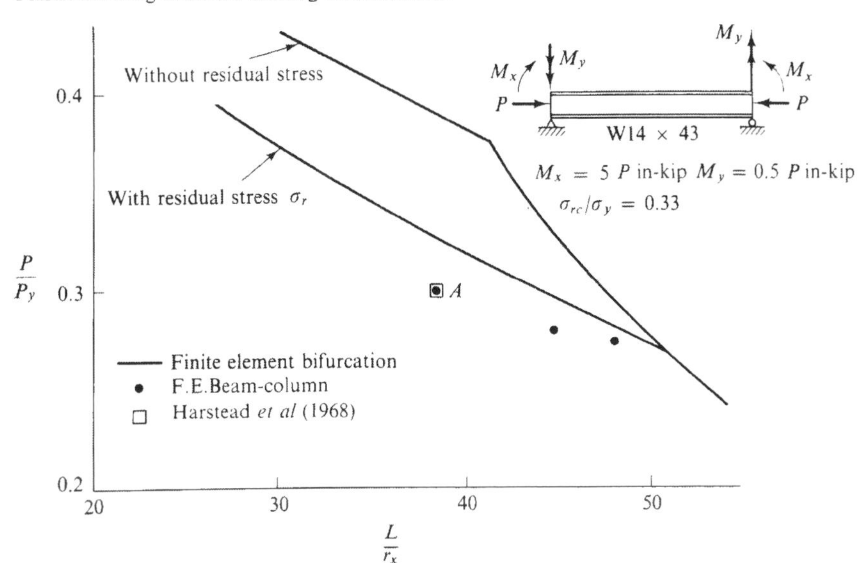

in Fig. 12.21. This figure shows a plot of critical loads for biaxial bending of the section determined from an eigenvalue analysis and neglecting prebuckling deformations. The curves are drawn for two cases: (a) without residual stresses, and (b) with residual stress. Again, for the same column, bending moment distribution is determined by load deflection approach for a particular load. Using the stress resultants obtained, buckling problem is solved as an eigenvalue problem. For a column of length of 221 in (5.6 m) ($L/r_x = 37.6$) and a load of 125 kip (556 kN) the critical length obtained is 283 in (7.2 m) showing that the load level is below the ultimate load. At $P = 134$ kip (596 kN), the critical length is found to be 226 in (5.7 m) indicating that 135 kip is very near to the ultimate load. However, the critical load based on the eigenvalue analysis and neglecting prebuckling deformation is 27 percent greater than the load-deformation approach. In such situations the effect of prebuckling displacements cannot be ignored and the stability problem should be solved only by load-deformation approach.

EXAMPLE 12-7 Yoshida *et al.*, (1975) applied finite element method to investigate the effects of beam slenderness on the plastic buckling behavior of the two-span continuous beam shown in Fig. 12.22. In that investigation, the value of the span ratio L_2/L_1 was kept constant at 1.5 while the span was decreased from 2.44 m to 1.83 m, 1.46 m, 1.22 m and 0.98 m. Nine different values of the load ratio P_1/P_2 between ranges 0 to ∞ were considered. Their finite element analysis consists of solving in-plane bending moment distribution by load-deflection approach for a particular load. Using the above stress resultants, buckling problem is solved as an

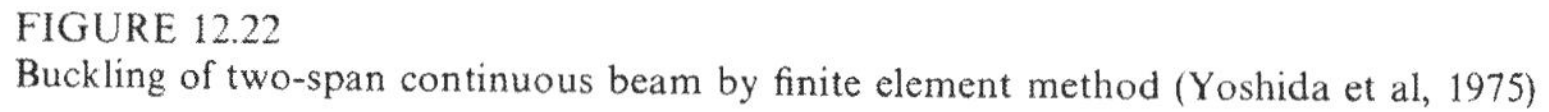
FIGURE 12.22
Buckling of two-span continuous beam by finite element method (Yoshida et al, 1975)

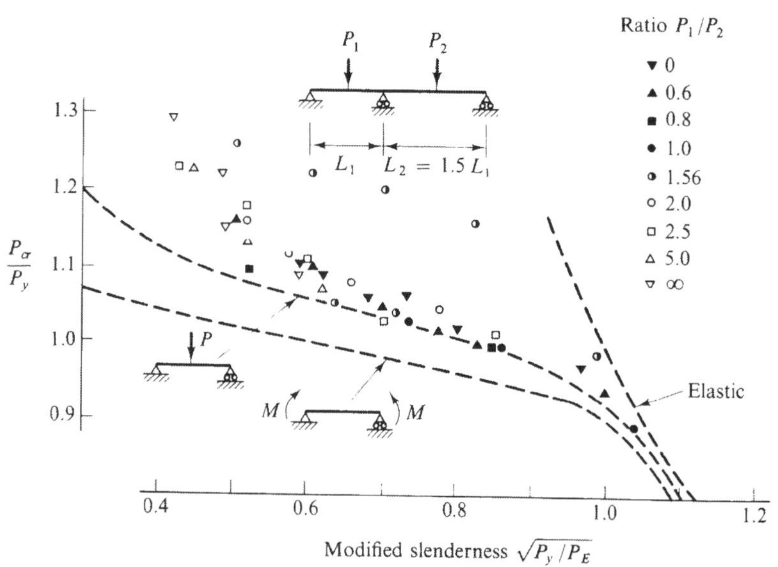

eigenvalue problem. This procedure is repeated until the value of $\lambda = 1$. Figure 12.22 shows the variation of the dimensionless critical load P_{cr}/P_y with a modified slenderness parameter $\sqrt{P_y/P_E}$ (P_E = theoretical elastic buckling load.) The results obtained show that until the beams are sufficiently stocky for the effects of strain-hardening, their plastic buckling loads are close to those for single span beam with similar patterns of in-plane moment.

12.9 PLASTIC STABILITY OF BEAM-COLUMNS

The columns in actual structures are frequently subjected to compression combined with biaxial bending. The realistic approach of determining the maximum strength of biaxially loaded columns is by load-deflection approach, since classical eigenvalue solution gives the upper bound for the ultimate load.

The solution procedure used for plastic response is iterative in nature and is based on the fact that Eq. (12.15a) and Eq. (12.16) and the associated natural boundary conditions of Table 12.1 are the requirements of equilibrium without regard to the origin of stress and the nature of the constitutive relations. For the assumption that no strain reversal occurs, the relationship between total strain and total stress is unique and any configuration which yields stress resultants that satisfy the equilibrium equations represent a solution for the given loading condition. It is necessary to carry out the total equilibrium check accurately at each loading.

Given an estimate of nodal displacement vectors, the longitudinal strain throughout an element may be computed from Eq. (12.23). After augmenting this strain with any set of residual strain, the total stress may be determined on the nodal cross section from the stress-strain relationship. The stress resultants at the nodal sections for each element may be obtained by numerical integration of normal stress. The stress resultants associated with each degree of freedom, which equilibrates the element may then be evaluated from the expressions given in Table 12.1. Summation of the element equilibrating forces and the generalized applied forces [see Eq. (12.110b)] at the nodes yields the out-of-balance or residual forces $\{E\}$, which the configuration does not equilibrate. A correction to the configuration $\{\dot{u}\}$ may then be evaluated by a further solution of Eq. (12.102) as

$$[K_S^T]\{\dot{u}\} + [K_G]\{\dot{u}\} = \{E\} \tag{12.110a}$$

where

$$\{E\} = \{\dot{F}\} \qquad \text{when} \qquad n = 1$$

and

$$\{E\} = \{F\} - [K_G]\{u\} - \text{static stress resultants} \tag{12.110b}$$

The process is continued until the norm of the unbalanced forces to that of the actual forces is less than allowable tolerance

$$\frac{\{E\}^T\{E\}}{\{F\}^T\{F\}} < \varepsilon = 0.001 \tag{12.111}$$

In the solution scheme $[K_S^T]$ is kept unchanged for some iterations (modified Newton-Raphson) as discussed in Sec. 12.6.1, until the rate of convergence begins to deteriorate, at which time the matrix is updated with properties based on current displacements. Since the nature of the problem is very sensitive to the applied forces and the type of stress-strain curve, an under-relaxation parameter has to be used to the residual force vector. The procedure may diverge if the matrix is not updated often enough.

12.9.1 Numerical Examples

EXAMPLE 12-8 *Plastic H-Columns Under Biaxial Bending* A W14 × 43 section with a doubly eccentric axial load, applied at eccentricities of $e_\xi = 0.5$ in (12.7 mm) and $e_\eta = 5$ in (127 mm) and $L/r_y = 117$ was analyzed and compared with the results obtained by Harstead, Birnstiel and Leu (1968). The beam was divided into four elements each of size 55.25 in (1403 mm). The load was incremented until the displacements no longer converged. Maximum load was reached in fifteen load increments. The procedure of incrementing the forces yielded results until a maximum load was attained, but numerical difficulties did not permit the evaluation in the unloading region. The results agree with Harstead *et al.* Figure 12.23 shows the load-midspan displacement for the beam-column.

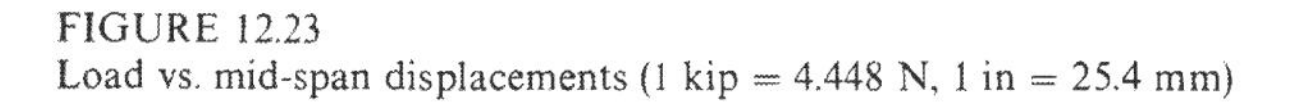
FIGURE 12.23
Load vs. mid-span displacements (1 kip = 4.448 N, 1 in = 25.4 mm)

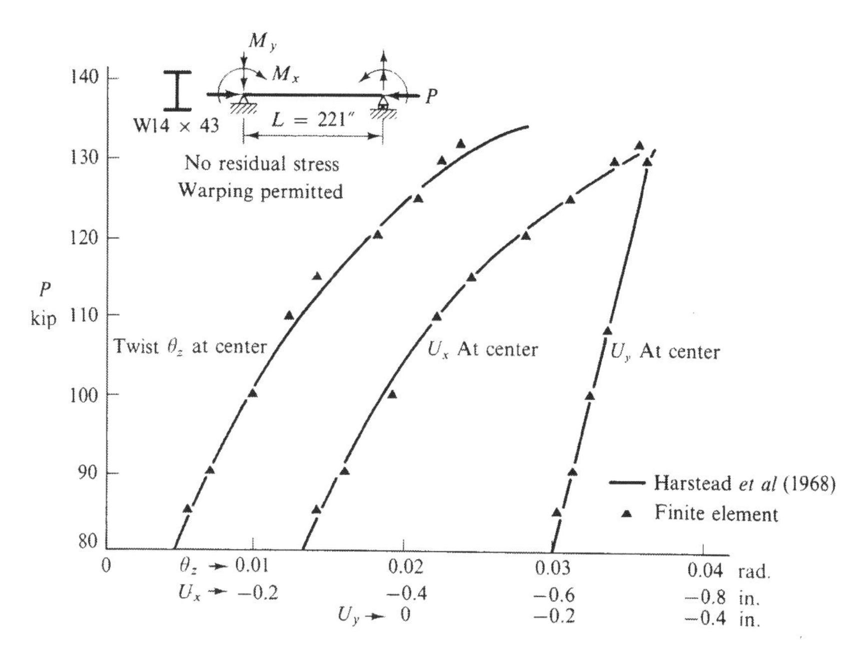

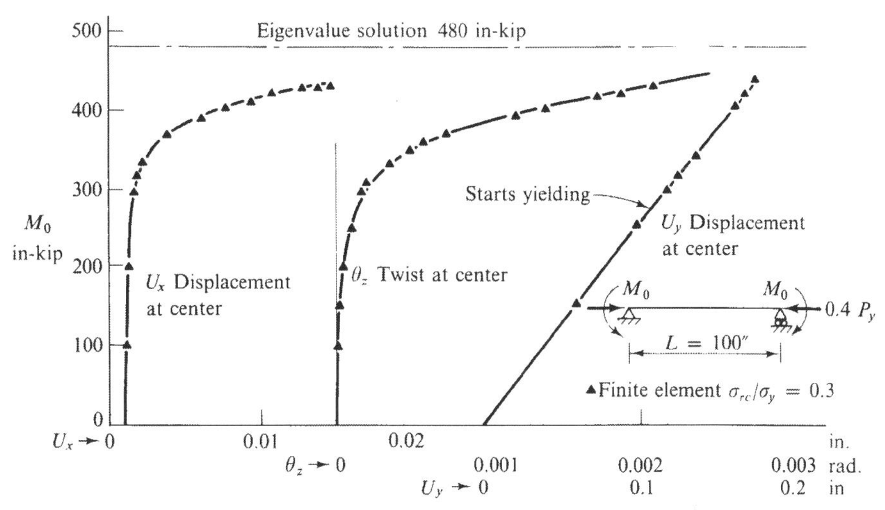

FIGURE 12.24
Moment vs. in and out of plane response (1 kip-in = 112 N-m, 1 in = 25.4 mm)

EXAMPLE 12-9 *Plastic Lateral Response with Initial Imperfection* A W8 × 31 beam-column with linear residual stress distribution over the cross-section was analyzed while subjected to a constant axial load of 120 kip (534 kN) ($P/P_y = 0.4$) and with an initial lateral displacement $U_x = 0.01 \sin(\pi z/L)$. A load deflection solution was carried out until with an end moment of 440 in-kip (49 kN-m), the residual force resultants no longer converged. Moment vs. in-plane and out-of-plane response is shown in Fig. 12.24. For the same problem the eigenvalue solution yields a critical length of 100 in (2.54 m) when an end moment of 480 in-kip (54 kN-m) and this is shown as an upper bound in the figure.

Efficient solution techniques such as those discussed in the paper by Rajasekaran and Murray (1975), may be used to increase the computational efficiency for the analysis of plastic beam-column problems.

12.10 COUPLED LOCAL AND MEMBER BUCKLING

In developing design procedures for thin-walled metallic members, consideration is generally given to local buckling as a phenomenon independent of, and separate from, member buckling. Observation of member failures, however, often indicates that failure involves both phenomena as shown in Fig. 12.25 and it is reasonable to investigate the conditions under which they interact. Przemieniecki (1972) presented a finite element analysis of local instability in plates, stiffened panels, and columns by representing the normal deflection u_η of the plate (see Fig. 12.26a) in the form of sine function in the longitudinal direction and cubic displacement function $\{n_3\}$ in the thickness (or transverse) direction as

$$u_\eta = Y(\beta) \sin(\pi\zeta/L), \quad \beta = \xi/b \tag{12.112a}$$

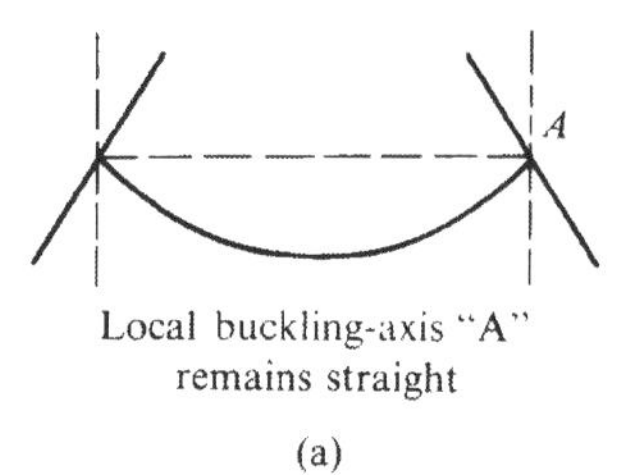

Local buckling-axis "A" remains straight

(a)

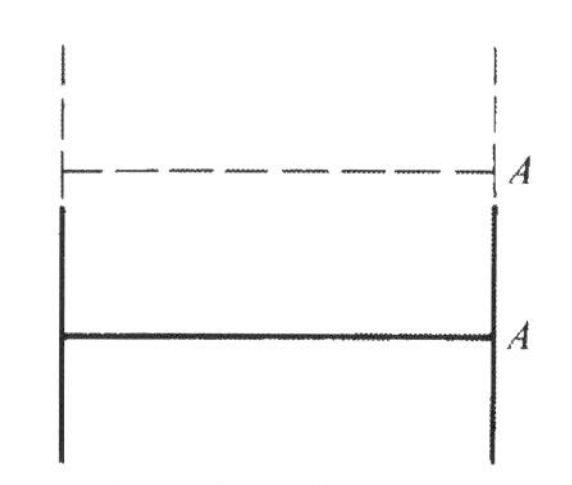

Column buckling-axis "A" deflects no distortion of cross section

(b)

A
A

Interaction of (a) and (b)

(c)

FIGURE 12.25
Interaction between column and local buckling

FIGURE 12.26(a)
Plate element for local instability

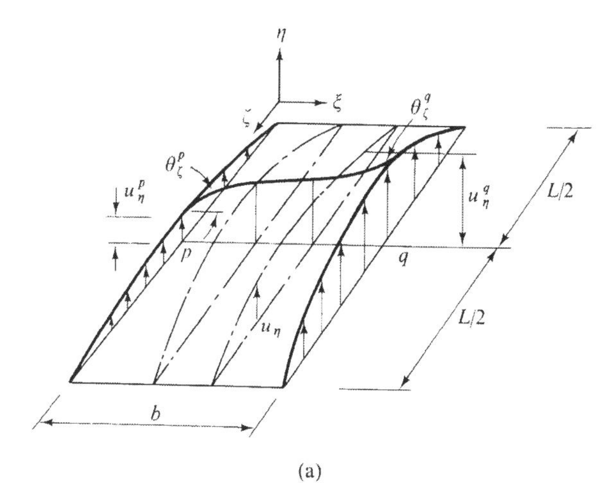

(a)

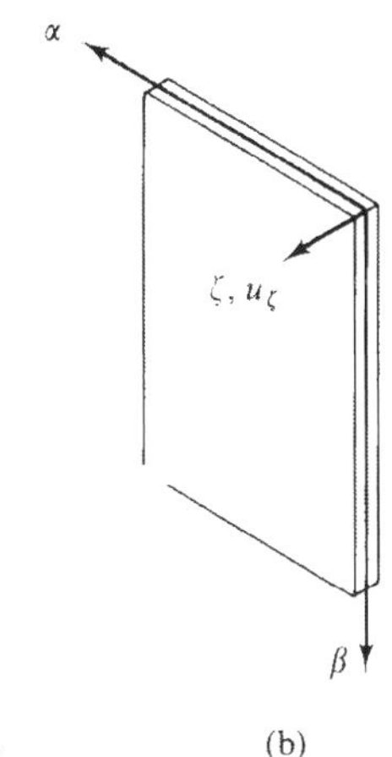

FIGURE 12.26(b)
Plate reference axis and displacements

and

$$Y(\beta) = \{n_3\}^T\{\underline{u}\} \tag{12.112b}$$

when $\{n_3\}^T$ is the same form as Eq. (12.38) and

$$\{\underline{u}\}^T = \langle u_\eta^p, b\theta_\zeta^p, u_\eta^q, b\theta_\zeta^q \rangle \tag{12.112c}$$

His analysis leads to the standard eigenvalue equation from which the buckling stress can be determined for any assumed wavelength of buckled pattern. The procedure is repeated until the value of the wavelength is found for the lowest buckling stress. Przemieniecki's analysis does not take into account the coupled behavior of the member and local buckling effects and also the variation of stress along the length of the thin-walled beam.

A finite element formulation of the virtual work equations combining the effects of local and member buckling is presented by Rajasekaran and Murray (1971, 1973b) for elastic and plastic cases. However, the element stiffnesses are derived from these equations for the particular case of elastic wide-flange sections and practical problems are investigated for the coupling behavior.

12.10.1 Formulation of the Problem

The assumptions associated with the formulation are those normally associated with an engineering theory for thin-walled beams, but out-of-plane flexure of the plate segments comprising the cross section is considered. Superposition of the two sets of displacements produces a beam-analysis which allows the cross section to distort. The basic assumptions may be listed as

1. The relative effect of shearing strain on the middle surface of the section is small in comparison to those arising from the flexural and torsional flexibility of the cross section and hence may be neglected.

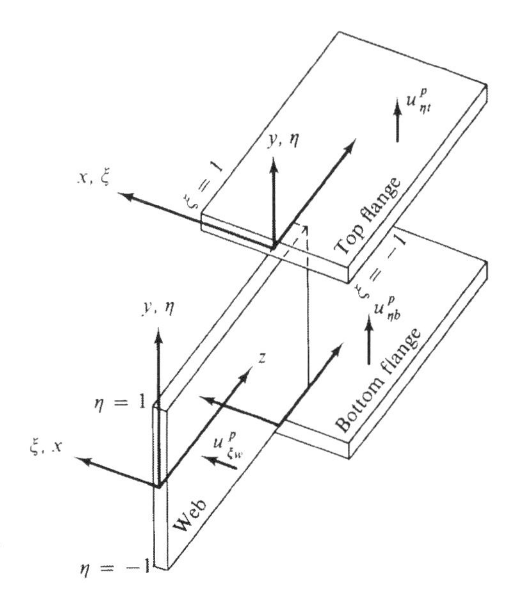

FIGURE 12.27
Local coordinate system for plate deformations

2. The angle between adjacent component plates of the cross section remains unchanged at their common edge during deformation.
3. Kirchhoff's plate assumptions remain valid for each plate segment.
4. Displacements of a point on the section may be obtained by superimposing displacements from an engineering theory for thin-walled beams and those from small deflection plate theory.
5. Initial curvatures and the effect of prebuckling displacements may be neglected.
6. The material responds in a linear elastic manner and residual stress may be neglected.

In the approximate incremental equilibrium equation, Eq. (12.21), $\dot{\sigma}$, $\dot{\tau}_{\xi\zeta}$ and $\dot{\tau}_{\eta\zeta}$ are the stress increments; σ, $\tau_{\xi\zeta}$ and $\tau_{\eta\zeta}$ are the stresses prior to the increment; and $\dot{t}_i$ is the increment in the surface loads.

Assuming the total displacement increments may be written as the sum of beam displacement increments $\dot{u}^b$, and plate flexure displacement increments $\dot{u}^p$, $\dot{u}$ may be expressed as

$$\dot{u} = \dot{u}^b + \dot{u}^p \tag{12.113}$$

The total stress increment may also be written as the sum of that due to beam displacements $\dot{\sigma}^b$ and that due to plate displacements $\dot{\sigma}^p$, so that

$$\dot{\sigma} = \dot{\sigma}^b + \dot{\sigma}^p \tag{12.114}$$

The expressions for the beam displacement increments are the same as given by

Eqs. (12.5) and (12.13). The normal expressions for plate displacements are [see Fig. 12.26(b)]

$$\begin{aligned} u_\alpha &= -\zeta u_{\zeta,\alpha} \\ u_\beta &= -\zeta u_{\zeta,\beta} \end{aligned} \tag{12.115}$$

in which α, β and ζ = local reference coordinates; and u_α, u_β and u_ζ = displacements in the directions of these local reference coordinates. For a symmetric wide flange shape the out of plane plate segment displacements are $u^p_{\eta t}$, $u^p_{\eta b}$ and $u^p_{\xi w}$ for top flange, bottom flange and web respectively (see Fig. 12.27). The approximate virtual work incremental equilibrium equation may be written in symbolic form as

$$\delta I_1^b + \delta I_1^p + \delta I_2^b + \delta I_2^p + \delta I_2^{bp} - \delta I_3 = 0 \tag{12.116}$$

The terms contained in the first bracket of Eq. (12.22) represents δI_1^b and the terms in the second bracket on the left hand side of Eq. (12.22) represents δ_2^b and the right hand side terms denote δI_3. When the equations are specialized to wide flange sections ($e_\xi = e_\eta = 0$), δI_1^b leads to usual elastic beam-column flexural stiffness matrix, δI_1^p is plate flexural stiffness matrix, δI_2^b is geometric stiffness matrix for the beam-column element, δI_2^p is geometric stiffness matrix for the plate element and δI_2^{bp} is coupling between member and plate displacements.

$$\delta I_1^p = -\sum_{i=1}^{N} \int_{A_i^*} (\dot{M}_\alpha\, \delta\dot{\kappa}_\alpha + \dot{M}_\beta\, \delta\dot{\kappa}_\beta + 2\dot{M}_{\alpha\beta}\, \delta\dot{\kappa}_{\alpha\beta})_i \, dA^* \tag{12.117}$$

where

A_i^* = the middle surface area of the plate i in the α, β coordinate system

N = number of plates

$\dot{M}_\alpha, \dot{M}_\beta, \dot{M}_{\alpha\beta}$ = increments in plate flexure stress resultants

$\dot{\kappa}_\alpha, \dot{\kappa}_\beta, \dot{\kappa}_{\alpha\beta}$ = increments in plate flexure curvatures

$$\dot{M}_\alpha = -D(\dot{\kappa}_\alpha + \nu\dot{\kappa}_\beta) \tag{12.118a}$$

$$\dot{M}_\beta = -D(\dot{\kappa}_\beta + \nu\dot{\kappa}_\alpha) \tag{12.118b}$$

and

$$\dot{M}_{\alpha\beta} = -D(1-\nu)\dot{\kappa}_{\alpha\beta} \tag{12.118c}$$

where

$$D = \frac{Et^3}{12}(1-\nu^2) \qquad \text{and} \qquad \nu = \text{Poisson's ratio} \tag{12.118d}$$

δI_2^p is given by

$$\delta I_2^p = \delta\left[\sum_{i=1}^{N}\left\{\frac{1}{2}\int_{A_i^*} \sigma t[(\dot{u}^p_{\alpha,\zeta})^2 + (\dot{u}^p_{\beta,\zeta})^2]\, dA^*\right\}_i\right] \tag{12.119}$$

and the coupling effect δI_2^{bp} is given by

$$\delta I_2^{bp} = \delta\left\{\sum_{i=1}^{N}\left[\int_{v_i} \sigma(\dot{u}^p_{\alpha,\zeta}\dot{u}^b_{\alpha,\zeta} + \dot{u}^p_{\beta,\zeta}\dot{u}^b_{\beta,\zeta})\right]\right\} dv_i \tag{12.120}$$

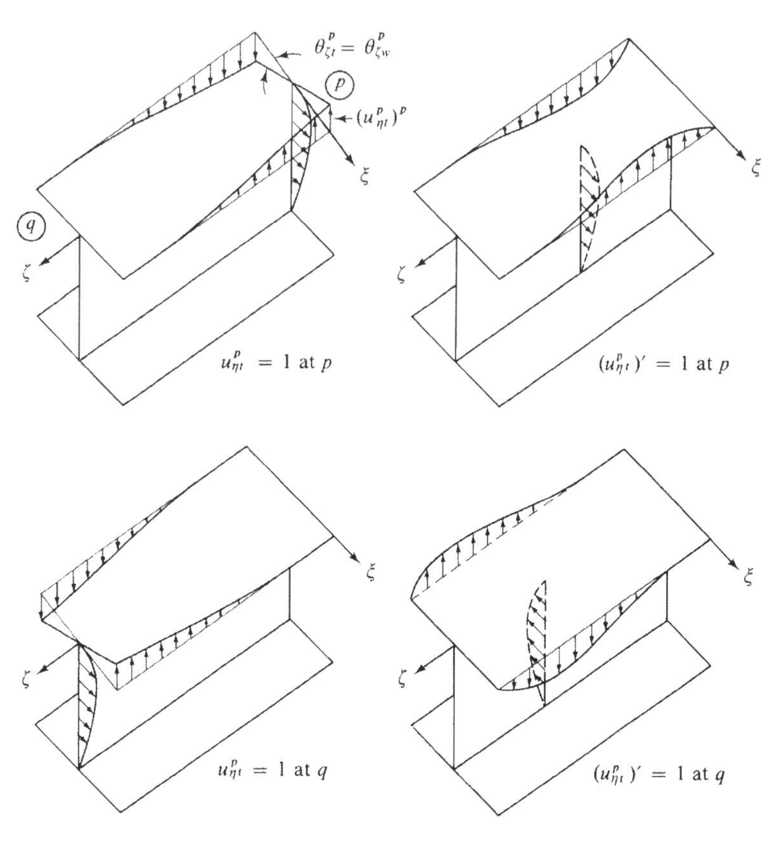

FIGURE 12.28
Deflected shape of the I-section due to unit value of the generalized displacement

12.10.2 Finite Element Model

The *shape functions* for beam displacements are the same as given by Eq. (12.41). The plate flexure displacements are assumed to occur as a linear combination of the mode shapes shown for the top flange only in Fig. 12.28. These mode shapes assume that a line on the flange normal to the longitudinal axis, remains straight as the cross section distorts; the displacements of a line parallel to the beam axis may be represented by cubic beam functions; that the web distortion may be represented on a normal section by cubic beam functions; and that the angle between the web and flange remains constant. Hence,

$$\begin{Bmatrix} u^p_{\eta t} \\ u^p_{\eta b} \end{Bmatrix} = \xi \begin{bmatrix} \langle n_3 \rangle & 0 \\ 0 & \langle n_3 \rangle \end{bmatrix} \begin{Bmatrix} \underline{u}^p_{\eta t} \\ \underline{u}^p_{\eta b} \end{Bmatrix} \tag{12.121a}$$

and

$$u^p_{\xi w} = \frac{(D-t)}{4B} \begin{Bmatrix} (1+\delta) & (1-\delta^2) \\ (\delta - 1) & (1-\delta^2) \end{Bmatrix}^T \begin{bmatrix} \langle n_3 \rangle & 0 \\ 0 & \langle n_3 \rangle \end{bmatrix} \begin{Bmatrix} \underline{u}^p_{\eta t} \\ \underline{u}^p_{\eta b} \end{Bmatrix} \tag{12.121b}$$

where D = section depth.
B = flange width.
t = flange thickness.
$u^p_{\eta t}, u^p_{\eta b}, u^p_{\xi w}$ = normal displacement of the top flange, normal displacement of the bottom flange and web of the wide flange beam, respectively.
$\delta = 2\eta/(D - t)$.

The nodal displacement

$$\begin{aligned} \langle \underline{u}^p_{\eta t} \rangle &= \langle (u^p_{\eta t})^p, l(u^p_{\eta t,\zeta})^p, (u^p_{\eta t})^q, l(u^p_{\eta t,\zeta})^q \rangle \\ \langle \underline{u}^p_{\eta b} \rangle &= \langle (u^p_{\eta b})^p, l(u^p_{\eta b,\zeta})^p, (u^p_{\eta b})^q, l(u^p_{\eta b,\zeta})^q \rangle \end{aligned} \tag{12.122}$$

Substituting the displacement assumptions of Eq. (12.121) into the appropriate terms of the incremental virtual work equation, and carrying out the integration, the element incremental equilibrium equation may be written as

$$([k_S]_E + [k_G]_E)\{\dot{u}\}_E = \{\dot{F}\}_E \tag{12.123}$$

and the nodal vector $\{\dot{u}\}_E$ is now defined as

$$\langle \dot{u} \rangle_E = \langle \langle \underline{\dot{u}}_{\xi o} \rangle, \langle \underline{\dot{u}}_{\eta o} \rangle, \langle \underline{\dot{\theta}}_{\zeta} \rangle, \langle \underline{\dot{u}}_{\zeta c} \rangle, \langle \underline{\dot{l}} \rangle \rangle \tag{12.124}$$

in which $\langle \underline{l} \rangle$ = the vector of nodal plate displacements given by

$$\langle \underline{l} \rangle = \langle \langle \underline{u}^p_{\eta t} \rangle, \langle \underline{u}^p_{\eta b} \rangle \rangle \tag{12.125}$$

Now each node of the beam-column element has eleven degrees of freedom if we include the plate deformations. Assembling the element equilibrium equations, the total set of incremental equilibrium equations may be written as

$$[K]\{\dot{u}\} + [K_G]\{\dot{u}\} = \{\dot{F}\} \tag{12.126}$$

The bifurcation condition may be written as

$$\frac{1}{\lambda}\{\dot{u}\} = -[K]^{-1}[K_G]\{\dot{u}\} \tag{12.127}$$

12.10.3 Numerical Examples

EXAMPLE 12-10 *Axially Loaded Column* Figure 12.29 shows the buckling load of an axially loaded wide flange column for the following conditions; (1) uncoupled local buckling, (2) uncoupled Euler buckling, (3) interactive member, and local buckling. The interaction curve between critical stress coefficient ($\sigma_c b_1^2 w/\pi^2 D$) and L/b_1, gives only a few percent lower than that of the Euler column buckling curve. These results agree with those of Bulson (1967). Curves are also drawn when the load is applied at an eccentricity of 5 in (127 mm).

EXAMPLE 12-11 *Simply Supported Beam with a Central Load* Figure 12.30 shows a comparison of (1) the (uncoupled) elastic local buckling load, (2) the

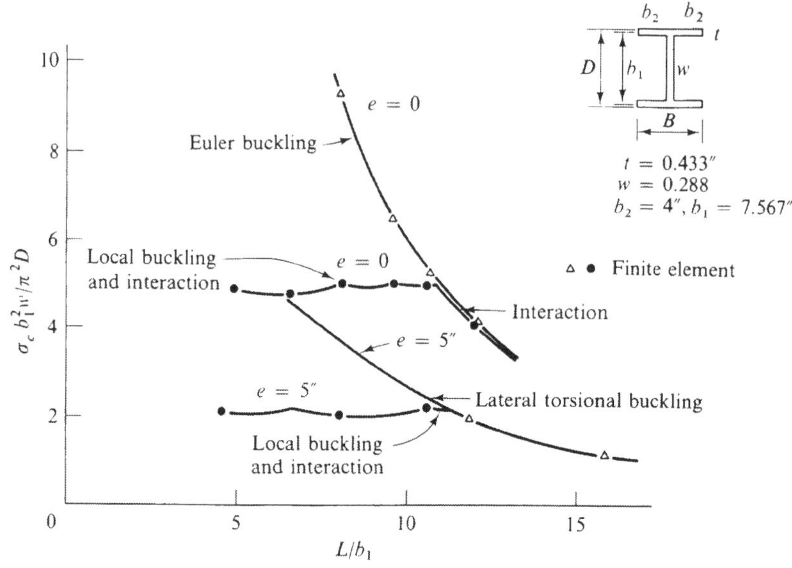

FIGURE 12.29
Column/local buckling interaction

(uncoupled) elastic lateral buckling load, and (3) the elastic critical load including interactive effects for a simply supported beam with central load.

The results of the examples verify the validity of the classical approach to local buckling by showing that the interactive effects normally small, and local and

FIGURE 12.30
Interaction and local buckling (1 kip = 4.448 N, 1 in = 25.4 mm)

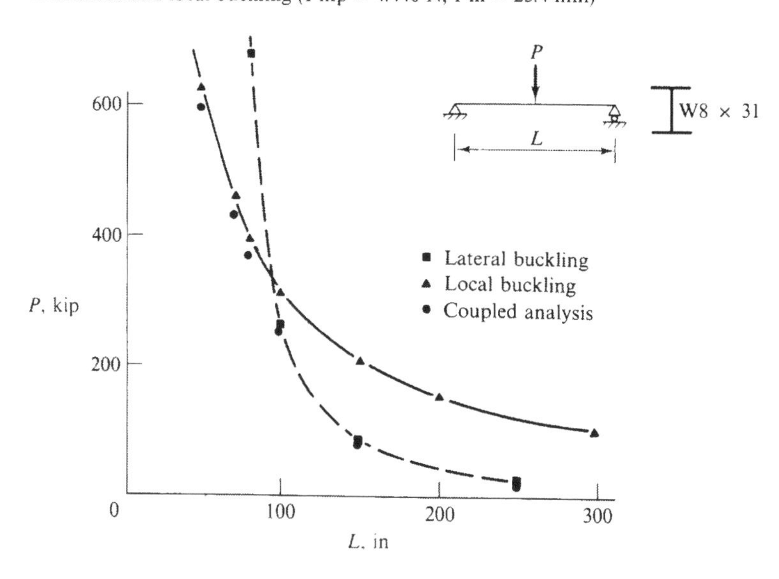

member buckling are initiated as a relatively detached phenomena. This conclusion may not be reached if initial deformations such as flange misalignment or plastic material were included in the analysis. However, the value of the finite element approach appears to be immensely high for further development on these lines.

12.11 SUMMARY

A direct virtual work formulation of the thin-walled beam equations is developed for arbitrary reference axes and is applicable to both elastic and plastic cases. Finite element total and incremental equilibrium equations are derived based on the assumed displacement model. This method is applied to stability of beam-column problems of arbitrary thin-walled open section by eigenvalue and load-deflection approaches for elastic and plastic cases. The technique has been illustrated by solving simple problems for which results by alternative methods are available. The systematic generality of the method makes it a powerful and versatile tool for a wide range of problems.

A most important question for the user of the finite element method is whether the method yields sufficiently accurate results for his purpose. If conforming elements are used incorporating displacement function satisfying completeness criteria and the element size is decreased according to minimizing sequence, it is possible to get the results to any desired degree of accuracy. As in any other approximate numerical method, engineering judgement should be exercised in interpreting the results obtained by the finite element method since the accuracy of results depends on the assumptions employed in the formulation and the limitations in the material idealization.

Since the finite element approach to beam-columns presented in this chapter is not restricted to simple cross sections or loading conditions, the most general types of beams and beam-column problems are now amenable to solution. It is therefore possible to carry out parametric studies for the codification of design procedures to include the effects of plastic member and plate instability, for realistic loading and material response.

12.12 PROBLEMS

12.1 Compute the elastic critical load on the tapered column shown in Fig. 12.31. The sectional properties are given by

$$I = I_C\left[1 - \frac{1}{2}(x/L)^2\right]$$

$$A = A_O(1 - x/2L)$$

Idealize the tapered column with (a) prismatic elements, (b) tapered elements, and obtain the critical load for one element, two elements and four elements. Draw a graph between percentage error and number of elements for the above two cases. Discuss your results.

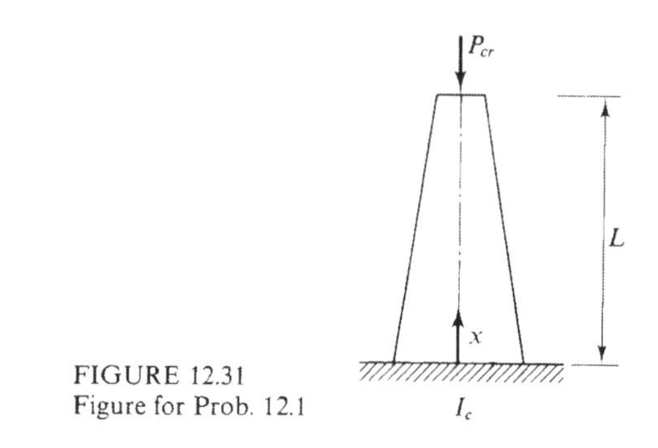

FIGURE 12.31
Figure for Prob. 12.1

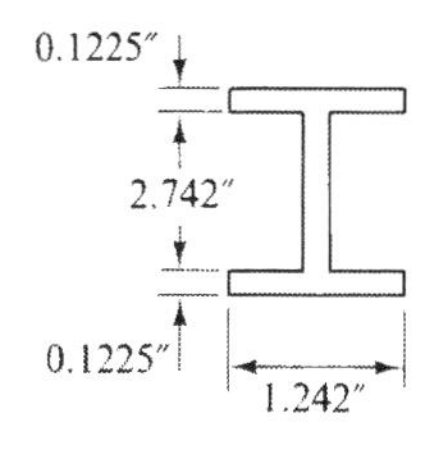

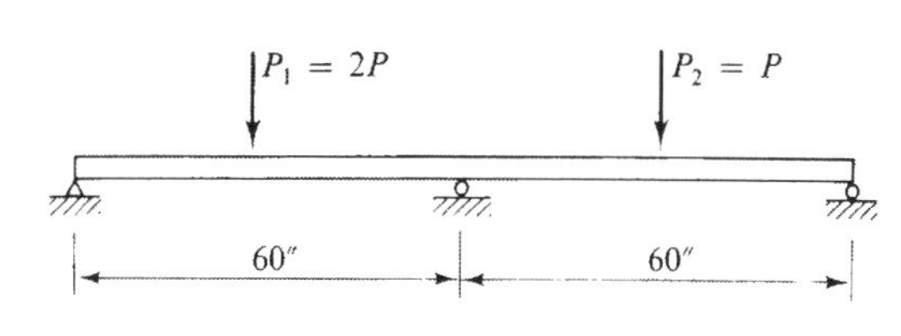

$EI_y = 372\ \text{in}^2\text{-lb}$
$EI_\omega = 764\ \text{in}^4\text{-lb}$
$GK_T = 7.98\ \text{in}^2\text{-lb}$

FIGURE 12.32
Figure for Prob. 12.2

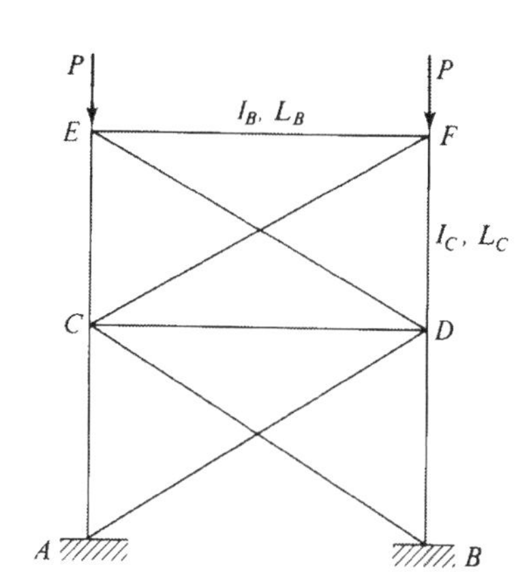

FIGURE 12.33
Figure for Prob. 12.3

12.2 Obtain the lateral buckling load for the elastic case using finite element method for a two-span continuous beam shown in Fig. 12.32 and compare your results with the results obtained by Tebedge and Tall (1973).

12.3 For the frame shown in Fig. 12.33 determine the critical load for the elastic case using the finite element method for $\lambda = 5$, 10 and 15 where

$$\lambda = \frac{I_B/L_B}{I_C/L_C}$$

Assume that transverse bracing system prevents sway.

12.4 Carry out finite element beam-column analysis to predict the axial load and lateral deformation response of a beam-column shown in Fig. 12.34 with initial imperfection as $U_{x0} = 0.01 \sin(\pi z/L)$ where U_0 is the initial lateral displacement. Compare the maximum load obtained from the analysis with that of eigenvalue solution which neglects prebuckling deformation. Use $e/r_x = 0, 0.73, 2, 4$ and 10 and assume material is linear elastic.

12.5 Hat-shaped section shown in Fig. 12.35 is used as a chord member of open web steel joist. Calculate the critical length using the finite element technique when the load is $0.9P_y$.

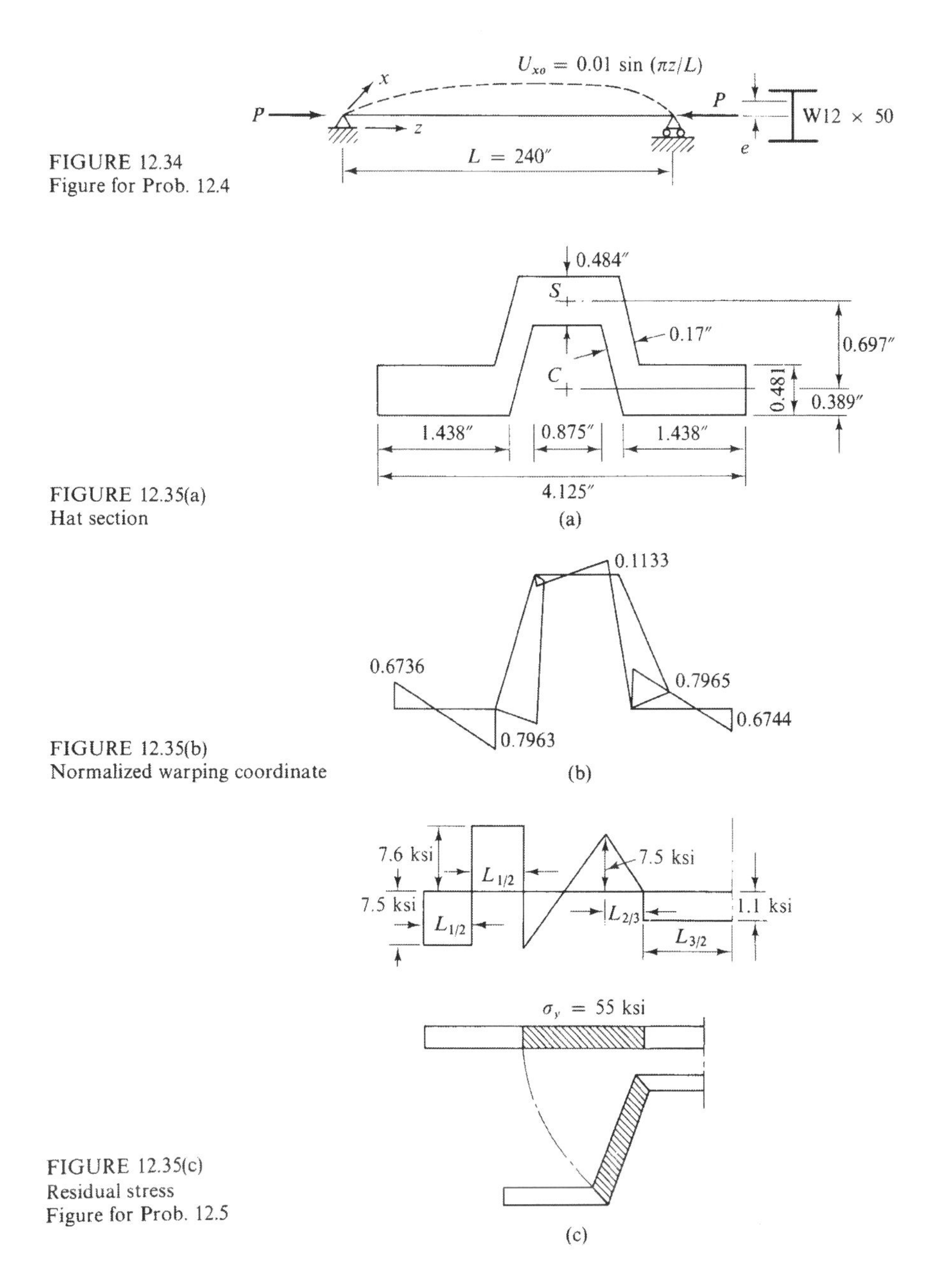

FIGURE 12.34
Figure for Prob. 12.4

FIGURE 12.35(a)
Hat section

FIGURE 12.35(b)
Normalized warping coordinate

FIGURE 12.35(c)
Residual stress
Figure for Prob. 12.5

The residual stress and the warping coordinate diagrams are shown in Fig. 12.35. [*Note:* compute tangent properties and stress resultants using expressions given in Sect. 12.7.3. The problem was solved by Heaton and Adams (1970) by finite difference approach and by Rajasekaran and Murray (1971) by finite element approach.]

12.6 Obtain the plastic beam buckling curve for a W8 × 31 section subjected to moment gradient as shown in Fig. 12.36 solving every time for critical length for a given moment. Assume a linear residual stress distribution as $\sigma_{rc}/\sigma_y = 0.3$. Assume $P/P_y = 0.6$. Compare your results with the results obtained by Fukumoto and Galambos (1966).

12.7 A column cross section is shown in Fig. 12.37. The individual plates are hot rolled of A36 steel. The plates are then welded together to form the cross section. Accounting for reasonable distribution of residual stress, determine the relation between the critical stress and slenderness ratio (i.e., develop column strength curves) using finite element method for the pin-ended columns. Check the curves using the SSRC(CRC) or AISC equation. Comment on your results.

Figure 12.36
Figure for Prob. 12.6

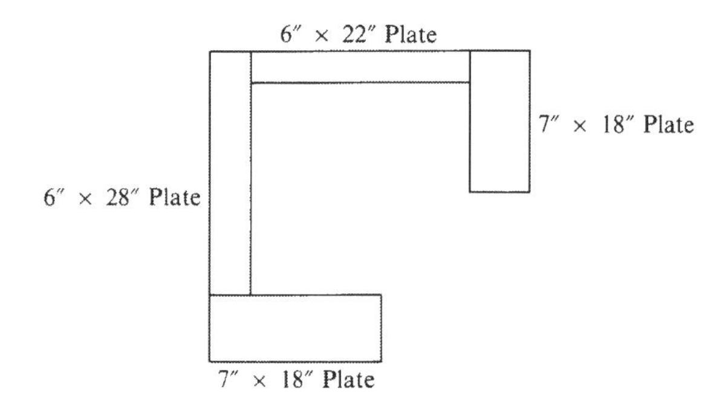

FIGURE 12.37
Figure for Prob. 12.7

12.13 REFERENCES

Arghyris, J. H., *Recent Advances in Matrix Methods of Structural Analysis*, Progress in Aeronautical Science, Pergamon Press, (1964).

Barsoum, R. S., "A Finite Element Formulation for the General Stability Analysis of Thin-Walled Members." Thesis presented to Cornell University in partial fulfillment of the requirements for the degree of Doctor of Philosophy, September, (1970).

Barsoum, R. S. and Gallagher, R. H., "Finite Element Analysis of Torsional-Flexural Stability Problems," *International Journal for Numerical Methods in Engineering*, Vol. 2, No. 1, (1970).

Bazant, Z. P., "Nonuniform Torsion of Thin-Walled Bars of Variable Cross Section," *International Association for Bridge and Structural Engineering, Publication*, Vol. 25, pp. 245–267, (1965).

Biot, M. A., *Mechanics of Incremental Deformation*, John Wiley and Sons, (1965).

Brebbia, C. A. and Connor, J. J., *Fundamentals of Finite Element Techniques for Structural Engineers*, Butterworth and Company, London, (1973).

Bulson, P. S., "Local Stability and Strength of Structural Sections," *Thin-Walled Structures* (A. H. Chilver, editor), Chatto and Windus, pp. 153–207, London, (1967).

Fukumoto, Y. and Galambos, T. V., "Inelastic Lateral-Torsional Buckling of Beam-Columns," *Journal of the Structural Division, ASCE*, Vol. 92, No. ST2, Proc. Paper 4770, April, pp. 41–61, (1966).

Galambos, T. V., "Inelastic Lateral Buckling of Beams," *Journal of the Structural Division, ASCE*, Vol. 89, No. ST5, Proc. Paper 3683, October, pp. 217–242, (1963).

Gallagher, R. H. and Padlog, J., "Discrete Element Approach to Structural Instability Analysis," *American Institute of Aeronautics and Astronautics Journal*, Vol. 1, No. 6, June, pp. 1437–1439, (1963).

Harstead, G. A., Birnstiel, C. and Leu, K. C., "Inelastic H-Columns Under Biaxial Bending," *Journal of the Structural Division, ASCE*, Vol. 94, No. ST10, Proc. Paper 6173, October, pp. 2371–2398, (1968).

Heaton, D. A. and Adams, P. F., "Buckling Strength of Hot-Rolled Hat Sections," *Structural Engineering Report No. 26*, University of Alberta, Canada, July, (1970).

Hsu, Y., "Elastic Buckling of Two-Component Columns," *Journal of Engineering Mechanics Division, ASCE*, Vol. 95, No. EM3, Proc. Paper 6612, June, pp. 611–628, (1969).

Krahula, J. L., "Analysis of Bent and Twisted Bars Using the Finite Element Method," *American Institute of Aeronautics and Astronautics Journal*, Vol. 5, No. 6, June, pp. 1194–1197, (1967).

Krajcinovic, D., "A Consistent Discrete Elements Technique for Thin-Walled Assemblages," *International Journal of Solids and Structures*, Vol. 5, No. 7, July, pp. 639–661, (1969).

Martin, H. C., "Finite Elements and the Analysis of Geometrically Nonlinear Problems," *Proceedings, Japan–U.S. Seminar on Recent Advances in Matrix Methods of Structural Analysis and Design*, Tokyo, Japan, (1969).

Massey, C. and Pitman, F. S., "Inelastic Lateral Stability Under a Moment Gradient," *Journal of the Engineering Mechanics Division, ASCE*, Vol. 92, No. EM2, Proc. Paper 4779, April, pp. 101–111, (1966).

Mei, C., "Coupled Vibration of Thin-Walled Beams of Open Section Using the Finite Element Method," *International Journal of Mechanical Sciences*, Vol. 12, pp. 883–891, (1970).

Murray, D. W. and Rajasekaran, S., "A Technique for Formulating Beam Equations," *Journal of the Engineering Mechanics Division, ASCE*, Vol. 101, No. EM10, Proc. Paper 11613, October, pp. 561–573, (1975).

Nayak, G. C. and Zienkiewicz, O. C., "Note on the 'Alpha' Constant Stiffness Method for Analysis of Nonlinear Problems," *International Journal for Numerical Methods in Engineering*, Vol. 4, No. 4, pp. 582–597, (1972).

Nuttal, N. J. and Adams, P. F., "Flexural and Lateral-Torsional Buckling Strength of Double Angle Struts," *Structural Engineering Report No. 30*, Department of Civil Engineering, University of Alberta, Canada, Fall, (1970).

Powell, G. H. and Klingner, R., "Elastic Lateral Buckling of Steel Beams," *Journal of the Structural Division, ASCE*, Vol. 96, No. ST9, Proc. Paper 7555, September, pp. 1919–1932, (1970).

Przemieniecki, J. S., "Discrete Element Methods for Stability Analysis of Complex Structures," *The Aeronautical Journal of the Royal Aeronautical Society*, Vol. 72, December, (1968).

Przemieniecki, J. S., "Matrix Analysis of Local Instability in Plates, Stiffened Panels and Columns," *International Journal for Numerical Methods in Engineering*, Vol. 5, pp. 209–216, (1972).

Rajasekaran, S., "Finite Element Analysis of Thin-Walled Members of Open Section," Ph.D. Thesis, University of Alberta, Edmonton, September, (1971).

Rajasekaran, S. and Murray, D. W., "Inelastic Buckling of Thin-Walled Members," First Speciality Conference on Cold-Formed Steel Structures, University of Missouri at Rolla, August, pp. 43–51, (1971).

Rajasekaran, S. and Murray, D. W., "Finite Element Solution of Inelastic Beam Equations," *Journal of the Structural Division, ASCE*, Vol. 99, No. ST6, June, Proc. Paper 9773, pp. 1024–1042, (1973a).

Rajasekaran, S. and Murray, D. W., "Coupled Local Buckling in Wide-Flange Beam Columns," *Journal of the Structural Division, ASCE*, Vol. 99, No. ST6, June, Proc. Paper 9774, pp. 1003–1023, (1973b).

Rajasekaran, S. and Murray, D. W., "Finite Element-Large Deflection Analysis of Thin-Walled Beams of Open Section," *Proceedings, First International Conference on Finite Element Methods in Engineering*, University of New South Wales, August 28–30, Sydney, Australia, (1974).

Rajasekaran, S. and Murray, D. W., "Solution Techniques to Nonlinear Problems," Proceedings, AICA International Symposium on Computer Methods for Partial Differential Equations, Lehigh University, Bethlehem, Pa., June 17–19, (1975).

Renton, J. D., "Buckling of Frames Composed of Thin-Walled Members," *Thin-Walled Structures* (A. H. Chilver, Editor), Chatto and Windus, London, pp. 1–59, (1967).

Soltis, L. A. and Christiano, P., "Finite Deformation of Biaxially Loaded Columns," *Journal of the Structural Division, ASCE*, Vol. 98, No. ST12, Proc. Paper 9407, pp. 2647–2661, (1972).

Strang, G. and Fix, G. I., *An Analysis of the Finite Element Method*, Prentice-Hall, Inc., Englewood Cliffs, New Jersey, (1973).

Tebedge, N., "Application of the Finite Element Method to Beam-Column Problems." Thesis presented to Lehigh University in partial fulfillment of the requirements for the degree of Doctor of Philosophy, Bethlehem, Pa., (1972).

Tebedge, N. and Tall, L., "Linear Stability Analysis of Beam-Columns," *Journal of the Structural Division, ASCE*, Vol. 99, No. ST12, Proc. Paper 10232, December, pp. 2439–2457, (1973).

Trahair, N. S., "Deformation of Geometrically Imperfect Beams," *Journal of the Structural Division, ASCE*, Vol. 95, No. ST7, Proc. Paper 6682, July, pp. 1475–1496, (1969).

Turner, M. J., Dill, E. H. and Martin, H. C., "Large Deflection of Structures Subjected to Heating and External Loads," *Journal of the Aerospace Sciences*, Vol. 27, February, pp. 97–106, (1960).

Vlasov, V. Z., "Thin-Walled Elastic Beams," English translation, Israel Program for Scientific Translations, Jerusalem, Israel, (1961).

Yoshida, H., Nethercot, D. A. and Trahair, N. S., "Analysis of Inelastic Buckling of Continuous Beams," *Research Report No. R65*, Department of Civil and Structural Engineering, University of Sheffield, England, August, (1975).

Zienkiewicz, O. C., *The Finite Element Method in Engineering Science*, McGraw-Hill Book Company, Inc., London, (1971).

13

DESIGN OF BIAXIALLY LOADED BEAM-COLUMNS

13.1 INTRODUCTION

The columns in an actual structure are frequently subjected to compression combined with biaxial bending, i.e., to bending moments acting in two principal axes directions in addition to an axial compressive load. The biaxial bending moments may result from the space action of the entire framing system, and/or from an axial load biaxially located with respect to the principal axes of the column cross section. All columns in an actual building framed structure should therefore be designed as a biaxially loaded beam-column. It is the purpose of this chapter to review the existing design specification requirements for biaxially loaded beam-columns, to present the new design procedures developed from a computer model and refined to achieve both simplicity in use and, as far as possible, a realistic representation of actual strength, to compare the size of beam-column section required by the existing and proposed new design rules for several practical design problems, and to provide final confirmation of the validity of the proposed methods by extensive comparisons with test results on actual beam-columns.

General approach

The design of a biaxially loaded beam-column may be approached from the standpoint of a precise computer solution. This is destined to be complex, since even for the design of a section for *strength* only, neither the location nor the angular disposition of the neutral axis of the fully plastic cross section subjected to a given axial load and biaxial bending moments is known. In consequence, an iterative procedure must be followed. In the case of *stability* analysis, the problem is far more complex. In addition to the problem of locating the neutral axis, the effects of torsional buckling and the lateral deflection of the beam-column, due to both end moments and the effect of the axial load acting on the deflected beam-column, must be determined to satisfy equilibrium and compatibility. Superimposed on this is the effect of residual stresses and initial out-of-straightness, so that once the actual stress induced by the applied loading and magnified by the initial out-of-straightness plus the residual stress exceeds the yield value of the material, the determination of the deflected shape must take account of the partially plastified section. Clearly, this is a difficult problem; demanding far too much computer time to be solved for each beam-column to be designed.

From the discussions in previous chapters, we know that even a reasonable approximate solution involves considerable labor and computer time in numerical calculations and therefore limits its practical use. Such a procedure may be used only to generate data from which dimensionless tables, charts, or nomographs for various shapes of beam-column cross sections and slenderness ratios can be prepared. In this way, the design of a complete beam-column can be made quite rapid. The only disadvantage of this approach is that an interpolation is required on each chart or table, followed sometimes by an additional interpolation between the charts or tables.

Many designers dislike graphical procedures, although the accuracy obtained thereby is probably within the limits with which we can predict the load carrying capacity of a beam-column under biaxial loading. For this and other reasons, specifications are generally expressed in terms of algebraic formulas, where possible. Such formulas are particularly suitable for incorporation into computer design programs. By this approach, a considerable amount of data on biaxially loaded beam-columns must be generated first using a computer model before simple design formulas are derived and developed.

The development of such design formulas is generally based on the concept of the so-called *interaction surface*. With this concept, an interaction surface can be constructed that relates the axial force and biaxial bending moments under the condition of the maximum load-carrying capacity of the member. For beam-columns with $L/r = 0$, or for "short" beam-columns for which the effect of lateral deflections on the magnitudes of bending moments is negligible, the interaction surface represents the condition when the entire cross section is fully plastic; whereas, for longer beam-columns it represents when the member reaches its maximum or buckling strength. Extensive work has been performed for beam-columns with $L/r = 0$ to establish the biaxial interaction relationship (Chap. 5). Sufficient information is now available on the maximum strength of biaxially loaded H-columns under symmetric loading conditions with simply supported ends (Chap. 7). Limited solutions are presently available for biaxially loaded plastic beam-columns under unsymmetric loading conditions (Chap. 7). From these interaction surfaces, simple empirical surfaces may be fitted in the derivation of design expressions (Tebedge and Chen, 1974). This method of approach for the design of beam-column subjected to biaxial bending has been recommended by Task Group 3 on "Columns with Biaxial Bending" of the Structural Stability Research Council (formerly the Column Research Council) for improved design techniques for the most commonly used type of steel column, namely the wide flange shape. This will be described in detail in this chapter. Studies of reinforced concrete columns made using this approach will also be included.

In England, Young (1973) has presented an alternative approach for the design of a biaxially loaded beam-column. In his approach, the section is selected such that its major axis fully plastic moment capacity, reduced by *influence coefficients* reflecting the presence of axial load, buckling about major and minor axes, and torsional buckling, fulfills the design requirements. The method, however, requires both formulas and design charts.

Scope

The beam-columns are assumed to be isolated, simply supported, warping permitted, prismatic, and made of as-rolled wide-flange shapes. The loading is considered to be fixed or to increase monotonically as failure of the beam-column progresses. Three classes of problems are considered: (1) short beam-columns; (2) long beam-columns loaded symmetrically at the ends; and (3) long beam-columns loaded unsymmetrically at the ends. For each case, the numerical results are given. The maximum load values are given for beam-columns with slenderness ratios ranging from 0 to 100 in increments of 10; and for major axis bending moment ratios $m_x = M_x/M_{px}$ from 0 to the maximum value in increments of 0.1. Here, all the axial force and bending moments are nondimensionalized with respect to quantities at the limit plastic state, and the basic section used in the study is W8 × 31. The spacing of the points is sufficiently close so that linear interpolation can be used for intermediate points within an acceptable margin of error. All computations are performed on mild steel columns ($\sigma_y = 36$ ksi), but the results can also be extrapolated for other beam-columns with different yield stress levels using the concept of equivalent slenderness ratio.

Since it is impractical to provide numerical results for all shapes of cross section currently used in practice, the question whether one particular shape can represent closely all the commonly used wide-flange beam-columns will be first investigated. The main characteristics taken into consideration in selecting the representative shape are the effects of the variations in shape, weight, and dimensions on the resulting interaction surfaces.

In the numerical solutions for the developing of interaction surfaces the following assumptions are made:

1. The stress-strain relationship of the beam-column material is elastic-perfectly plastic, and the effects of strain hardening are neglected.

2. Residual stresses are present in the beam-columns and an idealized pattern corresponding to hot-rolled wide-flange sections is used.
3. The deflections of the beam-column depend only on the final value of the applied loads and are independent of the actual history of loading. In other words, the irreversibility of plastic deformation is not considered.
4. The effects of shear on the yielding of material are neglected.
5. There will be no local buckling failure.
6. The initial imperfection of the beam-column is a half-sine curve.

In performing the numerical calculations, it is assumed that the axial force, P, is applied first and then the bending moment, M_x. These loads are maintained at constant values as the bending moment, M_y, is continuously increased from zero until the maximum load is reached. This assumption on the loading condition may be unnecessarily restrictive, but it does agree with the more recent findings that the plastic deformation of a biaxially loaded H-section is independent of loading path provided the loads are increased monotonically (Chen and Atsuta, 1973a; or Chap. 6). In presenting the families of interaction surfaces the parameters L/r_x, P/P_y, M_x/M_{px}, and M_y/M_{py} were chosen. The computed interaction surfaces are used as a basis for developing simple approximate formulas.

13.2 SHORT BEAM-COLUMN

In the design of a biaxially loaded beam-column, one of the two conditions will limit the load carrying capacity of a restrained long or intermediate length beam-column. These are *strength* of cross section at one of the restrained ends or at an intermediate braced location within the beam-column, or *instability* of the member arising from the magnification of the primary moments by the axial load acting on the laterally deflected beam-column. The term "short" beam-column used here implies that the effect of this lateral deflection upon the overall geometry can be ignored. Hence, the strength of a short beam-column is limited only by full plastic yielding of the material of the cross section, provided of course, that local buckling does not occur; that is, that the cross section is compact. Design of intermediate and slender columns will be considered in the next section.

13.2.1 Methods of Solution

For the uniaxial bending case, the determination of the location of the neutral axis in a cross section corresponding to a given pair of axial force and bending moment is relatively simple, and exact and approximate expressions for the simple *plastic-hinge* moment M_{pc} including the effect of axial compression P can be obtained directly for most shapes of cross section. In the case of a biaxially loaded cross section, however, both the location and angular disposition of the neutral axis corresponding to a given load has to be determined by an iteration procedure to satisfy the applied load P and moments M_x and M_y. Various procedures and approaches that have been made to solve this problem have been presented in Chap. 5. They may be categorized into three groups, as follows:

1. Given a predetermined series of locations and angular dispositions of the neutral axis, a corresponding series of P, M_x and M_y can be determined from simple statics. From these data, interaction diagrams can be constructed for a specific structural shape. Ringo, McDonough and Baseheart (1973) used this approach and found that the 88 common American column shapes could be represented by seven base section interaction curves and developed a proportionality function relating the sections (Table 5.5). They presented nomographs to facilitate the design use of the seven base section interaction curves (Figs. 5.36 and 5.37). Detail of this development is presented in Sec. 5.6.1, Chap. 5.

2. Deriving analytically the relations relating P, M_x, M_y to the location and disposition of the neutral axis for various possible cases for which neutral axis could be located, interaction curves can then be obtained numerically from these relations. The disadvantage of this approach is that the cases for the possible neutral axis location may be great (Fig. 5.12). For example, for the case of a wide-flange cross section, there exist at least six significant cases to be considered (see Table 5.2), and an iterative procedure must be followed in order to determine which one of these six cases controls corresponds to a given set of P, M_x and M_y. Morris and Fenves (1969) derived approximately the lower bound interaction equations in terms of axial force, torsional moment and biaxial bending moments (see Probs. 5.4 to 5.7, Chap. 5). Santathadaporn and Chen (1970) checked the accuracy of their solution using the upper and lower bound theorems of limit analysis (Table 5.2) and presented some interaction results from these equations (Table 5.3).

3. Using the interaction equations derived for a rectangular section, any other commonly used shape is then obtained by the method of *superposition*, i.e., by subtracting rectangular areas from the enclosing rectangular area. Using this concept, Chen and Atsuta (1972, 1974) derived general expressions (suitable for programming) for a double flanged, double web shape which can be used to solve most commonly occurring shapes (Fig. 5.18). The method is valid for both symmetric and asymmetric sections, interaction curves having been prepared for circular, wide-flange, angle, and channel sections (Figs. 5.31 to 5.35). Details of this development are presented in Sect. 5.3 and Sect. 5.4, Chap. 5. The method of superposition is found to be extremely powerful and efficient for computer solution, and is applied in the following to obtain various numerical results.

13.2.2 Numerical Results

Four standard W12 profiles were investigated to determine the effects of variations in shape as reflected by the flange width to depth ratio, B/D, varying from 0.540 to 0.975. The results are shown in Fig. 13.1 which are seen to form a rather narrow band of all values of $p = P/P_y$ varying from 0 to 0.9. Also shown in the figure are the interaction curves for the shape W8 × 31 ($B/D = 1.0$) which pass within the band. It may be concluded from these comparisons that the interaction curves are relatively insensitive to the variation of the B/D ratio and the particular shape W8 × 31 may be used as a representative shape for all the commonly used wide-flange shapes.

To determine the effects of the variation of the weight of shape on the interaction surfaces, two profiles were investigated: W8 × 31 to represent the light shapes

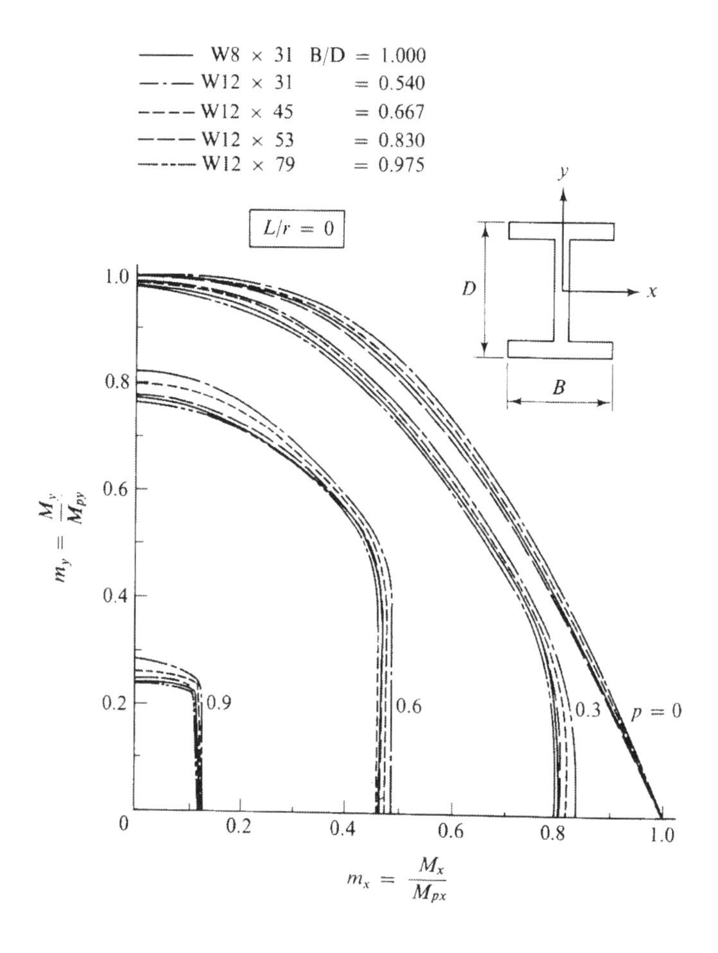

FIGURE 13.1
Comparison of interaction curves for various B/D ratios

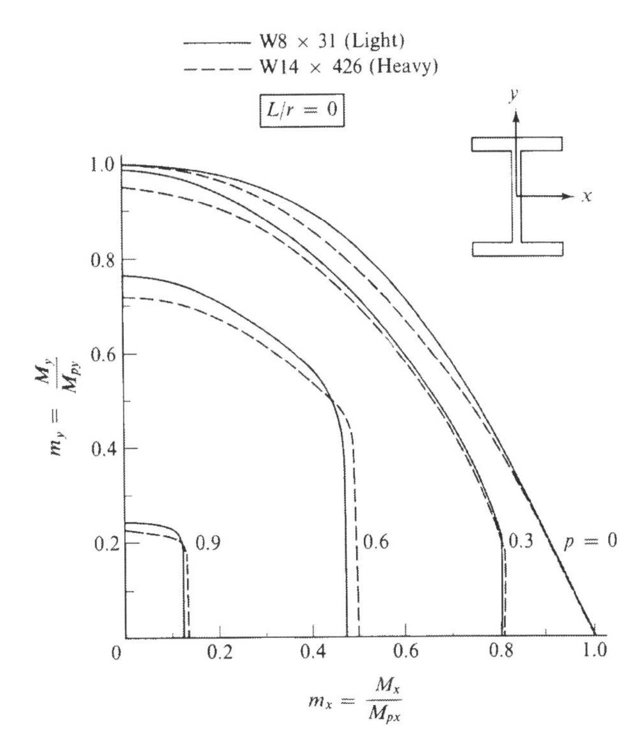

FIGURE 13.2
Comparison of interaction curves between heavy and light sections

and W14 × 426 for the heavy shapes. The interaction curves are shown in Fig. 13.2 for values of p ranging from 0 to 0.9. In this case also, the interaction curves are seen to be relatively insensitive to the variation in shape weight. Consequently, from considerations of cross sectional properties only, the profile W8 × 31 may be taken as a shape that can represent closely most commonly used wide-flange sections. This observation is in agreement with previous investigations on inplane bending beam-column problems (ASCE Manual 41, 1971). Furthermore, the W8 × 31 section has a low shape factor, 1.10, as compared to the average value of 1.14 for wide-flange sections. Therefore, the curves corresponding to the W8 × 31 section represent a good approximation for any other wide-flange section and are conservative for the sections in the ratio of their shape factor to 1.10. The validity for selecting this shape as a representative section for longer beam-columns will be examined later.

Based on the cross-sectional properties of W8 × 31, families of numerical interaction relations for the short column case ($L/r_x = 0$) are presented in column 3 of Table 13.1. This table gives a nondimensionalized relationship between the end moments, M_x/M_{px} and M_y/M_{py}, and the slenderness ratio, L/r_x, for constant values of the axial force, P/P_y.

Table 13.1 MAXIMUM STRENGTH TABLE FOR BIAXIALLY LOADED BEAM-COLUMNS (MAXIMUM VALUES OF M_y/M_{py}) (TEBEDGE AND CHEN, 1974) (W8 × 31, B/D = 1.0)

		Slenderness Ratio, $L/r_x(L/r_y)$										
P/P_y	M_x/M_{px}	0	10 (15)	20 (30)	30 (45)	40 (60)	50 (75)	60 (90)	70 (105)	80 (120)	90 (135)	100 (150)
0.0	0.0	1.000	1.000	1.000	1.000	1.000	1.000	1.000	1.000	1.000	1.000	1.000
	0.1	0.999	0.997	0.990	0.984	0.980	0.973	0.967	0.962	0.958	0.956	0.953
	0.2	0.991	0.986	0.972	0.956	0.940	0.924	0.909	0.894	0.880	0.868	0.857
	0.3	0.956	0.952	0.934	0.909	0.880	0.850	0.821	0.794	0.769	0.750	0.725
	0.4	0.897	0.889	0.867	0.834	0.795	0.752	0.708	0.663	0.620	0.585	0.545
	0.5	0.811	0.802	0.773	0.728	0.675	0.622	0.558	0.492	0.429	0.359	0.294
	0.6	0.699	0.689	0.651	0.595	0.528	0.449	0.356	0.286	0.228	0.169	0.095
	0.7	0.562	0.548	0.498	0.424	0.341	0.263	0.209	0.155	0.081	—	—
	0.8	0.398	0.379	0.318	0.241	0.177	0.131	—	—	—	—	—
	0.9	0.210	0.185	0.131	0.079	—	—	—	—	—	—	—
	1.0	0.000	—	—	—	—	—	—	—	—	—	—
0.1	0.0	0.999	0.971	0.931	0.882	0.824	0.761	0.698	0.626	0.554	0.481	0.396
	0.1	0.996	0.967	0.924	0.871	0.812	0.750	0.686	0.607	0.538	0.455	0.360
	0.2	0.983	0.951	0.902	0.844	0.782	0.710	0.638	0.558	0.476	0.390	0.304
	0.3	0.947	0.917	0.863	0.797	0.725	0.643	0.547	0.469	0.373	0.281	0.184
	0.4	0.880	0.856	0.799	0.727	0.643	0.550	0.448	0.347	0.241	0.130	0.047
	0.5	0.801	0.768	0.706	0.625	0.526	0.421	0.310	0.201	0.084	0.015	0.007
	0.6	0.688	0.652	0.585	0.488	0.381	0.281	0.175	0.058	—	—	—
	0.7	0.550	0.517	0.433	0.330	0.242	0.085	0.043	—	—	—	—
	0.8	0.387	0.342	0.262	0.180	0.126	—	—	—	—	—	—
	0.9	0.195	0.155	0.073	0.035	—	—	—	—	—	—	—
0.2	0.0	0.998	0.949	0.874	0.787	0.694	0.588	0.476	0.351	0.245	0.153	0.105
	0.1	0.991	0.943	0.869	0.779	0.681	0.573	0.469	0.332	0.234	0.123	0.011
	0.2	0.968	0.921	0.843	0.752	0.650	0.537	0.418	0.294	0.184	0.069	0.009
	0.3	0.927	0.879	0.798	0.699	0.589	0.465	0.325	0.216	0.150	0.019	0.004
	0.4	0.862	0.812	0.728	0.623	0.503	0.376	0.282	0.090	0.025	—	—
	0.5	0.772	0.719	0.629	0.518	0.393	0.260	0.113	0.004	—	—	—
	0.6	0.656	0.603	0.507	0.393	0.265	0.129	—	—	—	—	—
	0.7	0.516	0.459	0.363	0.250	0.129	—	—	—	—	—	—
	0.8	0.349	0.295	0.194	0.110	—	—	—	—	—	—	—
	0.9	0.046	—	—	—	—	—	—	—	—	—	—
0.3	0.0	0.986	0.922	0.811	0.693	0.561	0.409	0.267	0.160	0.080	0.005	—
	0.1	0.975	0.912	0.804	0.681	0.544	0.402	0.261	0.150	0.070	0.004	—

Table 13.1 Continued

P/P_y	M_x/M_{px}	Slenderness Ratio, $L/r_x\ (L/r_y)$										
		0	10 (15)	20 (30)	30 (45)	40 (60)	50 (75)	60 (90)	70 (105)	80 (120)	90 (135)	100 (150)
	0.2	0.940	0.880	0.778	0.649	0.472	0.364	0.225	0.094	0.030	—	—
	0.3	0.889	0.826	0.724	0.597	0.452	0.306	0.146	0.0323	—	—	—
	0.4	0.815	0.751	0.642	0.514	0.374	0.217	0.031	—			
	0.5	0.721	0.653	0.543	0.390	0.271	0.076	—	—	—	—	—
	0.6	0.600	0.532	0.427	0.292	0.120	—	—	—	—	—	—
	0.7	0.457	0.393	0.280	—	—	—	—	—	—	—	—
	0.8	0.001	—	—	—	—	—	—	—	—	—	—
0.4	0.0	0.944	0.883	0.744	0.582	0.403	0.246	0.147	0.070	0.010	—	—
	0.1	0.930	0.864	0.733	0.573	0.400	0.238	0.128	0.032	0.008	—	—
	0.2	0.893	0.821	0.697	0.569	0.361	0.203	0.078	0.029	—	—	—
	0.3	0.832	0.759	0.633	0.481	0.317	0.159	0.011	—	—	—	—
	0.4	0.752	0.674	0.549	0.404	0.231	0.040	—	—	—	—	—
	0.5	0.648	0.569	0.454	0.304	0.093	—	—	—	—	—	—
	0.6	0.524	0.451	0.333	—	—	—	—	—	—	—	—
	0.7	0.001	—	—	—	—	—	—	—	—	—	—
0.5	0.0	0.872	0.797	0.663	0.469	0.241	0.123	0.064	0.014	—	—	—
	0.1	0.857	0.783	0.641	0.455	0.231	0.122	0.012	0.013	—	—	—
	0.2	0.817	0.738	0.586	0.417	0.217	0.085	—	—	—	—	—
	0.3	0.753	0.669	0.523	0.357	0.169	0.012	—	—	—	—	—
	0.4	0.663	0.578	0.444	0.272	0.056	—	—	—	—	—	—
	0.5	0.555	0.475	0.339	—	—	—	—	—	—	—	—
0.6	0.0	0.761	0.678	0.535	0.314	0.134	0.102	0.051	—	—	—	—
	0.1	0.746	0.665	0.511	0.293	0.132	0.071	—	—	—	—	—
	0.2	0.707	0.623	0.467	0.274	0.100	0.012	—	—	—	—	—
	0.3	0.643	0.554	0.402	0.207	0.0016	—	—	—	—	—	—
	0.4	0.552	0.467	—	—	—	—	—	—	—	—	—
0.7	0.0	0.618	0.528	0.374	0.157	0.046	0.008	—	—	—	—	—
	0.1	0.604	0.515	0.361	0.153	0.040	0.009	—	—	—	—	—
	0.2	0.566	0.478	0.319	0.133	—	—	—	—	—	—	—
	0.3	0.499	0.416	—	—	—	—	—	—	—	—	—
0.8	0.0	0.445	0.346	0.222	0.052	0.0033	—	—	—	—	—	—
	0.1	0.420	0.300	0.205	—	—	—	—	—	—	—	—
	0.2	0.390	0.240	0.190	—	—	—	—	—	—	—	—
0.9	0.0	0.235	0.164	0.083	—	—	—	—	—	—	—	—
	0.1	0.205	—	—	—	—	—	—	—	—	—	—

13.2.3 Linear Interaction Equations

The strength criteria for a short steel beam-column may be defined in a simple manner by approximating the surface with straight line interaction equations such as the one shown in Fig. 13.3 for a steel H-column. For the particular case of an uniaxial bending of H-column subject to a constant axial compression P, the plastic-hinge moment, M_{pc}, modified to include the effect of this axial compression can be determined simply by using the superposition procedure developed by Chen and Atsuta (1972) for biaxially bent beam-columns. Equations for M_{pcx} and M_{pcy} for strong and weak axes bending respectively applicable for most wide-flange sections, are plotted non-dimensionally in Figs. 13.4 and 13.5 for two values of flange area to web area ratio $A_f/A_w = 1.0$ and 1.5. In Figs. 13.4 and 13.5 are also shown by dotted lines simple approximate equations to compute M_{pcx} and M_{pcy}. The limitations $M_x \leq M_{pcx}$ and $M_y \leq M_{pcy}$ lead to the interaction expressions as follows:

For strong axis bending moment, M_x,

$$\frac{P}{P_y} + 0.85\frac{M_x}{M_{px}} \leq 1.0, \qquad M_x \leq M_{px} \tag{13.1}$$

and for weak axis bending moment, M_y,

$$\left(\frac{P}{P_y}\right)^2 + 0.84\frac{M_y}{M_{py}} \leq 1.0, \qquad M_y \leq M_{py} \tag{13.2}$$

Pillai and Ellis (1974) have suggested that the nonlinear expression (13.2) may be linearized to the simple form

$$\frac{P}{P_y} + 0.6\frac{M_y}{M_{py}} \leq 1.0, \qquad M_y \leq M_{py} \tag{13.3}$$

From Fig. 13.5, it can be seen that this expression has the same limits as expression

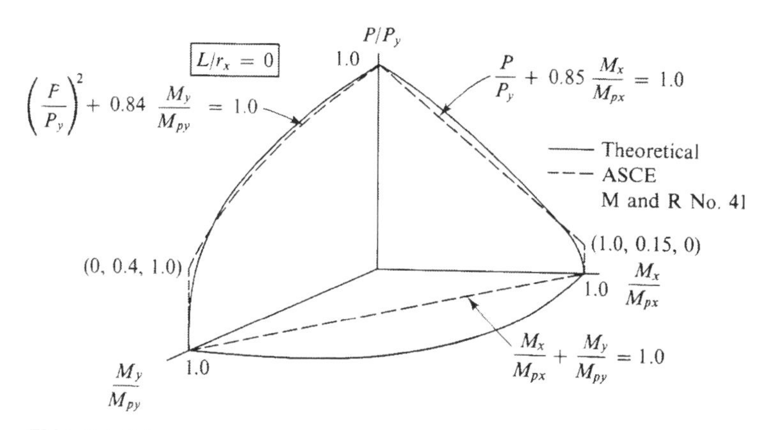

FIGURE 13.3
Interaction surfaces for short H-columns and ASCE M and R no. 41 equations

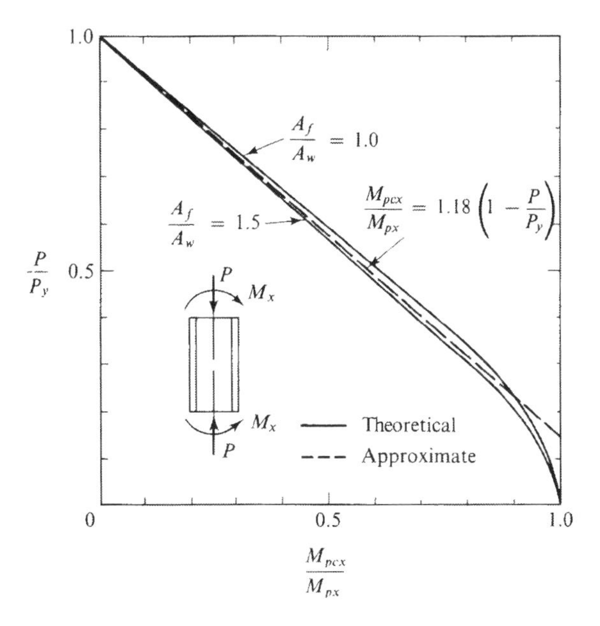

FIGURE 13.4
Approximate interaction equation for wide-flange section (strong axis bending, short column)

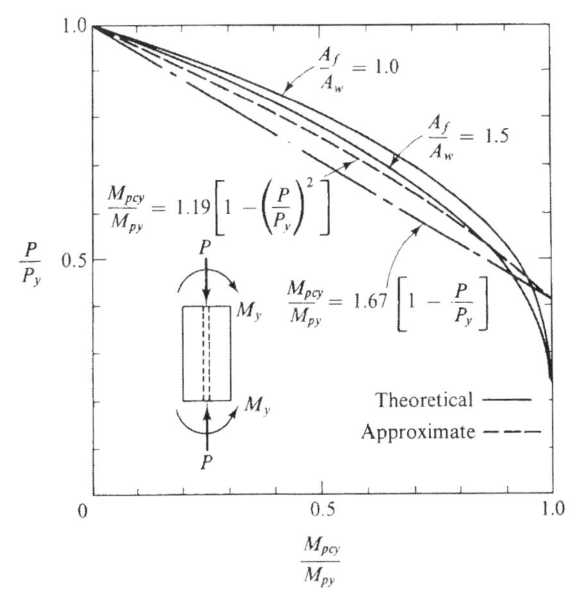

FIGURE 13.5
Approximate interaction equation for wide-flange section (weak axis bending, short column)

(13.2), and is only slightly more conservative in the middle range. In the following, it will be seen that the virtue of this linear form of the minor-axis interaction expression is the ease with which it can be incorporated into a biaxial linear interaction expression.

Expressions (13.1) and (13.2) approximately express mathematically the limit of interaction between bending moment and axial force and are the basis for the current AISC (1969) and CSA (1974) Specifications.

Expressions (13.1) and (13.3) can now be extended in a simple manner to the general case of biaxial bending. Figure 13.6 shows exact interaction curves corresponding to a constant value of P which may be thought of as "load contours." Also shown in Fig. 13.6 for comparison, is the load carrying capacity permitted by the CSA (1974) Specification:

$$\frac{P}{P_y} + 0.85\frac{M_x}{M_{px}} + 0.85\frac{M_y}{M_{py}} \leq 1.0 \tag{13.4}$$

and

$$\frac{M_x}{M_{px}} + \frac{M_y}{M_{py}} \leq 1.0 \tag{13.5}$$

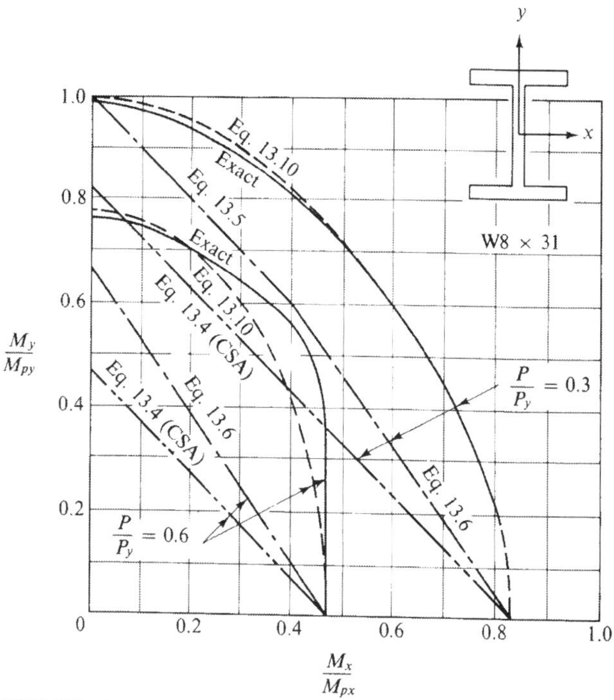

FIGURE 13.6
Interaction curves for zero length columns

Expressions (13.4) and (13.5) are seen to give a very conservative estimate of the biaxial strength of a cross section, especially with respect to the weak axis bending. Part of this extreme conservatism is due to the assumption of the same relation between P/P_y and M_y/M_{py} as exists in expression (13.4) as between P/P_y and M_x/M_{px}, whereas they are different as can be seen from Figs. 13.4 and 13.5 and Eqs. (13.1), (13.2) and (13.3). This can also be seen from the fact that the shape factor, or the ratio between fully plastic and initial yield moment capacity about weak axis bending, is approximately 1.5 which is roughly 1.5/1.12 = 1.34 times greater than that of strong axis bending 1.12. Pillai and Ellis (1974) suggested the following correction about the weak axis limit of expression (13.4), as follows:

$$\frac{P}{P_y} + 0.85\frac{M_x}{M_{px}} + 0.6\frac{M_y}{M_{py}} \leq 1.0 \tag{13.6}$$

provided that expression (13.5) is complied with. This expression, shown also on Fig. 13.6, is probably the limit of accuracy for a linear interaction. Nonlinear interaction equations will be presented in what follows.

The AISC Specification does not provide for the ultimate strength of beam-columns in biaxial bending, but the working stress expression

$$\frac{f_a}{0.6F_y} + \frac{f_{bx}}{F_{bx}} + \frac{f_{by}}{F_{by}} \leq 1.0 \tag{13.7}$$

factored by 1.67 is little different to the CSA expressions. In expression (13.7), f_a is axial stress, f_b is bending stress, F_y is yield stress, and F_b is permissible bending stress. AISC expression essentially places stress limits on the extreme corner fibers of a beam-column section. The limit for axial stress f_a was originally $0.6F_y$ giving a factor of safety against nominal first yield of material 1.67. Through successive editions of the AISC Specifications, minor modifications have been incorporated. The permissible bending stress has been increased from $0.6F_y$ to $0.66F_y$ in recognition of the shape factor of wide flange sections about their major axis. It is to be noted that the same relationship has been assumed between the fully plastic and initial yield moment capacity about the weak axis as about the strong axis.

For square hollow structural sections, Pillai and Ellis (1974) have proposed the following interaction expressions:

$$\frac{P}{P_y} + 0.85\left(\frac{M_x}{M_{px}} + 0.5\frac{M_y}{M_{py}}\right) \leq 1.0 \tag{13.8}$$

$$\frac{M_x}{M_{px}} + 0.5\frac{M_y}{M_{py}} \leq 1.0 \tag{13.9}$$

in which M_x is the numerically larger moment.

13.2.4 Nonlinear Interaction Equations

Both the AISC and CSA design expressions are straight line interaction equations.

Exact interaction curves shown in Fig. 13.6 are not linear; on the contrary, the interaction of biaxial bending moment resembles more closely the quadrant of a circle. Further, it is important to note that if a short beam-column is fully loaded under axial load and bending about one axis, then there is no spare capacity to accept bending moment about the other axes. However, as the loading decreases slightly below the maximum, moment capacity rapidly develops to accept bending about the other axis.

Since the end points of the exact curve are M_{pcx} and M_{pcy} where the curve crosses the M_x/M_{px} and M_y/M_{py} axes respectively, the exact interaction curves corresponding to four given values of axial load $p = P/P_y = 0$, 0.3, 0.6 and 0.9 are replotted on axes of M_x/M_{pcx} vs. M_y/M_{pcy} in Fig. 13.7. The general form of these curves can now be approximated by the nondimensional interaction equation:

$$\left(\frac{M_x}{M_{pcx}}\right)^{\alpha} + \left(\frac{M_y}{M_{pcy}}\right)^{\alpha} = 1.0 \tag{13.10}$$

Exact values of M_{pcx} and M_{pcy} for H-columns are listed in tabular form in Table 13.1.

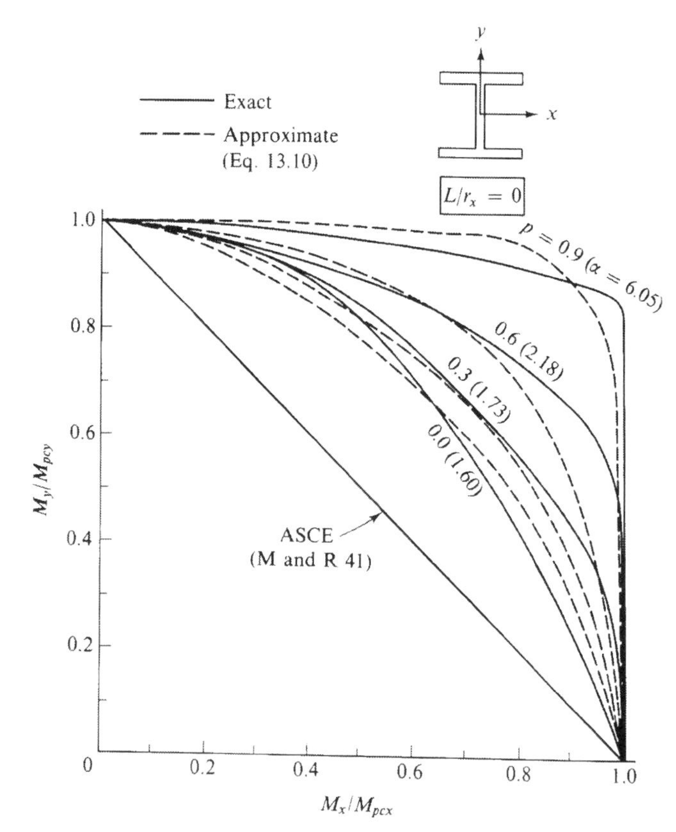

FIGURE 13.7
Comparison of interaction curves for short beam-columns

The values of M_{pcx} and M_{pcy} may also be computed with good accuracy from the formulas given in Figs. 13.4 and 13.5 for major and minor axes bending of short columns.

$$M_{pcx} = 1.18\, M_{px}\left(1 - \frac{P}{P_y}\right) \leq M_{px} \tag{13.11a}$$

$$M_{pcy} = 1.19 M_{py}\left[1 - \left(\frac{P}{P_y}\right)^2\right] \leq M_{py} \tag{13.11b}$$

The exponent, α, is a numerical factor whose value depends on the shape of a particular cross section and the magnitude of the axial load. By deriving the correct exponents, an interaction curve could be defined which would fit the actual strength curve of a biaxially loaded beam-column of a particular cross section, such as a wide flange shape. The curve-fitting values of α for H-column can be determined from the numerical interaction relations (Table 13.1) and are found to vary from 1.58 when $p = P/P_y = 0.0$ to 6.05 when $p = 0.9$.

The variation of these calculated values of α for H-column can be represented by the approximate expression

$$\alpha = 1.60 - \frac{p}{2 \ln p} \tag{13.12}$$

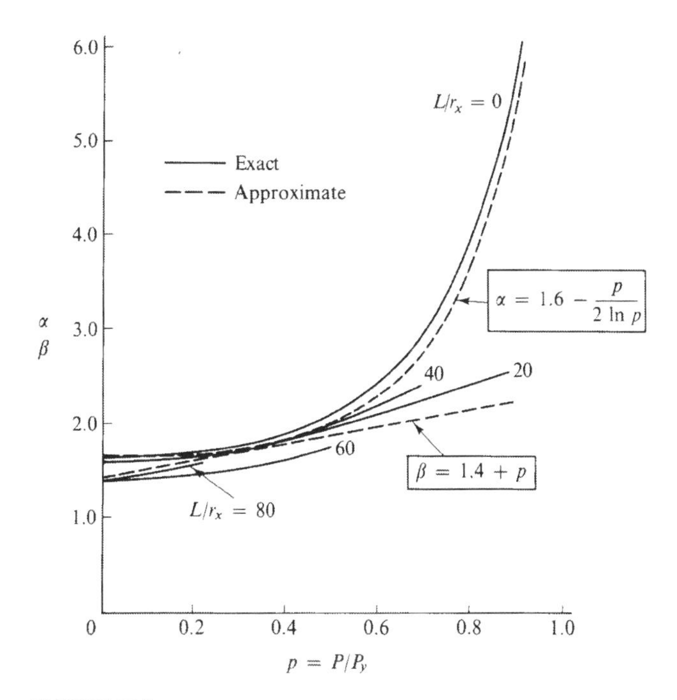

FIGURE 13.8
Values of the exponent α, β

where ln is the natural logarithm. A comparison of the "exact" curve-fitting values of α and those given by Eq. (13.12) is shown in Fig. 13.8.

Using the approximate values of α given in Eq. (13.12), the approximate nonlinear interaction curves for all values of p can now be determined in a simple manner from Eq. (13.10). A comparison between the approximate equation and the exact numerical results for H-column is also shown in Figs. 13.6 and 13.7. Good agreement is observed. Note that the current ASCE design formulas (Manual No. 41, 1971) governing the strength of biaxially loaded short columns (Fig. 13.3) is equivalent to Eq. (13.10) with $\alpha = 1.0$ valid for all values of p. Clearly, the formula gives a conservative estimate of the strength of biaxially loaded short columns, especially for the cases of higher values of p (Fig. 13.7).

For the case of a square box column the following expression for the exponent α is suggested by Chen (1977)

$$\alpha = 1.7 - \frac{p}{\ln p} \tag{13.13}$$

From the foregoing discussion, we can conclude here that for short beam-columns, considerable advantage may be gained by considering the ultimate capacity of the fully plastified section. Aids are now available, either in the form of tables such as Table 13.1 or charts as shown in Fig. 13.6, to enable designers to tackle this problem readily. The linear and nonlinear interaction equations developed here appear to be especially appropriate for incorporation into column-design programs such as those currently sponsored by the AISC (1969) and CISC (1967).

13.3 LONG BEAM-COLUMNS UNDER SYMMETRIC LOADING

When a beam-column is subjected to a compressive load combined with biaxial bending, it will usually deflect and twist simultaneously at any load. The essential feature of the beam-column due to the space action is that the lateral displacement is always accompanied by a twist of the beam-column cross sections. The importance of this twisting lies in the fact that the ultimate load-carrying capacity of such beam-columns, especially for beam-columns with open thin-walled sections that have low torsional rigidity, may be much less than the maximum load carrying capacity for inplane loading.

13.3.1 Method of Solution

The analysis of such long or intermediate length of beam-column is much more complex than that of the short beam-column. The solutions are dependent on the assumptions made in the analytical treatment. The methods of solution to the problem may be categorized into three groups as follows, although they are all closely related:

1. The governing differential equilibrium equations are solved rigorously by the use of formal mathematics. Closed form solutions are possible only for simple *elastic*

beam-column problems (Chap. 3 and Chap. 4). In the plastic or nonlinear range, numerical methods must be used to obtain solutions. The numerical methods which have been used by various investigators include the finite integral method (Chap. 3), numerical integration (Chap. 9), and the finite difference method (Chap. 10). The disadvantage of these methods is that they are efficient and successful only for isolated members or small structures.

2. The governing differential equilibrium equations are solved approximately by introducing additional assumptions. The assumptions which have been used by various investigators include establishing equilibrium only at midheight of the beam-column or at a number of stations along the beam-column length (Chaps. 7 and 8), assuming the three components of displacement of a cross section to be given by known functions such as half waves of a sine curve (Chap. 7), and idealizing the shape of the biaxial moment-curvature-thrust relationships (Chap. 7). These methods are developed only for isolated members and usually give accurate solutions. The methods are found quite powerful and efficient for computer solution, and are used in the following to generate numerical results.

3. The beam-column is assumed to consist of a number of finite segments or elements, and the assemblage of these elements is then treated as a space structure. The matrix stiffness method, which has been widely used for large structures, is applied to obtain solutions. In this method, the complete stiffness matrix of the beam-column is constantly modified for each load increment to account for the change of lateral displacements, as well as the deterioration of the segment stiffness caused by the plastic yielding of the cross section. Because of its capability in dealing with complex problems along with the appeal of a systematic approach of computation, the method is especially useful, for example, for solving restrained type of beam-column problems for which the beam-column must be considered as a part of the entire rigid framework. Details of the method can be found in Chap. 3 for the analysis of elastic flexural-torsional buckling type of problems, and in Chaps. 11 and 12 for the analysis of elastic-plastic biaxially loaded beam-columns and space frame type of structures.

The tangent stiffness method developed in Sect. 7.3, Chap. 7 is used here for numerical studies. The method is based on the analytical development of the linear relationship between the incremental changes of the generalized forces $\{\dot{W}\}$ and displacements $\{\dot{\Delta}\}$. The mathematical derivation of this relationship is based on the assumption that the beam-column has initial imperfections of a half-sine wave and the subsequent deflected shapes under increasing loads can be described by the sum of sine and parabolic curves. Equilibrium equations can then be established at the midheight section of the member.

The incremental force-deformation relationship at the beam-column midheight has the simple form

$$\{\dot{W}\} = [K]\{\dot{\Delta}\} \tag{13.14a}$$

in which

$\{\dot{W}\}$ = the rate of change of the external force vector

$\{\dot{\Delta}\}$ = the incremental deformation vector

$[K]$ = the tangent stiffness matrix.

In this procedure the load is applied as a sequence of sufficiently small increments. During the application of each increment the beam-column has essentially to behave linearly. Thus, the nonlinear behavior of the beam-column is determined by solving a sequence of linearized equations.

$$\{\dot{\Delta}\} = [K]^{-1}\{\dot{W}\} \tag{13.14b}$$

An improved solution may be obtained by starting with an initial estimate of the displacement solution. This solution is then back substituted into the equations and the procedure is repeated until an accepted convergence or a prescribed tolerance is obtained. The iterational scheme is similar to that of the Newton-Raphson method. In general, the solution will converge within a few cycles of iteration even for larger load increments.

13.3.2 Numerical Results

In constructing the interaction surface the maximum load carrying capacities of a given member corresponding to various combinations of applied loads must be determined. Unlike short columns, where the limit plastic state can be evaluated in a single step, the maximum loading condition for a long beam-column cannot readily be determined before establishing the complete load-deflection relationship. This requires an enormous amount of computation effort to furnish sufficient points that describe smoothly the families of interaction surfaces. A computer program was developed at Lehigh University (Santathadaporn, 1970) that computes the relationship between the applied loads and the three generalized displacements for biaxially loaded H-columns. The program can also handle residual stress and yield strength variations including variations through the thickness of the component plates, end warping restraint, end bending restraint and initial imperfections.

Although the numerical procedure was developed to solve exclusively load-deflection problems, it was also employed in solving the inherent eigenvalue problems, i.e., the maximum axial loads, P_u, when $M_x = M_y = 0$ or the critical moment, $(M_x)_{cr}$, when $M_y = 0$. Herein, the load-deflection method was utilized by assuming sufficiently small imperfections ($\delta_{x0} = \delta_{y0} = 1.0 \times 10^{-6}\ L$, $\theta_{im} = 1.0 \times 10^{-6}$ rad. in which δ = lateral deflection at midspan; θ_{im} = twist angle at midspan; and L = length). The results were compared with those obtained by the eigenvalue approach and good correlations were observed. The validity of the program was also verified by comparing the predictions with the experimental results of a recently tested heavy column, H23 × 681 (Tebedge, Chen and Tall, 1972). The failure of the column was observed to be in biaxial bending although the column was centrally loaded. The comparison of the theoretical predictions based on the biaxial column analysis and the column test results showed good agreement.

To confirm the selection of the shape W8 × 31 as a representative shape of cross section for long beam-columns, the effects of the variation in shape and weight were investigated. Four different shapes ranging from W8 × 31 to W14 × 87 were studied to determine the effects of the variation of column sizes on the interaction curves. The sections were chosen to have equal flange width to depth ratios ($B/D \simeq 1.0$). Figure 13.9 shows the nondimensionalized interaction curves for $p = P/P_y = 0.3$. It is observed that the curves for all slenderness ratios indicate small differences. To determine the effects of the variation of column weight on the interaction curves, the shapes W8 × 31 and W14 × 426 were used. Figure 13.10 shows the results for four different slenderness ratios when $p = 0.3$. Again, the interaction curves are seen to be relatively insensitive to the variations in column weights. Here, as in the previous

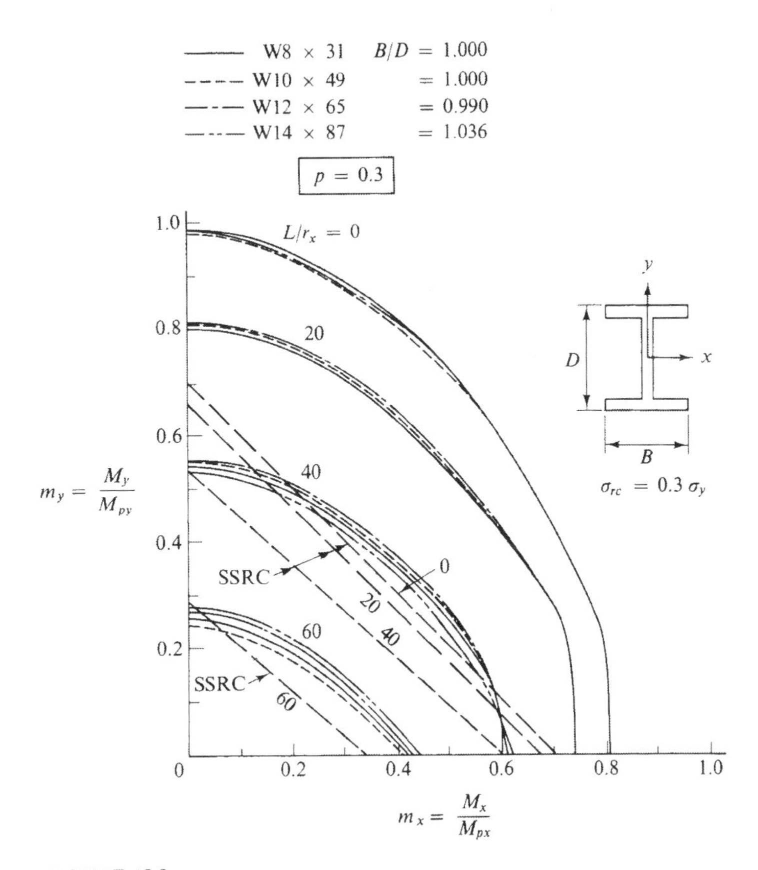

FIGURE 13.9
Comparison of maximum strength interaction curves for various depths of light shapes ($B/D \cong 1.0$)

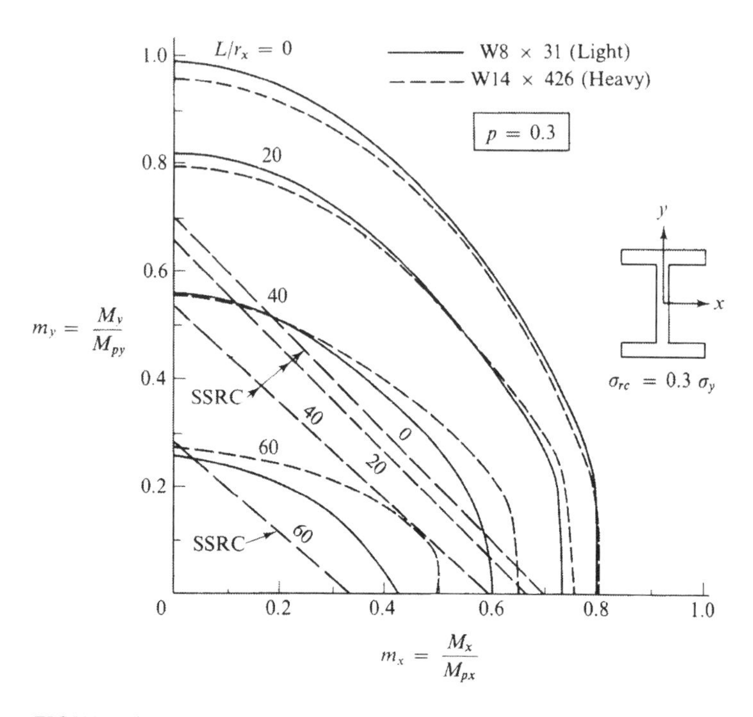

FIGURE 13.10
Comparison of maximum strength interaction curves of a light and a heavy shape

case of short columns, the particular shape W8 × 31 is found to be a suitable section representing most commonly used wide-flange sections.

Based on the cross sectional properties of W8 × 31, with linear distribution of residual stresses due to rolling $\sigma_{rc} = 0.3\ \sigma_y$ (residual stress at flange tip) and with $\sigma_y = 36$ ksi (248 N/mm^2), families of numerical interaction relations are prepared and presented in Table 13.1. To minimize the required computational efforts in determining maximum values, different curvature increments were used for each slenderness ratio: the increments varied from $\dot{u}''_m = 0.0001$ for $L/r_x = 10$ to $\dot{u}''_m = 0.000001$ for $L/r_x = 100$.

Although the interaction curves were prepared for mild steel with a nominal yield strength of 36 ksi (248 N/mm^2), the results may also be applied to steels of other yield stress levels by entering the tables with a modified or equivalent slenderness ratio:

$$\left(\frac{L}{r}\right)_{\text{equ}} = \frac{L}{r}\sqrt{\frac{\sigma_y}{36}} \tag{13.15}$$

A few examples were solved to determine the discrepancy that may result when using Eq. (13.15) for the same members with different steel grades. Figure 13.11 shows a

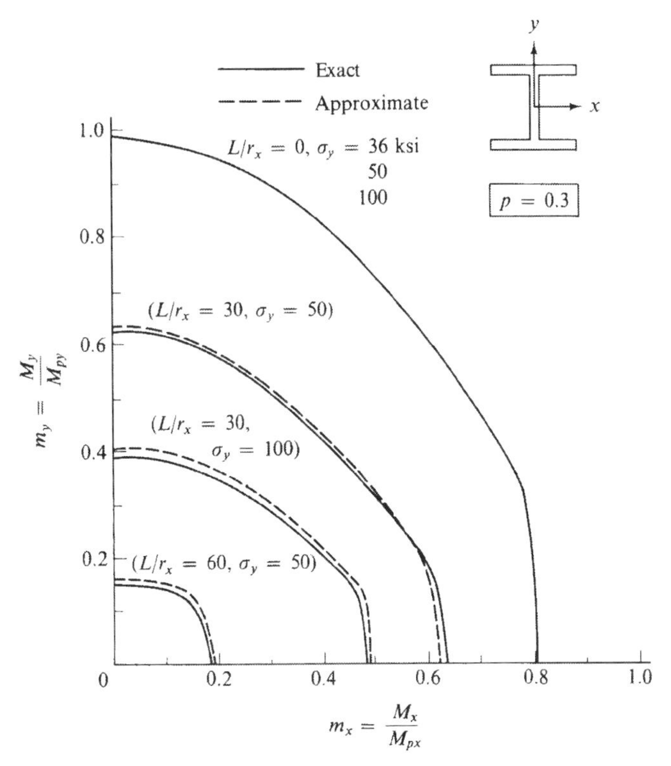

FIGURE 13.11
Interaction curves for different steel grades

comparison of the exact interaction curves for members of high grade steels [$\sigma_y = 50$ ksi (345 N/mm^2) and 100 ksi (689 N/mm^2)] and the approximate curves obtained through Eq. (13.15). For the beam-columns with $L/r_x = 0$ the results were seen to be identical, as expected. For the long beam-columns, small differences were observed but, in general, the correlations are good.

13.3.3 Linear Interaction Equations

In the current AISC Specification (1969), the expression governing the working stress design of long beam-columns subject to biaxial bending has the form

$$\frac{f_a}{F_a} + \frac{f_{bx}}{F_{bx}\left(1 - \frac{f_a}{F'_{ex}}\right)} + \frac{f_{by}}{F_{by}\left(1 - \frac{f_a}{F'_{ey}}\right)} \leq 1.0 \tag{13.16}$$

where

f_a = computed axial stresses

f_{bx}, f_{by} = computed compressive bending stresses at the point under consideration

F_a = axial stress that would be permitted for the column if axial force alone existed

F_{bx}, F_{by} = compressive bending stresses that would be permitted for the beam if bending moment alone existed. The stresses are determined considering all failure modes including plastic yielding, lateral torsional buckling, and local buckling.

F'_{ex}, F'_{ey} = $\frac{12}{23}$ times the Euler buckling stresses in the plane of bending.

The term F_a is based on the ultimate load-carrying capacity of an axially loaded column P_u modified by a variable safety factor which increases from $5/3 = 1.67$ to $23/12 = 1.92$ as the slenderness ratio increases. The term F'_e is based on the Euler buckling load modified by a constant value of safety factor 23/12. The term F_{bx} is based on a 10 percent increase over the initial yield moment to recognize that the fully plastic moment capacity of compact sections is greater than the nominal initial yield moment, that is, the permissible bending stress F_{bx} is $1.1(0.6\,F_y) = 0.66\,F_y$. The term F_{by} is treated similarly to F_{bx}, but in wide flange sections the shape factor about the minor axis is approximately 1.55 versus 1.12 about the major axis. Therefore, when the moment about the minor axis of the wide flange shape is significant, there may be a marked difference between design by the AISC working strength method and by the ultimate strength method as predicted by Eq. (6.19) of the 2nd edn (1966) of the Column Research Council's *Guide to Design Criteria for Metal Compression Members* [or Eq. (8.29) of the 3rd edn (1976)]. The SSRC (formerly CRC) expression for the ultimate strength of beam-columns subject to biaxial bending, for a stability type failure, has the form

$$\frac{P}{P_u} + \frac{M_x}{M_{ux}\left(1 - \frac{P}{P_{ex}}\right)} + \frac{M_y}{M_{uy}\left(1 - \frac{P}{P_{ey}}\right)} \leq 1.0 \tag{13.17}$$

where

M_{ux}, M_{uy} = ultimate bending moments of a beam about the x- and y-axis, respectively, reduced for the possible presence of lateral torsional buckling, if necessary, when $P = 0$

P_{ex}, P_{ey} = Euler buckling loads

P_u = ultimate load of an axially loaded column.

For wide flange shapes, M_{uy} would be the plastic moment capacity of the section M_{py}. M_{ux} would be the plastic moment capacity of the beam about the major axis, reduced as may be necessitated by lateral torsional buckling considerations, $M_{ux} = M_m \leq M_{px}$. Therefore, the conservatism mentioned previously with respect to the different shape factors about the major and minor axes is eliminated in this expression, since correct ultimate moment capacity about each respective axis is used.

The two interaction expressions, stated above, are both straight line interaction of moments about the orthogonal axes, i.e., for a given ratio of axial load to axial capacity, the variation in moment about the two orthogonal axes is linear. Figure 13.12 shows the three-dimensional interaction surfaces for long H-columns along with SSRC expressions. Note that the interaction curves at the $p = 0$ plane pass through the point $M_y/M_{py} = 1.0$ as there is no likelihood of lateral buckling due to minor axis bending.

By algebraic transposition, Eq. (13.17) may be written in the following linear form:

$$\frac{M_x}{M_{ucx}} + \frac{M_y}{M_{ucy}} \leq 1.0 \tag{13.18}$$

where M_{ucx} and M_{ucy} are the ultimate bending moment capacities of an axially loaded beam-column about the x- and y-axis respectively, when there is zero moment about the other axis ($P \neq 0$). Thus for a given compression, P, the values of M_{ucx} and M_{ucy} are given on the $M_y = 0$ and $M_x = 0$ axes of Figs. 13.9 to 13.11. The values for H-columns may be obtained from Table 13.1. The values of M_{ucx} and M_{ucy} may also be computed with good accuracy from the formulas given in Part 2 of the AISC Specifications (1969).

$$M_{ucx} = M_m\left(1 - \frac{P}{P_u}\right)\left(1 - \frac{P}{P_{ex}}\right) \tag{13.19a}$$

$$M_{ucy} = M_{py}\left(1 - \frac{P}{P_u}\right)\left(1 - \frac{P}{P_{ey}}\right) \tag{13.19b}$$

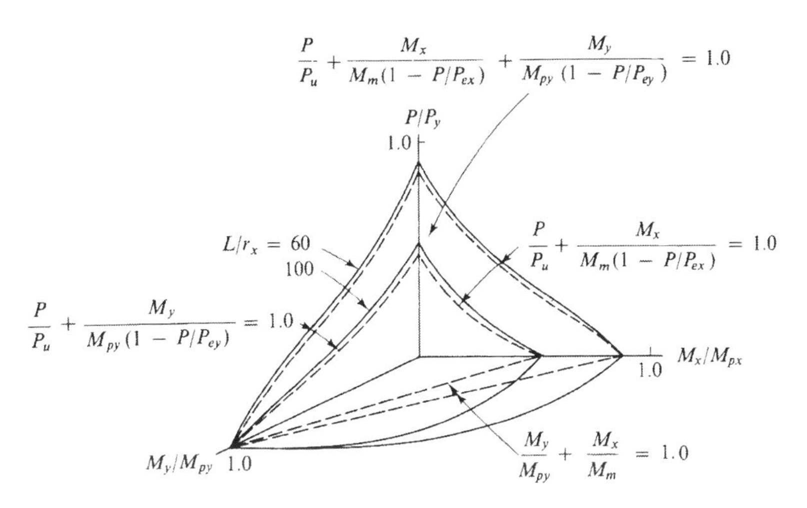

FIGURE 13.12
Interaction surfaces for long H-columns and SSRC equations

where

M_m = maximum moment that can be resisted by the member in strong axis bending in the absence of axial load and weak axis bending moment.

In lieu of a more precise calculation, M_m may be taken as

$$M_m = \left[1.07 - \frac{\left(\frac{L}{r_y}\right)\sqrt{\sigma_y}}{3160} \right] M_{px} \le M_{px} \tag{13.20}$$

in which σ_y is in ksi.
It should be noted that slenderness ratios greater than $\sqrt{2\pi^2 E/\sigma_y}$ do not appear to have been considered in the development of Eq. (13.20). Beyond this ratio, values of M_m could be based on 1.67 times the allowable stress given in Sec. 1.5.1.4.6a of the AISC Specification. The effect of lateral torsional buckling generally is to reduce the axial load capacity only slightly. The reader is referred to Chap. 3 for further discussion on this subject.

Based on the properties of the section W8 × 31, the SSRC curves were computed using Eq. (13.17) for the cases $L/r_x = 0$ to 60 and $p = 0.3$, and these are compared with the theoretical interaction curves in Figs. 13.9 and 13.10. The SSRC formula is seen to be over-conservative for short columns, conservative for intermediate columns and less conservative for long columns.

The SSRC curves in Figs. 13.9 and 13.10 were replotted on axes of M_y/M_{ucy} vs. M_x/M_{ucx} in the form of expression (13.18) and are compared to the exact numerical results in Fig. 13.13. It should be noted that the value of M_{uc} used in Fig. 13.13 was the exact value from Table 13.1, whereas the end points of SSRC lines were obtained approximately from Eq. (13.19). Hence the intercept on the axes is not at unity.

Although a short beam-column is a special case of a long beam-column when L/r_x approaches zero, the SSRC formula, Eq. (13.17), or the formula (13.19) for the end points is not applicable to the case $L/r_x = 0$. In this special case, one should turn to Eqs. (13.11). Relative to Eqs. (13.11), the formula (13.19) would give much less value. The SSRC interaction lines marked $L/r_x = 0$ in Figs. 13.9, 13.10 and 13.13 should therefore be viewed as only providing the limiting lines of Eq. (13.17) when L/r_x becomes very small.

Pillai and Ellis (1971) proposed a modified SSRC interaction expression in the form

$$\frac{P}{P_u} + C\left[\frac{M_x}{M_{ux}\left(1 - \frac{P}{P_{ex}}\right)} + \frac{M_y}{M_{uy}\left(1 - \frac{P}{P_{ey}}\right)} \right] \le 1.0 \tag{13.21}$$

for hollow, circular, and square box section beam-columns. In expression (13.21), the coefficient C is defined in terms of biaxial eccentricities e_x and e_y as

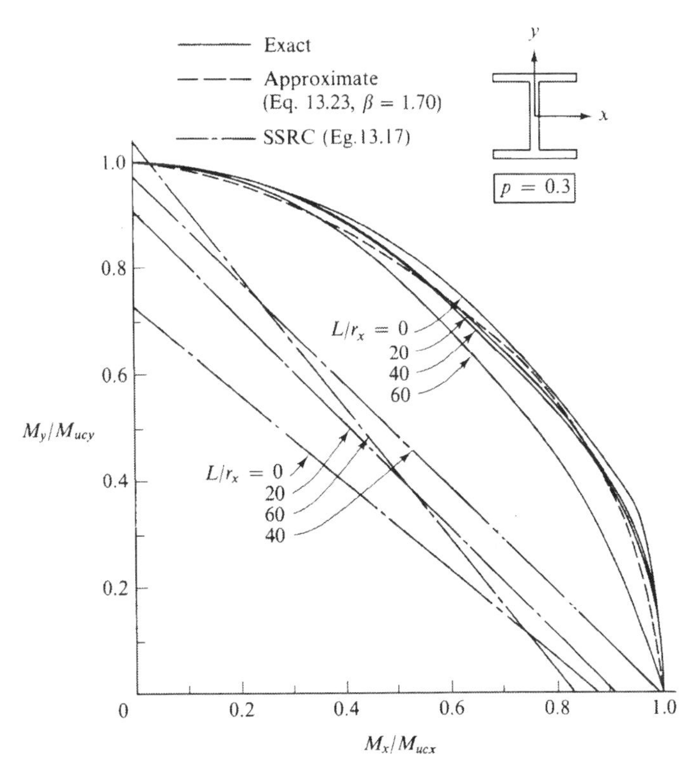

FIGURE 13.13
Comparison of interaction curves for long columns

$$C = \frac{\sqrt{e_x^2 + e_y^2}}{e_x + e_y} \tag{13.22}$$

and all other terms are the same as in Eq. (13.17). On comparing with results of 21 tests, Eq. (13.21) gave consistent results with an average ratio of test load to predicted load of 1.11. Numerical interaction diagrams for hollow square tubular sections subject to biaxial bending are also available (Ellis, Jury and Kirk, 1964; Marshall and Ellis, 1970, or see Sect. 9.2.4, Chap. 9, page 420)

13.3.4 Nonlinear Interaction Equations

From Fig. 13.13, it is clear that the interaction of moments about the orthogonal axes is not linear; on the contrary, the interaction curve more closely resembles the quadrant of a circle. In parallel with their proposal for the case of short columns, Eq. (13.10), Tebedge and Chen (1974) proposed the same form of interaction expression for long beam-columns, that is,

$$\left(\frac{M_x}{M_{ucx}}\right)^{\beta} + \left(\frac{M_y}{M_{ucy}}\right)^{\beta} \leq 1.0 \tag{13.23}$$

which is a simple modification of expression (13.18).

The appropriate values of β in Eq. (13.23) determined from the numerical interaction relations (Table 13.1), are found to vary from 1.35 when $p = 0$ and $L/r_x = 80$ to 2.54 when $p = 0.9$ and $L/r_x = 20$. The calculated values of β for different values of L/r_x are shown in Fig. 13.8. It is evident that the variation of these calculated values may be approximated by the simple equation

$$\beta = 1.40 + p \tag{13.24a}$$

valid for all values of slenderness ratios greater than 10 (Fig. 13.8).

Using the approximate values of β given by Eq. (13.24a), the approximate interaction curves for all values of p are determined from Eq. (13.23). A comparison between the approximate equation and the exact numerical results is shown in Fig. 13.13. Good agreement is again observed.

Since the values of the exponents α and β determined from Eqs. (13.12) and (13.24a) are very close to each other for small values of $p \leq 0.6$ (Fig. 13.8), it follows that the approximate interaction Eq. (13.23) proposed herein as the strength criteria for long steel H-columns can also be used with good accuracy for estimating the strength of a short beam-column when the axial compression is moderate ($p \leq 0.6$). This is demonstrated in Fig. 13.13.

The development of Eq. (13.24a) is based essentially on a particular wide flange shape, i.e., W8 × 31, for which the flange width B to section depth D is almost equal. Two pertinent questions might then be asked: Does the exponent β have any shape dependence, and, if so, how should Eq. (13.24a) be modified to account for this? In the following, the question of shape dependence is investigated and a simple modification to Eq. (13.24a) is then suggested.

Two narrower I-shapes in the lightweight range were adopted for this investigation: one with $B/D \approx 0.5$ (W14 × 38), and the other with $B/D \approx 0.3$ (W21 × 44). The axial load cases considered are $p = 0.1$, 0.3, 0.5 and 0.7, and the numerical results are summarized in Table 13.2. Some of these results are shown in Figs. 13.14 and 13.15. In view of the results obtained, a simple modification is suggested to Eq. (13.24a) (Ross and Chen, 1976)

$$\beta = 0.4 + p + B/D \geq 1 \qquad \text{for} \qquad B/D \geq 0.3$$

and (13.24b)

$$\beta = 1.0 \qquad \text{for} \qquad B/D < 0.3$$

Using Eq. (13.24b), the nonlinear interaction of moments calculated from expression (13.23) is also plotted in each of Figs. 13.14 and 13.15, dashed curves.

Figures 13.14 and 13.15 illustrate the applicability of the expression (13.24b). When $B/D \approx 1.0$, this expression becomes identical to Eq. (13.24a). The equation was found to give remarkably accurate predictions of maximum strength interaction of moments for sections with $B/D \approx 1.0$. For sections which deviate markedly from being nearly square (i.e., $B/D \neq 1.0$), Eq. (13.24b) is found to be adequate, but to give more conservative predictions.

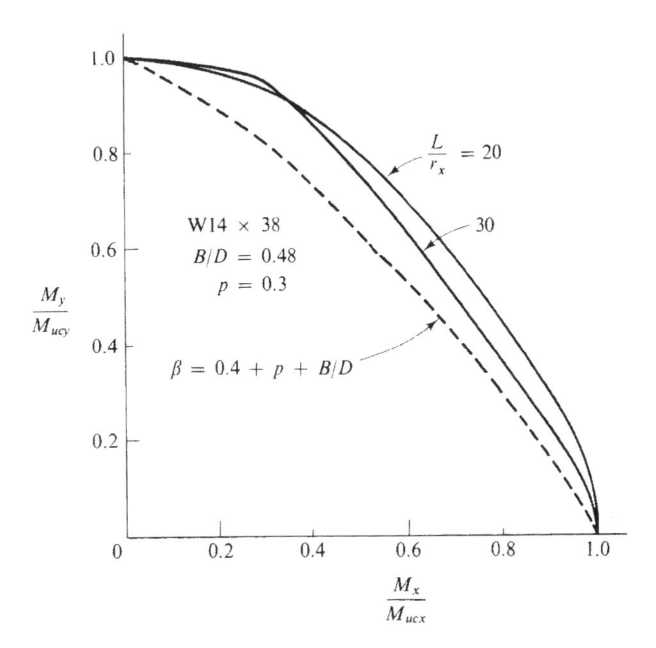

FIGURE 13.14
Maximum strength interaction curves for W14 × 38 at $p = 0.3$

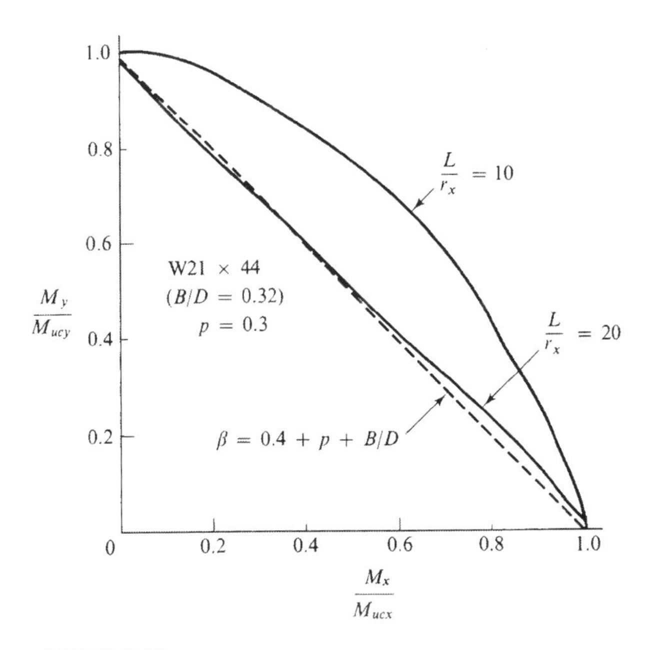

FIGURE 13.15
Maximum strength interaction curves for W21 × 44 at $p = 0.3$

Table 13.2 MAXIMUM STRENGTH TABLE FOR BIAXIALLY LOADED BEAM-COLUMNS (MAXIMUM VALUES OF M_y/M_{py}) (ROSS AND CHEN, 1976)

$\frac{P}{P_y}$	$\frac{M_x}{M_{px}}$	W14 × 38 $(B/D = 0.48)$			$\frac{M_x}{M_{px}}$	W21 × 44 $(B/D = 0.32)$		
		$L/r_x = 30$	$L/r_x = 50$			$L/r_x = 5$	$L/r_x = 20$	$L/r_x = 30$
0.1	0.0	0.648	0.282		0.0	0.885	0.58	0.267
	0.10		0.243		0.05			0.262
	0.15		0.181		0.10			0.199
	0.20	0.541	0.092		0.12			0.163
	0.22		0.070		0.15			0.098
	0.40	0.276			0.17			0.068
	0.50	0.105			0.20	0.869		0.025
	0.53	0.050			0.40	0.828		
					0.60	0.668		
					0.70	0.514		
					0.80	0.311		
					0.85	0.194		
					0.90	0.075		
		$L/r_x = 20$	$L/r_x = 30$	$L/r_x = 40$		$L/r_x = 10$	$L/r_x = 20$	
0.3	0.0	0.493	0.179	0.070	0.0	0.583	0.109	
	0.10		0.168		0.10	0.572	0.068	
	0.20	0.448	0.110		0.20		0.033	
	0.30		0.037		0.26		0.011	
	0.40	0.290			0.30	0.527		
	0.45	0.250			0.50	0.341		
	0.52	0.160			0.70	0.060		
	0.56	0.096						
		$L/r_x = 10$	$L/r_x = 20$	$L/r_x = 25$		$L/r_x = 8$	$L/r_x = 12$	$L/r_x = 15$
0.5	0.0	0.640	0.165	0.078	0.0	0.507	0.164	0.071
	0.10		0.158		0.20	0.477	0.150	
	0.15		0.156		0.40	0.335	0.027	
	0.20	0.590	0.151		0.60	0.074		
	0.30		0.105					
	0.35		0.064					
	0.40	0.436						
	0.50	0.333						
	0.52	0.307						
	0.56	0.255						
	0.60	No Sol.						
		$L/r_x = 5$	$L/r_x = 10.7$	$L/r_x = 15$		$L/r_x = 5$	$L/r_x = 12$	$L/r_x = 15$
0.7	0.0	0.579	0.369	0.105	0.0	0.567	0.055	0
	0.10		0.339		0.20	0.447		
	0.20	0.512	0.287		0.25	0.400		
	0.30	0.430	0.207		0.30	0.349		
	0.32	0.411	0.186		0.35	0.294		
	0.34	0.390	No Sol.		0.40	No Sol.		
	0.36	No Sol.						

It is noted also that Eq. (13.24b) proposed that $\beta = 1.0$ be adopted as a lower limit. The curve $\beta = 1.0$ is the straight line representing the design expressions assumed in traditional code requirements. There is evidence that for $B/D \leq 0.3$ the true value of β is close to $\beta = 1.0$. As a check on the viability of this lower limit, an interaction curve was derived for the W21 × 44 section with an axial load ratio of $p = 0.1$. Thus, by calculation, the corresponding value of $\beta = 0.82$. But we have stipulated a β_{min} of 1.0. It can be seen from Fig. 13.16 that there is not any concavity in the predicted strength curve. The adoption of the lower limit of $\beta = 1.0$ is recommended. In design practice, rolled sections with $B/D < 0.3$ are very rare. Further, the formula is not recommended for use with non-standard sections.

For the case of a square box column the following expression for the exponent β is suggested by Chen (1977)

$$\beta = 1.3 + \frac{1000p}{(L/r)^2} \geq 1.4 \tag{13.25}$$

The advantages of using the nonlinear interaction equations for design are that both strength and stability are covered by the same basic equation and the equation has universal applicability, for if the exponent for the particular shape of cross section being designed has not been determined, it may conservatively be assumed to be unity. In using the proposed interaction equation, however, the determination of axial load capacity of the column should be as realistic as possible, such as using the multiple column curves presented previously in Sect. 14.4, Chap. 14, Vol. 1 of this book.

13.4 LONG BEAM-COLUMNS UNDER UNSYMMETRIC LOADING

The required analysis for the determination of the maximum strength of compression members subjected to unequal biaxial bending moments or transverse loads, is often

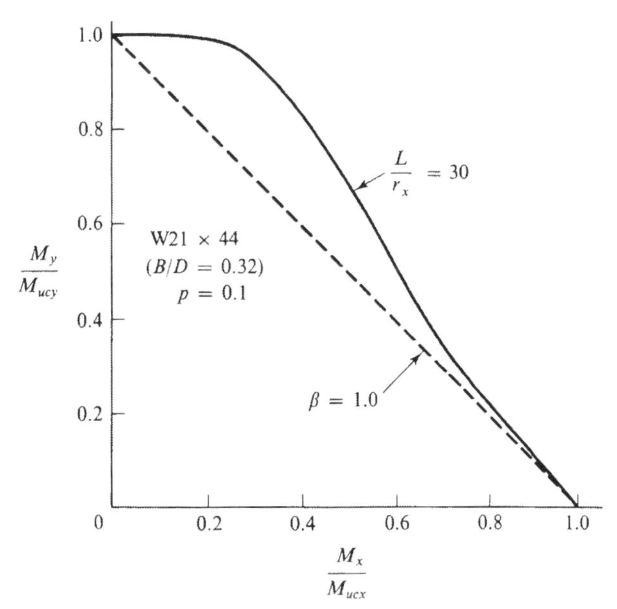

FIGURE 13.16
Strength of beam-column in region where $\beta = 1.0$

too complicated for practical applications. Limited solutions are presently available for biaxially loaded plastic beam-columns subjected to unequal end moments. The possibility of extending the use of the interaction curves obtained for the case of symmetrically loaded beam-columns to the case of unsymmetrically loaded beam-columns using the C_m factor method, will be discussed here. According to the concept, a *reduction factor* or *equivalent uniform moment factor* termed C_m will be introduced. The C_m factor has been used extensively in the in-plane beam-column design. Direct modification of the equation recommended by AISC (1969) for in-plane cases is now extended to the biaxial cases in the following manner

$$C_{mx} = 0.6 + 0.4\frac{M_{xl}}{M_{xo}} \geq 0.4 \tag{13.26a}$$

$$C_{my} = 0.6 + 0.4\frac{M_{yl}}{M_{yo}} \geq 0.4 \tag{13.26b}$$

in which M_{xo}, M_{yo}, M_{xl} and M_{yl} are the applied end moments, and M_{xo} and M_{yo} being the larger ones. M_{xl}/M_{xo} and M_{yl}/M_{yo} are positive when the member is bent in single curvature and negative when it is bent in reverse curvature. The equivalent uniform bending moments in the two directions are

$$(M_x)_{\text{equ}} = C_{mx} M_{xo} \tag{13.27a}$$

$$(M_y)_{\text{equ}} = C_{my} M_{yo} \tag{13.27b}$$

these equivalent moment values should be used when entering Tables 13.1 and 13.2.

Five maximum axial loads corresponding to various end eccentricities are computed from the interaction relations (Table 13.1) using the C_{mx} and C_{my} factors given in Eqs. (13.26) and these are compared with experimental results and theoretical values in Table 13.3. In all cases satisfactory agreement is observed. Sharma and Gaylord (1969) also reported briefly on the accuracy of the C_m factor method as a means of extending their design charts (Fig. 7.17, Chap. 7) to encompass all conditions met in practice (see Sec. 7.5, Chap. 7). In a further investigation by Keramati, Gaylord, and Robinson (1972), the ultimate loads for 15 cases of unsymmetrical biaxial bending were computed and compared with predictions using the C_m factor method with the Sharma and Gaylord's charts. One case involved single-curvature bending in one principal plane and double-curvature bending in the other; another involved double-curvature bending in both principal planes, while the remaining 13 were cases of unsymmetrical single curvature in each principal plane. The largest difference between computed values and those given by the interaction curves was 5 percent. It may therefore be concluded from these studies that the strength of unsymmetrically loaded biaxial beam-columns may be obtained directly from either Table 13.1, interaction charts presented in previous chapters, or interaction equations presented here using the equivalent uniform moment concept, that is, the C_m factor method.

Table 13.3 COMPARISON OF THEORETICAL SOLUTIONS WITH EXPERIMENTS AND C_m FACTOR METHOD

Specimen number (1)	Column shape (2)	Length in (m) (3)	Yield stress ksi (N/mm^2) (4)	Eccentricities of Loading in (mm)				Reduction Factor		Maximum Axial Load, $p = P/P_y$		
				e_{xo} (5)	e_{xl} (6)	e_{yo} (7)	e_{yl} (8)	C_{mx} (9)	C_{my} (10)	Theory[a] (11)	Experiment[b] (12)	C_m Factor (13)
9	I4 × 8	96 (2.43)	36 (248)	0.38 (10)	0.39 (10)	1.72 (44)	−1.66 (−42)	0.40	0.99	0.373	0.387	0.40
11	H5 × 5	96 (2.43)	36 (248)	0.83 (21)	−0.78 (−20)	1.73 (44)	−1.74 (−44)	0.40	0.40	0.565	0.562	0.50
15	H5 × 5	120 (3.19)	65 (448)	1.75 (44)	−1.47 (−37)	2.96 (75)	−2.96 (−75)	0.40	0.40	0.305	0.312	0.30
16	H5 × 5	120 (3.19)	65 (448)	0.14 (4)	0.35 (9)	−0.11 (−3)	−2.80 (−71)	0.62	0.75	0.308	0.309	0.30
H1	W14 × 43	264.6 (6.94)	33 (228)	0.50 (13)	0.00 (0)	5.00 (127)	0.00 (0)	0.60	0.60	0.317	0.334	0.35

[a] Theoretical results reported by Chen and Atsuta (1973b); and Sharma and Gaylord (1969).
[b] Experimental results reported by Birnstiel (1968).

13.5 METHOD OF K COEFFICIENTS

The greatest major axis moment M_x which could possibly be applied to a *short* beam-column is the full plastic hinge moment M_{px}. The presence of axial load P in the cross section has the immediate effect of reducing this simple plastic hinge moment from its full value M_{px} to the lower value M_{pcx} [see Eq. (13.11a)]. The addition of minor axis moment M_y in addition to the axial load P, further reduces this major axis moment carrying-capacity of the short beam-column. For this case, we may write the reduced major axis moment M_x in the form

$$M_x = K_h M_{px} \tag{13.28}$$

where K_h is the dimensionless *reduction coefficient* which accounts for the combined reducing effect of axial load and minor axis moment on the major plastic hinge moment capacity M_{px}. K_h is obviously less than, or at most equal to, the ratio M_{pcx}/M_{px}.

The values of K_h can be determined directly from the existing biaxial moment interaction diagrams. Figure 13.1 shows a typical biaxial plastic hinge interaction diagram developed previously for H-columns (Chap. 5). The horizontal axis of the interaction diagram gives the ratio of the maximum major axis moment M_x under the combined effect of axial load and minor axis moment to the full plastic moment M_{px}. Reference to Eq. (13.28) will confirm that this ratio $m_x = M_x/M_{px}$ is the definition of K_h. The coefficient may thus be obtained from an interaction diagram with the horizontal axis relabelled to read K_h directly.

The problem becomes far more complex in the case of a *long* beam-column. In addition to the problem of strength reduction of the cross section due to axial load and minor axis bending moment, the effects of torsional buckling and flexural buckling about major and minor axes must also be considered. Superimposed on this are the effects of residual stresses and initial curvatures. Each one of these factors will have the effect of reducing the ideal major axis bending moment value M_{px}.

Young (1973) proposes that the effect of these factors may be assessed separately and introduces the following three additional coefficients K_x, K_y, K_t to account for the length effect of the beam-column so that the product of the four coefficients and the full major axis bending moment M_{px} provides the actual major axis moment carrying-capacity of the long beam-column, thus

$$M_x = K_x K_t K_y K_h M_{px} \tag{13.29}$$

in which K_x represents the proportion of the major axis plastic hinge moment which may be attained at one end of the beam-column, if axial load and major axis moment alone are considered. K_t accounts for the proportion of the plastic moment which can be carried by the member if it is treated as a beam susceptible to torsional-flexural failure solely under major axis moment loading. K_y is similar to K_x, except that only the effects of axial load and minor axis moment are considered. The calculation of these coefficients is presented in the following.

The K_x coefficient

The K_x coefficient accounts for the effect of major axis deflection on the major axis moment capacity of an uniaxially loaded beam-column. Two important length parameters are required in this calculation: one is the limiting length of a long column L_c under axial load alone, and the other is the limiting length for a short, or stocky column, L_s. This means that only when the beam-column is shorter than its critical length for buckling about the major axis as an axially loaded column L_{cx} and longer than its stocky column length L_{sx}, need the effect of this major axis deflection on the major axis moment be considered. When the beam-column reaches its major axis critical length, the major axis moment capacity falls immediately to zero. For a short beam-column the effect of this deflection upon the major axis moment capacity can be ignored. This simple approach leads to the following expressions for K_x:

$$
\begin{aligned}
&L \le L_{sx}, && K_x = 1.0 \\
&L_{sx} < L < L_{cx}, && 0 < K_x = \frac{C_{mx} M_{ucx}}{M_{pcx}} < 1.0 \qquad (13.30) \\
&L_{cx} \le L, && K_x = 0
\end{aligned}
$$

in which M_{ucx} is given by Eq. (13.19a) with $M_m = M_{px}$ and C_{mx} is defined in Eq. (13.26a).

The values of the limiting column length L_c under axial load alone can be determined from existing column strength curves. Studies of axially loaded column strength have been made in Sect. 14.4, Chap. 14, Vol. 1 from the standpoint of deflections considering the effect of residual stresses and initial curvature. Probabilistic analysis of axially loaded column strength has also been made in Chap. 15, Vol. 1. The end result of these studies was the recommendation of three multiple column curves. With the aid of a column curve selection chart, a wide range of typical, axially loaded steel column sections of various material strengths, can be designed for failure about major or minor axes. The value of L_c for a section appropriate to the axial load, includes implicitly the imperfections mentioned above.

A stocky beam-column is one which may be designed for a plastic hinge to appear at one end without the need to consider stability effects. In this case, the major axis moment capacity is $M_x = M_{pcx}$, that is $K_x = 1.0$. This is particularly useful since many practical beam-column lengths fall into this category.

The coefficient K_x may be obtained directly from a beta-chart of Chap. 13, Vol. 1. A typical beta-chart for a universal column section in major axis bending is shown in Fig. 13.17. Three parameters are used in the figure, the ratio of the greater end moment to the plastic moment reduced in the presence of axial load, M_x/M_{pcx}, the ratio of the end moments, β_x, and the ratio of the actual column length to limiting column length, L/L_{cx}. From the definition of K_x coefficient it follows that $K_x = M_x/M_{pcx}$.

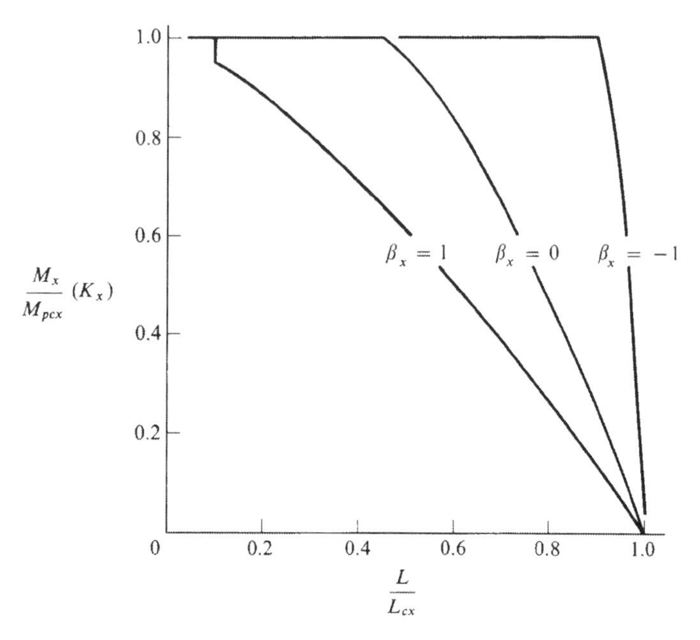

FIGURE 13.17
Beta chart for a universal column section in major axis bending (Young, 1973)

The K_t Coefficient

The K_t coefficient accounts for the effect of torsional and warping stiffness on the major axis moment capacity of an I-section beam-column. The coefficient K_t is therefore derived by consideration of the pure beam behavior of the member and represents the proportion of the full plastic moment of the section which can be developed at one end when account is taken of the St. Venant torsional stiffness and differential warping of the flanges.

If M_{cr} denotes the critical moment for a beam in pure bending, M_{cr} is then related to the flexural-torsional buckling coefficient, K_t, by the equation

$$M_{cr} = K_t M_{px} \tag{13.31}$$

In lieu of a more precise calculation, M_{cr} may be taken directly from the AISC Specification (1969) as $M_{cr} = M_m$ [Eq. (13.20)] or $K_t = M_m/M_{px}$. The reader is referred to Chap. 3 for the detailed development of more precise design formulas for elastic and plastic lateral torsional buckling of beams.

Like the development of K_x coefficient, the strength of a laterally unsupported long beam is determined by elastic behavior, while a short or stocky beam is expected to develop the full yield strength of the cross section, M_{px}. For an intermediate length beam, like that of a corresponding column, the critical lateral buckling moment is determined by plastic behavior. A simple approximate method of estimating the effect of plastic action on the buckling strength of beams, is to assume that the relationship between elastic and plastic buckling strength is the same for beams as it is for columns. It follows that the plastic buckling strength of beams can then be estimated from an existing column strength curve. Thus by entering an appropriate column curve a slenderness ratio L/r_y, a value of P/P_y is found which could be interpreted directly as K_t for a real beam with residual stresses and initial curvature.

In the case of a beam subjected to unequal end moments, the concept of equivalent uniform moment can be applied. In this case the value of $K_t(=M_{cr}/M_{px})$ from a column curve should be multiplied by $(1/C_{mx})$ but with the restriction that (K_t/C_{mx}) must not be greater than unity. The limiting length of a stocky beam is defined as $K_t/C_{mx} = 1.0$.

The coefficient K_t may be obtained directly from a prepared beam design curve which is a modification of a column curve, taking into account the right shape for beam instead of column. Two British beam design curves proposed by Young (1973) are shown in Fig. 13.18. The shape of these curves is established empirically with test results. The upper one is for beams in Grade 55 steel and the lower curve is for Grade 43 steel. Test results by Dibley (1969) for I-sections in Grade 55 steel subjected to uniform bending ($\beta_x = +1.0$) are also shown in Fig. 13.18. In Fig. 13.18, $(L/r_y)_{equ}$ is the "equivalent slenderness ratio". For British universal sections it can be expressed

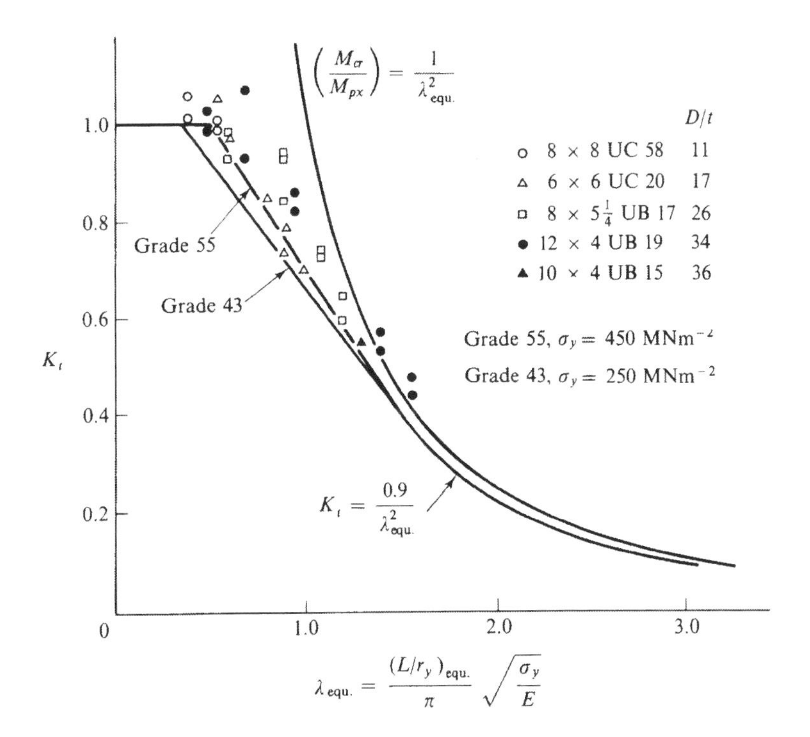

FIGURE 13.18
Dibley's experimental results for flexural-torsional buckling of grade 55 beams under uniform moment (Young, 1973)

in terms of the actual slenderness ratio and the ratio of the section depth, D, to the flange thickness, t, by

$$\left(\frac{L}{r_y}\right)_{\text{equ}} = \left(\frac{L}{r_y}\right)\left[\frac{14(D/t)^2}{(L/r_y)^2 + 15.5(D/t)^2 + 0.23(D/t)^3}\right]^{1/4} \tag{13.32}$$

The K_y coefficient

The K_y coefficient accounts for the effect of minor axis deflection on the major axis moment carrying-capacity of a beam-column. The major axis moment capacity of a biaxially loaded beam-column must obviously be zero when the member is on the point of failing by flexural buckling about the minor axis under the action of axial load and minor axis moments alone. This limiting length L_{cmy} is defined here as the greatest length the beam-column can have to withstand a given combination of axial load and minor axis moment alone ($L_{cmy} = L_{cy}$ when no minor axis moments act). For a short or stocky beam-column, L_{slat}, the effect of this minor deflection on the major axis moment capacity can be ignored. This approach leads to the following upper and lower bound expressions for K_y:

$$\begin{aligned} &L \le L_{\text{slat}}, && K_y = 1.0 \\ &L_{\text{slat}} < L < L_{cmy}, && 0 < K_y < 1.0 \\ &L \ge L_{cmy}, && K_y = 0 \end{aligned} \tag{13.33}$$

Since the ratio L_{cmy}/L_{cy} is identical to the ratio L/L_{cy} in Eq. (13.29), Chap. 13, Vol. 1, for hot-rolled I-sections, the value for L_{cmy} can be obtained directly from a beta-chart in minor axis bending with the symbol L replaced by L_{cmy} in Fig. 13.11, Chap. 13, Vol. 1.

The stockiness limit for no lateral buckling (L_{slat}) has a similarity with previously defined stockiness limits (L_{sx} and L_{sy}) for in-plane flexural buckling. Here we are concerned, however, with the limiting length of a column for which lateral-flexural buckling need not be considered. L_{sx} and L_{sy} were limiting column lengths for which in-plane flexural buckling could be ignored.

By comparing with tests on hot-rolled column sections at Liége (Campus and Massonnet, 1956), Young (1973) proposed the following approximate expressions for K_y

$$L_{\text{slat}} = 0.4\, L_{cmy}(1 - \beta_{\text{eff}}) \tag{13.34}$$

$$K_y = 1.0 \qquad \text{when} \quad \begin{cases} L \le L_{\text{slat}} \\ L \le 0.1\, L_{cmy} \end{cases} \tag{13.35}$$

and K_y takes the lesser value from:

$$K_y = 1 - \frac{1}{5}\left(\frac{L - L_{\text{slat}}}{L_{cmy} - L_{\text{slat}}}\right)$$

or

$$K_y = 5\left(1 - \frac{L}{L_{cmy}}\right) \tag{13.36}$$

when

$$\begin{cases} L > L_{slat} \\ L > 0.1\, L_{cmy} \end{cases}$$

In Eq. (13.34), β_{eff} denotes the *effective end moment ratio* which takes account of major and minor axis moment ratios β_x and β_y. The value for β_{eff} is given in Fig. 13.19. Two extreme combinations of β_x and β_y are represented by the boundaries AB and BC in Fig. 13.19. Intermediate combinations of β_x and β_y are easily obtained from the chart. Similar procedure for K_y may be developed for box or other shapes of cross section.

13.6 DESIGN EXAMPLES

For a given problem case, the design values of L, P, M_x and M_y are assumed known and the main purpose here is to select the most economical section that can carry these loads. This may be performed by using Table 13.1. The use of the table can be illustrated by the following typical example:

Example 13.1 Stability design using Table 13.1

GIVEN

Design a W10 column section in a building for an effective length of 15 ft-0 in story height to support an ultimate axial load $P = 175$ kip, subjected to equal end moments, $M_{xo} = -M_{xl} = 870$ kip-in and $M_{yo} = M_{yl} = 95$ kip-in, inducing antisymmetric

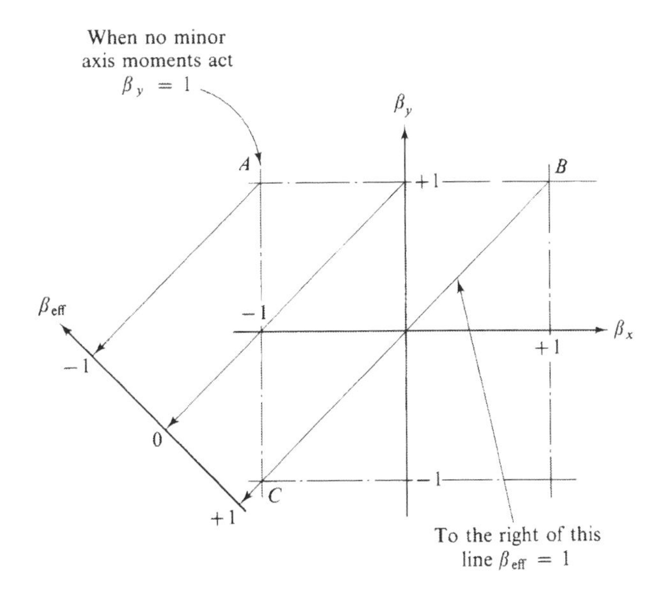

FIGURE 13.19
β_{eff} chart for combining the effect of β_x and β_y in biaxially loaded beam columns (Young, 1973)

double curvature about x-axis and symmetric single curvature about y-axis. Use steel with $\sigma_y = 50$ ksi.

SOLUTION

The equivalent moments and the equivalent slenderness ratio are determined from Eq. (13.27) and (13.15), respectively. Thus,

$$(M_x)_{\text{equ}} = 0.4\,M_{xo} = 348\,\text{kip-in};\quad (M_y)_{\text{equ}} = M_{yo} = 95\,\text{kip-in};$$
$$(L/r)_{\text{equ}} = L/r\sqrt{50/36} = 1.18(L/r).$$

Trial 1: W10 × 29

For this section: $P/P_y = 0.41$; $M_x/M_{px} = 0.20$; $M_y/M_{py} = 0.22$; $r_x = 4.30$ in, $r_y = 1.38$ in, $(L/r_x)_{\text{equ}} = 50$; $(L/r_y)_{\text{equ}} = 154$. From Table 13.1, for $P/P_y = 0.4$; $M_x/M_{px} = 0.20$; $L/r_y = 150$, it is found that $M_y/M_{py} = 0 < 0.22$ and therefore the section W10 × 29 is inadequate.

Trial 2: W10 × 33

For this section: $P/P_y = 0.36$; $M_x/M_{px} = 0.18$; $M_y/M_{py} = 0.14$; $r_x = 4.20$ in, $r_y = 1.94$ in, $(L/r_x)_{\text{equ}} = 51$; $(L/r_y)_{\text{equ}} = 110$. From Table 13.1, for $P/P_y = 0.30$; $M_x/M_{px} = 0.20$; $L/r_y = 105$, it is found that $M_y/M_{py} = 0.094 \approx 0.14$.

Therefore, the section *W10 × 33 is adequate.*

13.6.1 Comparative Strength Designs

The following example has been chosen to illustrate the use of various design methods for strength-type analysis and to demonstrate the difference in column section required by these methods.

Example 13.2 Strength design using equations and charts

Design a W12 section in A36 steel, effective length 15 ft-0 in, to resist an ultimate axial load of 375 kip, subject to equal and opposite end moments of 160 and 60 kip-ft about the x- and y-axes, respectively, inducing antisymmetric double curvature about both axes.

For this case, since reverse curvature is induced by the end moments, maximum bending stresses will occur at the ends and will be unaffected by the axial load acting upon the deflected shape of the column. The strength-type analysis is found to control in this example.

The column section required for strength according to each design method is shown in Table 13.4, along with a factor indicating the relative economy of each section.

Table 13.4 COMPARISON OF STRENGTH SOLUTIONS TO DESIGN EXAMPLE 13.2

Method	Column Section	Economy Factor
(a) AISC (Eq. 13.7)	W12 × 106	1.00
(b) CSA-S16 (Eqs. 13.5, 13.6)	W12 × 85	0.80
(c) Nonlinear Interaction Equations (Eqs. 13.10–13.12)	W12 × 72	0.68
(d) Ringo's Charts (Figs. 5.36, 5.37, Table 5.5)	W12 × 72	0.68
(e) Sharma and Gaylord's Charts (Fig. 7.17)	W12 × 65	0.61
(f) *K* Coefficients (Eq. 13.29)	W12 × 72	0.68

The example has been solved on the basis of the following design methods:

(a) *AISC Specification* (*Working Stress Design*)

To use expression (13.7) for strength-type analysis, we use a load factor of $1/0.6 = 1.667$, permissible bending stress $0.66\ F_y$ for strong axis bending, and $0.75\ F_y$ for weak axis bending, respectively. The corresponding expression (13.16) for stability analysis is found not critical in this example.

AISC Solution

Try W12 × 106

$$A = 31.19 \text{ in}^2, \quad S_x = 144.5 \text{ in}^3, \quad S_y = 49.2 \text{ in}^3$$

From formula (13.7) for strength, we have,

$$\frac{375/1.667}{31.19 \times 0.6 \times 36} + \frac{160 \times 12/1.667}{144.5 \times 0.66 \times 36} + \frac{60 \times 12/1.667}{49.2 \times 0.75 \times 36}$$

$$= 0.334 + 0.336 + 0.325 = 0.995 < 1.0 \qquad \text{O.K.}$$

(b) *CSA S16.1—1974* (*Limit States Design*)

Canadian Standards Association Code is based on the ultimate load formulas (13.5) and (13.6) for strength-type analysis using an overall load factor of 1.667. The corresponding stability formula (13.17) is found not critical in this example.

CSA-S16 Solution

Try W12 × 85

$$A = 24.98 \text{ in}^2, \quad Z_x = 129 \text{ in}^3, \quad Z_y = 59.1 \text{ in}^3$$

From formula (13.5) for strength we have

$$\frac{160 \times 12}{36 \times 129} + \frac{60 \times 12}{36 \times 59.1} = 0.413 + 0.338 = 0.751 < 1.0 \qquad \text{O.K.}$$

From formula (13.6) for strength we have

$$\frac{375}{36 \times 24.98} + \frac{0.85 \times 160 \times 12}{36 \times 129} + \frac{0.6 \times 60 \times 12}{36 \times 59.1} = 0.417 + 0.351 + 0.203$$

$$= 0.971 < 1.0 \qquad \text{O.K.}$$

(c) *Nonlinear interaction equations* (13.10) *to* (13.12)

The corresponding nonlinear interaction formula (13.23) together with (13.24) for stability-type of analysis is found not critical in this example.

Try W12 × 72

$$A = 21.16 \text{ in}^2, \quad Z_x = 108 \text{ in}^3, \quad Z_y = 49.2 \text{ in}^3$$

$$p = \frac{P}{P_y} = \frac{375}{21.16 \times 36} = 0.492 \text{ ksi}$$

From Eqs. (13.11) and (13.12), we find

$$M_{pcx} = 1.18 \times 108 \times 36(1 - 0.492) = 2330 \text{ k-in}$$

$$M_{pcy} = 1.19 \times 49.2 \times 36(1 - 0.492^2) = 1598 \text{ k-in}$$

$$\alpha = 1.60 - \frac{0.492}{2 \times \ln 0.492} = 1.95$$

From Eq. (13.10) for strength, we have,

$$\left(\frac{160 \times 12}{2330}\right)^{1.95} + \left(\frac{60 \times 12}{1598}\right)^{1.95} = 0.686 + 0.211 = 0.897 < 1.0 \qquad \text{O.K.}$$

(d) *Nomographs developed by Ringo, et al. (1973)—Figs. 5.36 and 5.37, and Table 5.5, Chap. 5*

Try W12 × 72

From Table 5.5 (page 245), the W12 × 72 is a member of Group 3 with $P/P_b = 2.332$, $M_x/M_{bx} = 0.35$ and $M_y/M_{by} = 3.50$. Thus, we have

$$M_{bx} = \frac{160 \times 12}{0.35} = 5486 \text{ k-in}, \quad M_{by} = \frac{60 \times 12}{3.5} = 205.7 \text{ k-in}$$

From Group 3 interaction curves of Fig. 5.37 (page 247), we find $P_b = 180$ or $P = 2.332 \times 180 = 420 \text{ kip} > 375 \text{ kip}$ O.K.

(e) *Charts developed by Sharma and Gaylord (1969)—Fig. 7.17, Chap. 7*

Try W12 × 65

$$B = 12.00 \text{ in}, \quad D = 12.12 \text{ in}, \quad A = 19.11 \text{ in}^2, \quad r_x = 5.20 \text{ in}, \quad r_y = 3.02 \text{ in}$$

From Eqs. (13.26) and (13.27), we find

$$C_{mx} = C_{my} = 0.4$$

$$(M_x)_{\text{equ}} = 0.4 \times 160 = 64 \text{ k-ft}, \quad (M_y)_{\text{equ}} = 0.4 \times 60 = 24 \text{ k-ft}$$

From which we obtain

$$e_x = \frac{24 \times 12}{375} = 0.77 \text{ in}, \quad e_y = \frac{64 \times 12}{375} = 2.05 \text{ in}$$

$$\frac{e_x B}{2r_y^2} = \frac{0.77 \times 12.00}{2(3.02)^2} = 0.51$$

$$\frac{e_y D}{2r_x^2} = \frac{2.05 \times 12.12}{2(5.20)^2} = 0.46$$

From Fig. 7.17 (page 344), we find $P/P_y = 0.58$ corresponding to $L/r_y = 59$. Thus

$$P = 0.58 \times 19.11 \times 36 = 399 \text{ kip} > 375 \text{ kip} \qquad \text{O.K.}$$

(f) *K coefficients method*

Try W12 × 72

$A = 21.16 \text{ in}^2$, $D = 12.25$ in, $B = 12.04$ in, $t = 0.671$ in

$r_x = 5.31$ in, $r_y = 3.04$ in, $Z_x = 108 \text{ in}^3$, $Z_y = 49.2 \text{ in}^3$

$\beta_x = -1$, $C_{mx} = 0.4$, $L/r_y = 59$, $L/r_x = 34$

Calculation of K_x

$$p = \frac{P}{P_y} = \frac{375}{21.16 \times 36} = 0.49$$

From multiple column curve 2, Fig. 14.10, Chap. 14, Vol. 1, we find

$$\lambda = \frac{1}{\pi}\sqrt{\frac{\sigma_y}{E}}\left(\frac{L_{cx}}{r_x}\right) = 1.15$$

corresponding to $p = 0.49$. Thus

$$L_{cx} = 45 \text{ ft} \quad \text{or} \quad \frac{L}{L_{cx}} = 0.33$$

From the Beta-Chart of Fig. 13.17, we find $K_x = 1.0$.

Calculation of K_t

From Eq. (13.32), we obtain

$$\left(\frac{L}{r_y}\right)_{\text{equ}} = 59 \times \left[\frac{14(12.25/0.671)^2}{(59)^2 + 15.5(12.25/0.671)^2 + 0.23(12.25/0.671)^3}\right]^{1/4} = 48.7$$

From which we find

$$\lambda_{\text{equ}} = \frac{1}{\pi}\sqrt{\frac{\sigma_y}{E}}\left(\frac{L}{r_y}\right)_{\text{equ}} = 0.55$$

The value of K_t obtained from the curves of Fig. 13.18 should be multiplied by $1/C_{mx} = 1/0.4$ for the nonuniform moment case which is found to be greater than unity. Hence $K_t = 1.0$.

Calculation of K_y

From Fig. 13.19 with $\beta_x = -1.0$ and $\beta_y = -1.0$, we find $\beta_{\text{eff}} = +1.0$. Substituting this into Eq. (13.34), we obtain $L_{\text{slat}} = 0$. From calculation of K_x, we know $\lambda = 1.15$, which gives

$$L_{cy} = 26 \text{ ft}$$

From the Beta-Chart of Fig. 13.11, Vol. 1, for minor axis bending with $M_y/M_{pcy} = 60 \times 12/1598 = 0.45$, we find

$$\frac{L_{cmy}}{L_{cy}} = 0.92 \qquad \text{or} \qquad L_{cmy} = 24 \text{ ft}$$

Substituting this into Eq. (13.36), we obtain

$$K_y = 1 - \frac{1}{5}\left(\frac{15}{24}\right) = 0.875$$

Calculation of K_h

From Fig. 5.31(c), Chap. 5, with

$$p = \frac{P}{P_y} = \frac{375}{21.16 \times 36} = 0.49$$

$$m_y = \frac{M_y}{M_{py}} = \frac{60 \times 12}{49.2 \times 36} = 0.41$$

we find $K_h = m_x = 0.59$

From Eq. (13.29), we have

$$\begin{aligned} M_x &= K_x K_t K_y K_h M_{px} \\ &= 1.0 \times 1.0 \times 0.875 \times 0.59 \times 108 \times 36/12 \\ &= 167 \text{ k-ft} > 160 \text{ k-ft} \end{aligned}$$

O.K.

13.6.2 Comparative Stability Designs

Example 13.3 Stability design using equations and charts

Same as Example 2, except that the end moments induce single curvature about both axes.

For this case, since single curvature is induced by the end moments, maximum bending stresses will occur between the ends of the column, as a result of the axial load acting upon the deflected shape. Therefore, a stability type analysis must be made.

The column section required for stability according to each design method is shown in Table 13.5, along with a factor indicating the relative economy of each section.

Table 13.5 COMPARISON OF STABILITY SOLUTIONS TO DESIGN EXAMPLE 13.3

Method	Column Section	Economy Factor
(a) AISC (Eq. 13.16)	W12 × 133	1.00
(b) CSA-S16 (Eq. 13.17)	W12 × 120	0.90
(c) Nonlinear Interaction Equations (Eqs. 13.23, 13.24b)	W12 × 99	0.74
(d) Ringo's Charts	not applicable	—
(e) Sharma and Gaylord's Charts (Fig. 7.17)	W12 × 92	0.69
(f) *K* Coefficients (Eq. 13.29)	W12 × 92	0.69

To illustrate the use of various methods, following are the required calculations for solving Example 13.3.

(a) *AISC specification (working stress design)*

Strength is O.K. if the section is greater than W12 × 106. We therefore try W12 × 133

$$A = 39.11 \text{ in}^2, \quad S_x = 182.5 \text{ in}^3, \quad S_y = 63.1 \text{ in}^3$$

$$E = 29\,000 \text{ ksi}, \quad r_x = 5.59 \text{ in}, \quad r_y = 3.16 \text{ in}$$

Euler buckling stress $= \pi^2 E/(L/r)^2$

$$L/r_x = 32.2, \quad F'_{ex} = 144 \text{ ksi}$$

$$L/r_y = 57, \quad F'_{ey} = 46 \text{ ksi}$$

From the AISC column curve, $F_a = 17.71$ ksi. Using a load factor of 1.667, we have

$$f_a = \frac{375}{1.667 \times 39.11} = 5.75 \text{ ksi}, \qquad \frac{f_a}{F'_{ex}} = 0.04, \qquad \frac{f_a}{F'_{ey}} = 0.125$$

$$f_{bx} = \frac{160 \times 12}{1.667 \times 182.5} = 6.31 \text{ ksi}, \qquad f_{by} = \frac{60 \times 12}{1.667 \times 63.1} = 6.85 \text{ ksi}$$

From expression (13.16) for stability analysis, we find

$$\frac{5.75}{17.71} + \frac{6.31}{0.66 \times 36 \times (1 - 0.04)} + \frac{6.85}{0.75 \times 36 \times (1 - 0.125)}$$

$$= 0.325 + 0.277 + 0.290 = 0.892 < 1.0 \qquad \text{O.K.}$$

(b) *CSA S-16.1-1974 (limit state design)*

Strength is O.K. if the section is W12 × 85 or larger. We therefore try W12 × 120

$$A = 35.31 \text{ in}^2, \quad Z_x = 186 \text{ in}^3, \quad Z_y = 85.5 \text{ in}^3$$

$$E = 29\,000 \text{ ksi}, \quad r_x = 5.51 \text{ in}, \quad r_y = 3.13 \text{ in}$$

Euler buckling stress $= \pi^2 E/(L/r)^2$

$$L/r_x = 32.7, \quad F_{ex} = 268 \text{ ksi}, \quad P_{ex} = 9463 \text{ kip}$$

$$L/r_y = 57.5, \quad F_{ey} = 86.5 \text{ ksi}, \quad P_{ey} = 3054 \text{ kip}$$

Using a load factor of 1.667, we calculate P_u from CSA axial compression formula for column which results $P_u = 1032$ kip.
From Eq. (13.20) we obtain

$$M_{ux} = M_{px}\left[1.07 - \frac{(L/r_y)\sqrt{\sigma_y}}{3160}\right] = (186 \times 36)\left[1.07 - \frac{57.5 \times 6}{3160}\right]$$

$$= 6434 \text{ k-in} = 536 \text{ k-ft} < 558 \text{ k-ft } (= M_{px})$$

$$M_{uy} = M_{py} = 85.5 \times 36/12 = 256.5 \text{ k-ft}$$

From expression (13.17) for stability analysis, we find

$$\frac{375}{1032} + \frac{160}{536\left(1 - \dfrac{375}{9463}\right)} + \frac{60}{256.5\left(1 - \dfrac{375}{3054}\right)}$$

$$= 0.363 + 0.311 + 0.267 = 0.941 < 1.0 \qquad \text{O.K.}$$

(c) *Nonlinear interaction equations (13.23) and (13.24b)*

Strength is O.K. if section W12 × 72 or larger is used. We therefore try W12 × 99:

$$A = 29.09 \text{ in}^2, \quad Z_x = 152 \text{ in}^3, \quad r_x = 5.43 \text{ in}$$

$$E = 29\,000 \text{ ksi}, \quad Z_y = 69.5 \text{ in}^3, \quad r_y = 3.09 \text{ in}$$

Euler buckling stress $= \pi^2 E/(L/r)^2$

$$L/r_x = 33, \quad F_{ex} = 260 \text{ ksi}, \quad P_{ex} = 7563 \text{ kip}$$

$$L/r_y = 58.2, \quad F_{ey} = 84.3 \text{ ksi}, \quad P_{ey} = 2452 \text{ kip}$$

From the AISC column curve, $F_a = 17.60$ ksi.
Using a load factor of 1.667, $P_u = 853$ kip.

$$M_{ux} = M_{px}\left[1.07 - \frac{(L/r_y)\sqrt{\sigma_y}}{3160}\right] = (152 \times 36)\left[1.07 - \frac{58.2 \times 6}{3160}\right]$$

$$= 5250 \text{ kip-in} = 438 \text{ kip-ft} < 456 \text{ k-ft } (= M_{px})$$

From Eqs. (13.19), we find

$$M_{ucx} = M_{ux}(1 - P/P_u)(1 - P/P_{ex}) = 438(1 - 375/853)(1 - 375/7563)$$

$$= 2796 \text{ kip-in} = 233 \text{ kip-ft}$$

$$M_{ucy} = M_{py}(1 - P/P_u)(1 - P/P_{ey}) = 69.5 \times 36(1 - 375/853)(1 - 375/2452)$$

$$= 1189 \text{ kip-in} = 99 \text{ kip-ft}$$

$$\frac{C_{mx}M_x}{M_{ucx}} = \frac{1.0 \times 160}{233} = 0.687$$

$$\frac{C_{my}M_y}{M_{ucy}} = \frac{1.0 \times 60}{99} = 0.606$$

$$P/P_y = 375/(29.09 \times 36) = 0.358$$

Since $B = D$, find $\beta = 1.76$ from Fig. 13.20; alternatively, from Eq. (13.24b), $\beta = 0.4 + P/P_y + B/D = 1.758$.

From Fig. 13.21, enter with appropriate moment ratios, finding that the required

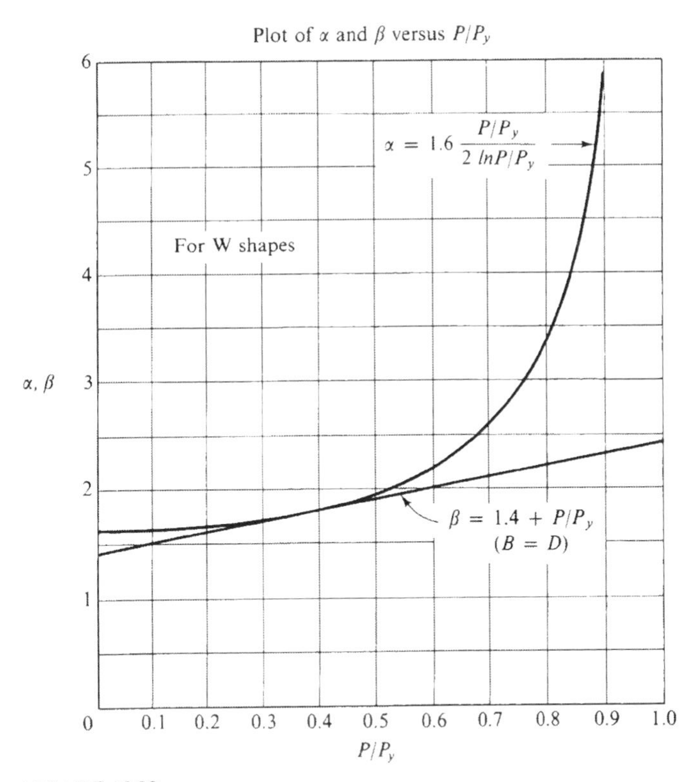

FIGURE 13.20
Plot of α and β vs. P/P_y

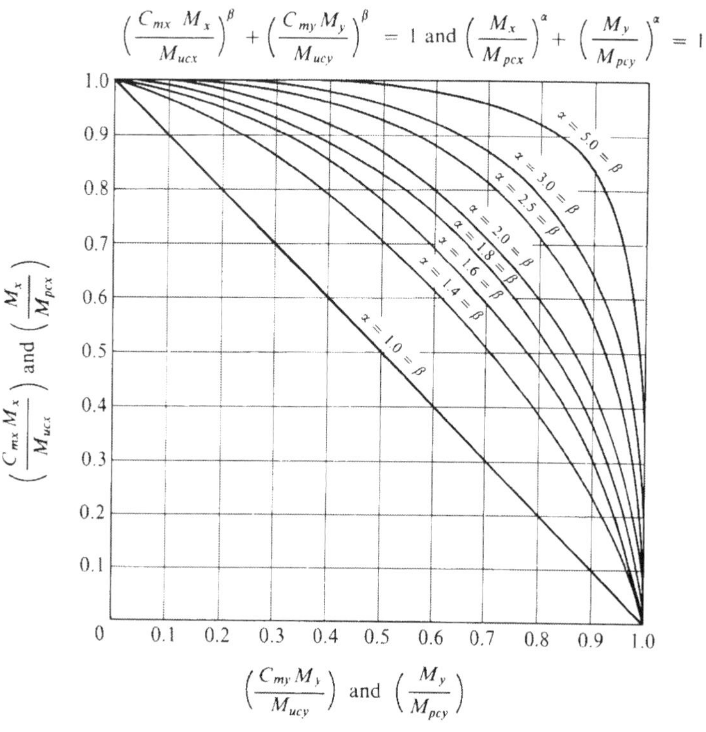

FIGURE 13.21
Plot of interaction Equations (13.10) and (13.23)

value of β is $1.6 < 1.76$; therefore, the W12 × 99 section is adequate. This may be verified by solution of the interaction equation:

$$0.687^{1.76} + 0.606^{1.76} = 0.93 < 1.0 \qquad \text{O.K.}$$

(d) *Nomographs developed by Ringo, et al., (1973)*

Nomographs are developed for braced points only. They are not applicable here.

(e) *Charts developed by Sharma and Gaylord (1969)*

Strength is O.K. if the section is W12 × 65 or larger. We therefore try W12 × 92

$$A = 27.06 \text{ in}^2, \quad D = 12.62 \text{ in}, \quad B = 12.155 \text{ in}$$

$$r_x = 5.40 \text{ in}, \qquad r_y = 3.08 \text{ in}$$

$$e_x = \frac{60 \times 12}{375} = 1.92 \text{ in}, \quad e_y = \frac{160 \times 12}{375} = 5.12 \text{ in}$$

From which we can calculate

$$\frac{e_x B}{2r_y^2} = \frac{1.92 \times 12.155}{2 \times (3.08)^2} = 1.23$$

$$\frac{e_y D}{2r_x^2} = \frac{5.12 \times 12.62}{2 \times (5.40)^2} = 1.11$$

From Fig. 7.17 (page 344), we find $P/P_y = 0.40$ for $L/r_y = 58$. Thus,

$$P = 0.40 \times 27.06 \times 36 = 390 \text{ kip} > 375 \text{ kip} \qquad \text{O.K.}$$

(f) *K coefficients*

Strength is O.K. if the section is W12 × 72 or greater. We therefore try W12 × 92

$A = 27.06 \text{ in}^2$, $D = 12.62$ in, $B = 12.155$ in, $t = 0.856$ in

$r_x = 5.4$ in, $r_y = 3.08$ in, $Z_x = 140 \text{ in}^3$, $Z_y = 64.2 \text{ in}^3$

$\beta_x = 1.0$, $C_{mx} = 1.0$, $L/r_y = 58.4$, $L/r_x = 33.3$

Calculation of K_x

$$p = \frac{P}{P_y} = \frac{375}{27.06 \times 36} = 0.385$$

From multiple column curve 2, Fig. 14.10(a), page 440, Vol. 1, we find

$$\lambda = \frac{1}{\pi}\sqrt{\frac{\sigma_y}{E}}\left(\frac{L_{cx}}{r_x}\right) = 1.38$$

corresponding to $p = 0.385$. Thus,

$$L_{cx} = 55.4 \text{ ft} \qquad \text{or} \qquad \frac{L}{L_{cx}} = 0.271$$

From the Beta-Chart of Fig. 13.17 with $\beta_x = 1.0$, we find $K_x = 0.82$.

Calculation of K_t

From Eq. (13.32), we obtain

$$\left(\frac{L}{r_y}\right)_{equ} = 58.4 \times \left[\frac{14(12.62/0.856)^2}{(58.4)^2 + 15.5(12.62/0.856)^2 + 0.23(12.62/0.856)^3}\right]^{1/4} = 46.6$$

From which we find

$$\lambda_{equ} = \frac{1}{\pi}\sqrt{\frac{\sigma_y}{E}}\left(\frac{L}{r_y}\right)_{equ} = 0.52$$

Entering $\lambda_{equ} = 0.52$ into Fig. 13.18 and interpolating the curves for A36 steel, we find $K_t = 0.86$.

Calculation of K_y

From Fig. 13.19 with $\beta_x = 1.0$ and $\beta_y = 1.0$, we find $\beta_{eff} = 1.0$. Substituting this into Eq. (13.34), we obtain $L_{slat} = 0$.

From the calculation of K_x, we know $\lambda = 1.38$ which gives

$$L_{cy} = 31.6 \text{ ft}$$

From Eq. (13.11b), we have

$$M_{pcy} = 1.19 \times 64.2 \times 36[1 - (0.385)^2] = 2343 \text{ k-in}$$

From the Beta-Chart of Fig. 13.11, (page 400), Vol. 1, for minor axis bending with $M_y/M_{pcy} = 60 \times 12/2343 = 0.31$, we find

$$\frac{L_{cmy}}{L_{cy}} = 0.63 \qquad \text{or} \qquad L_{cmy} = 19.9 \text{ ft}$$

Substituting this into Eq. (13.36), we find

$$K_y = 1 - \frac{1}{5}\left(\frac{15}{19.9}\right) = 0.85$$

Calculation of K_h

From Fig. 5.31(c), (page 238), with $p = 0.385$,

$$m_y = \frac{M_y}{M_{py}} = \frac{60 \times 12}{64.2 \times 36} = 0.31$$

we find $K_h = m_x = 0.73$. From Eq. (13.29), we have

$$\begin{aligned} M_x &= K_x K_t K_y K_h M_{px} \\ &= 0.82 \times 0.86 \times 0.85 \times 0.73 \times 140 \times 36/12 \\ &= 184 \text{ k-ft} > 160 \text{ k-ft} \qquad \text{O.K.} \end{aligned}$$

13.6.3 Comparisons of Results

In Tables 13.4 and 13.5, the most striking features of the comparative designs are the remarkable reduction in column section required by methods (c) through (f) for both strength and stability, and the agreement of their results.

In Table 13.6, comparison is made with an example published by Young (1973). The ultimate major axis moment capacity predicted by various methods is given, assuming fixed values for axial load and minor axis moment.

Table 13.6 COMPARISON OF MAJOR AXIS BIAXIAL MOMENT CAPACITIES (KIP-FT) PREDICTED BY VARIOUS DESIGN METHODS[a]

C_m	1.0	0.6	0.4
AISC[b]	43[c]	43[c]	43[c]
CSA S16.1	117	157[c]	157[c]
Nonlinear Interaction	167	209[c]	209[c]
K Coefficients	167	183	198

[a] Based on Young's example (1973): W12 × 65, $L = 118''$, $P = 0.3\,P_y$, $M_{y(\text{top})} = 0.4\,M_{py}$, $M_{y(\text{bottom})} = 0$.
[b] Determined by applying 0.6 times ultimate thrust and minor axis moment, and multiplying resulting major axis working stress design moment by 1.67 for direct comparison.
[c] Strength controls ("stocky" column). 1 kip-ft = 1.36 kN-m.

It is most encouraging to observe that for the single curvature case the K coefficients method proposed by Young and the nonlinear interaction equations proposed by Chen are in very close agreement in their predictions even though their origins are very different.

Similar comparisons are made in Table 13.7 for uniaxial bending cases.

Table 13.7 COMPARISON OF MAJOR AXIS UNIAXIAL MOMENT CAPACITIES (KIP-FT) PREDICTED BY VARIOUS DESIGN METHODS[a]

C_m	1.0	0.6	0.4
AISC[b]	93	116[c]	116[c]
AISC (Plastic)	93	136	136
CSA S16.1	93	137[c]	137[c]
Nonlinear Interaction	93	137[c]	137[c]
K Coefficients	101	130	132

[a] Based on Young's example (1973): W12 × 65, $L = 118''$, $P = 0.6\,P_y$.
[b] Determined by applying 0.6 times ultimate thrust and minor axis moment and multiplying resulting major axis working stress design moment by 1.67 for direct comparison.
[c] Strength controls ("stocky" column). 1 kip-ft = 1.36 kN-m.

13.6.4 Limitations of the Ultimate Strength Methods

Before adopting designs according to the ultimate strength methods, some general thought should be given to the extent to which yielding under service load is likely and, in view of this, whether further restrictions on the design are necessary. In Table 13.8, for each of two ratios of actual axial load to yield axial load and three ratios of M_x/M_{pcx}, the corresponding ratios of M_y/M_{pcy} have been determined from the interaction equation. (13.10). In evaluating M_x and M_y, a W12 × 65 section of A36 steel has been assumed and the extreme corner fiber stress has been calculated elastically and divided by a load factor of 1.7. It will be noted that, for the lower ratio of axial stress, the yield stress of 36 ksi has been exceeded by an appreciable margin.

This rather simple calculation demonstrates that the onset of yielding under service loads is a real probability for sections designed according to the proposed interaction equation.

Table 13.8 STRESS IN A BIAXIALLY LOADED SECTION*

P/P_y	$\frac{M_x}{M_{pcx}}$	$\frac{M_y}{M_{pcy}}$	Stress in ksi f_a	f_{bx}	f_{by}	$\frac{f}{1.7}$
0.3	0.2	0.963	10.8	6.55	52.6	41.1
	0.5	0.812	10.8	16.4	44.3	42.0
	0.8	0.516	10.8	26.2	28.2	38.3
0.7	0.2	0.994	25.2	2.81	32.9	35.8
	0.5	0.932	25.2	7.02	30.9	35.1
	0.8	0.726	25.2	11.24	24.1	35.6

* Based on a W12 × 65 section, A36 steel, Eq. (13.10) 1 ksi = 6.9 MPa.

As pointed out by Springfield (1975), three precautions are desirable when designing beam-columns by the ultimate strength methods. These are:

1. Sections should be proportioned so that influences such as wind or earthquake, which are reversible, should not, when taken alone, cause stresses in extreme fibers which exceed the nominal yield strength.
2. Sections should be proportioned so that influences which are variable, such as combined wind plus live load (taken at a load factor of 1.5 reduced to account for probability of occurrence, say by 0.7), should not in themselves cause extreme fiber stresses that exceed the nominal yield strength of the material.
3. Keep in mind the fact that the procedure was developed on the assumption that the section is compact.

It appears that further advances in the design of biaxially loaded beam-columns must recognize shape dependency. From a practical viewpoint, this creates no problem, since a very large proportion of the columns being designed in buildings are of wide flange shape. If we are able to develop the exponent required for Chen's interaction equation for all wide flange shapes and for square and rectangular tubular sections, almost the whole practical field will have been covered.

13.7 TESTS ON BIAXIALLY LOADED STEEL WIDE-FLANGE BEAM-COLUMNS

Experimental investigations concerning the strength and behavior of biaxially loaded H- and I-columns are reviewed here. Comparisons are made between the results of 125 tests on beam-columns subjected to symmetrical biaxial eccentric loads, and the ultimate load-carrying capacities predicted by the linear and nonlinear interaction equations (13.17) and (13.23) described in Sec. 13.3.

The basic interaction equation under evaluation is the nonlinear equation (13.23) from which typical interaction curves are shown in Fig. 13.21. Both the AISC design expressions and SSRC Eq. (13.17) are straight line interaction equations corresponding to Eq. (13.23) with the exponent $\beta = 1.0$. For H-sections for which the flange width B is approximately equal to the depth D, the exponent β of Eq. (13.23) has the value $\beta = 1.4 + p$ or Eq. (13.24a). For all narrow I-sections, the more conservative exponent $\beta = 0.4 + p + B/D \geq 1.0$ or Eq. (13.24b) should be used. The interaction curves for I-sections lie between the straight line ($\beta = 1.0$) and the interaction curves for H-sections ($\beta = 1.4 + p$). In the following evaluation, these two extreme values of β are used in Eq. (13.23) for comparisons with test results regardless of the actual B/D ratio of the test specimen.

13.7.1 Klöppel and Winkelmann's Tests (1962)

Klöppel and Winkelmann (1962) conducted a series of tests on 74 rolled steel wide-flange beam-columns. The end conditions were such that the beam-columns were essentially pinned against rotation, and warping was restrained by heavy end plates. The specimens investigated were sections IP 16/16/0 (16 × 16 cm) and IP 10/10/0 (10 × 10 cm), for various biaxial eccentricities. In the series of tests with section IP 16/16/0, the strong axis slenderness ratio was maintained at 34, while the weak axis slenderness ratio varied from 57 to 114. The successively more slender sections were obtained by trimming the tips of the flanges. Similar tests were carried out for the section IP 10/10/0 with the strong axis slenderness ratio maintained at 49, while the weak axis slenderness ratio varied from 83 to 121. Figure 13.22 summarizes these tests in outline form and the details were given in a report by Galambos (1963). Reasonable agreement was found between the observed beam-column strength and the values calculated from Sharma's approximate solutions (1965).

Table 13.9 COMPARISON OF INTERACTION EQUATIONS WITH KLÖPPEL AND WINKELMANN'S TESTS (1962) (SEE FIG. 13.22)

Group	Size $B \times D$ (cm)	Number of Test	Mean Value, $P_{test}/P_{formula}$	
			Linear Interaction Formula (13.17)	Nonlinear Interaction Formulas (13.23) & (13.24a)
1	16 × 16	12	1.044	0.911
	13 × 16	9	1.029	0.874
	11 × 16	7	0.975	0.827
	10 × 16	8	1.026	0.879
	9 × 16	6	1.139	0.988
2	10 × 10	15	1.053	0.931
	8 × 10	8	1.019	0.917
	7 × 10	6	1.051	0.958
Overall Average			1.04	0.92

The ultimate load carrying capacities of the 71 beam-column specimens shown in Fig. 13.22 were compared with the values predicted from the linear interaction equation (13.17) and the nonlinear interaction equation (13.23) with β given by Eq. (13.24a). The observed ultimate loads are denoted by P_{test} and listed in Table 13.9 together with the values computed from the interaction equations, $P_{formula}$. The computations of ultimate axial load capacity P_u were based on the SSRC column strength equation

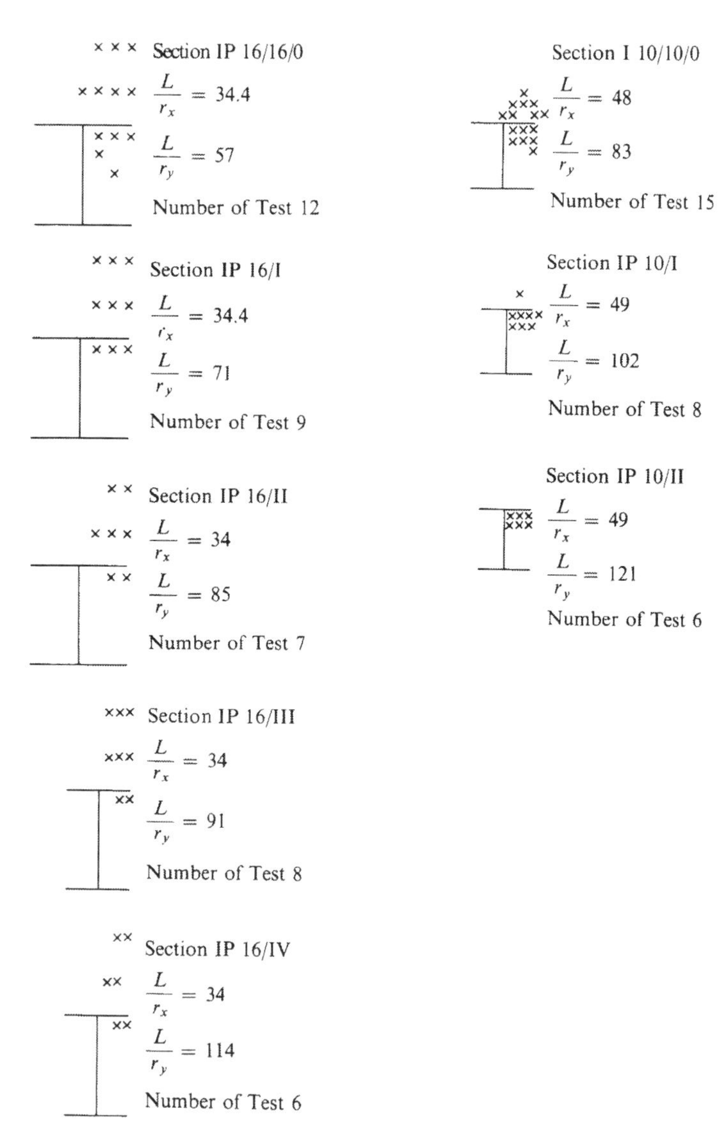

FIGURE 13.22
Outline of Klöppel and Winkelman's tests (X denotes the location for the various biaxial eccentricities used in their tests)

$$\sigma_c = \sigma_y - \frac{\sigma_{rc}}{\pi^2 E}(\sigma_y - \sigma_{rc})\left(\frac{kL}{r}\right)^2 \tag{13.37}$$

in which the residual stress σ_{rc} was set at 0.5 of the reported yield stress σ_y of the section under test for all yield strengths up to 38 ksi. For the Birnstiel tests on V65 steels which will be compared later, the residual stress was set at 18 ksi. Equation (13.37) depicts a curve that meets the Euler curve at the stress $(\sigma_y - \sigma_{rc})$ and is tangent to it when $\sigma_{rc} = 0.5\sigma_y$. If we let $\sigma_{rc} = 0.5\sigma_y$, Eq. (13.37) provides a suitable compromise between strong- and weak-axis buckling of wide flange sections that actually have a maximum compressive residual stress of $0.3\sigma_y$.

For almost all the beam-columns, the observed ultimate load P_{test} exceeded that computed from the linear interaction formula P_{formula}. The ratio of observed ultimate load to that computed from the linear interaction formula varied between 0.975 and 1.139 with an average value of 1.04. The nonlinear interaction formula, however, overpredicted all the ultimate load carrying capacities. The corresponding ratios varied from 0.827 to 0.988 with an average value of 0.92.

From Table 13.9, it would be concluded that the linear interaction equation (13.17) predicts the load-carrying capacity better than the nonlinear equation (13.23) with exponent β given by Eq. (13.24a). Although the exponent β used in the comparison is the upper bound limit for nearly square column sections, the difference between the linear and nonlinear equations is fairly constant as the flanges become more narrow, being 15 percent for Group 1 and 10 percent for Group 2 and this difference will become smaller for narrow I-sections when the more conservative exponent β given by Eq. (13.24b) is used. Further, it should be kept in mind that the machining down of the beam-column flanges induced another variable into the tests other than the reduction in L/r_y while L/r_x was constant. The above conclusion is not supported by the test results of Chubkin (1959) and Birnstiel (1968).

13.7.2 Chubkin's Tests (1959)

Chubkin reported on another series of extensive tests. Two hundred eighty-one steel beam-columns were tested with various conditions of eccentricity (axial, uniaxial, biaxial) and end conditions (warping restrained and warping free). In one series, the specimens were I-type (9.4 × 18 cm) rolled sections with the weak axis slenderness ratios 50, 100 and 150, respectively. In the other series, 12 × 12 cm welded built-up H-shapes were used with weak axis slenderness ratios of 50 and 100 respectively. The details were also summarized by Galambos (1963). Interaction curves obtained by Sharma and the test results are compared in Fig. 13.23. The observed values are in reasonable agreement with the calculated results.

The load carrying capacities predicted by the linear and nonlinear interaction equations for the Chubkin's tests are listed in Table 13.10. In the case of Group 1 tests (rolled sections) the nonlinear interaction equation (13.23) with exponent β given by Eq. (13.24a) is seen to give a much better prediction than that given by the linear

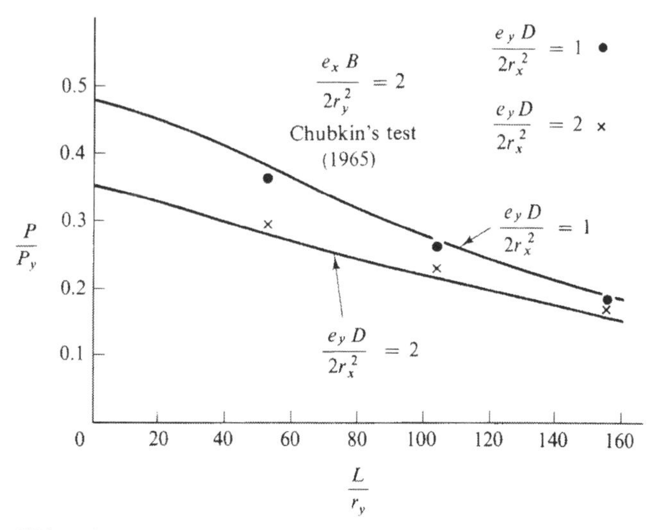

FIGURE 13.23
The comparison of Sharma's interaction curves with Chubkin's tests (1965)

equation (13.17), even though the specimens were I-shaped. The mean ratios $P_{test}/P_{formula}$ of 1.05 and 1.003 for slenderness ratios of 50 and 100 are extremely accurate. For the extreme slenderness ratio of 150, both linear and nonlinear interaction equations give more conservative predictions (1.315 and 1.219).

Table 13.10 COMPARISON OF INTERACTION EQUATIONS WITH CHUBKIN'S TESTS (1959)

Group	Size $B \times D$ (cm)	Number of Test	$\frac{L}{r_y}$	Mean Value, $P_{test}/P_{formula}$: Linear Interaction Formula (13.17)	Nonlinear Interaction Formulas (13.23) & (13.24a)
1	9.4 × 18	13	50	1.213	1.050
	9.4 × 18	13	100	1.129	1.003
	9.4 × 18	13	150	1.315	1.219
2*	12 × 12	2	50	1.170	0.970
	12 × 12	2	100	0.880	0.765
Overall Average				1.22	1.09

* Welded built-up H-shapes

In the case of Group 2 tests (welded built-up H-shapes), Chubkin's tests in Table 13.10 indicate the reduced load-carrying capacity of beam-columns when warping free. A reduction in capacity of the order of 12 percent results from the removal of warping restraint. It has been shown in Chap. 7 that the presence of warping restraint may increase theoretically the ultimate load-carrying capacity by 10 percent over that for the same beam-column with ends free to warp. Chubkin's test results are consistent with this theoretical conclusion. For slenderness ratio of 100, both linear and nonlinear interaction equations overpredict the load carrying capacities of welded built-up H-shapes.

13.7.3 Birnstiel's Tests (1968)

Recently conducted tests on 16 H-shaped beam-columns by Birnstiel provide further experimental information on biaxially loaded beam-columns. All test specimens were loaded through heavy end plates, and thus warping was restrained. The tests were made on beam-columns of various square H-shapes and one I-shape made of three types of steel; A7, A36 and V65. The biaxial eccentricities varied from -0.89 in (-22.6 mm) to 2.38 in (60.5 mm) for e_x and 1.87 in (47.5 mm) to 3.21 in (81.5 mm) for e_y.

Analytical studies were also made using the procedure outlined in Chap. 8 and assuming an elastic-perfectly plastic material behavior. Typical results of the experimental and analytical study are compared in Fig. 13.24. The shapes of the computed load P versus lateral displacements u and v curves are found to be similar to those formed by the plotted experimental points for most beam-columns. The computed and measured values of u and v are in adequate agreement at loads less than 90 percent of the measured ultimate load for most specimens. However, the load P versus twist θ curves for some beam-columns are found to differ considerably from the curves formed by the corresponding experimental points. Figure 13.24 is representative of the correspondence between computed and observed midheight displacements. The

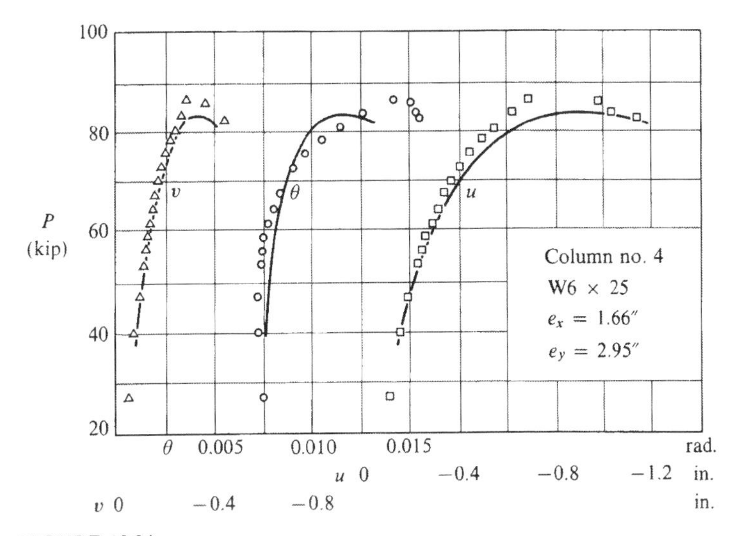

FIGURE 13.24
Comparison of Birnstiel and Harstead's analytical curves (Chap. 8) with their tests (Birnstiel, 1968). (1 in = 25.4 mm) (1 kip = 4.45 kN)

discrepancy between some observed and computed values of twist is believed to be due to the initial twist of the column specimens. Overall, the agreement between numerical and experimental results is found to be satisfactory.

The comparison of the ultimate load-carrying capacities of the beam-column specimens with the theoretical calculations, and linear and nonlinear interaction equations is given in Table 13.11 for four types of specimens. These are specimens fabricated by machining all surfaces of a stress-relieved weldment, specimens made by machining the edges of the flanges of rolled wide flange shapes which had been stress-relieved, as rolled shapes that were not straightened, and shaped specimens as delivered which had been rotarized. The load carrying capacities predicted by the nonlinear interaction equation for Birnstiel's tests are seen extremely good. The overall mean value of the capacity is within 1 percent and 20 percent of the test value for nonlinear and linear equations, respectively. The extreme predictions, expressed by $P_{test}/P_{formula}$, range from 0.89 to 1.14 for the nonlinear equation (13.23) and 1.08 to 1.30 for the linear SSRC equation (13.17).

The load-carrying capacities predicted by the nonlinear equation (13.23) are found to be extremely close to the respective capacities predicted by the Birnstiel indirect incremental-load computer analysis described in Chap. 8. With the exception of Tests 1, 10, 13 and 14, the difference between nonlinear interaction prediction and "exact" theoretical calculation does not exceed 3 percent. Test 10 was on a 4 × 8I. Since Tests 13 and 14 were on V65 steels, a discrepancy may have arisen from the prediction of axial load. Thus, the only significant difference is one of 7 percent for Test 1. For Test 10, the nonlinear equation prediction of the ratio $P_{test}/P_{calculate}$ is 1.14. The nonlinear interaction equation (13.23) with exponent β for H-shapes underestimates the capacity of Birnstiel's test of an I-shaped column. If the β equation (13.24b) for I-shapes is used, the ratio $P_{test}/P_{calculate}$ should lie between the value 1.14 and 1.29. It is noted that maximum error in predicted capacity by computer solution also occurs with this test, the ratio being 1.05. Tests 10, 13 and 14 are the only tests in Table 13.11 where there is a significant difference between M_{px} and M_{ux} taking account of lateral torsional buckling about the major axis. When the lateral torsional buckling reduction is suppressed, the prediction is found to be improved for both linear and nonlinear equations, except for the last test of Table 13.11. This is on a V65 steel.

From Table 13.11, the value $P_{test}/P_{calculate}$ is seen to be reduced below unity either by high moment ratio M_y/M_x or a high ratio of r_x/r_y. While Birnstiel's tests are not so numerous as the previous one discussed, Table 13.11 shows clear groupings above and below unity for $P_{test}/P_{calculate}$, as determined by the nonlinear interaction equation (13.23). The lower values of Tests 5 and 6 may be attributed to high applied moments (P/P_y is low). For Tests 7 and 8, the ratio r_x/r_y is higher. Test 10 appears to contradict Tests 7 and 8 since r_x/r_y is 3.57, but it must be remembered that this was a 4 × 8 in fabricated I-shape with much higher slenderness ratio (107 vs. 78). Tests 13 and 14 cannot be used properly for comparisons since the high yield strength precludes an identical determination of axial load capacity.

Table 13.11 COMPARISON OF INTERACTION EQUATIONS WITH BIRNSTIEL'S TESTS AND THEORETICAL PREDICTIONS (1968)

Test No.	Size (in)	Type	$\frac{r_x}{r_y}$	$\frac{M_y}{M_x}$	P/P_y Test	$P_{test}/P_{calculated}$: Exact Theoretical Calculation	$P_{test}/P_{calculated}$: Linear Interaction Formula (13.17)	$P_{test}/P_{calculated}$: Nonlinear Interaction Formulas (13.23) & (13.24a)
1[a]	6 × 6	Stress Relieved Weldment	1.63	1.54	0.329	1.00	1.30	1.07
2	5 × 5		1.67	1.38	0.274	1.03	1.28	1.05
3	5 × 5	As Rolled	1.66	0.83	0.319	1.00	1.24	1.02
4	6 × 6		1.78	1.69	0.327	1.03	1.28	1.06
5	5 × 5	Stress Relieved	1.62	1.96	0.196	0.94	1.13	0.94
6	5 × 5	Weldment	1.62	2.49	0.188	0.97	1.11	0.95
7	5 × 6	Stress Relieved	2.11	1.12	0.306	0.96	1.12	0.94
8	5 × 6	WF	2.11	0.64	0.437	0.99	1.14	0.98
10	4 × 8		3.57	0.47	0.424	1.05	1.29	1.14
12	5 × 5	As Rolled	1.65	0.75	0.273	0.91	1.08	0.89
13[b]	4 × 4	As Rolled	1.70	0.44	0.193	1.02	1.29	1.10
14[b]	4 × 4	Rotarized	1.70	1.00	0.160	0.94	1.16	1.00
Oveall Average						0.99	1.20	1.01
Total Average of Tables 13.9, 13.10 and 13.11							1.15	1.00

[a] A7 steel. [b] V65 steel. All others A36 steel. 1 in = 25.4 mm.

The variation in $P_{test}/P_{formula}$ for all the tests described above has been evaluated statistically by Springfield and Hegan (1973). While the differences between the linear and nonlinear equations are not great, the nonlinear equation (13.23) results in an improvement over the linear SSRC equation (13.17), which in turn is found to be a significant improvement over the AISC working stress design interaction equation (13.16).

It is worth noting here that the Pillai and Ellis's modified equation (13.21) to the SSRC equation (13.17) is also found to be applicable to square H-columns, even though it was proposed only for square hollow shapes. However, this modification has not been proven by extensive theoretical comparisons, as has the nonlinear interaction equation (13.23).

13.7.4 Milner's Tests (1965)

Milner reported on tests of 10 H-shaped elastically restrained beam-columns which were also restrained against sway. The experimental work was concentrated on a series of exploratory tests, in which significant parameters were varied. The specimens are shown diagrammatically in Fig. 13.25. Loads were applied to the beams by means of turnbuckles. In the first three tests, major axis beam loads and axial load were applied. In the second seven tests, both the major and minor axis beam loads and axial load were applied. In all tests, beam loads were first applied and then held constant and the beam-column was finally loaded to failure by axial force.

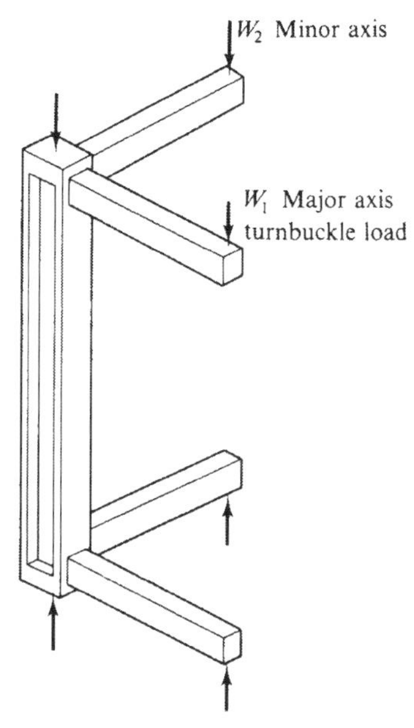

FIGURE 13.25
Milner's tests (1965)

Milner's numerical solutions of the maximum load-carrying capacity of his experimental sub-assemblages were within 8 percent of his test results, except for one case, where the difference was 11 percent.

13.8 REINFORCED CONCRETE COLUMNS UNDER BIAXIAL LOADING

The mathematics involved in the analysis of reinforced concrete columns subjected simultaneously to bending moments about both major axes in addition to an axial compressive load is quite involved, even for the special case of relatively short columns for which the effect of lateral deflections on the magnitudes of bending moments is negligible. For the most part, analysis and design of such columns have in the past been directed toward the study of the ultimate strength of reinforced concrete short columns (Sect. 5.7, Chap. 5). For square columns, detailed ultimate strength interaction diagrams relating axial, biaxial bending moments are given in Figs. 5.40 and 5.41, Chap. 5. In the case of long columns, the present design procedure of biaxially loaded columns does not differ from uniaxially loaded columns. The 1971 ACI Building Code, for example, recommends calculating the moment magnifier separately and applying it to the moment about each axis independently. The long columns are then designed according to the given axial compressive load and the magnified biaxial moments. For a short column, the moment magnification factor is taken as unity.

Although this procedure has been used extensively in design computations, it does not give accurate indications of the true load-carrying capacity of a biaxially loaded beam-column. To determine the ultimate strength of such a beam-column, it is necessary to perform a stability analysis that considers the entire range of loading up to ultimate load. Since the required analysis is too complex for practical applications, simple procedures to enable designers to assess the ultimate strength of long columns must be developed or simple approximate formulas must be proposed as an alternate method for design. The work described in this section attempts to achieve this goal. Details of this development are given elsewhere (Chen and Shoraka, 1974).

The first part of the section discusses a rigorous method for performing elastic-plastic stability analysis of reinforced concrete columns subject to axial load combined with biaxial bending. Three classes of problems are considered: short columns, long columns under symmetrical loading, and long columns under unsymmetrical loading conditions. The ultimate strength interaction curves for symmetrical and unsymmetrical loading cases are presented in forms of charts suitable for direct analysis and design. The important factors influencing the behavior of these curves are discussed such as strength of materials, percentage of reinforcement and the magnitude of compression load.

A design method, based on the C_m factor method, is then developed. Preliminary verification indicates that maximum load-carrying capacity prediced by C_m factor method for unsymmetrically loaded cases from the exact symmetrically loaded solutions, is in good agreement with calculated exact unsymmetric theoretical solutions.

A nonlinear interaction formula for symmetrically loaded columns is also proposed as an alternate method for design. It is concluded that the method is satisfactory.

The analytical results are also compared with the 1971 ACI design method for the biaxial bending case. It is found that the ACI method in general gives overconservative results.

13.8.1 Scope and Assumptions

The columns are assumed to be isolated, simply supported, prismatic and made of rectangular cross section as shown in Fig. 6.26. The range of variables considered in the numerical solutions are summarized as follows:

A square cross section with $a = b = 24$ in (610 mm) as shown in Fig. 6.26(b) is considered. The positioning of the reinforcement is chosen as being close to a practical average as shown in the inset of Fig. 6.28. Three percentages of reinforcement, $A_s/ab = 1.25$ percent, 3.25 percent and 8.33 percent are used. The stress-strain relationship is assumed to be linearly elastic-perfectly plastic [Fig. 6.27(b)] and Young's modulus E_s is taken at 29 000 000 psi (200 000 MPa). Three types of steel, $f_y = 40$, 60 and 80 ksi (276, 414, 552 MPa), are used.

Three types of concrete are considered, having compressive strengths $k_1 f_c' =$ 3, 4.2 and 5 ksi (21, 29, 35 MPa). The characteristic stress-strain curve assumed for the concrete in compression is shown in Fig. 6.27(a). The tensile strength and creep effect of concrete are neglected. The initial modulus of elasticity is taken as $E_c = 57\,600\sqrt{f_c'}$ where f_c' is in psi. The concrete strain ε_c' when concrete stress is $k_1 f_c'$ is taken at 0.002. The values of $\gamma_2 = 4$ and $\gamma_1 = E_c \varepsilon_c'/k_1 f_c'$ are used [see Fig. 6.27(a)]. Details of the description of the stress-strain curve are given in Sec. 6.8, Chap. 6.

Analyses are carried out for two values of crushing strain $\varepsilon_0 = 0.003$ (ACI 1971) and $\varepsilon_0 = 0.004$. (*Note*: CEB (1970) recommends 0.0035.) The column is subjected to three levels of axial compressive load, $P/f_c'ab = 0.1$, 0.5 and 1.0. The slenderness ratios, L/a considered are 0, 10, 20, 30 and 40. The total number of separate analyses performed may be seen from Table 13.12.

The following two additional assumptions are made in the solutions: (1) plane sections remain plane after bending; and (2) lateral-torsional twisting of the column is neglected. The failure is always caused by the crushing of concrete due to excessive bending curvature. In performing numerical calculations, it is further assumed that the axial compressive load P is applied first and maintained at a constant value as the biaxial bending moments increase or decrease proportionally.

Table 13.12 COMPUTED VALUES OF β [FOR EQ. (13.23)]

$\frac{P}{f'_c ab}$	$k_1 f'_c$ (ksi)	f_y (ksi)	$\frac{A_s}{ab}$	$\varepsilon_0/\varepsilon'_c = 1.5$ (ACI) L/a					$\varepsilon_0/\varepsilon'_c = 2.0$ L/a					Note
				0	10	20	30	40	0	10	20	30	40	
1.0	4.2	60	0.0325	1.70	2.00	—	—	—	2.40	2.50	—	—	—	Effect of
0.5				1.30	1.40	1.50	—	—	1.45	1.40	1.50	—	—	axial
0.1				1.40	1.50	1.70	1.90	1.80	1.60	1.70	1.70	1.90	1.80	load
0.5	3.0	60	0.0325	1.25	1.25	1.35	—	—	1.40	1.30	1.60	—	—	Effect of
	4.2			1.30	1.40	1.50	—	—	1.45	1.40	1.50	—	—	strength of
	5.0			1.35	1.35	1.55	—	—	1.60	1.50	1.55	—	—	concrete
0.5	4.2	40	0.0325	1.40	1.70	1.60	—	—	1.50	2.00	1.60	—	—	Effect of
		60		1.30	1.40	1.50	—	—	1.45	1.40	1.50	—	—	strength of
		80		1.20	1.25	1.45	—	—	1.30	1.27	1.40	—	—	steel
														Effect of
0.5	4.2	60	0.0125	1.40	1.60	1.30	—	—	1.45	1.60	1.30	—	—	percent-
			0.0325	1.30	1.40	1.50	—	—	1.45	1.40	1.50	—	—	age of
			0.0833	1.10	1.20	1.25	—	—	1.30	1.35	1.50	—	—	reinforce-
														ment

1 ksi = 6.9 MPa.

In presenting the families of interaction curves the parameters L/a, $P/f'_c ab$, $M_x/f'_c ab^2$ and $M_y/f'_c a^2 b$ are chosen. The computed interaction curves are used as a basis for (1) developing a simple approximate formula; (2) comparing with the C_m method; and (3) comparing with 1971 ACI moment magnifier method.

13.8.2 Moment-Curvature-Thrust Relationships

The relations between the bending moments M_x, M_y and axial force P, and the bending curvatures Φ_x, Φ_y and axial strain ε_0 at corner 0 [Fig. 6.26(a)] are of prime importance in the analysis of a long reinforced concrete column. The relationship used in the present calculations was determined by a separate program using the formulation developed in Sect. 6.8, Chap. 6. In this program a moment vs. curvature curve was developed for a constant axial compressive load with the other moment being held constant. Typical moment-curvature relations for a square section with $P/f'_c ab = 0.5$ are shown in Fig. 6.28, which has been computed from the stress-strain curves given in Fig. 6.27. The moment-curvature curve is obtained numerically by dividing the cross section into a large number of finite elements as shown in Fig. 6.26(d). Assuming linear strain distribution over the section, the strains of the elements are related to the curvatures of the section. The stresses are related to the applied bending moments through the condition of equilibrium. The relationship between the applied moments and the resulting curvatures can therefore be found through the stress-strain relations shown in Fig. 6.27. The details of the method and the computer solution are described in Sec. 6.8, Chap. 6.

In the numerical analysis, the square cross section was divided into 100 (10×10) and 400 (20×20) elements. The increase in accuracy obtained by using the finer grids was only 0.1 percent. A partitioning of the concrete cross section into 100 elements, and steel areas into 12 elements, distributed uniformly around the sides of the section, was used here. The strain and stress in each element were computed as the average value at its centroid. The allowable error in $P/f_c'ab$ was 0.002.

13.8.3 Method of Solution

The desired response of a given column (with known length and cross sectional properties) subjected to a specified axial compressive load, is the relation between an end moment and a lateral deflection. Once the complete load-deflection curve is obtained, the maximum biaxial moment that can be carried by the reinforced concrete column can be easily determined from the peak of the curve.

The numerical integration process used here was the Newmark method (1943). Details of the Newmark method has been presented in Sect. 13.3, Chap. 13, Vol. 1. The column lengths were divided into nine segments. This degree of subdivision of column length was checked against a subdivision of 15 and 20 segments. The refined analysis only led to an improvement of less than 1 percent in the results, when compared with the 9-segment solution, while the computational time was increased by more than twice.

The starting point in Newmark's method is to assume a reasonable initial deflected shape. Here, the elastic deflection is used with the flexural rigidity EI being computed from the approximate EI formula given by 1971 ACI Code [Eq. (14.42), Chap. 14, Vol. 1].

13.8.4 Numerical Results

The interaction diagrams giving the combinations of axial compressive load and biaxial moment corresponding to maximum or ultimate load conditions, are shown in Figs. 13.26 through 13.34 for symmetrically loaded cases and in Figs. 13.35 to 13.38 for unsymmetrical loading cases. The interaction curves shown in Fig. 13.26 and the moment curvature relations shown in Fig. 6.28 are considered here as the standard case. Each diagram is for a particular reinforced concrete column with a given compression load. Since these interaction curves are nondimensionalized, they can be directly used in analysis and design computations and also in checking the validity of the existing design approximations. This will be described later.

Referring now to the symmetric loading cases (Figs. 13.26 to 13.34), plotted on each interaction diagram for a slender ratio are two sets of interaction curves corresponding to two different values of concrete crushing strain $\varepsilon_0/\varepsilon_c' = 1.5$ (solid lines) and 2.0 (dashed curve). As can be seen, an increase in concrete crushing strain from $\varepsilon_0 = 0.003$ to 0.004 significantly increases the ultimate strength of a biaxially loaded column, when the slenderness ratio L/a is less than 20. This is almost true for all the variables investigated here.

The important factors influencing the ultimate strength of a long column are the magnitude of compression load P (Figs. 13.26, 13.27 and 13.28), concrete quality $k_1 f_c'$ (Figs. 13.29 and 13.30), steel quality f_y (Figs. 13.31 and 13.32), and percentage of reinforcement A_s/ab (Figs. 13.33 and 13.34). The variations of these factors with respect to the standard case may be obtained by comparing Figs. 13.27 to 13.34 with Fig. 13.26. The shape of the interaction curves is obviously a function of these factors considered.

13.8.5 Comparison with Equivalent Moment Method (C_m-Method)

Since the determination of the ultimate strength of reinforced concrete columns subjected to biaxial bending requires lengthy computations, design aids in the form of

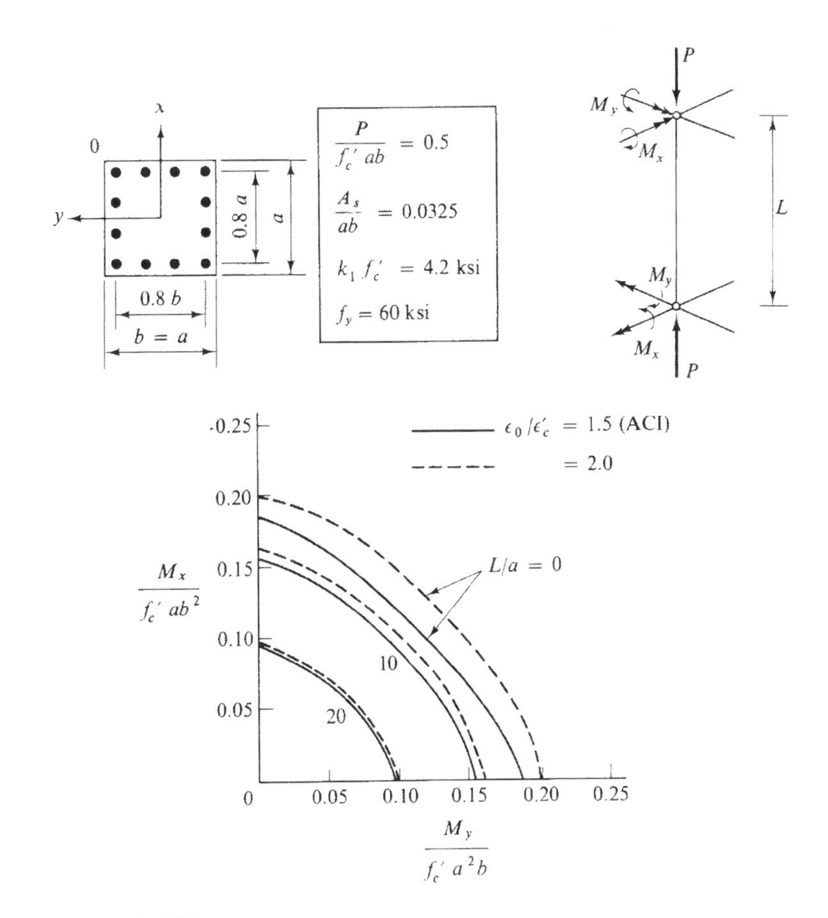

FIGURE 13.26
Interaction curves: standard case, long column (1 ksi = 6.9 MPa)

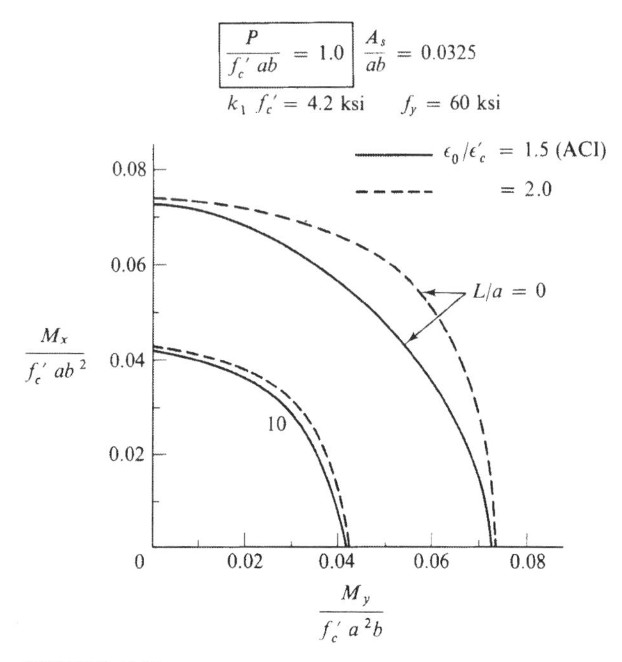

FIGURE 13.27
Interaction curves: maximum axial compression force effect (1 ksi = 6.9 MPa)

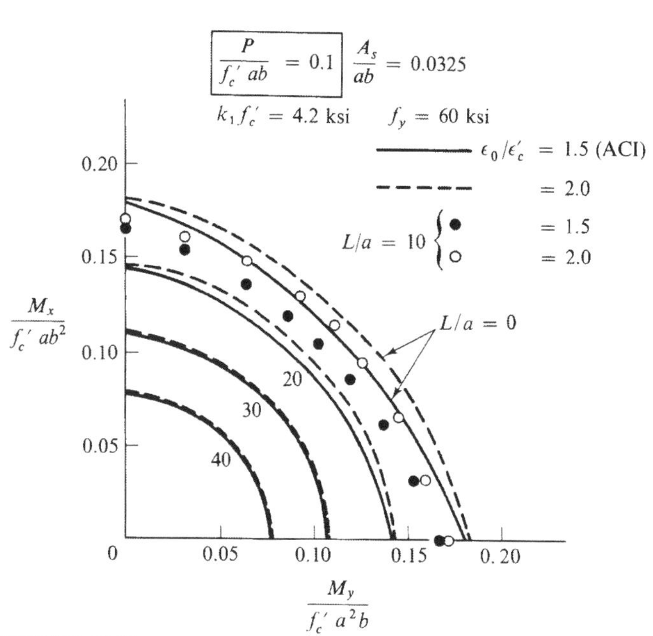

FIGURE 13.28
Interaction curves: minimum axial compression force effect (1 ksi = 6.9 MPa)

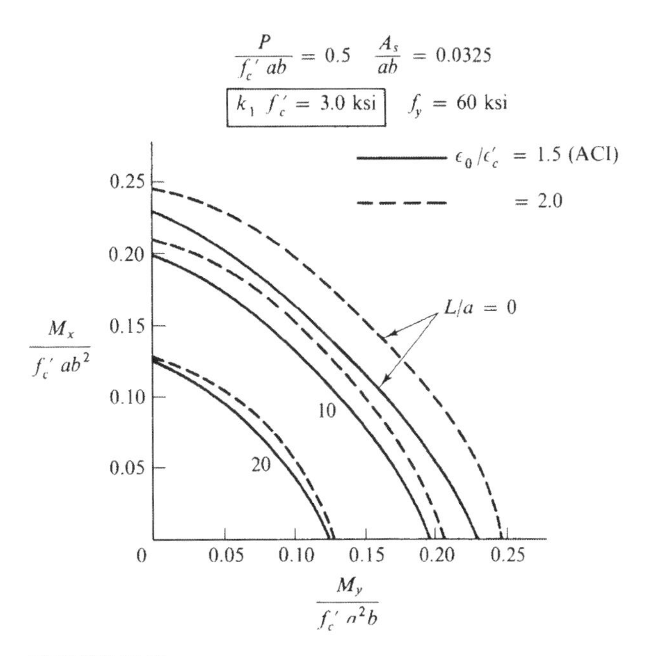

FIGURE 13.29
Interaction curves; minimum concrete quality effect
(1 ksi = 6.9 MPa)

interaction curves such as those shown in Figs. 13.26 to 13.38, or tables are needed for design in practice. However, even for the symmetrically loaded case, the range of important variables which could be considered is very large, and some restriction is necessary to keep the total number of design aids to a practical level. It is obvious that further presentation of interaction curves for various ratio of combinations of unsymmetric biaxial bending moments about each axis is too large in number to be practical. Hence, to cover adequately and comprehensively unsymmetric biaxial bending within a manageable compilation, the ultimate strength of columns to unsymmetric biaxial bending conditions must be related to its symmetric counterpart. The equivalent moment or C_m method, as given by Eqs. (13.26) and (13.27) is adopted herein to achieve this.

The equivalent moment values should be used in Figs. 13.26 to 13.34 when the interaction curves for the case of unsymmetrically loaded beam-columns are sought from the corresponding interaction curves developed for the case of symmetrically loaded conditions. These results (dotted curves) are compared with the exact solutions in Figs. 13.35 to 13.38 for several ratios of combinations of unsymmetric biaxial bending moments $\kappa_x = M_{xl}/M_{xo} = 0$, 1/2 and 1 and $\kappa_y = M_{yl}/M_{yo} = 0$, 1/2 and 1. In all cases excellent agreement is observed. It may be concluded from this study that the strength of unsymmetrically loaded biaxial beam-columns may be obtained directly from Figs. 13.26 to 13.34 using the equivalent moment concept. This observation is in agreement with previous investigations on steel beam-columns (Sect. 13.4).

13.8.6 Comparison With Moment Magnifier Method (δ-Method)

The formula which will be used in the comparison is formula (14.39) given in Chap. 14, Vol. 1 [or formula (10-4) given in Chap. 10 of the 1971 ACI Code]. In formula (14.39), the capacity reduction factor ϕ is taken as 1.0 and the creep reduction factor β_d is taken as 0. The value EI is computed from formula (14.42), Chap. 14, Vol. 1. The C_m factor as given in Eqs. (13.26), is used for the case of unsymmetrical loading conditions. The maximum loads determined by the ACI formula (14-39), Vol. 1, are compared in Figs. 13.35 to 13.38 (open circles) with the present "exact" solutions.

The moment magnifier method combined with the C_m factor is seen to give extremely conservative results for all the cases investigated. This is probably due to the fact that a more precise formula for estimating the initial stiffness EI of the column is required, when a column is subjected to biaxial bending conditions.

13.8.7 Simple Interaction Equation

The ultimate strength interaction curves shown in Figs. 13.26 to 13.34 for reinforced concrete columns subjected to combined axial load and symmetric biaxial bending may be approximated by the nonlinear interaction expression (13.23). In expression (13.23) M_{ucx} and M_{ucy} represent the load-carrying capacities of the column under compression and uniaxial bending moment about x and y axes respectively. Thus, for a given compression P, the values of M_{ucx} and M_{ucy} are the values given

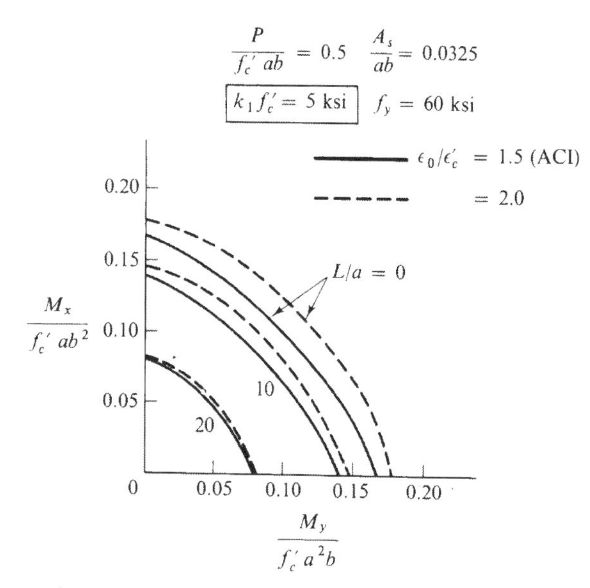

FIGURE 13.30
Interaction curves: maximum concrete equality effect (1 ksi = 6.9 MPa)

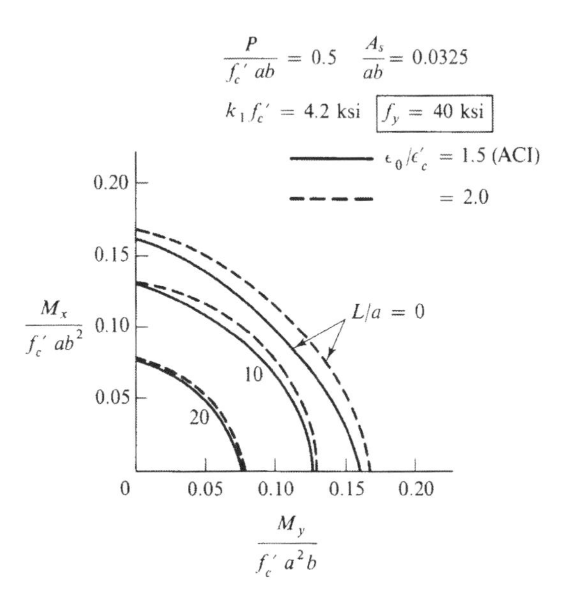

FIGURE 13.31
Interaction curves: minimum steel quality effect
(1 ksi = 6.9 MPa)

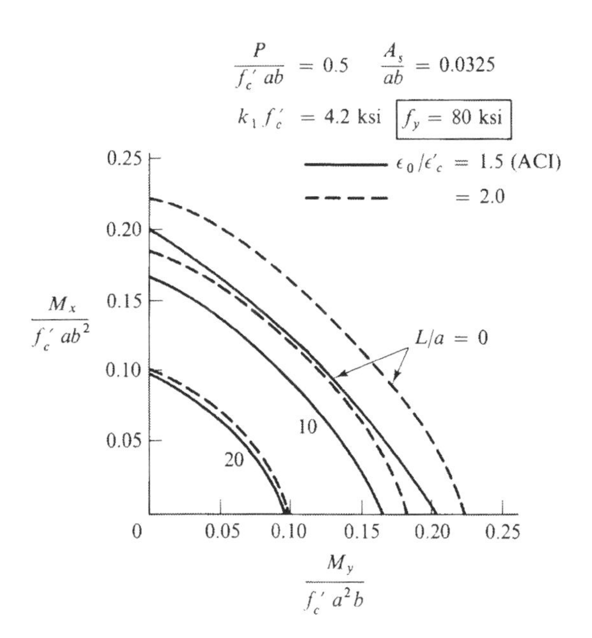

FIGURE 13.32
Interaction curves: maximum steel quality effect
(1 ksi = 6.9 MPa)

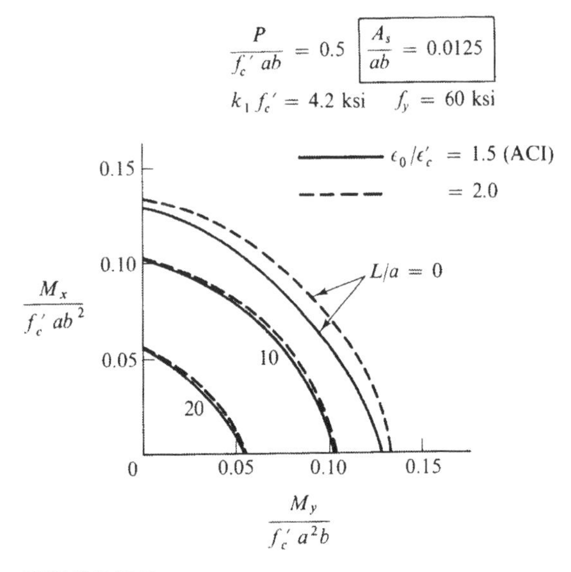

FIGURE 13.33
Interaction curves: minimum percentage of steel
(1 ksi = 6.9 MPa)

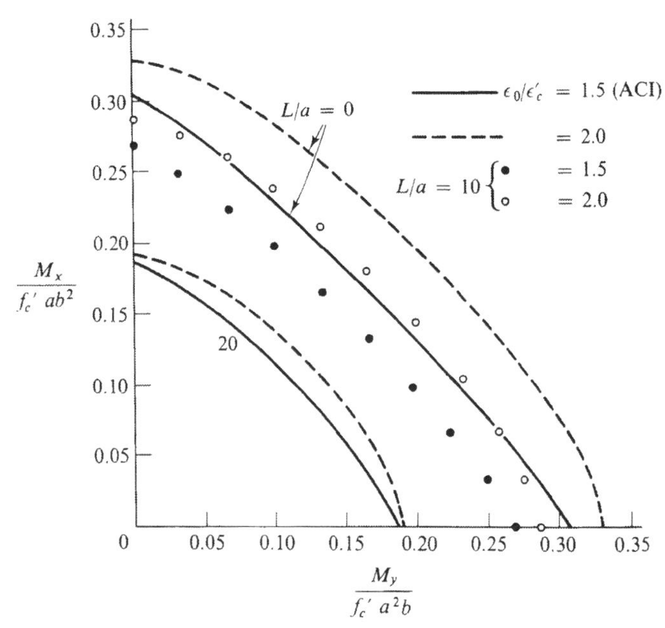

FIGURE 13.34
Interaction curves: maximum percentage of steel (1 ksi = 6.9 MPa)

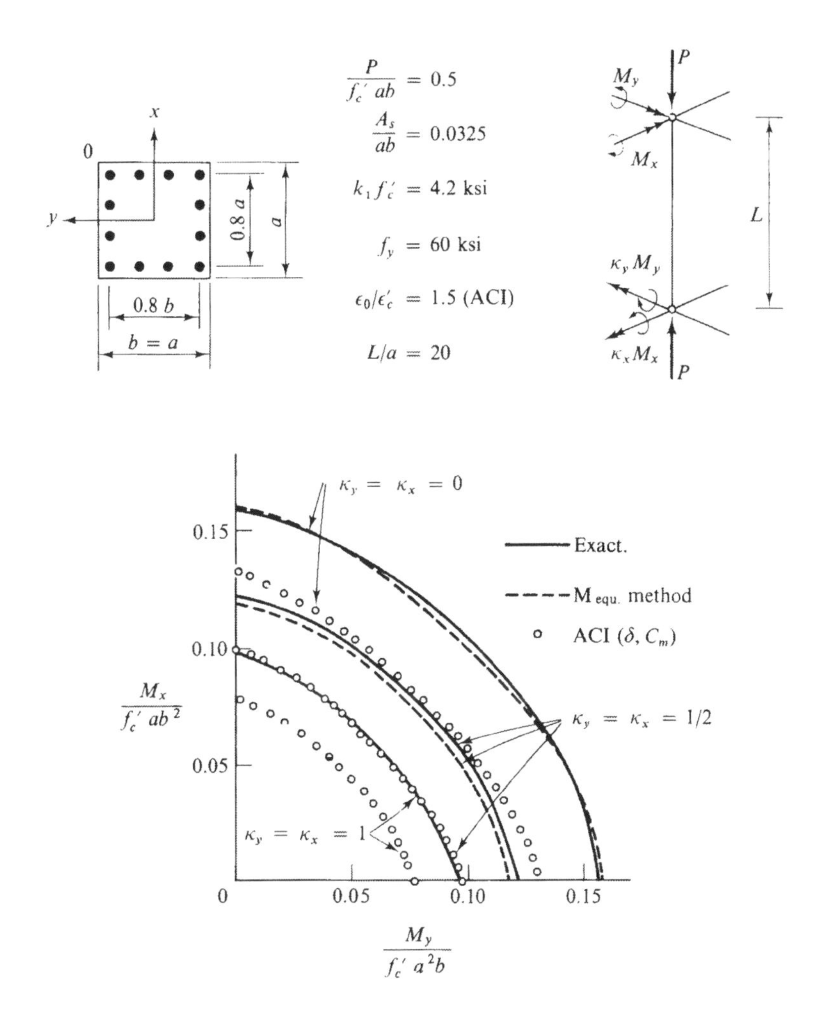

FIGURE 13.35
Comparison of exact solutions with equivalent moment method and ACI moment magnifier method (1 ksi = 6.9 MPa)

on the $M_y = 0$ and $M_x = 0$ axes shown in Figs. 13.26 to 13.34. Similar expression for the case of short reinforcement concrete columns is given by Eq. (5.125), Chap. 5.

The exponent β is a numerical factor whose value depends on, among other variables, the magnitude of the axial load, strength of materials, and distribution and percentage of reinforcement. For a given compression and a given column characteristic, the value of β is a numerical constant.

The interaction curves given previously in Fig. 13.26 for the standard case are now nondimensionalized by the values M_{ucx} and M_{ucy} and plotted in Fig. 13.39. The maximum and minimum values of β are also marked on the curves. Figures 13.40 to 13.43 shows the values of β affected by the magnitude of axial compression ($P/f_c'ab = 0.1$, 0.5 and 1.0 in Fig. 13.40), the strength of concrete ($k_1 f_c' = 3$, 4.2

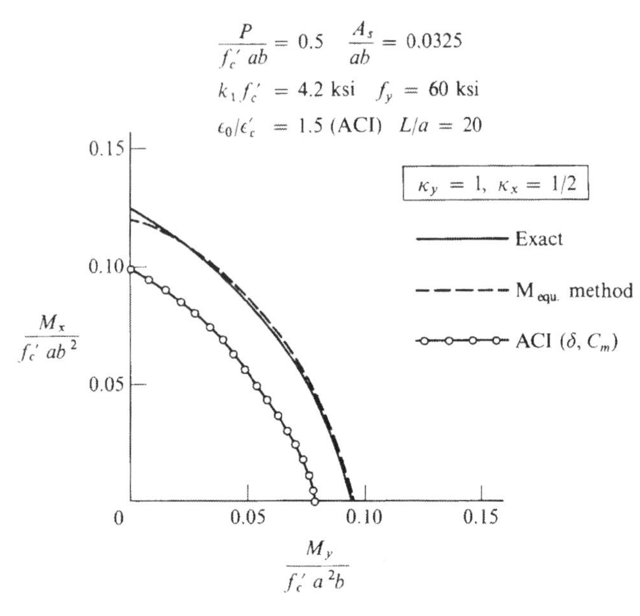

FIGURE 13.36
Comparison of exact solutions with equivalent moment method and ACI moment magnifier method (1 ksi = 6.9 MPa)

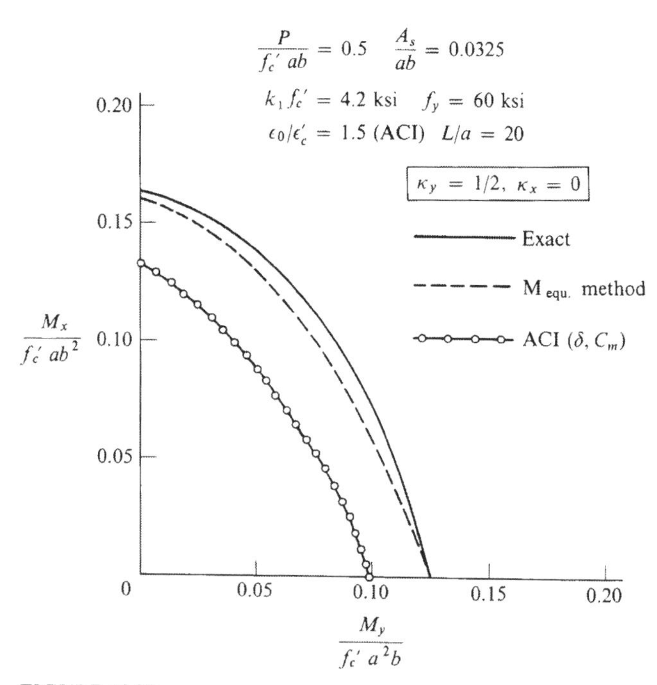

FIGURE 13.37
Comparison of exact solutions with equivalent moment method and ACI moment magnifier method (1 ksi = 6.9 MPa)

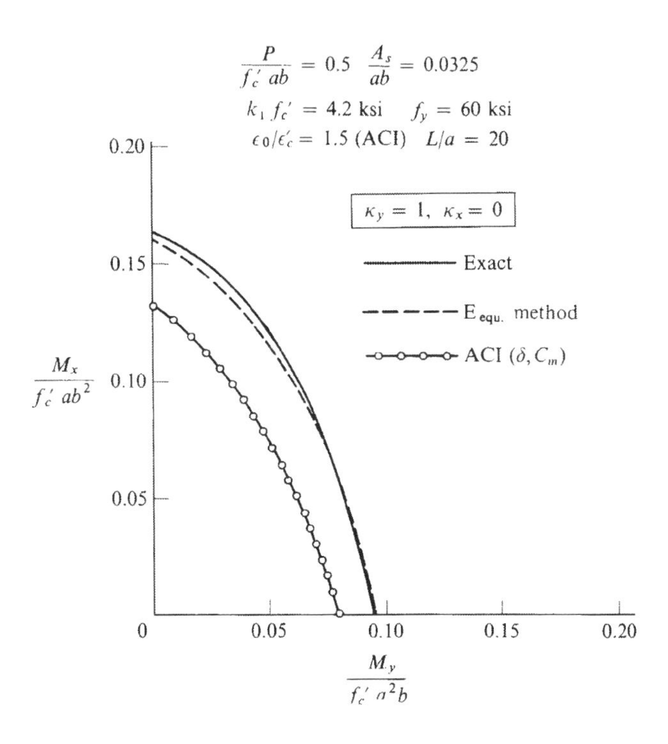

FIGURE 13.38
Comparison of exact solutions with equivalent moment method and ACI moment magnifier method (1 ksi = 6.9 MPa)

and 5 ksi in Fig. 13.41), strength of reinforcement ($f_y = 40$, 60 and 80 ksi in Fig. 13.42), and percentage of reinforcement ($A_s/ab = 0.0125$, 0.0325 and 0.0833 in Fig. 13.43). The values of β for columns with wide range of variation in values of f_c', f_y, A_s/ab, L/a and P are summarized in Table 13.12 for two strain ratios $\varepsilon_0/\varepsilon_c' = 1.5$ and 2.0.

It can be seen from Table 13.12 that the value of β is dependent primarily on the magnitude of axial compression and to a lesser, though still significant extent, on the slenderness ratio L/a, the percentage of steel, A_s/ab, and the strength of the materials f_c' and f_y. In general, the values of β in the range 1.1 to 1.6 give a good approximation to the exact curves for all the cases investigated in the lower and moderate axial compression range, but large variation in values of β is observed for columns with high axial compression.

13.9 SUMMARY

This chapter reviews the existing specification design requirements for biaxially loaded beam-columns, presents various ultimate strength design procedures, com-

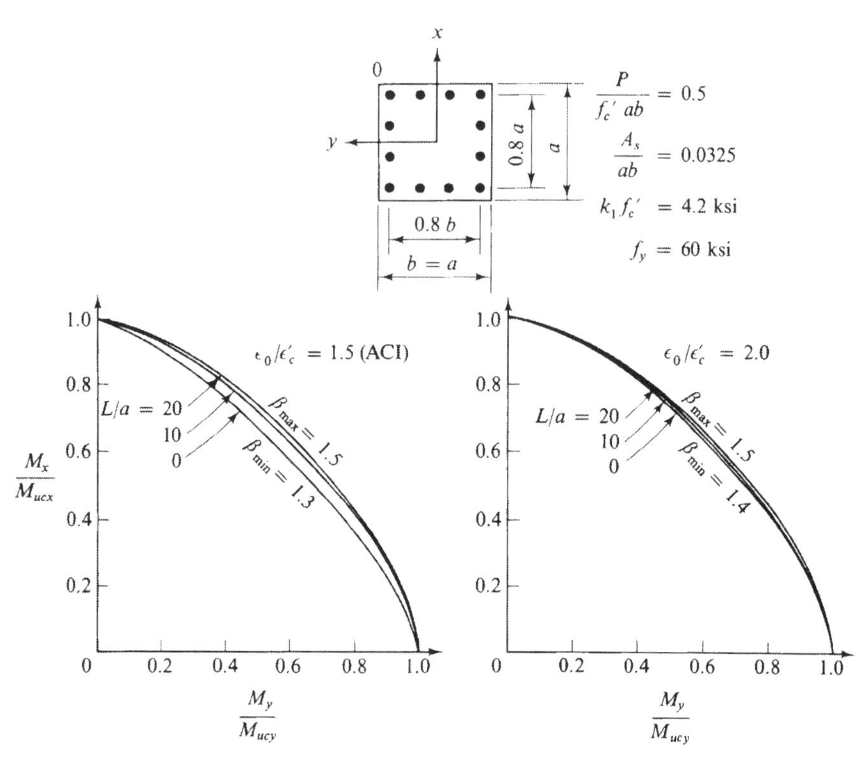

FIGURE 13.39
Nondimensionalized interaction curves for standard section (1 ksi = 6.9 MPa)

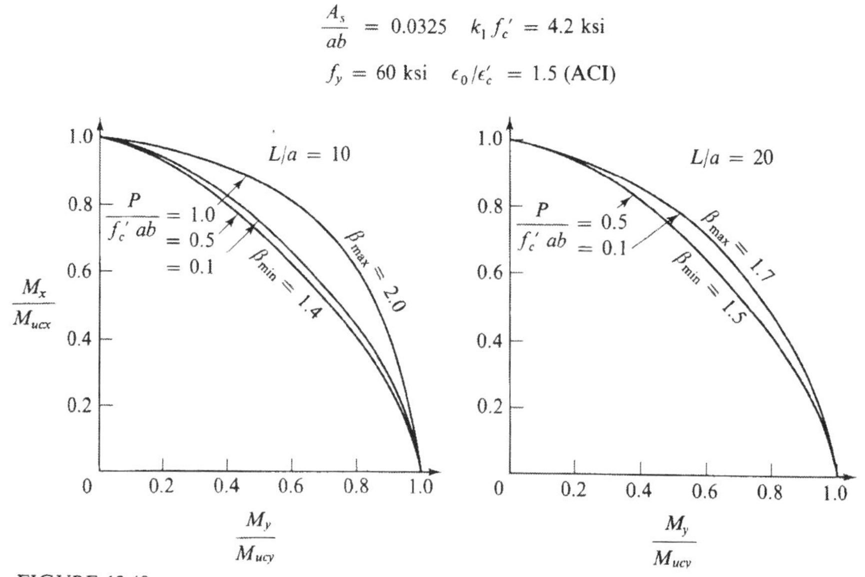

FIGURE 13.40
Nondimensionalized interaction curves: axial force effect (1 ksi = 6.9 MPa)

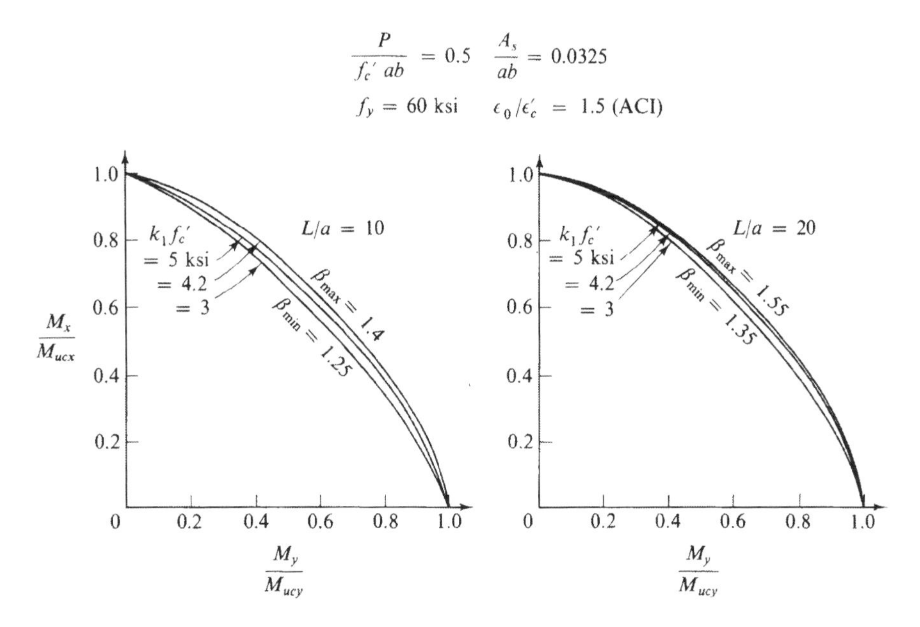

FIGURE 13.41
Nondimensionalized interaction curves: concrete quality effect (1 ksi = 6.9 MPa)

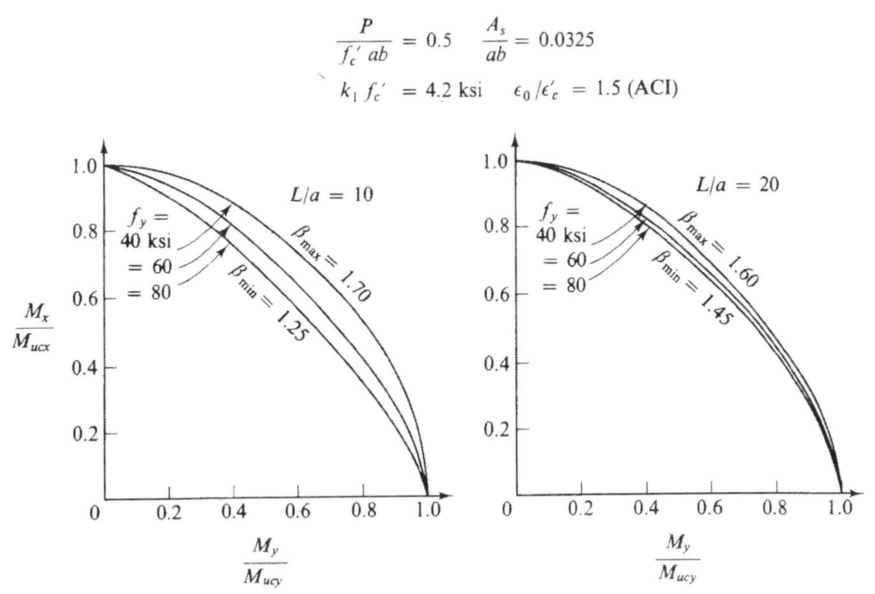

FIGURE 13.42
Nondimensionalized interaction curves: steel quality effect (1 ksi = 6.9 MPa)

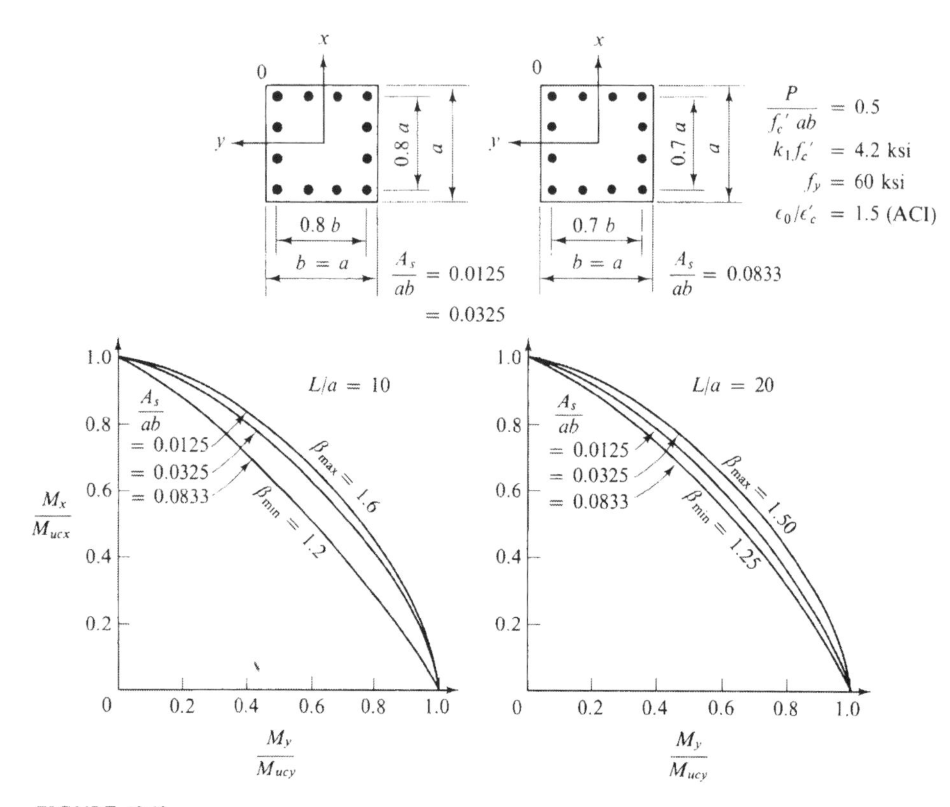

FIGURE 13.43
Nondimensionalized interaction curves: percentage of reinforcement effects (1 ksi = 6.9 MPa)

pares the size of beam-column section required by the existing and recently developed design rules for several practical design problems, and verifies the proposed interaction equation method with extensive test results. Similar attempts have also been made in the case of reinforced concrete beam-columns.

Several acceptable methods of design for biaxially loaded steel beam-columns have been discussed here. Ringo, McDonough and Basehart (1973) have studied the beam-column at a brace point. They were able to classify the 77 commonly occurring wide flange column sections into seven groups and to provide convenient design aids for these sections in the form of seven direct reading nomographs (Figs. 5.36 and 5.37, pp. 246–248).

Sharma and Gaylord (1969) developed a series of five design charts for biaxially loaded beam-columns (Fig. 7.17, pp. 344–347), plotting dimensionless parameters against slenderness ratio. In this way they were able to offer design of the complete column, both for strength at the brace point and for stability. Design using the Sharma and Gaylord charts is quite rapid; the only disadvantage is that interpolation is required on each chart followed by interpolation between the charts.

In England, Young (1973) has recently published a design method called K-coefficients, complete with formulas and design aids, in which the section selected is such that its major axis plastic moment, reduced by coefficients reflecting the presence of axial load, buckling about major and minor axes, and torsional buckling, fulfills the design requirements.

Many designers dislike graphical procedures, although the accuracy obtained thereby is probably within the limits with which we can predict the capacity of a beam-column under biaxial loading. For this and other reasons, specifications are generally expressed in terms of algebraic formulas, where possible. Such formulas are particularly suitable for incorporation into computer design programs.

By algebraic transposition, Eq. (13.17) may be written in the following form:

$$\frac{M_x}{M_{ucx}} + \frac{M_y}{M_{ucy}} \leq 1.0 \tag{13.18}$$

where M_{ucx} and M_{ucy} are the ultimate symmetrical single curvature moment capacities of an axially loaded beam-column, about the x- and y-axes, respectively, when there is zero moment about the other axis. These denominator terms may be expressed as follows:

At braced points:

$$M_{ucx} = M_{pcx} = 1.18\,M_{px}[1 - (P/P_y)] \leq M_{px} \tag{13.11a}$$

$$M_{ucy} = M_{pcy} = 1.19\,M_{py}[1 - (P/P_y)^2] \leq M_{py} \tag{13.11b}$$

Considering stability:

$$M_{ucx} = M_{ux}[1 - (P/P_u)][1 - (P/P_{ex})] \tag{13.19a}$$

$$M_{ucy} = M_{py}[1 - (P/P_u)][1 - (P/P_{ey})] \tag{13.19b}$$

in the above equations,

M_{pcx}, M_{pcy} = plastic moment capacity about the x- or y-axis, respectively, reduced for the presence of axial load

M_{px}, M_{py} = plastic moment capacities

M_{ux} = plastic moment capacity about the x-axis, reduced for the presence of lateral torsional buckling, if necessary

$$= M_m = M_{px}\left[1.07 - \frac{(L/r_y)\sqrt{\sigma_y}}{3160}\right] \leq M_{px} \tag{13.20}$$

P_y = axial load at full yield condition.

Both the AISC design expressions (13.7) and (13.16) and SSRC equations (13.17) are straight line interaction equations. Research has shown that the interaction of moments about the orthogonal axes is not linear; on the contrary, the interaction curve resembles more closely the quadrant of a circle (see Fig. 13.13). The following nonlinear interaction equation was suggested by Tebedge and Chen (1974).

$$\left(\frac{M_x}{M_{ucx}}\right)^{\text{exponent}} + \left(\frac{M_y}{M_{ucy}}\right)^{\text{exponent}} \leq 1.0 \tag{13.23}$$

By deriving the correct exponents, an interaction curve could be defined which would fit the actual strength curve of a biaxially loaded beam-column of a particular cross section, such as a wide flange shape.

At a braced section, the capacity was determined by a superposition technique (Chen and Atsuta, 1972, Chap. 5) and interaction curves plotted. For stability, Tebedge and Chen (1974) determined empirical interaction curves for wide flange beam-columns at various levels of axial load. Exponent expressions giving a good fit to these curves were developed. The following design interaction formulas are suggested:

At a braced location, the following equation should be satisfied:

$$\left(\frac{M_x}{M_{pcx}}\right)^{\alpha} + \left(\frac{M_y}{M_{pcy}}\right)^{\alpha} \leq 1.0 \tag{13.10}$$

For wide flange shapes having a flange width to web depth ratio from 0.5 to 1.0:

$$\alpha = 1.6 - \frac{P/P_y}{2ln(P/P_y)} \tag{13.12}$$

where ln indicates the natural logarithm.

To check stability between braced points, the following equation should be satisfied:

$$\left(\frac{C_{mx}M_x}{M_{ucx}}\right)^{\beta} + \left(\frac{C_{my}M_y}{M_{ucy}}\right)^{\beta} \leq 1.0 \tag{13.23}$$

in which,

C_{mx}, C_{my} = equivalent moment factors used in the AISC Specification interaction formula,

i.e.,

$$C_m = 0.6 + 0.4\frac{M_1}{M_2} \geq 0.4 \tag{13.26}$$

where M_1/M_2 is the ratio of the smaller to larger moments at the ends of that portion of the member unbraced in the plane of bending under consideration. M_1/M_2 is positive when the member is bent in single curvature and negative when it is bent in reverse curvature.

M_x, M_y = the greater of the moments applied at one or the other end of the beam-column

$$\beta = 0.4 + P/P_y + B/D \geq 1.0 \quad \text{when} \quad B/D \geq 0.3 \tag{13.24b}$$

$$= 1.0 \quad \text{when} \quad B/D < 0.3$$

B = flange width of W or I section

D = depth of W or I section

and in which M_{ucx} and M_{ucy} can be determined from Eqs. (13.19a) and (13.19b), respectively.

The advantages of the design interaction equation suggested by Chen *et al.*, are that both strength and stability are covered by the same basic equation and the equation has universal applicability, for if the exponent for the particular section being designed has not been determined, it may conservatively be assumed to be unity.

The assumptions made in all the design procedures are:

1. The design is for an isolated member, not for a part of a framework in which the forces redistribute as ultimate load is approached.
2. The design method is valid for beam-columns in which the axial load and end moments are known, being determined by a first or second order elastic analysis, as appropriate to braced or unbraced frames. The adoption of biaxial plastic design is not suggested.
3. The sections are compact, i.e., premature failure will not occur due to local buckling of flanges or webs prior to the member attaining its ultimate strength.
4. The column is initially twisted and out-of-straight, contains residual stresses, and the material is elastic-perfectly plastic.

In using these methods, the determination of axial load capacity of the beam-column should be as realistic as possible, such as the use of Multiple Column Curves described in Chap. 14, Vol. 1, rather than the unmodified tangent modulus critical load given by SSRC Eq. (13.37), which is based on the concept of buckling of a perfectly straight column using the tangent modulus dependent on residual stresses, whereas the fundamental analysis of beam-column is based on the load-deflection approach to the determination of ultimate strength considering geometrical and material imperfections.

In an extension of an earlier evaluation of the SSRC linear interaction equation (13.17) by Galambos (1963), Springfield and Hegan (1973) included the Birnstiel tests (Table 13.11) and compared also the nonlinear interaction equation (13.23) of Chen with exponent given by (13.24), and of Pillai, Eq. (13.21). Although Pillai (1970) did not propose Eq. (13.21) for wide-flange sections, when flange to depth ratios were equal, the results showed a marked improvement over current specification procedures in predicted capacity, as did Eq. (13.23). The latter equation was superior in predicting capacities of sections with varying flange width to web depth. For the Birnstiel's tests, Chen's nonlinear equation (13.23) was found extremely reliable (Mean 1.01, S.D. 0.074), while the SSRC linear interaction equation (13.17) was conservative (Mean 1.20, S.D. 0.085). A further verification of Chen's nonlinear equation (13.23) is its good agreement with Birnstiel's incremental analytical procedure (Chap. 8). Aside from one result, in which the error was 7 percent conservative, all the other values agree to within 3 percent.

Ultimate strength interaction relations for reinforced concrete columns subjected to compression combined with biaxial bending, have been developed for short as well as long columns under symmetric and unsymmetric loading conditions. The results are presented in Figs. 13.26 to 13.38 in the form of interaction curves relating the axial compression, maximum biaxial moment and slenderness ratio.

The maximum biaxial moments determined by the ACI moment magnifier method have been compared with the present analytical results and are found to give over-conservative results for all the cases investigated. It is also found that maximum load-carrying capacity predicted by C_m factor method for unsymmetrically loaded cases from the symmetric cases is in excellent agreement with calculated theoretical values.

Nonlinear interaction equation as given by Eq. (13.23) provides an alternate method for design. The values of β varying from 1.1 to 1.6 (see Table 13.12) are found to provide a good approximation of analytical results when the axial compression is moderate.

13.10 PROBLEMS

13.1 Check Example 1 in Sec. 13.6 using
 (a) Sharma and Gaylord's Charts (Fig. 7.17)
 (b) Nonlinear interaction equations (13.23) and (13.25).

13.2 Consider a beam-column of W8 × 31 section in A36 steel which is subjected to a constant axial load of $P = 0.3P_y$ combined with a symmetric biaxial moment M_x and M_y. The residual stresses are assumed to be the linear type with maximum compressive stresses at flange tip of $\sigma_{rc} = 0.3\sigma_y$. Construct the maximum biaxial moment strength interaction curves for the two beam-column lengths; $L = 30r_x$ and $L = 50r_x$
 (a) Use the computer solution of Table 13.1 with initial imperfections: $\delta = \delta_{xo} = \delta_{yo} = 1.0 \times 10^{-6}\ L$ and $\theta_{im} = 1.0 \times 10^{-6}$ rad.
 (b) Use the computer solution of Table 10.6 with initial imperfections: $\delta = \delta_{xo} = \delta_{yo} = L/5000$ and $\theta_{im} = 0$.
 (c) Use the nonlinear interaction equation (13.23) along with Eqs. (13.19), (13.20) and (13.24). Use the Multiple Column Curve 2 for the determination of axial load capacity P_u of the column [Eq. (14.18) or Fig. 14.10, Chap. 14, Vol. 1].
 (d) Use the nonlinear interaction equation (13.23) along with Eqs. (13.19), (13.20) and (13.24). Use the SSRC Basic Column Strength Equation (13.37) for the determination of axial load capacity P_u of the column.
 (e) Use the modified interaction equation

$$\left(\frac{M_x}{M_{ucx}}\right)^{\beta} + \frac{M_y}{M_{ucy}} = 1.0 \tag{13.38}$$

 along with Eqs. (13.19), (13.20) and (13.24). Use the SSRC Basic Column Strength Equation (13.37) for the determination of axial load capacity P_u of the column. [Note: A generalization of Eq. (13.38) has been reported by Djalaly (1975)].

13.11 REFERENCES

ACI 318-71, "Building Code Requirements for Reinforced Concrete," American Concrete Institute, (1971).

AISC, "Computer Program for Steel Column Design," American Institute of Steel Construction, CDA 1-69, New York, (1969).

AISC, "Specification for the Design, Fabrication and Erection of Structural Steel for Buildings," American Institute of Steel Construction, New York, February, (1969).

ASCE-WRC, "Plastic Design in Steel—A Guide and Commentary," *Manual 41*, Welding Research Council and the American Society of Civil Engineers, (1971).

Birnstiel, C., "Experiments on H-Columns Under Biaxial Bending," *Journal of the Structural Division, ASCE*, Vol. 94, No. ST10, Proc. Paper 6186, October, pp. 2429–2448, (1968).

Campus, F. and Massonnet, C., "Recherches sur le flambement de colonnes en acier A37, à profil en double Té, Sollicitées obliquement," *Comptes Rendus de Recherches, Bulletin* No. 17, Institut de Recherche Scientifique, Industrielle et Agriculturelle, April, (1956).

CEB, Comité Européen du Beton, "International Recommendations for the Design and Construction of Concrete Structures," Paris, June, (1970).

Chen, W. F., "Design of Box Columns under Biaxial Bending," Proceedings, International Colloquium on Stability of Structures, European Convention for Constructional Steelwork, Liege, Belgium, April 13–15, pp. 355–361, (1977).

Chen, W. F. and Atsuta, T., "Interaction Equations for Biaxially Loaded Sections," *Journal of the Structural Division, ASCE*, Vol. 98, No. ST5, Proc. Paper 8902, May, pp. 1035–1052, (1972).

Chen, W. F. and Atsuta, T., "Inelastic Response of Column Segments Under Biaxial Loads," *Journal of the Engineering Mechanics Division, ASCE*, Vol. 99, No. EM4, Proc. Paper 9957, August, pp. 685–701, (1973a).

Chen, W. F. and Atsuta, T., "Ultimate Strength of Biaxially Loaded Steel H-Columns," *Journal of the Structural Division, ASCE*, Vol. 99, No. ST3, Proc. Paper 9613, March, pp. 469–489, (1973b).

Chen, W. F. and Atsuta, T., "Interaction Curves for Steel Sections Under Axial Load and Biaxial Bending," *Engineering Journal*, Vol. 17, No. A-3, Engineering Institute of Canada, March/April, pp. I–VIII, (1974).

Chen, W. F. and Santathadaporn, S., "Review of Column Behavior Under Biaxial Loading," *Journal of the Structural Division, ASCE*, Vol. 94, No. ST12, Proc. Paper 6316, December, pp. 2999–3021, (1968).

Chen, W. F. and Shoraka, M. T., "Analysis and Design of Reinforced Columns Under Biaxial Loading," *Proceedings, Symposium on Design and Safety of Reinforced Concrete Compression Members*, Publications, International Association for Bridge and Structural Engineering, Quebec, pp. 179–195, (1974).

Chubkin, G. M., "Experimental Research on Stability of Thin Plate Steel Members with Biaxial Eccentricity," Paper No. 6 in, *Analysis of Spatial Structures*, Vol. 5, Moscow, GILS, (1959).

CISC, "Column Selection Program," Canadian Institute of Steel Construction, Toronto, Canada, (1967).

CSA, "Steel Structures for Buildings—Limit States Design," *Canadian Standards Association, CSA Standard* S16, (1974).

Dibley, J. E., "Lateral-Torsional Buckling of I-Sections in Grade 55 Steel," Proceedings, Institute of Civil Engineers, Vol. 43, August, pp. 599–627, (1969).

Djalaly, H, "Calcul de la Résistance Ultime des Barres Comprimées et Fléchies," Construction Métallique, No. 4, pp. 17–47, (1975).

Ellis, J. S., Jury, E. J. and Kirk, D. W., "Ultimate Capacity of Steel Columns Loaded Biaxially," *Transactions, Engineering Institute of Canada*, Vol. 7, No. A-2, 64-BR, STR2, February, pp. 3–11, (1964).

Galambos, T. V., "Review of Tests on Biaxially Loaded Steel Wide-Flange Beam-Columns," Report to the Column Research Council Task Group No. 3, *Fritz Engineering Laboratory Report* No. 287.4, Lehigh University, Bethlehem, Pa., April, (1963).

Johnston, B. G., *Guide to Design Criteria for Metal Compression Members*, 2nd edn, (1966), Guide to Stability Design Criteria for Metal Structures, 3rd edn, John Wiley, New York, pp. 76–83, (1976).

Keramati, S., Gaylord, E. H. and Robinson, A. R., "Ultimate Strength of Eccentrically Loaded Columns," *Proceedings, International Colloquium on Colun.n Strength*, European Convention for Constructional Steelwork-Column Research Council-CRC Japan, Paris, November 23–24, IABSE publication, pp. 171–181, (1972).

Klöppel, K. and Winkelmann, E., "Experimentelle und Theoretische Untersuchungen über die Traglast von Zweiachsig Aussermittig Gedruckten Stahlstaben," *Der Stahlbau*, Vol. 31, No. 2, February, March, April, p. 33, (1962).

Marshall, P. J. and Ellis, J. S., "Ultimate Biaxial Capacity of Box Steel Columns," *Journal of the Structural Division, ASCE*, Vol. 96, No. ST9, Proc. Paper 7515, September, pp. 1873–1887, (1970).

Milner, H. R., "The Elastic Plastic Stability of Stanchions Bent About Two Axes." Dissertation presented to the College at the University of London, in partial fulfillment of the requirements for the degree of Doctor of Philosophy, December, (1965).

Morris, G. A. and Fenves, S. J., "Approximate Yield Surface Equations," *Journal of the Engineering Mechanics Division, ASCE*, Vol. 95, No. EM4, Proc. Paper 6741, August, pp. 937–954, (1969).

Newmark, N. M., "Numerical Procedure for Computing Deflections, Moments and Buckling Loads," *Transactions, ASCE*, Vol. 108, p. 1161, (1943).

Pillai, S. U., "Review of Recent Research on the Behavior of Beam-Columns Under Biaxial Bending,"

Civil Engineering Research Report No. CE70-1, Royal Military College of Canada, Kingston, Canada, January, (1970).

Pillai, S. U., Ellis, J. S., "Beam Columns of Hollow Structural Sections," *Canadian Journal of Civil Engineering*, Vol. 1, No. 4, December (1974).

Pillai, S. U. and Ellis, J. S., "Hollow Tubular Beam-Columns in Biaxial Bending," *Journal of the Structural Division, ASCE*, Vol. 97, No. ST5, Proc. Paper 8094, May, pp. 1399–1406, (1971).

Ringo, B. C., McDonough, J. F. and Baseheart, T. M., "Plastic Design of Biaxially Loaded Beam-Columns," *Engineering Journal*, American Institute of Steel Construction, Vol. 10, No. 1, First Quarter, pp. 6–18, (1973).

Ross, D. A. and Chen, W. F., "Design Criteria for Steel I-Columns under Axial Load and Biaxial Bending," *Canadian Journal of Civil Engineering*, Vol. 3, No. 3, pp. 466–473, (1976).

Santathadaporn, S., "Analysis of Biaxially Loaded Columns." Thesis presented to Lehigh University, Bethlehem, Pa., in partial fulfillment of the requirements for the degree of Doctor of Philosophy (University Microfilms, Ann Arbor, Michigan), (1970).

Santathadaporn, S. and Chen, W. F., "Interaction Curves for Sections Under Combined Biaxial Bending and Axial Force," *Welding Research Council Bulletin* No. 148, February, (1970).

Santathadaporn, S. and Chen, W. F., "Tangent Stiffness Method for Biaxial Bending," *Journal of the Structural Division, ASCE*, Vol. 98, No. ST1, Proc. Paper 8637, January, pp. 153–163, (1972).

Santathadaporn, S. and Chen, W. F., "Analysis of Biaxially Loaded H-Columns," *Journal of the Structural Division, ASCE*, Vol. 99, No. ST3, Proc. Paper 9621, March, pp. 491–509, (1973).

Sharma, S. S., "Ultimate Strength of Steel Columns Under Biaxial Eccentric Load." Thesis presented to the University of Illinois at Urbana, in partial fulfillment of the requirements for the degree of Doctor of Philosophy, (1965).

Sharma, S. S. and Gaylord, E. H., "Strength of Steel Columns with Biaxially Eccentric Load," *Journal of the Structural Division, ASCE*, Vol. 95, No. ST12, Proc. Paper 6960, December, pp. 2797–2812, (1969).

Springfield, J., "Design of Columns Subject to Biaxial Bending," *Engineering Journal, American Institute of Steel Construction*, Third Quarter, pp. 73–81, (1975).

Springfield, J. and Hegan, B., "Comparison of Test Results with Design Equations for Biaxially Loaded Steel Wide Flange Beam-Columns," *Report to the Canadian Steel Industries Construction Council*, Toronto, Canada, under the auspices of Carruthers and Wallace, Consultants, Ltd., joint investigation, (1973).

Syal, I. C. and Sharma, S. S., "Biaxially Loaded Beam-Column Analysis," *Journal of the Structural Division, ASCE*, Vol. 97, No. ST9, Proc. Paper 8384, September, pp. 2245–2259, (1971).

Tebedge, N., Chen, W. F. and Tall, L., "On the Behavior of a Heavy Steel Column," *Proceedings, International Colloquium on Column Strength*, European Convention for Constructional Steelwork-Column Research Council-CRC Japan, Paris, France, November 23–24, IABSE publication, pp. 9–24, (1972).

Tebedge, N. and Chen, W. F., "Design Criteria for Steel H-Columns Under Biaxial Loading," *Journal of the Structural Division, ASCE*, Vol. 100, No. ST3, Proc. Paper 10400, March, pp. 579–598, (1974).

Young, B. W., "Steel Column Design," *Journal of the Institute of Structural Engineering, The Structural Engineer*, Vol. 51, No. 9, September, pp. 323–336, (1973).

ANSWERS TO SOME SELECTED PROBLEMS

CHAPTER 2

Problem 2.1

$$A = \int t\, ds = 1500 \text{ in}^2$$

$$I_x = \int \eta^2 t\, ds = 5.04 \times 10^5 \text{ in}^4$$

$$I_y = \int \xi^2 t\, ds = 2.97 \times 10^6 \text{ in}^4$$

$$\omega = \int_0^s \rho\, ds \qquad \text{[Fig. 2.26(a)]}$$

$$I_{\omega x} = \int \omega \xi t\, ds = -51.9 \times 10^6 \text{ in}^5$$

$$I_{\omega y} = \int \omega \eta t\, ds = 0$$

$$K_T = \frac{1}{3} \sum bt^3 = 12\,500 \text{ in}^4$$

shear center $\xi_0 = \frac{I_{\omega y}}{I_x} = 0, \quad \eta_0 = -\frac{I_{\omega x}}{I_y} = 17.5 \text{ in}$

$$\omega_0 = \omega + \eta_0(\xi - \xi_A) - \xi_0(\eta - \eta_A) = \omega + 17.5(\xi - 90) \qquad \text{[Fig. 2.26(b)]}$$

$$\omega_n = \frac{1}{A}\int \omega_0 t \, ds - \omega_0 = -495 - \omega_0 \qquad \text{[Fig. 2.26(c)]}$$

$$S_\omega = \int_0^s \omega_n t \, ds \qquad \text{[Fig. 2.26(d)]}$$

$$I_\omega = \int \omega_n^2 t \, ds = 5.59 \times 10^8 \text{ in}^6,$$

$$\lambda = \sqrt{\frac{GK_T}{EI_\omega}} = \sqrt{\frac{11.5 \times 10^3 \times 12\,500}{30 \times 10^3 \times 5.59 \times 10^8}} = 2.93 \times 10^{-3} \text{ in}^{-1}$$

Differential equation $EI_\omega \theta''' - GK_T\theta' = -\frac{1}{2}M_z$

Solution $$\theta = \frac{M_z}{2\lambda^3 EI_\omega}\left[\frac{1 - \cosh\dfrac{\lambda L}{2}}{\sinh\dfrac{\lambda L}{2}}(1 - \cosh\lambda z) - \sinh\lambda z + \lambda z\right]$$

$$\theta'' = 1.27 \times 10^{-10}\, Q \text{ in}^{-2} \quad \text{at} \quad z = \frac{L}{2}, \quad \theta'' = 0 \quad \text{at} \quad z = \frac{L}{4}$$

$$\theta''' = -8.15 \times 10^{-13}\, Q \text{ in}^{-3} \quad \text{at} \quad z = \frac{L}{4}, \quad \theta''' = -8.95 \times 10^{-13}\, Q \text{ in}^{-3} \quad \text{at} \quad z = \frac{L}{2}$$

$$\sigma_w = E\omega_n\theta'' = 3.81 \times 10^{-6}\, \omega_n Q \quad \text{at} \quad z = \frac{L}{2}, \quad \sigma_w = 0 \quad \text{at} \quad z = \frac{L}{4}$$

$$\tau_w = -\frac{1}{t}ES_\omega\theta''' = 4.90 \times 10^{-9}\, S_\omega Q \quad \text{at} \quad z = \frac{L}{4}, \quad \tau_w = 5.38 \times 10^{-9}\, S_\omega Q \quad \text{at} \quad z = \frac{L}{2}$$

$$\sigma_b = \frac{M_\xi}{I_x}\eta = 9.92 \times 10^{-7}\, Qz\eta \qquad \text{[Fig. 2.26(f)]}$$

$$\tau_b = -\frac{V_\eta}{I_x}\int_0^s \eta t \, ds = 4.96 \times 10^{-6} Q \int_0^s \eta \, ds \qquad \text{[Fig. 2.26(e)]}$$

$$\max \tau_{sv} = \frac{t}{K_T}T_{sv} = 0.006Q \qquad \text{[Fig. 2.26(g)]}$$

Deflection at the midspan due to bending,

$$\delta_B = \frac{QL^3}{48EI_x} = \frac{600^3}{48 \times 30 \times 10^3 \times 5.04 \times 10^5} Q = 2.98 \times 10^{-4}\, Q$$

Rotation angle at the midspan due to torsion

$$\theta_{L/2} = \frac{30Q}{2\lambda^3 EI_\omega}\left[\frac{\left(1-\cosh\frac{\lambda L}{2}\right)^2}{\sinh\frac{\lambda L}{2}} - \sinh\frac{\lambda L}{2} + \frac{\lambda L}{2}\right]$$

$$= \frac{30Q}{2\times 2.93^3\times 10^{-9}\times 30\times 10^3\times 5.59\times 10^8}\left[\frac{0.413^2}{0.998} - 0.998 + 0.880\right]$$

$$= 1.89\times 10^{-6}Q$$

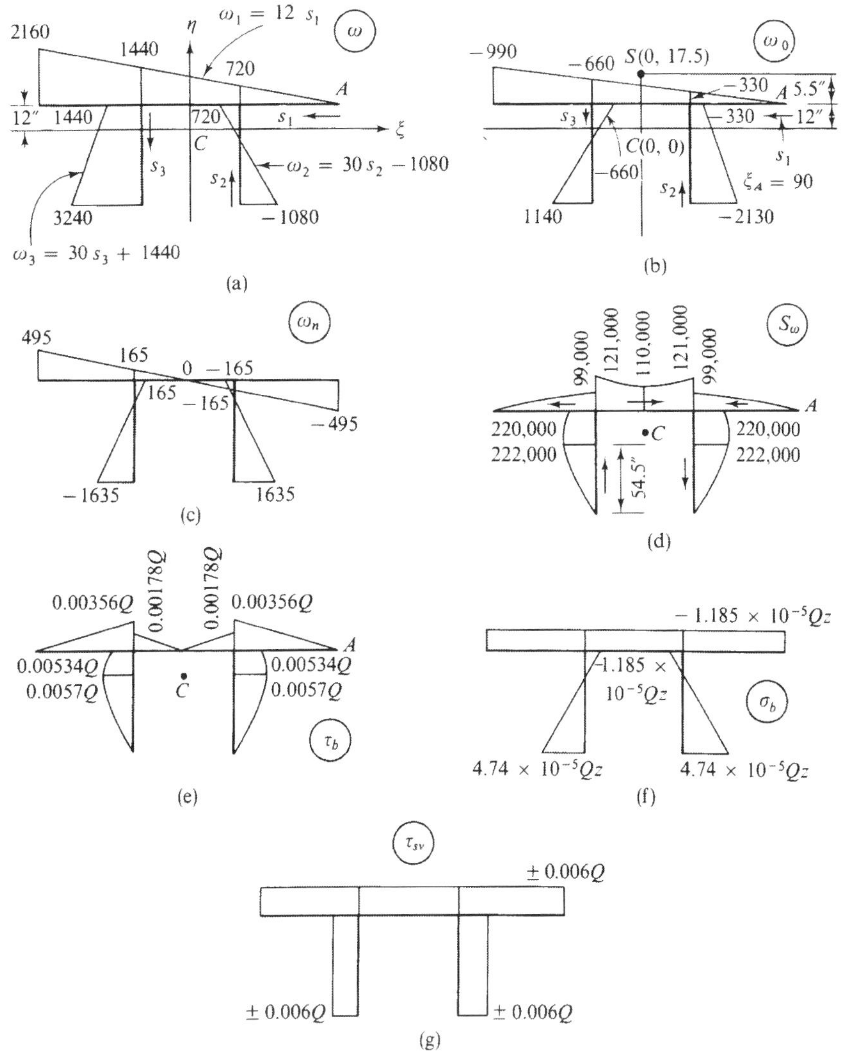

FIGURE 2.26
Figures for solution to Prob. 2.1

Deflection of point A due to torsion,

$$\delta_T = \theta_{L/2}90 = 1.89 \times 90 \times 10^{-6} Q = 1.70 \times 10^{-4} Q$$

Total deflection of point A is

$$\delta_A = \delta_B + \delta_T = 2.98 \times 10^{-4} Q + 1.70 \times 10^{-4} Q = 4.68 \times 10^{-4} Q$$

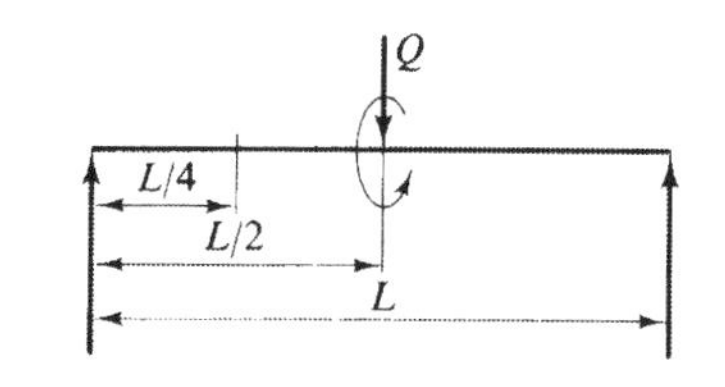

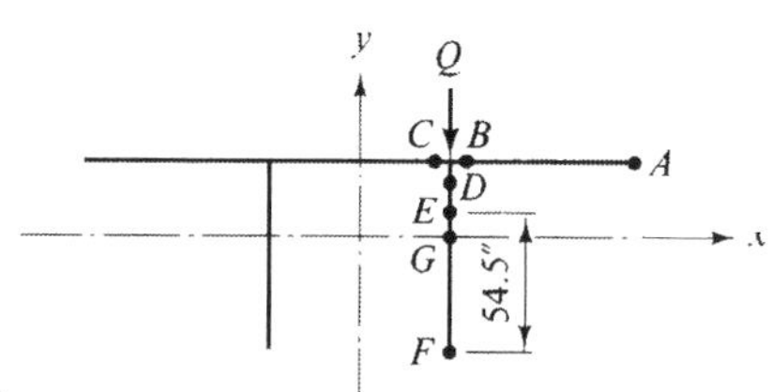

Fig. for Table 2.2.

Table 2.2 NORMAL STRESS ($\times 10^{-3}Q$)

	$z = L/4$					
	A	B	C	D	E	F
σ_b	−1.78	−1.78	−1.78	−1.78	−0.97	7.12
σ_w	0	0	0	0	0	0
Total	−1.78	−1.78	−1.78	−1.78	−0.97	7.12

	$z = L/2$					
	A	B	C	D	E	F
σ_b	−3.56	−3.56	−3.56	−3.56	−1.93	14.23
σ_w	−1.89	−0.63	−0.63	−0.63	0	6.23
Total	−5.45	−4.19	−4.19	−4.19	−1.93	20.46

Table 2.3 SHEAR STRESS ($\times 10^{-3}Q$)

	$z = L/4$					
	A	B	C	D	E	G
τ_b	0	3.56	1.78	5.34	5.58	5.70
τ_{sv}	± 6.0	± 6.0	± 6.0	± 6.0	± 6.0	± 6.0
τ_w	0	0.49	0.59	1.08	1.09	1.07
Total	6.0 -6.0	10.05 -1.95	8.37 -3.63	12.42 0.42	12.67 0.67	12.77 0.77

	$z = L/2$					
	A	B	C	D	E	G
τ_b	0	3.56	1.78	5.34	5.58	5.70
τ_{sv}	± 6.0	± 6.0	± 6.0	± 6.0	± 6.0	± 6.0
τ_w	0	0.53	0.65	1.18	1.19	1.18
Total	6.0 -6.0	10.09 -1.91	8.43 -3.57	12.52 0.52	12.77 0.77	12.88 0.88

Maximum stresses: $\sigma_{max} = 2.05 \times 10^{-2}Q$ (at point F), $\tau_{max} = 1.29 \times 10^{-2}Q$ (at point G)
Deflection: $\delta_A = \delta_{bending} + \delta_{torsion} = 2.98 \times 10^{-4}Q + 1.70 \times 10^{-4}Q = 4.68 \times 10^{-4}Q$ in

Problem 2.2

$$A = 32ta$$

$$I_x = 85.33ta^3$$

$$I_y = 309.33ta^3$$

$$K_T = 192ta^3$$

Make a cut as shown in Fig. 2.27(a).

Shear flow $q = \dfrac{V_\eta}{I_x}\left(7.0a^2t - \displaystyle\int_0^s \eta t\, ds\right)$ [Fig. 2.27(b)]

Shear center $\xi_0 = -0.25a$, $\eta_0 = 0$ [Fig. 2.27(d)]

Shear stress due to V_η [Fig. 2.27(c)]

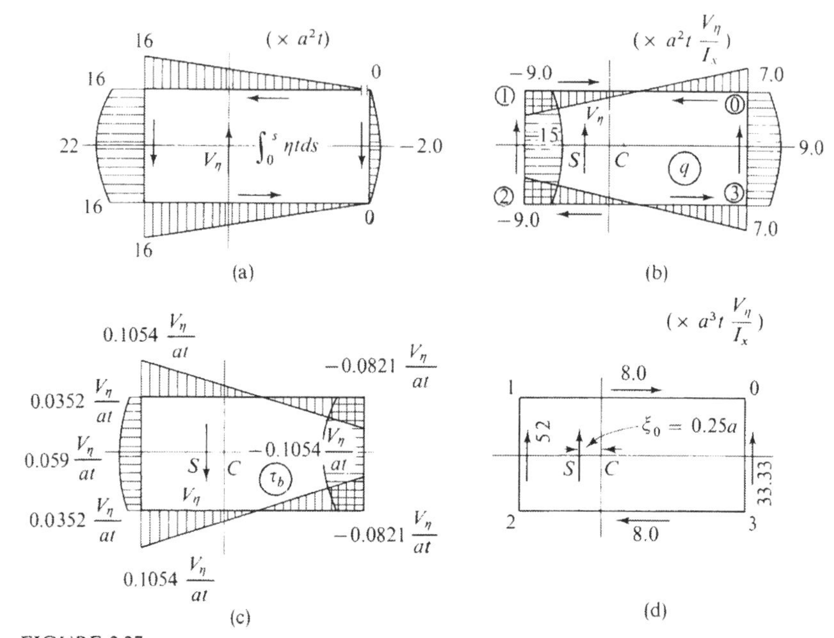

FIGURE 2.27
Figures for solution to Prob. 2.2

Problem 2.3

Assume the particular solution $\theta = A \sin(\pi/L)z$,

$$\theta = \frac{m_0 L^4}{\pi^3 EI_\omega + \pi GK_T L^2}\left[\frac{1}{\pi}\sin\frac{\pi}{L}z + \frac{\sinh\lambda z - \tanh\lambda L\left(\dfrac{z}{L} + \cosh\lambda z - 1\right)}{\tanh\lambda L - \lambda L}\right]$$

Problem 2.4

Denote the pole for the unit warping ω by $P(\xi_p, \eta_p)$, then the unit warping with the pole at the shear center $S(\xi_0, \eta_0)$ is given by the expression similar to that of Eq. (2.94) as

$$\omega_0 = \omega + (\eta_0 - \eta_p)(\xi - \xi_A) - (\xi_0 - \xi_p)(\eta - \eta_A) \tag{a}$$

in which $A(\xi_A, \eta_A)$ is the reference point ($s = 0$) which is common for both unit warpings, ω and ω_0. The warping normal strain is from Eq. (2.101), using Eq. (a),

$$\begin{aligned}\varepsilon_w = \theta''\left(\frac{1}{A}\int_0^E \omega_0\, dA - \omega_0\right) &= \theta''\left[\frac{S_\omega}{A} + (\eta_0 - \eta_p)\left(\frac{S_x}{A} - \xi_A\right) - (\xi_0 - \xi_p)\left(\frac{S_y}{A} - \eta_A\right)\right.\\ &\quad \left. - \omega - (\eta_0 - \eta_p)(\xi - \xi_A) + (\xi_0 - \xi_p)(\eta - \eta_A)\right] \qquad (b)\\ &= \theta''\left[\left(\frac{S_\omega}{A} - \omega\right) + (\eta_0 - \eta_p)\left(\frac{S_x}{A} - \xi\right) - (\xi_0 - \xi_p)\left(\frac{S_y}{A} - \eta\right)\right]\end{aligned}$$

Adding the bending strain ε_b which is referred to the coordinates ξ, η with its origin at center of the cross section

$$\varepsilon_b = \bar{\varepsilon}_0 + \eta\overline{\Phi}_\xi - \xi\overline{\Phi}_\eta \tag{c}$$

to the warping strain, the total normal strain is obtained

$$\begin{aligned}\varepsilon = \varepsilon_b + \varepsilon_w = \bar{\varepsilon}_0 + \left[\frac{S_\omega}{A} + (\eta_0 - \eta_p)\frac{S_x}{A} - (\xi_0 - \xi_p)\frac{S_y}{A}\right]\theta'' \\ + \eta[\overline{\Phi}_\xi + (\xi_0 - \xi_p)\theta''] - \xi[\overline{\Phi}_\eta + (\eta_0 - \eta_p)\theta''] - \omega\theta''\end{aligned} \tag{d}$$

Denoting

$$\begin{aligned}\varepsilon_0 &= \bar{\varepsilon}_0 + \left[\frac{S_\omega}{A} + (\eta_0 - \eta_p)\frac{S_x}{A} - (\xi_0 - \xi_p)\frac{S_y}{A}\right]\theta'' \\ \Phi_\xi &= \overline{\Phi}_\xi + (\xi_0 - \xi_p)\theta'' \\ \Phi_\eta &= \overline{\Phi}_\eta + (\eta_0 - \eta_p)\theta''\end{aligned} \tag{e}$$

Then, one obtains

$$\varepsilon = \varepsilon_0 + \eta\Phi_\xi - \xi\Phi_\eta - \omega\theta'' \tag{f}$$

Problem 2.5

From Eqs. (2.116) and (2.119), we obtain

$$\sigma = Ew = E\varepsilon_0 + E\eta\theta'_\xi - E\xi\theta'_\eta - E\omega\theta''_\zeta \tag{a}$$

Thus, from Eqs. (2.118) and (2.120), we have

$$M_\omega = \int_0^E \sigma\omega\, dA = ES_\omega\varepsilon_0 + EI_{\omega y}\theta'_\xi - EI_{\omega x}\theta'_\eta - EI_\omega\theta''_\zeta \tag{b}$$

While from Eq. (2.15), using Eq. (a), we find

$$\begin{aligned}\tau t &= -\int_0^s \frac{\partial\sigma t}{\partial z}\, ds + \tau_0 t_0 \\ &= -E\left[\varepsilon'_0\int_0^s t\, ds + \theta''_\xi\int_0^s \eta t\, ds - \theta''_\eta\int_0^s \xi t\, ds - \theta'''_\zeta\int_0^s \omega t\, ds\right] + \tau_0 t_0\end{aligned} \tag{c}$$

Thus, the twisting moment due to warping shear stress is

$$\begin{aligned}T_w &= \int_0^E \rho\tau t\, ds \\ &= -E\left[\varepsilon'_0\int_0^E \rho\left(\int_0^s t\, ds\right) ds + \theta''_\xi\int_0^E \rho\left(\int_0^s \eta t\, ds\right) ds\right. \\ &\quad \left. - \theta''_\eta\int_0^E \rho\left(\int_0^s \xi t\, ds\right) ds - \theta'''_\zeta\int_0^E \rho\left(\int_0^s \omega t\, ds\right) ds\right] + \tau_0 t_0\int_0^E \rho\, ds\end{aligned} \tag{d}$$

Integrating by parts, for example

$$\int_0^E \rho \left(\int_0^s \eta t \, ds \right) ds = \left[\left(\int_0^s \rho \, ds \right) \left(\int_0^s \eta t \, ds \right) \right]_0^E - \int_0^E \left(\int_0^s \rho \, ds \right) \eta t \, ds$$

$$= \omega_E \int_0^E \eta t \, ds - \int_0^E \omega \eta t \, ds = \omega_E S_y - I_{\omega y} \tag{e}$$

Then, Eq. (*d*) can be written as

$$\begin{aligned} T_w &= -E[\varepsilon_0'(\omega_E A - S_\omega) + \theta_\xi''(\omega_E S_y - I_{\omega y}) - \theta_\eta''(\omega_E S_x - I_{\omega x}) - \theta_\zeta'''(\omega_E S_\omega - I_\omega)] \\ &\quad + \omega_E \tau_0 t_0 \\ &= ES_\omega \varepsilon_0' + EI_{\omega y}\theta_\xi'' - EI_{\omega x}\theta_\eta'' - EI_\omega \theta_\zeta''' - \omega_E(EA\varepsilon_0' + ES_y\theta_\xi'' - ES_x\theta_\eta'' \\ &\quad - ES_\omega \theta_\zeta''' - \tau_0 t_0) \\ &= M_\omega' - \omega_E(EA\varepsilon_0' + ES_y\theta_\xi'' - ES_x\theta_\eta'' - ES_\omega\theta_\zeta''' - \tau_0 t_0) \end{aligned} \tag{f}$$

which is the expression given in Eq. (2.128*a*). It is known from Eq. (*f*) that, if the pole for ω is appropriately selected such that shear flow at the reference point A ($s = 0$) has the value

$$\tau_0 t_0 = EA\varepsilon_0' + ES_y\theta_\xi'' - ES_x\theta_\eta'' - ES_\omega\theta_\zeta''' \tag{g}$$

then, the twisting moment T_w is obtained directly from the bimoment M_ω by

$$T_w = M_\omega' \tag{h}$$

Problem 2.6

From Fig. 2.28

$$\begin{aligned} \varepsilon_0 &= \frac{\sqrt{(1 + w')^2 + (u')^2 + (v')^2}\, dz - dz}{dz} \\ &\approx \frac{1}{2}[1 + 2w' + (w')^2 + (u')^2 + (v')^2 - 1] \\ &= w' + \frac{1}{2}(u')^2 + \frac{1}{2}(v')^2 \end{aligned}$$

Note

$$w' \ll u'$$

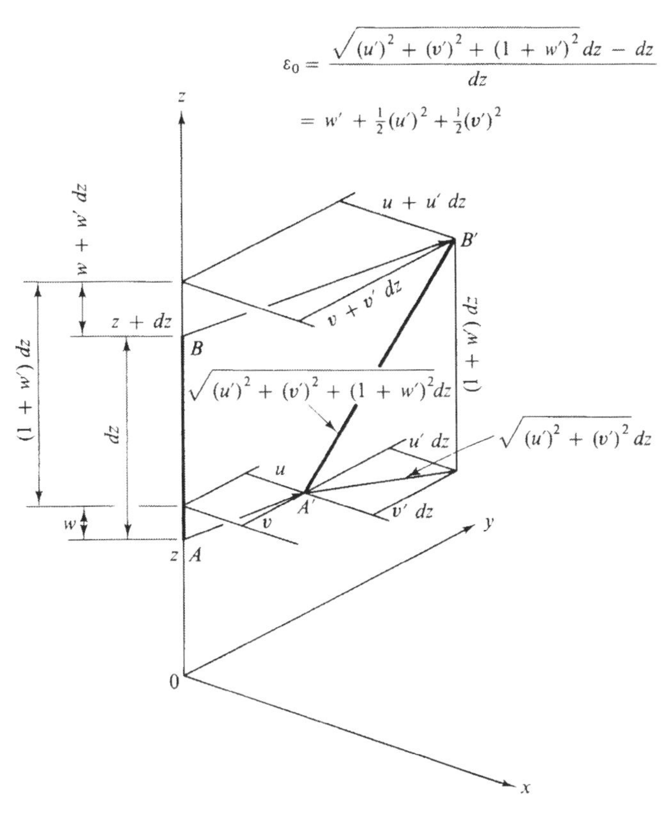

FIGURE 2.28
Normal strain

CHAPTER 3

3.1 Note that u_0 and u always satisfy the boundary conditions $(u)_0 = (u'')_0 = (u)_l = (u'')_l = 0$, and that θ_0 and θ always satisfy Eq. (3.8). The differential equations of minor axis bending and torsion are

$$-EI_y u'' = M(\theta + \theta_0)$$

$$GK_T \theta' - EI_\omega \theta''' = M(u' + u_0')$$

If u is eliminated from these, then

$$(M^2/EI_y)\theta + GK_T\theta'' - EI_\omega\theta'''' = Mu_0'' - (M^2/EI_y)\theta_0$$

Substituting for θ and θ_0 and dividing by $\sin \pi z/L$,

$$\left\{\frac{M^2}{EI_y} - \frac{\pi^2}{L^2}GK_T\left(1 + \frac{\pi^2 EI_\omega}{L^2 GK_T}\right)\right\}\theta_m = -\frac{\pi^2}{L^2}Mu_{mo} - \frac{M^2}{EI_y}\theta_{mo}$$

which can be rearranged as

$$\left(1-\frac{M^2}{M_e^2}\right)\theta_m=\frac{M^2}{M_e^2}\left(\theta_{mo}+\frac{\pi^2 EI_y/L^2}{M}u_{mo}\right)$$

If Eq. (3.205) is substituted, then this becomes

$$\left(1-\frac{M^2}{M_e^2}\right)\theta_m=\frac{M^2}{M_e^2}\theta_{mo}\left(1+\frac{M_e}{M}\right)$$

whence

$$\frac{\theta_m}{\theta_{mo}}=\frac{M/M_e}{1-M/M_e}$$

which is the second of Eq. (3.207).

If this and Eq. (3.205) are substituted in the minor axis bending equation, then, after some rearrangement, the first of Eq. (3.207) can be obtained.

Equations (3.207) are plotted in Fig. 3.5 (page 77).

3.2 Because lateral deflection and twist are prevented at midspan, each halfspan buckles independently as a simply supported beam with end moments $M = PL/4$ and $\kappa M = 0$. Using Eq. (3.31), $\beta = 1.75 - 1.05 \times 0 + 0.3 \times 0^2 = 1.75$. Using Eq. (3.30),

$$M_e=\frac{1.75\pi\sqrt{30\,000\times 57.5\times 12\,000\times 1.97}}{35\times 12/2}$$

$$\times\sqrt{1+\frac{\pi^2\times 30\,000\times 5968}{(35\times 12/2)^2\times 12\,000\times 1.97}}$$

$$= 8679 \text{ kip in}$$

$$P_e=\frac{8679\times 4}{35\times 12}=82.7\text{ kip}$$

3.3 The section rigidities are

$$EI_y = 30\,000\times 1^3\times 10/12 = 25\,000\text{ kip in}^2$$

$$GK_T = 12\,000\times 10\times 1^3/3 = 40\,000\text{ kip in}^2$$

$$EI_\omega = 0 \text{ (narrow rectangular section)}$$

Therefore

$$\text{beam parameter} = \sqrt{\pi EI_\omega/L^2 GK_T} = 0$$

and

$$\delta=\frac{5}{5\times 12}\sqrt{\frac{25\,000}{40\,000}}=0.066$$

Interpolating in Fig. 3.8 (page 81),

$$\frac{P_e L^2}{\sqrt{EI_y GK_T}} \approx 15$$

and so

$$P_e = \frac{\sqrt{25\,000 \times 40\,000 \times 15}}{(5 \times 12)^2} = 131.8 \text{ kip.}$$

3.4 Using Fig. 3.10 (page 84) with $B_2 = 0$

$$h = 20.99 - 2 \times 0.615/2 = 20.375 \text{ in}$$

$$\bar{y} = \frac{0 + (20.99 - 2 \times 0.615)(20.99 - 0.615) \times 0.400/2}{(8.24 \times 0.615) + 0 + (20.99 - 2 \times 0.615) \times 0.400} = 6.21 \text{ in}$$

$$\alpha = 0$$

$$y_0 = 0 - 6.21 = -6.21 \text{ in}$$

$$I_x = 8.24 \times 0.615^3/12 + 8.24 \times 0.615 \times 6.21^2$$
$$+ (20.99 - 2 \times 0.615)^3 \times 0.4/12$$
$$+ (20.99 - 2 \times 0.615) \times 0.4 \times [(20.99 - 0.615)/2 - 6.21]^2 = 577.8 \text{ in}^4$$

$$\beta_x = \{0 - 6.21(8.24^3 \times 0.615/12 + 8.24 \times 0.615 \times 6.21^2)$$
$$+ (20.375 - 6.21 - 0.615/2)^4 \times 0.4/4 - (6.21 - 0.615/2)^4 \times 0.4/4\}/$$
$$577.8 - 2 \times (-6.21) = 16.19 \text{ in}$$

$$I_{w} = 0$$

$$I_y = 8.24^3 \times 0.615/12 + (20.99 - 2 \times 0.615) \times 0.4^3/12 = 28.78 \text{ in}^4$$

$$K_T \approx 8.24 \times 0.615^3/3 + (20.99 - 2 \times 0.615) \times 0.4^3/3 = 1.06 \text{ in}^4$$

Using Eq. (3.45)

$$\gamma_m = \pm \frac{16.19}{12 \times 12}\sqrt{\frac{30\,000 \times 28.78}{12\,000 \times 1.06}} = \pm 0.926$$

Using Eq. (3.44) for the case when the flange is in compression ($\gamma_m = +0.926$),

$$M_e = \frac{\pi\sqrt{30\,000 \times 28.78 \times 12\,000 \times 1.06}}{12 \times 12}\left\{\sqrt{1 + 0 + \left(\frac{\pi \times 0.926}{2}\right)^2}\right.$$
$$\left. + \frac{\pi \times 0.926}{2}\right\} = 7361 \text{ kip in}$$

When the flange is in tension, $\gamma_m = -0.926$, and so $M_e = 710$ kip in. An approximate solution can be obtained from Fig. 3.11 (page 85).

$$\text{Beam parameter} = \sqrt{\frac{\pi^2 \times 30\,000 \times 28.78 \times 20.375^2}{4 \times (12 \times 12)^2 \times 12\,000 \times 1.06}} = 1.83$$

When the flange is in compression, $\rho = 1.0$, and $M_e L/\sqrt{EI_y GK_T} \approx 9.3$. Thus

$$M_e \approx \frac{\sqrt{30\,000 \times 28.78 \times 12\,000 \times 1.06}}{12 \times 12} \times 9.3 = 6800 \text{ kip in}$$

When the flange is in tension, $\rho = 0$, and $M_e L/\sqrt{EI_y GK_T} \approx 1.0$. Thus

$$M_e \approx 730 \text{ kip in.}$$

3.5
$$\sqrt{\frac{\pi^2 EI_\omega}{L^2 GK_T}} = \sqrt{\frac{\pi^2 \times 30\,000 \times 5968}{(12 \times 12)^2 \times 12\,000 \times 1.97}} = 1.90$$

Using the curve in Fig. 3.13 (page 88) for top flange loading.

$$P_e L^2/\sqrt{EI_y GK_T} \approx 3$$

and so

$$P_e = \frac{3 \times \sqrt{30\,000 \times 57.5 \times 12\,000 \times 1.97}}{(12 \times 12)^2} = 29.2 \text{ kip.}$$

3.6 Simplify the proof outlined in Sect. 3.8.3 for beam-columns by putting $P = 0$.

3.7 Note that Eqs. (3.209) satisfy the end boundary conditions $(u)_o = (u'')_0 = (\theta)_0 = (\theta'')_0 = 0$ and the midspan symmetry conditions $(u')_m = (\theta')_m = 0$. The differential equation of minor axis bending is

$$-EI_y u'' = M_e \theta - (\alpha_t z/2)(u_m + \bar{b}\theta_m)$$

If Eqs. (3.209) are substituted into this, it becomes

$$-EI_y \frac{\pi^2}{k^2 L^2} \frac{\sin \pi z/kL}{(\pi/k) \cos \pi/2k} = \frac{\pi^2 EI_y}{k^2 L^2}\left(\frac{z}{L} - \frac{\sin \pi z/kL}{(\pi/k) \cos \pi/2k}\right)$$
$$- \frac{\alpha_t z}{2}\left(1 + \frac{\bar{b}\pi^2 EI_y/k^2 L^2}{M_e}\right)\left(\frac{1}{2} - \frac{\tan \pi/2k}{\pi/k}\right)$$

which is simplified to

$$0 = \frac{\pi^2 EI_y}{k^2 L^2} - \frac{\alpha_t L}{2}\left(1 + \frac{\bar{b}\pi^2 EI_y/k^2 L^2}{M_e}\right)\left(\frac{1}{2} - \frac{\tan \pi/2k}{\pi/k}\right)$$

When Eq. (3.70) is substituted, this becomes Eq. (3.68).

The differential equation of torsion is

$$GK_T \theta' - EI_\omega \theta''' = M_e u' - (\alpha_t \bar{b}/2)(u_m + \bar{b}\theta_m) - (\alpha_r/2)\theta_m$$

If Eqs. (3.209) are substituted into this, it becomes

$$\frac{\pi^2 EI_y/k^2L^2}{M_e}\left\{GK_T\left(\frac{1}{L}-\frac{\pi}{kL}\frac{\cos \pi z/kL}{(\pi/k)\cos \pi/2k}\right)-\frac{\pi^3 EI_\omega}{k^3L^3}\frac{\cos \pi z/kL}{(\pi/k)\cos \pi/2k}\right\}$$

$$= M_e\left\{\frac{1}{L}-\frac{\pi}{kL}\frac{\cos \pi z/kL}{(\pi/k)\cos \pi/2k}\right\}-\left\{\frac{\alpha_t \bar{b}}{2}\left(1+\frac{\bar{b}\pi^2 EI_y/k^2L^2}{M_e}\right)\right.$$

$$\left.+\frac{\alpha_r}{2}\frac{\pi^2 EI_y/k^2L^2}{M_e}\right\}\left(\frac{1}{2}-\frac{\tan \pi/2k}{\pi/k}\right)$$

and when Eq. (3.70) is substituted, this simplifies to

$$0=\frac{\pi^2 EI_y/k^2L^2}{M_e}\frac{\pi^2 EI_\omega}{k^2L^3}-\left\{\frac{\alpha_t}{2}\bar{b}\left(1+\frac{\bar{b}}{b_0}\right)+\frac{\alpha_r}{2b_0}\right\}\left(\frac{1}{2}-\frac{\tan \pi/2k}{\pi/k}\right)$$

If Eqs. (3.68) and (3.70) are substituted, then this becomes

$$\frac{\alpha_r}{2b_0}\left(\frac{1}{2}-\frac{\tan \pi/2k}{\pi/k}\right)=\frac{\pi^2}{k^2L^2}\left\{\frac{EI_\omega}{b_0 L}-\frac{EI_y\bar{b}}{L}\right\}$$

and if $I_y = 4I_\omega/h^2$ is substituted, then this can be rearranged as Eq. (3.69).

3.8 From Prob. 3.2, $\kappa = 0$ and $M_e = 8679$ kip in. From Fig. 3.50, $M_p = \sigma_y Z_x = 45.0 \times 144.1 = 6485$ kip in. Using Eq. (3.80),

$$\frac{M_i}{M_p} = 0.70 + 0.30\,\frac{(1 - 0.70 \times 6485/8679)}{(0.61 + 0.30 \times 0 + 0.07 \times 0)} = 0.935$$

and so

$$M_i = 0.935 \times 6485 = 6063 \text{ kip in}$$

Thus

$$P_i = 4 \times 6063/(35 \times 12) = 57.7 \text{ kip.}$$

3.9 Using $\kappa = 0$ in Eq. (3.31), $C_b = 1.75$

$$\sqrt{\frac{\sigma_y(L/r_t)^2}{286\,200 C_b}} = \sqrt{\frac{45 \times (35 \times 12)/(2 \times 1.93)^2}{286\,200 \times 1.75}} = 1.031 < 4/3$$

Therefore, use Eqs. (3.85) and (3.88). After inserting C_b into Eq. (3.85),

$$\sigma_{bc} = \frac{12\,000 \times 1.75}{(35 \times 12/2) \times 20.99/(8.24 \times 0.615)} = 24.1 \text{ ksi}$$

After inserting C_b into Eq. (3.88),

$$\sigma_{bc} = \frac{2 \times 45.0}{3}\left\{1 - \frac{1}{2}\left(\frac{3}{4}\right)^2 \times \frac{45 \times (35 \times 12/2 \times 1.93)^2}{286\,200 \times 1.75}\right\}$$

$$= 30.0\left\{1 - \frac{1}{2}\left(\frac{3}{4}\right)^2 \times 1.031^2\right\} = 21.0 \text{ ksi} < 24.1$$

Therefore, $\sigma_{bc} = 24.1$ ksi. Checking Eq. (3.89), σ_{bc} must not exceed $0.60\sigma_y = 0.60 \times 45.0 = 27.0$ ksi, and this is so. Hence

$$P_d = \frac{4\sigma_{bc}S_x}{L} = \frac{4 \times 24.1 \times 126.4}{35 \times 12} = 29.0 \text{ kip.}$$

3.10 (a) Method of Taylor, Dwight and Nethercot (1974). In this method, M_e is determined for the uniform moment condition (for which $\beta = 1.00$), instead of for the actual moment gradient (for which $\beta = 1.75$). Thus, adapting the solution of Prob. 3.2,

$$M_e = \frac{8679 \times 1.00}{1.75} = 4959 \text{ kip in}$$

From the solution of Prob. 3.8, $M_p = 6485$ kip in. Substituting in Eq. (3.94),

$$\eta = 0.007\sqrt{\frac{\pi^2 \times 30\,000}{45.0}}\left(\sqrt{\frac{6485}{4959}} - 0.4\right) = 0.422$$

and so

$$\frac{1 + (1 + \eta)M_e/M_p}{2} = \frac{1 + (1 + 0.422) \times 4959/6485}{2} = 1.044$$

Substituting in Eq. (3.93) and inserting $\beta = 1.75$,

$$\frac{M_u}{M_p} = 1.75\left\{1.044 - \sqrt{1.044^2 - \frac{4959}{6485}}\right\} = 0.830$$

and so

$$P_u = \frac{4(M_u/M_p)M_p}{L} = \frac{4 \times 0.830 \times 6485}{35 \times 12} = 51.3 \text{ kip.}$$

(b) Method of Nethercot and Trahair (1976b). From the solution of Prob. 3.8,

$$M_c/M_p = M_i/M_p = 0.935$$

and so

$$\sqrt{M_p/M_i} = 1.034 < 1.1$$

Therefore, using Eq. (3.110),

$$M_u/M_p = (1.57 \times 0.935 - 0.57) \times 0.935 = 0.840$$

and so

$$P_u = \frac{4 \times 0.840 \times 6485}{35 \times 12} = 51.9 \text{ kip.}$$

3.11 $M_1 = 0, M_2 = 75P$ kip in, $M_3 = -60P$ kip in, $\kappa_{12} = 0$, $\kappa_{23} = -60/75 = -0.800$, $\beta_{12} = 1.75$, $\beta_{23} = 2.56$ using Eq. (3.31). Assume $k_{12} = k_{23} = 1.00$ as a first approximation. Making use of the solution of Prob. 3.2,

$$M_{e12} = 8679 \text{ kip in}$$

and

$$M_{e23} = 8679 \times 2.56/1.75 = 12\,696 \text{ kip in}$$

$$P_{s12} = \frac{8679}{75} = 115.7 \qquad P_{s23} = \frac{12\,696}{75} = 169.3$$

and so $P_{ms} = 115.7$ and critical segment is 12.

$$\alpha_{12} = \frac{2 \times 30\,000 \times 57.5}{17.5 \times 12} = 16\,429 \text{ kip in}$$

$$\alpha_{23} = \frac{3 \times 30\,000 \times 57.5}{17.5 \times 12}\left(1 - \frac{115.7}{169.3}\right) = 7801 \text{ kip in}$$

$$\therefore G_2 = \frac{16\,429}{7801} = 2.11, \quad G_1 = \infty$$

Using Fig. 3.16 (page 92), $k_{12} = 0.92$

$$\therefore M_{e12} = \frac{1.75\pi\sqrt{30\,000 \times 57.5 \times 12\,000 \times 1.97}}{17.5 \times 0.92 \times 12} \times \sqrt{1 + \frac{\pi^2 \times 30\,000 \times 5968}{(17.5 \times 0.92 \times 12)^2 \times 12\,000 \times 1.97}}$$

$$= 9957 \text{ kip in}$$

$$\therefore P_e = \frac{9957}{75} = 132.8 \text{ kip}$$

3.12 $P_{ex} = \pi^2 \times 30\,000 \times 1330/(17.5 \times 12)^2 = 8930$ kip

$P_{ey} = \pi^2 \times 30\,000 \times 57.5/(17.5 \times 12)^2 = 386$ kip

$r_0 = \sqrt{(1330 + 57.5)/18.35} = 8.7$ in

$$P_z = \frac{12\,000 \times 1.97}{8.7^2}\left(1 + \frac{\pi^2 \times 30\,000 \times 5968}{(17.5 \times 12)^2 \times 12\,000 \times 1.97}\right) = 843 \text{ kip}$$

$M_e = 8679$ kip in from the solution of Prob. 3.2. An approximate solution can be obtained from Eq. (3.135). To solve this, first ignore the $(1 - P/P_{ex})$ term, whence

$$\frac{P}{386} + \frac{20P}{8679} = 1 \qquad \text{whence} \qquad P \approx 204 \text{ kip}$$

Using this to approximate the $(1 - P/P_{ex})$ term,

$$\frac{P}{386} + \frac{20P}{(1 - 204/8930) \times 8679} = 1$$

whence $P = 202$ kip, approximately. A more accurate solution can be obtained from Eq. (3.133) which becomes

$$\frac{(20P)^2}{8679^2} = \left(1 - \frac{P}{8930}\right)\left(1 - \frac{P}{386}\right)\left(1 - \frac{P}{843}\right)$$

This cubic equation can be solved iteratively starting with the approximate value of $P = 202$ kip. The final solution is $P = 231$ kip.

3.13 (a) $$L/r_y = 17.5 \times 12/1.77 = 118.6$$

Using this and $\sigma_y = 45.0$ ksi, the maximum working load stress is $\sigma_a = 10.6$ ksi. Thus

$$P_u \approx 1.7 \times 10.6 \times 18.35 = 331 \text{ kip.}$$

Using the solution of Prob. 3.9,

$$\sigma_{bc} = 24.1 \text{ ksi}$$

Thus,

$$M_{ux}/C_m \approx 1.7 \times 24.1 \times 126.4 = 5179 \text{ kip in}$$

Using the solution of Prob. 3.12,

$$P_{ex} = 8930 \text{ kip.}$$

Thus, the modified form of Eq. (3.151) becomes

$$\frac{P}{331} + \frac{20P}{(1 - P/8930) \times 5179} = 1$$

An approximate solution [ignoring the $(1 - P/8930)$ term] is $P = 145$ kip. The accurate solution is $P = 144$ kip.

(b) $$L/r_x = 17.5 \times 12/8.54 = 24.6$$

Using this and the same approximations for P_u and M_{ux}/C_m as those used in (b) above, then Eq. (3.154) becomes

$$\frac{P}{331} + \frac{20P(1 - 100/24.6^2)}{(1 - P/8930) \times 5179} = 1$$

An approximate solution [ignoring the $(1 - P/8930)$ term] is 160 kip. The accurate solution is $P = 159$ kip.

(c) $$M_{yy} = \sigma_y S_x = 45.0 \times 126.4 = 5688 \text{ kip in}$$

$$P_y = \sigma_y A = 45.0 \times 18.35 = 826 \text{ kip.}$$

Thus, Eq. (3.152) becomes

$$P_{fy}/826 = 1 - 20P_{fy}/5688$$

the solution of which is $P_{fy} = 212$ kip. Using the solution of Prob. 3.12, the approximate elastic critical load is 202 kip. Thus, Eq. (3.153) becomes

$$\frac{P_i}{212} = 1 - \frac{1}{4} \times \frac{212}{202}$$

whence $P_i = 156$ kip.

CHAPTER 4

4.1 For a cylindrical tubular beam column,

$$P_x = P_y$$

Thus from Eq. (4.103),

$$C = 0$$

which means no torsional deformation.

4.2 It is clear from the first two equations of Eq. (4.104).

4.3 The condition $P_x = P_z$ is given

$$\frac{\pi^2 EI_x}{L^2} = \left(\frac{\pi^2 EI_\omega}{L^2} + GK_T\right) \Big/ r_0^2$$

from which

$$L = \pi \sqrt{\frac{E}{G} \frac{r_0^2 I_x - I_\omega}{K_T}}$$

For section W14 × 43,

$$\frac{E}{G} = 2.6, \quad I_x = 429 \text{ in}^4, \quad I_\omega = 1950 \text{ in}^6$$

$$r_0 = 6.15 \text{ in}, \quad K_T = 1.05 \text{ in}^4$$

thus

$$L = \pi \sqrt{(2.6) \frac{(6.15)^2(429) - 1950}{1.05}} = 591 \text{ in}$$

The common buckling load is

$$P_{cr} = \frac{\pi^2 EI_x}{L^2} = \frac{\pi^2(29\,000)(429)}{(591)^2} = 351 \text{ kip}$$

CHAPTER 5

5.1 (omitted)
5.2 (omitted)
5.3 (omitted)
5.4 (omitted)
5.5 (omitted)
5.6 (omitted)
5.7 (omitted)

CHAPTER 6

6.1 Take one of the generalized stresses, say, M_x, which is a function of the generalized strains,

$$M_x = M_x(\varepsilon_o, \Phi_x, \Phi_y, \Phi_\omega) \tag{6.73}$$

This relation in the rate form can be written using the Taylor series expansion,

$$\dot{M}_x = \frac{\partial M_x}{\partial \varepsilon_o}\dot{\varepsilon}_o + \frac{\partial M_x}{\partial \Phi_x}\dot{\Phi}_x + \frac{\partial M_x}{\partial \Phi_y}\dot{\Phi}_y + \frac{\partial M_x}{\partial \Phi_\omega}\dot{\Phi}_\omega \tag{6.74}$$

which gives the second row of the tangent stiffness matrix, and the other rows can be obtained analogously.

6.2 Assuming that a plane in each leg, AB or BC, remains plane after deformation, the normal strains at ends of the legs, ε_A, ε_B, ε_C can be a set of the generalized strains and the corresponding generalized stresses are forces at these points, F_A, F_B, F_C. The usual notations for the generalized stresses (P, M_x, M_y) and the generalized strains $(\varepsilon_o, \Phi_x, \Phi_y)$ can be obtained as the linear combinations of $(\varepsilon_A, \varepsilon_B, \varepsilon_C)$ and (F_A, F_B, F_C), respectively.

6.3 (omitted)

CHAPTER 7

7.1 (omitted)

7.2 The exact maximum moment is given by Eq. (7.61) of Vol. 1, which can be expanded into Taylor series as

$$M_{\max} = \bar{M} \sec\left(\frac{\pi}{2}\sqrt{\frac{P}{P_E}}\right) = \bar{M}\left\{1 + \frac{\pi^2}{8}\frac{P}{P_E} + \frac{5\pi^4}{384}\left(\frac{P}{P_E}\right)^2 + \cdots\right\}$$

From this it is known that Eq. (7.42) gets close to the exact value when the axial force is small comparing to the buckling load P_E or P_{ex}. Accuracy of the approximate deflection method depends on whether the deflection function is close to column deflection or beam deflection and on the magnitude of the axial force P/P_y.

7.3 When an energy method is applied, the assumed deflection functions are preferably orthogonal as was discussed in Chap. 4 of Vol. 1. When the equilibrium equations are applied only at some specific locations along the beam-column, the assumed deflection can be any function, orthogonal or not.

7.4 (omitted)

CHAPTER 8

8.1 From Eqs. (8.53), (8.54) and (8.55), the relationship between the internal forces and the twisting center and centroidal strain to be adjusted are given by

$$\begin{Bmatrix} \Delta \dot{N}_w \\ \Delta \dot{M}_{\xi w} \\ -\Delta \dot{M}_{\eta w} \end{Bmatrix} = \begin{bmatrix} \int E_t\, dA & \int E_t \eta\, dA & \int E_t \xi\, dA \\ \int E_t \eta\, dA & \int E_t \eta^2\, dA & \int E_t \xi\eta\, dA \\ \int E_t \xi\, dA & \int E_t \xi\eta\, dA & \int E_t \xi^2\, dA \end{bmatrix} \begin{Bmatrix} \Delta \dot{\varepsilon}_{w0} \\ \dot{\theta}'' \Delta\xi_0 \\ -\dot{\theta}''\Delta\eta_0 \end{Bmatrix} + \dot{\theta}'' \begin{Bmatrix} \int E_t \xi\eta\, dA \\ \int E_t \xi\eta^2\, dA \\ \int E_t \xi^2 \eta\, dA \end{Bmatrix}$$

Assuming elastic section properties, i.e.,

$$EA = \int E_t\, dA, \quad EI_x = \int E_t \eta^2\, dA, \quad EI_y = \int E_t \xi^2\, dA$$

and

$$\int E_t \eta\, dA = \int E_t \xi\, dA = \int E_t \xi\eta\, dA = \int E_t \xi^2 \eta\, dA = \int E_t \xi\eta^2\, dA = 0$$

Thus, Eqs. (8.56) are derived.

8.2 $GK_T = \dfrac{G}{3}\sum_k b_k t_k^3$ (for elastic portion only)

$$= \frac{1}{3}\Sigma G t^2 \Delta A$$

$$= \frac{1}{3}\int G_t t^2\, dA \quad \text{(for the entire cross section)}$$

For an I-shaped section, we can assume the warping function in Eq. (2.102) as

$$\omega_n = \frac{1}{A}\int_0^E \omega_0\, dA - \omega_0 = -\omega_0$$

and from Eq. (2.94) with the reference point A at the centroid, $\xi_A = \eta_A = 0$,

$$\omega_0 = -\xi\eta + \xi\eta_0 - \xi_0\eta$$

Further the warping strain is from Eq. (8.53a),

$$\varepsilon_w = \varepsilon_{w0} - \theta''\omega_0$$

Next, assume the center of torsion on the web plate, $\xi_0 = 0$ and $\varepsilon_{w0} = 0$. Thus

$$\omega_0 = -\xi(\eta - \eta_0)$$
$$\varepsilon_w = -\theta''\omega_0 = \theta''\xi(\eta - \eta_0)$$

Now the bimoment is from Eq. (2.118),

$$\begin{aligned} M_\omega &= \int \sigma_w \omega_n \, dA \\ &= -\int E_t \varepsilon_w \omega_0 \, dA \\ &= \theta'' \int E_t \xi^2 (\eta - \eta_0)^2 \, dA \end{aligned}$$

This is the expression given in the problem.

While the bimoment can be expressed in another way,

$$\begin{aligned} M_\omega &= \int \sigma_w \omega_n \, dA \\ &= -\int \sigma_w \omega_0 \, dA \\ &= \int \sigma_w \xi (\eta - \eta_0) \, dA \end{aligned}$$

Since no warping is assumed on the web plate,

$$M_\omega = (\eta - \eta_0)_{\mathrm{UF}} \int_{\mathrm{UF}} \sigma_w \xi \, dA + (\eta - \eta_0)_{\mathrm{LF}} \int_{\mathrm{LF}} \sigma_w \xi \, dA$$

in which the integral parts are the in-plane bending moments of the upper and lower flange plates, respectively.

$$M_{\omega\eta\mathrm{UF}} = \int_{\mathrm{UF}} \sigma_w \xi \, dA \quad \text{on} \quad \eta = \frac{D - t}{2}$$

$$M_{\omega\eta\mathrm{LF}} = -\int_{\mathrm{LF}} \sigma_w \xi \, dA \quad \text{on} \quad \eta = -\frac{D - t}{2}$$

thus

$$M_\omega = \left(\frac{D - t}{2} - \eta_0\right) M_{\omega\eta\mathrm{UF}} + \left(\frac{D - t}{2} + \eta_0\right) M_{\omega\eta\mathrm{LF}}$$

This is the expression given in Eq. (8.60).

8.3 From Eq. (8.69),

$$\{\xi''\} = -P[G(n \times n)]^{-1}\{\xi\}$$

or

$$\begin{Bmatrix} \xi_1'' \\ \xi_2'' \\ \xi_3'' \\ \cdot \\ \cdot \\ \xi_{n-1}'' \\ \xi_n'' \end{Bmatrix} = -\frac{1}{\lambda^2} \begin{bmatrix} 2 & -1 & & & & & \\ -1 & 2 & -1 & & (0) & & \\ & -1 & 2 & -1 & & & \\ & & & \cdots & & & \\ & & & \cdots & & & \\ & (0) & & & -1 & 2 & -1 \\ & & & & & -2 & 2 \end{bmatrix} \begin{Bmatrix} \xi_1 \\ \xi_2 \\ \xi_3 \\ \cdot \\ \cdot \\ \xi_{n-1} \\ \xi_n \end{Bmatrix}$$

Take the inverse of $[G(n \times n)]^{-1}$ and get $[G(n \times n)]$ which is identical to Eq. (8.71).

CHAPTER 9

9.1 For a loading condition which is near the ultimate state of a beam-column, there exist two different deflection configurations. One corresponds to a loading or stable state and the other to an unloading or unstable state as can be seen in Fig. 9.12 (page 428). Thus, at the ultimate or collapse state, the difference between the two deflection configurations must be infinitesimally small.

9.2 The strain distribution on the plate element can be expressed by

$$\varepsilon = \varepsilon_o + \Phi_\xi \eta$$

from which the boundary of elastic and plastic zones is known

$$\eta_y = \frac{1}{\Phi_\xi}(\varepsilon_y - \varepsilon_o) \tag{a}$$

and the normal stress distribution is

$$\sigma = \begin{cases} E(\varepsilon_o + \Phi_\xi \eta) & \left(-\dfrac{B}{2} \le \eta \le \eta_y\right) \\ E\varepsilon_y & \left(\eta_y \le \eta \le \dfrac{B}{2}\right) \end{cases} \tag{b}$$

Thus, the normal force P on the plate element is

$$\begin{aligned} P &= \int \sigma\, dA = t \int_{-B/2}^{\eta_y} E(\varepsilon_o + \Phi_\xi \eta)\, d\eta + t \int_{\eta_y}^{B/2} E\varepsilon_y\, d\eta \\ &= Et\varepsilon_o\left(\eta_y + \frac{B}{2}\right) + \frac{1}{2}Et\Phi_\xi\left(\eta_y^2 - \frac{B^2}{4}\right) + Et\varepsilon_y\left(\frac{B}{2} - \eta_y\right) \end{aligned} \tag{c}$$

Substitute Eq. (a) into Eq. (c) and rearrange

$$P = \frac{Et}{2\Phi_\xi}\left[3\varepsilon_o^2 + 2\left(\varepsilon_y + \frac{B}{2}\Phi_\xi\right)\varepsilon_o - \left(\varepsilon_y - \frac{B}{2}\Phi_\xi\right)^2\right] \tag{d}$$

from which Eq. (9.32) is obtained.

CHAPTER 10

10.1 (omitted)

10.2 Designate the symmetric, single curvature, major-axis end-moments by M_0. As there are no transverse loads ($q_x = q_y = 0$) and no axial load ($N = 0$) acting on the beam, Eq. (10.51) reduces to

$$M_Y = 0 \qquad M_X = M_0$$
$$M''_\omega + [(GK_T + \bar{K})\theta']' - u''M_0 = 0 \tag{a}$$

Taking the shear center of the section as the pole and the principal axes as the axes ξ, η, we have the following orthogonal conditions:

$$\int \xi\, dA = \int \eta\, dA = \int \xi\eta\, dA = 0$$
$$\int \omega\, dA = \int \omega\xi\, dA = \int \omega\eta\, dA = 0 \tag{b}$$

where the integration is carried for the entire section that remains elastic. With the help of these conditions and making use of the relations (10.26), Eqs. (10.67) could be simplified to

$$S_Y = \int X\, dA = \cos\theta \int \xi\, dA - \sin\theta \int \eta\, dA = 0$$
$$S_X = \int Y\, dA = 0 \qquad S_\omega = \int \omega\, dA = 0$$
$$I_Y = \int X^2\, dA = I_y \cos^2\theta + I_x \sin^2\theta$$
$$I_X = \int Y^2\, dA = I_y \sin^2\theta + I_x \cos^2\theta$$
$$I_{XY} = \int XY\, dA = (I_y - I_x)\sin\theta\cos\theta \tag{c}$$
$$I_\omega = \int \omega^2\, dA$$

$$I_{\omega X} = \int \omega X \, dA = \cos\theta \int \omega\xi \, dA - \sin\theta \int \omega\eta \, dA = 0$$

$$I_{\omega Y} = \int \omega Y \, dA = 0$$

where

$$I_y = \int \xi^2 \, dA \qquad I_x = \int \eta^2 \, dA \tag{d}$$

By substituting these relations in Eq. (10.68), we have

$$\begin{aligned} \tilde{I}_{\omega X} = \tilde{I}_{\omega Y} = 0 \qquad \tilde{I}_\omega = I_\omega \\ \tilde{I}_Y = I_Y \qquad \tilde{I}_X = I_X \qquad \tilde{I}_{XY} = I_{XY} \end{aligned} \tag{e}$$

With the help of Eqs. (e) and (c), the generalized stresses M_Y, M_X and M_ω become [see Eq. (10.70)]:

$$\begin{aligned} M_Y &= Eu''(I_y \cos^2\theta + I_x \sin^2\theta) + Ev''(I_y - I_x)\sin\theta\cos\theta \\ M_X &= -Eu''(I_y - I_x)\sin\theta\cos\theta - Ev''(I_y\sin^2\theta + I_x\cos^2\theta) \\ M_\omega &= -EI_\omega\theta'' \end{aligned} \tag{f}$$

Also from its definition given in Eq. (10.45), we have

$$\bar{K} = 0 \tag{g}$$

for the doubly symmetric section under consideration.

With the help of relations (f), (g) and (a), we obtain an equation which is only a function of θ

$$EI_\omega\theta'''' - GK_T\theta'' - \frac{M_0^2}{E}\left(\frac{1}{I_y} - \frac{1}{I_x}\right)\sin\theta\cos\theta = 0 \tag{h}$$

With the simply supported end conditions ($\theta = \theta'' = 0$), and assuming that θ is small, the characteristic value of the above differential equation furnishes the lateral buckling moment of the beam

$$M_{0\,cr} = \frac{\pi}{L}\sqrt{\frac{1}{(1 - I_y/I_x)} EI_y GK_T\left(1 + \frac{\pi^2 EI_\omega}{GK_T L^2}\right)} \tag{i}$$

The corresponding critical or buckling moment obtained by assuming a linear deformation field defined by Eq. (10.26) is (see Chap. 3 for details):

$$M_{0\,cr} = \frac{\pi}{L}\sqrt{EI_y GK_T\left(1 + \frac{\pi^2 EI_\omega}{GK_T L^2}\right)} \tag{j}$$

Equation (i) unlike Eq. (j) explicitly shows that the phenomenon of lateral buckling of I-beams is possible only for moments acting about the major axis ($I_y < I_x$). When $I_y \ll I_x$ the two equations yield the same value for the elastic critical moment; such is the case for thin rectangular sections. For I sections and more so for wide flange sections, Eq. (i) based on the non-linear deformation field results in a higher critical moment than that given by Eq. (j) based on linear deformation field (see also Fig. 10.31 and 10.32, pages 494–495).

10.3 (omitted)

10.4 From Fig. 10.38, we have

$$\begin{Bmatrix} X \\ Y \end{Bmatrix} = \begin{bmatrix} \cos\theta & -\sin\theta \\ \sin\theta & \cos\theta \end{bmatrix} \begin{Bmatrix} \xi \\ \eta \end{Bmatrix} \tag{a}$$

$$\begin{Bmatrix} X_0 \\ Y_0 \end{Bmatrix} = \begin{bmatrix} \cos\theta & -\sin\theta \\ \sin\theta & \cos\theta \end{bmatrix} \begin{Bmatrix} \xi_0 \\ \eta_0 \end{Bmatrix} \tag{b}$$

Hence, their difference has the relation

$$\begin{Bmatrix} X - X_0 \\ Y - Y_0 \end{Bmatrix} = \begin{bmatrix} \cos\theta & -\sin\theta \\ \sin\theta & \cos\theta \end{bmatrix} \begin{Bmatrix} \xi - \xi_0 \\ \eta - \eta_0 \end{Bmatrix} \tag{c}$$

or

$$\begin{Bmatrix} \xi - \xi_0 \\ \eta - \eta_0 \end{Bmatrix} = \begin{bmatrix} \cos\theta & -\sin\theta \\ \sin\theta & \cos\theta \end{bmatrix}^{-1} \begin{Bmatrix} X - X_0 \\ Y - Y_0 \end{Bmatrix}$$

$$= \begin{bmatrix} \cos\theta & \sin\theta \\ -\sin\theta & \cos\theta \end{bmatrix} \begin{Bmatrix} X - X_0 \\ Y - Y_0 \end{Bmatrix} \tag{d}$$

Referring again to the figure shown, we have

$$\begin{aligned} \overrightarrow{PP'} &= \overrightarrow{OO'} - \overrightarrow{OP} + \overrightarrow{O'P'} \\ &= \overrightarrow{OO'} - (\overrightarrow{CP} - \overrightarrow{CO}) + [\theta](\overrightarrow{CP} - \overrightarrow{CO}) \\ &= \overrightarrow{OO'} - ([I] - [\theta])(\overrightarrow{CP} - \overrightarrow{CO}) \end{aligned} \tag{e}$$

where $[\theta]$ is rotation matrix defined in Eq. (c). Using Eq. (c), Eq. (e) can be rewritten in the form

$$\begin{Bmatrix} u_P \\ v_P \end{Bmatrix} = \begin{Bmatrix} u \\ v \end{Bmatrix} - \left(\begin{bmatrix} 1 & 0 \\ 0 & 1 \end{bmatrix} - \begin{bmatrix} \cos\theta & -\sin\theta \\ \sin\theta & \cos\theta \end{bmatrix} \right) \begin{Bmatrix} \xi - \xi_0 \\ \eta - \eta_0 \end{Bmatrix}$$

$$= \begin{Bmatrix} u \\ v \end{Bmatrix} - \begin{bmatrix} 1 - \cos\theta & \sin\theta \\ -\sin\theta & 1 - \cos\theta \end{bmatrix} \begin{Bmatrix} \xi - \xi_0 \\ \eta - \eta_0 \end{Bmatrix} \tag{f}$$

which is Eq. (10.27).

Substituting Eq. (d) into Eq. (f) and differentiate once with respect to z, we have,

$$\begin{Bmatrix} u_P \\ v_P \end{Bmatrix}' = \begin{Bmatrix} u \\ v \end{Bmatrix}' - \begin{bmatrix} \cos\theta - 1 & \sin\theta \\ -\sin\theta & \cos\theta - 1 \end{bmatrix}' \begin{Bmatrix} X - X_0 \\ Y - Y_0 \end{Bmatrix}$$

$$= \begin{Bmatrix} u \\ v \end{Bmatrix}' - \begin{bmatrix} -\theta'\sin\theta & \theta'\cos\theta \\ -\theta'\cos\theta & -\theta'\sin\theta \end{bmatrix} \begin{Bmatrix} X - X_0 \\ Y - Y_0 \end{Bmatrix}$$

$$\approx \begin{Bmatrix} u \\ v \end{Bmatrix}' - \begin{bmatrix} 0 & \theta' \\ -\theta' & 0 \end{bmatrix} \begin{Bmatrix} X - X_0 \\ Y - Y_0 \end{Bmatrix} \tag{g}$$

which is Eq. (10.29).

Noting that $\overrightarrow{O'P'} = [\theta]'\overrightarrow{OP}$, we obtain

$$\begin{Bmatrix} X - X_0 \\ Y - X_0 \end{Bmatrix}' = \begin{bmatrix} \cos\theta & -\sin\theta \\ \sin\theta & \cos\theta \end{bmatrix}' \begin{Bmatrix} X - X_0 \\ Y - Y_0 \end{Bmatrix}$$

$$\approx \begin{bmatrix} 0 & -\theta' \\ \theta' & 0 \end{bmatrix} \begin{Bmatrix} X - X_0 \\ Y - Y_0 \end{Bmatrix} \tag{h}$$

Using Eq. (h), Eq. (g) can be reduced to the form

$$\begin{Bmatrix} u_P \\ v_P \end{Bmatrix}' = \begin{Bmatrix} u \\ v \end{Bmatrix}' + \begin{Bmatrix} X - X_0 \\ Y - Y_0 \end{Bmatrix}'$$

$$= \begin{Bmatrix} u + X - X_0 \\ v + Y - Y_0 \end{Bmatrix}'$$

which is Eq. (10.30).

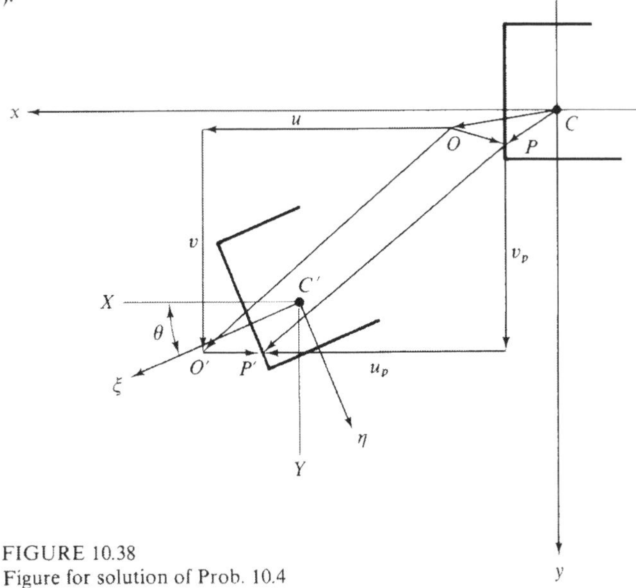

FIGURE 10.38
Figure for solution of Prob. 10.4

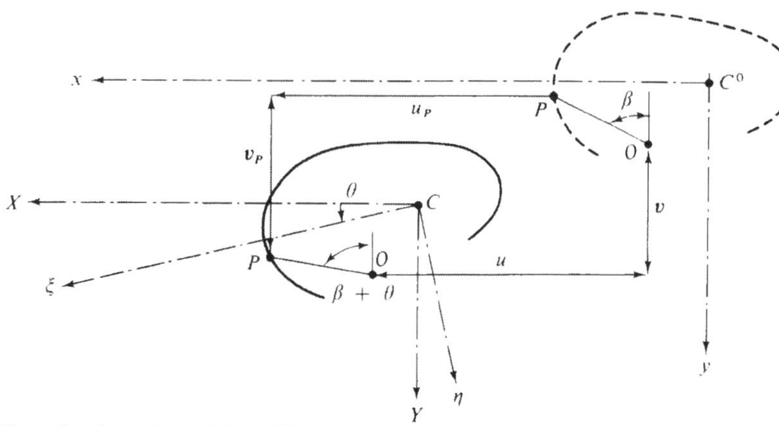

Figure for alternative solution of Prob. (10.4)

Alternative approach to the solution of Prob. 10.4

Let β be an angle between the vector $\overline{OP}$ and the y axis. Due to the fact that the shape of the section is indeformable, it follows that the vector $\overline{OP}$ does not change its length during the deformation of the bar. We have, from the figure shown:

$$\begin{aligned} u_P &= u - \overline{OP} \sin \beta + \overline{OP} \sin (\beta + \theta) \\ &= u - \overline{OP} \sin \beta + \overline{OP} (\sin \beta \cos \theta + \cos \beta \sin \theta) \\ &= u + \overline{OP} \cos \beta \sin \theta - \overline{OP} \sin \beta (1 - \cos \theta) \end{aligned} \tag{j}$$

$$\begin{aligned} v_P &= v + \overline{OP} \cos \beta - \overline{OP} \cos (\beta + \theta) \\ &= v + \overline{OP} \cos \beta - \overline{OP} (\cos \beta \cos \theta - \sin \beta \sin \theta) \\ &= v + \overline{OP} \sin \beta \sin \theta + \overline{OP} \cos \beta (1 - \cos \theta) \end{aligned} \tag{k}$$

Noting that

$$\overline{OP} \cos \beta = (\eta_0 - \eta); \quad \overline{OP} \sin \beta = (\xi - \xi_0) \tag{l}$$

we get

$$\begin{aligned} u_P &= u - (\eta - \eta_0) \sin \theta - (\xi - \xi_0)(1 - \cos \theta) \\ v_P &= v + (\xi - \xi_0) \sin \theta - (\eta - \eta_0)(1 - \cos \theta) \end{aligned} \tag{m}$$

which is Eq. (10.27).

Differentiating these relations once with respect to z, we obtain, with the help of Eq. (10.26)

$$\begin{aligned} u'_P &= u' - (\eta - \eta_0)\theta' \cos\theta - (\xi - \xi_0)\theta' \sin\theta \\ &= u' - [(\xi - \xi_0)\sin\theta + (\eta - \eta_0)\cos\theta]\theta' \qquad \text{(n)} \\ &= u' - [(Y - Y_0)\theta'] \end{aligned}$$

$$\begin{aligned} v'_P &= v' + (\xi - \xi_0)\theta' \cos\theta - (\eta - \eta_0)\theta' \sin\theta \\ &= v' + [(\xi - \xi_0)\cos\theta - (\eta - \eta_0)\sin\theta]\theta' \qquad \text{(o)} \\ &= v' + (X - X_0)\theta' \end{aligned}$$

which are the relations given in Eq. (10.29).

Differentiating once the relations for X and Y given in Eq. (10.26) we obtain

$$\begin{aligned} X' &= -\xi\theta' \sin\theta - \eta\theta' \cos\theta = -Y\theta' \\ Y' &= \xi\theta' \cos\theta - \eta\theta' \sin\theta = X\theta' \end{aligned} \qquad \text{(p)}$$

The relations for u'_P and v'_P could therefore be written as:

$$\begin{aligned} u'_P &= (u + X - X_0)' \\ v'_P &= (v + Y - Y_0)' \end{aligned} \qquad \text{(q)}$$

which are the relations (10.30).

CHAPTER 11

11.1 Use Taylor series expansion

$$C_\xi = \cos k_\xi L = 1 - \frac{1}{2}(k_\xi L)^2 + \frac{1}{24}(k_\xi L)^4$$

$$S_\xi = \sin k_\xi L = k_\xi L - \frac{1}{6}(k_\xi L)^3$$

in Eq. (11.56) through Eq. (11.59).

11.2 (omitted)

11.3 From the definition of principal axes of a cross section,

$$\begin{bmatrix} EI'_\xi & 0 \\ 0 & EI_\eta \end{bmatrix} = \begin{bmatrix} \cos\beta & \sin\beta \\ -\sin\beta & \cos\beta \end{bmatrix} \begin{bmatrix} EI'_\xi & -EI'_{\xi\eta} \\ -EI'_{\xi\eta} & EI'_\eta \end{bmatrix} \begin{bmatrix} \cos\beta & -\sin\beta \\ \sin\beta & \cos\beta \end{bmatrix}$$

from which Eqs. (11.10) are obtained together with the condition

$$0 = I'_\xi \cos\beta \sin\beta - I'_{\xi\eta} \sin^2\beta + I'_{\xi\eta} \cos^2\beta - I'_\eta \sin\beta \cos\beta$$

Eq. (11.7) is the solution for β of this equation or use Mohr's circle to check this result.

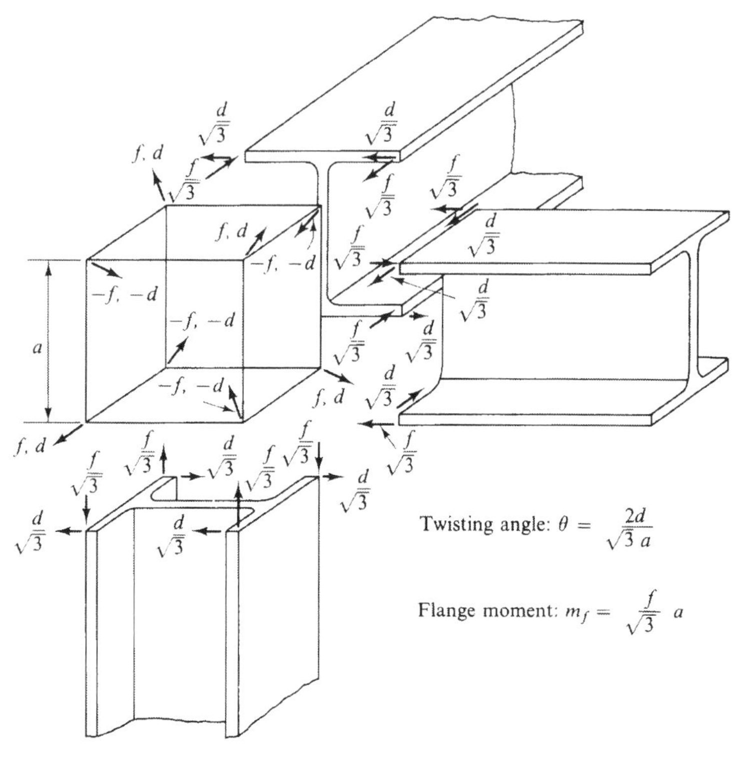

FIGURE 11.17
Scaler warping cube (Figure for solution of Prob. 11.4)

11.4 Imagine a cube of size "a" which is subjected to constant forces "f" and displacements "d" outward or inward at each corner point as shown in Fig. 11.17. Then, the warping and bimoment are the same in all faces on the cube,

$$\text{warping:} \qquad \omega = \frac{\partial \theta}{\partial \zeta} = \frac{d/\sqrt{3}}{a/2}\left(\frac{2}{a}\right) = \frac{4d}{\sqrt{3}\,a^2}$$

$$\text{bimoment:} \quad m_\omega = m_f\, a = \frac{1}{\sqrt{3}} f a^2$$

CHAPTER 12

12.1 Using the actual variation of I as (Fig. 12.38)

$$I = \langle (1 - 2\beta)(1 - \beta), 4\beta(1 - \beta), \beta(2\beta - 1) \rangle \begin{Bmatrix} I_1 \\ I_2 \\ I_3 \end{Bmatrix}$$

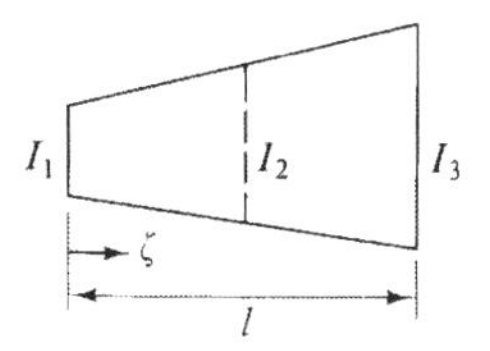

FIGURE 12.38
Figure for Prob. 12.1

Flexural stiffness $[k_S]$ for the nonprismatic element is

$$[k_S] = \frac{E}{l^3}\left\{ I_1 \begin{bmatrix} 3.6 & 2.8 & -3.6 & 0.8 \\ 2.8 & 2.0666 & -2.8 & 0.7334 \\ -3.6 & -2.8 & 3.6 & -0.8 \\ 0.8 & 0.7334 & -0.8 & 0.0666 \end{bmatrix} \right.$$

$$+ I_2 \begin{bmatrix} 4.8 & 2.4 & -4.8 & 2.4 \\ 2.4 & 1.8668 & -2.4 & 0.5332 \\ -4.8 & -2.4 & 4.8 & -2.4 \\ 2.4 & 0.5332 & -2.4 & 1.8668 \end{bmatrix}$$

$$\left. + I_3 \begin{bmatrix} 3.6 & 0.8 & -3.6 & 2.8 \\ 0.8 & 0.0666 & -0.8 & 0.7334 \\ -3.6 & -0.8 & 3.6 & -2.8 \\ 2.8 & 0.7334 & -2.8 & 2.0666 \end{bmatrix} \right\}$$

Flexural stiffness matrix $[k_S]$ for the prismatic element with I_2 as average moment of inertia is

$$[k_S] = \frac{EI_2}{l^3} \begin{bmatrix} 12 & 6 & -12 & 6 \\ 6 & 4 & -6 & 2 \\ -12 & -6 & 12 & -6 \\ 6 & 2 & -6 & 4 \end{bmatrix}$$

Geometric stiffness $[k_G]$ for an element is

$$[k_G] = \frac{P_{cr}}{30l} \begin{bmatrix} 36 & 3 & -36 & 3 \\ 3 & 4 & -3 & -1 \\ -36 & -3 & 36 & -3 \\ 3 & -1 & -3 & 4 \end{bmatrix}$$

Number of Elements	Prismatic Element		I-Parabolic Variation	
	P_{cr}	% Error	P_{cr}	% Error
1	2.17521	−4.8	2.31117	1.2
2	2.24917	−1.5	2.28955	0.3
4	2.272	−0.5	2.28378	0

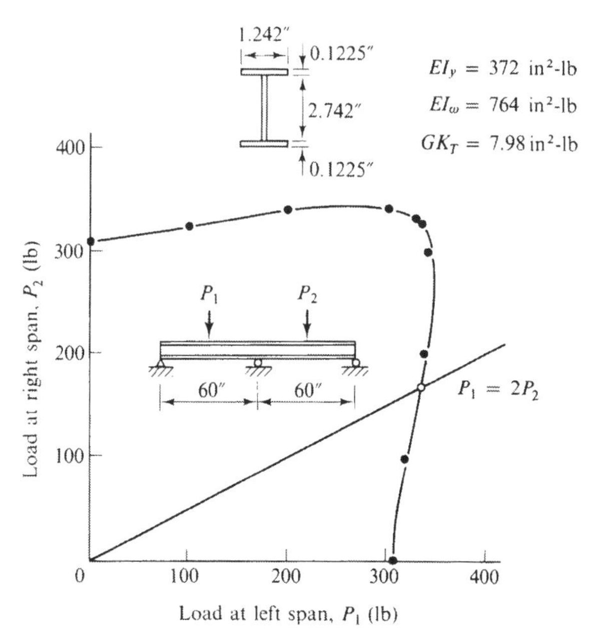

FIGURE 12.39
Buckling load for two-span continuous beam—solution to Prob. 12.2 (1 lb = 4.448 N, 1 in = 25.4 mm)

If a graph is drawn between percentage error and number of elements, it can be seen that accuracy at any level of refinement is strikingly less for uniform element than for the tapered element. Rather than converging upon the solution from the upper side, the results for the prismatic element converge from the lower side. The increased flexibility due to geometric approximation far outbalances the approximation in the element displacement function representation.

12.2 Divide the continuous beam into eight elements. Formulate element flexural and geometric stiffness matrices (see Table 12.3). Assemble element stiffness matrices to form the stiffness matrices for the entire beam. Apply boundary conditions and carry out the eigenvalue solution.

$$P = 185 \text{ kip}$$

Buckling load for two-span continuous beam is shown in Fig. 12.39.

12.3 When transverse bracing system prevents sways, the frame buckles in a symmetric node. The problem may be converted to that of beam-column with rotational springs at joints C and E rotational stiffness $K_\theta = 2EI_B/l_B$ or $K_\theta = 2\lambda I_c/l_c$ (see Fig. 12.40). Assume $I_c = 1$ and the length of the column is 1. Hence, $K_\theta = 2\lambda$. Divide the beam-column into a number of elements and assemble flexural stiffness $[k_S]$ and geometric stiffness $[k_G]$. Add the stiffness of the spring corresponding to appropriate degree of freedom

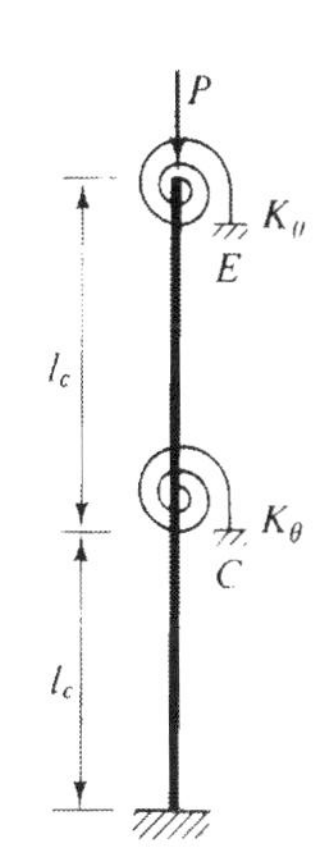

FIGURE 12.40
Figure for Prob. 12.3

$\lambda = \dfrac{I_B/l_B}{I_c/l_c}$	P_{cr} Actual	Four Elements
5	2.73762	2.74597
10	3.19192	3.21365
15	3.6635	3.7143

12.4
$$[K_S]\{u\} + [K_G]\{u\} + [K_G]\{u\}^* = \{0\}$$

or

$$\{E\}_{n+1} = -[K_S]\{u\}_n - [K_G]_n\{u\}_n - [K_G]_n\{u\}_n^*$$

use incremental method combined with either Newton-Raphson or modified Newton-Raphson method to arrive at the solution. See Fig. 12.41 for the load and the lateral deformation response.

12.5 Compute the tangent properties and stress resultants with respect to instantaneous centroid and shear center using the expressions given in Sec. 12.7.3. Formulate $[k_S^T]$ and $[k_G]$ matrices for the element (see Table 12.6). Assemble the element stiffness matrices to form the global stiffness matrices for the entire column. Solve this as an eigenvalue problem for critical length. Figure 12.42 shows the plastic column buckling curve.

12.6 Figure 12.43 shows the finite element results for plastic lateral torsional buckling of a beam-column subjected to moment gradient. The results agree with Fukumoto and Galambos (1966).

12.7 Assume a reasonable residual stress distribution shown in Fig. 12.44. The variation of critical stress with slenderness ratio is shown in Fig. 12.44 and compared with basic SSRC (or CRC) column strength curve.

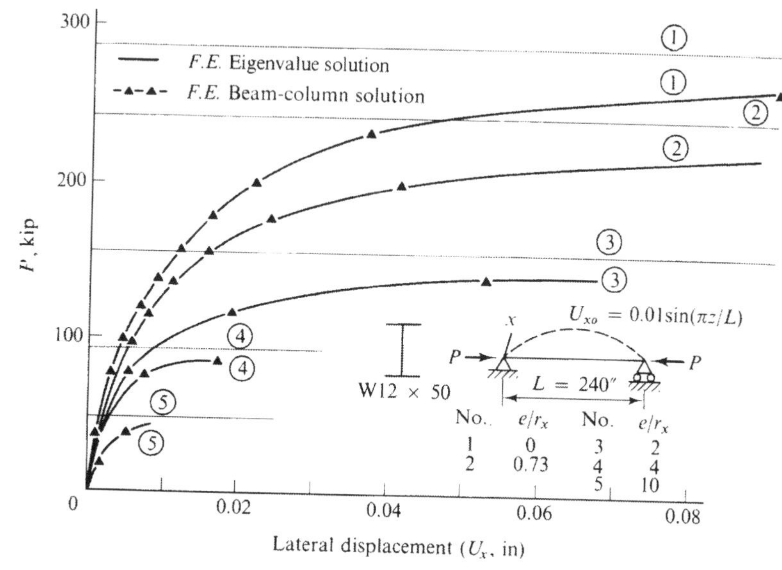

FIGURE 12.41
Beam-column with sinusoidal imperfection solution to Prob. 12.4 (1 kip = 4.448 N, 1 in = 25.4 mm)

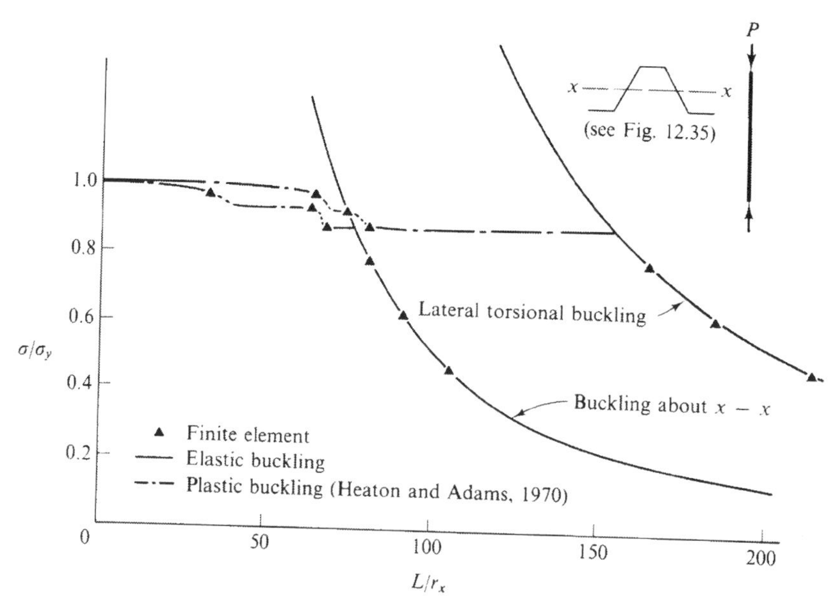

FIGURE 12.42
Plastic buckling of a column (hat section)—solution to Prob. 12.5

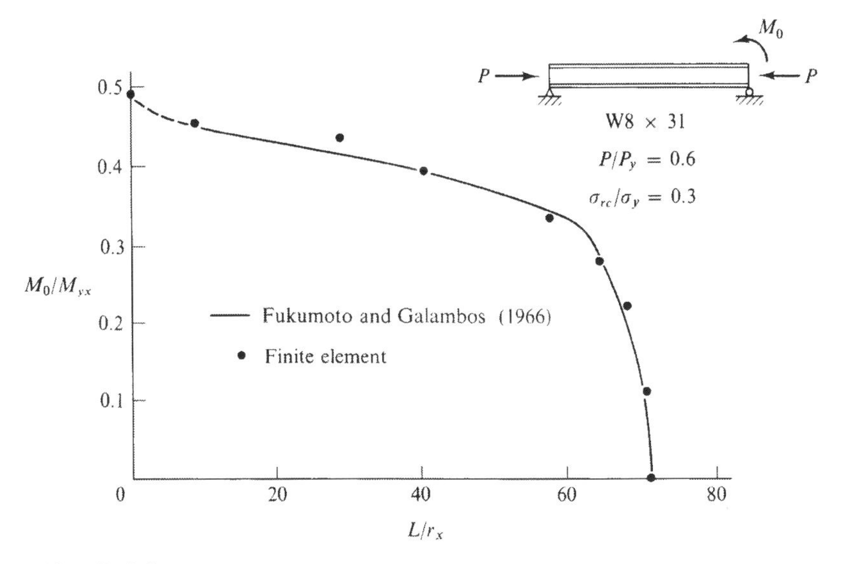

FIGURE 12.43
Plastic lateral torsional buckling of a beam-column with moment gradient—solution to Prob. 12.6

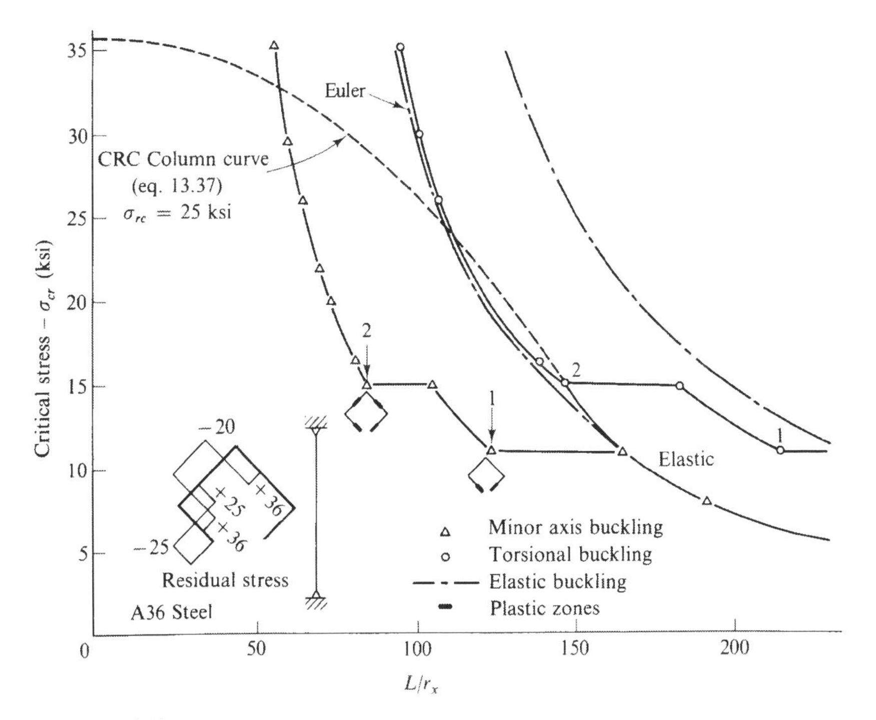

FIGURE 12.44
Column strength curves—solution to Prob. 12.7 (1 ksi = 6.89 MN/m²)

CHAPTER 13

13.1(a) Try W10 × 33

$$A = 9.71 \text{ in}^2, \quad D = 9.75 \text{ in}, \quad B = 7.964 \text{ in}$$

$$r_x = 4.20 \text{ in}, \quad r_y = 1.94 \text{ in}$$

$$P = 175 \text{ kip}, \quad C_{mx} = 0.4, \quad C_{my} = 1.0$$

$$(M_x)_{\text{equ}} = 0.4 \times 870 = 348 \text{ k-in}$$

$$(M_y)_{\text{equ}} = 1.0 \times 95 = 95 \text{ k-in}$$

$$e_x = \frac{95}{175} = 0.54 \text{ in}, \quad e_y = \frac{348}{175} = 2.00 \text{ in}$$

$$\frac{e_x B}{2r_y^2} = \frac{0.54 \times 7.964}{2(1.94)^2} = 0.57$$

$$\frac{e_y D}{2r_x^2} = \frac{2 \times 9.75}{2(4.20)^2} = 0.55$$

$$\left(\frac{L}{r_y}\right)_{\text{equ}} = \frac{15 \times 12}{1.94}\sqrt{\frac{50}{36}} = 109$$

From Fig. 7.17 (pages 345 and 346), we find $P/P_y = 0.43$ for $(L/r_y)_{\text{equ}} = 109$. Thus

$$P = 0.43 \times 9.71 \times 50 = 210 \text{ kip} > 175 \text{ kip}$$

O.K. but too conservative.

Try a smaller section, W10 × 29

$$A = 8.53 \text{ in}^2, \quad D = 10.22 \text{ in}, \quad B = 5.799 \text{ in}$$

$$r_x = 4.29 \text{ in}, \quad r_y = 1.34 \text{ in}$$

$$\frac{e_x B}{2r_y^2} = \frac{0.54 \times 5.799}{2(1.34)^2} = 0.87$$

$$\frac{e_y D}{2r_x^2} = \frac{2 \times 10.22}{2(4.29)^2} = 0.56$$

$$\left(\frac{L}{r_y}\right)_{\text{equ}} = \frac{15 \times 12}{1.34}\sqrt{\frac{50}{36}} = 158$$

From Fig. 7.17, we find $P/P_y = 0.42$ for $(L/r_y)_{\text{equ}} = 158$. Thus,

$$P = 0.42 \times 8.53 \times 50 = 179 \text{ kip} > 175 \text{ kip} \qquad \text{O.K.}$$

13.1(b) Try W10 × 33

$$A = 9.71 \text{ in}^2, \quad Z_x = 38.8 \text{ in}^3, \quad r_x = 4.20 \text{ in}$$

$$E = 29\,000 \text{ ksi}, \quad Z_y = 14.0 \text{ in}^3, \quad r_y = 1.94 \text{ in}$$

Euler buckling stress $\sigma_e = \pi^2 E/(L/r)^2$

$$L/r_x = 42.9, \quad \sigma_{ex} = 155.8 \text{ ksi}, \quad P_{ex} = 1513 \text{ kip}$$

$$L/r_y = 92.8, \quad \sigma_{ey} = 33.3 \text{ ksi}, \quad P_{ey} = 323 \text{ kip}$$

From SSRC Basic Column Strength Equation

$$P_u = P_y - \frac{A\sigma_y^2}{4\pi^2 E}\left(\frac{kL}{r_y}\right)^2 = 485.5 - 182.6 = 303 \text{ kip}$$

$$M_{ux} = M_{px}\left[1.07 - \frac{(L/r_y)\sqrt{\sigma_y}}{3160}\right]$$

$$= (38.8 \times 50)\left[1.07 - \frac{92.8 \times \sqrt{50}}{3160}\right] = 1673 \text{ kip-in}$$

$$M_{ucx} = M_{ux}(1 - P/P_u)(1 - P/P_{ex})$$

$$= 1673(1 - 175/303)(1 - 175/1513) = 625 \text{ kip-in}$$

$$M_{ucy} = M_{py}(1 - P/P_u)(1 - P/P_{ey})$$

$$= (14 \times 50)(1 - 175/303)(1 - 175/323) = 135.5 \text{ kip-in}$$

$$\frac{C_{mx} M_x}{M_{ucx}} = \frac{0.4 \times 870}{625} = 0.557$$

$$\frac{C_{my} M_y}{M_{ucy}} = \frac{1.0 \times 95}{135.5} = 0.701$$

$$p = \frac{P}{P_y} = \frac{175}{9.71 \times 50} = 0.361$$

$$\frac{B}{D} = \frac{7.964}{9.75} = 0.817$$

From Eq. (13.24b) we have

$$\beta = 0.4 + p + B/D = 1.58$$

Substituting into Eq. (13.23), we find

$$0.557^{1.58} + 0.701^{1.58} = 0.97 < 1.0 \qquad \text{O.K.}$$

13.2(a) and (b) see the curves labelled (a) and (b) in Figs. 13.44(a) and 13.44(b).

13.2(c) Multiple column strength curve 2 [equation (14.18), page 441, Vol. 1] with $\lambda = (1/\pi)\sqrt{\sigma_y/E}(L/r_y)$ has the form

$$\sigma_u = \sigma_y(1.035 - 0.202\lambda - 0.222\lambda^2) \quad \text{for} \quad 0.15 \le \lambda \le 1.0$$

Euler buckling stress

$$\sigma_e = \frac{\pi^2 E}{(L/r)^2}$$

For W8 × 31,

$$r_x = 3.47 \text{ in}, \quad r_y = 2.01 \text{ in}, \quad r_x/r_y = 1.73$$

	$L/r_x = 30$, $L/r_y = 52$	$L/r_x = 50$, $L/r_y = 86$
λ	0.583	0.965
σ_u, ksi	30.30	22.80
σ_{ey}, ksi	105.85	38.70
σ_{ex}, ksi	318.00	114.50
M_m, Eq. (13.20)	$0.971M_{px}$	$0.907M_{px}$
M_{ucx}, Eq. 13.19(a)	$0.602M_{px}$	$0.430M_{px}$
M_{ucy}, Eq. 13.19(b)	$0.576M_{py}$	$0.373M_{py}$

Solution to interaction equation (13.23)

$$\left(\frac{M_x}{M_{ucx}}\right)^{1.7} + \left(\frac{M_y}{M_{ucy}}\right)^{1.7} = 1.0$$

expressed in terms of

$$\frac{M_x}{M_{px}} \quad \text{and} \quad \frac{M_y}{M_{py}}$$

is tabulated below.

		$L/r_y = 52$, $L/r_x = 30$		$L/r_y = 86$, $L/r_x = 50$	
$\frac{M_x}{M_{ucx}}$	$\frac{M_y}{M_{ucy}}$	$\frac{M_y}{M_{py}}$	$\frac{M_x}{M_{px}}$	$\frac{M_y}{M_{py}}$	$\frac{M_x}{M_{px}}$
0	1.0	0.58	0.0	0.37	0.0
0.2	0.96	0.55	0.12	0.36	0.086
0.4	0.87	0.50	0.24	0.33	0.17
0.6	0.73	0.42	0.36	0.27	0.26
0.8	0.51	0.29	0.48	0.19	0.34
0.9	0.35	0.20	0.54	0.13	0.39
1.0	0.0	0.0	0.60	0.0	0.43

The two interaction curves plotted from the above table are labelled (c) in Fig. 13.44(a) for $L/r_x = 30$ and in Fig. 13.44(b) for $L/r_x = 50$, respectively.

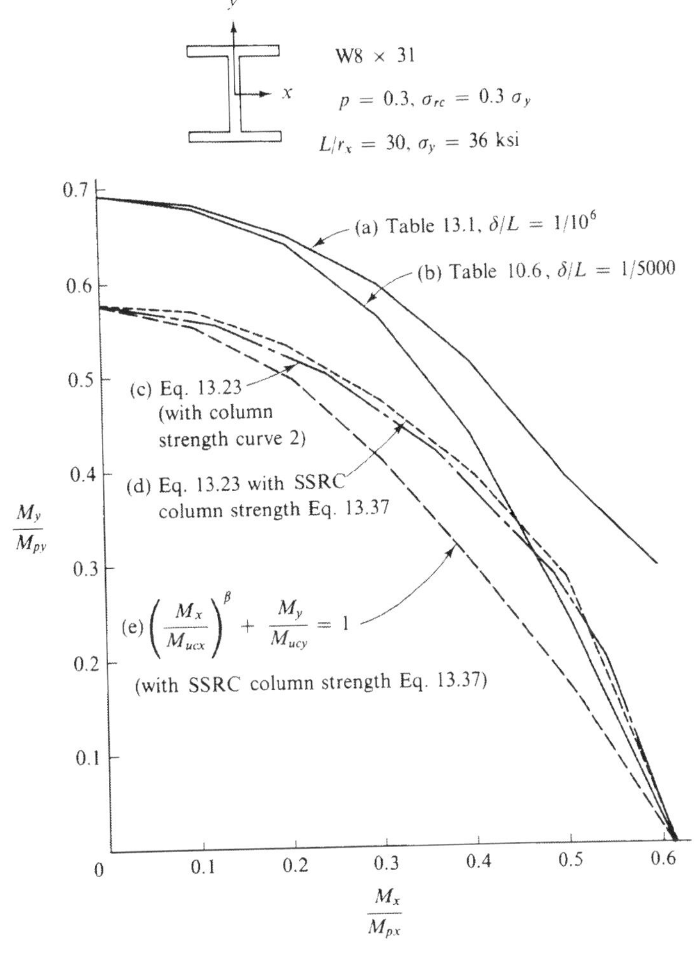

FIGURE 13.44(a)
Figure for Prob. 13.2 with $L/r_x = 30$

13.2(d) and (e)

SSRC Basic Column Strength Equation (13.37) with $\sigma_{rc} = 0.5\sigma_y$ has the form

$$\sigma_c = \sigma_y - \frac{\sigma_y^2}{4\pi^2 E}\left(\frac{kL}{r_y}\right)^2$$

which results in

$$\sigma_c = 32.94 \text{ ksi} \quad \text{for} \quad L/r_x = 30, \quad L/r_y = 52$$

and

$$\sigma_c = 27.63 \text{ ksi} \quad \text{for} \quad L/r_x = 50, \quad L/r_y = 86$$

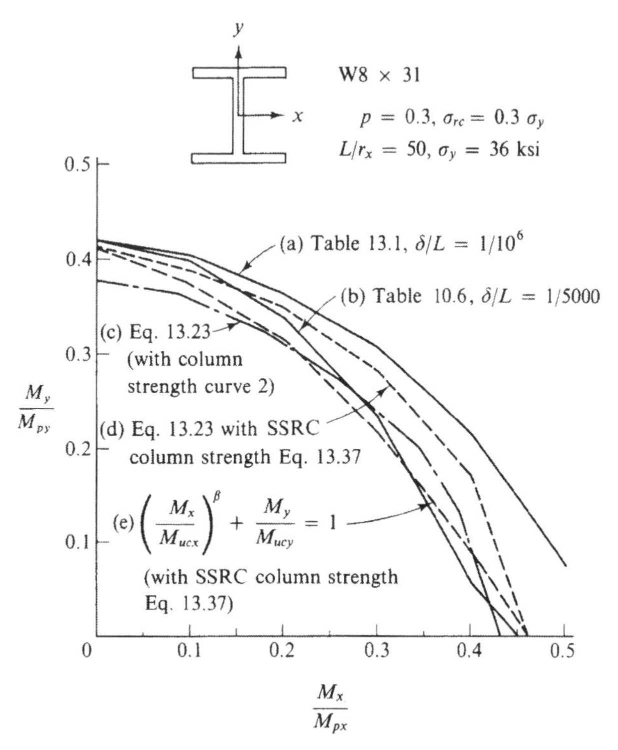

FIGURE 13.44(b)
Figure for Prob. 13.2 with $L/r_x = 50$

Using σ_c instead of σ_u, similar procedure as Prob. 13.2(c) can be followed and the corresponding interaction curves, labelled (d) and (e) in the figures, are plotted in Fig. 13.44(a) for $L/r_x = 30$ and in Fig. 13.44(b) for $L/r_x = 50$, respectively.

AUTHOR INDEX

Note: Page numbers in bold-face type indicate references at the end of chapters

SUBJECT INDEX